Lecture Notes in Artificial Intelligence 1424

Subseries of Lecture Notes in Computer Science
Edited by J. G. Carbonell and J. Siekmann

Lecture Notes in Computer Science

Edited by G. Goos, J. Hartmanis and J. van Leeuwen

Springer-Verlag Berlin Heidelberg GmbH

Lech Polkowski   Andrzej Skowron (Eds.)

Rough Sets and Current Trends in Computing

First International Conference, RSCTC'98
Warsaw, Poland, June 22-26, 1998
Proceedings

 Springer

Series Editors
Jaime G. Carbonell, Carnegie Mellon University, Pittsburgh, PA, USA
Jörg Siekmann, University of Saarland, Saarbrücken, Germany

Volume Editors

Lech Polkowski
Institute of Mathematics, Warsaw University of Technology
Pl. Politechniki 1, 00-665 Warszawa, Poland
E-mail: polk@mimuw.edu.pl

Andrzej Skowron
Institute of Mathematics, Warsaw University
Banacha 2, 02-097 Warszawa, Poland
E-mail: skowron@mimuw.edu.pl

Cataloging-in-Publication data applied for

Die Deutsche Bibliothek - CIP-Einheitsaufnahme

Rough sets and current trends in computing : first international
conference ; proceedings / RSCTC'98, Warsaw, Poland, June 22 -
26, 1998. Lech Polkowski ; Andrzej Skowron (ed.). - Berlin ;
Heidelberg ; New York ; Barcelona ; Budapest ; Hong Kong ;
London ; Milan ; Paris ; Santa Clara ; Singapore ; Tokyo : Springer,
1998
  (Lecture notes in computer science ; Vol. 1424 : Lecture notes in artificial
  intelligence)
  ISBN 3-540-64655-8

CR Subject Classification (1991): I.2, F.1, F.4.1, I.5.1, I.4
ISBN 978-3-540-64655-6   ISBN 978-3-540-69115-0 (eBook)
DOI 10.1007/978-3-540-69115-0

Typesetting: Camera-ready by author
SPIN 10637485   06/3142 – 5 4 3 2 1 0   Printed on acid-free paper

Preface

The papers collected in this volume were presented at the International Conference on Rough Sets and Current Trends in Computing – RSCTC'98 held in Warsaw, June 22-26, 1998. The conference was intended as a forum for exchanging ideas among experts in various areas of soft computing and researchers in rough set theory and its applications – a branch of soft computing showing a significant growth in years – in order to stimulate mutual understanding and cooperation.

It is our pleasure to dedicate this volume to Professor Zdzisław Pawlak who created the theory of rough sets. The present state of this theory and its applications owes very much to Professor Pawlak's enthusiasm and support.

The conference was held under the patronage of the Committee for Computer Science of the Polish Academy of Sciences as one of the events celebrating the 50th Anniversary of starting Computer Science in Poland and for this we wish to express our thanks to Professor Stefan Węgrzyn, chairman of the committee.

We wish to express our gratitude to Professors Edward A. Feigenbaum, Zdzisław Pawlak, Carl A. Petri, and Lotfi A. Zadeh who accepted our invitation to act as honorary chairs of the conference.

The conference was held in the Barnabite Cultural Center in Warsaw and it was sponsored by grants from the Polish State Committee for Scientific Research (KBN) and by the Institute of Mathematics at Warsaw University, the Department of Electronics and Information Techniques at Warsaw University of Technology, the Polish-Japanese Institute of Computer Techniques and the Japan International Cooperation Agency (JICA).

The following major areas were selected for RSCTC'98:

- Rough Set Theory and Applications
- Fuzzy Set Theory and Applications
- Knowledge Discovery and Data Mining
- Decision Support Systems
- Evolutionary Computations
- Neural Networks
- Computing with Words and Granular Computations
- Grammar Systems and Molecular Computations
- Petri Nets and Concurrency
- Complexity Aspects of Soft Computing
- Pattern Recognition and Image Processing
- Statistical Inference
- Logical Aspects of Soft Computing
- Applications of Soft Computing

We wish to express our thanks to members of the advisory board: M. Drzewiecki, J. W. Grzymala-Busse, T.Y. Lin, K. Malinowski, T. Munakata, A. Nakamura, S. Ohsuga, F. Petry, Z. W. Raś, G. Rozenberg, S. Shimada, R. Słowiński, H. Tanaka, B. Trakhtenbrot, S. Tsumoto, P.P. Wang, S. Węgrzyn, W. Ziarko for their contribution to the scientific program of the conference.

The accepted papers selected from 90 draft papers were divided into regular communications (allotted 8 pages in this volume) and short communications (allotted 4 pages in this volume) on the basis of reviewer grades. The process of reviewing rested with members of the program committee: R. Agrawal, Shun-ichi Amari, Th. Baeck, M.C.F. Fernández-Baizán, J. Bielecki, H.-D. Burkhard, G. Cattaneo, M. Chakraborty, N. Cercone, J.-C. Cubero, A. Czyżewski, J. Doroszewski, D. Dubois, I. Duentsch, M. Grabish, J. Kacprzyk, T. Kaczorek, W. Kloesgen, J. Komorowski, J. Koronacki, W. Kowalczyk, M. Kryszkiewicz, P. Lingras, T. Luba, S. Marcus, V. W. Marek, Z. Michalewicz, R. Michalski, M. Moshkov, Son H. Nguyen, M. Novotny, E. Orłowska, K. Oshima, P. Pagliani, S. K. Pal, Gh. Paun, W. Pedrycz, J. F. Peters, A. Pettorossi, Z. Piasta, F. G. Pin, L. Polkowski, H. Prade, V. Raghavan, B. Reusch, H. Rybiński, R. Schaefer, W. Skarbek, A. Skowron (chair), K. Słowiński, J. Stefanowski, J. Stepaniuk, R. Swiniarski, A. Szałas, R. Tadeusiewicz, H. Thiele, W. Traczyk, D. Vakarelov, H.-M. Voigt, A. Wasilewska, A. Wierzbicki, S.K.M. Wong, S. Yamane, Y. Y. Yao, J. Zabrodzki, Ning Zhong, J. Żytkow.

We would like to acknowledge help in reviewing from J. Dassow, C. Jędrzejek, O. Pons Capote, L. Rudak, D. Ślęzak, Z. Suraj, M. Szczuka, S. Wierzchoń, P. Wojdyłło, J. Wróblewski.

Invited lectures were presented at the conference by Professors Rakesh Agrawal, Edward A. Feigenbaum, Wolfgang Haerdle, Willi Kloesgen, Solomon Marcus, V. Wiktor Marek, Zdzisław Pawlak and Lotfi A. Zadeh. We wish to express our thanks to them.

The organizing committee members: A. Jankowski, M. Toho (JICA), D. Gello (secretary), Son Hung Nguyen, Sinh Hoa Nguyen, Tuan Trung Nguyen (conference database, papers processing), D. Ślęzak (contact person), M. Szczuka (finances), P. Synak, P. Wojdyłło, J. Wróblewski, P. Ejdys, G. Góra, M. Kryszkiewicz (deputy chair), L. Polkowski (chair) wish to express their thanks to all partners for their kind cooperation.

Our special thanks go to all individuals who submitted papers for the conference and to all conference participants.

We wish to express our thanks to Alfred Hofmann of Springer-Verlag for his support and cooperation.

April 1998   Lech Polkowski and Andrzej Skowron

Table of Contents

Invited Talks (received in text form)

Communications

Session RSM1 Rough Set Methods

Session ST1 Statistical Inference

Deviation and Association Patterns for Subgroup Mining in Temporal, Spatial, and Textual Data Bases

Willi Klösgen

German National Research Center
for Information Technology – GMD
D-53757 St. Augustin, Germany
email:kloesgen@gmd.de

[Abstract.] Data mining is usually introduced as search for interesting patterns in data. It is often an explorative step iteratively performed within a process of knowledge discovery in data bases (KDD). A mining step typically relies on strategies for systematic search in large hypotheses spaces guided by the autonomous evaluation of statistical tests. We describe the subgroup mining approach that is based on deviation and association patterns. A typical database contains values of attributes for many objects (persons, transactions, documents). Interpretable subgroups of these objects are searched that deviate from a designated expected behavior. Many types of data analysis questions can be answered by subgroup mining with diverse specializations of general deviation and association patterns. Tests measure the statistical interestingness of subgroup deviations. After summarizing the approach by discussing the fundamental components of subgroup pattern classes concerning validation, search and interactive presentation of pattern instances, we explain how deviation patterns of subgroup mining are applied for temporal, spatial and textual databases.

1 Introduction: The pattern paradigm

The paradigm of large scale and systematic search for patterns serves as a fundamental idea for introducing data mining. So most informal definitions describe data mining as that step in a KDD process in which valid, novel, potentially useful, and understandable patterns are searched in data (Fayyad et al., 1996). A pattern is seen as a statement S in a high level language that describes a subset $D(S)$ of a data base D with a quality $q(S)$ (Klösgen and Żytkow, 1996). Since a tremendous number of subsets exists in a data base for which a huge collection of quite different types of statements can be checked, a problem oriented focusing is necessary when setting a data mining step. Typically the user of a data mining system, when directing a mining step, specifies the type of the statements and several constraints on the data subsets. Thus the search space

L. Polkowski and A. Skowron (Eds.): RSCTC'98, LNAI 1424, pp. 1–18, 1998.
© Springer-Verlag Berlin Heidelberg 1998

of a mining step contains similar statements of a selected type on each of the subsets implicitly selected by the specification of constraints.

Similar statements can be considered as instances of a special pattern class. A pattern class encompasses the generic properties of its instances. Therefore, the definition of the generic properties is constitutive for a pattern class providing the methods by which all the instances of the class are constructed and processed. Three main properties are treated in this paper that determine how pattern instances are validated, searched and presented.

A pattern class is defined by its meaning, i.e. the statistical content or analytic question, given for instance by a model or hypothesis type. An abstract analytic question is operationalized by the validation method including a verification and a quality computation part. A pattern instance is statistically verified in an associated subset of data, e.g. by testing an hypothesis, and its quality is computed by optionally considering also non-statistical aspects of interestingness.

Other generic properties refer to the presentation schemes, the search dimensions of the instance space, and the search control and pruning possibilities. Presentation schemes especially must determine an appropriate visualization of a statement showing its significance and quality figures and their components. Moreover, presentation of a set of interdependent statements and the exploratory, interactive operations that a user can perform on these presentations have to be designed. Based on a description of the search dimensions and pruning options of a pattern class, a general search algorithm can efficiently organize and process the search space. For a given analytic question, several options can be selected for verification and quality computation, presentation in natural language or graphical form, and search control. For instance, diverse statistical tests can check, if the probability of a binary event deviates in a subgroup.

Based on the generic properties defining a pattern class, a pattern instance is treated in a mining system as a node in a search space which is processed by a search algorithm, a hypothesis or a model on a subset of the data base tested and evaluated by a verification and a quality method, and a statement which is presented to the user.

Since the pattern paradigm refers to a very broad definition, many mining tasks can be captured within this framework. Some simple patterns illustrating the definition are included in Table 1, and some model-type patterns are shown in Table 2. An important KDD pattern class, the subgroup paradigm, is then treated in the rest of this paper, where we will discuss in detail the constitutive aspects of subgroup mining given by the generic properties of this pattern class.

2   Interesting subgroups: Two general pattern classes

The language for building pattern instances has not been elaborated in the introductory pattern definition. Hence many specializations of this definition are possible, as the examples in Table 1 and Table 2 show. KDD usually deals with given data bases that contain data on objects such as clients, transactions,

	Covering sets of attributes	Key attributes in a database
Meaning	At least c records of database assume the (positive) values for all (binary) attributes of the set.	There are no different records in data base that assume the same values for all key attributes.
Validation	simple arithmetic calculation	simple arithmetic calculation
Presentation	symptom1, symptom9, symptom27, appear together in 23% of cases	$\{att1, att5, att13\}$ is a set of key attributes
Instance space	all subsets of attributes	all subsets of attributes
Search control	set covering $\rightarrow$ each subset covering	set is key $\rightarrow$ each superset is key

Table 1. Generic properties for two simple patterns

	Regression model	Conceptual cluster
Meaning	Describes a linear relation between a dependent and a set of independent variables.	A set of similar cases described by the predictive and predictable values of normative attributes.
Validation	verification: F-test quality: adjusted R square	category utility measure e.g. Cobweb, see (Fisher, 1987)
Presentation	log Salary = 3.85 + 0.001 Age + 0.103 Sex + 0.3 Education Level	Toxic Waste - yes (0.81,0.90) Budget Cuts - yes (0.81,0.81) SDI Reduction - no (0.93,0.88)
Instance space	all subsets of independent attributes	all subsets of cases
Search control	forward selection, backward elimination, stepwise regression	hill climbing

Table 2. Generic properties for two model-type patterns

documents. Therefore, one obvious pattern language describes subgroups of these objects. A subgroup is an interpretable and analysis relevant subset of the object population. This specialization of the general pattern approach can informally be characterized as searching for interesting subgroups. A first generation subgroup mining system is the Explora system (Klösgen; 1992, 1996). The subgroup mining task has more formally been introduced by Siebes (1995) as a basis for the development of the *Data Surveyor* mining system.

The motivation for the subgroup approach is given by a frequent data analytic goal where an analyst is interested in a special property or behavior of one or several selected target variables. Those regions of the input variable space are searched where the target variables show this behavior. The analyst could e.g. be interested in regions with a high average value of a selected continuous target variable, or with a high share of one value of a binary target, or in regions for which this share has significantly more increased between two years than in the complementary region, or in regions that show a special time trend of a target variable. Thus many different data analytic questions can be represented as special behavior types of target variables.

The usual approach applied for many of these data analysis relies on an approximation of the unknown function that describes the dependency between the target variable(s) and the input (independent) variables. The function derived as an approximation is then studied to identify the interesting regions. However, a good approximation is often difficult to find, especially when there are many input variables, a global approximation (in the whole input space) has to be derived, and a probabilistically based approximation must be found with a sample of noisy data. So often the direct approach of searching for the interesting subgroups without relying on an intermediary functional approximation is more efficient (Friedman and Fisher, 1997).

To formalize the subgroup approach, a specification of the description language to build subgroups (Section 3) and an operationalization of behavior patterns are needed. The behavior of target variables $\mathbf{y} = (y_1, ..., y_k)$ is captured by assuming a probabilistic approach and referring to their joint distribution with the input variables $\mathbf{x} = (x_1, ..., x_m)$. One is interested in some designated property of the unknown joint distribution, respectively of the density function $p(\mathbf{y}, \mathbf{x})$. The interesting behavior is now defined by a statistical test. In the null hypothesis of such a test, the assumption (the property or behavior) for the distribution of the target variables in the subgroup is specified that is regarded as expected or uninteresting. The alternative hypothesis defines the deviating or interesting subgroup referring to the property of interest for the distribution. When a given data set spots the null hypothesis for a subgroup as very unlikely (under a given confidence threshold), the subgroup (i.e. the behavior of the target variable(s) in the subgroup given by the distribution property) is identified as interesting.

The test approach has three advantages for subgroup mining. It allows a broad spectrum of data analytic questions to be treated, offers intelligent solutions to balance the trade-off between diverse criteria for assessing the statistical interestingness of a deviation, e.g. size of subgroup versus amount of deviation, and finally mitigates the problem of just discovering random fluctuations of the target variables in a given noisy sample data base as interesting.

Besides searching for interesting subgroups, a second general mining task consists of searching for interesting pairs of subgroups. Interestingness is then evaluated by association measures for the two subgroups of a pair. Association rules (Agrawal et al., 1996) or sequences (Mannila et al., 1997) are typical examples of this second general subgroup mining pattern. Mostly these association measures rely on an evaluation of the 2×2 cross table constructed for the two subgroups. In case of time stamped objects, time relations between objects (e.g. successor) can underlie the cross table calculation and special association measures (e.g. for sequences).

The generic components of subgroup patterns include a description language to construct subgroups, a verification method to test the statistical significance of a subgroup, quality functions to measure the interestingness of a single subgroup and of a set of subgroups, constraints limiting the space of admissible subgroups, and search goals and controls defining additional properties of the subgroups to be found. Interactive visualization of individual subgroups and sets of interdependent subgroups has to be fixed in the presentation component of a pattern. The mining task consists in applying a search strategy to discover the interesting subgroups which is appropriate for the given search goals and constraints and to present the results to the user so that she can operate on the presentation. In a very similar way, the generic properties can be introduced for pattern classes that deal with searching for interesting pairs of subgroups.

3   Description languages to build subgroups

Description languages for subgroup mining are mostly conjunctive, propositional languages. So we assume that the data base consists of one or several relations, each with a schema $\{A_1, A_2, ..., A_n\}$ and associated domains D_i for the attributes A_i. A conjunctive, propositional description of a subgroup is then given by: $(A_1 \in V_1) \wedge ... \wedge (A_n \in V_n)$ with $V_i \subseteq D_i$. E.g. $age \in [18, 25] \wedge region \in \{BO, DO, EN, RE\}$. Conjunctive selectors with $V_i = D_i$ can of course be omitted in a description.

Single relation languages allow only to analyze a single relation, possibly constructed in a preprocessing phase by a join of several relations. Then only the attributes belonging to the schema of this single relation can be used for subgroup descriptions. Multirelational languages do not require a preprocessing join operation and allow to build descriptions with attributes from several relations. They are especially useful for applications that require different target object classes with flexible joins to be analyzed. In the *Kepler* system, we use the *MIDOS* subgroup miner (Wrobel, 1997) to discover subgroups of objects of a selected target relation. A description may include selectors from several relations, which are linked by foreign link attributes.

A data base, for instance, includes relations on hospitals, patients, diagnoses of patients, and therapies for patient-diagnoses with obvious foreign links such as patient-id linking patient-diagnoses and patients. The analyst chooses a target object class, e.g. patients, and can decide on the other object classes, e.g. hospitals and patient-diagnoses, and their attributes to be used for building subgroups of patients. Such a subgroup could e.g. be described by *male patients with a cancer diagnosis treated in small hospitals*.

By these foreign links and the implicit existential quantifier used for linking relations (e.g. patients with at least one diagnosis of a type), a very limited Inductive Logic Programming approach is applied, extending the simple one relational propositional approach. The full ILP approach has not (yet) been used for subgroup mining.

A next aspect for specializations of description languages refers to taxonomies that can be used for subgroup descriptions. To restrict the number of descriptions, usually not every subset of the domain of an attribute A_i is allowed in a description. A taxonomy H_i consisting of a set of subsets of D_i holds the allowed subsets. A taxonomy is hierarchically arranged, i.e. the subsets are partially ordered by inclusion. Usually H_i will be much smaller than the power set of D_i. A taxonomy for an attribute can explicitly and statically be selected by the user for the description language of a mining task, or dynamically and implicitly determined by a special subsearch process that generates and evaluates certain subsets of attribute values during a mining task.

Restricting the description space by taxonomies is not only important to reduce the search effort for a mining task, but also to produce descriptions which are simpler, present the subgroups on an appropriate hierarchical level, and are relevant for the application domain by avoiding faked or nonsense subsets V_i . With a taxonomy, implicit internal disjunctions are introduced. For attributes with many values, elementary selectors built with a single value and added as a further conjunct to a subgroup will often lead to a small resulting subgroup which possibly has a too small statistical significance, so that results can only be found on more general levels. And finally taxonomies are also important for the effectiveness of search algorithms: they can avoid too greedy algorithms and allow more patient search strategies (Section 5).

User defined statical taxonomies usually are global and not adjusted to an analytic question or pattern type. To generate taxonomies dynamically for a subgroup mining task, appropriate subsets of values must be found. Dependent on the type of an attribute A_i, diverse methods are used for automatically generating value subsets for A_i. Nominal, ordinal, or continuous attributes are distinguished. Sometimes background knowledge on the domain D_i can be used for a taxonomy construction (see Section 8). Most work in taxonomy construction relates to the continuous case and the discretization problem; see (Dougherty et al., 1995) for an overview.

In subgroup mining, we mainly need supervised methods that exploit the joint distribution with the target variables selected for a special mining task. Supervised discretization methods have been developed for the classification pattern type and usually deal with a binary target variable. Then discretizations shall be generated that optimize the classification accuracy of e.g. a decision tree. Unsupervised methods usually rely only on the univariate distribution of the attribute for which a taxonomy shall be derived. Quantiles and other simple methods, but also density based or clustering approaches which may also exploit all other attributes (not particularly the target attributes) are used.

Next, we distinguish global and local methods. Global methods find a taxonomy independent from a subgroup which is being expanded by a conjunctive selector. In the context of subgroup mining and mostly not homogeneous data bases, local methods are preferable. For example, a taxonomy derived in the context of expanding the local subgroup *males* could be different from a taxonomy for the subgroup *females*.

Top down methods generate taxonomies by recursive specializations of already found sets on each hierarchical level. For ordinal (and continuous) attributes, a best splitting point into two (or more) intervals is found on each level. However, recursive splitting does not necessarily find the best multi-interval split.

Bottom-up methods generalize the sets found already on a hierarchical level to create the sets on the next higher level, e.g. by merge techniques.

For subgroup mining, we are mainly interested in supervised, local methods that are adjusted to the pattern type. Bottom-up methods can be problematic, because of the small significance of the lower levels in a local context. Time efficiency is another criterium. Local taxonomy derivation is of course extremely time-consuming. Further, it is preferable to use the same evaluation framework for taxonomy finding as for the pattern type dependent evaluation of subgroups. A too sophisticated mixture of methods might be difficult to understand for an user. We have considered these requirements when implementing the taxonomy finder in *Data Surveyor*.

4   Validation of subgroup patterns

The general subgroup pattern (Section 2) can be specialized in various ways. These special analytic questions can be classified with two dimensions. At first, the type of the target variable is important. If this is a binary variable, a single percentage is analyzed for a subgroup, e.g. the percentage of good productions (or of complementary not good productions). In case of a nominal variable, a whole vector is studied, e.g. the percentages of bad, medium, and good productions. When the target variable is ordinal, the probability of a better value in the subgroup than in a reference group can be analyzed. E.g. the probability that a subgroup of productions has a better quality than all the productions. Finally, the target variable can be continuous (interval or ratio type). Then statements on the median or mean value of the variable can be inferred. If several target variables are selected, their joint distribution in a subgroup is analyzed.

The second dimension for classifying analytic questions refers to the number of studied populations which can e.g. relate to several time points or countries. When the database contains one cross section (i.e. one population of objects), the subgroup is usually compared for deviations with the whole population or some root or parent subgroup, resp. with its complementary subgroup. In case of two independent cross sections or of a panel of a population, the latter including the same objects for two (or more) time points, the change of the distribution of the target variables is analyzed for a subgroup. Next, k cross sections can be analyzed, for instance data for k time points or countries. Finally, a database includes a time series of populations when the segmentation attribute that generates the k cross sections is ordinal with equidistant values.

A pair of verification and quality functions is used to evaluate the deviation of a subgroup. The verification method operationalizes a special analytic question by a statistical test. Table 3 lists some tests according to the above classification of analytic questions that are offered in *Explora, Kepler* and *Data Surveyor*. When relying on parametrical tests in subgroup mining, a property of one of the distribution parameters of the target variable(s) in the subgroup determines the meaning (semantics) of an analytical question. For continuous target variables, nonparametrical tests are appropriate in the data mining context, because the smaller test power (adhering longer to the null hypothesis) is mostly not a problem, and the modest distribution assumptions and calculation efforts of these tests are preferable. Also the usually large sample and explorative situation in data mining favors non-parametrical tests. The verification method is used as a filter constraint to subselect pattern instances. Thus only deviations are selected that have a very low probability of being generated just by random fluctuations of the target variables.

Type of dependent variable(s)	One cross section	Two independent cross sections	k independent cross sections and time series
Binary	binomial test chi square test confidence intervals information gain	bin.test:pooled variance chi square test log odds ratio: z-scores (each with absolute / relative version)	chi square tests trend test

Nominal	chi square:
	  goodness of fit
	  independence test
	Gini diversity index
	information gain
	twoing criterium
		chi square tests
		Gini diversity index
			chi square test
			trend analysis

Ordered	ridit analysis	ridit analysis	ridits & trend analysis
Continuous	median test median-quantile test U-test H-test 1 or 2 sample t-test	median test median-quantile test U-test H-test two sample t-test	analysis of variance

Table 3. Some statistical verification tests for subtypes of the subgroup pattern

The quality function is used by the search algorithm to rank the instances. For instance, in a beam search strategy (Section 5), only the best n pattern instances according to their quality value are further expanded. The quality computation can relate on statistical and other interestingness aspects such as simplicity, usefulness, novelty. It can directly be given by the significance value or test statistic of the verification method, or by a function exploiting this significance value as one component for the final quality. A typical statistical quality function (e.g. defined by z-scores) combines several aspects of interestingness such as *strength* (deviation of parameter from a-priori value) and *generality* (size of the subgroup).

5 Search

Large scale and mostly brute force search is the core of data mining algorithms. In a sense, this approach mimics the procedure of data analysts when looking at a set of cross-tabulations to find interesting cells or when generating a sequence of statistical (e.g. regression) models. Due to the limited manual analysis capacities, a computerized brute force search can be organized more systematically and more completely to cover large parts of hypotheses spaces.

Search can be exhaustive or heuristic. An exhaustive strategy prunes only hypotheses that can not belong to the solutions, whereas heuristic strategies like hill-climbing and its extensions (tabu search, simulated annealing), beam-, tree-, or stepwise search aspire to process prospective regions of the search space, but cannot exclude that there are (better) solutions in the pruned subspaces. Often the description language cannot be restricted in a way that allows exhaustive search (e.g. by limiting the maximal number of conjunctions, applying only coarse discretizations and taxonomies). Another possibility to avoid combinatorial explosion is to apply constraints on the search space. Ideally, such constraints represent domain knowledge, preventing also from discovering many uninteresting descriptions.

Many search strategies exploit the partial order that is associated to a space of descriptions. Descriptions can be partially ordered by the generality of the intensional descriptions or by the subset relation of their extensions (set of all objects that satisfy the description). Genetic search strategies apply genetic operators like mutation and cross-over to produce a new generation of descriptions which iterates in an evolutionary process to a set of high quality descriptions. Parallel approaches distribute search onto subsearches that can be scheduled in parallel.

Search can be scheduled in several phases. In a first brute force phase, all subgroups are determined that satisfy the given constraints and search goals. In a second refinement phase, redundancies are eliminated and selected subgroups are elaborated. Redundancies relate mainly to the correlation between subgroups which may be responsible for spurious effects. Elaborations analyze the homogeneity of subgroups to avoid e.g. that not the subgroup as a whole is relevant but a subset of the subgroup. Brute force and refinement phases can be scheduled iteratively.

Search strategies iterate over two main steps: validating hypotheses and generating new hypotheses. Operating on a current population of hypotheses, neighborhood operators generate the neighbors of hypotheses, e.g. by expanding hypotheses with additional conjunctive terms, or genetic operators create a next generation of hypotheses by mutation and crossover operations. Both the validation and the generation step consist of four substeps. At first promising hypotheses are selected from the list of not yet validated, newly generated offsprings, respectively from the list of not yet further expanded, newly validated hypotheses. Then the selected hypotheses are validated by applying a statistical verification and quality computation module, resp. expanded by applying neighborhood or genetic operators creating new hypotheses. In the third substep, the newly validated or generated hypotheses are jointly evaluated, e.g. to identify solutions or check pruning possibilities. Finally, the populations of hypotheses are updated.

Search Step	Beam Search	Broad View	Best n	Patient
Select hypos for validation from list of generated, not yet validated hypotheses	all	all	all	all
validate	apply verification test and quality computation			
evaluation of validated hypos	sort successfully verified, not prunable hypos (cover constr.) by quality and put best n on list of hypos to be expanded	put not successfully verified, not prunable hypotheses (cover constr.) on list of hypos to be expanded put successfully verified hypos on result list	update list of best n hypos with successfully verified hypos. Put not prunable hypos (cover constrain, optimistic estimate) on list of hypos to be expanded	sort successfully verified, not prunable hypos by quality and put best one on list of hypos to be expanded. If no better hypo, repeat process, but eliminate all cases covered by found subgroups
update list of hypos not yet validated	not applicable: all have been validated			
select hypos to expand	all	all	all	all
expand hypos	dependent on type of expansion attribute: discretization regional clustering			eliminate 1 internal disjunction/quantile
evaluation of expanded hypos		eliminate successors of results		
update list of hypos to be expanded	not applicable: all have been expanded			

Table 4. Four simple brute force search strategies for subgroup mining

The validation and generation steps iterate until the solutions are found or the space of hypotheses is exploited. Search strategies fix the details within this general search frame, e.g. the order in which the hypotheses are evaluated, expanded and validated, the selection and pruning criteria, and the iteration, recursion or backtracking of the search. In Table 4, these steps are summarized for search strategies implemented in *Data Surveyor*. These strategies take simple decisions in most of the steps. All these strategies perform a brute force search to identify a set of hypotheses with high quality. Whereas beam search at each step only expands the best hypotheses to find more specialized, better subgroups, the broad view strategy is complementary. If a high quality subgroup is found, it is not further expanded. So subgroups can be identified that consist of a conjunction of selectors, where each selector alone is not interesting. The "best n" strategy is exhaustive, so that an efficient pruning is necessary for large hypothesis spaces. This can be achieved by a restrictive cover constraint (requiring a relatively large size of a subgroup). The optimistic estimate evaluation of a subgroup checks, if any specialization (expansion) can have a higher quality than the worst of the currently best n hypotheses.

Another aspect of a search strategy relates to its greediness. The usual general to specific search over subgroups realized by successively adding further conjuncts is very greedy, i.e. the size of the next subgroup is much reduced by a further conjunct. Especially for hill climbing strategies, this is a problem. Friedman and Fisher (1997) therefore propose a patient strategy based on a description language offering all internal disjunctions for categorical variables and (high) quantiles for continuous variables. At each specialization step, one internal disjunction is eliminated or one upper or lower quantile is taken away from the current interval for the variable. So only a small part of the objects of a current subgroup is reduced in a specialization step.

6   Navigation and Visualization

The kind and extent of user involvement into a data mining step variates dependent on applications and user preferences. Subgroup mining systems differ in the degree of autonomy that is incorporated in the system by the parameterization of decision processes and treatment of trade-offs between evaluations aspects. Involving the user interactively and iteratively in the mining process is often necessary to ensure that the mining results best serve the particular data analytic goals. Then a user centered search is incrementally scheduled, where the user assesses individual results by judging various trade-offs and compiles more or less manually a consolidated set of subgroups. These judgments on trade-offs are best made by an user if they are highly application dependent. Visualizations can help to compare diverse alternatives so that the user can select the most appropriate ones for the current application. This incremental search is directed by intermediate results and by operations of the user on interactive visualizations.

User directed incremental and iterative search can be supported by navigation operators to specify search processes that are run in subspaces of a multidimensional hypothesis space, compare their results, and redefine search tasks. Visualization of search results is important for these navigation operators. The analyst should be able to operate on the presented results to perform comparison, focusing, explanation, browsing and scheduling operations.

When relying on visualization approaches only, the user must identify the patterns (regional clusters, concurrencies of lines, emergent groups of associations) in the presentations. Because of well-established visual capabilities, it is of course much easier for an analyst to detect these patterns in the presented visualizations than in the numerical raw data. Data mining methods detect these patterns more autonomously, e.g. by searching and evaluating clusters of neighboring regions. A statistical test ensures that such a cluster is not a random result. Although the eye is quite efficient in detecting any regularities, the situation is not quite as easy. Often the user sees patterns in the visualizations that are not really statistically valid, or ignores existing patterns. Therefore a combination of data mining and visualization approaches is important.

Presentation issues deal with four aspects: how can a single pattern instance be appropriately visualized, how can a set of interconnected patterns be visualized, which interactive operations can be performed on a presentation graph, and which additional visualizations are important to explain the results, compare trade-offs and support an explorative analysis.

Patterns and sets of patterns must be presented in textual and graphical form to the user. A set of patterns can often be represented as a graph referring to the partial ordering of the subgroups (ordered by generality). Associations between subgroups imply a graph structure as well. Various operations on these graphs allow the user of a mining system to redirect a mining task, to filter or group mining results, and to browse into the data base. Thus these graphs provide an interaction medium for the user based on interactive visualization techniques.

Besides the interactivity of operations on the visualized subgroup results, the pattern specific presentation of a single instance has to be designed. Text presentations of subgroups can be simply arranged with presentation templates (compare *Explora*; Klösgen, 1992). Additionally, appropriate graphical presentations of subgroups and their deviation figures must be designed. For example, a simple share pattern (binary dependent variable, one population) can be graphically represented as a fourfold display including the confidence intervals of the share. In case of a nominal dependent variable, the set of percentages could be represented as a pie chart. However, already the application of pie charts to illustrate a single frequency distribution is heavily discussed among visualization experts, because of the limited capacities of humans to compare a set of angles. Using many pie charts to compare several subgroups and their frequency distributions for the values of a nominal variable is even more doubtful.

Additional visualizations explaining mining results support the analyst in selecting subgroups by assessing the trade-off between generality and strength. Friedman and Fisher (1997) propose a trajectory visualization of subgroups in a two dimensional generality vs. strength space. Another example relates to the multicollinearity problem, a frequency distribution of the values of an input variable for a subgroup can help to identify the correlations between subgroups. Other visualizations can uncover the overlapping degree between subgroups and explain a suppression refinement. Friedman and Fisher (1997) also propose sensitivity plots that can be used to judge the sensitivity of the hypothesis (subgroup) quality to the values of the subgroup description variables. With these plots, the overfitting problem is addressed.

7 Analysis of Change and of Time Stamped Data

Three measurements of change can be distinguished: individual, absolute, and relative change. When the data base includes the same objects for different time points (panel), individual change can be studied. Partially this can be reduced to the one cross section case by simply deriving variables that have the difference between the two time points as values. However when analyzing subgroups, a three dimensional cross table is studied (time x dependent variable x subgroup & complementary group). Thus more analysis options are available compared to the two dimensional cross tabulations used for the one cross section case. In the following we regard the case, when the cross sections are independent, i.e. include different objects.

We first consider two independent cross sections, e.g. samples for two time points. Since the two cross sections do not include the same objects, change cannot be analyzed on an individual level. For panel data (same objects), more elaborated analysis methods can be used. Again we refer to one (or several) dependent variables, for which changes in their distribution shall be found. Two approaches are possible. The first one finds subgroups, for which the distributions of the two time points are different (absolute change). A relative approach compares the differences of the distributions for the subgroup with a reference group (e.g. the whole population or the complementary subgroup). Different tests depend on the types of the dependent variable.

For a binary dependent variable we want to find subgroups, for which the shares, i.e. Bernoulli parameter $P(Y = 1)$, are different for the two cross sections. Under the null hypothesis of equal shares, the quality function based on z-scores (difference of shares divided by estimation of its standard error; pooled estimator for variance) is asymptotically $N(0, 1)$ distributed. Subgroups are selected as statistically interesting for which this test statistic rejects the null hypothesis. With this approach, we refer to absolute change, i.e. we look for subgroups that have changed. Additionally, relative change can be analyzed. Then we relate the change of a subgroup to the change of a reference group, e.g. the complementary group.

For relative change, we use as quality a z-score based test statistic that is $AN(0, 1)$ distributed under the null hypothesis of equal change in the subgroup and its complementary subgroup. We can also regard a fixed change (measured in the root group) and compare the change in the subgroup with this fixed change. Additionally, we can rely on confidence intervals. For the subgroup and its parent group, resp. its complementary group, confidence intervals for change (difference of shares) are computed and checked for overlapping.

Various other tests (and quality functions) are possible. For analyzing absolute change in a subgroup, we can apply a chi-square test of homogeneity for the cross table ($Y \times Time$) and get a quality that is approximatively chi-square distributed. A next quality is based on the odds values and the computation of z-scores for the log odds ratio. The odds values for the two time points are compared (odds-ratio). An odd value is given by the quotient of the probabilities of positive and negative values of the binary dependent variable. Finally, we analyse for a subgroup the relative difference of the shares for the two time points, e.g. $(p_2 - p_1)/(1 - p_1)$. Under the null hypothesis of no change, this relative difference is asymptotically normal distributed. The z-score is calculated as usual based on an estimation of the standard error.

For ordered dependent variables and analyzing absolute change in a subgroup, we refer to ridits to make a statement on the probability of a larger value of the ordered dependent variable for the second time point. Quality functions can be defined similarly as above. In case of continuous variables, methods based on median tests, order statistics, t-tests, and analysis of variance are used. For very skewed distributions, the mean-based approaches may cause problems. However, large samples (subgroups) allow an approximation with a normal distribution.

We finally consider the case of more than two independent cross sections, e.g. samples for several time points. Some basic questions are: Is there a variability in time? Is there a positive trend in time? A possible approach is, first to find subgroups with variability in time, and then to elaborate this pattern by subsequent more specialized analyzes for trends, etc. Variability of the shares for the m time points are identified with a chi-square test. Special tests are used to derive quality functions that measure the degree of an increase (e.g. gradient test), or of a positive trend. For ordered variables, ridits are analyzed with more complex statistical methods to deal with this case. For instance, a quality function can be defined that measures the degree of a monotonous trend for ridits. More complex methods apply GLIMS for ridits.

Another type of time dependent data is given when each object has a separate time reference, e.g. a time stamp. In this case, the time attribute is not categorical with only a few values, but continuous. A typical example is a data base with error or action logs. For instance, Web log-files can be preprocessed to three relations on users, sessions, and actions. Each action has a time stamp. Multi-relational subgroup mining methods can be run on these data to identify deviating subgroups of users, sessions, or actions.

In the context of time-stamped data, we finally shortly refer to the second general subgroup mining pattern directed to pairs of subgroups. The identification of interesting pairs of action subgroups is based on an association measure. See (Feldman et al., 1997) for a comparison of several association measures. Whereas these association measures rely on a two dimensional cross table (number of objects in the intersection of the two subgroups and their complementary parts), for time stamped data another association option is important to identify rules:

if subgroup A **then** subgroup B ($p\%$, within n time units)

This rule states that $p\%$ of the actions in subgroup A are succeeded by at least one action in subgroup B in the same session. For instance: 87% *of upload ac-*

tions submitted in long sessions by experienced users are succeeded by a comment action within an average time of 123 seconds. Several types of "successor" definitions are possible, e.g. immediate successors with no other actions in between. The time information can relate to given fixed time windows, or express an overall measure for the actions such as the mean successor time.

8   Spatial Mining

Spatial Mining is necessary when each object has a space reference captured within a spatial attribute. Several types of spatial references can be distinguished. The objects can directly represent spatial entities such as points, lines or areas, or they can indirectly be related to such entities such as persons living in areas. Specifically we deal now with the latter case where many objects are related to each spatial entity and the problem of how to use the spatial attribute for a subgroup description. For spatial attributes with a few nominal, ordered or equidistant values, the k-population patterns (Section 4) can be applied to compare k areas. We assume now, that the spatial attribute represents many non-overlapping, contiguous areas defined as polygons that cover a whole region (e.g. a country).

The background knowledge on polygons can be exploited to construct taxonomies for the spatial attribute. An admissible value subset V_i of the spatial attribute can be defined by some conditions on the neighborhood structure of regional clusters given by a subset of neighbored values. Such a geographical taxonomy finder is implemented in the *Data Surveyor* system.

To construct regional clusters, a bottom up, supervised and local subsearch for value subsets of the spatial attribute is scheduled, conditional on the outcome of a global spatial autocorrelation test. When the standard statistic for testing the target variable on independence in areas (Moran's I) indicates a departure from independent observations, the clustering subsearch is run; see (Gebhardt, 1997) for the theoretical foundations of this solution that overcomes the combinatorial explosion and randomness of regional clusters.

The separate clustering search is necessary, because the global test does not tell where and in which direction (positive or negative) the spatial deviations occur. The merge operation used in the bottom-up search relies on merging compact triplets of neighbored areas in each recursive search step. These compact triplets are derived in a preprocessing GIS operation using the polygons. The polygons are further used for geographic visualizations of the resulting clusters within a geographical map showing the spatial deviations of the target variable distribution.

9   Text Mining

Document Explorer (Feldman et al., 1997) is a system searching for patterns in document collections based on subgroup mining. Patterns relate to subgroups of documents and pairs of subgroups (Section 2). A subgroup is described by a set of terms or concepts (concept set). Each document described in the subgroup has to include all concepts of the concept set. These patterns provide knowledge on the application domain that is represented by the collection. A pattern can also be seen as a query or implying a query that, when addressed to the collection, retrieves a set of documents. Thus the data mining tools also identify interesting queries which can be used to browse the collection. The system searches for interesting concept sets and associations between concept sets, using explicit bias for capturing interestingness. The bias is provided by the user specifying syntactical, background, quality and redundancy constraints to direct the search in the vast implicit spaces of pattern instances which exist in the collection. The patterns which have been verified as interesting are structured and presented in a visual user interface allowing the user to operate on the results to refine and redirect search tasks or to access the associated documents. The system offers preprocessing tools to construct or refine a knowledge base of domain concepts and to create an internal representation of the collection which will be used by all subsequent mining operations. The source documents can be of text or multimedia type and be distributed, e.g. in Internet or Intranet.

A knowledge base includes domain knowledge about the document area. It includes a *concept DAG* (directed acyclical graph) of the relevant concepts for the domain. Several *categories* of concepts are hierarchically arranged in this *DAG*. For the application domain of Reuters newswire collection, e.g. categories correspond to countries, persons, topics, etc. with subcategories like European Union, politicians, economic indicators. Additionally, the knowledge base contains *background relations*. These are binary relations between categories such as nationality (relation between persons and countries) or export partners (between countries). In preprocessing, the knowledge base and a target database are constructed. The *target database* contains binary tuples. A tuple represents a document and the concepts being relevant for the document. All data mining operations in *Document Explorer* are operated on a derived trie structure, that is an efficient data structure to manage all aggregates existing in the target database.

A *concept set* is simply a set of concepts. A set of concepts can be seen as an intermediate concept that is given by the conjunction of the concepts of the set. E.g. the concepts "data mining" and "text analysis" define a joint concept which can be interpreted as "data mining in text data". *Frequent concept* sets are sets of concepts with a minimal *support*, i.e. all the concepts of the set must appear together in at least s documents. A *context* is given by a concept set and is used as a subselection of the document collection. Then only the documents in this subcollection are analyzed in a search task. The system derives, for example, patterns "in the context of crude oil" for the documents that contain crude oil as a phrase or are annotated by crude oil using text categorization algorithms.

A *binary relation* between concept sets is a subset of the crossproduct of the set of all concept sets. An *association* is a binary relation given by a *similarity function*. To measure the degree of connection (similarity) between two sets of concepts, we usually rely on the support of the documents in the collection, that include all the concepts of the two sets. If there is no document that contains all the concepts, then the two concept sets will have no connection (similarity = 0). If all the concepts of the two sets always appear together, the strongest connection measurable by the document collection (similarity = 1) is given. An *association rule* is a special association, defined as usual by a minimal support and *confidence*. Furthermore, the similarity of two concept sets relative to a category can be measured by comparing the conditional distributions of the concepts of the category with respect to the two concept sets.

A *keyword graph* is a pair consisting of a set of nodes and a set of edges. Each edge connects two nodes. Quality measures are calculated for each node and each edge. A node corresponds to a concept set and an edge to an element of a binary relation. Special subsets of nodes and connections can be defined, e.g. a *clique* is a subset of nodes of a keyword graph, for which all pairs of its elements are connected by an edge. A *path* connects two nodes of a keyword graph by a chain of connected nodes.

A *search task* is specified in *Document Explorer* by syntactical, background, quality and redundancy constraints for searching spaces of concept sets or of associations. The result of a search task is a group of concept sets or associations satisfying the specified constraints. These groups of results are arranged in keyword graphs offering to the user interactive operations on the nodes and edges of the graph.

10 Conclusion

Subgroup mining is a pragmatic data exploration approach that can be applied for various data analytic questions. Although subgroup mining has already reached a quite impressive development status, it is an evolving area for which a lot of problems must still be solved. These problems mainly relate to statistical validity of subgroup mining results, robustness of results, advanced description languages for time and space related variables, evaluating sets of subgroups, non-statistical interestingness evaluations, second order mining to combine subgroups found for different pattern types, and scalability to very large data bases.

References

1. R. Agrawal, H. Mannila, R. Srikant, H. Toivonen, I. Verkamo 1996. Fast Discovery of Association Rules. In Advances in Knowledge Discovery and Data Mining, eds. U. Fayyad, G. Piatetsky–Shapiro, P. Smyth, R. Uthurusamy, Cambridge, MA: MIT Press.
2. J. Dougherty, R. Kohavi, M. Sahami 1995. Supervised and unsupervised discretization of continuous features. Proceedings of 12th Internat. Conference on Machine Learning.
3. U. Fayyad, G. Piatetsky–Shapiro, P. Smyth, R. Uthurusamy (eds) 1996. Advances in Knowledge Discovery and Data Mining, Cambridge, MA: MIT Press.

4. R. Feldman, W. Klösgen, A. Zilberstein 1997. Visualization Techniques to Explore Data Mining Results for Document Collections. Proceedings of Third International Conference on Knowledge Discovery and Data Mining (KDD-97), eds. D. Heckerman, H. Mannila, D. Pregibon, Menlo Park: AAAI Press.

5. D. Fisher 1987. Knowledge Acquisition via Incremental Conceptual Clustering. Machine Learning 2.

6. J. Friedman, N. Fisher 1997. Bump Hunting in High-Dimensional Data. http://stat.stanford.edu/~jhf/ftp/prim.ps.Z

7. F.Gebhardt 1991. Choosing among Competing Generalizations. Knowledge Acquisition 3.

8. F. Gebhardt 1997. Finding Spatial Clusters. In Proceedings of the First European Symposium on Principles of KDD, eds. J. Komorowski and J. Zytkow. Berlin: Springer.

9. W. Klösgen 1992. Problems for Knowledge Discovery in Databases and their Treatment in the Statistics Interpreter Explora, Internat. Journal for Intelligent Systems vol 7(7).

10. W. Klösgen 1996. Explora: A Multipattern and Multistrategy Discovery Assistant. In Advances in Knowledge Discovery and Data Mining, eds. U. Fayyad, G. Piatetsky-Shapiro, P. Smyth, and R. Uthurusamy, Cambridge, MA: MIT Press.

11. W. Klösgen, J. Zytkow 1996. Knowledge Discovery in Databases Terminology. In Advances in Knowledge Discovery and Data Mining, eds. U. Fayyad, G. Piatetsky-Shapiro, P. Smyth, and R. Uthurusamy, Cambridge, MA: MIT Press.

12. H. Mannila, H. Toivonen, I. Verkamo 1997. Discovery of Frequent Episodes in Event Sequences. Data Mining and Knowledge Discovery, Vol. 1, No. 3.

13. A. Siebes 1995. Data Surveying: Foundations of an Inductive Query Language. Proceedings of the First International Conference on Knowledge Discovery and Data Mining (KDDM95), eds. U. Fayyad and R. Uthurusamy, Menlo Park, CA: AAAI Press.

14. S. Wrobel 1997. An Algorithm for Multirelational Discovery of Subgroups. Proceedings of the First European Symposium on Principles of KDD, eds. J. Komorowski and J. Zytkow. Berlin: Springer.

The Paradox of the Heap of Grains in Respect to Roughness, Fuzziness and Negligibility

Solomon Marcus

Romanian Academy, Mathematics
Calea Victoriei 125
Bucuresti, Romania
e-mail: smarcus@stoilow.imar.ro

In a first step, roughness and fuzziness fail to account for the type of graduality (vagueness) involved in the concept of a heap, as it is conceived in the famous *Eubulides' paradox*. One can partially bridge this gap by means of tolerance rough sets. Even in this case, a non-concordance persists between the empirical finiteness and the theoretical infinity of a heap. Another way to approach this problem could be via negligibility (be it cardinal, measure-theoretic or topological).

The paradox of the heap of grains due to Eubulides (a pupil of Euclid) consists of the impossibility to answer the question: Which is the smallest number of grains making a heap of grains ? If such a number n exists, then $n-1$ grains no longer form a heap, in contradiction with the empirical fact that a heap remains still a heap when only one grain is eliminated from it. Correspondingly, there is another empirical fact, according to which a non-heap cannot become a heap by adding to it only one grain. This is a particular way to assert that the switch from non-heap to heap is gradual. The concept of a heap does not have a sharp boundary. Eubulides' paradox is due to the fact that the question looking for an answer is based on a wrong presupposition: the concept of a heap has a sharp boundary.

Formally, given a universal set U, its subsets are of two kinds: heaps and non-heaps. These two classes are not defined intrinsically, but by their behavior in respect to some operations. As a matter of fact, it is enough to define one of them and the definition of the other follows immediately.

A heap is a non-empty subset X of U such that for any $x \in X$ the set $X - \{x\}$ is still a heap. A subset Y of U is said to be a non-heap if it is not a heap.

Proposition 1. *Given a non-heap Y and an element x in $U - Y$, the union Y' of Y and $\{x\}$ is still a non-heap.*

Proof. Accepting, by contradiction, that Y' is a heap, it follows that $Y' - \{x\} = Y$ is still a heap, in conflict with the hypothesis.

Proposition 2. *Any heap is an infinite subset of U.*

Proof. Let us suppose the existence of a finite heap X. There exists a natural number n and n elements $x(1)$, $x(2),..., x(n)$ in U, such that $X = \{x(1), x(2),...,$

L. Polkowski and A. Skowron (Eds.): RSCTC'98, LNAI 1424, pp. 19–22, 1998.
© Springer-Verlag Berlin Heidelberg 1998

$x(n)$}. Applying $n - 1$ times the definition of a heap, we infer that the sets

$$X(1) = X - \{x(1)\} \tag{1}$$
$$X(2) = X(1) - \{x(2)\} \tag{2}$$
$$\text{...} \tag{3}$$
$$X(n - 1) = X(n - 2) - \{x(n - 1)\} \tag{4}$$

are all heaps. But we have

$$X(1) = \{x(2), x(3)..., x(n)\} \tag{5}$$
$$X(2) = \{x(3), ..., x(n)\} \tag{6}$$
$$\text{...} \tag{7}$$
$$X(n - 1) = \{x(n)\} \tag{8}$$

Applying to $X(n-1)$ the definition of a heap, the set $X(n) = X(n-1) - \{x(n)\}$ should be still a heap, in contradiction with the fact that $X(n)$ is empty.

Corollary 3. *Any finite subset of U is a non-heap.*

Corollary 4. *A necessary condition for the existence of a heap is the infinity of U*

Proposition 5. *Any infinite subset of U is a heap.*

Proof. Obvious, because any infinite set remains infinite when one element is eliminated from it.

Proposition 6. *A subset of U is a heap if and only if it is infinite.*

Proof. Follows from Propositions 2 and 5.

Corollary 7. *A subset of U is a non-heap if and only if it is finite.*

Proof. Follows from Proposition 6.

Proposition 6 and Corollary 7 contradict the empirical-intuitive base of the concept of a heap; empirically, a heap is finite, although very large. So far, we have no therapy for this illness.

Given a set Z contained in U, let us try to approximate Z by means of various equivalence relations between subsets of U. Define r as a binary relation between subsets of U, such that two subsets are in relation r either if they are both finite or if they are both infinite. Obviously, r is an equivalence relation.

Following Pawlak [4], a subset of U is said to be totally e-unobservable, where e is an equivalence relation in the set $P(U)$ of all subsets of U, if its e-lower approximation is empty, while its e-upper approximation is U.

Proposition 8. *Every proper subset of U is totally r-unobservable.*

Proof. Excepting the trivial case when $Z = U$, no r-equivalence class is such that the union of its sets is contained in Z; so, the r-lower approximation of Z is the empty set. On the other hand, the r-upper approximation of Z is given by U, because among the infinite subsets of U which meet Z is U itself. Since there are only two e-quivalence classes, we reach a situation of total r-unobservability.

We will consider now the binary relation p defined as follows: two subsets A and B of U are in relation p if their symmetric difference is finite.

Proposition 9. *The binary relation p is an equivalence relation in $P(U)$.*

Proof. The relation p is obviously reflexive and symmetric. Let A p B and B p C. It follows that $A - B$, $B - A$, $B - C$, $C - B$ are all finite sets. The set $A - C$ is the union of a part contained in B, i.e., in $B - C$, which is finite, and a part disjoint from B, so contained in $A - B$, which is again finite; it follows that $A - C$ is finite. The set $C - A$ is the union of a part contained in B,i.e., in $B - A$, which is finite, and a part disjoint from B, so contained in $C - B$, which is also finite; it follows that $C - A$ is finite and the symmetric difference of A and C is finite and we have A p C, proving so the transitivity of p.

Proposition 10. *If A is a heap and B is p-equivalent to A, then B is still a heap.*

Proof. Since A is a heap, it follows from Proposition 6 that A is infinite. Since $A - B$ is finite, it follows that the common part of A and B is infinite, so B is infinite and again, in view of Proposition 6, B is a heap.

Proposition 11. *There exist sometimes two heaps that are not p-equivalent.*

Proof. If U is the set of real numbers, the set of rationals and the set of irrationals are both heaps, although they are disjoint.

Proposition 12. *Any set Z strictly contained in U is totally p-unobservable.*

Proof. Since all finite subsets of U belong to the same p-equivalence class and $U - Z$ is not empty, it follows that the p-lower approximation of Z is empty. On the other hand, the union of subsets p-equivalent to U intersects Z, because Z is contained in U, so the p-upper approximation of Z is U and Z is totally p-unobservable.

Another possibility is to introduce the similarity relation s; two subsets A and B of U are in relation s if one of the following conditions is satisfied:

1. $A = B$;
2. there is an element a in U such that either $A - \{a\} = B$ or $B - \{a\} = A$.

The relation s is reflexive and symmetric in $P(U)$, but not transitive. It is a tolerance relation in $P(U)$. We associate to each subset X of U its tolerance class $s(X)$, i.e., the union of all subsets Y of U such that X s Y. Given a set Z contained in U, let us call the s-lower approximation of Z the union $m(Z, s)$ of all $s(X)$ contained in Z; let us call the s-upper approximation of Z the union $u(Z, s)$

of all $s(X)$ intersecting Z. For details of this approach, see Marcus [1], where it is shown how an infinite sequence of approximations of Z can be obtained. See also, in this respect, Nieminen [3],Polkowski et al. [5] and Pomykala [6]. In respect to the paradox of the heap, the relevant case for a tolerance approach is obtained when U is a countable infinite set. The corresponding infinite sequence of approximations is represented by the successive finite sections. They can be organized as a Cech topology [1].

A fuzzy approach to the concept of a heap could start in the following way (in view of the already obtained results). We assimilate a heap with a mapping from $P(U)$ into $[0,1]$, where for any finite subset X of U we put $f(X) = 0$ and for any infinite subset X of U we put $f(X) = 1$ (in view of Proposition 6). In this way, we get the special case of a crisp set, in conflict with the empirical vagueness (graduality) of a heap. The only way to avoid this failure is to follow an itinerary parallel to the tolerance rough set approach. For instance, let us put $f(\emptyset) = 0$, $f(X) = n/(n+1)$ when X is a finite subset of U of cardinal n and $f(X) = 1$ when X is an infinite subset of U.

Another fuzzy approach could follow the way proposed by Mares [2] namely by means of what he calls a trapezoidal fuzzy quantity and a triangular fuzzy quantity; maybe in this way we could bridge the gap between the theoretical and the empirical aspects of our problem, by avoiding to work with infinite sets.

The idea of a heap exploits the elementary fact that finite sets are negligible in respect to infinite sets, in the same sense in which zero is negligible in respect to the addition of real numbers. We can extend this perspective, by defining cardinally generalized heaps. Any set of transfinite cardinal a is a heap in respect to all sets of cardinal smaller than a. In a measure-theoretic perspective, sets of strictly positive measure are heaps in respect to sets of measure zero, while in a topological perspective sets of second Baire category are heaps in respect to sets of first Baire category.

The baldness paradox could be investigated following similar methods, but the main difficulties still remain unsolved.

References

1. Marcus, Solomon. Tolerance rough sets, Cech topology, learning processes. Bulletin of the Polish Acad. Sci. Technical Sci. 42(1994)3, 471–487.
2. Mares, Milan. Fuzzy zero, algebraic equivalence: yes or not ? Kybernetika 32(1996)4, 343–351.
3. Nieminen, J. Rough tolerance equality. Fundam. Inf. 11(1988), 289–296.
4. Pawlak, Zdzislaw. Vagueness - a rough set view. Lecture Notes in Computer Science 1261 (J. Mycielski, G. Rozenberg, A. Salomaa, eds.) "Structures in Logic and Computer Science - A Selection of Essays in Honor of A. Ehrenfeucht. Berlin-New York: Springer, 1997, 106–117.
5. Polkowski, L., A. Skowron, J. Zytkov. Tolerance based rough sets, in "Soft Computing (T.Y.Lin, A.M. Wildberger, eds.), "Simulations Councils", San Diego, 1995, 55–58.
6. Pomykala, J.A. On definability in the non-deterministic information systems. Bulletin of the Polish Acad. Sci., Math. 36(1988), 193–210.

Rough Sets - What Are They About?

V. Wiktor Marek and Mirosław Truszczyński

Department of Computer Science
University of Kentucky
Lexington, KY 40506-0046, USA
{marek, mirek}@cs.engr.uky.edu, Phone: +(606) 257-3961,
Fax: +(606) 323-1971.

We discuss philosophical and metamathematical origins of rough sets and their fundamental properties. We argue that rough sets are necessary in the light of the Platonian concept of ideal mathematical objects. We show how rough sets have been present in concept formation, diagnosis, classification and other reasoning tasks. We present examples indicating that the intuitive idea of a rough set has been used (under various names) by physicians, engineers and philosophers as a basic tool to classify and utilize concepts in their respective domains of activity. We discuss the differences between rough sets and other approaches to incomplete and imprecise information such as fuzzy logic and logics that formalize the process of "jumping to conclusions".

In a more formal part of out presentation, we discuss the connection of rough sets with three- and four- valued logics (relevance logics). We show how equivalence relations and related notions of rough sets generate three-valued and four-valued approximations of relational systems. We prove monotonicity results for such approximations as well as preservation theorems.

We discuss computational tasks associated with rough sets such as minimal discerning set of attributes selection and its weighted version. We present complexity results and some algorithms.

L. Polkowski and A. Skowron (Eds.): RSCTC'98, LNAI 1424, pp. 24–24, 1998.
© Springer-Verlag Berlin Heidelberg 1998

Reasoning about Data – A Rough Set Perspective

Zdzisław Pawlak

Institute of Theoretical and Applied Informatics
Polish Academy of Sciences
ul. Baltycka 5, 44 000 Gliwice, Poland
e-mail:zpw@ii.pw.edu.pl

[**Abstract.**] **The paper contains some considerations concerning the relationship between decision rules and inference rules from the rough set theory perspective. It is shown that decision rules can be interpreted as a generalization of the** *modus ponens* **inference rule, however there is an essential difference between these two concepts. Decision rules in the rough set approach are used to describe dependencies in data, whereas** *modus ponens* **is used in general to derive conclusions from premises.**

1 Introduction

Data analysis, recently known also as data mining, is, no doubt, a very important and rapidly growing area of research and applications. Historically, data mining methods were first based on statistics, but it is worth mentioning that their origin can be traced back to some ideas of Bertrand Russell and Karl Popper concerning reasoning about data. Recently machine intelligence and machine learning contributed essentially to this domain. Particularly fuzzy sets, rough sets, genetic algorithms, neural networks, cluster analysis and other branches of AI can be considered as a basic tools for knowledge discovery in databases, nowadays.

Main objective of data analysis is finding hidden patterns in data. More specifically, data analysis is about searching for dependencies, or in other words, pursuing "cause-effect" relations, in data.

From logical point of view, data analysis can be perceived as a part of inductive reasoning, and therefore it can be understood as a kind of reasoning about data methods, with specific inference tools.

Reasoning methods are usually classified into three classes: deductive, inductive and common sense reasoning.

Deductive methods are based on axioms and deduction rules, inductive reasoning hinges on data and induction rules, whereas common sense reasoning is based on common knowledge and common sense evident inferences from the knowledge.

Deductive methods are used exclusively in mathematics, inductive methods – in natural sciences, e.g., physics, chemistry etc., while common sense reasoning

L. Polkowski and A. Skowron (Eds.): RSCTC'98, LNAI 1424, pp. 25–34, 1998.

is used in human sciences, e.g., politics, medicine, economy, etc. but mainly this method is used almost everywhere in every day life debates, discussions and polemics.

This paper shows that the rough set approach to data analysis bridges somehow the deductive and inductive approach in reasoning about data. The rough set reasoning is also, to some extent, related to common sense reasoning.

Rough set theory gave rise to extensive research in deductive logic, and various logical systems, called rough logics, have been proposed and investigated (see e.g., [3, 6, 7, 10, 11, 18, 20, 23]). However, the basic idea of rough set based reasoning about data is rather of inductive than deductive character. Particularly interesting in this context is the relationship between an implication in deductive logic and a decision rule in the rough set approach.

In deductive logic basic rule of inference, *modus ponens* (*MP*) is based on implication, which can be seen as counterpart of a decision rule in decision rule based methods of data analysis. Although formally decision rules used in the rough set approach are similar to *MP* rule of inference, they play different role to that of *MP* inference rule in logical reasoning. Deduction rules are used to derive true consequences from true premises (axioms), whereas decision rules are description of total or partial dependencies in databases. Besides, in inductive reasoning optimization of decision rules is of essential importance, but in deductive logic we don't need to care about optimization of implications used in reasoning. Hence, implications and decision rules, although formally similar, are totally different concepts and play various roles in both kinds of reasoning methods. Moreover decision rules can be also understood as exact or approximate description of decisions in terms of conditions.

It is also interesting to note a relationship between rough set based reasoning and common sense reasoning methods. Common sense reasoning usually starts from common knowledge shared by domain experts. In the rough set based reasoning the common knowledge is not assumed but derived from data about the domain of interest. Thus the rough set approach can be also seen as a new approach to (common) knowledge acquisition. Also the common rules of inference can be understood in our approach as data explanation methods. Note, that qualitative reasoning, part of common sense reasoning, can be also explained in the rough set philosophy.

Summing up, rough set based reasoning has an overlap with deductive, inductive and common sense reasoning, however it has its own specific features and can be considered in its own right.

2   Data, Information Systems and Decision Tables

Starting point of rough set theory is a set of data (information) about some objects of interest. Data are usually organized in a form of a table called *information system* or *information table*.

A very simple, fictitious example of an information table is shown in Table 1. The table describes six cars in terms of their (attributes) features such as *fuel consumption* (F), *perceived quality* (Q), *selling price* (P) and *marketability* (M).

Table 1. An example of information system

Car	F	Q	P	M
1	high	fair	med.	poor
2	v. high	good	med.	poor
3	high	good	low	poor
4	med.	fair	med.	good
5	v. high	fair	low	poor
6	high	good	low	good

Our main problem can be characterized as determining the nature of the relationship between selected features of the cars and their marketability. In particular, we would like to identify the main factors affecting the market acceptance of the cars.

Information systems with distinguished decision and condition attributes are called *decision tables*.

Each row of a decision table determines a *decision rule*, which specifies *decisions* (*actions*) that should be taken when conditions pointed out by *condition* attributes are satisfied. For example in Table 1 the condition $(F,high)$, $(Q,fair)$, (P,med) determines uniquely the decision $(M,poor)$. Decision rules 3) and 6) in Table 1 have the same conditions but different decisions. Such rules are called *inconsistent* (*nondeterministic, conflicting, possible*); otherwise the rules are referred to as *consistent* (*certain, deterministic, nonconflicting, sure*). Decision tables containing inconsistent decision rules are called *inconsistent* (*nondeterministic, etc*); otherwise the table is *consistent* (*deterministic, etc*).

The number of consistent rules to all rules in a decision table can be used as *consistency factor* of the decision table, and will be denoted by $\gamma(C, D)$, where C and D are condition and decision attributes respectively. Thus if $\gamma(C, D) < 1$ the decision table is consistent and if $\gamma(C, D) \neq 1$ the decision table is inconsistent. For example for Table 1 $\gamma(C, D) = 4/6$.

In what follows information systems will be denoted by $S = (U, A)$, where U – is *universe*, A is a set of *attributes*, such that for every $x \in U$ and $a \in A$, $a(x) \in V_a$, and V_a is the domain (set of values of a) of a.

3   Decision Rules and Certainty Factor

Decision rules are often presented as implications and are called "*if... then...*" rules. For example, Table 1 determines the following set of implications:

1) *if (F,high) and (Q,fair) and (P,med) then (M,poor)*,

2) *if (F,v.high) and (Q,good) and (P,med) then (M,poor)*,

3) *if (F,high) and (Q,good) and (P,low) then (M,poor)*,

4) *if (F,med.) and (Q,fair) and (P,med.) then (M,good)*,

5) *if (F,v.high) and (Q,fair) and (P,low.) then (M,poor)*,

6) *if (F,high) and (Q,good) and (P,low) then (M,good)*,

In general decision rules are implications built up from elementary formulas (attribute name, attribute value) and combined together by means of propositional connectives "and", "or" and "implication" in a usual way.

Let Φ and Ψ be logical formulas representing conditions and decisions, respectively and let $\Phi \to \Psi$ be a decision rule, where Φ_S denote the meaning of Φ in the system S, i.e., the set of all objects satisfying Φ in S, defined in a usual way.

With every decision rule $\Phi \to \Psi$ we associate a number, called a *certainty factor* of the rule, and defined as

$$\mu_S(\Phi, \Psi) = \frac{|\Phi_S \cap \Psi_S|}{|\Phi_S|},$$

where $|\Phi|$ denotes the cardinality of Φ. Of course $0 \le \mu_S(\Phi, \Psi) \le 1$; if the rule $\Phi \to \Psi$ is consistent then $\mu_S(\Phi, \Psi) = 1$, and for inconsistent rules $\mu_S(\Phi, \Psi) < 1$. For example, the certainty factor for decision rule 2) is 1, and for decision rule 3) is 0.5.

The certainty factor can be interpreted as a conditional probability of a decision Ψ given the probability of the condition Φ.

It is worth mentioning that association of conditional probability with implication first was proposed by J. Lukasiewicz in the context of multivalued logic and probabilistic logic [4]. This idea has been pursued by other logicians years after [1]. In the rule based knowledge systems many authors also proposed using conditional probability to characterize certainty of the decision rule [2]. In particular in the rough set approach association of condition probabilities with decision rules have been pursued e.g., in [21, 24, 27].

Now the difference between use of implications in classical logic and in data analysis can be clearly seen, particularly in the rough set framework. Implication in deductive logic is used to draw conclusions from premises, by means of *modus ponens* rule of inference. In reasoning about data implications are decision rules used to describe patterns in data. Hence, the role of implications in both cases is completely different. Besides, *modus ponens* is an universal rule of inference valid in any logical system, but decision rules are strictly associated with a specific data and are not valid universally.

However in the rough set approach decision rules can be also exploited in a similar way as *modus ponens* in logic. Let us consider the following formula:

$$\pi_S(\Psi) = \Sigma(\pi_S(\Phi) \cdot \mu_S(\Phi, \Psi)) = \Sigma\pi_S(\Phi \wedge \Psi) \qquad (*)$$

where Σ is taken over all conditions Φ associated with the decision corresponding to Ψ, and $\pi_S(\Phi) = \frac{|\Phi_S|}{|U|}$.

$\pi_S(\Phi)$ is a probability that the condition Φ is satisfied in S. Thus formula (*) shows the relationship between the probability of conditions, certainty factor of a decision rule and the probability of decisions.

Hence the formula (*) allows to compute probability that the decision Ψ is satisfied in S, in terms of the probability of condition Φ and conditional probability of the decision rule $\Phi \Rightarrow \Psi$.

This is a kind of analogous structure to *modus ponens* inference rule and can be treated as its generalization, called *rough modus ponens* (*RMP*) [15]. The certainty factor of a decision rule can be seen as generalization of the rough membership function. It can be also understood as a rough inclusion factor in rough mereology [16, 17] or as a degree of truth of the implication associated with the inclusion.

4 Approximations of Sets

The main problem discussed in the previous section can be also formulated as follows: can we uniquely describe well (poorly) selling cars in terms of their features. Of course, as before, this question cannot be answered uniquely, since cars 3 and 6 have the same features but car 3 sells poorly whereas car 6 sells well, hence we are unable to give unique description of cars selling well or poorly.

But one can observe that in view of the available information we can state that cars 1, 2 and 5 *surely* belong to the set of cars which are selling poorly, whereas cars 1, 2, 3, 5 and 6 *possibly* belong, to the set of cars selling poorly, i.e. cannot be excluded as cars selling poorly. Similarly car 4 surely belongs to well selling cars, whereas cars 3, 4 and 6 possible belong to well selling cars. Hence, because we are unable to give an unique characteristic of cars selling well (poorly), instead we propose to use of two sets, called the *lower* and the *upper approximation* of the set of well (poorly) selling cars.

Now, let us formulate the problem more precisely.

Any subset B of A determines a binary relation I_B on U, which will be called an *indiscernibility relation*, and is defined as follows: xI_By if and only if $a(x) = a(y)$ for every $a \in B$, where $a(x)$ denotes the value of attribute a for element x. Obviously I_B is an equivalence relation. The family of all equivalence classes of I_B, i.e., the partition determined by B, will be denoted by U/I_B, or simply U/B; an equivalence class of I_B, i.e., the block of the partition U/B, containing x will be denoted by $B(x)$.

If (x, y) belongs to I_B we will say that x and y are *B-indiscernible*. Equivalence classes of the relation I_B (or blocks of the partition U/B) are referred to as *B-elementary concepts* or *B-granules*. As mentioned previously in the rough set approach the elementary concepts are the basic building blocks (concepts) of our knowledge about reality.

The indiscernibility relation will be used next to define basic concepts of rough set theory. Let us define now the following two operations on sets

$$B_*(X) = \{x \in U : B(x) \subseteq X\},$$

$$B^*(X) = \{x \in U : B(x) \cap X \neq \emptyset\},$$

assigning to every subset X of the universe U two sets $B_*(X)$ and $B^*(X)$ called the B-lower and the B-*upper approximation* of X, respectively. The set

$$BN_B(X) = B^*(X) - B_*(X)$$

will be referred to as the B-*boundary* region of X. If the boundary region of X is the empty set, i.e., $BN_B(X) = \emptyset$, then the set X is *crisp* (exact) with respect to B; in the opposite case, i.e., if $BN_B(X) \neq \emptyset$, the set X is referred to as *rough* (*inexact*) with respect to B.

For example, the lower approximation of the set $\{1,2,3,5\}$, of poorly selling cars, is the set $\{1,2,5\}$, whereas the upper approximation of poorly selling cars is the set $\{1,2,3,5,6\}$. The boundary region is the set $\{3,6\}$. That means that cars 1, 2 and 5 can be surely classified, in terms of their features, as poorly selling cars, while cars 3 and 6 cannot be characterized, by means of available data, as selling poorly or not. Rough sets can be also defined using a *rough membership function*, defined as

$$\mu_X^B(x) = \frac{|X \cap B(x)|}{|B(x)|}, \text{ and } \mu_X^B(x) \in [0,1].$$

Value of the membership function $\mu_X(x)$ is kind of conditional probability, and can be interpreted as a degree of *certainty* to which x belongs to X (or $1 - \mu_X(x)$, as a degree of *uncertainty*).

For example, car 1 belongs to the set $\{1,2,3,5\}$ of cars selling poorly with the conditional probability 1, whereas car 3 belongs to the set with conditional probability 0.5.

5   Dependency of Attributes

Our main problem can be rephrased as whether there is a functional dependency between the attribute M and attributes F, Q and P. In other words we are asking whether the value of the decision attribute is determined uniquely by the values of the condition attributes. It is easily seen that this is not the case for the example since cars 3 and 6 have the same values of condition attributes but different value of decision attribute. The consistency factor $\gamma(C, D)$ can be also interpreted as a *degree of dependency* between C and D. We will say that D *depends on* C *in a degree* k $(0 \leq k \leq 1)$, denoted $C \Rightarrow_k D$, if $k = \gamma(C, D)$.

If $k = 1$ we say that D *depends totally* on C, and if $k < 1$, we say that D *depends partially* (in a *degree k*) on C.

For example, for dependency $\{F, P, Q\} \Rightarrow \{M\}$ we get $k = 4/6 = 2/3$.

Dependency of attributes can be also defined using approximations as shown below.

We will say that D depends on C in a degree $k(0 \leq k \leq 1)$, denoted $C \Rightarrow_k D$, if

$$k = \gamma(C, D) = \frac{|POS_C(D)|}{|U|}, \text{ where } POS_C(D) = \bigcup_{X \in U/D} C_*(X),$$

called a *positive region* of the partition U/D with respect to C, is the set of all elements of U that can be uniquely classified to blocks of the partition U/D, by means of C. Obviously

$$\gamma(C, D) = \sum_{X \in U/D} \frac{|C_*(X)|}{|U|}.$$

If $k = 1$ we say that D *depends totally* on C, and if $k < 1$, we say that D *depends partially* (in a *degree k*) on C.

The coefficient k expresses the ratio of all elements of the universe, which can be properly classified to block of the partition U/D, employing attributes C.

6   Reduction of Attributes

We often face a question whether we can remove some data from a data table preserving its basic properties, that is – whether a table contains some superfluous data. This can be formulated as follows.

Let $C, D \subseteq A$, be sets of condition and decision attributes, respectively. We will say that $C' \subseteq C$ is a *D-reduct* (reduct with *respect* to D) of C, if C' is a minimal subset of C such that

$$\gamma(C, D) = \gamma(C', D).$$

Thus reduct enables us to make decisions employing minimal number of conditions.

For example, for Table 1 we have two reducts F, Q and F, P. It means that instead of Table 1 we can use either Table 2 or Table 3, shown below.

These simplifications yield to the following sets of decision rules. For Table 2 we get

   1) *if (F,high) and (Q,fair) then (M,poor)*,

   2) *if (F,v.high) and (Q,good) then (M,poor)*,

Table 2. Reduced information system

Table 3. Another reduced information system

Car	F	Q	M
1	high	fair	poor
2	v. high	good	poor
3	high	good	poor
4	med.	fair	good
5	v. high	fair	poor
6	high	good	good

Car	F	P	M
1	high	med.	poor
2	v. high	med.	poor
3	high	low	poor
4	med.	med.	good
5	v. high	low	poor
6	high	low	good

3) *if* (*F,high*) *and* (*Q,good*) *then* (*M,poor*),

4) *if* (*F,med.*) *and* (*Q,fair*) *then* (*M,good*),

5) *if* (*F,v.high*) *and* (*Q,fair*) *then* (*M,poor*),

6) *if* (*F,high*) *and* (*Q,good*) *then* (*M,good*),

and for Table 3 we have

7) *if* (*F,high*) *and* (*P,med*) *then* (*M,poor*),

8) *if* (*F,v.high*) *and* (*P,med.*) *then* (*M,poor*),

9) *if* (*F,high*) *and* (*P,low*) *then* (*M,poor*),

10) *if* (*F,med.*) *and* (*P,med.*) *then* (*M,good*),

11) *if* (*F,v.high*) *and* (*P,low.*) *then* (*M,poor*),

12) *if* (*F,high*) *and*
(*P,low*) *then* (*M,good*).

Hence, employing the notion of the reduct we can simplify the set of decision rules.

7   Conclusions

Using rough sets to reason about data hinges on three basic concepts of rough set theory: approximations, decision rules and dependencies. All these three notions are strictly connected and are used to express our imprecise knowledge about reality, represented by data obtained from measurements, observations or from knowledgeable expert.

The rough set approach to reasoning about data bridges to some extent the deductive and inductive way of reasoning. Decision rules in this approach can be understood as implications, whose degree of truth is expressed by the certainty factor. Consequently, this leads to generalization of the *modus ponens* inference rule, which in the rough set framework has a probabilistic flavor. It is interesting that the certainty factor of a decision rule is closely related to the rough membership function and to rough inclusion of sets, basic concept of rough mereology.

8 Acknowledgments

Thanks are due to Prof. Andrzej Skowron and Dr. Marzena Kryszkiewicz for critical remarks.

References

1. Adams, E. W.: The logic of Conditionals, An Application of Probability to Deductive Logic. D. Reidel Publishing Company, Dordrecht, Boston (1975)
2. Bandler, W. Kohout, L.: Fuzzy power sets and fuzzy implication operators. Fuzzy Sets and Systems **4** (1980) 183–190
3. Banerjee, M., Chakraborty, M.K.: Rough logics: A survay with further directions. In: E. Orłowska (ed.): Incomplete information: Rough set analysis. Physica–Verlag, Heidelberg (1997) 579–600
4. Borkowski, L. (ed.): Jan Łukasiewicz – Selected Works. North Holland Publishing Company, Amsterdam, London, Polish Scientific Publishers, Warszawa (1970)
5. Dempster, A. P.: Upper and lower probabilities induced by induced by the multiplevalued mapping. Ann. Math. Statistics **38** (1967) 325–339
6. Demri, S., Orłowska, E.: Logical analysis of indiscernibility. Institute of Computer Science, Warsaw University of Technology, ICS Research Report **11/96** (1996); see also: E. Orłowska (ed.): Incomplete information: Rough set analysis. Physica–Verlag, Heidelberg (1997) 347–380
7. Gabbay, D., Guenthner, F.: Handbook of Philosophical Logic Vol.1, Elements of Classical Logic, Kluwer Academic Publishers, Dordrecht, Boston, London (1983)
8. Łukasiewicz, J.: Die logischen Grundlagen der Wahrscheinlichkeitsrechnung. Krakow (1913)
9. Magrez, P., Smets, P.: Fuzzy modus ponens: A new model suitable for applications in knowledge-based systems. Information Journal of Intelligent Systems **4** (1975) 181–200
10. Orłowska, E.: Modal logics in the theory of information systems. Zeitschrift für Mathematische Logik und Grundlagen der Mathematik **30** (1984) 213–222
11. Pagliani, P.: Rough set theory and logic–algebraic structures. In: E. Orłowska (ed.): Incomplete information: Rough set analysis. Physica–Verlag, Heidelberg (1997) 109–190
12. Pawlak, Z.: Rough probability. Bull. Polish Acad., Sci. Tech. **33(9-10)** (1985) 499–504
13. Pawlak, Z.: Rough set theory and its application to data analysis. Systems and Cybernetics (to appear)

14. Pawlak, Z.: Granularity of knowledge, indiscernibility and rough sets. IEEE Conference on Evolutionary Computation (1998) 100–103
15. Pawlak, Z.: Rough Modus Ponens. IPMU'98 Conference, Paris (1998)
16. Polkowski, L., Skowron, A.: Rough Mereology. Proc. of the Symphosium on Methodologies for Intelligent Systems **869** (1994) 85–94, Charlotte, N.C., Lecture Notes in Artificial Intelligence, Springer Verlag
17. Polkowski, L., Skowron, A.: Rough Mereology: A New Paradigm for Approximate Reasoning. Journ. of Approximate Reasoning **15(4)** (1996) 333–365
18. Rasiowa, H., Marek, W.: Approximating sets with equivalence relations. Theoret. Comput. Sci. **48** (1986) 145–152
19. Rasiowa, H., Skowron, A.: Rough concepts logic. In: A. Skowron (ed.), Computation Theory, Lecture Notes in Computer Science **208** (1985) 288–297
20. Rauszer, C.: A logic for indiscernibility relations. In: Proceedings of the Conference on Information Sciences and Systems, Princeton University (1986) 834–837
21. Skowron, A.: Management of uncertainty in AI: A rough set approach. In: V. Alagar, S. Belgrer and F.Q. Dong (eds.) Proc. SOFTEKS Workshop on Incompleteness and Uncertainty in Information Systems, Springer Verlag and British Computer Society (1994) 69–86
22. Trillas, E., Valverde, L.: On implication and indistinguishability in the setting of fuzzy logic. Management Decision Support Systems Using Fuzzy Sets and Possibility Theory, Verlag TÜ (1985) 198–212
23. Vakarelov, D.: A modal logic for similarity relations in Pawlak knowledge representation systems. Fundamental Informaticae **15** (1991) 61–79
24. Wong, S.K.M., Ziarko, W.: On learning and evaluation of decision rules in the context of rough sets. Proceedings of the International Symposium on Methodologies for Intelligent Systems (1986) 308–224
25. Zadeh, L.: Fuzzy sets as a basis for a theory of possibility. Fuzzy Sets and Systems **1** (1977) 3–28
26. Zadeh, L.: The role of fuzzy logic in in the management of uncertainty in expert systems. Fuzzy Sets and Systems **11** (1983) 199–277
27. Ziarko, W., Shan, N.: KDD-R: A comprehensive system for knowledge discovery using rough sets. Proceedings of the International Workshop on Rough Sets and Soft Computing (RSSC'94) 164–173, San Jose (1994); see also: T. Y. Lin and A. M. Wildberger (eds.), Soft Computing, Simulation Councils, Inc. (1995) 298–301

Information Granulation and its Centrality in Human and Machine Intelligence

Lotfi A. Zadeh

Professor in the Graduate School and Director
Berkeley Initiative in Soft Computing (BISC).
Computer Science Division and the Electronics Research Laboratory
Department of EECS, University of California, Berkeley, CA 94720-1776
Telephone: 510-642-4959; Fax: 510-642-1712
E-mail: zadeh@cs.berkeley.edu

Abstract

In our quest for machines which are capable of performing non-trivial human tasks, we are developing a better understanding of the centrality of information granulation in human cognition, human reasoning and human decision-making. In many contexts, information granulation is a reflection of the finiteness of human ability to resolve detail and store information. In many other contexts, granulation is employed to solve a complex problem by partitioning it into simpler subproblems. This is the essence of the strategy of divide and conquer. What is remarkable is that humans are capable of performing a wide variety of tasks without any measurements and any computations. A familiar example is the task of parking a car. For a human it is an easy task so long as the final position of the car is not specified precisely. In performing this and similar tasks, humans employ their ability to exploit the tolerance for imprecision to achieve tractability, robustness and low solution cost. What is important to recognize is that this essential ability is closely linked to the modality of granulation and, more particularly, to information granulation.

In a very broad sense, granulation involves partitioning of whole into parts. In more specific terms, granulation involves partitioning a physical or mental object into a collection of granules, with a granule being a clump of objects (points) drawn together by indistinguishability, similarity, proximity or functionality. Granulation may be physical or mental; dense or sparse; and crisp or fuzzy, depending on whether the boundaries of granules are or are not sharply defined.

Modes of information granulation (IG) in which granules are crisp play important roles in a wide variety of methods, approaches and techniques. Among them are: interval analysis, quantization, chunking, rough set theory, diakoptics, divide and conquer, Dempster-Shafer theory, machine learning from examples, qualitative process theory, decision trees, semantic networks, analog-to-digital

---

* Research supported in part by NASA Grant NCC 2-275, ONR Grant N00014-96-1-0556, LLNL Grant 442427-26449, ARO Grant DAAH 04-961-0341, and the BISC Program of UC Berkeley.

L. Polkowski and A. Skowron (Eds.): RSCTC'98, LNAI 1424, pp. 35–36, 1998.

conversion, constraint programming, image segmentation, cluster analysis and many others.

Important though it is, crisp IG has a major blind spot. More specifically, it fails to reflect the fact that in much – perhaps most – of human reasoning and concept formation the granules are fuzzy rather than crisp. For example, the fuzzy granules of a human head are the nose, ears, forehead, hair, cheeks, etc. Each of the fuzzy granules is associated with a set of fuzzy attributes, e.g., in the case of hair, the fuzzy attributes are color, length, texture, etc. In turn, each of the fuzzy attributes is associated with a set of fuzzy values. For example, in the case of the fuzzy attribute Length(hair), the fuzzy values are long, short, not very long, etc. The fuzziness of granules, their attributes and their values is characteristic of the ways in which human concepts are formed, organized and manipulated. In effect, fuzzy information granulation (fuzzy IG) may be viewed as a human way of employing data compression for reasoning and, more particularly, making rational decisions in an environment of imprecision, uncertainty and partial truth.

In fuzzy logic, the machinery of fuzzy information granulation – based on the concepts of a linguistic variable, fuzzy if-then rule and fuzzy graph has long played a key role in most of its applications. However, what is emerging now is a much more general theory of information granulation which goes considerably beyond its place in fuzzy logic. This more general theory leads to two linked methodologies – granular computing (GrC) and computing with words (CW).

In CW, words play the role of labels of granules and the initial and terminal datasets are assumed to consist of propositions expressed in a natural language. The input interface serves to translate from a natural language (NL) to a generalized constraint language (GCL), while the output interface serves to re-translate from GLC to NL. Internally, granular computing is employed to propagate constraints from premises to conclusions.

The importance of the methodologies of granular computing and computing with words derives from the fact that they make it possible to conceive and design systems which achieve high MIQ (Machine Intelligence Quotient) by mimicking the remarkable human ability to perform complex tasks without any measurements and any computations.

Although GrC and CW are intended to deal with imprecision, uncertainty and partial truth, both are well-defined theories built on a mathematical foundation. In coming years, they are likely to play an increasingly important role in the conception, design, construction and utilization of information/intelligent systems.

Classification Strategies
Using Certain and Possible Rules *

Jerzy W. Grzymala-Busse and Xihong Zou

Department of Electrical Engineering and Computer Science
University of Kansas
Lawrence, KS 66045, USA
jerzy@eecs.ukans.edu

Abstract. A typical real-life data set is affected by inconsistencies—cases characterized by the same attribute values are classified as members of different concepts. The most apparent methodology to handle inconsistencies is offered by rough set theory. For every concept two sets are computed: the lower approximation and the upper approximation. From these two sets a rule induction system induces two rule sets: certain and possible.

The problem is how to use these two sets in the process of classification of new, unseen cases. For example, should we use only certain rules (or only possible rules) for classification? Should certain rules be used first and, when a case does not match any certain rule, should possible rules be used later? How to combine certain and possible rules with complete and partial matching of rules by a case? This paper presents experiments that were done to answer these questions. Different strategies were compared by classifying ten real-life data sets, using the error rate as a criterion of quality.

1   Introduction

The main idea of knowledge discovery is to look for regularities in the raw data describing some real-life phenomena. Such regularities are often presented in the form of if-then rules [8]. For example, a data set describes patients diagnosed by a clinician on the basis of many attributes (tests done in a laboratory, questions asked by a physician, etc.). The resulting rule set is put into an expert system and is used for diagnosis of new patients.

In experiments presented in this paper a rule induction system called LERS (Learning from Examples based on Rough Sets) was used. LERS was studied in many papers, see, e.g., [3]. Other rule induction systems, based on rough set theory, were described in [11], [12], [14], and [15]. Usually, training data sets, i.e., data sets used for rule induction, are inconsistent. For example, two patients described by the same values of all attributes are diagnosed as members

---

* This work was supported in part by the Committee for Scientific Research, Warsaw, Poland, Grant No. 8T11C00512

L. Polkowski and A. Skowron (Eds.): RSCTC'98, LNAI 1424, pp. 37–44, 1998.

of two different concepts (e.g., one may be sick and the other may be healthy). LERS uses an approach to inconsistent data sets based on rough sets [9] and [10]. First, LERS checks training data for consistency. If data are inconsistent, for every concept two sets are computed: lower approximation and upper approximation [2] and [3]. Then rule sets are induced separately from both sets. In our experiments, option LEM2 (Learning from Examples, version 2) was chosen to induce rule sets. Rules induced from lower approximations are called *certain*, while rules induced from upper approximations are called *possible*. The terminology, introduced in [2], is based on the following observation: if a case is a member of the lower approximation of the concept, it is *certainly* a member of the concept. Similarly, if a case is a member of the upper approximation, it is only *possibly* a member of the concept.

The question is how to use both rule sets for classification of new cases, members of a testing data set. For example, how to use these two rule sets for classification of new patients, which rule induction system was not aware of. Besides, a classification system of LERS has four parameters that may be set up by the user. Thus the user may select many strategies for classification. Our objective was to determine the ranking of these strategies.

The standard process of LERS classification system may use four parameters: strength, specificity, matching factor and support. In our experiments, classification was performed using four options for choosing rule sets, two options for using specificity, and two options for using matching. Based on these different combinations, sixteen classification strategies were developed. A set of experiments was conducted: the sixteen strategies were tested on ten real-world data sets. The performance of the sixteen strategies were measured by the error rates for each strategy and each data set.

2   LERS classification scheme

The process of classification used in LERS has four factors: *Strength, Specificity, Matching_factor,* and *Support.* The original approach was introduced under the name of *bucket brigade algorithm,* see [1] and [7]. In this approach, the classification of a case is based on three factors: strength, specificity, and support. The additional factor, used for partial matching, was added to LERS [4]. In the bucket brigade algorithm partial matching is not used at all. These four factors are defined as following:

Strength is a measure of how well the rule performed during training. It is the number of cases correctly classified by the rule in training data. The bigger strength is, the better. Strength was used in all sixteen strategies, following a recommendation from [5].

Specificity is a measure of completeness of a rule. It is the number of conditions (attribute-value pairs) of a rule. It means a rule with a bigger number of attribute-value pairs is more specific. Specificity may or may not be used to classify cases. In our experiments, both options were used: using specificity and not using specificity.

For a specific case, if complete matching, where all attribute-value pairs of at least one rule match all attribute-value pairs of a case is impossible, LERS tries partial matching. During partial matching all rules with at least one match between the attribute-value pairs of a rule and the attribute-value pairs of a case are identified. *Matching_factor* is a measure of matching of a case and a rule. Matching_factor is defined as the ratio of the number of matched attribute-value pairs of a rule with a case to the total number of attribute-value pairs of the rule. In our experiments, two options of matching were used: using complete matching first, then using partial matching if necessary and using both complete matching and partial matching at the same time. Thus, in partial matching, Matching_factor was always used.

Support is related to a concept C. It is the sum of scores of all matching rules from C. Support is defined as follows:

$$\sum_{\substack{partially\ matching \\ rules\ R\ describing\ C}} Strength(R) * Specificity(R) * Matching\_factor(R)$$

The concept with the largest score wins the contest and the case is classified as belonging to this concept. If there is a tie among concepts, the strongest rule determines a concept. Support was used for classifying in all sixteen strategies, again, following an advice from [5]. In the above formula any factor may be equal to one, for example we may set specificity as equal to one. We say then that the corresponding classification strategy does not use specificity. During complete matching the value of Matching_factor is always equal to one. Obviously, during complete matching, in the above formula, *partially matching rules R describing C* should be interpreted as *completely matching rules R describing C.*

3 Sixteen classification strategies

In our experiments, we used four combinations for using rule sets: using only certain rules, using only possible rules, using certain rules first, then possible rules if necessary, and using both certain and possible rules. Since we used four options for choosing rule sets, two options for specificity, and two options for matching, sixteen different strategies were tested in our experiments:

1. Using only certain rules, specificity, complete matching, then partial matching if necessary,
2. Using only certain rules, specificity and both complete matching and partial matching,
3. Using only certain rules and complete matching, then partial matching if necessary, not using specificity,
4. Using only certain rules and both complete matching and partial matching, not using specificity,
5. Using only possible rules, specificity and complete matching, then partial matching if necessary,

6. Using only possible rules, specificity, both complete matching and partial matching,
7. Using only possible rules, complete matching, then partial matching if necessary, not using specificity;
8. Using only possible rules and both complete matching and partial matching, not using specificity,
9. Using certain rules first, then using possible rules if necessary, specificity, and complete matching, then partial matching if necessary,
10. Using certain rules first, then using possible rules if necessary, specificity and both complete matching and partial matching,
11. Using certain rules first, then using possible rules if necessary and complete matching, then partial matching if necessary, not using specificity,
12. Using certain rules first, then using possible rules if necessary and both complete matching and partial matching, not using specificity,
13. Using both certain rules and possible rules, specificity, complete matching, then partial matching if necessary,
14. Using both certain rules and possible rules, specificity, and both complete matching and partial matching,
15. Using both certain rules and possible rules, complete matching, then partial matching if necessary, not using specificity,
16. Using both certain rules and possible rules and both complete matching and partial matching, not using specificity.

4   Experiments

An overview of the ten real-life data sets is presented in Table 1. For each data set 16 experiments were performed, using all 16 different classification strategies. For each data set the performance of each classification strategy was measured in terms of the error rate. The error rate was estimated using *n-fold-cross-validation* [13]. The training data and testing data were generated from an available data set. During the process, first each data set was re-shuffled (the order of cases was randomly changed), then was divided into n subsets with approximately equal size. Each such data subset was used for testing exactly once. Each time, the remaining $n - 1$ subsets were used as training data (to induce rules by LERS). The average error rate for all n iterations was defined as the final error rate for the data set and the classification strategy. For moderate and large sets (more than or equal to 100 cases), n is chosen to be equal to 10; for small sets (less than 100 cases), n was the total number of cases. The later method is called a *leaving-one-out* approach. The average error rates for the ten data sets are presented in Tables 2 and 3.

The purpose of our experiments was to compare the strategies and find the best strategy. In order to compare the overall performance of all strategies, the Wilcoxon Signed Ranks Test [6], a nonparametric test for significant differences between paired observations, was used. All sixteen strategies were compared and ordered, from the best to the worst, see Figure 1. Note that some strategies are

Data set	Number of cases	Number of attributes	Number of concepts
Thesaurus	129,797	3	5
Wisconsin	625	9	9
Breast cancer	286	9	2
Nursing	90	8	3
HSV	122	12	4
Primary tumor	339	17	21
Mammography	1284	12	2
Luktrain	1654	13	2
Bupa	345	6	2
Iris	150	4	3

Table 1. Data sets

	Thesaurus	Wisconsin	Breast	Nursing	HSV
1	44.19	21.76	32.52	36.67	50.00
2	44.69	17.92	29.72	28.89	34.43
3	44.01	21.76	33.22	35.56	50.00
4	44.51	17.92	29.72	28.89	33.61
5	32.34	22.08	31.47	38.89	45.90
6	44.29	17.92	29.72	28.89	33.61
7	36.33	21.76	31.12	38.89	45.90
8	44.53	17.92	29.72	28.89	33.61
9	32.34	22.56	31.12	36.67	50.00
10	44.23	17.92	29.72	28.89	34.43
11	36.33	22.56	31.82	35.56	50.00
12	44.51	17.92	29.72	28.89	33.61
13	32.34	21.76	30.07	37.78	47.54
14	44.25	17.92	29.72	28.89	33.61
15	36.33	21.60	31.12	35.56	48.36
16	44.52	17.92	29.72	28.89	33.61

Table 2. Error rate

	Tumor	Mammography	Luktrain	Bupa	Iris
1	61.95	31.54	8.04	48.70	26.00
2	74.93	38.55	10.40	47.54	30.00
3	61.65	32.40	7.92	52.75	26.00
4	75.22	39.95	10.40	51.59	28.67
5	59.88	23.13	7.62	38.26	11.33
6	73.16	23.60	10.40	42.03	30.67
7	59.00	22.43	7.38	41.45	26.67
8	74.93	25.00	10.40	41.16	29.33
9	57.52	23.52	8.22	38.55	11.33
10	74.93	38.55	10.40	47.54	30.00
11	57.52	22.74	8.10	41.74	26.67
12	75.22	39.95	10.40	51.59	28.67
13	56.34	23.05	8.16	38.55	11.33
14	75.22	24.07	10.40	41.74	34.00
15	55.46	22.12	7.74	41.45	26.67
16	75.22	27.80	10.40	40.87	29.33

Table 3. Error rate

presented by the same node of the diagram in Figure 1, e.g., strategies 5 and 9. This means that the corresponding strategies are not significantly different. On the other hand, some strategies, e.g., 2 and 3, are not connected by any arc in Figure 1. This means that these strategies are statistically incomparable.

5   Conclusions

Our research focused on the classification process using rule sets induced by LERS. The following conclusions may be drawn from our experiments.

Choosing different matching options has no significant impact on the classification results. It means that there is no significant difference between using complete matching first, then partial matching if necessary and using both, complete and partial matching. As follows from the diagram in Figure 1, the even numbered strategies have about the same performance as the odd numbered strategies do. Thus, different options of matching were not important to the classification process.

When using the same rule set option with matching option: using complete matching first, then partial matching if necessary (strategies 1, 3, 5, 7, 9, 11, 13 and 15) there is no significant difference between using specificity and not using specificity.

The following matching option: using complete matching first, then partial matching if necessary, using both certain and possible rules yields the best results (strategies 13 and 15), using only certain rules yields the worst results (strategies 1 and 3).

Using the same rule set option with matching option: using both complete and partial matching, there is no significant difference between using specificity and not using specificity. The corresponding strategies are: 2 vs. 4, 6 vs. 8, 10 vs. 12, and 14 vs. 16.

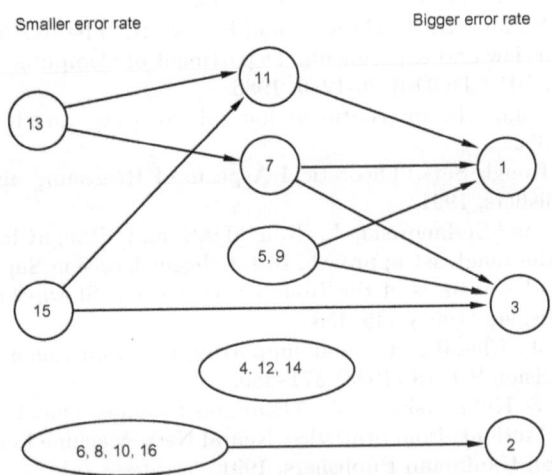

Fig. 1. Comparison of 16 strategies for classification

Choosing between the certain rules and possible rules is an important factor for classification. The option with only certain rules (strategies 1, 2, and 3) results in a bigger error rate, and the option with using both certain rules and possible rules (strategies 13, 14, 15, and 16) yields a smaller error rate.

Summarizing, there is no one single best strategy among 16 strategies used for classification. Thus, a choice of successful strategy depends on kind of input data as well.

References

1. Booker, L. B., Goldberg, D. E., and Holland, J. F.: Classifier systems and genetic algorithms. In Machine Learning. Paradigms and Methods. Carbonell, J. G. (ed.), The MIT Press (1990) 235–282.
2. Grzymala–Busse, J. W.: Knowledge acquisition under uncertainty—A rough set approach. Journal of Intelligent & Robotic Systems 1 (1988) 3–16.
3. Grzymala–Busse, J. W.: LERS—A system for learning from examples based on rough sets. In Intelligent Decision Support. Handbook of Applications and Advances of the Rough Sets Theory. Slowinski, R. (ed.), Kluwer Academic Publishers (1992) 3–18.
4. Grzymala–Busse, J. W.: Managing uncertainty in machine learning from examples. Proc. of the Third Intelligent Information Systems Workshop, Wigry, Poland, June 6–11, 1994, 70–84.
5. Grzymala–Busse, J. W. and Wang, C. P. B.: Classification and rule induction based on rough sets. Proc. of the 5th IEEE International Conference on Fuzzy Systems FUZZ-IEEE'96, New Orleans, Louisiana, September 8–11, 1996, 744–747.

6. Hamburg, M.: Statistical Analysis for Decision Making. Harcourt Brace Jovanovich Inc., 1983, Third Edition.

7. Holland, J. H., Holyoak K. J., and Nisbett, R. E.: Induction. Processes of Inference, Learning, and Discovery. The MIT Press, 1986.

8. Michalski, R. S., Mozetic, I., Hong, J. and Lavrac, N.: The AQ15 inductive learning system: An overview and experiments. Department of Computer Science, University of Illinois, Rep. UIUCDCD-R-86-1260, 1986.

9. Pawlak, Z.: Rough sets. International Journal Computer and Information Sciences 11 (1982) 341–356.

10. Pawlak, Z.: Rough Sets. Theoretical Aspects of Reasoning about Data. Kluwer Academic Publishers, 1991.

11. Slowinski, R. and Stefanowski, J.: 'RoughDAS' and "RoughClass' software implementations of the rough set approach. In Intelligent Decision Support. Handbook of Applications and Advances of the Rough Sets Theory. Slowinski, R. (ed.), Kluwer Academic Publishers (1992) 445–456.

12. Stefanowski, J.: Classification and supporting based on rough set theory, Found. Computing Decision Sci. 18 (1993) 371–380.

13. Weiss, S. M. & Kulikowski, C. A.: Computer Systems That Learn: Classification and Prediction Methods from Statistics, Neural Nets, Machine Learning, and Expert Systems. Morgan Kaufmann Publishers, 1991.

14. Ziarko, W. P.: Acquisition of control algorithms from operation data. In Intelligent Decision Support. Handbook of Applications and Advances of the Rough Sets Theory. Slowinski, R. (ed.), Kluwer Academic Publishers (1992) 61–75.

15. Ziarko, W.: Analysis of uncertain information in the framework of variable precision rough sets. Found. Computing Decision Sci. 18 (1993) 381–396.

Well-Behaviored Operations for Approximate Sets

Jose Miro–Julia and José Miró

Departament de Matemàtiques i Informàtica. Universitat de les Illes Balears, Campus UIB, 07071 Palma de Mallorca, SPAIN. joe@ipc4.uib.es, dmijmn0@clust.uib.es

[Abstract.] The prime ingredients of the operations of the human cognitive mind are descriptions. Descriptions may be approximate in the sense that imprecision may not allow the construction of a set from a word description. In a previous paper this type of imprecision was introduced with the name of *approximate sets*. The operations on approximate sets defined there were as precise as possible, but they had some difficulties, one of them being that the union operation defined was not associative. It was foreseen that precision in the operations might be traded for operational convenience. In this paper such a possibility is investigated, less precise operations are offered, and their convenience is studied.

1 Introduction

The prime ingredients of the operations of the human cognitive mind are descriptions. The objects about whose descriptions the human mind deal with may be considered from many points of view, and consequently may admit different classifications. At this moment we wish to bear in mind a particular classification according to which an object may be

considered an *individual* object or a *set* object. We admit that an individual object is described by the set of its features. We pay attention now to the set objects. One possible way of describing a set object is by means of the list of all of its components. Another way is by means of sentences relating the features of the components without referring to any individual object in particular.

This has been recognized for a long time, and used to establish two kinds of definitions for sets: (1) the definition by enumeration of its components, and (2) the definition of a set by a property.

Descriptions may be approximate in the sense that imprecision does not allow the construction of a set from a word description. An example will help to clarify this discussion. Suppose a soccer team A. A has 25 registered players $p_1, p_2, \ldots, p_{25}$. A has to play an official tournament game, say in Warsaw, and a friendly invited game in Cracow. The Coach has decided that players p_1 to p_{16} will be available to play in Warsaw and players p_8 to p_{25} will travel to Cracow. It is known that a soccer team consists of 11 players and that in a particular official game only three players may be substituted. This means that at least 11

L. Polkowski and A. Skowron (Eds.): RSCTC'98, LNAI 1424, pp. 45–51, 1998.
© Springer-Verlag Berlin Heidelberg 1998

and at most 14 of players p_1 to p_{16} will play in Warsaw. It has been a greed that in Cracow four players may be substituted, thus in Cracow from 11 to 15 of players p_8 to p_{25} will play. This type of imprecise knowledge may be described by a formal statement.

In a previous paper [2] this type of imprecision was introduced with the name of *approximate sets*, and the way of handling these descriptions to compute some possible inferences was studied. Operations of union and intersection were defined. In the above example the union wo uld determine the approximate set describing the players that would play in either Warsaw or Cracow or both, while the intersection would establish those who would play in both cities.

The operations on approximate sets defined in our previous paper we re as precise as possible, but they had some difficulties, to be discussed later in more detail, one of them being that the union as defined was not associative. It was foreseen that precision in the operations might be traded for operational conveni ence. In this paper such a possibility is investigated, less precise operations are offered, and their convenience is studied.

2   Approximate sets

In this section we present succintly the definition of approximate sets and the main results that were presented in our previous paper [2].

Let $X = \{a, b, \ldots, n\}$ be a finite, extensionally defined, set of cardinal $|X|$, and let x_1, x_2 be two natural numbers. We define *approximate set* $\mathbf{X}$ as a three-tuple $\langle x_1, x_2 \rangle X$, where $0 \leq x_1 \leq x_2 \leq |X|$. The meaning of $\mathbf{X} = \langle x_1, x_2 \rangle X$ is that $\mathbf{X}$ contains between x_1 and x_2 elements of X. We call x_1 the lower bound of the approximate set, and x_2, the upper bound; X is called the *base set*. An approximate set describes approximately a subset of X. It is not known which elements constitute it, not even its cardinality. If we apply this definition to the example presented above, the players playing the game in Warsaw could be defined by the approximate set $\langle 11, 14 \rangle \{p_1, \ldots, p_{16}\}$, while the players in the Cracow game could be described by $\langle 11, 15 \rangle \{p_8, \ldots, p_{25}\}$.

The usual mathematical quantifiers $\forall$ and $\exists$ define approximate sets. For instance, given a universe X $\forall x$ defines the approximate set $\langle |X|, |X| \rangle X$, and $\exists x$ defines $\langle 1, |X| \rangle X$. Quantifiers used in more sofist icate logic models, like the *counting quantifiers* [1] also define approximate sets. The counting quantifier $\exists^{\geq l}$ with the interpretation 'there are at least l', defines the approximate set $\langle l, |X| \rangle X$.

Two approximate sets $\mathbf{X} = \langle x_1, x_2 \rangle X$, $\mathbf{Y} = \langle y_1, y_2 \rangle Y$ are said to be equal iff $x_1 = y_1$, $x_2 = y_2$, $X = Y$. It must be noted that even though $\mathbf{X}$ and $\mathbf{Y}$ are both subsets of the same set,

$\mathbf{X} = \mathbf{Y}$ does not imply we are referring to the same subset. Not only that, it is entirely possible that the two subsets be disjoint.

Approximate sets show a partial order. We shall say that given two approximate sets $\mathbf{X} = \langle x_1, x_2 \rangle X$, $\mathbf{Y} = \langle y_1, y_2 \rangle Y$, $\mathbf{X}$ is included in $\mathbf{Y}$ and represent it as

$\mathbf{X} \sqsubseteq \mathbf{Y}$ iff $x_1 \leq y_1$, $x_2 \leq y_2$, $X \subseteq Y$. As before, the two subsets defin ed by $\mathbf{X}$ and $\mathbf{Y}$ are not necesarily included one in the other.

There are two operations defined over approximate sets. Given the sets $X, Y, X \cup Y, X \cap Y$, where $\cup$, $\cap$ represent the union of sets and intersection of s ets respectively, we define the *union*, $\mathbf{X} \sqcup \mathbf{Y}$, and *intersection*, $\mathbf{X} \sqcap \mathbf{Y}$ of approximate sets $\mathbf{X} = \langle x_1, x_2 \rangle X$, $\mathbf{Y} = \langle y_1, y_2 \rangle Y$ as

$$\mathbf{X} \sqcup \mathbf{Y} = \langle a, b \rangle (X \cup Y)$$
$$\mathbf{X} \sqcap \mathbf{Y} = \langle c, d \rangle (X \cap Y)$$

where

$$a = \max(x_1, y_1) \qquad\qquad b = \min((x_2 + y_2), |X \cup Y|)$$
$$c = \max(0, (x_1 + y_1 - |X \cup Y|)) \quad d = \min(x_2, y_2, |X \cap Y|)$$

These operations are commutative, neither is idempotent, the intersection is associative, and the union $\mathbf{X} \sqcup \mathbf{Y} \sqcup \mathbf{Z}$ is not associative.

These operations were defined so as to have the best *computational behavior*, that is, to lose the least precision possible when operating with approximate sets. Unfortunately, as the last paragraph shows, the *operational behavior*, the algebraic characteristics of the operations, is very poor. In the next section we present new operations that, although may decrease the precision while computing, do have much better algebraic characteristics.

3 New definition of operations

In this section we will define the new union and intersection operations on approximate sets. From now on the operations defined in Section 2 will be referred to as square-union and square-intersection, and the new operations to be defined below will be referred to simply as union and intersection.

3.1 Union

The union operation over two approximate sets $\mathbf{X} = \langle x_1, x_2 \rangle X$ and $\mathbf{Y} = \langle y_1, y_2 \rangle Y$ is defined as

$$\mathbf{X} \vee \mathbf{Y} = \bigsqcup_{n=1}^{\infty} (\mathbf{X} \sqcup \mathbf{Y})^{(n)}$$

that is, first a square-union is done between $\mathbf{X}$ and

$\mathbf{Y}$, and the result is operated with itself through square-unions infinite times. Once $\mathbf{X} \sqcup \mathbf{Y}$ is performed, on each operation of this square-union with itself, the lower bound remains unchanged while the upper bound, if it is not already $|X \cup Y|$, will strictly increase until reaching this value. Therefore, it is easy to show that the final result of this operation is

$$\mathbf{X} \vee \mathbf{Y} = \langle \max(x_1, y_1), |X \cup Y| \rangle (X \cup Y) \tag{1}$$

Comparing this operation to the square-union, we see that there might be a loss in precision, as the upper bound grows goes from $\min((x_2 + y_2), |X \cup Y|)$ to $|X \cup Y|$. That is, $\mathbf{X} \sqcup \mathbf{Y} \sqsubseteq \mathbf{X} \vee \mathbf{Y}$.

3.2 Intersection

The intersection operation i s defined analogously to the union as

$$\mathbf{X} \wedge \mathbf{Y} = \prod_{n=1}^{\infty} (\mathbf{X} \sqcap \mathbf{Y})^{(n)}$$

After the first square-intersection is performed, on each operation of the result with itself, the upper bound will remain unchanged, while the lower bound will strictly decrease, unless it is zero. Therefore the final result is

$$\mathbf{X} \wedge \mathbf{Y} = \langle 0, \min(x_2, y_2, |X \cap Y|)\rangle(X \cap Y) \tag{2}$$

Again there might be a loss of precision respect to the square-intersection as the lower bound diminishes from $\max(0, (x_1 + y_1 - |X \cup Y|))$ to 0. That is, $\mathbf{X} \sqcap \mathbf{Y} \sqsubseteq \mathbf{X} \wedge \mathbf{Y}$.

As we shall see in Section 4, this loss of precision is compensated by a better behavior of these operations.

3.3 Complement of an approximate set

In our previous paper in which we introduced approximate sets there was no definition of the complement of an approximate set. We proceed to introduce the definition here.

An approximate set is an imprecise description of a concept, so the complement of an approximate set should be an imprecise description of the complement of the concept. If given a set S of 30 students, we know that between 2 and 12 are bright, we can express this concept by $\mathbf{B} = \langle 2, 12\rangle S$. Suppose we want to know how many students are *not* bright. It is easy to see that this concept is described as $\mathbf{NB} = \langle 18, 28\rangle S$. From here we establish a definition of complement of an approximate set.

[**Definition 1.**] *Given an approximate set* $\mathbf{X} = \langle x_1, x_2\rangle X$, *we define the complement of this set,* $\overline{\mathbf{X}}$ *as*

$$\overline{\mathbf{X}} = \langle |X| - x_2, |X| - x_1\rangle X$$

It must be stressed that $\mathbf{X}$ and $\overline{\mathbf{X}}$ are both subsets of the same base set X.

4   Operation behavior

[**Theorem 1.**] *The union and intersection operations are commutative.*

This property follows from the definition of the operations as shown in (1) and (2).

[**Theorem 2.**] *The intersection operation is associative.*

This result follows directly from the associativity of the square-intersection. □

[**Theorem 3.**] *The union operation is associative.*

[*Proof.*]

$$
\begin{aligned}
(\mathbf{X} \vee \mathbf{Y}) \vee \mathbf{Z} &= \langle \max(x_1, y_1), |X \cup Y| \rangle (X \cup Y) \vee \mathbf{Z} \\
&= \langle \max(\max(x_1, y_1), z_1), |(X \cup Y) \cup Z| \rangle ((X \cup Y) \cup Z) \\
&= \langle \max(x_1, y_1 z_1), |X \cup Y \cup Z| \rangle (X \cup Y \cup Z) \\
&= \langle \max(x_1, \max(y_1, z_1)), |X \cup (Y \cup Z)| \rangle (X \cup (Y \cup Z)) \\
&= \mathbf{X} \vee \langle \max(y_1, z_1), |Y \cup Z| \rangle (Y \cup Z) \\
&= \mathbf{X} \vee (\mathbf{Y} \vee \mathbf{Z})
\end{aligned}
$$

□

[**Theorem 4.**] *The intersect ion is distributive over the union*

[*Proof.*]

$$
\begin{aligned}
\mathbf{X} \wedge (\mathbf{Y} \vee \mathbf{Z}) &= \mathbf{X} \wedge \langle \max(y_1, z_1), |Y \cup Z| \rangle (Y \cup Z) \\
&= \langle 0, \min(x_2, |Y \cup Z|, |X \cap (Y \cup Z)| \rangle (X \cap (Y \cup Z)) \\
&= \langle 0, \min(x_2, |(X \cap Y) \cup (X \cap Z)| \rangle ((X \cap Y) \cup (X \cap Z))
\end{aligned}
$$

Given that the upper bound of a union does not depend on the upper bounds of its operands, we can now divide this into the union of two approximate sets with the upper bounds set to $\min(x_2, y_2, |X \cap Y|)$ and $\min(x_2, z_2, |X \cap Z|)$ respectively, and the proof is almost complete

$$
\begin{aligned}
&= \langle 0, \min(x_2, y_2, |X \cap Y|) \rangle (X \cap Y) \vee \langle 0, \min(x_2, z_2, |X \cap Z|) \rangle (X \cap Z) \\
&= (\mathbf{X} \wedge \mathbf{Y}) \vee (\mathbf{X} \wedge \mathbf{Z})
\end{aligned}
$$

□

In likewise fashion, it is easily seen that the union is *not* distributive over the intersection.

[**Theorem 5.**] *There is no zero element for the union operation*

[*Proof.*] Let $\mathbf{X} = \langle x_1, x_2 \rangle X$ be an approximate set, and $\mathbf{N} = \langle n_1, n_2 \rangle N$ be the candidate to zero element, that is, $\mathbf{X} \vee \mathbf{N} = \mathbf{X}$. From the definition of the union operation, it must be that $X \cup N = X$, so $N \subseteq X$, but then the upper bound of the union will be $|X|$, and this is in general different from x_2, so the result of $\mathbf{X} \vee \mathbf{N}$ is in general different from $\mathbf{X}$. □

[**Theorem 6.**] *There is no unit element for the intersection operation*

[*Proof.*]From the definition of the intersection operation, the lower bound must be 0, and that is different from x_1, so there is no unit element.   □

With the theorems proven above we establish that the set of all approximate sets of a given universe with the union and intersection have the structure of a semi-ring without a unit element. This is clearly an improvement over the algebraic properties of the square-union and square-intersection, that presented no definite structure.

Following we present some results showing the characteristics with respect to sets of the new operations.

[**Theorem 7.**]$\mathbf{X} \wedge \mathbf{Y} \sqsubseteq \mathbf{X} \sqsubseteq \mathbf{X} \vee \mathbf{Y}$

[*Proof.*]We apply the definition of the operations and notice that

$$0 \leq x_1 \leq \max(x_1, y_1)$$
$$\min(x_2, y_2, |X \cap Y|) \leq x_2 \leq |X \cup Y|$$
$$(X \cap Y) \subseteq X \subseteq (X \cup Y)$$

and by the definition of inclusion of approximate sets given in Section 2, the result is proven.   □

[**Theorem 8.**]*Neither operation is idempotent*

[*Proof.*]Applying the definition of the operations, same as above, we obtain

$$0 \leq x_1 = \max(x_1, x_1)$$
$$\min(x_2, x_2, |X \cap X|) = x_2 \leq |X \cup X|$$
$$(X \cap X) = X = (X \cup X).$$

Given the inequalities that remain, the proof is complete.   □

[**Theorem 9.**]*Although not idempotent, we could say that they are* quasi-idempotent, *because* $(\mathbf{X} \vee \mathbf{X}) \vee \mathbf{X} = (\mathbf{X} \vee \mathbf{X})$ *and* $(\mathbf{X} \wedge \mathbf{X}) \wedge \mathbf{X} = (\mathbf{X} \wedge \mathbf{X})$. *begintheorem* $\overline{\overline{\mathbf{X}}} = \mathbf{X}$

[*Proof.*]As usual, $\mathbf{X} = \langle x_1, x_2 \rangle X$.

$$\overline{\overline{\mathbf{X}}} = \overline{\langle |X| - x_2, |X| - x_1 \rangle X} = \langle |X| - (|X| - x_1), |X| - (|X| - x_2) \rangle X = \langle x_1, x_2 \rangle X$$

     □

When trying to establish the operational behavior of the complement we quickly run into a problem due to the lack of a common framework: the complement of $\langle x_1, x_2 \rangle X$ belongs to X, the complement of $\langle y_1, y_2 \rangle Y$ belongs to Y, and therefore there is no way in general to relate one to the other. In set theory this is solved through the Universe, that is, all sets are supposed to be subsets of a common frame. If we establish a universe U, we can prove the following theorem.

[**Theorem 10.**] *Within a common framework U both de Morgan's Laws hold for approximate sets with the $\vee, \wedge$ operations and the complement.*

[*Proof.*]For t he first de Morgan Law. Let $\mathbf{X} = \langle x_1, x_2 \rangle U$ and $\mathbf{Y} = \langle y_1, y_2 \rangle U$. On the one hand

$$\overline{\mathbf{X} \vee \mathbf{Y}} = \overline{\langle \max(x_1, y_1), |U| \rangle U}$$
$$= \langle 0, |U| - \max(x_1, y_1) \rangle U$$

And on the other hand

$$\overline{\mathbf{X}} \wedge \overline{\mathbf{Y}} = \langle |U| - x_2, |U| - x_1 \rangle U \wedge \langle |U| - y_2, |U| - y_1 \rangle U$$
$$= \langle 0, \min(|U| - x_1, |U| - y_1, |U|) \rangle U$$
$$= \langle 0, |U| - \max(x_1, y_1) \rangle U$$

For the second de Morgan's Law we proceed in the same fashion.

$$\overline{\mathbf{X} \wedge \mathbf{Y}} = \overline{\langle 0, \min(x_2, y_2, |U|) \rangle U}$$
$$= \langle |U| - \min(x_2, y_2), |U| \rangle U$$

And

$$\overline{\mathbf{X}} \vee \overline{\mathbf{Y}} = \langle |U| - x_2, |U| - x_1 \rangle U \vee \langle |U| - y_2, |U| - y_1 \rangle U$$
$$= \langle \max(|U| - x_2, |U| - y_2), |U| \rangle U$$
$$= \langle |U| - \min(x_2, y_2), |U| \rangle U$$

$\square$

5 Conclusion

We have presented new union and intersection operations on approximate sets. These operations present better algebraic behavior than the original ones, although some loss of precision might occur. We have also introduced a complement operation and proven that under certain circumstances de Morgan's Laws hold for approximate sets.

This work is just a first approach in trying to establish well-behaviored operations on approximate sets. Further research on exactly what loss of precision must be incurred to obtain a good algebraic behavior is warranted.

References

1. Ebbinghaus, H-D., Flum, J.: Finite Model Theory. Springer-Verlag, Berlin Heidelberg New York (1995)
2. Miró, J., Miro–Julia, J.: Uncertainty and Inference Through Approximate Sets. In: Bouchon–Meunier, B., Valverde, L., Yager, R.R. (eds.): Uncertainty in Intelligent Systems. North-Holland, Amsterdam London New York Tokyo (1993) 203–214
3. Yao, Y.Y., Noroozi, N. A Unified Framework for Set-Based Computations. In: Lin, T.Y. (Ed.): Proceedings of the 3rd International Workshop on Rough Sets and Soft Computing, San Jose, California, (1994) 236–243.

Searching for Frequential Reducts in Decision Tables with Uncertain Objects *

Dominik Ślęzak

Institute of Mathematics
University of Warsaw
Banacha 2, 02-097 Warsaw, Poland
email: slezak@alfa.mimuw.edu.pl

Abstract. We discuss an uncertainty representation based on rough membership functions in inconsistent decision tables. We propose a reasoning model for dealing with objects having probability distributions instead of concrete values on attributes. We prove discernibility characteristics for minimal boolean implicants in inconsistent decision tables with indeterministically defined objects.

1 Introduction

Theory of rough sets ([4]) gave the origin to many methods of reasoning about data. One of them is how to search for optimal sets of features for classification of new cases. By adopting the principles of boolean reasoning, a lot of algorithmic approaches to the above task have been developed (see e.g. [3] for further references), with the notion of consistent decision table as the starting point. However, since inconsistency has been introduced by handling lower and upper set approximations ([4]), rough sets based research has been providing more and more strategies of dealing with inconsistency (see e.g. [9], [10]).

We begin from recalling the rough set approach to data analysis (see e.g. [2], [7]). Then we discuss uncertainty representation in inconsistent decision tables with respect to rough membership functions ([5]), which have a strong support in probability theory and statistics (see [6] for further references). Most of the definitions and results presented in Sections 2 and 3 can be found in rough set literature (see e.g. [5], [8]), assumed that attributes remain discrete, with values uniquely determined over considered objects. In Section 4 we let objects be indeterministically defined, with probability distributions instead of concrete values on attributes. Decision tables begin to be a basis for reasoning about new cases remaining indeterministic as well. Such generalization requires verification of fundamental notions and results. Here we focus especially on the problem of finding minimal frequential reduct in a decision table with uncertain objects, providing discernibility characteristics analogous to that for classical model.

---

* This work was supported by KBN Research Grant No. 8T11C01011 and ESPRIT project CRIT-2 No. 20288 of European Union.

L. Polkowski and A. Skowron (Eds.): RSCTC'98, LNAI 1424, pp. 52–59, 1998.

2   Rough sets based approach to data

While reasoning about a domain specified by our needs, we are usually forced to base just on information gathered by the analysis of some sample of objects. The main paradigm of rough sets theory ([4]) states that such a universe of known objects, stored within an information system, is assumed to be the only source of knowledge able to be used for classifications of cases outside the sample.

An information system is a tuple $\mathbf{A} = (U, A)$, where each attribute $a \in A$, corresponding to some feature which may be important with respect to object classification, is identified with function $a : U \to V_a$, from the universe U of objects, onto the set V_a of all possible values on a.

While reasoning about new objects outside U we refer to equivalence classes of indiscernibility relation, defined, for arbitrary subset $B \subseteq A$, as

$$IND\,(B) = \{(u_1, u_2) \in U \times U : Inf_B\,(u_1) = Inf_B\,(u_2)\} \tag{1}$$

where information function Inf_B, such that $Inf_B\,(u) = \left(a_{i_1}\,(u), .., a_{i_{card(B)}}\,(u)\right)$, is consistent with fixed linear ordering $A = < a_1, .., a_{card(A)} >$. One can see that there is a one-to-one correspondence between equivalence classes of $IND\,(B)$ and elements from $V_B \subseteq \times_{a \in B} V_a$ - the set of all vector values occurring on B in $\mathbf{A}$, i.e. such that $w_B \in V_B$ iff there is at least one $u \in U$ satisfying $Inf_B\,(u) = w_B$.

In applications, reasoning is usually stated as a classification problem, concerning distinguished decision attribute to predict under given conditions. By a decision table we understand a triple $\mathbf{A} = (U, A, d)$, where $d \notin A$ corresponds to partition of U onto pairwise disjoint decision classes denoted by $d^{-1}\,(v_d)$, $v_d \in V_d$. The very initial model of decision table is the following:

Definition 1. *Decision table* $\mathbf{A} = (U, A, d)$ *is called consistent iff for each* $u \in U$ *its indiscernibility class* $[u]_A = \{u' \in U : Inf_A\,(u') = Inf_A\,(u)\}$ *is included in one of decision classes* $d^{-1}\,(v_d)$, $v_d \in V_d$.

In case of such consistency we classify new cases by analogy with those from the universe, i.e., given some *new* $\notin U$ indiscernible from $u \in U$, we predict that *new* has decision value $d\,(u)$, since this is the only supported choice. If $Inf_A\,(u) = w_A$ and $d\,(u) = v_d$, then we can say that our table generates decision rule $A = w_A \Rightarrow d = v_d$, stating that if any object is equal to w_A on A, then it is going to have decision value v_d. We can also introduce boolean implication $A \Rightarrow d$ which holds iff each $w_A \in V_A$ implies some $v_d \in V_d$ in the above sense. One can easily see that decision table $\mathbf{A} = (U, A, d)$ is consistent iff boolean implication $A \Rightarrow d$ is satisfied. In other words, the decision table is consistent iff decision d can be completely determined from the conditions.

Definition 2. *Given consistent decision table* $\mathbf{A} = (U, A, d)$, *subset* $B \subseteq A$ *is called a decision implicant iff it satisfies boolean implication* $B \Rightarrow d$. *If* B *is minimal in the sense of inclusion, i.e. such that there is no its proper subset* $C \subset B$ *holding implication* $C \Rightarrow d$, *then we call* B *a decision reduct for* $\mathbf{A}$.

The above definition is connected with the fundamental question whether we do need all conditions to determine the decision. Let us give an exemplar argumentation for searching for decision implicants with possibly small number of attributes. We say that decision table $\mathbf{A}$ is applicable to an object $new \notin U$ under $B \subseteq A$ iff new fits some indiscernibility class of $IND(B)$. Since for any $w_A \in \times_{a \in A} V_a$ we have implication $w_A \in V_A \Rightarrow w_A^{\downarrow B} \in V_B$ (where $w_A^{\downarrow B} \in V_B$ is the projection of w_A onto B), we know that applicability to new objects is potentially more probable under smaller $B \subseteq A$, unless it does not preserve precision of decision classification.

Proposition 1. *Given consistent decision table* $\mathbf{A} = (U, A, d)$, *subset* $B \subseteq A$ *is a decision implicant iff it discerns objects belonging to different decision classes.*

As a corollary, we obtain that searching for minimal decision reducts is complex enough to force us to base it on some heuristics. One can prove that the problem of finding decision implicant with minimal number of attributes in a consistent decision table is NP-hard. On the other hand, Proposition 1 enables to implement some random heuristics efficiently enough in order to obtain approximately optimal solutions in a relatively short time (see e.g. [3]). In fact, this advantage suggested us to verify the power of discernibility characteristics with respect to more general rough set models.

3   Frequential reasoning

In classical model of consistent decision table $\mathbf{A} = (U, A, d)$, where each indiscernibility class of $IND(A)$ is contained in one of the decision classes, preserving an information about decision is equivalent with its complete determination. In case of any inconsistencies, however, we must settle the way of representing indeterministic knowledge. For any fixed $B \subseteq A$, let us introduce rough membership function $\mu_{d/B} : V_d \times V_B \to [0, 1]$, defined by the following formula

$$\mu_{d/B}(v_d/w_B) = \frac{|w_B, v_d|}{|w_B|} \tag{2}$$

where $|w_B|$ is the number of objects with vector value w_B on B, and $|w_B, v_d|$ - the number of objects which additionally have the decision value equal to v_d. We propose, given linear ordering $V_d = \left\langle v_d^1, ..., v_d^{|d|} \right\rangle$ (where $|d|$ denotes the number of values possible for d), to consider rough membership distribution $\mu_{d/B} : V_B \to \triangle_{|d|-1}$ defined by

$$\mu_{d/B}(w_B) = \left(\mu_{d/B}(v_d^1/w_B), ..., \mu_{d/B}(v_d^{|d|}/w_B) \right) \tag{3}$$

where $\triangle_{|d|-1}$ denotes $(|d|-1)$-dimensional simplex. Rough membership distributions correspond to frequencies of putting indiscernibility classes of $IND(B)$ into particular decision classes. They are widely studied in rough sets as well as in statistics (see e.g. [5], [6]). We have to remember that $\mu_{d/B}$ is just one of the examples of a decision function which, for arbitrary $B \subseteq A$, specifies conditional information about the decision attribute. On the other hand, however, it expresses the whole knowledge about dependencies of the decision on conditions, unless some additional information outside the decision table is provided (compare with [10]).

Given the above representation, one begins to handle uncertain decision rules of the form $B = w_B \Rightarrow d = v_d$ with the probability $\mu_{d/B}(v_d/w_B)$, for any $B \subseteq A$, $w_B \in V_B$. Thus, the notion of decision reduct is reformulated as follows.

Definition 3. *Given decision table* $\mathbf{A} = (U, A, d)$*, subset* $B \subseteq A$ *is called a* μ*-implicant iff for each* $w_A \in V_A$

$$\mu_{d/B}\left(w_A^{\downarrow B}\right) = \mu_{d/A}(w_A) \tag{4}$$

If B is minimal in sense of inclusion, satisfying (4) for all $w_A \in V_A$, we call it a μ-decision reduct.

For consistent decision tables, the above is equivalent to Definition 2. Indeed, in such a case, distribution $\mu_{d/A}(w_A)$, for each particular $w_A \in V_A$, corresponds to a unique vertex of $\triangle_{|d|-1}$ and thus it must be also the case for $\mu_{d/B}\left(w_A^{\downarrow B}\right)$, if (4) satisfied. Characteristics provided by Proposition 2 for consistent tables can be generalized as well.

Proposition 2. *Given decision table* $\mathbf{A} = (U, A, d)$*, subset* $B \subseteq A$ *is a* μ*-implicant iff it is an implicant for a consistent decision table* $\mathbf{A}_\mu = (U, A, \mu_{d/A})$*, or, equivalently iff implication* $B \Rightarrow \mu_{d/A}$ *holds; This means that each two* $w_A^1, w_A^2 \in V_A$*, such that* $\mu_{d/A}\left(v_d/w_A^1\right) \neq \mu_{d/A}\left(v_d/w_A^2\right)$ *for at least one* $v_d \in V_d$*, must be discerned by B.*

Proof We have to prove that equivalence $\mu_{d/B} = \mu_{d/A}$, understood in terms of (4), holds iff $B \Rightarrow \mu_{d/A}$. ($\Rightarrow$) From left to right, by converse, let us assume that there are $w_A^1, w_A^2 \in V_A$, such that $\mu_{d/A}\left(v_d/w_A^1\right) \neq \mu_{d/A}\left(v_d/w_A^2\right)$ for some $v_d \in V_d$, with the same projection onto B, denoted by w_B. Then condition (4) must fail for at least one of w_A^i, $i = 1, 2$, because if not, then we would have $\mu_{d/B}(v_d/w_B) = \mu_{d/A}(v_d/w_A^i)$ for both $i = 1, 2$ and as a result - equality $\mu_{d/A}\left(v_d/w_A^1\right) = \mu_{d/A}\left(v_d/w_A^2\right)$ contradictive to that above. ($\Leftarrow$) Now, let us assume that $B \Rightarrow \mu_{d/A}$ is satisfied. Let us consider arbitrary $w_B \in V_B$ and subset $V_A(w_B) = \left\{w_A \in V_A : w_A^{\downarrow B} = w_B\right\}$. To finish the proof it is enough to show that if for arbitrary $v_d \in V_d$ values $\mu_{d/A}(v_d/w_A)$ are the same for all $w_A \in V_A(w_B)$, then they are equal to $\mu_{d/B}(v_d/w_B)$. Let us enumerate members of $V_A(w_B)$ from w_A^1 to w_A^n for appropriate n. Since $\mu_{d/A}\left(v_d/w_A^i\right)$ is constant, we have that for each $i = 1, ..., n$ there is such real number k_i that $\left|w_A^i, v_d\right| = k_i \cdot \left|w_A^1, v_d\right|$ and $\left|w_A^i\right| = k_i \cdot \left|w_A^1\right|$, where $k_1 = 1$.

We can rewrite $\mu_{d/B}(v_d/w_B)$ as

$$\frac{|w_B, v_d|}{|w_B|} = \frac{\sum_{w_A \in V_A(w_B)} |w_A, v_d|}{\sum_{w_A \in V_A(w_B)} |w_A|} = \frac{\sum_{i=1,..,n} \left(k_i \cdot |w_A^1, v_d|\right)}{\sum_{i=1,..,n} \left(k_i \cdot |w_A^1|\right)} = \frac{|w_A^1, v_d|}{|w_A^1|}$$

(5)

So $\mu_{d/B}(v_d/w_B)$ is equal to $\mu_{d/A}(v_d/w_A^1)$ and thus to $\mu_{d/A}(v_d/w_A^i)$ for any other $i = 2, ..., n$.

The above result can be interpreted in both pessimistic and optimistic ways. The first of them is due to the corollary that for any inconsistent decision table the problem of finding minimal μ-implicant is NP-hard. The optimistic interpretation is that we could not expect lower time complexity, since the above problem had already turned out to be NP-hard for the special case of consistent decision tables. We should rather be happy that complexity does not increase; indeed, we managed to reformulate the search problem for inconsistent tables to consistent ones. The cost of such reformulation, for each particular $\mathbf{A}$, is negligible with respect to further computations. Moreover, space complexity remains the same and, finally, we can apply algorithmic tools based on discernibility for $\mathbf{A}_\mu$.

4   A generalization of the frequential model

So far, we assumed that the source of our knowledge consists of objects which had deterministically defined attribute values. In many applications, however, we cannot observe concrete values of attributes from $A \cup \{d\}$. Instead, for each $u \in U$, we have a probability distribution, denoted here by $u_a = \left(u_a\left(v_a^1\right), ..., u_a\left(v_a^{|a|}\right)\right)$. From now on we will write $\mathbf{A} = (U_\mu, A, d)$ while talking about decision tables with objects remaining uncertain in the above sense.

One must realize that once we begin to deal with uncertain objects inside the universe, we cannot expect certain values from new objects occurring at the input of reasoning process. Thus each *new* takes the form of vector of new_a, $a \in A$, of simplex elements, where $new_a(v_a)$ is the chance that it has value v_a on a. To obtain probability that given *new* has decision value $v_d \in V_d$, by referring it to U_μ with respect to $B \subseteq A$, we first compute distribution concerning its possible vector values on B, and then combine it with estimated conditional probabilities of v_d under particular elements of $\times_{a \in B} V_a$. Starting from distribution over vector values, let us put

$$new_B(w_B) = \prod_{a \in B} new_a\left(w_B^{\downarrow\{a\}}\right)$$

(6)

as the expected chance that *new* has w_B on B. By $new_B^* \subseteq \times_{a \in B} V_a$ we denote the subset_B of all vector values such that $new_B(w_B) > 0$. Further, in order to extract conditional information contained in $\mathbf{A} = (U_\mu, A, d)$, let us introduce an expected membership function $\mu_{d/B}^* : V_d \times V_B^* \to [0, 1]$ defined by

$$\mu_{d/B}^*(v_d/w_B) = \frac{\mu_{B,d}^*(w_B, v_d)}{\mu_B^*(w_B)}$$

(7)

where

$$\mu_{B,d}^* (w_B, v_d) = \sum_{u \in U_\mu} \prod_{a \in B \cup \{d\}} u_a \left((w_B, v_d)^{\downarrow \{a\}} \right) \qquad (8)$$

and $V_B^* \subseteq \times_{a \in B} V_a$ is the set of all vector values w_B such that $\mu_B^* (w_B)$ (defined analogously to (8)) is greater than zero. Formula (7) generalizes rough membership function, since it is equivalent to (2) in case of deterministically defined attribute values over the universe. Obviously, in case of both (6) and (7), one could wonder whether we should be allowed to use multiplications, as if assuming that attributes from B correspond to pairwise independent random variables. However, on the other hand, this is the only way to combine probabilistic knowledge within a product space, unless some additional information concerning dependencies among attributes is given. We would like to use the following estimator

$$\mu_{d/B} (v_d/new) = \sum_{w_B \in V_B^*} new_B (w_B) \mu_{d/B}^* (v_d/w_B) \qquad (9)$$

corresponding to probability of v_d for new under B. It is analogous to the formula for the total probability, where, under fixed w_B, $\mu_{d/B}^* (v_d/w_B)$ is the chance that $d = v_d$ and $new_B (w_B)$ corresponds to the chance that new has particular w_B on B. Let us generalize the notion of applicability introduced in Section 2 as follows.

Definition 4. *We say that decision table* $\mathbf{A} = (U_\mu, A, d)$ *is applicable to new object* $new \notin U_\mu$ *with respect to* $B \subseteq A$ *iff there is inclusion* $new_B^* \subseteq V_B^*$.

From now on we would like, for fixed $B \subseteq A$, to reason only about new objects satisfying the above inclusion. Obviously, one could argue that such a restriction is too strong, since it would be better to take decision under at least a part of information, corresponding to subdomain of possible vector values from $new_B^* \cap V_B^*$, than to do nothing. However, we must remember that such an approach may lead to wrong predictions - here we are trying to deal with probabilities as concrete real numbers and thus, for any new, the loss of any information may result with completely irrelevant conclusions.

Consideration of only these new objects which satisfy conditions of Definition 4 for given $B \subseteq A$ has also one additional advantage. Let us note that to obtain well defined distribution estimator $\mu_{d/B} (new)$, defined in a standard way, we have to check whether $\sum_{v_d \in V_d} \mu_{d/B} (v_d/new) = 1$ and normalize formula (9) if it is not the case. It turns out that such equality is satisfied iff $new_B^* \subseteq V_B^*$. This fact shows that indeed only in case of objects to which $\mathbf{A}$ is applicable there is no loss of information along the reasoning procedure, i.e. the whole knowledge about a new object can be used for its probabilistic classification.

Definition 5. *Given decision table* $\mathbf{A} = (U_\mu, A, d)$, *subset* $B \subseteq A$ *is called an expected μ-implicant iff for each* $new \notin U_\mu$, *such that* $new_A^* \subseteq V_A^*$, *we have equality*

$$\mu_{d/B} (new) = \mu_{d/A} (new) \qquad (10)$$

If B is minimal in the sense of inclusion, satisfying (10), we call it an expected μ-decision reduct.

Given the generalization of the notion of applicability onto decision tables with uncertain objects, we can repeat the argument for searching for minimal implicants from Section 2. Inclusion $new_A^* \subseteq V_A^*$ in the above definition means that we can be initially interested only in new cases for which given decision table is applicable with respect to the whole set of conditional attributes. For fixed new, it is easy to prove that this inclusion implies $new_B^* \subseteq V_B^*$. It lets us claim that, just like before, handling smaller subset of conditions gives a chance of applying gathered information to potentially larger number of new objects. Let us conclude with the following discernibility characteristics, generalizing Proposition 2 from Section 3.

Proposition 3. *Given decision table* $\mathbf{A} = (U_\mu, A, d)$, *subset* $B \subseteq A$ *is an expected μ-implicant iff it satisfies boolean implication* $B \Rightarrow \mu_{d/A}^*$, *i.e. discerns all pairs* $w_A^1, w_A^2 \in V_A^*$ *with different uncertain membership distributions.*

Proof ($\Rightarrow$) Just like before, let us start with proving that condition connected with (10) implies $B \Rightarrow \mu_{d/A}^*$. By converse, let us assume that there are $w_A^1, w_A^2 \in V_A^*$, such that $\mu_{d/A}^* \left(v_d / w_A^1 \right) \neq \mu_{d/A}^* \left(v_d / w_A^2 \right)$ for some $v_d \in V_d$, with the same projection onto B, denoted by $w_B \in V_B^*$. Then, for at least one $i = 1, 2$ we must have $\mu_{d/B}^* \left(v_d / w_B \right) \neq \mu_{d/A}^* \left(v_d / w_A^i \right)$ and it remains to take as a counterexample object new indicating w_A^i, i.e. such that $new_A \left(w_A^i \right) = 1$, for which both equalities $\mu_{d/B} \left(v_d / new \right) = \mu_{d/B}^* \left(v_d / w_B \right)$ and $\mu_{d/A} \left(v_d / new \right) = \mu_{d/A}^* \left(v_d / w_A^i \right)$ are satisfied. ($\Leftarrow$) To opposite direction, let us just note that computations analogous to (5) result with the fact that $B \Rightarrow \mu_{d/A}^*$ implies equality $\mu_{d/B}^* \left(w_B \right) = \mu_{d/A}^* \left(w_A \right)$, for any $w_B \in V_B^*$ and $w_A \in V_A^*$, belonging to $V_A^* \left(w_B \right) = \left\{ w_A \in V_A^* : w_A^{\downarrow B} = w_B \right\}$. Further, let us consider arbitrary new such that $new_A^* \subseteq V_A^*$. For any $B \subseteq A$ and $v_d \in V_d$ we can write

$$\mu_{d/A} \left(v_d / new \right) = \sum_{w_B \in V_B^*} \sum_{w_A \in V_A^*(w_B)} new_A \left(w_A \right) \mu_{d/A}^* \left(v_d / w_A \right) \qquad (11)$$

However, we already know that if $B \Rightarrow \mu_{d/A}^*$, then for any fixed $w_B \in V_B^*$ values $\mu_{d/A}^* \left(v_d / w_A \right)$ are equal to $\mu_{d/B}^* \left(v_d / w_B \right)$ for all $w_A \in V_A^* \left(w_B \right)$. Thus we obtain

$$\mu_{d/A} \left(v_d / new \right) = \sum_{w_B \in V_B^*} \left(\sum_{w_A \in V_A^*(w_B)} new_A \left(w_A \right) \right) \mu_{d/B}^* \left(v_d / w_B \right) \qquad (12)$$

Now it is enough to realize that inclusion $new_A^* \subseteq V_A^*$ implies that we can replace $\sum_{w_A \in V_A^*(w_B)} new_A \left(w_A \right)$ by $new_B \left(w_B \right)$, which finally leads to the equality $\mu_{d/A} \left(v_d / new \right) = \mu_{d/B} \left(v_d / new \right)$.

5   Conclusions

We presented a rough set approach to searching for minimal sets of features classifying objects in terms of rough membership information. We also proposed a reasoning model for dealing with objects indeterministically defined on attributes. Due to discernibility characteristics for minimal frequential implicant search problem, attributes preserving frequential decision information can be found using algorithms implemented for consistent decision tables, where complexity remains similar. Still, requirements corresponding to such preservation are too rigorous in practice and thus we may need some approximations (compare with [10]). Another direction for further research is to reason about indeterministic cases under "almost" applicable information, by introducing a kind of rough inclusion ([1]) to definition of applicability.

References

1. Komorowski, J., Polkowski, L., Skowron, A.: Towards a rough mereology–based logic for approximate solution synthesis Part 1. Studia Logica **58/1** (1997) 143–184
2. Nguyen, S. Hoa, Skowron, A., Synak, P.: Rough sets in data mining: approximate description of decision classes. In: Proceedings of the Fourth European Congress on Intelligent Techniques and Soft Computing (EUFIT'96), September 2–5, Aachen, Germany, Varlag Mainz (1996) 149–153
3. Øhrn, A., Komorowski, J., Skowron, A., Synak, P.: A software system for rough data analysis. Bulletin of the International Rough Set Society 1/ 2 (1997) 58–59
4. Pawlak, Z.: Rough sets - Theoretical aspects of reasoning about data. Kluwer Academic Publishers, Dordrecht (1991)
5. Pawlak, Z., Skowron, A.: Rough membership functions. In: R.R. Yaeger, M. Fedrizzi, and J. Kacprzyk (eds.), Advances in the Dempster Shafer Theory of Evidence, John Wiley & Sons, Inc., New York, Chichester, Brisbane, Toronto, Singapore (1994) 251–271
6. Pearl, J.: Probabilistic reasoning in intelligent systems: Networks of plausible inference. Morgan Kaufmann (1988)
7. Skowron, A.: Extracting laws from decision tables. Computational Intelligence 11/2 (1995) 371–388
8. Skowron, A., Rauszer, C.: The discernibility matrices and functions in information systems. In: R. Słowiński (ed.), Intelligent Decision Support. Handbook of Applications and Advances of the Rough Set Theory, Kluwer Academic Publishers, Dordrecht (1992) 311–362
9. Ślęzak, D.: Approximate reducts in decision tables. In: Proceedings of the Sixth International Conference on Information Processing and Management of Uncertainty in Knowledge-Based Systems (IPMU'96), July 1–5, Granada, Spain (1996) 3 1159–1164
10. Ślęzak, D.: Searching for dynamic reducts in inconsistent decision tables. Accepted to the Seventh International Conference on Information Processing and Management of Uncertainty in Knowledge-Based Systems (IPMU'98), July 6–10, Paris, France (1998)

A New Rough Set Approach to Multicriteria and Multiattribute Classification

Salvatore Greco[1], Benedetto Matarazzo[1], and Roman Slowinski[2]

[1] Faculty of Economics, University of Catania,
55, Corso Italia, I-95129 Catania, Italy
[2] Institute of Computing Science, Poznan University of Technology,
3a, Piotrowo 60-965 Poznan, Poland

1 Introduction

As pointed out by Greco, Matarazzo and Slowinski [1] the original rough set approach does not consider *criteria*, i.e. attributes with ordered domains. However, in many real problems the *ordering properties* of the considered attributes may play an important role. E.g. in a bankruptcy evaluation problem, if firm A has a low value of the debt ratio (Total debt/Total assets) and firm B has a large value of the same ratio, within the original rough set approach the two firms are just discernible, but no preference is established between them two with respect to the attribute "debt ratio". Instead, from a decisional point of view, it would be better to consider firm A as preferred to firm B, and not simply "discernible", with respect to the attribute in question.

Motivated by the previous considerations, Greco, Matarazzo and Slowinski [2] proposed a new rough set approach to take into account the ordering properties of criteria. Similarly to the original rough set analysis, the proposed approach is based on approximations of a partition of objects in some pre-defined categories. However, differently from the original approach, the categories are ordered from the best to the worst and the approximations are built using dominance relations, being specific order binary relations, instead of indiscernibility relations, being equivalence relation. The considered dominance relations are built on the basis of the information supplied by condition attributes which are all criteria. In this paper we generalize this approach considering a set of condition attributes which are not all criteria.

The paper is organized in the following way. In the second section, the main concepts of the rough approximation based on criteria and attributes are introduced. In section 3 we apply the proposed approach to a didactic example to compare the results with the original rough set approach. Final section groups conclusions.

2 Multicriteria and multiattribute rough approximation

As usual, by an *information table* we understand the 4-tuple $S = \langle U, Q, V, f \rangle$, where U is a finite set of objects, Q is a finite set of *attributes*, $V = \bigcup_{q \in Q} V_q$

L. Polkowski and A. Skowron (Eds.): RSCTC'98, LNAI 1424, pp. 60–67, 1998.
© Springer-Verlag Berlin Heidelberg 1998

and V_q is a domain of the attribute q, and $f : U \times Q \to V$ is a total function such that $f(x,q) \in V_q$ for every $q \in Q$, $x \in U$, called an *information function* (cf. Pawlak [4]).

Moreover, an information table can be seen as *decision table* assuming the set of attributes $Q = C \cup D$ and $C \cap D = \emptyset$, where set C contains so called *condition attributes*, and D, *decision attributes*.

In general, the notion of attribute differs from that of criterion, because the domain (scale) of a criterion has to be ordered according to a decreasing or increasing preference, while the domain of the attribute does not have to be ordered. We will use the notion of criterion only when the preferential ordering of the attribute domain is important in a given context. Formally, for each $q \in C$ which is a criterion there exists an *outranking* relation (Roy [6]) S_q on U such that xS_qy means "x is at least as good as y with respect to attribute q". We suppose that S_q is a total preorder, i.e. a strongly complete and transitive binary relation on U. Instead, for each attribute $q \in C$ which is not a criterion, there exists an *indiscernibility* relation I_q on U which, as usual in rough sets theory, is an equivalence binary relation, i.e. reflexive, symmetric and transitive. We denote by $C^>$ the subset of attributes being criteria in C and by $C^=$ the subset of attributes which are not criteria, such that $C^> \cup C^= = C$ and $C^> \cap C^= = \emptyset$. Moreover, for each $P \subseteq C$ we denote by $P^>$ the set of criteria contained in C, i.e. $P^> = P \cap C^>$, and by $P^=$ the set of attributes which are not criteria contained in C, i.e. $P^= = P \cap C^=$.

Let R_P be a reflexive and transitive binary relation on U, i.e. R_P is a a partial preorder on U, defined on the basis of the information given by the attributes in $P \subseteq C$. More precisely, for each $P \subseteq C$ we can define R_P as follows: $\forall x, y \in U$, xR_Py if xS_qy for each $q \in P^>$ (i.e. x outranks y with respect to all the criteria in P) and xI_qy for each $q \in P^=$ (i.e. x is indiscernible with y with respect to all the attributes which are not criteria in P). If $P \subseteq C^>$ (i.e. if all the attributes in P are criteria) and xR_Py, then x outranks y with respect to each $q \in P$ and therefore we can say that x dominates y with respect to P. Let us observe that in general $\forall x, y \in U$ and $\forall P \subseteq C$, xR_Py if and only if x dominates y with respect to $P^>$ and x is indiscernible with y with respect to $P^=$.

Furthermore let $\mathbf{Cl} = \{Cl_t, t \in T\}$, $T = \{1, \cdots, n\}$, be a set of classes of U, such that each $x \in U$ belongs to one and only one $Cl_t \in \mathbf{Cl}$. We suppose that $\forall r, s \in T$, such that $r > s$, the elements of Cl_r are preferred (strictly or weakly (Roy [6])) to the elements of Cl_s. More formally, if S is a comprehensive outranking relation on U, i.e. if $\forall x, y \in U$ xSy means "x is at least as good as y", we suppose

$$[x \in Cl_r, y \in Cl_s, r > s] \Rightarrow [xSy \text{ and not } ySx].$$

In simple words the classes $\mathbf{Cl}$ represent a comprehensive evaluation of the objects in U: the worst objects are in Cl_1, the best objects are in Cl_n, the other objects belong to the remaining classes Cl_r, according to an evaluation improving with the index $r \in T$. E.g. considering a credit evaluation problem we can have $T = \{1, 2, 3\}$, $\mathbf{Cl} = \{Cl_1, Cl_2, Cl_3\}$ and Cl_1 represents the class of the "unacceptable" firms, Cl_2 represents the class of "uncertain" firms, Cl_3 represents the class of "acceptable" firms.

Starting from the classes in $\mathbf{Cl}$, we can define the following sets:

$$Cl_t^{\geq} = \bigcup_{s \geq t} Cl_s,$$

$$Cl_t^{\leq} = \bigcup_{s \leq t} Cl_s.$$

Let us remark that $Cl_1^{\geq} = Cl_n^{\leq} = U$, $Cl_n^{\geq} = Cl_n$ and $Cl_1^{\leq} = Cl_1$. Furthermore $\forall t = 2, \cdots, n$ we have:

$$Cl_{t-1}^{\leq} = U - Cl_t^{\geq} \tag{1}$$

and

$$Cl_t^{\geq} = U - Cl_{t-1}^{\leq}. \tag{2}$$

For each $P \subseteq C$, let be

$$R_P^+(x) = \{y \in U : y R_P x\},$$

$$R_P^-(x) = \{y \in U : x R_P y\}.$$

Let us observe that, given $x \in U$, $R_P^+(x)$ represents the set of all the objects $y \in U$ which dominates x with respect to $P^>$ (i.e. the criteria of P) and are indiscernible with x with respect to $P^=$ (i.e. the attributes of P). Analogously $R_P^-(x)$ represents the set of all the objects $y \in U$ which are dominated by x with respect to $P^>$ and are indiscernible with x with respect to $P^=$.

We say that, with respect to $P \subseteq C$ and $t \in T$, $x \in U$ belongs to $Cl_t^{\geq}$ *without any ambiguity* if $x \in Cl_t^{\geq}$ and $y \in Cl_t^{\geq}$ for all the objects $y \in U$ dominating x with respect to $P^>$ and indiscernible with x with respect to $P^=$.

Formally, remembering the reflexivity of R_P, we can say that $x \in U$ belongs to $Cl_t^{\geq}$ without any ambiguity if $R_P^+(x) \subseteq Cl_t^{\geq}$. Furthermore we say that, with respect to $P \subseteq C$ and $t \in T$, $y \in U$ *could belong* to $Cl_t^{\geq}$ if there exists at least one object $x \in Cl_t^{\geq}$ such that y dominates x with respect to $P^>$ and y is indiscernible with x with respect to $P^=$, i.e. $y \in R_P^+(x)$. Our definitions of lower and upper approximation are based on the previous ideas. Thus, with respect to $P \subseteq C$, the set of all the objects belonging to $Cl_t^{\geq}$ without any ambiguity constitutes the lower approximation of $Cl_t^{\geq}$, while the set of all the objects which could belong to $Cl_t^{\geq}$ constitutes the upper approximation of $Cl_t^{\geq}$.

Formally, $\forall t \in T$ and $\forall P \subseteq C$ we define the lower approximation of $Cl_t^{\geq}$ with respect to P, denoted by $\underline{P}Cl_t^{\geq}$, and the upper approximation of $Cl_t^{\geq}$ with respect to P, denoted by $\overline{P}Cl_t^{\geq}$, as:

$$\underline{P}Cl_t^{\geq} = \{x \in U : R_P^+(x) \subseteq Cl_t^{\geq}\},$$

$$\overline{P}Cl_t^{\geq} = \bigcup_{x \in Cl_t^{\geq}} R_P^+(x).$$

We say that, with respect to $P \subseteq C$ and $t \in T$, $x \in U$ belongs to $\text{Cl}_t^{\leq}$ without any ambiguity if $x \in \text{Cl}_t^{\leq}$ and $y \in \text{Cl}_t^{\leq}$ for all the objects $y \in U$ dominated by x with respect to $P^>$ and indiscernible with x with respect to $P^=$.

Formally, remembering the reflexivity of R_P, we can say that $x \in U$ belongs to $\text{Cl}_t^{\leq}$ without any ambiguity if $R_P^-(x) \subseteq \text{Cl}_t^{\geq}$. Furthermore we say that with respect to $P \subseteq C$, $y \in U$ could belong to $\text{Cl}_t^{\leq}$ if there exists at least one object $x \in \text{Cl}_t^{\leq}$ such that x dominates y with respect to $P^>$ and y is indiscernible with x with respect to $P^=$, i.e. $y \in R_P^-(x)$. Thus, with respect to $P \subseteq C$, the set of all the objects belonging to $\text{Cl}_t^{\leq}$ without any ambiguity constitutes the lower approximation of $\text{Cl}_t^{\leq}$, while the set of all the objects which could belong to $\text{Cl}_t^{\leq}$ constitutes the upper approximation of $\text{Cl}_t^{\leq}$.

Formally, $\forall t \in T$ and $\forall P \subseteq C$, we define the lower approximation of $\text{Cl}_t^{\leq}$ with respect to P, denoted by $\underline{P}\text{Cl}_t^{\leq}$, and the upper approximation of $\text{Cl}_t^{\leq}$ with respect to P, denoted by $\overline{P}\text{Cl}_t^{\leq}$, as:

$$\underline{P}\text{Cl}_t^{\leq} = \{x \in U : R_P^-(x) \subseteq \text{Cl}_t^{\leq}\},$$

$$\overline{P}\text{Cl}_t^{\leq} = \bigcup_{x \in \text{Cl}_t^{\leq}} R_P^-(x).$$

The *P-boundary* (doubtful region) of $\text{Cl}_t^{\geq}$ and $\text{Cl}_t^{\leq}$ are respectively defined as

$$\text{Bn}_P(\text{Cl}_t^{\geq}) = \overline{P}\text{Cl}_t^{\geq} - \underline{P}\text{Cl}_t^{\geq},$$

$$\text{Bn}_P(\text{Cl}_t^{\leq}) = \overline{P}\text{Cl}_t^{\leq} - \underline{P}\text{Cl}_t^{\leq}.$$

$\forall t \in T$ and $\forall P \subseteq C$ we define the *accuracy* of the approximation of $\text{Cl}_t^{\geq}$ and $\text{Cl}_t^{\leq}$ as the ratios:

$$\alpha_P(\text{Cl}_t^{\geq}) = \frac{\text{card}(\underline{P}\text{Cl}_t^{\geq})}{\text{card}(\overline{P}\text{Cl}_t^{\geq})},$$

$$\alpha_P(\text{Cl}_t^{\leq}) = \frac{\text{card}(\underline{P}\text{Cl}_t^{\leq})}{\text{card}(\overline{P}\text{Cl}_t^{\leq})},$$

respectively. The coefficient

$$\gamma_P(\text{Cl}) = \frac{\text{card}(U - ((\bigcup_{t \in T} \text{Bn}_P(\text{Cl}_t^{\leq})) \cup (\bigcup_{t \in T} \text{Bn}_P(\text{Cl}_t^{\geq}))))}{\text{card}(U)}$$

is called the *quality of approximation of partition* **Cl** by set of attributes P, or in short, *quality of classification*. It expresses the ratio of all P-correctly classified objects to all objects in the table. Each minimal subset $P \subseteq C$ such that $\gamma_P(\text{Cl}) = \gamma_C(\text{Cl})$ is called a *reduct* of **Cl** and denoted by RED_{Cl}. Let us remark that an information table can have more than one reduct. The intersection of all reducts is called the *core* and denoted by CORE_{Cl}.

3   An example

The following example (based on a previous example proposed by Pawlak [5]) illustrates the concepts introduced above. In Table 3, twelve warehouses are described by means of five attributes:

- A_1, capacity of the sales staff,
- A_2, perceived quality of goods,
- A_3, high traffic location,
- A_4, geographical region,
- A_5, warehouse profit or loss.

In fact, A_1, A_2 and A_3 are criteria, because their domains are ordered, A_4 is an attribute, whose domain is not ordered, and A_5 is a decision attribute, defining two ordered decision classes. More in detail we have that

- with respect to A_1 "high" is better than "medium" and "medium" is better than "low",
- with respect to A_2 "good" is better than "medium",
- with respect to A_3 "yes" is better than "no",
- with respect to A_5 "profit" is better than "loss".

Table 1. Example of an information table.

Warehouse	A_1	A_2	A_3	A_4	A_5
1	High	Good	no	A	Profit
2	Medium	Good	no	A	Loss
3	Medium	Good	no	A	Profit
4	Low	Medium	no	A	Loss
5	Medium	Medium	yes	A	Loss
6	High	Medium	yes	A	Profit
7	Medium	Medium	no	A	Profit
8	High	Good	no	B	Profit
9	Medium	Good	no	B	Profit
10	Low	Medium	no	B	Loss
11	Medium	Medium	yes	B	Profit
12	High	Medium	yes	B	Profit

3.1   The results from classical rough set approach

By means of the classical rough set approach we approximate the class Cl_1 of the warehouses making loss and the class Cl_2 of the warehouses making profit. It is clear that $C = \{A_1, A_2, A_3, A_4\}$ and $D = \{A_5\}$. The C-lower approximations, the C-upper approximations and the C-boundaries of sets Cl_1 and Cl_2 are respectively:

$\underline{C}Cl_1 = \{4, 5, 10\}$, $\overline{C}Cl_1 = \{2, 3, 4, 5, 10\}$, $Bn_C(Cl_1) = \{2, 3\}$, $\underline{C}Cl_2 = \{1, 6, 7, 8, 9, 11, 12\}$, $\overline{C}Cl_2 = \{1, 2, 3, 6, 7, 8, 9, 11, 12\}$, $Bn_C(Cl_2) = \{2, 3\}$. Therefore the accuracy of the approximation is 0.6 for the class of warehouses making loss and 0.78 for the class of warehouses making profit and the quality of classification is equal to 0.83. There is only one reduct which is also the core, i.e. $Red(C) = Core(C) = \{A_1, A_2, A_3, A_4\}$.

Using the algorithm LERS (Grzymala-Busse [3]) the following set of decision rules is obtained from the considered decision table 3 (Table 1) (within brackets there are the objects supporting the corresponding rules):

1. if $f(x, A_1) =$ high, then $x \in Cl_2$ (1, 6, 8, 12)
2. if $f(x, A_1) =$ medium and $f(x, A_4) = B$, then $x \in Cl_2$ (9, 11)
3. if $f(x, A_1) =$ medium and $f(x, A_2) =$ medium and $f(x, A_3) =$ no, then $x \in Cl_2$ (7)
4. if $f(x, A_1) =$ medium and $f(x, A_2) =$ good and $f(x, A_4) = A$, then $x \in Cl_1$ or $x \in Cl_2$ (2, 3)
5. if $f(x, A_1) =$ low, then $x \in Cl_1$ (4, 10)
6. if $f(x, A_1) =$ medium and $f(x, A_3) =$ yes and $f(x, A_4) = A$, then $x \in Cl_1$ (5)

3.2 The results from approximations by dominance and indiscernibility relations

With this approach we approximate the class $Cl_1^{\leq}$ of the warehouses at most making loss and the class $Cl_2^{\geq}$ of the warehouses at least making profit. Since only two classes are considered, we have $Cl_1^{\leq} = Cl_1$ and $Cl_2^{\geq} = Cl_2$. When a larger number of classes is considered this equalities are not satisfied.

The C-lower approximations, the C-upper approximations and the C-boundaries of sets $Cl_1^{\leq}$ and $Cl_2^{\geq}$ are respectively: $\underline{C}Cl_1^{\leq} = \{4, 10\}$, $\overline{C}Cl_1^{\leq} = \{2, 3, 4, 5, 7, 10\}$, $Bn_C(Cl_1^{\leq}) = \{2, 3, 5, 7\}$, $\underline{C}Cl_2^{\geq} = \{1, 6, 8, 9, 11, 12\}$, $\overline{C}Cl_2^{\geq} = \{1, 2, 3, 5, 6, 7, 8, 9, 11, 12\}$, $Bn_C(Cl_2^{\geq}) = \{2, 3, 5, 7\}$. Therefore, the accuracy of the approximation is 0.33 for $Cl_1^{\leq}$ and 0.6 for $Cl_2^{\geq}$ while the quality of classification is equal to 0.67. There is only one reduct, which is also the core, i.e. $RED_{Cl}(C) = CORE_{Cl}(C) = \{A_1, A_4\}$.

The following minimal set of decision rules can be obtained from the considered decision table (within parentheses there are the objects supporting the corresponding rules):

1. if $f(x, A_1)$ is high, then $x \in Cl_2^{\geq}$ (1, 6, 8, 12)
2. if $f(x, A_1)$ is at least medium and $f(x, A_4)$ is B, then $x \in Cl_2^{\geq}$ (8, 9, 11, 12)
3. if $f(x, A_1)$ is low, then $x \in Cl_1^{\leq}$ (4, 10)
4. if $f(x, A_1)$ is medium and $f(x, A_4)$ is A, then $x \in Cl_1^{\leq}$ or $x \in Cl_2^{\geq}$ (2, 3, 5, 7).

3.3   Comparison of the results

The advantages of the rough set approach based on dominance and indiscernibility relations over the original rough set analysis, based on the indiscernibility
relation, can be summarized in the following points.

The results of the approximation are more satisfactory. This improvement
is represented by a smaller reduct ($\{A_1, A_4\}$ against $\{A_1, A_2, A_3, A_4\}$). Let us
observe that even if the quality of the approximation is deteriorated (0.67 vs.
0.83), this is another point in favour of the proposed approach. In fact, this
difference is due to the warehouses 5 and 7. Let us notice that with respect to
the attributes A_1, A_2, A_3, which are criteria, warehouse 5 dominates warehouse
7 and with respect to the attribute A_4, which is not a criterion, the warehouse
5 and 7 are indiscernible. However warehouse 5 has a comprehensive evaluation
worse than warehouse 7. Therefore, this can be interpreted as an inconsistency
revealed by the approximation by dominance and indiscernibility that cannot be
pointed out when we consider the approximation by indiscernibility only.

From the viewpoint of the quality of the set of decision rules extracted from
the information table by the two approaches, let us remark that the decision rules
obtained from the approximation by dominance and indiscernibility relations
give a more synthetic representation of knowledge contained in the information
table. The minimal set of decision rules obtained from the new approach has a
smaller number of rules (4 against 6), uses a smaller number of attributes and
descriptors than the set of the decision rules obtained from the classical rough set
approach, obtains rules supported by a larger number of objects. Furthermore,
let us observe that the rules obtained from the original rough sets approach
present some problems with respect to their interpretation. E.g. rule 3 obtained
by the original rough set approach says that if the capacity of the sale staff
is medium, the perceived quality of goods is medium and if the warehouse is
not in a high traffic location then the warehouse makes profit. One can expect
that improving the quality of the warehouse, e.g. considering a warehouse with
the same capacity of the sales staff and the same quality of goods but located
in a high traffic location the warehouse should also make profit. Surprisingly,
the warehouse 5 of the considered decision table has these characteristics but
it makes loss. Finally, let us remark that rule 4 from the new approach is an
approximate rule, as well as rule 4 from the classical approach. However, rule 4
from the new approach is based on a small number of descriptors and supports
a greater number of actions.

4   Conclusion

We presented a new rough set approach whose purpose is to approximate sets
of objects divided in ordered predefined categories considering criteria, i.e. attributes with ordered domains, jointly with attributes which are not criteria.
We showed that the basic concepts of the rough sets theory can be restored in
the new context. We also applied the proposed methodology to an exemplary

problem approached also with the classical rough set analysis. The comparison of the results proved the usefulness of the new approach.

Acknowledgments

The research of the first two authors has been supported by grant no. 96.01658. *ct*10 from Italian National Research Council (CNR). The third author wishes to acknowledge financial support from State Committee for Scientific Research, KBN research grant no. 8 T11C 013 13, and from CRIT 2 - Esprit Project no. 20288. For the task of typing this paper, we are indebted to the high qualification of Ms Silvia Angilella.

References

1. Greco S., Matarazzo, B., Slowinski, R.: Rough approximation of a preference relation by dominance relations, ICS Research Report **16**, Warsaw University of Technology, Warsaw, (1996) and to be published on European Journal of Operational Research.
2. Greco, S., Matarazzo, B. Slowinski, R.: A new rough set approach to evaluation of bankruptcy risk, in C. Zopounidis (ed.), Operational Tools in the Management of Financial Risks, Kluwer, Dordrecht, (1998), 121–136.
3. Grzymala–Busse, J.W: LERS - a system for learning from examples based on rough sets, in R. Slowinski, (ed.), Intelligent Decision Support. Handbook of Applications and Advances of the Rough Sets Theory, Kluwer Academic Publishers, Dordrecht, (1992), 3–18.
4. Pawlak, Z.: Rough Sets. Theoretical Aspects of Reasoning about Data, Kluwer Academic Publishers, Dordrecht, (1991).
5. Pawlak, Z.: Rough set approach to knowledge-based decision support, European Journal of Operational Research, **99**, (1997), 48–57.
6. Roy, B.: Méthodologie Multicritère d'Aide à la Décision, Economica, Paris, (1985).

A Heuristic Method of Model Choice for Nonlinear Regression

J. Ćwik and J. Koronacki

Institute of Computer Science, Polish Academy of Sciences,
ul. Ordona 21, 01-237 Warsaw, Poland
jc, korona@ipipan.waw.pl

[Abstract.] A heuristic method of model choice for a nonlinear regression problem on real line, based on the Equation Finder (EF) of Zembowicz and Żytkow (1992), is proposed and discussed. In our implementations of the EF we use a new, actually a three-stage, procedure for stabilizing model selection. First, a set of pseudosamples is obtained from the original sample by resampling in some way. Second, for each pseudosample, a family of acceptable models is found by a clustering-like algorithm performed on models with largest (adjusted) coefficients of determination. And third, the final selection is made from among the models which appear most often in the families obtained in the second stage.

1 Introduction and Outline of the Model Selection Procedure

Discovering equations or functional relationships from data in presence of random errors should be an inherent capability of every data mining and, more generally, knowledge discovery system. Such discoveries, in statistical terms, amount to choosing a model in a regression set-up. That is, given a random sample of pairs of variables, $(x_1, y_1), \ldots, (x_n, y_n)$, where $y_i = f(x_i) + \varepsilon_i$, $i = 1, 2, \ldots, n$, ε_i's are zero-mean random variables and f is an unknown function, the task is to estimate f by a member of a given family of nonlinear functions and to assess validity of the model thus obtained.

In this report, we present a new application of Equation Finder (EF), a discovery system of Zembowicz and Żytkow (see Zembowicz and Żytkow (1992) for a detailed description and discussion of the system and Moulet (1997) for the latter). The system finds "acceptable" models by means of a systematic search among polynomials of transformed data, when transformations – such as, e.g., logarithm, exponent and inverse – of both the independent and response variable are allowed. The family of all models to be considered is decided upon in advance by listing possible transformations of data and choosing the highest possible order of the polynomials. In the original version of EF, possible acceptance of a particular model is based on weighted least-squares (WLS) estimation of the model parameters and on the χ^2 test of fit.

L. Polkowski and A. Skowron (Eds.): RSCTC'98, LNAI 1424, pp. 68–74, 1998.
© Springer-Verlag Berlin Heidelberg 1998

Although EF uniquely determines the true model for samples from the same true model with sufficiently small error or large range of x values, it fails to provide sufficiently stable results for samples that carry larger error. As will be seen from simulation results in Section 3, for different samples from the same distribution the system provides results that may differ substantially from sample to sample. (Admittedly, this sort of instability is common to practically all the well-established methods of model choice in nonlinear regression.) It follows that some sort of stabilization of the original model selection procedure is needed, at least for regression problems with "larger errors".

In our implementations of the system, both the regression models and their parameters are provided by EF, the latter in accordance with the WLS method. But our choice of acceptable models (from those provided by EF) is different. It consists of three stages. First, a set of pseudosamples is obtained from the original sample by resampling combined with leave-many-out procedure: the original data are randomly permuted and the first pseudosample is obtained by leaving-out the first 10% of the data (for simplicity, we assume here that the sample size is divisible by 10); then the second 10% is left-out to obtain the second pseudosample, again of the size equal to 90% of the size of the original data, and so on, each time giving a pseudosample of the same size; once the last 10% of the data is left-out, the whole process, starting with permuting the original data, is repeated several times (e.g., if the sample size is 100, each permutation enables one to obtain 10 pseudosamples of size 90). Second, for each pseudosample, a family of acceptable models is found by a clustering-like algorithm performed on models with largest (adjusted) coefficients of determination. Loooosely speaking, clustering rests on two-dimensional scaling based on distances between the models. And third, the final selection is made from among the models which appear most often in the families obtained in the second stage.

The rationale behind the method of choice of acceptable models is simple. Had we many independent sets of data, it would be reasonable to find a regression model for each set independently of the others, and then to aggregate the models obtained into one. Given just one set, we use it to produce many pseudosamples and proceed as if they were independent sets of data. In fact, our method of model selection borrows from stabilization ideas of Breiman (1996a) (see also Breiman (1996b)), with aggregation of multiple versions of a predictor[1] essentially made by a plurality vote: the final selection is made from among the models which appear most often in clusters of models obtained for subsequent pseudosamples. It is believed, and has been confirmed by simulations, that predictors which are relatively close to the true model underlying the data are likely to form clusters in a properly defined space of models.

In the next section, more details on the method will be given. Simulation results will be discussed in Sect. 3. Although we shall confine ourselves to discussing just one example, several more have been thoroughly investigated, leading to essentially the same conclusions. All in all, the simulations show that the method

---

[1] Borrowing from machine learning terminology, we use the terms regression model and predictor interchangeably.

proposed can be considered a useful tool for model selection in the nonlinear regression context.

In this report, we deal with discovering functions of one variable only, in accordance with the current capability of EF. In a separate report by the authors and J. Żytkow, possible extensions to discovering functions of several variables will be discussed.

2   More on the Method

Assume that the true model has the form

$$Y = f(X) + \varepsilon(X), \tag{1}$$

where f is a regression function (i.e., $E(\varepsilon(X)|X) = 0$) and $Var(\varepsilon(X)|X)$ is a function of X. Unlike Zembowicz and Żytkow (1992) we do not assume the variances $Var(\varepsilon(X)|X)$ to be known (in EF only the WLS method of estimation is implemented and therefore the system requires that the weights for the WLS equation be provided). Instead, given a sample $(x_1, y_1), (x_2, y_2), \ldots, (x_n, y_n)$, we apply the Additive and Variance Stabilizing Transformation (Tibshirani's (1988) AVAS with the Supersmoother of Friedman and Stuetzle (1982)) to obtain an auxiliary estimate $\tilde{f}$ of f and we use

$$w_i \equiv (y_i - \tilde{f}(x_i))^{-2}$$

as the weights in the WLS equation for computing parameters of any particular model provided by EF (this simple choice of weights has proved to work surprisingly well in the simulations). The EF is applied separately to each pseudosample generated as described in the previous section.

For each fixed pseudosample (by abuse of notation to be denoted identically as the original sample), the models obtained are ranked by the adjusted coefficient of determination

$$R^2_{adj} = 1 - \frac{\text{WSSR}/(n-p)}{\text{WSST}/(n-1)},$$

where, as usual, p is the number of parameters in the model, WSSR is the weighted sum of squares of residuals (with $\hat{f}$ denoting the model obtained),

$$\text{WSSR} = \sum_{i=1}^{n} w_i (y_i - \hat{f}(x_i))^2,$$

and WSST is the weighted total sum of squares,

$$\text{WSST} = \sum_{i=1}^{n} w_i (y_i - \bar{y})^2,$$

with $\bar{y} = n^{-1} \sum y_i$. Now, a fraction of models with largest R^2_{adj} is chosen for later investigation; either the fraction value can be determined a priori or a threshold value of R^2_{adj} can be predetermined or the threshold value can be found by inspection: as a rule, "large" values of R^2_{adj} appear in a clear-cut cluster. In turn, the matrix of "empirical squared distances" between all pairs of the models chosen is formed, with the "empirical squared distance" between models $\hat{f}_j$ and $\hat{f}_k$ defined as

$$n^{-1} \sum_{i=1}^{n} (\hat{f}_j(x_i) - \hat{f}_k(x_i))^2.$$

In this way, a matrix of dissimilarities between the models is formed, which is then used to represent the models in a two-dimensional space via classical metric scaling (see, e.g., Krzanowski (1988); in our computations, S-Plus's cmdscale function is used). The "most outlying" models in the 2D-representation obtained are removed and a new representation of the remaining models is obtained, again by classical metric scaling. This process of removing the "most outlying" models and building a 2D-representation of the models still in the family is performed repeatedly. In our implementations of such a clustering-like algorithm, we proceeded in one of two ways. One was automatic: the number of repetitions of the process was determined in advance and, at each repetition, a fixed number of models was retained for further analysis. For instance, for $n = 100$, the process was repeated 3 times, with the numbers of models retained equal, respectively, to 20, 15 and 10. The most outlying models were defined as those farthest from the models' center of gravity in the 2D-representation. Surprisingly, despite obvious drawbacks of this criterion, it proved to work generally well in the simulations. The other way was subjective: the outlying models were found by inspection, and the process of removing models and re-representing them in a 2D-space was repeated until no obviously "outlying models" could be found (usually, the process was again repeated thrice). Subjective detection of "outliers" was facilitated by constructing also a dendrogram of models, again based on the same dissimilarities between the models (S-Plus's hclust function for complete linkage clustering was used).

Equipped with the outcomes of applying the clustering-like algorithm to all the pseudosamples, the final selection of acceptable models can be made from among the models which appear most often in the families obtained for the pseudosamples. That is, each model is ranked according to the number of pseudosamples for which the model has been retained after applying the clustering-like algorithm and the final selection is made from among models with highest ranks (the given way of ranking is referred to as the plurality vote).

3   Simulation Results

In our example, the regression function was of the form

$$f(x) = 1 + \frac{1}{x},$$

$x \in [1, 10]$, and the ε's were normally distributed with homogeneous variance, $Var(\varepsilon(X)|X) = 0.02$. Equation Finder was set at search depth equal to one, with transformations SQR, SQRT, EXP, LOG, INV, MULTIPLICATION and DIVISION, and with maximum polynomial degree equal to three. Fifty i.i.d. random samples $(X_1, Y_1), (X_2, Y_2), \ldots, (X_{100}, Y_{100})$ of size 100 were generated from model (1).

The EF was first used on all 50 samples and stability of solutions obtained was assessed. For each sample, models provided by EF were ranked by their WSSR's and the best 40 were subjected to two additional rankings, one based on the empirical mean squared error,

$$\text{EMSE} = \frac{1}{n} \sum_{i-1}^{n} (\hat{f}(x_i) - f(x_i))^2,$$

and another on the mean squared residual error,

$$\text{MSRE} = \frac{\text{SSR}}{n} = \frac{1}{n} \sum_{i=1}^{n} (y_i - \hat{f}(x_i))^2;$$

that is, the 40 models were ranked by their EMSE's and MSRE's. Then, only the models which appeared for all 50 or for 49 samples were retained for stability assessment. For each such model, two histograms were constructed: one for the model's EMSE ranks in all samples and another for the model's MSRE ranks in all samples.

In figure 1, histograms for EF models: No. 64 $(\hat{f}(x) = x(a + b/x + c/x^2))$, No. 113 $(\hat{f}(x) = a + b/x)$, No. 158 $(\hat{f}(x) = (a + b\sqrt{x} + cx)/x)$, No. 262 $(\hat{f}(x) = x/(a + b\log x + c\log^2 x + d\log^3 x))$, No. 359 $(\hat{f}(x) = \log(a + bx^{1/2} + cx + dx^{3/2}))$ and No. 464 $(\hat{f}(x) = \sqrt{a + b/x + c/x^2})$ are given for the sake of illustration (for each model, the two histograms are depicted together, with the adjacent bars belonging alternately to the first and the second histogram: in the leftmost pair of bars, ranks 1–3 according to EMSE and to MSRE, respectively, are grouped, in the second leftmost pair, ranks 4–6, etc.). We note that models 64, 113, 158 and 464 are all equal to f provided their parameters assume suitable values.

It is apparent from all the histograms obtained that, whatever the form of a particular model, the MSRE ranks hardly follow those based on EMSE. While the EMSE rankings for models 64, 113, 158 and 464 point to the possibility of dealing with data from f, it is by far not the case for the MSRE rankings. Rather, by the "MSRE standards" it is model 262 which should be considered most likely to be the true one. Generally, these are strong indications that not only the solutions are unstable in the given sense but we cannot rely unconditionally on the LS methodology when choosing a model (we do not have to refer here specifically to the WLS methodology since the error variance is homogeneous in the example under scrutiny). On the other hand, the plurality vote performed after applying the (automatic) clustering-like algorithm to all the 50 independent cases ranked model 158 as the first one, model 464 as the second, model 64 as the fourth and model 113 as the seventh one.

In reality, of course, we are most often given just one sample. Table 1 provides a summary of results obtained for three example samples (actually, samples No.

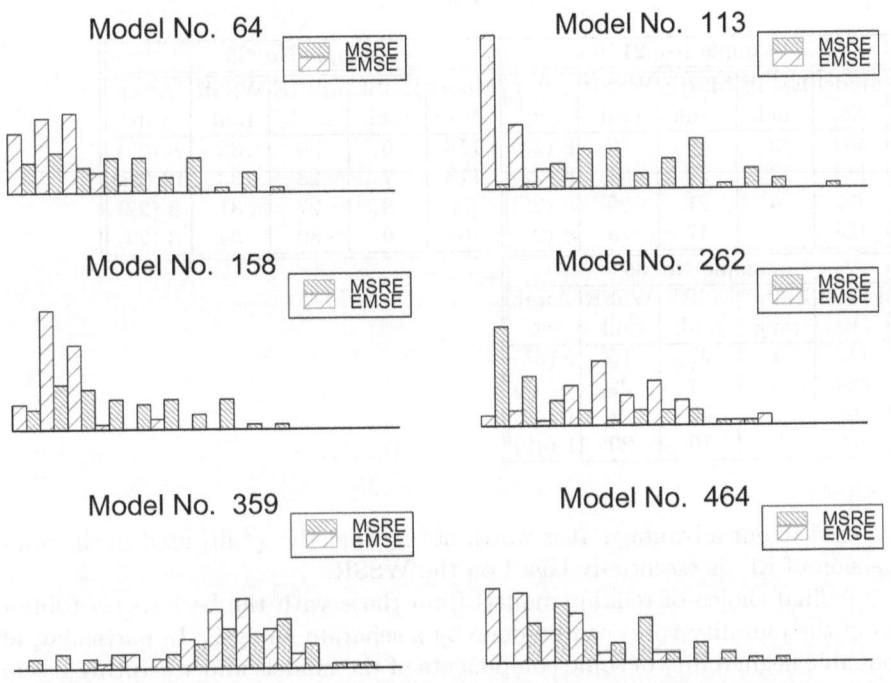

Fig. 1. Histograms of EMSE and MSRE for six example models

21, 25 and 36), when each sample was used to generate 50 pseudosamples (as described in Sect. 1) and the automatic clustering-like algorithm from Sect. 2 was applied to the pseudosamples. In the first three columns, results of rankings based on the EMSE, MSRE and WSSR are given for the four models mentioned above. The models' ranks which resulted from the plurality vote are given in the last column along with the number of pseudosamples for which the model was retained by the clustering-like algorithm (the number given in brackets). Judged by the concordance/discordance between the EMSE and MSRE, samples 21 and 36 can be considered a medium-hard basis for estimating the regression function sought, while sample 25 can be considered a hard one. For sample 21, the best model according to the plurality vote scored 36 votes, and the subsequent scores were: 33, 32, 28, 27, 27, 26, 22 six times, etc. For sample 25, the best model scored 28 votes and the subsequent scores were: 27, 26 twice, 25 twice, 23, 22 four times, etc. For sample 36, the three best models scored 49 votes and the subsequent votes were: 47 twice, 44, 41, 35, 31, 28, 20, 16, 14, 5, etc. Results reported for the three samples are rather typical when compared with those for other samples in their categories.

As the results summarized in the table are rather typical for other samples (and other examples studied), it is clearly seen that the stabilizing method can

Table 1.

sample No. 21					sample No. 25				
Model No.	EMSE rank	MSRE rank	WSSR rank	After vote	Model No.	EMSE rank	MSRE rank	WSSR rank	After vote
464	2	16	16	8 (22)	158	6	30	32	8 (22)
113	3	22	21	8 (22)	113	7	28	33	12 (21)
64	4	21	20	8 (22)	64	8	27	31	8 (22)
158	5	17	18	8 (22)	464	9	30	34	8 (22)

sample No. 36			
Model No.	EMSE rank	MSRE rank	WSSR rank	After vote
113	1	22	19	9 (31)
464	3	21	22	4 (47)
158	5	20	21	1 (49)
64	7	19	20	1 (49)

be used to our advantage. It is worth noting that the χ^2 fit, used in the original version of EF, is essentially based on the WSSR.

A final choice of reliable models from those with the best scores (obtained from the plurality vote) can be done by a separate analysis. In particular, after possible negligibility of some components of the models under scrutiny has been taken into account, such an analysis can be based on the distances between those models, their plausibility in view of some other information, etc.

Acknowledgment

We are grateful to Jan Żytkow for comments and discussion.

References

[1996a]Breiman, L.: Heuristics of instability and stabilization in model selection. Ann. Statist. **24** (1996a) 2350–2393

[1996b]Breiman, L.: Bagging predictors. Machine Learning **26** (1996b) 123–140

[1982]Friedman, J.H. and Stuetzle, W.: Smoothing of scatterplots. Technical Report, Stanford University, Dept. of Statist., 1982

[1988]Krzanowski, W.J.: Principles of multivariate analysis. Oxford University Press, 1988

[1997]Moulet, M.: Comparison of three inductive numerical law discovery systems. In: Machine learning and statistics, G. Nakhaeizadeh and C.C. Taylor (eds.), Wiley, 1997, 293–317

[1988]Tibshirani, R.: Estimating transformations for regression additivity and variance stabilization. J. American Statist. Assoc. **83** (1988) 394–405

[1992]Zembowicz, R. and Żytkow, J.M.: Discovery of equations: Experimental evaluation of convergence. In: Proceedings of the 10th National Conference on Artificial Intelligence, AAAI-92, AAAI Press, 1992, 70-75

How a New Statistical Infrastructure Induced a New Computing Trend in Data Analysis

A. Ciok, T. Kowalczyk, and E. Pleszczyńska

Institute of Computer Science, Polish Academy of Sciences
01-237 Warsaw, Ordona 21, Poland
eple@ipipan.waw.pl

Abstract. A statistical infrastructure based on the concentration measures is presented. It aims at being a consistent system of descriptive parameters with a clear interpretation, summarizing knowledge on distributions of variables in populations of objects. The system requires unified formulas for sets of mixed continuous-categorical variables. These demands are met e.g. by the formulas given in Sec. 3 for the bivariate dependence measures, presented as suitably weighted averages of concentration indices for pairs of conditional distributions. Further, in Sec. 4, a concentration oriented modification of the statistical procedure called correspondence analysis is mentioned and exemplified by applying it to data from the Polish Parliament Elections in 1993 and 1997. Sec. 5 contains remarks on creating an inference and computing system adjusted to this new concentration - based approach to statistical descriptive parameters.

1 Introduction

Statistical infrastructure appearing in the title of this paper refers to descriptive parameters which summarize our knowledge about distributions of sets of variables in populations of objects. These descriptive parameters are then exploited in various decision making schemes. It is well known that very simple data models, the multinormal model in particular, can be easily described and led usually to easy and consistent patterns of inference and computing. However, models too simplified are not helpful in solving real world problems. Many efforts have been done to construct models more complicated than multinormal but retaining a similar simple ("linear") inference and computing pattern. The most popular among them is the log-linear model applicable to sets of ordinal and nominal categorical variables (and thus also to discretized continuous ones) (cf [1]). The log-linear inference and computing system is very similar to that in the multinormal case. Computing is well-organized; it enables dealing simultaneously with many inference problems. This is achieved due to a special log-linear infrastructure which aims at transforming sets of categorical variables onto "linear-like" models. But neither descriptive parameters nor inference have a convincing interpretation and justification.

L. Polkowski and A. Skowron (Eds.): RSCTC'98, LNAI 1424, pp. 75–82, 1998.

Some recently developed computational systems refer to rather narrowly designed specific problems and admit specific families of decision rules (see e.g. [8] a comparative study of exploratory methods related to the DEDICOM system meaning DEcomposition into DIrectional COMponents). Various approaches to a data set may provide helpful complementary information but may also lead to a total mess.

Let us stress that it is the computing system which forms an interface between a specific inference system and its users. Such a system plays the role of a specific language. Users often rely too strongly on loose interpretation of the results provided by the computing system such as "significant", "strongly dependent", "redundant", "outlier", etc. They become accustomed to a particular computing system and are reluctant to make a change. This is why new ideas are not easily conveyed to practitioners. To have any chance, new ideas should be incorporated into a ready-made computing system. In the sequel we tell about a new computing trend in data analysis which is now *in statu nascendi*. This new trend is induced by statistical infrastructure based on concentration measures.

It is astonishing that statistical infrastructure based on concentration measures is being developed so slowly although basic ideas were formulated by Gini, Kakwani and Fogelson as early as the beginning of the 20th century (cf. e.g. [12]). Obviously, we will not even try to trace here the main publications in this area. Our own contribution has been reached in a team (consisting of the authors of this paper together with T. Bromek, W. Szczesny, M. Niewiadomska-Bugaj and W. Wysocki). The team's results can be found in the bibliography at the end of the present paper. The important starting points were: [5], [10], [11], [13], [21], [23]. We aim at constructing a consistent system of descriptive tools based on concentration measures, defined in a unified way for mixed data (i.e. for sets of variables measured on various measurement scales). We tried to show how this system of descriptive tools is linked with important inference problems and how this worked in a number of case studies [2], [8], [9], [10], [19], [22], [24].

In the present paper we restrict ourselves to sketch the descriptive tools used to compare one probability distribution with another and to describe bivariate dependence. This allows to indicate our main inference and computing procedures and the corresponding visualization tools. A few illustrations are made using data which concern two last elections (1993 and 1997) to the Polish Parliament.

2   A look on concentration measures

The vague idea of concentration is a fundamental informal statistical notion. Formal definitions of concentration measures serve to create the basic structure of important descriptive tools.

Let us start with two categorical variables X and Y valued $1, \ldots, k$ with probabilities $p_1, \ldots, p_k$ and $q_1, \ldots, q_k$, respectively. The concentration curve of Y w.r.t. X related to the natural order of points in $\{1, \ldots, k\}$, consists of segments which join points $(0,0)$, (p_1, q_1), $(p_1 + p_2, q_1 + q_2)$, $\ldots, (1, 1)$. To illustrate, let X and Y be two nominal categorical variables with probabilities equal to fractions of votes gained by 10 political parties in the Polish Parliament Elections in 1993 (for X) and in 1997 (for Y) (cf [24]). To be precise, there are 7 parties and 3 "quasi parties" corresponding respectively to non-voters, those who issued formally invalid votes, and those who voted in 1993 on a party which vanished in 1997 (in case of X) or vice versa (in case of Y). Suppose that the parties and quasi parties are labeled so that the likelihood ratios q_i/p_i are increasing (i.e. the parties are ordered from that which had the maximal loss to that with the maximal gain). The respective concentration curve is shown in Fig. 1. It describes changes in relative concentration of votes fractions. It is convex and consists of 6 longer segments (1, 2, 6, 7, 8, 9) and four very short ones corresponding to less popular parties. Number 7 refers to non-voters. Comparing the slopes of particular segments with the diagonal $y = x$ we see that the fraction of non-voters slightly increased in 1997, parties 8 and 9 also belong to "winners" with increased fractions of voters, the fraction corresponding to party 6 is practically unchanged, parties 1-5 are "losers".

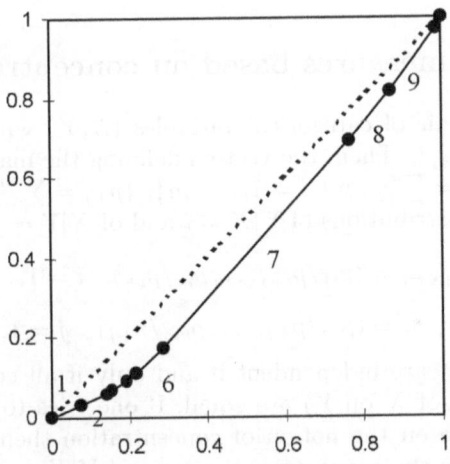

Fig. 1. The absolute concentration curve of the distribution of votes in the 1997 election to that in 1993 election.

For nominal variables the ordering related to increasing likelihood ratios seems to be the only one worth consideration. For ordinal variables it may be not the case. For instance, let X and Y be two income distributions for k ordered income categories selected a priori. Then one may be interested in the concentration curve related to the order $1, \ldots, k$ which may be different from that corresponding to increasing likelihood ratios. Thus we may draw both curves; the second one will evidently lie under the first.

For continuous variables X and Y with distribution functions F and G the situation is similar; the ratios q_i/p_i are replaced by the ratio of densities $h(x) = \frac{g(x)}{f(x)}$. The concentration curve corresponding to h will be called the *absolute concentration curve* (cf [3]) and denoted $C_{max}(Y : X)$. It lies under the *directed concentration curve* $C(Y : X; \varphi)$ for any $\varphi : \Omega \to R$ which orders the set of elementary events on which X and Y are defined.

In the general case, distributions of X and Y may be replaced by any probability measures P and Q defined on $(\Omega, \mathcal{A})$. Curves $C(Y : X; \varphi)$ and $C_{max}(Y : X)$ are function-valued measures of *differentation* between P and Q. This differentation is reflected by the position of a particular curve with respect to the diagonal $y = x$ which corresponds to equal P and Q. Departures of a concentration curve from the diagonal, positive under $y = x$, negative over it, serve to construct summary measures of differentation between P and Q. Thus we have

$$ar(Y : X; \varphi) = \int_0^1 (t - C(Y : X; \varphi)(t)) \, dt,$$

$$ar_{max}(Y : X) = \int_0^1 (t - C_{max}(Y : X)(t)) \, dt.$$

If X and Y are at least ordinal and $\varphi(x) = x$, we omit φ and write $ar(Y : X)$.

3 Dependence measures based on concentration indices

Let us start with a pair of categorical variables (X, Y) with probabilities p_{ij}, $i = 1, \ldots, m$, $j = 1, \ldots, k$. Then, the vectors defining the marginal distributions of X and Y are $\{p_{i+} = \sum_{j=1}^k p_{ij}, i = 1, \ldots, m\}$, $\{p_{+j} = \sum_{i=1}^m p_{ij}, j = 1, \ldots, k\}$; and the conditional distributions of $Y|X = i$ and of $X|Y = j$ are:

$$P_i = P_{Y|X=i} = (p_{i1}/p_{i+}, \ldots, p_{ik}/p_{i+}), \quad i = 1, \ldots, m,$$

$$Q_j = P_{X|Y=j} = (p_{1j}/p_{+j}, \ldots, p_{mj}/p_{+j}), \quad j = 1, \ldots, k.$$

Variables X and Y are independent if and only if all conditional distributions (of Y on X and of X on Y) are equal. If one tries to ground the notion of positive dependence on the notion of concentration then, the further down the diagonal $y = x$ are the concentration curves of $Y|X = x_2$ w.r.t. $Y|X = x_1$ for any $x_1 < x_2$, the stronger should be the positive dependence of Y on X. This introduces a suitable ordering in the set of pairs (X, Y) related to positive dependence. A global measure of this dependence could be based on the concentration indices of $Y|X = x_2$ w.r.t. $Y|X = x_1$ directed according to Y. Therefore, for X valued $1, ..., m$ and Y valued $1, ..., k$, we introduce summary measures of positive dependence of Y on X in the form of linear combinations of $ar(P_t : P_s)$, $s < t$, $s, t = 1, ..., m$. The leading role of concentration measures in creating statistical descriptive tools is illustrated by the fact that (cf [15]) a popular measure of positive dependence, known as Spearman's *rho* and denoted ρ^*, is representable in this form. Due to generality of the definition of the concentration index, the definition of ρ^* is immediately generalized to the whole family of pairs of continuous or categorical-continuous variables.

To mention another link between ρ^* and ar (cf e.g. [11]), we note that the distribution of any pair (X, Y) may be mapped, by means of the so called grade transformations performed simultaneously on each marginal distribution, into a continuous distribution (with uniform marginals) on the unit square. Let (X^*, Y^*) denote the respective "grade representation" of (X, Y). We have $\rho^*(X, Y) = \rho^*(X^*, Y^*) = cor(X^*, Y^*)$, and ρ^* is therefore called the grade correlation between X and Y. What is essential here, $\rho^*(X, Y) = 3ar(r(X^*) : X^*)$ where r is the regression function of Y^* on X^*.

4   A look at the grade correspondence-cluster analysis

We remind that for any $X, Y : \Omega \to R$ the concentration index $ar(Y : X)$ is accompanied by the absolute concentration index ar_{max} which is the maximum value of $ar(Y : X; \varphi)$ over all possible orderings φ in Ω. Similarly, for a pair $(X, Y) : \Omega_1 \times \Omega_2 \to R^2$, $\rho^*(X, Y)$ is accompanied by $\rho^*_{max}(X, Y)$ which is the maximum value of $\rho^*(f(X), g(Y))$ over all possible pairs of orderings in Ω_1 and Ω_2, introduced respectively by f and g. Transformations f^*, g^* of X and Y which achieve the maximum value of ρ^* provide a "ρ^* best" approximation to positive dependence. This resembles an important statistical procedure called the correspondence analysis which is based on transformations maximizing the Pearson's correlation coefficient. Our procedure has been called in [5] the *grade correspondence analysis* (GCA). According to Sec.3, ρ^*_{max} is the concentration index of the regression function of $g^*(Y^*)$ on $f^*(X^*)$, multiplied by 3.

GCA proved to be a very useful exploratory procedure, apt to a convenient visualization ([7], [9], [22]). To illustrate, we will use once more the Parliament elections data. Let $\{p_{ij}, i = 1, \ldots, 52, j = 1, \ldots, 10\}$ be the two-way table with rows corresponding to election regions (voivodships) and columns corresponding to the parties and quasi parties, while p_{ij} is the fraction of votes gained in region i by party j in the 1993 election. A similar table $\{q_{ij}\}$ refers to the 1997 election. Each table was first considered separately, but a more interesting result was obtained when the two tables were combined into one, containing still 52 rows (regions) but 20 columns (10 parties, each appearing twice, once in 1993 and once in 1997; parties corresponding to 1997 are denoted $1, 2, \ldots, 10$ as introduced in Sec. 2, their counterparts in 1993 are denoted $11, 12, \ldots, 20$, respectively). The combined 52×20 table has been transformed by the GCA procedure onto a table $\{\pi_{ij}\}$ with suitably permuted rows and columns. This allows tracing and interpreting latent traits which implied the orderings of voivodships and parties obtained due to GCA. We will analyse this latent structure looking at the results of GCA visualized in Fig. 2.

Table $\{\pi_{ij}\}$ is mapped in Fig. 2 into a unit square with 52×20 rectangles. The width and length of rectangle (i, j) is π_{i+} and π_{+j}. The product $\pi_{i+}\pi_{+j}$ is equal to the fraction of votes expected under "fair representation" related to fraction of votes ascribed to region i and total gain of votes by party j in the first or second election. The quotient $\pi_{ij}/\pi_{i+}\pi_{+j}$ is the overrepresentation index. Four thresholds of this index have been chosen: $2/3, 99/100, 100/99, 3/2$. The respective intervals of the index were called: strong underrepresentation, weak underrepresentation, almost fair representation, weak overrepresentation, strong overrepresentation. Strong and weak underrepresentation were marked white and light grey, weak and strong overrepresentation were marked dark grey and black, almost fairness was marked as white with vertical lines. The GCA transformation forced a concentration of black and dark grey rectangles in the upper left and lower right corners of the unit square and along the diagonal which joins these corners. This concentration indicates the trend of overrepresentation as a relation between parties and election regions.

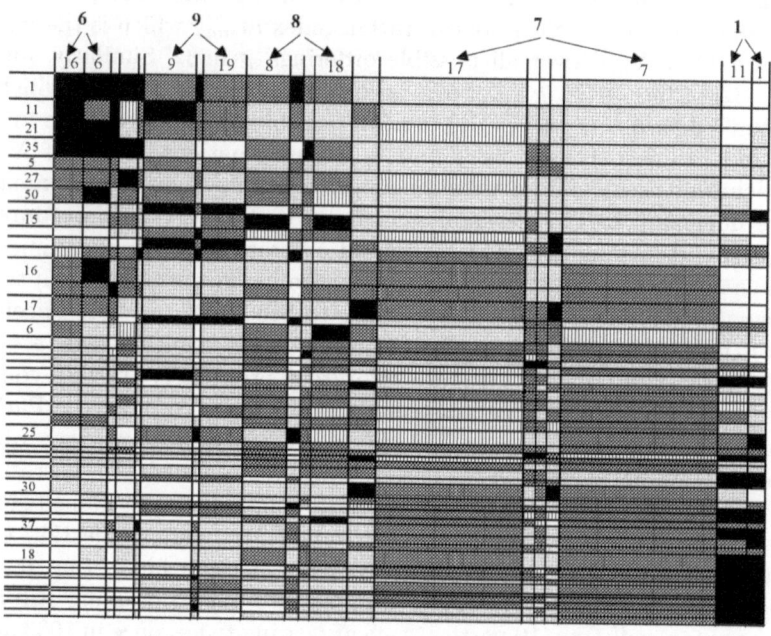

Fig. 2. A graphical presentation of the table containing votes fractions gained by 10 political parties in 1993 and 1997 elections to the Polish Parliament in 52 voivodships. Rows and columns are transformed due to the grade correspondence analysis.

The map in Fig. 2 provides much information. It is clear that the pairs of columns which refer to the same party or quasi party in both elections tend to keep together: 6 with 16, 9 with 19, 8 with 18, 7 with 17, 1 with 11. The conditional distributions of columns, referring to the same party are rather similar (the 1993 and 1997 columns of party 8 are even amazingly similar!).The voivodships are ordered starting from such big towns in the west and central Poland as: Warsaw(1), Gdańsk(11), Kraków(21), Poznań(35), Lódź(27), Wroclaw(50), through Katowice(16), Gliwice(17), Bydgoszcz(6), to smaller rural districts in the eastern part of Poland. This seems to meet the ordering of parties, from the most liberal ones (6) to party 1 representing mainly the inhabitants of rural areas. Two very large columns 17 and 7 refer to non-voters; it is seen that the percent of non-voters tends to concentrate rather in the rural voivodships, especially in 1997.

Usually, the GCA should be followed by a simultaneous clustering of rows and columns. The idea of clustering following GCA is to divide rows as well as columns into *non-overlapping* clusters which will retain positive dependence after aggregation; the aggregated table transformed by the GCA will remain unchanged. These problems are tackled in the papers [4], [5], [16], [20] but no general optimal solutions are yet available.

5   Closing remarks

The previous sections gave only a superficial insight into the construction of descriptive parameters and inference procedures induced by the concentration measures. In particular, we didn't mention the computer-intensive tests introduced to make a suitable reference distribution to a considered descriptive parameter (cf [7], [9], [10]). This is specially important in case of ρ^*_{max} since it is necesary to check whether the GCA transformations and trends are meaningful. It is also possible to introduce a computer-intensive estimation, using a sufficiently regular one-parameter model as reference (cf [2]).

It follows that there are three directions of computing which have to be considered. The main one concerns procedures concerning GCA and inference based on it (e.g. variables selection, outliers detection, discrimination). Two other directions concern visualization and computer-intensive testing and estimating.

References

1. Agresti, A. Analysis of ordinal categorical data. (1990) Wiley, N. York.
2. Bondarczuk, B., Kowalczyk, T., Pleszczyńska, E., Szczesny, W.: Evaluating departures from fair representation. Applied Stochastic Models and Data Analysis **10** (1994) 169–185.
3. Cifarelli, D. M., Regazzini, E.: On a general definition of concentration function. Snakhya B **49** (19987) 307–319.
4. Ciok, A.: Discretization as a tool in cluster analysis. Sixth Conference of the International Federation of Classification Societies, Rome, July 21–24, 1998.
5. Ciok, A., Kowalczyk, T., Pleszczyńska, E., Szczesny, W.: Algorithms of grade correspondence-cluster analysis. The Collected Papers of Theoretical and Applied Computer Science **7** (1995) 5–22.
6. Ciok, A., Kowalczyk, T., Pleszczyńska, E., Szczesny, W.: Inequality measures in data analysis. The Collected Papers of Theoretical and Applied Computer Science **6** (1994) 3–20.
7. Ciok, A.: Generating random MxK tables when row and column totals and cell ranges are given. ICS PAS Reports, No 844 (1997).

8. Ciok, A.: A comparative study of exploratory methods applied to car switching data. Bulletin of the Polish Academy of Sciences - Technical Sciences (1998).

9. Ciok, A., Bulhak, W., Skoczylas, R.: Exploration of control-experimental data by means of grade corresponding analysis. Biocybernetics and Biomedical Engineering **17** (1997).

10. Ciok, A., Kowalczyk, T., Pleszczyńska, E., Szczesny, W. On comparing departures from proportionality with application to the Polish parliament elections of 1991 and 1993. In: R. Dutter and W. Grossmann (eds.), Short Communications in Computational Statistics (1994) 223–224.

11. Ciok, A., Kowalczyk, T., Pleszczyńska, E., Szczesny, W.: On a system of grade dependence measures. ICS PAS Reports (1994) No 741.

12. Fogelson, S.: Measures of concentration and their applications (in Polish), Kwart. Statyst. **10** (1933) 149–197.

13. Kowalczyk, T.: A unified Lorenz-type approach to divergence and dependence, Dissert. Math. **CCXXXV** (1994) 1–54.

14. Kowalczyk, T., Ciok, A., Pleszczyńska, E., Szczesny, W.: Grade dependence in bivariate mixed data: a uniform approach based on concentration curves, submitted (also ICS PAS Reports (1996) No 815).

15. Kowalczyk, T.: Link between grade measures of dependence and of separability in pairs of conditional distributions, submitted (also ICS PAS Reports (1998) No 849).

16. Kowalczyk, T.: Decomposition of the concentration index and its implications for the Gini Inequality and Dependence Measures (submitted).

17. Kowalczyk, T.: Properties of the standard concepts of monotone dependence based on separation curves. Discussiones Mathematicae - Algebra and Stochastic Methods (1997) (to appear).

18. Kowalczyk, T., Niewiadomska-Bugaj, M.: Grade correspondence analysis based on Kendall's tau. Sixth Conference of the International Federation of Classification Societies, Rome, July 21–24, 1998.

19. Pleszczyńska, E., Ciok, A., Kowalczyk, T., Szczesny, W.: New tools of statistical data analysis. Biocybernetics and Biomedical Engineering **15** (1995) 121–137.

20. Pleszczyńska, E., Ciok, A., Kowalczyk, T.: On parameter decomposition in statistical data analysis. Intelligent Information Systems VI (1997) Proceedings of the Workshop held in Zakopane, Poland, 9-13 June, 1997, 74–85.

21. Szczesny, W.: On the Performance of a Discriminant Function. J. Classif. **8** (1991) 201–215.

22. Szczesny, W., Pleszczyńska, E.: A grade statistics approach to exploratory analysis of the HSV data. Biocybernetics and Biomedical Engineering **17** (1997).

23. Szczesny, W., Grabowski, T., Ciok, A., Kowalczyk, T., Pleszczyńska, E.: A mixed-data grade dependence system and its applications in selection and clustering. In: R. Dutter and W. Grossmann (eds.), COMPSTAT'94, Short Communications in Computational Statistics (1994) 53–54.

24. Szczesny, W., Ciok, A., Kowalczyk, T., Pleszczyńska, E., Wysocki, W.: A statistical comparative study of 1993 and 1997 Elections to the Polish Parliament (in Polish). ICS PAS Reports (1998) No 851.

Some Remarks on Networks
of Parallel Language Processors

Jürgen Dassow

Otto-von-Guericke-Universität Magdeburg
Fakultät für Informatik
PSF 4120, D-39016 Magdeburg, Germany

[Abstract.] Networks of parallel language processors form a
special type of grammar systems where derivation steps in the
mode of L systems and communication steps alternate. We
prove that, given an arbitrary system, an equivalent system
with at most three components can be constructed. This improves a result of [2]. Furthermore we consider the function
$f(m)$ which gives the number of words generated in m steps.
We relate this function to growth functions of D0L systems
and prove the undecidability of the equivalence with respect to
this function.

1 Introduction and Definitions

Grammar systems have been intensively studied in the last decade (see [1], [4]).
Essentially, there are two variants of grammar systems. The components of the
system rewrite a common sentential form, or the components have their own
sentential forms and some communication steps are performed.

In [2] grammar systems of the second type called networks of parallel language
processors (NLP for short) have been introduced. The process of generation consists of derivation and communication steps which alternate. The derivation step
is a parallel rewriting of all letters of the current sentential form as in Linden-
mayer systems. In a communication step, any component sends its sentential
forms which satisfy its exit condition to any other component, and any component receives only those words which satisfy its entrance condition and adds
these words to its sentential forms.

In [2] it is shown that the corresponding families of generated languages
coincide with some well-known families of languages generated by ET0L systems
with a control of the application of tables. We shall improve these results for some
control mechanisms. We show that systems with three components are sufficient
to generate any language which can be generated by such systems (with an
unbounded number of components).

Further, in [2] some results are given on the cardinality of the multiset of
words generated by NLP systems in a certain number of steps. We consider the
function $f(m)$ which gives the number of words generated in m steps. We relate
this function to growth functions of D0L systems and prove that it is undecidable
whether or not the functions of two systems are equal.

L. Polkowski and A. Skowron (Eds.): RSCTC'98, LNAI 1424, pp. 83–90, 1998.

We now give the formal definitions and some notation. Throughout the paper we assume that the reader is familiar with the basic notions of the theory of formal languages especially of L languages (see [5], [6]).

By $l(w)$ and $|M|$ we denote the length of a word w and the cardinality of the set M, respectively.

A *conditional* ET0L system (with n tables, $n \geq 1$) is a construct

$$G = (V, V', \varrho_1 : P_1, \varrho_2 : P_2, \ldots, \varrho_n : P_n, w)$$

where

- $(V, V', P_1, P_2, \ldots, P_n, w)$ is a usual ET0L system with the alphabet V, the set $V' \subseteq V$ of terminals, the production sets $P_1, P_2, \ldots, P_n$ and the axiom w, and

- $\varrho_1, \varrho_2, \ldots, \varrho_n$ are mappings from V^* into $\{\underline{true}, \underline{false}\}$.

The language generated by a conditional ET0L system as above consists of all words $z \in (V')^*$ such that there is a derivation

$$w = w_0 \Longrightarrow_{P_{i_1}} w_1 \Longrightarrow_{P_{i_2}} w_2 \Longrightarrow_{P_{i_3}} \cdots \Longrightarrow_{P_{i_{m-1}}} w_{m-1} \Longrightarrow_{P_{i_m}} w_m = z$$

for some $m \geq 0$ and $\varrho_{i_j}(w_{j-1}) = \underline{true}$ for $1 \leq j \leq m$.

Mostly we are only interested in some special conditions.

We say that a condition ϱ is a *regular context condition*, if there is a regular language R such that $\varrho(w) = \underline{true}$ holds if and only if $w \in R$.

We say that a condition ϱ is a *random context condition*, if there are finite sets Q and R such that $\varrho(w) = \underline{true}$ holds if and only if any letter of Q occurs in w and no letter of R occurs in w.

We say that a condition ϱ is a *forbidden context condition*, if it is a random context condition where the set Q of required letters is empty.

We shall write $\varrho = R$ and $\varrho = (Q, R)$ in case of a regular and random context condition, respectively.

By $E(reg)T0L$, $E(rc)T0L$ and $E(for)T0L$ we denote the families of languages generated by ET0L systems with regular, random and forbidden context conditions, respectively.

An NPL_F0L system (of degree n, $n \geq 1$) is a construct

$$\Gamma = (V, (P_1, F_1, \varrho_1, \sigma_1), (P_2, F_2, \varrho_2, \sigma_2), \ldots, (P_n, F_n, \varrho_n, \sigma_n))$$

where

- V is an alphabet,
- for $1 \leq i \leq n$, P_i is a finite subset of $V \times V^*$ such that $V = \{a \mid (a, v) \in P_i\}$,
- for $1 \leq i \leq n$, F_i is a finite subset of V^*,
- for $1 \leq i \leq n$, ϱ_i and σ_i are mappings from V^* to $\{\underline{true}, \underline{false}\}$.

The quadruples $(P_i, F_i, \varrho_i, \sigma_i)$, $1 \leq i \leq n$, are called the components or nodes of the system, and P_i, F_i, ϱ_i and σ_i are called their set of productions, set of axioms, exit filter and entrance filter, respectively. Note that any P_i, $1 \leq i \leq n$, is a usual set of productions of an 0L system.

A configuration C of an NPL_F0L system Γ as above is an n-tuple of languages over V, i.e. $C = (L_1, L_2, \ldots, L_n)$ where $L_i \subseteq V^*$ for $1 \leq i \leq n$.

Let $C = (L_1, L_2, \ldots, L_n)$ and $C' = (L'_1, L'_2, \ldots, L'_n)$ be two configurations of an NPL_F0L system Γ as above. Then we say that

- C' is obtained from C by a derivation step, written as $C \Longrightarrow C'$, if for $1 \leq i \leq n$

$$L'_i = \{w : w' \Longrightarrow_{P_i} w \text{ for some } w \in L_i\}$$

- C' is obtained from C by a communication step, written as $C \vdash C'$, if for $1 \leq i \leq n$

$$L'_i = L_i \cup \bigcup_{j=1}^n \{w \mid w \in L_j, \ \varrho_j(w) = \sigma_i(w) = \underline{true}\}.$$

With any NPL_F0L system Γ as above we associate a sequence of configurations

$$C_m = (L_{m,1}, L_{m_2}, \ldots, L_{m,n}), \qquad m \geq 0$$

such that

$$C_0 = (F_1, F_2, \ldots, F_n) \quad \text{and} \quad C_{2k} \Longrightarrow C_{2k+1} \text{ and } C_{2k+1} \vdash C_{2k+2} \text{ for } k \geq 0$$

and define the language $L(\Gamma)$ generated by Γ by

$$L(\Gamma) = \bigcup_{k \geq 0} L_{2k+1,1}.$$

We call an NLP_F0L system an NLP_0L system if, for $1 \leq i \leq n$, any set F_i contains exactly one element. Moreover, we call an NLP_F0L system propagating if, for $1 \leq i \leq n$, P_i is a subset of $V \times V^+$. Furthermore, an NLP_F0L system is called deterministic if, for $1 \leq i \leq n$, $a \to \alpha \in P_i$ and $a \to \beta \in P_i$ imply $\alpha = \beta$. If the system under consideration is propagating or deterministic we add the letter P or D, respectively. Thus we get NLP_PF0L, NLP_DF0L, NLP_DP0L etc. systems.

By NLP_n_F0L and NLP_F0L we denote the families of all languages generated by NLP_F0L systems with n components and of all languages generated by NLP_F0L systems (without a restriction concerning the degree). NLP_n_0L, NLP_n_PF0L, NLP_DF0L etc. are defined analogously.

Again, we are mostly interested in filters which are regular or random or forbidden context conditions. The corresponding language families are denoted by $(reg)NLP_n\_F0L$, $(rc)NLP_n\_F0L$ and $(for)NLP_n\_F0L$, and accordingly the notation is used in case of deterministic, propagating and/or single axiom systems.

[*Example 1.*] The NLP_0L system

$$\Gamma = (\{a, A, F\}, (P_1, \{a\}, \varrho_1, \sigma_1)(P_2, \{A\}, \varrho_2, \sigma_2))$$

with

$$P_1 = \{a \to a^2\}, \qquad P_2 = \{A \to A^3, A \to a, a \to F\}$$

and the filters (regular context conditions)

$$\varrho_1 = \emptyset, \ \sigma_1 = \{a\}^*, \ \varrho_2 = \{a\}^*, \ \sigma_2 = \emptyset$$

generates the language

$$L(\Gamma) = \{a^{2^n \cdot 3^m} \mid n \geq 1, m \geq 0\}.$$

2   The Number of Components

We start with a simple observation.

[**Lemma 1.**] *For* $n \geq 1$, $Y \in \{F0L, 0L, PF0L, P0L\}$ *and* $x \in \{(reg), (rc), (for)\}$,

$$xNLP_n\text{-}Y \subseteq xNLP_{n+1}\text{-}Y.$$

[*Proof.*]Let $L \in (reg)NLP_n\text{-}Y$ be generated by an NLP_Y system Γ with n components, with the alphabet V and regular context conditions. Then we define the NLP_Y system Γ' by adding the component $(\{x \to x \mid x \in V\}, \emptyset, \emptyset, \emptyset)$ which does not obtain words from other components and does not send words to other components by the definition of its exit and entrance filter. Thus $L(\Gamma) = L(\Gamma')$ which proves $(reg)NLP_n\text{-}Y = (reg)NLP_{n+1}\text{-}Y$.

The modifications for random and forbidden context conditions are left to the reader.

We now investigate the hierarchy given in Lemma 1 in more detail.

[**Theorem 1.**] *For* $x \in (reg), (rc), (for)$ *and* $Y \in \{F0L, 0L, PF0L, P0L\}$,

$$Y = xNLP_1\text{-}Y \subset xNLP_2\text{-}Y.$$

[*Proof.*]The equality $Y = NLP_1\text{-}Y$ holds by definition.

By Example 1 $L = \{a^{2^n \cdot 3^m} \mid n \geq 1, m \geq 0\} \in (reg)NLP_2\text{-}Y$. On the other hand, $L \notin F0L$ can be shown by standard methods. This proves that the inclusion $(reg)NLP_1\text{-}Y \subseteq (reg)NLP_2\text{-}Y$ from Lemma 1 is proper in the case of regular context conditions.

For the case of random and forbidden context conditions as filters, we modify the filters in Example 1 such that they are random or forbidden context conditions and have the same effect as the regular context conditions.

[**Theorem 2.**] *For* $n \geq 3$, $Y \in \{0L, F0L\}$ *and* $x \in \{(reg), (for)\}$,

$$ExT0L = xNLP_n\text{-}Y \qquad and \qquad ExPT0L = xNLP_n\text{-}PY.$$

[*Proof.*]We only give the proof for $Y = F0L$ and $x = (reg)$. The necessary modifications for the other cases are left to the reader (and refer to [3]).

By [2] Theorem 4.3 and Lemma 1,

$$(reg)NLP_3\text{-}F0L \subseteq (reg)NLP_4\text{-}F0L \subseteq \ldots \subseteq (reg)NLP\text{-}F0L \subseteq E(reg)T0L.$$

Hence it is sufficient to show that $E(reg)T0L \subseteq (reg)NPL_3\text{-}F0L$.

Let $L \in E(reg)T0L$ and let $L = L(G)$ for some ET0L system

$$G = (V, V', R_1 : P_1, R_2 : P_2, \ldots, R_n : P_n, w)$$

with n tables and regular languages R_i, $1 \leq i \leq n$, as conditions. For $1 \leq i \leq n$, we set

$$V^{(i)} = \{x^{(i)} : x \in V\}.$$

Moreover, for a word $y = x_1 x_2 \ldots x_m$ with $x_j \in V$ for $1 \leq j \leq m$, a set $R \subseteq V^*$ and $1 \leq i \leq n$, we set $y^{(i)} = x_1^{(i)} x_2^{(i)} \ldots x_m^{(i)}$ and $R^{(i)} = \{y^{(i)} \mid y \in R\}$.

We now consider the NPL system

$$\Gamma = (W, (Q_1, F_1, \varrho_1, \sigma_1), (Q_2, F_2, \varrho_2, \sigma_2), (Q_3, F_3, \varrho_3, \sigma_3))$$

where

$$W = V \cup \bigcup_{i=1}^{n} V^{(i)} \cup \{S\} \quad (S \text{ is an additional letter}),$$

$$Q_1 = \{z \to z \mid z \in W\},$$

$$F_1 = \emptyset, \qquad \varrho_1 = W^*, \qquad \sigma_1 = (V')^*,$$

$$Q_2 = \{x \to x^{(1)} \mid x \in V\} \cup \{x^{(i)} \to x^{(i+1)} \mid x \in V, 1 \leq i \leq n - 1\}$$

$$\cup \{x^{(n)} \to x : x \in V\} \cup \{S \to w\},$$

$$F_2 = \{S\}, \qquad \varrho_2 = W^*, \qquad \sigma_2 = V^*,$$

$$Q_3 = \{x \to x \mid x \in V\} \cup \{x^{(i)} \to w_x : x \to w_x \in P_i, 1 \leq i \leq n\} \cup \{S \to S\},$$

$$F_3 = \emptyset, \qquad \varrho_3 = \emptyset, \qquad \sigma_3 = \bigcup_{i=1}^{n} R_i^{(i)}.$$

Now assume that there is a derivation

$$w \Longrightarrow_{P_{i_1}} w_1 \Longrightarrow_{P_{i_2}} w_2 \Longrightarrow_{P_{i_3}} \cdots \Longrightarrow_{P_{i_m}} w_m \tag{1}$$

in G with $m \geq 1$. By induction on m it is easy to show that

$$w_m \in L_{2i_1 + 2i_2 \ldots 2i_m + 2m + 1, 3}. \tag{2}$$

Now let $z \in L(G)$. Then $z = w \in (V')^*$ or there is a derivation (1) with $m \geq 1$ and $z = w_m \in (V')^*$. In the former case $w \in L_{1,2}$, $w \in L_{2,1}$ by a communication step and $w \in L_{3,1}$ by a derivation step, and in the latter case $w_m \in L_{2i_1 + 2i_2 \ldots 2i_m + 2m + 2, 1}$ by a communication step and $w_m \in L_{2i_1 + 2i_2 \ldots 2i_m + 2m + 3, 1}$ by a derivation step. Thus $z \in L(\Gamma)$ and hence $L(G) \subseteq L(\Gamma)$.

Conversely, by induction on the number of steps it is easy to prove that, for $m \geq 1$,

- if $w \in L_{m,2} \cup L_{m,3}$, then $w \in S(G)$ or $w = z^{(i)}$ for some $z \in S(G)$ and $1 \leq i \leq n$,

- $L_{m,1} \subseteq L(G)$.

Therefore $L(\Gamma) \subseteq L(G)$.

Thus $L(G) = L(\Gamma)$ and $E(reg)T0L \subseteq (reg)NPL_3\_F0L$.

Combining these results, for $x \in (reg), (for)\}$ and $Y \in \{F0L, 0L\}$, we obtain the hierarchy

$$Y = xNLP_1\_Y \subset xNLP_2\_Y \subseteq xNLP_3\_Y = xNLP_4\_Y = \cdots$$

$$\cdots = xNLP\_Y = ExT0L.$$

The only problem remaining open is the question whether or not two components are as powerful as three (or more) components.

3   Growth Functions of NPL Systems

[**Definition 1.**]*Let $x \in \{(reg), (rc), (for)\}$. For an $xNPL\_F0L$ system Γ with n components and $1 \le i \le n$ we define*
  (i) the growth functions $f_{i,\Gamma} : \mathbf{N} \to \mathbf{N}$ of Γ at node i by $f_{i,\Gamma}(m) = |L_{m,i}|$
and
  (ii) the growth function $f_\Gamma : \mathbf{N} \to \mathbf{N}$ of Γ by $f_\Gamma(m) = \sum_{i=1}^{n} f_{i,\Gamma}(m)$.

In [2] growth functions of NLP_F0L have been introduced. However, there $L_{m,i}$ is considered as a multiset whereas we consider it as a set.

[**Theorem 3.**] *Let $x \in \{(reg), (rc), (for)\}$. Then it is undecidable whether or not the growth functions or the growth functions at some node coincide for two given $(x)NPL\_P0L$ systems with at least three components.*

[*Proof.*]We only give the proof for $x = (reg)$. The obvious modifications for $x \in \{(rc), (for)\}$ are left to the reader.

For an arbitrary number $n \ge 1$, we consider the (reg)NPL_P0L system

$$\Gamma = (V, (P_1, \{c\}, V^*, V^* \setminus V^*\{d\}V^*), (P_2, \{d\}, V^*, V^* \setminus V^*\{c\}V^*), (P_3, \{c\}, \emptyset, V^*)$$

where

$$
\begin{aligned}
V &= \{a, b, c, d, [1], [2], \ldots [n]\}, \\
P_1 &= \{c \to [i]ca \mid 1 \le i \le n\} \cup \{x \to x \mid x \in V, x \ne c\}, \\
P_2 &= \{d \to [i]db \mid 1 \le i \le n\} \cup \{x \to x \mid x \in V, x \ne d\}, \\
P_3 &= \{c \to d\} \cup \{x \to x \mid x \in V \setminus \{c\}\}.
\end{aligned}
$$

Then we obtain

$$
\begin{aligned}
&f_{1,\Gamma}(2m) = f_{1,\Gamma}(2m - 1) = f_{2,\Gamma}(2m) = f_{2,\Gamma}(2m - 1) = n^m \quad \text{for } m \ge 0, \\
&f_{3,\Gamma}(0) = f_{3,\Gamma}(1) = 1,
\end{aligned}
$$

$$f_{3,\Gamma}(2m) = f_{3,\Gamma}(2m + 1) = 1 + 2 \cdot \sum_{k=0}^{m} n^k \quad \text{for } m \ge 1.$$

With an instance $I = \{(u_1, v_1), (u_2, v_2), \ldots, (u_n, v_n)\}$ of the Post Correspondence Problem over some alphabet U we associate the (reg)NPL_0L system

$$
\begin{aligned}
\Delta_I = (W, (Q_1, \{c\}, W^*, W^* \setminus W^*\{d\}W^*), (Q_2, \{d\}, W^*, W^* \setminus W^*\{c\}W^*), \\
(Q_3, \{c\}, \emptyset, W^*))
\end{aligned}
$$

where

$$
\begin{aligned}
W &= U \cup \{c, d, [1], [2], \ldots, [n]\}, \quad U \cap \{c, d, [1], [2], \ldots, [n]\} = \emptyset, \\
Q_1 &= \{c \to [i]cu_i \mid 1 \le i \le n\} \cup \{x \to x \mid x \in W, x \ne c\}, \\
Q_2 &= \{d \to [i]dv_i \mid 1 \le i \le n\} \cup \{x \to x \mid x \in W, x \ne d\}, \\
Q_3 &= \{c \to d\} \cup \{x \to x \mid x \in W \setminus \{c\}\}.
\end{aligned}
$$

Obviously,
$$f_{1,\Delta_I}(m) = f_{2,\Delta_I}(m) = f_{1,\Gamma}(m) \qquad \text{for} \quad m \geq 0.$$

Moreover, if the instance I has no solution of the Post Correspondence Problem, then
$$f_{3,\Delta_I}(m) = f_{3,\Gamma}(m) \qquad \text{for} \quad m \geq 0.$$

However, if I has a minimal solution $i_1 i_2 \dots i_r$ (with respect to the length of the index sequence), then
$$f_{3,\Delta_I}(m) < f_{3,\Gamma}(m) \qquad \text{for} \quad m \geq r.$$

Hence $f_{3,\Delta_I} = f_{3,\Gamma}$ and $f_{\Delta_I} = f_\Gamma$ hold if and only if I has no solution.

The components NLP_0L systems constructed in the proof of Theorem 3 are nondeterministic. The following statement is the deterministic version (which we present without proof, a proof analogous to the above one is given in [3]).

[**Theorem 4.**] *Let $x \in \{(reg), (rc), (for)\}$. Then it is undecidable whether or not the growth functions or the growth functions at some node coincide for two given (x)NLP_PD0L systems with the same number of components.*

A D0L system $G = (V, P, w)$ generates a unique sequence
$$w = w_0 \Longrightarrow w_1 \Longrightarrow w_2 \Longrightarrow \dots \Longrightarrow w_m \Longrightarrow \dots$$

of words. The growth function $g_G : \mathbf{N} \to \mathbf{N}$ of G is defined by $g_G(m) = l(w_m)$. g is called a D0L growth function if there is a D0L system such that $g = g_G$.

We mention that the problem whether or not the growth functions of two NLP_DF0L systems coincide is decidable if we consider growth function on the basis of multisets. This follows from the facts that such growth functions are D0L growth functions (see [2], Theorem 5.1) and that the growth equivalence of D0L systems is decidable (see [6], Theorem 3.3).

We now present a relation between D0L growth functions and growth functions of NLP_F0L systems (with one component).

[**Theorem 5.**] *For any D0L growth function g, there is an $(NLP\_)F0L$ system Γ such that*
$$f_\Gamma(0) = f_{1,\Gamma}(0) = g(0),$$
$$f_\Gamma(2i-1) = f_{1,\Gamma}(2i-1) = g(i) \quad \text{for } i \geq 1.$$

[*Proof.*]Let $G = (V, P, w)$ be a D0L system with
$$V = \{a_1, a_2, \dots, a_n\},$$
$$P = \{p_1, p_2, \dots, p_n\},$$
$$p_i = a_i \to x_{i1} x_{i2} \dots x_{ir_i} \quad \text{for } 1 \leq i \leq n,$$
$$w = y_1 y_2 \dots y_m \quad \text{with } y_j \in V \text{ for } 1 \leq j \leq m,$$
$$g(G) = g.$$

First let us assume that $y_i \neq y_j$ for $1 \leq i < j \leq n$. We construct the (NLP-)F0L system (with one component) $(V', (P', F, \varrho, \sigma)$ with

$$V' = V \cup \{a_{i,j} \mid 1 \leq i \leq n, 1 \leq j \leq r_i\},$$

$$P' = \bigcup_{i=1}^{n} \bigcup_{j=1}^{r_i} \{a_i \rightarrow x_{ij}(a_i, j), (a_i, j) \rightarrow (a_i, j)\},$$

$$F = \{y_1, y_2, \ldots, y_m\}$$

and arbitrary filters. Since we have only one component, any communication step does not change the language, i.e. $L_{2i-1,1} = L_{2i,1}$ for $i \geq 1$. It is easy to prove by induction that any word $z \in L_{2i-1,1}$ has the form

$$z = b_i(b_{i-1}, j_{i-1})(b_{i-2}, j_{i-2}), \ldots, (b_0, j_0)$$

where, for $1 \leq k \leq i$, b_k is the j_{k-1} letter of the right hand side of the rule in P with the left hand side b_{k-1}. Therefore $z \in L_{2i-1,1}$ starts with a letter b_i of V and the remaining $i - 1$ letters store the "derivation" of b_i in the D0L system G. In the sequel let $b_i' = (b_{i-1}, j_{i-1})(b_{i-2}, j_{i-2}), \ldots, (b_0, j_0)$.

If the D0L system G generates the word $u_1 u_2 \ldots u_{t(i)}$ in i steps, then

$$L_{2i-1,1} = \{u_1 u_1', u_2, u_2', \ldots u_{t(i)} u_{t(i)}'\}.$$

Moreover, all words of $L_{2-i-1,1}$ are pairwise different. Hence

$$f_\Gamma(2i - 1) = |L_{2i-1,1}| = t(i) = l(u_1 u_2 \ldots u_{t(i)}) = g(i).$$

By slight modifications (introduce primed versions of letters) we can prove the statement for the case that some letters occur a number of times in the axiom of the D0L system.

References

1. E. Csuhaj-Varju, J. Dassow, J. Kelemen and Gh. Paun, *Grammar Systems: A Grammatical Approach to Distribution and Cooperation*. Gordon and Breach Science Publisher, Yverdon, Switzerland, 1994.
2. E. Csuhaj-Varju and A. Salomaa, Networks on parallel language processors. In: Gh. Paun and A. Salomaa (eds.), *New Trends in Formal Languages*. LNCS 1218, Springer-Verlag, 1997, 299–318.
3. J. Dassow, Some remarks on networks of parallel language processors. Techn. report, University of Magdeburg, 1998.
4. J. Dassow, Gh. Paun and G. Rozenberg, Grammar systems. In: G. Rozenberg and A. Salomaa (eds.), *Handbook on Formal Languages*, Vol. 2, Springer-Verlag, Berlin, 1997, 155–213.
5. G. T. Herman and G. Rozenberg, *Developmental Systems and Languages*. North-Holland Publ. Co., Amsterdam, 1975.
6. G. Rozenberg and A. Salomaa, *The Mathematical Theory of L Systems*. Academic Press, New York, 1980.

Molecular Computation for Genetic Algorithms

J. Castellanos, S. Leiva, J. Rodrigo, A. Rodríguez-Patón

Dpto. de Inteligencia Artificial. Facultad de Informática
Universidad Politécnica de Madrid
Campus de Montegancedo. Boadilla del Monte 28660. Madrid (Spain)
E-mail: jcastellanos@fi.upm.es

[Abstract.] In this paper we present a new computational model based on DNA molecules and genetic operations. This model incorporates the theoretic simulation of the main genetic algorithms operations like: selecting individuals from the population to create a new generation, crossing selected individuals, mutating crossed individuals, evaluating fitness of generated individuals, and introducing individuals in the population. This is a first step that will permit the resolution of larger instances of search problems far beyond the scope of exact and exponentially sized DNA algorithms like the proposed by Adleman [1] and Lipton [2].

1   Introduction

In a short period of time great advances have been reached in DNA based computations. With a large number of DNA strands and with some biological operations it is possible to obtain an universal model able to resolve any given decidable problem. Adleman [1] began this field describing an abstract model that he applied to the resolution of a $\mathcal{NP}$–complete problem, the Directed Hamiltonian Path. Then, Lipton [2] showed how to resolve more general problems through finding a solution to the SAT problem.

In this paper, we show the possibility of solving optimization problems without generating or exploring the complete search space. In the second section of the paper the steps of the genetic algorithms as well as their main operations are explained. In the third, we describe the principles of molecular computation and the main molecular operations. In the fourth section, we detail the simulation of genetic algorithms employing DNA molecules.

2   Genetic algorithms

Genetic algorithms are adaptative searching techniques inspired on the Darwin's evolution theory. According to Darwin's theory, individuals who are better adapted to the environment, survive and transfer some kind of information to their descendants through their genetic code. Genetic algorithms follow the same principle: potential solutions for a problem are coded by generating an initial random population that will develops thanks to recombination and mutation

L. Polkowski and A. Skowron (Eds.): RSCTC'98, LNAI 1424, pp. 91–98, 1998.
© Springer-Verlag Berlin Heidelberg 1998

operations. The following generations are formed by evolution so that in time the population comes to consist of better individuals (solutions). In this way, the best solutions to the problem survive and transfer their internal information to their descendants.

The structure of a basic genetic algorithm include the following steps: (1) Generate an initial random population evaluating the fitness for each individual, (2) Extract individuals, (3) Cross and mutate extracted individuals, (4) Evaluate and introduce the new created individuals, looping steps 2 to 4 until population has converged.

3   Molecular Computation

Molecular computation, or DNA-computation encodes solutions of complex problems in DNA strands and applies conventional techniques from molecular biology to obtain the right solution filtering out the wrong candidates.

Biological computation presents a set of advantages and disadvantages with regard to conventional computation. Some advantages of molecular computation are massive parallelism (10^{20} molecules of DNA in a tube), high storage density and low energy consumption.

Among disadvantages that can be found in molecular computers are the limited number of biological operations available as well as the complexity and experimental errors of these biological operations.

3.1   Main Molecular Operations

We present the main biological operations that we employ to manipulate DNA strands.

 - Strands separation according to their length using gel electrophoresis [8].
 - Strands separation according to a determinated subchain s using complementary probes anchored to magnetic beads.
 - Denaturation of DNA strands.
 - Strands Separation with a certain symmetry forming palindromes: An initial tube T containing DNA chains is separated in two tubes T_1 and T_2. T_1 will contains the chains that have a palindrome symmetry and the tube T_2 the rest of chains. A strand has palindrome symmetry when contains complementary substrands at both sides of a central point. If we denature the double strands, the following strands will remain having the following geometry due to the complementary nucleotides in the symmetry zone (figure 1). To separate the chains forming palindromes from the ones that do not, we will use gel electrophoresis [8].
 - Append a sequence of nucleotides in a free end of a strand [3].
 - Site directed mutagenesis [4].
 - Cut on strands: Restriction enzymes cut the DNA strands in a sequence specific form. The cut produced by the restriction enzymes may leave sticky ends or blunt ends in the DNA fragments. DNA fragments with complementary sticky ends may reanneal forming a new double strand.
 - Chain duplication through Polymerase Chain Reaction (PCR): With this operations, two identical tubes T_1 and T_2 are obtained from a first tube T.

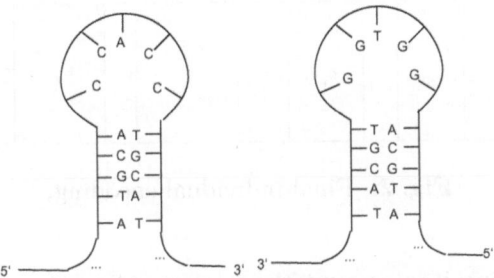

Fig. 1. Palindrome loop.

4 Genetic Algorithm Simulation with Biological Computation.

The simulation of genetic algorithms with DNA molecules require a set of test tubes where each one has a well defined utility: (1) Population tubes (2) Mating pool tubes and (3) Temporary tubes. Individuals from different tubes have different format.

4.1 Individual Encoding

The choice of a codification is a key point for the correct evolution of the population towards the final solution. Individuals need to be coded so that it can be made combinations, duplications, copies, quick fitness evaluation and selection of a specific individual inside the population or mating pool without the need of a complete sequencing.

- The Lipton encoding [2] is used to obtain each individual coded by a sequence of ones and zeros, where the $ENC(b, i)$ function returns the codification of b (0 or 1) at the ith bit place. The codification returned from the ENC function is unique for each value of b and i. ¿From now on, $ENC(b, i)$ will be represented by b_i.
- Between the DNA code belonging to bits places i and $i + 1$, a cutting or cleavage site for a restriction enzyme will be inserted.
- To allow a quick fitness evaluation, a field with the adaptation grade of the individual is included in the DNA strand. The fitness may be coded so that its length is proportional to the value it represents.
- A field indicating the numbering of that individual in the population will be included to locate it and it will be inserted at both sides of the DNA strand with the peculiarity that in one side (sequence N_p') it is symmetrically complementary with a palindrome structure (sequence N_p) that allow a later separation operation.
  Also one more cleavage site RE_p for a restriction enzyme will be included to separate its numbering.

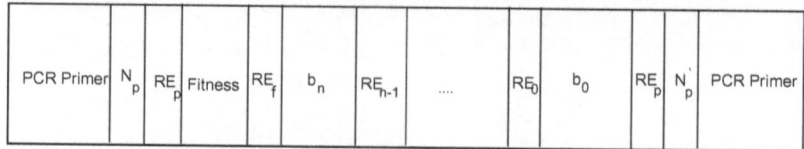

Fig. 2. Final individual encoding.

Taking everything into account the final encoding of an individual has the following format (figure 2).

It must be considered that the individuals of the population may go to the mating pool from where they will be selected for later genetic operations, so some identification is needed in the usual codification to select an specific individual inside the mating pool. That identificator will be the sequence N_m. This value will be introduced in the fitness value field so that mating pool individuals have the same format as population ones with N_m values instead of the fitness field.

4.2 Generation of the initial population

For a generic population with m individuals, in which each individual represents his genotype with n bits, $n-1$ steps are needed to generate all the initial population. Also, in each step we would need to sequence $2m$ molecules. The steps of the initial population generation algorithm are the following.

Synthesizing half of the population with a value 0 in the first bit place and with N having values for numbering half of the population. These sequences will have the following format: $RE_0-0_0-RE_p-N_p'$–PCR Primer, and the other half with a 1 in the first bit place with N having values for numbering the rest of the population. These sequences have the format: $RE_0-1_0-RE_p-N_p'$–PCR Primer.

It is created also the sequences $RE_1-1_1-RE_0$ and $RE_1-0_1-RE_0$. The previous chains are put together in a temporal test tube. The restriction enzyme corresponding to the site RE_0 is applied. The strands are cut and DNA ligase is applied to rejoin the fragments. Supposing that there are not mistakes in the operations of joining m molecules, the following sequences are obtained: $RE_1-b_1-RE_0-b_0-RE_p-N_p'$–PCR Primer with different values for b and N.

Until arrive at the step $n-1$, in a generic step i, the half of the individuals will be created with encoding format $RE_i-0_i-RE_{i-1}$ and the other half with value 1: $RE_i-1_i-RE_{i-1}$. They will be joined in a temporal test tube with strands obtained in step $i-1$. The restriction enzyme corresponding to site RE_{i-1} and DNA ligase will be applied to obtain resultant strands for step i. Strands will have the following format: $RE_i-b_i-RE_{i-1}-\ldots-RE_0-b_0-RE_p-N_p'$–PCR Primer.

When $i = n-1$, RE_f should be used instead of RE_i.

Once we get the individuals with the last format it is necessary to evaluate the fitness function to append its value [3] and join the number N_p. This number is encoded with a symmetric sequence of N_p'. We need $\log_2 m$ steps of extraction and append to write the $\log_2 m$ bits of N_p.

4.3   Individual selection for mating pool

Length separation is used to select individuals to fill the mating pool assuming that better adapted individuals are longer than worst adapted individuals. There are different methods for creating the mating pool.

Method of the best ones The purpose is to take the best n individuals of the population. Length separation will be used so that the n best individuals are selected. For viewing individuals with its respective length we add a radioactive marker corresponding for each individual number, so that applying X-rays to the gel electropheresis all individuals sorted by fitness will be seen and the n longer will be taken.

Roulette method This method will be carried out using a scale function. The referred function obtains the strands fitness contribution—its length—to global fitness. Starting from scale function we obtain the inverse function that will return the individual identification through the fitness degree aportation to global fitness in that particular population. Then, numbers will be generated randomly between one and hundred; and using the inverse function we get the individual identifications for these numbers. These individuals will be taken off the population separating the chains starting from the sequence corresponding to the individual number using magnetic beads.

Crowding method This method consist on taking individuals randomly from the population; then, those individuals are introduced in a temporal test tube and the best of them will be taken to introduce it in the mating pool. The last step will be repeated until the mating pool is filled.

For taking individuals randomly, a random number must be generated and obtain the individual from population as it has been described previously. For getting individuals randomly, instead of generate a random number, a little quantity of chains from the population tube could be taken and the best element is chosen separating it by length.

Once individuals selected using one of the previous explained methods are introduced in the mating pool test tube, it is necessary to change chain format to introduce individual numbers for the mating pool. This mating pool numbering will replace fitness field in the population format and it will be introduced applying site directed mutagenesis [4] over the mating pool tube individuals.

4.4   Individual Crossing

For crossing operation the mating pool individuals will be taken by pairs or the crowding method will be used for selecting an individual from the population who will be crossed with one from the mating pool. For getting a pair of individuals from the mating pool, we use strands separation according to the subchain of the two mating pool numbers selected to be crossed. In case that the crowding method is used, two individuals with different format will be crossed, due to the different chain format between population and mating pool individuals. For solving this, it is enough to place in the individual of the population a mating pool individual number 0 never used by a mating pool individual.

If a one point crossing will be made, a crossing point will be chosen randomly. It will be equivalent to choose a restriction enzyme randomly among $RE_0 \ldots RE_{n-1}$. Then, apply the chosen restriction enzyme and the chains are cut in two parts. It will be possible that original individuals will be generated again; to avoid this, chains having palindrome structure will be separated from the rest of the resultant chains, so that chains that have palindrome structure are the chains for which the crossing has not been made and these chains will have to be removed. The several points crossing can be made in several one point crossing steps.

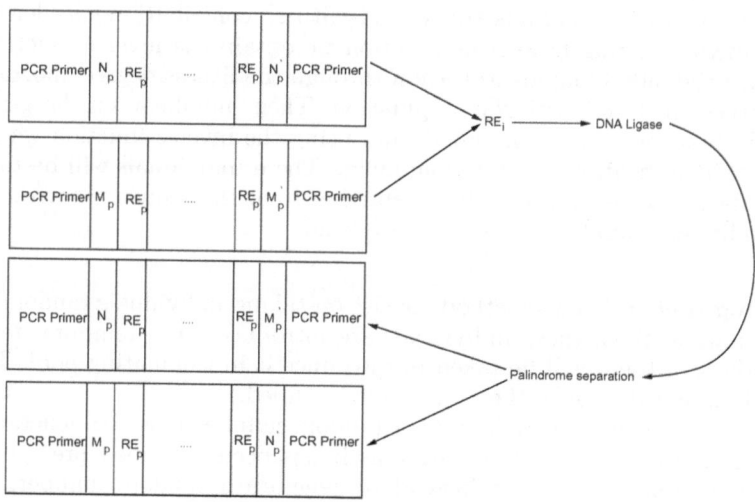

Fig. 3. Individual crossing.

4.5   Mutation

Once the individuals have been crossed, mutation operation should be applied. This mutation can be achieved employing site directed mutagenesis [9]. Previously to the substitution operation it is necessary to apply the probability of the mutation for each bit. To do so if the generic bit i is pretended to be mutated the following pieces will be generated in two different temporary tubes: $RE_i - 0_i - RE_{i-1}$ and $RE_i - 1_i - RE_{i-1}$. One of the last short strands will be taken randomly, so that depending on the value of the bit i in the individual strand and on the value of the short strand chosen the mutation will be made or not. This mutation will be repeated for each bit.

4.6   Introducing new individuals in the population

Recombined and mutated strands are evaluated and its fitness value are inserted. Once obtained the final chains to be introduced in the new population, it is necessary to extract the individuals from the last generation. When those individuals have been selected, one by one are taken from the mating pool by their identification number, so that for each individual we have to make this operation: take the individual to be extracted and the individual from the mating pool, introduce them in separated temporary tubes and apply the restriction enzyme corresponding to RE_p. Strands are separated by length getting the longer one from the tube containing the mating pool individual and the shorter one from the tube containing the individual to be extracted. They are introduced in a temporary tube and DNA ligase is applied (figure 4).

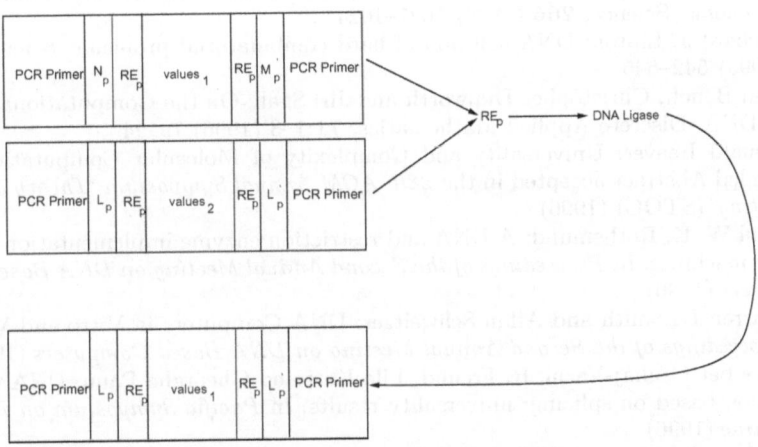

Fig. 4. Individual replacement.

4.7   Evolutionary computation

In this subsection, we propose an alternative evolutionary computation based on restrictions. We have an initial test tube with individuals and a list of restrictions that the individuals must verify to become a right solution. The fitness function of an individual is the grade of 'accepted' restrictions and we can evaluate these restrictions parallely for the whole population. We can crossover strands of individuals of a tube that agree with one restriction with strands of individuals of another tube that verify two or three restrictions. Then, we are mixing the Adleman's style of computation based on restrictions [1] with evolutionary search techniques.

5 Conclusions

We have simulated with DNA molecules and recombinant DNA technology the main genetic algorithms operations. Until now the molecular computation has been used to solve $\mathcal{NP}$–complete problems with exact and 'brute force' algorithms. It is necessary for DNA-computation to expand its algorithmic techniques to incorporate aproximative and probabilistics algorithms and heuristics so the resolution of large instances of $\mathcal{NP}$–complete problems will be possible. The simulation of concepts of genetic evolution with DNA that we have presented will help DNA-computation to resolve more complex problems because introduce genetic and evolutive search in the list of molecular computation techniques availables.

References

1. Leonard M. Adleman: Molecular Computation of Solutions to Combinatorial Problems. Science, **266** (1994) 1021–1024
2. Richard J. Lipton: DNA solution of hard combinatorial problems. Science, **268** (1995) 542–545
3. Dan Boneh, Christopher Dunworth and Jiří Sgall: On the Computational Power of DNA. Discrete Applied Mathematics, **71**:1–3 (1996) 79–94
4. Donald Beaver: Universality and Complexity of Molecular Computation. Extended Abstract accepted in the *28th ACM Annual Symposium 'Theory of Computing'* (STOC) (1996)
5. Paul W. K. Rothemund: A DNA and restriction enzyme implementation of Turing machines. In *Proceedings of the Second Annual Meeting on DNA Based Computers* (1996)
6. Warren D. Smith and Allan Schweitzer: DNA Computers in Vitro and Vivo. In *Proceedings of the Second Annual Meeting on DNA Based Computers* (1996)
7. Erzs bet Csuhaj-Varjú, R. Freund, Lila Kari and Gheorghe Paun: DNA computation based on splicing: universality results. In *Pacific Symposium on Biocomputing* (1996)
8. Richard R. Sinden: *DNA Structure and Functions.* Academic Press (1996)
9. J. Sambrook, E. F. Fritsch, and T. Maniatis: *Molecular Cloning: A Laboratory Manual.* Cold Spring Harbor Press, NY, 2nd edition, (1989)

Marcus Contextual Languages and their Cut-and-Paste Properties

Rodica Ceterchi

Faculty of Mathematics, Bucharest University
Academiei 14, R-70109, Bucureşti, Romania
e-mail: rc@funinf.math.unibuc.ro

Abstract. We explore the relationship between Marcus contextual languages and CP-languages. We prove that external and internal contextual languages without choice are 2CP- and 3CP-languages. We extend these results to contextual languages with choice, by appropriately defining a concept of selective CP-language.

1 Introduction

The class of Marcus contextual languages, and especially their generative mechanisms, the contextual grammars, are, since their introduction in 1969, both objects of intensive studies and helpful tools, in various areas of Linguistics, mathematical or not. The pioneering paper [3] introduces contextual grammars in their *external* variant, and [6] introduces the *internal* variant. The mathematical study of contextual grammars developed quickly, and is now a very large field, despite this, still rich in open and challenging problems. We refer the reader to the two monographs [7], and especially [8] to convince himself.

Cut-and-Paste languages were recently introduced [1], and are still trying to find their place. All the necessary definitions are given in this paper. Their "generative device" is Kleene's fix-point theorem. Their name is a natural consequence of the way "monomials" and "polynomials" are defined: they "cut" a word into n pieces, then "glue" them with coefficients, and finally "paste" the pieces. Cut and paste operations are central in DNA computing, molecular computing, see for instance [9] for overwhelming proof. The kind of cut-and-paste we describe by means of the concept of nCP-language is probably the most rudimentary one. Still, there are some very pregnant analogies with the operations performed by contextual grammars, which motivate the beginning of a comparative study, of which our present paper is nothing but a very small step. It can be also seen as an attempt to generate contextual languages via a fix-point mechanism, attempt which has been successfully made by others too, see [2].

We have considered here only the most simple types of contextual grammars, external and internal, first without choice, then with a choice function of the simplest type $\phi : V^* \to \mathcal{P}(C)$. As described in paragraphs 3 and 4, the passage from contextual grammars to nCP-polynomials with $n = 2$ and 3 presents no problems. The reverse passage, when one takes into account the most general form of an nCP-polynomial, does not "fall back precisely" on the same spot, as illustrated at the end of paragraph 3. Languages generated by contextual

L. Polkowski and A. Skowron (Eds.): RSCTC'98, LNAI 1424, pp. 99–106, 1998.
© Springer-Verlag Berlin Heidelberg 1998

grammars with shuffled contexts, a concept introduced in [5], could be considered as candidates for successfully completing this passage.

We introduce in 5 the concept of a ϕ-selective nCP-language attached to a contextual grammar with selection function ϕ, and afterwards the concept of selective nCP-language attached to an insertion grammar. We think that the mechanism described there is crucial for understanding the interplay between strings and contexts, interplay embedded in the choice operation, and not only. The typology generated by this interplay, as presented in [4], can be fully covered, and even enriched, by considering a more general concept of selective nCP-language.

2   Definitions and notations

Let V be an alphabet, V^* the free monoid generated by V, and let $1 \in V^*$ denote the empty word.

For a natural number $n \geq 2$, consider the product monoid

$$(V^*)^n = \underbrace{V^* \times V^* \times \ldots \times V^*}_{n \text{ times}}$$

with concatenation defined component wise and $(1, 1, \ldots, 1)$ the neutral element.

Let $c_n : (V^*)^n \to V^*$ be the *n-concatenation function*, defined by

$$c_n(x_1, x_2, ..., x_n) := x_1 x_2 ... x_n \in V^*,$$

and consider its extension to the image function $c_n : \mathcal{P}((V^*)^n) \to \mathcal{P}(V^*)$.

Let $c_n^{-1} : \mathcal{P}(V^*) \to \mathcal{P}((V^*)^n)$ be the pre-image function, i.e. for $w \in V^*$

$$c_n^{-1}(w) = \{(x_1, x_2, \ldots, x_n) \in (V^*)^n \mid x_1 x_2 \ldots x_n = w\},$$

and for an $L \subseteq V^*$, let $c_n^{-1}(L) = \bigcup_{w \in L} c^{-1}(w)$.

$\mathcal{P}((V^*)^n)$ is a monoid with the usual product of sets. For fixed $(a_1, \ldots, a_n)$, $(b_1, \ldots, b_n)$ in $(V^*)^n$, identified with their corresponding singletons, we have the product

$$(a_1, a_2, \ldots, a_n) c_n^{-1}(w)(b_1, b_2, \ldots, b_n) =$$
$$\{(a_1 x_1 b_1, a_2 x_2 b_2, \ldots, a_n x_n b_n) \mid x_1 x_2 \ldots x_n = w\}.$$

Definition 1. *We call* first degree n-cut-and-paste monomial *(nCP-monomial for short), an expression of the form*

$$M(X) = c_n((a_1, a_2, \ldots, a_n) c_n^{-1}(X)(b_1, b_2, \ldots, b_n)).$$

We will call first degree nCP-polynomial *an expression*

$$P(X) = \bigcup_{i=1}^{k} M_i(X) \bigcup L,$$

where $M_i(X)$, $i = \overline{1, k}$, are first-degree nCP-monomials, and $L \subset V^*$ is finite and is called the free term *of P.*

The *monomial function* $M : \mathcal{P}(V^*) \to \mathcal{P}(V^*)$, is defined by the composition of functions $c_n \circ (a_1, a_2, \ldots, a_n)(\cdot) \circ (\cdot)(b_1, b_2, \ldots, b_n) \circ c_n^{-1}$ and all these functions commute with arbitrary unions. It follows that any monomial function and also the polynomial functions commute with arbitrary unions. In particular, they will be ω-continuous functions from $\mathcal{P}(V^*)$ to itself.

Applying Kleene's fix-point theorem we obtain the following result:

Theorem 1. *Let P be a first-degree nCP-polynomial function. Then, the fix-point equation $P(X) = X$ has a least solution, L, obtained as the limit of the Kleene sequence:*

$$L = \bigcup_{m \geq 0} P^m(\emptyset).$$

L is also the smallest $L' \in \mathcal{P}(V^*)$ such that $P(L') \subseteq L'$.

Definition 2. *A language $L \subseteq V^*$ which is the solution of a fix-point equation $P(X) = X$, with P a first-degree nCP-polynomial, will be called an nCP-language.*

3 External contextual languages without choice are 2CP-languages

A *contextual grammar without choice* is a triple $G = (V, A, C)$, where V is a finite alphabet, A is a finite language over V (the set of axioms) and C is a finite subset of $V^* \times V^*$ (the contexts).

With respect to such a grammar, the *external derivation relation* on V^* is defined by:

$$x \underset{ex}{\Longrightarrow} y \text{ iff } y = uxv, \text{ for a context } (u, v) \in C,$$

and by $\underset{ex}{\overset{*}{\Longrightarrow}}$ we denote as usual the reflexive and transitive closure of $\underset{ex}{\Longrightarrow}$.

The *external contextual language* generated by G is:

$$L_{ex}(G) := \{y \in V^* \mid x \underset{ex}{\overset{*}{\Longrightarrow}} y, x \in A\}.$$

It can also be defined as the smallest language $L \subseteq V^*$ such that:

1. $A \subseteq L$;
2. if $x \in L$ and $(u, v) \in C$, then $uxv \in L$.

Theorem 2. *$L_{ex}(G)$ is a first-degree 2CP-language. More precisely, $L_{ex}(G)$ is the least solution of the fix-point equation attached to the first-degree 2CP-polynomial:*

$$P_G(X) = \bigcup_{(u,v) \in C} c_2((u, 1)c_2^{-1}(X)(1, v)) \cup A.$$

The proof is straightforward, and is based on identifying each member of P_G's Kleene sequence, $P_G^m(\emptyset)$, with the set $\{y \in V^* \mid x \overset{m}{\underset{ex}{\Longrightarrow}} y,\, x \in A\}$ of words obtained from axioms by applying at most k derivations (we start the Kleene sequence with $X_{-1} = \emptyset$, $X_0 = A$, etc.).

This first result shows that we can make a "canonical" passage from a contextual grammar G to a first-degree 2CP-polynomial, P_G, and that the languages generated by the two distinct mechanisms coincide.

The next result shows that we can also go ("canonically") in the reverse direction, from 2CP-polynomials of a certain type, to contextual grammars.

Theorem 3. *Let*

$$P(X) = \bigcup_{i=1}^{n} c_2((a_i, 1)c_2^{-1}(X)(1, b_i)) \cup A$$

be a 2CP-polynomial, and let $\overline{X}$ be the least solution of the fix-point equation attached to it.

With $C := \{(a_i, b_i) \mid i = \overline{1, n}\}$, $G_P = (V, A, C)$ is a contextual grammar such that $\overline{X} = L_{ex}(G_P)$.

The proof is similar to that of the preceding theorem.

Note that, in order to fall precisely on external contextual languages, we had to confine ourselves to 2CP-polynomials of a very particular type.

We are going to consider now 2CP-polynomials of the most general form, and construct a more general type of contextual grammar.

Let $P(X) = \bigcup_{i=1}^{n} c_2((a_i, b_i)c_2^{-1}(X)(a_i', b_i')) \cup A$ be a 2CP-polynomial, in its most general form. Consider the finite subsets of $V^* \times V^*$

$$C_{pref} := \{(a_i, a_i') \mid i = \overline{1, n}\}$$
$$C_{suf} := \{(b_i, b_i') \mid i = \overline{1, n}\}$$

and call them, respectively, *prefix-contexts* and *suffix-contexts*. We associate to P the quadruple:

$$G_P = (V, A, C_{pref}, C_{suf}).$$

For $x, y \in V^*$ define the (one-step) derivation in G_P as follows:

$$x \underset{G_P}{\Longrightarrow} y \text{ iff } \forall (x_1, x_2) \in V^* \times V^* \text{ such that } x_1 x_2 = x,$$

$$\exists (u, u') \in C_{pref},\ \exists (v, v') \in C_{suf} \text{ such that } y = u x_1 u' v x_2 v'.$$

Define the language generated by G_P as

$$L(G_P) := \{y \mid x \underset{G_P}{\overset{*}{\Longrightarrow}} y, x \in A\}.$$

It can alternatively be described as the smallest language $L \subseteq V^*$ such that:

1. $A \subseteq L$;
2. $x \in L$ with $x = x_1 x_2 \Rightarrow u x_1 u' v x_2 v' \in L$ for any $(u, u') \in C_{pref}$ and $(v, v') \in C_{suf}$.

Theorem 4. $L(G_P)$ *is precisely the least solution of the fix-point equation attached to* P.

As we can see, the grammar $G_P = (V, A, C_{pref}, C_{suf})$ attached to a 2CP-polynomial in its most general form, is *not* a contextual grammar without choice in the classical sense, but a much richer structure. The richness comes not only from the "multiplication" of the set of contexts, but essentially from the fact that the second component of a prefix-context, concatenated with the first component of a suffix-context, appears as an *internal context* in a derived word, thus blurring the distinction between *external* and *internal* languages generated by such a grammar. Only if $C_{pref} \subseteq V^* \times \{1\}$ and $C_{suf} \subseteq \{1\} \times V^*$, then, by taking $C = \pi_1(C_{pref}) \times \pi_2(C_{suf})$, can we fall back again on a contextual grammar without choice $G = (V, A, C)$.

4   Internal contextual languages without choice are 3CP-languages

Let $G_P = (V, A, C)$ be a contextual grammar without choice. Consider the *internal derivation* relation defined by:

$$x \underset{in}{\Longrightarrow} y \text{ iff } x = x_1 x_2 x_3, y = x_1 u x_2 v x_3 \text{ for any } x_1, x_2, x_3 \in V^*, (u, v) \in C$$

The *internal contextual language* generated by G is:

$$L_{in}(G) := \{ y \in V^* \mid x \underset{in}{\overset{*}{\Longrightarrow}} y, x \in A \}.$$

It can be written as the union $L_{in}(G) = \bigcup_{k \geq 0} L_{in}^k(G)$ where $L_{in}^k(G)$ is the set of words obtained applying at most k derivations to the axioms in A.

Consider now the 3CP-polynomial:

$$P_G(X) = \bigcup_{(a,b) \in C} c_3((1, a, b) c_3^{-1}(X)) \cup A.$$

Theorem 5. $L_{in}(G)$ *is a* 3CP-*language. More precisely,* $L_{in}(G)$ *is the least solution of the fix-point equation attached to the above polynomial* P_G.

The proof is again straightforward, and is based on identifying each $L_{in}^k(G)$ with the k term of the Kleene sequence, $P_G^k(\emptyset)$, starting the sequence with $P_G^{-1}(\emptyset) = \emptyset$, $P_G^0(\emptyset) = A$, etc.

As in the previous paragraph, we will now go from 3CP-polynomials towards internal contextual languages, via grammars.

For 3CP-polynomials of a particular form, the passage is easy:

Theorem 6. *Let* $P(X) = \bigcup\limits_{i=1}^{n} c_3((1, a_i, b_i)c_3^{-1}(X)) \cup A$ *be a* $3\,CP$-*polynomial, and denote by* $\overline{X}$ *its least fix-point solution. With* $C := \{(a_i, b_i) \mid i = \overline{1,n}\}$, $G_P = (V, A, C)$ *is a contextual grammar such that* $L_{in}(G_P) = \overline{X}$.

5   Contextual languages with choice and selective nCP-Languages

A contextual grammar with choice is a quadruple $G = (V, A, C, \phi)$, where (V, A, C) is a contextual grammar without choice as before, to which a **choice** (or **selection**) **function** $\phi : V^* \rightarrow \mathcal{P}(C)$ was added. With respect to such a grammar the **external** and **internal derivation** relations on V^* are defined by:

$$x \underset{ex}{\Longrightarrow} y \text{ iff } y = uxv, \text{ for a context } (u, v) \in \phi(x),$$

$$x \underset{in}{\Longrightarrow} y \text{ iff } x = x_1 x_2 x_3, y = x_1 u x_2 v x_3 \text{ for any } x_1, x_2, x_3 \in V^* \text{ and } (u, v) \in \phi(x_2)$$

Their respective reflexive and transitive closures define the **external** and the **internal contextual language with choice** generated by G, by:

$$L_\alpha(G) := \{y \in V^* \mid x \overset{*}{\underset{\alpha}{\Longrightarrow}} y, x \in A\} , \text{ with } \alpha \in \{ex, in\}.$$

We intend to prove that, as in the case without choice, contextual languages with choice can be obtained as nCP-languages, with $n = 2$ for external ones and $n = 3$ for internal ones. In order to achieve this goal, we will have to generalize the concept of nCP-polynomial, to that of **selective nCP-polynomial function**.

For a fixed $(u, v) \in C$, consider the partial function $M_{(u,v)} : V^* \rightarrow V^*$ with domain $Dom(M_{(u,v)}) = \{x \in V^* \mid (u, v) \in \phi(x)\}$, and defined for $x \in Dom(M_{(u,v)})$ by $M_{(u,v)}(x) = uxv = c_2((u, 1)c_2^{-1}(x)(1, v))$. We recognize in the last term of these equalities a 2CP-monomial function acting on a word $x \in V^*$. We can extend $M_{(u,v)}$ to a total function, $\overline{M}_{(u,v)} : \mathcal{P}(V^*) \rightarrow \mathcal{P}(V^*)$, by defining it on singletons as

$$\overline{M}_{(u,v)}(\{x\}) = \begin{cases} \{uxv\} & \text{, if } x \in Dom(M_{(u,v)}) \text{ iff } (u, v) \in \phi(x) \\ \emptyset & \text{, otherwise.} \end{cases}$$

and extending it naturally to a monotonous function by making it commute with arbitrary unions.

Definition 3. *A function* $\overline{M}_{(u,v)} : \mathcal{P}(V^*) \rightarrow \mathcal{P}(V^*)$, *defined as above will be called a* ϕ-*selective 2CP-mononomial function. A function* $P : \mathcal{P}(V^*) \rightarrow \mathcal{P}(V^*)$ *defined as*

$$P(X) := \bigcup_{(u,v) \in C} \overline{M}_{(u,v)}(X) \bigcup A$$

will be called a ϕ-*selective 2CP-polynomial function. A* ϕ-*selective 2CP-language will be a language* $L \subset V^*$ *which is the least solution of the fix-point equation* $P(X) = X$ *attached to a* ϕ-*selective 2 CP-polynomial function* P.

We now have the following result.

Theorem 7. *The external contextual language with choice $L_{ex}(G)$ is a ϕ-selective 2CP-language. More precisely, given the external contextual grammar G, we can canonically associate to it a ϕ-selective 2CP-polynomial function P_G as above, such that $L_{ex}(G)$ is the least solution of its attached fix-point equation.*

We will now consider the case $n = 3$ in order to obtain a similar result for internal contextual languages with choice.

Consider the partial function $N_{(u,v)} : V^* \rightarrow V^*$ obtained by the following recipe: restrict the codomain of c_3^{-1} to $V^* \times Dom(M_{(u,v)}) \times V^*$, and it will become a partial function, then extend it monotonously on $\mathcal{P}(V^*)$, and make it total by letting singletons $\{x\}$ go to $\emptyset$ for x's not in its domain. We obtain $\overline{N}_{(u,v)} : \mathcal{P}(V^*) \rightarrow \mathcal{P}(V^*)$, which is, with the considerations just made, precisely $c_3((1, u, v)c_3^{-1}(\cdot))$.

Definition 4. *A function $\overline{N}_{(u,v)} : \mathcal{P}(V^*) \rightarrow \mathcal{P}(V^*)$, defined as above will be called a ϕ-selective 3CP-mononomial function. A function $P : \mathcal{P}(V^*) \rightarrow \mathcal{P}(V^*)$ defined as*

$$P(X) := \bigcup_{(u,v) \in C} \overline{N}_{(u,v)}(X) \bigcup A$$

will be called a ϕ-selective 3CP-polynomial function. A ϕ-selective 3CP-language will be a language $L \subset V^*$ which is the least solution of the fix-point equation $P(X) = X$ attached to a ϕ-selective 3CP-polynomial function P.

Theorem 8. *The internal contextual language with choice $L_{in}(G)$ is a ϕ-selective 3CP-language. More precisely, given the external contextual grammar G, we can canonically associate to it a ϕ-selective 3CP-polynomial function P_G as in definition 4 above, such that $L_{in}(G)$ is the least solution of its attached fix-point equation.*

After having exemplified the construction of ϕ-selective nCP-polynomials for contextual languages with the simplest type of selection function $\phi : V^* \rightarrow \mathcal{P}(C)$, there is no problem to generalize it for $\phi : (V^*)^3 \rightarrow \mathcal{P}(C)$, thus obtaining the **total contextual languages with choice as ϕ-selective 3CP-languages**, and, for $\phi : (V^*)^n \rightarrow \mathcal{P}(C)$, the **$n$-contextual languages as ϕ-selective nCP-languages**.

Consider now an **insertion grammar** $G = (V, A, P)$, where V, A are as above, and the finite set $P \subset (V^*)^3$ is called the **set of productions** of G. The corresponding language $L(G)$ is defined as usual, using as one-step derivation the following relation:

$$x \underset{G}{\Longrightarrow} y \text{ iff } x = x_1 u v x_2, y = x_1 u x v x_2 \text{ for any } x_1, x_2 \in V^* \text{ and } (u, x, v) \in P$$

For fixed $(u, x, v) \in P$ consider the pre-image of two-concatenation with restricted codomain $c_2^{-1} : V^* \rightarrow V^* u \times v V^*$, and apply the recipe: make it total by letting singletons $\{x\}$ go to $\emptyset$ for x not in its domain, and then extend it ω-continuously to get a monomial function $Q_{(u,x,v)} : \mathcal{P}(V^*) \rightarrow \mathcal{P}(V^*)$, defined by $Q_{(u,x,v)}(X) = c_2((1, x)c_2^{-1}(X))$.

Theorem 9. *The language generated by an insertion grammar G is a selective* 2*CP-language. More precisely, it is the least solution of the fix-point equation attached to the selective* 2*CP-polynomial*

$$P(X) = \bigcup_{(u,x,v) \in P} Q_{(u,x,v)}(X) \bigcup A.$$

6   Some concluding remarks

The construction of different types of selective nCP-polynomials done in the last section illustrates the whole delicate interplay between strings and contexts. In particular, the selective 2CP-polynomials attached to an insertion grammar open the way towards an alternative formalism for the splicing operation, based on CP-polynomials.

We hope that, completing the research started here with the reverse passage, from selective nCP-languages to contextual ones, will contribute to one of the goals proposed by S. Marcus in [4], namely "to enrich the combinatorics of distinctions left-right, internal-external ..." providing at the same time an extremely powerful tool for their unified study.

References

1. R. Ceterchi, Cut-and-Paste Languages (submitted)
2. S. Istrail, A fixed-point approach to contextual languages, *Rev. Roum. Math. Pures Appl.*, 25 (1980), 861–869
3. S. Marcus, Contextual grammars, *Rev. Roum. Math. Pures Appl.*, 14(1969), 1525–1534
4. S. Marcus, Contextual grammars and natural languages, chapter 5 in vol.2 of [10], 215–235
5. A. Mateescu, Marcus contextual grammars with shuffled contexts, in *Mathematical Aspects of Natural and Formal Languages*, (Gh. Paun, ed.), World Sci. Publ., Singapore, 1994, 275–284
6. Gh. Paun, X. M. Nguyen, On the inner contextual grammars, *Rev. Roum. Math. Pures Appl.*, 25(1980), 641–651
7. Gh. Paun, *Contextual Grammars*, The Publ. House of the Romanian Academy, Bucharest, 1982 (in Romanian; 144 pages)
8. G h. Paun, *Marcus Contextual Grammars*, Kluwer Academic Publishers, Dordrecht, Boston, london, 1997
9. Gh. Paun, G. Rozenberg, A. Salomaa, *DNA Computing. New Computing Paradigms*, Springer-Verlag, Berlin, Heidelberg, 1998
10. G. Rozenberg, A. Salomaa (eds.), *The Handbook of Formal Languages*, 3 volumes, Springer-Verlag, Berlin, Heidelberg, 1997

Contextual Multilanguages: A Learning Method

Radu Gramatovici

Faculty of Mathematics, Bucharest University
Academiei 14, R-70109, Bucureşti, Romania
e-mail: radu@moisil.math.ro

Abstract. In this paper, we deal with a class of contextual languages defined in the frame of order-sorted algebra. We define a new way to describe contextual (multi)languages, using constraints. For this class of constraint contextual languages a learning method with motivations from natural language processing is provided.

1 Introduction

It is a reality that precise grammars for natural language processing are very hard to describe. At the same time, learning such grammars is usually a slow process and the result is difficult to use in practice. A comfortable compromise between these directions might be to write some general rules and then use a learning method to achieve other specific informations.

Contextual grammars ([6],[5]) are a very interesting model for describing non-context-free constructions in natural languages. Order-sorted algebra ([1]) is a powerful algebraic approach to computational semantics. At the intersection of these areas, we have developed a generative formalism called *order sorted multilanguages* ([3]), where syntactical categories are described as sorts and derivation rules as operations. We have also developed an extended stack automaton called *operatorial automaton* to deal with order-sorted multilanguages.

The link between order-sorted multilanguages and operatorial automata is established by a set of *operatorial relations* which are a generalization of operatorial precedence relations. A very interesting way to define *constraint* order-sorted multilanguages can be described using operatorial relations: define a "maximal" multilanguage and then constrain it by selecting a subset of operatorial relations. Only those words which are consistent with the remaining set of operatorial relations will be kept.

The learning algorithm developed in this paper for constraint contextual multilanguages splits the paradigm writing/learning in the following way: the "maximal" contextual multilanguage must be written and the actual subset of operatorial relations constraining the language may be learned from examples.

We believe that derivation rules are not as hard to write as it is to describe the precise application of these rules. Our learning algorithm covers two aspects: how the derivation rules apply and to which arguments may they be applied. Considering the case of natural languages, the first aspect concerns the order of words in a sentence, while the second aspect includes a refinement of sorts and operations.

L. Polkowski and A. Skowron (Eds.): RSCTC'98, LNAI 1424, pp. 107–110, 1998.

2   Preliminaries

We denote by $[n]$ the set of the first n natural numbers, not equal to 0. A *partially ordered set* or *poset* is a set S together with a binary relation $\leq$ on S that is reflexive, transitive and antisymmetric.

Let V be a set. We denote by 2^V the set of all finite subsets of V. We denote by V^+ the set of strings $v_1 \ldots v_n$, with $v_i \in V$, for any $i \in [n]$ and we denote $V^* = V^+ \cup \{\lambda\}$, where λ is the empty string. Let $w \in V^*$. We denote by $|w|$ the length of w and by $|w|_a$ the number of occurrences of the character a in w. We say that $v = v_1 \ldots v_n$ is a prefix of $w = w_1 \ldots w_m$ iff $n \leq m$ and $v_i = w_i$, for any $i \in [n]$. Also, λ is a prefix of any string of characters.

The *shuffle* operation between words, denoted $ш$, is defined recursively by $av \: ш \: bw = a(v \: ш \: bw) \cup b(av \: ш \: w)$ and $w \: ш \: \lambda = \lambda \: ш \: w = w$, where $v, w \in V^*$, $a, b \in V$.

Definition 1. *A* (simple) order-sorted signature *is a triple* $(S, \leq, \Sigma)$ *such that* $(S, \leq)$ *is a non-empty poset, called the* sort set *and* Σ *is an* $S^* \times S$-*sorted family* $\{\Sigma_{w,s}/w \in S^*, s \in S\}$ *such that* $\Sigma_{w_1,s_1} \cap \Sigma_{w_2,s_2} = \emptyset$, *for any* $w_1, w_2 \in S^*$, $s_1, s_2 \in S$, *with* $(w_1, s_1) \neq (w_2, s_2)$.

Definition 2. *Let* $(S, \leq, \Sigma)$ *an order-sorted signature. An* order-sorted multi-algebra *over* Σ *is a pair* $A = ((A_s)_{s \in S}, (\sigma_A)_{\sigma \in \Sigma})$, *where* A_s *is a* S-*sorted family of sets such that* $s \leq s'$ *implies* $A_s \subseteq A_{s'}$, *and the operations are defined as:*

i) if $\sigma \in \Sigma_{\lambda,s}$, *then* $\sigma_A \in 2^{A_s}$;
ii) if $\sigma \in \Sigma_{s_1 \ldots s_n, s}$, *then* $\sigma_A : A_{s_1} \times \ldots \times A_{s_n} \to 2^{A_s}$.

We denote $\mid A \mid = \cup_{s \in S} A_s$.

Definition 3. *Let* A *be an order-sorted multialgebra over* Σ *and* $s \in S$ *be a sort. An element* $a \in A_s$ *is called* reachable *iff:*

– *there is* $\sigma \in \Sigma_{\lambda,s}$ *such that* $a \in \sigma_A$, *or*
– *there are* $\sigma \in \Sigma_{s_1 \ldots s_n, s}$ *and* $a_i \in A_{s_i}$, $i \in [n]$, *reachable elements, such that* $a \in \sigma_A(a_1, \ldots, a_n)$.

A is called reachable *iff it has only reachable elements.*

3   Contextual multilanguages

Let Σ be an order-sorted signature and V a finite, non-empty set, called *alphabet*.

Definition 4. *A reachable order-sorted multialgebra* A *is an* order-sorted multilanguage *over* Σ *and* V *iff* $\mid A \mid \subseteq V^+$.

Definition 5. *An operation* $\sigma_A : A_{s_1} \to 2^{A_{s_2}}$ *is a* contextual rule *(respectively a* shuffle rule*) iff there exists an word* $mask_A(\sigma) \in V^+$ *such that* $\sigma_A(\alpha) \subseteq mask_A(\sigma) \: ш \: \alpha$ *(respectively* $\sigma_A(\alpha) = mask_A(\sigma) \: ш \: \alpha$).

Definition 6. *An order-sorted multilanguage is a contextual multilanguage (respectively a contextual language with shuffle) iff it contains only constants and contextual (respectively shuffle) rules.*

We call a *derivation of sort* s_2 of α in A a construction $\overset{\sigma}{\Rightarrow}_A \alpha$ iff $\sigma \in \Sigma_{\lambda,s_2}$ and $\alpha \in \sigma_A$, or a construction $\beta \overset{\sigma}{\Rightarrow}_A \alpha$ iff $\sigma \in \Sigma_{s_1,s_2}$, $\beta \in A_{s_1}$, $\alpha = b_1$ and $\alpha \in \sigma_A(\beta)$.

Example 1. We consider (S, Σ) an order-sorted signature, with $S = \{s\}$, $\Sigma = \{\sigma^1 :\to s, \sigma^2 : s \to s\}$ and $V = \{a, b, c\}$ an alphabet. Then, we consider a contextual language with shuffle $A = (A_s, \sigma_A^1, \sigma_A^2)$, where A_s is the set of all words $w \in V^*$, such that $|w|_a = |w|_b = |w|_c$ and for any prefix x of w, $|x|_a \geq |x|_b \geq |x|_c$ and $\sigma_A^1 = \{abc\}$, $\sigma_A^2(\alpha) = abc \amalg \alpha$, for any $\alpha \in A_s$.

4   Constraint contextual multilanguages

We will examine in the sequel a way to generate multilanguages using contextual operatorial relations which are formally defined in [2].

Definition 7. *Let A be a contextual multilanguage and R_A the set of operatorial relations generated by A. Let R be a subset of R_A. The constraint contextual multilanguage A/R is a contextual multilanguage defined by[1]:*

1. $\sigma_{A/R} = \{\alpha/\exists d :\overset{\sigma}{\Rightarrow}_A \alpha, R_A^{d,s_2}(\alpha) \subseteq R\}$, *for any* $\sigma \in \Sigma_{\lambda,s}$;
2. $\sigma_{A/R}(\beta) = \{\alpha/\exists d : \beta \overset{\sigma}{\Rightarrow}_A \alpha, R_A^{d,s_2}(\alpha) \subseteq R\}$, *for any* $\sigma \in \Sigma_{s_1,s_2}$, $\beta \in (A/R)_{s_1}$.

Example 2. We consider the contextual language with shuffle A defined in Example 1. We have $R_A = \{(a, a, 2), (a, a, 3), (a, b, 1), (a, b, 3), (a, c, 3), (b, a, 3), (b, b, 3), (b, c, 1), (b, c, 3), (c, a, 3), (c, b, 3), (c, c, 3)\}$.If $R = \{(a, a, 3), (a, b, 1), (b, b, 3), (b, c, 1), (c, c, 3)\} \subseteq R_A$, then $A/R = ((A/R)_s, \sigma_{A/R}^1, \sigma_{A/R}^2)$, with $(A/R)_s = \{a^n b^n c^n | n \geq 1\}$ and $\sigma_{A/R}^1 = \{abc\}$, $\sigma_{A/R}^2(a^n b^n c^n) = \{a^n ab^n bc^n c\}$, for any $n \geq 1$.

5   A learning method

A problem when writing down a contextual multilanguage is to establish the precise definition of derivation rules. Even with constraint contextual multilanguages we still have a problem: how many operatorial relations must be kept in order to obtain the desired constraint multilanguage.

The learning method we introduce here uses an operatorial automaton which is formally described in [3]. The learning algorithm has the following steps:

1. Define a "maximal" contextual multilanguage A.
2. Generate the corresponding set of operatorial relations R_A.

---

[1] $R_A^{d,s_2}(\alpha)$ is the set of operatorial relations corresponding to α with respect to the derivation d of sort s_2 in A.

3. If $M = (V, S, R_A, \delta)$ is an operatorial automaton which computes the (least) sort of any word in A , define a learning operatorial automaton $M' = (V, S, R, R_A, \delta)$ and set $R = \emptyset$.

4. Run the automaton M' on a set of examples $E \subseteq |A|$ in the following way:
   - if an operatorial relation between two characters a and b is required, take this relation from R;
   - if R doesn't contain relations between a and b or all the relations that R contains between a and b fail, search a relation between a and b in R_A;
   - if such a relation exists add this relation to R and continue the execution.

Example 3. With the above notations, let us consider A the contextual language with shuffle from Example 1, with the set of operatorial relations R_A, defined in Example 2. We take the set of positive examples $E = \{a^2b^2c^2\}$. The learning method has a degree of randomness, depending on the choices we make, when selecting an operatorial relation from R_A. The learning operatorial automaton may run in the following way:

$$(aabbcc\$, (\$)., \_) \quad \vdash \quad (abbcc\$, (\$)(a)., \_) \quad \overset{(a,a,3)\in R}{\vdash} \quad (abbcc\$, (\$).(a), \_)$$

$$\vdash (bbcc\$, (\$)(a)(a)., \_) \quad \overset{(a,b,1)\in R}{\vdash} \quad (bcc\$, (\$)(a)(ab)., \_) \quad \overset{(b,b,3)\in R}{\vdash} \quad (bcc\$, (\$)(a).(ab), \_)$$

$$\vdash (cc\$, (\$)(ab)(ab)., \_) \quad \overset{(b,c,1)\in R}{\vdash} \quad (c\$, (\$)(ab)(abc)., \_) \quad \overset{(c,c,3)\in R}{\vdash} \quad (c\$, (\$)(ab).(abc), \_)$$

$$\vdash (\$, (\$)(abc)(abc)., \_) \vdash \quad (\$, (\$)(abc)., s) \quad \vdash \quad (\$, (\$)., s)$$

and we obtain a set of operatorial relations $R = \{(a, a, 3), (a, b, 1), (b, b, 3), (b, c, 1),$ $(c, c, 3)\}$, hence the procedure finds exactly the last constraint contextual multi-language from Example 2.

Theorem 1. *The learning method described above is an algorithm. It learns a constraint contextual multilanguage, included in the initial "maximal" contextual multilanguage and consistent with the given set of examples.*

Proof. The learning procedure ends every time because the set of operatorial relations R_A (hence also R) is finite.   □

References

1. J. Goguen, J. Meseguer, Order-sorted algebra I: equational deduction for multiple inheritance, overloading, exceptions and partial operations, *Theoretical Computer Science* 105(1992), 217–273.
2. R. Gramatovici, An efficient parser for a class of contextual languages, *Fundamenta Informaticae*, to appear.
3. R. Gramatovici, Contextual multilanguages and operatorial automata, submitted *MFCS'98.*
4. M. Kudlek, S. Marcus, A. Mateescu, Contextual Grammars with Distributed Catenation and Shuffle, *Technical Report* FBI-HH-B-200/97, University of Hamburg, 1997.
5. Gh. Păun, *Marcus Contextual Grammars*, Kluwer, 1997.
6. S. Marcus, Contextual grammars, *Rev. Roum. Math. Pures Appl.*, 14, 10(1969), 69–74.

On Recognition of Words from Languages Generated by Linear Grammars with One Nonterminal Symbol

Alexander Knyazev

Faculty of Calculating Mathematics and
Cybernetics of Nizhni Novgorod State University
23, Gagarina Av., Nizhni Novgorod, 603600, Russia

1 Introduction

In this paper we consider the classification of languages generated by linear grammars with one nonterminal symbol depending on the minimal depth of decision trees for language word recognition.

Assume that L is a language in a finite alphabet, n is a natural number and $L(n)$ is the set of all words from L for which the length is equal to n. We denote by $h_L(n)$ the minimal depth of a decision tree which recognizes words from $L(n)$ and uses only such checks each of which determines i-th letter of a word, $i \in \{1, \ldots, n\}$. If $L(n) = \emptyset$, then $h_L(n) = 0$. (Note: the belonging recognition problem a word to $L(n)$ is not explored here, but any word must be determined in assumption that one belongs to $L(n)$).

Instead of the function $h_L(n)$ we will consider the following function:

$$H_L(n) = \max\{h_L(m) : m \leq n\}.$$

In [2, 3, 4] it was shown that for an arbitrary regular language L either $H_L(n) = O(1)$, or $H_L(n) = \Theta(\log n)$, or $H_L(n) = \Theta(n)$. This paper deals with the investigation of the function H_L behavior for an arbitrary language L generated by a linear grammar with one nonterminal symbol. We show that either $H_L(n) = O(1)$, or $H_L(n) = \Theta(\log n)$, or $H_L(n) = \Theta(n)$. Further this results will be used for investigation of linear grammars with many nonterminal symbols and context-free grammars with one nonterminal symbol.

In proofs methods of test theory [1, 2] and rough set theory [5, 6] are used. For a set of words $L(n)$ we construst corresponding decision table, and use lower and upper bounds of minimal depth of decision tree for this one.

One can interpret a word from $L(n)$ as a description of an image on the screen with n cells: the i-th letter of the word defines the color of the i-th cell of the screen. In this case a decision tree which recognizes words from $L(n)$ may be interpreted as an algorithm for the recognition of images which are defined by words from $L(n)$.

* This work was partially supported by Russian Federal Program "Integration" (project # 473 "Educational-Research Center "Methods of Discrete Mathematics for New Information Technologies").

L. Polkowski and A. Skowron (Eds.): RSCTC'98, LNAI 1424, pp. 111–114, 1998.

2   Definitions

Let E be a finite nonempty set (alphabet) and E^* be a set of all finite words in the alphabet E, including the empty word λ.

The root of a word $\alpha \in E^*$ is the word $\beta \in E^*$ of minimal length such that for some natural number t the equality $\alpha = \beta^t$ holds. The root of the word α will be denoted by $rt(\alpha)$.

Denote by $\pi_k(\alpha)$ the cyclic permutation of k letters from the begining to the end of α. Words α and β will be called similar if there exists k such that $rt(\beta) = \pi_k(rt(\alpha))$. The minimal k such that $rt(\beta) = \pi_k(rt(\alpha))$ will be denoted by $RT(\alpha, \beta)$. If the words α and β are not similar then $RT(\alpha, \beta) = \infty$.

Denote by $l(\alpha)$ the length of the word α.

Let S be a symbol which does not belong to E. We will consider arbitrary grammar Γ of the following form:

$$S \to \alpha_1 S \beta_1, \ldots, S \to \alpha_p S \beta_p, S \to \varepsilon_1, \ldots, S \to \varepsilon_q$$

where $p \geq 1$, $q \geq 1$ and $\alpha_i, \beta_i, \varepsilon_j$ are words from E^*, $i = 1, \ldots, p$, $j = 1, \ldots, q$.

Denote by $L(\Gamma)$ the language in the alphabet E generated by the grammar Γ.

With the grammar Γ we will associate the following two sets $R_1(\Gamma) = \{rt(\alpha_i) : i = 1, \ldots, p, \alpha_i \neq \lambda\}$ and $R_2(\Gamma) = \{rt(\beta_i) : i = 1, \ldots, p, \beta_i \neq \lambda\}$, and also the matrix $\Delta(\Gamma)$ with p rows and 2 columns. Elements of this matrix Δ_{i1} and Δ_{i2}, $i = 1, \ldots, p$, are determined as follows:

a) $\Delta_{i1} = \frac{l(\alpha_i)}{l(rt(\alpha_i))}$ if $\alpha_i \neq \lambda$, and $\Delta_{i1} = 0$ otherwise;

b) $\Delta_{i2} = \frac{l(\beta_i)}{l(rt(\beta_i))}$ if $\beta_i \neq \lambda$, and $\Delta_{i2} = 0$ otherwise.

Denote by $rank\Delta(\Gamma)$ the rank of the matrix $\Delta(\Gamma)$.

For $j = 1, \ldots, q$ denote by Γ_j the following grammar:

$$S \to \alpha_1 S \beta_1, \ldots, S \to \alpha_p S \beta_p, S \to \varepsilon_j.$$

It is clear that

$$L(\Gamma) \subseteq \bigcup_{j=1}^{q} L(\Gamma_j).$$

One can show that

$$\max\{H_{L(\Gamma_j)}(n) : j = 1, \ldots, q\} \leq H_{L(\Gamma)}(n) \leq \sum_{j=1}^{q}(H_{L(\Gamma_j)}(n) + 1) - 1.$$

Therefore $H_{L(\Gamma)}(n) = \Theta(\max\{H_{L(\Gamma_j)}(n) : j = 1, \ldots, q\})$. Consequently we can consider only the case when $q = 1$.

3   Main Result

[**Theorem 1.**]*Let Γ be linear grammar with one nonterminal symbol*

$$S \to \alpha_1 S \beta_1, \dots, S \to \alpha_p S \beta_p, S \to \varepsilon.$$

Then the following statements hold:

1. *If $|R_1(\Gamma)| \geq 2$ or $|R_2(\Gamma)| \geq 2$ then $H_{L(\Gamma)}(n) = \Theta(n)$;*
2. *If $|R_1(\Gamma)| \leq 1$, $|R_2(\Gamma)| \leq 1$ and $rank\Delta(\Gamma) \leq 1$ then $H_{L(\Gamma)}(n) = 0$ for any n.*
3. *Let $R_1(\Gamma) = \{\alpha\}$, $R_2(\Gamma) = \{\beta\}$ and $rank\Delta(\Gamma) = 2$. Then*
   a) if $\alpha\varepsilon = \varepsilon\beta$ then $H_{L(\Gamma)}(n) = 0$ for any n;
   b) if $\alpha\varepsilon \neq \varepsilon\beta$ and $RT(\alpha, \beta) = \infty$ or $RT(\alpha, \beta) \not\equiv l(\varepsilon)\mathrm{mod}(l(\alpha))$ then $H_{L(\Gamma)}(n) = \Theta(\log n)$.
   c) if $\alpha\varepsilon \neq \varepsilon\beta$ and $RT(\alpha, \beta) \equiv l(\varepsilon)\mathrm{mod}(l(\alpha))$ then $H_{L(\Gamma)}(n) = \Theta(n)$;

4   Examples

[*Example 1.*]Consider the following grammar Γ_1:

$$S \to 1S0, S \to 0S1, S \to 0110.$$

One can show that $R_1(\Gamma_1) = R_2(\Gamma_1) = \{0, 1\}$ and $|R_1(\Gamma_1)| = 2$. Therefore $H_{L(\Gamma_1)}(n) = \Theta(n)$.

[*Example 2.*]Consider the following grammar Γ_2:

$$S \to 1010S01010101, S \to 101010S010101010101, S \to \lambda.$$

One can show that $R_1(\Gamma_2) = \{10\}$, $R_2(\Gamma_2) = \{01\}$, $|R_1(\Gamma_2)| = |R_2(\Gamma_2)| = 1$, $\Delta(\Gamma_2) = \begin{bmatrix} 2 & 4 \\ 3 & 6 \end{bmatrix}$. Therefore $H_{L(\Gamma_2)}(n) = 0$ for any n.

[*Example 3.*]Consider the following grammar Γ_3:

$$S \to 10S, S \to S01, S \to 1.$$

One can show that $R_1(\Gamma_3) = \{10\}$, $R_2(\Gamma_3) = \{01\}$, $|R_1(\Gamma_3)| = |R_2(\Gamma_3)| = 1$, $\Delta(\Gamma_3) = \begin{bmatrix} 1 & 0 \\ 0 & 1 \end{bmatrix}$, $\underbrace{10 \ \overbrace{1}} = \overbrace{1} \ \underbrace{01}$. Therefore $H_{L(\Gamma_3)}(n) = 0$ for any n.

[*Example 4.*]Consider the following grammar Γ_4:

$$S \to 1S, S \to S0, S \to \lambda.$$

One can show that $R_1(\Gamma_4) = \{1\}$, $R_2(\Gamma_4) = \{0\}$, $|R_1(\Gamma_4)| = |R_2(\Gamma_4)| = 1$, $\Delta(\Gamma_4) = \begin{bmatrix} 1 & 0 \\ 0 & 1 \end{bmatrix}$, $1\lambda \neq \lambda 0$, $RT(1, 0) = \infty$. Therefore $H_{L(\Gamma_4)}(n) = \Theta(\log n)$.

[*Example 5.*]Consider the following grammar Γ_5:

$$S \to 10S, S \to S01, S \to \lambda.$$

One can show that $R_1(\Gamma_5) = \{10\}$, $R_2(\Gamma_5) = \{01\}$, $|R_1(\Gamma_5)| = |R_2(\Gamma_5)| = 1$, $\Delta(\Gamma_5) = \begin{bmatrix} 1 & 0 \\ 0 & 1 \end{bmatrix}$, $10\lambda \neq \lambda01$, $RT(10,01) = 1 \not\equiv l(\lambda)\mathrm{mod}(2)$. Hence $H_{L(\Gamma_5)}(n) = \Theta(\log n)$.

[*Example 6.*]Consider the following grammar Γ_6:

$$S \to 10S, S \to S01, S \to 0.$$

One can show that $R_1(\Gamma_6) = \{10\}$, $R_2(\Gamma_6) = \{01\}$, $|R_1(\Gamma_6)| = |R_2(\Gamma_6)| = 1$, $\Delta(\Gamma_6) = \begin{bmatrix} 1 & 0 \\ 0 & 1 \end{bmatrix}$, $\underbrace{10}\ \overbrace{0} = \overbrace{0}\ \underbrace{01}$, $RT(10,01) = 1 \equiv l(0)\mathrm{mod}(2)$. Therefore $H_{L(\Gamma_6)}(n) = \Theta(n)$.

References

1. Chegis, I.A., Yablonskii, S.V.: Logical methods for electric circuit control. Trudy MI AN SSSR **51** (1958) 270–360 (in Russian)
2. Moshkov, M.Ju.: Decision Trees. Theory and Applications. Nizhni Novgorod University Publishers, Nizhni Novgorod (1994) (in Russian)
3. Moshkov, M.Ju.: Complexity of decision trees for regular language word recognition. Preproceedings of the Second International Conference Developments in Language Theory, Magdeburg (1995)
4. Moshkov, M.Ju.: Complexity of deterministic and nondeterministic decission trees for regular language word recognition. Proceedings of the Interlational Conference Developments in Language Theory, Thessaloniki (1997) 343–349
5. Pawlak, Z.: Rough Sets - Theoretical Aspects of Reasoning about Data. Kluwer Academic Publishers, Dordrecht (1991)
6. Skowron, A., Rauszer, C.: The discernibility matrices and functions in information systems, in Intelligent Decision Support. Handbook of Applications and Advances of the Rough Set Theory. Kluwer Academic Publishers, Dordrecht (1992) 331–362

Approximation Spaces and Definability for Incomplete Information Systems

Wojciech Buszkowski

Faculty of Mathematics and Computer Science
Adam Mickiewicz University
Poznań Poland

[**Abstract.**] **Incomplete information systems are approached here by general methods of rough set theory (see Pawlak [6,7]). We define approximation spaces of incomplete information systems and study definability and strong definability of sets of objects.**

1   Basic notions

An information system is a quadruple $I = (\mathcal{O}, \mathcal{A}, \mathcal{V}, \rho)$ such that $\mathcal{O}, \mathcal{A}, \mathcal{V}$ are nonempty finite sets, and $\rho : \mathcal{O} \times \mathcal{A} \mapsto \mathcal{V}$. Elements of $\mathcal{O}$, $\mathcal{A}$ and $\mathcal{V}$ are called *objects*, *attributes* and *values*, respectively, and $\rho(x, A)$ is *the value* of attribute A for object x. For instance, let $\mathcal{O}$ consist of some persons and $\mathcal{A}$ contain three attributes A, B, C, interpreted as Sex, Age and Nationality, respectively. Then, $\rho(x, A)$ is F or M, $\rho(x, B)$ is a nonnegative integer, and $\rho(x, C)$ is English or German or Polish etc. The set $\mathcal{V}$ consists of all values of attributes appearing in the system.

The above notion of an information system can be used to express a complete knowledge about some piece of reality. In this paper, we are concerned with incomplete information systems, corresponding to an incomplete knowledge. Then, we admit ρ be a partial mapping from $\mathcal{O} \times \mathcal{A}$ to $\mathcal{V}$; $\rho(x, A)$ is undefined, if one does not know the value of attribute A for object x. For different models of incomplete information systems see Orłowska and Pawlak [5]. Hereafter, by an information system we mean an incomplete information system. We always assume *the nonempty knowledge condition*:

$$(\text{NKC}) \ (\forall x \in \mathcal{O})(\exists A \in \mathcal{A}) \, \rho(x, A) \text{ is defined.}$$

An information system $I = (\mathcal{O}, \mathcal{A}, \mathcal{V}, \rho)$ is said to be *complete*, if ρ is a total mapping.

To simplify the framework we often consider (incomplete) two-valued information systems in which $\mathcal{V}$ consists of truth values 1 and 0 (truth and falsehood). Attributes are partial unary predicates on the set of objects: attribute A holds for object x, if $\rho(A, x) = 1$, attribute A does not hold for object x, if $\rho(A, x) = 0$, and A is undetermined on x, otherwise. Every information system $I = (\mathcal{O}, \mathcal{A}, \mathcal{V}, \rho)$

L. Polkowski and A. Skowron (Eds.): RSCTC'98, LNAI 1424, pp. 115–122, 1998.
© Springer-Verlag Berlin Heidelberg 1998

can be represented by the two-valued information system $I' = (\mathcal{O}', \mathcal{A}', \mathcal{V}', \rho')$ such that $\mathcal{O}' = \mathcal{O}$, $\mathcal{A}' = \mathcal{A} \times \mathcal{V}$, $\mathcal{V}' = \{0, 1\}$, and ρ' is defined as follows:

$\rho'(x, (A, v)) = 1$ if $\rho(x, A) = v$,

$\rho'(x, (A, v)) = 0$ if $\rho(x, A) \neq v$ and $\rho(x, A)$ is defined,

$\rho'(x, (A, v))$ is undefined if $\rho(x, A)$ is undefined.

The system I' will be called *the two-valued representation* of I and denoted by $T(I)$.

With any two-valued information system $I = (\mathcal{O}, \mathcal{A}, \{0, 1\}, \rho)$ we associate a propositional language L_I whose atomic formulas are symbols $\overline{A}$, for all $A \in \mathcal{A}$, and complex formulas are built by means of logical connectives: $\neg$ and $\wedge$. Disjunction $\vee$, implication $\rightarrow$ and equivalence $\leftrightarrow$ are defined as in classical logic.

For any formula φ, we define two sets $P(\varphi), N(\varphi) \subseteq \mathcal{O}$, called *the positive extension* and *the negative extension*, respectively, of formula φ in system I:

(PN0) $P(\overline{A}) = \{x \in \mathcal{O} : \rho(x, A) = 1\}$, $N(\overline{A}) = \{x \in \mathcal{O} : \rho(x, A) = 0\}$,

(PN1) $P(\neg\varphi) = N(\varphi)$, $N(\neg\varphi) = P(\varphi)$,

(PN2) $P(\varphi \wedge \psi) = P(\varphi) \cap P(\psi)$, $N(\varphi \wedge \psi) = N(\varphi) \cup N(\psi)$.

Observe that $P(\varphi) \cap N(\varphi) = \emptyset$, for every formula φ, but $P(\varphi) \cup N(\varphi) = \mathcal{O}$ need not hold.

We briefly recall some notions of partial logic (or: Kleene 3-valued logic with strong connectives [1,4,2]) which is a standard logical formalism for describing incomplete information. The truth values are 0 (falsehood), u (truth value gap) and 1 (truth), and one stipulates the ordering $0 < u < 1$. An *assignment* is a mapping α from the set of atomic formulas to $\{0, u, 1\}$, and it is defined for all formulas, by setting:

(a1) $\alpha(\neg\varphi) = 1 - \alpha(\varphi)$,

(a2) $\alpha(\varphi \wedge \psi) = \min(\alpha(\varphi), \alpha(\psi))$.

Let Γ be a set of formulas, and let φ be a formula. Then, $\Gamma \vdash_3 \varphi$ holds, if, for every assignment α, if $\alpha(\psi) = 1$, for all $\psi \in \Gamma$, then $\alpha(\varphi) = 1$. $\vdash_3$ is *the consequence relation* of partial logic. By $\vdash_2$ we denote the consequence relation of classical logic (one considers classical assignments only, i.e. mappings from the set of atomic formulas to $\{0, 1\}$).

Let $I = (\mathcal{O}, \mathcal{A}, \{0, 1\}, \rho)$ be a two-valued information system. For any $x \in \mathcal{O}$, we define the set:

$$D_I(x) = \{\overline{A} : \rho(x, A) = 1\} \cup \{\neg\overline{A} : \rho(x, A) = 0\},$$

called *the description of x in I*. By $\delta_I(x)$ we denote the conjunction of all formulas from $D_I(x)$. Notice that $D_I(x) \neq \emptyset$, by (NKC).

[**Proposition 1**]*For all objects $x \in \mathcal{O}$ and formulas φ of L_I, there hold the following equivalences:*

(D1) $x \in P(\varphi)$ if, and only if, $D_I(x) \vdash_3 \varphi$,

(D2) $x \in N(\varphi)$ if, and only if, $D_I(x) \vdash_3 \neg\varphi$.

(D1) and (D2) can be proved by induction on φ, using (PN0)-(PN2) and obvious properties of $\vdash_3$. For complete information systems I, $\vdash_3$ can be replaced by $\vdash_2$.

2   Approximation spaces

Given a complete information system $I = (\mathcal{O}, \mathcal{A}, \mathcal{V}, \rho)$, one defines *the indiscernibility relation* $\sim_I$ on the set $\mathcal{O}$ in the following way: $x \sim_I y$ if, for all $A \in \mathcal{A}$, $\rho(x, A) = \rho(y, A)$. For $X \subseteq \mathcal{O}$, $C_L X$ is the join of all equivalence classes of $\sim_I$ which are totally contained in X, and $C_U X$ is the join of all equivalence classes of $\sim_I$ which are not disjoint with X. $C_L X$ and $C_U X$ are called *the lower approximation* and *the upper approximation*, respectively, of the set X. A set X is said to be *definable* in I, if $X = C_L X$ (equivalently: $X = C_U X$). These notions constitute a foundation of Pawlak's rough set theory [6,7,3].

One can prove that approximation operations C_L and C_U satisfy the following conditions, for all $X, Y \subseteq \mathcal{O}$:

(C0) $-C_L X = C_U(-X)$, $-C_U X = C_L(-X)$.
(C1) $C_L X \subseteq X$, $X \subseteq C_U X$,
(C2) if $X \subseteq Y$, then $C_L X \subseteq C_L Y$ and $C_U X \subseteq C_U Y$,
(C3) $C_L C_U X = C_U X$, $C_U C_L X = C_L X$,

Notice that (C3) implies the idempotence condition:

(C4) $C_L C_L X = C_L X$, $C_U C_U X = C_U X$,

since $C_L C_L X = C_L C_U C_L X = C_U C_L X = C_L X$, and similarly for C_U.

In fact, $(\mathcal{O}, C_L, C_U)$ is a 0-dimensional topological space, that is, a space based on a family of clopen subsets of $\mathcal{O}$; C_U is the closure operation, and C_L is its dual, i.e. the interior operation, of this space.

Incomplete information systems give rise to more general approximation operations which need not fulfill (C0). In this section, we briefly outline basic properties of these generalized notions.

We define *an approximation space* as a triple $S = (\mathcal{O}, C_L, C_U)$ such that $\mathcal{O}$ is a set, and C_L, C_U are mappings from $\mathrm{Pow}(\mathcal{O})$ to $\mathrm{Pow}(\mathcal{O})$, satisfying (C1), (C2) and (C3), for all $X, Y \subseteq \mathcal{O}$. Since (C4) follows from (C3), (C4) must also hold in every approximation space.

A set $X \subseteq \mathcal{O}$ is said to be *definable* in an approximation space S, if $X = C_L X$ (by (C3), $X = C_L X$ iff $X = C_U X$). $\mathrm{Def}(S)$ denotes the family of all sets definable in S.

One easily shows that $\mathrm{Def}(S)$ is closed under arbitrary joins and meets, hence it is a complete lattice of sets. Conversely, every complete lattice L of subsets of a set $\mathcal{O}$ equals $\mathrm{Def}(S)$, for the approximation space $S = (\mathcal{O}, C_L, C_U)$ such that $C_L X$ is the largest set in L contained in X, and $C_U X$ is the smallest set in L containing X.

Another characterization of approximation spaces can be given in terms of preordered sets. Let $P = (\mathcal{O}, \leq)$ be a preordered set, that means, $\leq$ is a reflexive and transitive relation on $\mathcal{O}$. A set $X \subseteq \mathcal{O}$ is called *a positive cone* in P, if, for all $x, y \in \mathcal{O}$, if $x \in X$ and $x \leq y$ then $y \in X$. $\mathrm{Con}(P)$ denotes the family of all positive cones in P. Since $\mathrm{Con}(P)$ is a complete lattice of sets, then, by the preceding paragraph, $\text{Con}(P) = \text{Def}(S)$, for some approximation space $S = (\mathcal{O}, C_L, C_U)$. One proves that operations C_L, C_U can be defined as follows:

$$C_L X = \{x \in X : (\forall y \in \mathcal{O})(x \leq y \Rightarrow y \in X)\} \tag{1}$$

$$C_U X = \{y \in \mathcal{O} : (\exists x \in X)(x \leq y)\}. \tag{2}$$

Conversely, every approximation space $S = (\mathcal{O}, C_L, C_U)$ determines a preordered set $P = (\mathcal{O}, \leq)$ such that, for $x, y \in \mathcal{O}$, $x \leq y$ iff, for all $X \in \text{Def}(S)$, $x \in X$ entails $y \in X$. It is easy to show $\text{Def}(S) = \text{Con}(P)$. The relation $\leq$ is called *the specialization preorder* of the approximation space S.

[**Proposition 2**]*For every approximation space S, the following conditions are equivalent:*
(i) S fulfills (C0),
(ii) the specialization preorder of S is symmetric.

PROOF. Assume (i). Then, for all $X \subseteq \mathcal{O}$, $X \in \text{Def}(S)$ iff $-X \in \text{Def}(S)$, which yields: $x \leq y$ iff $y \leq x$, for all $x, y \in \mathcal{O}$. Assume (ii). Then, the relation $\leq$ is an equivalence relation on $\mathcal{O}$. By (1), (2), $C_L X$ is the join of all equivalence classes of $\leq$ totally contained in X, and $C_U X$ is the join of all equivalence classes of $\leq$ which are not disjoint with X, and these operations obviously fulfil (C0).

Given an information system $I = (\mathcal{O}, \mathcal{A}, \mathcal{V}, \rho)$, one defines the relation $\sqsubseteq$: for $x, y \in \mathcal{O}$, $x \sqsubseteq y$ iff, for all $A \in \mathcal{A}$, if $\rho(x, A)$ is defined then $\rho(x, A) = \rho(y, A)$ (that means, object x is not more specified than object y). A set $X \subseteq \mathcal{O}$ is said to be *definable* in I, if X is a positive cone in the preordered set $(\mathcal{O}, \sqsubseteq)$. $\text{Def}(I)$ denotes the family of all sets definable in I. Clearly, $\text{Def}(I) = \text{Def}(S)$, for the approximation space $S = (\mathcal{O}, C_L, C_U)$ in which C_L, C_U are defined by (1), (2) with $\leq$ interpreted as $\sqsubseteq$. This approximation space is denoted by $S(I)$. One proves that $\sqsubseteq$ is precisely the specialization preorder of $S(I)$. For $x \in \mathcal{O}$, we define $[x] = \{y \in \mathcal{O} : x \sqsubseteq y\}$, i.e. $[x]$ is the principal positive cone determined by object x.

The label 'definable sets' for positive cones with respect to $\sqsubseteq$ will be better justified in section 3, where we show that these sets are precisely the extensions of formulas of L_I.

At the end of this section, we observe that every nonempty finite approximation space $S = (\mathcal{O}, C_L, C_U)$ equals $S(I)$, for some two-valued information system I, satisfying (NKC). Define $I = (\mathcal{O}, \mathcal{A}, \{0, 1\}, \rho)$ by setting $\mathcal{A} = \text{Def}(S)$ and, for all $x \in \mathcal{O}, A \in \mathcal{A}$:
$\rho(x, A) = 1$ iff $x \in A$,
$\rho(x, A)$ is undefined, otherwise.

This information system I satisfies (NKC), since $\mathcal{O} \in \text{Def}(S)$ (use (C1)), and consequently, $\rho(x, \mathcal{O}) = 1$, for all $x \in \mathcal{O}$.

[**Proposition 3**]$S = S(I)$.

PROOF. We prove $\text{Def}(S) = \text{Def}(I)$ (that entails $S = S(I)$). First, we show $\subseteq$. Let $A \in \text{Def}(S)$. Assume $x \in A$ and $x \sqsubseteq y$. Then, $\rho(x, A) = 1$, and consequently,

$\rho(y, A) = 1$, which yields $y \in A$. We have shown that A is a positive cone with respect to $\sqsubseteq$, hence $A \in \mathrm{Def}(I)$. Now, we show the converse inclusion. Let $A \in \mathrm{Def}(I)$. Then, A is a positive cone with respect to $\sqsubseteq$. Clearly, A is the join of all $[x]$, for $x \in A$. Since $\mathrm{Def}(S)$ is closed under arbitrary (also empty) joins, in order to prove $A \in \mathrm{Def}(S)$ it suffices to show $[x] \in \mathrm{Def}(S)$, for any $x \in \mathcal{O}$. Fix $x \in \mathcal{O}$. Let $A_1, \ldots, A_n$ be all $A \in \mathrm{Def}(S)$ such that $x \in A$. We have $[x] = A_1 \cap \ldots \cap A_n$, and consequently, $[x] \in \mathrm{Def}(S)$.

Actually, system I, defined above, is one-valued. One may extend it to a properly two-valued system, by setting: $\rho(x, A) = 0$, if $x \in C_L(-A)$, and proposition 3 for the extended system can be proved in a similar way.

3   Definability

In this section, we characterize sets definable and strongly definable in information systems in terms of propositional definability. Main results are first proved for two-valued information systems and, then, generalized for arbitrary information systems.

Let $I = (\mathcal{O}, \mathcal{A}, \{0, 1\}, \rho)$ be an information system. A set $X \subseteq \mathcal{O}$ is said to be *propositionally definable* in I, if $X = P(\varphi)$, for some formula φ of L_I. By (PN0), X is propositionally definable in I iff $X = N(\psi)$, for some formula ψ of L_I.

[**Theorem 1**]*For any $X \subseteq \mathcal{O}$, X is propositionally definable in I if, and only if, $X \in Def(I)$.*

PROOF. By induction on φ, we prove $P(\varphi), N(\varphi) \in \mathrm{Def}(I)$. Since $P(\overline{A}) = \{x : \rho(x, A) = 1\}$, then $P(\overline{A})$ is a positive cone with respect to $\sqsubseteq$, and similarly for $N(\overline{A})$. For $\varphi \equiv \neg\psi$, we apply (PN1) and the induction hypothesis. For $\varphi \equiv \psi \wedge \chi$, we apply (PN2), the induction hypothesis and the fact that $\mathrm{Def}(I)$ is closed under joins and meets.

Now, assume $X \in \mathrm{Def}(I)$. If $X = \emptyset$, then $X = P(\overline{A} \wedge \neg\overline{A})$, for any $A \in \mathcal{A}$. So, assume $X \neq \emptyset$. Let φ denote the disjunction of all $\delta_I(x)$, for $x \in X$. We have $P(\psi \vee \chi) = P(\psi) \cup P(\chi)$, by (PN1), (PN2), hence:

$$P(\varphi) = \bigcup_{x \in X} P(\delta_I(x)) = X,$$

since $P(\delta_I(x)) = [x]$ and $X \in \mathrm{Def}(I)$.

Theorem 1 generalizes for non-two-valued information systems: for any information system I, $\mathrm{Def}(I)$ is precisely the family of sets propositionally definable in $T(I)$. That follows from theorem 1 and the equality $\mathrm{Def}(I) = \mathrm{Def}(T(I))$.

Let $I = (\mathcal{O}, \mathcal{A}, \mathcal{V}, \rho)$, $I' = (\mathcal{O}', \mathcal{A}', \mathcal{V}', \rho')$ be information systems. System I' is called *an informational extension* of system I (write: $I \sqsubseteq I'$), if $\mathcal{O} = \mathcal{O}'$, $\mathcal{A} = \mathcal{A}'$, $\mathcal{V} = \mathcal{V}'$ and $\rho \subseteq \rho'$, that means, for all $x \in \mathcal{O}$ and $A \in \mathcal{A}$, if $\rho(x, A)$ is defined, then $\rho'(x, A) = \rho(x, A)$. One easily shows: $I \sqsubseteq I'$ entails $P(\varphi) \subseteq P'(\varphi)$ and $N(\varphi) \subseteq N'(\varphi)$, for all formulas φ of $L_I = L_{I'}$. By $\mathrm{Com}(I)$ we denote the set of all complete informational extensions of I.

A set $X \subseteq \mathcal{O}$ is said to be *strongly definable* in I, if $X \in \mathrm{Def}(I')$, for all $I' \in \mathrm{Com}(I)$. Accordingly, the sets strongly definable in an information system are the sets definable in all complete informational extensions of this system. Evidently, the notion of a strongly definable set deserves a close attention as another natural 'incomplete' variant of the notion of a definable set in complete information systems. Below we characterize this notion in terms of propositional definability.

A formula φ of L_I is said to be *determined* in I, if $P(\varphi) \cup N(\varphi) = \mathcal{O}$. It is said to be *positively determined* in I, if $P(\varphi) = P'(\varphi)$, for all informational extensions I' of I. Every formula determined in I is also positively determined in I, but the converse does not hold (take $\varphi \equiv \psi \wedge \neg\psi$; $P(\varphi) = \emptyset$ holds in all information systems, but $P(\varphi) \cup N(\varphi) = N(\varphi)$ does not contain all objects, if ψ is not determined). We say that object x is *incompatible* with object y in an information system I, if there is $A \in \mathcal{A}$ such that $\rho(x, A) \neq \rho(y, A)$ (both defined).

[**Theorem 2**] *For any information system $I = (\mathcal{O}, \mathcal{A}, \{0, 1\}, \rho)$, and all $X \subseteq \mathcal{O}$, the following conditions are equivalent:*
(i) X *is strongly definable in* I,
(ii) *for all* $x \in X$, $y \notin X$, x *is incompatible with* y *in* I,
(iii) $X = P(\varphi)$, *for some formula* φ *determined in* I,
(iv) $X = P(\varphi)$, *for some formula* φ *positively determined in* I,
(v) *there is a formula* φ *such that* $X = P'(\varphi)$, *for all* $I' \in \mathrm{Com}(I)$.

PROOF. Implications (iii)⇒(iv), (iv)⇒(v) and (v)⇒(i) are obvious. We prove (i)⇒(ii). Assume (ii) do not hold. Then, there exist $x \in X$, $y \notin X$ such that, for all attributes A, if both $\rho(x, A)$ and $\rho(y, A)$ are defined, then $\rho(x, A) = \rho(y, A)$. Clearly, there exists $I' \in \mathrm{Com}(I)$ such that $\rho'(x, A) = \rho'(y, A)$, for all attributes A, and consequently $x \sim_{I'} y$. Therefore, X is not a join of equivalence classes of $\sim_{I'}$, hence X is not definable in I', and (i) fails. We prove (ii)⇒(iii). Assume (ii). We consider two cases. (I) $X = \emptyset$. Then, $X = P(\varphi)$, where φ is the conjunction of all $\overline{A}, \neg\overline{A}$, for $A \in \mathcal{A}$, and φ is determined in I, since $N(\varphi) = \mathcal{O}$, by (NKC). (II) $X \neq \emptyset$. Let φ be the disjunction of all formulas $\delta_I(x)$, for $x \in X$. We show $X \subseteq P(\varphi)$. Let $x \in X$. Since $D_I(x) \vdash_3 \delta_I(x)$ and $\delta_I(x) \vdash_3 \varphi$, then $D_I(x) \vdash_3 \varphi$, hence $x \in P(\varphi)$, by proposition 1. We show $-X \subseteq N(\varphi)$. Let $y \notin X$. By (ii), $D_I(y) \vdash_3 \neg\delta_I(x)$, for all $x \in X$, hence by proposition 1, $y \in N(\delta_I(x))$, for all $x \in X$. Accordingly, $y \in N(\varphi)$ (use $N(\psi \vee \chi) = N(\psi) \cap N(\chi)$). Since $P(\varphi) \cap N(\varphi) = \emptyset$, it follows that $X = P(\varphi)$ and $-X = N(\varphi)$, and φ is determined in I.

It is not that easy to generalize this theorem for non-two-valued information systems, since not every complete informational extension of $T(I)$ is the two-valued representation of a system from $\mathrm{Com}(I)$. Let $I = (\mathcal{O}, \mathcal{A}, \mathcal{V}, \rho)$ be an information system. One easily shows that a system $I' \in \mathrm{Com}(T(I))$ equals $T(J)$, for some $J \in \mathrm{Com}(I)$, if, and only if, for all $A \in \mathcal{A}$, the following formulas are valid in I' (i.e. their positive extensions equal $\mathcal{O}$):

$$\neg(\overline{(A, v_i)} \wedge \overline{(A, v_j)}) \text{ for } i \neq j,$$

$$\overline{(A, v_1)} \vee \ldots \vee \overline{(A, v_n)},$$

where $v_1, \ldots, v_n$ are all values in $\mathcal{V}$. The set of all formulas of the above form will be denoted by com(I).

This observation leads us to the following definitions. Let Γ be a set of formulas of L_I, where I is two-valued. By Com$_\Gamma(I)$ we denote the set of all $I' \in$ Com(I) such that every formula from Γ is valid in I'. A set $X \subseteq \mathcal{O}$, where $\mathcal{O}$ is the set of objects of I, is said to be *strongly definable* in I *with regarding* Γ, if X is definable in all systems $I' \in$ Com$_\Gamma(I)$. For a formula φ of L_I, by $P_\Gamma(\varphi)$ (resp. $N_\Gamma(\varphi)$) we denote the meet of all sets $P'(\varphi)$ (resp. $N'(\varphi)$), for $I' \in$ Com$_\Gamma(I)$.

[**Proposition 4**] *For all sets Γ, formulas φ and objects x, there hold the following equivalences:*
(SD1) $x \in P_\Gamma(\varphi)$ *iff* $\Gamma \cup D_I(x) \vdash_2 \varphi$,
(SD2) $x \in N_\Gamma(\varphi)$ *iff* $\Gamma \cup D_I(x) \vdash_2 \neg\varphi$.

A formula φ is said to be *determined* in I *with regarding* Γ, if $P_\Gamma(\varphi) \cup N_\Gamma(\varphi) = \mathcal{O}$. Object x is said to be *incompatible* with object y in system I *with regarding* Γ, if the set $\Gamma \cup D_I(x) \cup D_I(y)$ is unsatisfiable in the sense of classical logic.

[**Theorem 3**] *For all information systems $I = (\mathcal{O}, \mathcal{A}, \{0, 1\}, \rho)$, all sets $X \subseteq \mathcal{O}$ and all sets Γ such that $Com_\Gamma(I) \neq \emptyset$ the following conditions are equivalent:*
(i) X is strongly definable in I with regarding Γ,
(ii) for all $x \in X$, $y \notin X$, x is incompatible with y in I with regarding Γ,
(iii) $X = P_\Gamma(\varphi)$, for some formula φ determined in I with regarding Γ,
(iv) there is a formula φ such that $X = P'(\varphi)$, for all $I' \in Com_\Gamma(I)$.

We omit proofs of proposition 4 and theorem 3. As an easy corollary from theorem 3, we obtain: for all information systems $I = (\mathcal{O}, \mathcal{A}, \mathcal{V}, \rho)$ and sets $X \subseteq \mathcal{O}$, X is strongly definable in I iff there is a formula φ (of $L_{T(I)}$) determined in $T(I)$ with regarding com(I) such that $X = P_{com(I)}(\varphi)$ iff there is a formula φ (of $L_{T(I)}$) such that $X = P'(\varphi)$, for all systems $I' \in$ Com$_{com(I)}(T(I))$. Notice that $I' \in$ Com$_{com(I)}(T(I))$ iff there exists $J \in$ Com(I) such that $I' = T(J)$.

References

1. S. Blamey, Partial logic, in: D. Gabbay and F. Guenthner (eds.), *Handbook of Philosophical Logic*, vol. 3, D. Reidel, Dordrecht, 1986, 1–70.
2. L. Bolc and P. Borowik, *Many-Valued Logics. 1. Theoretical Foundations*, Springer, Berlin, 1992.
3. S. Demri and E. Orłowska, Logical Analysis of Indiscernibility, in: E. Orłowska (ed.), *Incomplete Information: Rough Set Analysis*, Physica-Verlag, Heidelberg, 1998, 347–380.
4. S.C. Kleene, *Introduction to metamathematics*, North-Holland, Amsterdam, 1952.
5. E. Orłowska and Z. Pawlak, *Logical foundations of knowledge representation*, ICS PAS Reports, 537, Warszawa, 1984.
6. Z. Pawlak, Information systems - theoretical foundations, *Information Systems* 6 (1981), 205–218.
7. Z. Pawlak, *Rough Sets. Theoretical Aspects of Reasoning about Data*, Kluwer, Dordrecht, 1991.

Intrinsic Co-Heyting Boundaries and Information Incompleteness in Rough Set Analysis

Piero Pagliani

FINSIEL-TELECOM
Telecommunication Business Unit
Via Bona, 65, 00156 Roma, Italy
p.pagliani@agora.stm.it

Abstract. Probably the distinguishing concept in incomplete information analysis is that of "boundary": in fact a boundary is precisely the region that represents those doubts arising from our information gaps. In the paper it is shown that the rough set analysis adequately and elegantly grasps this notion via the algebraic features provided by co-Heyting algebras.

1 Algebraic Views of Rough Set Systems

Any Rough Set Systems, that is the family of all the rough sets induced by an Approximation Space over a set U (see the definition below), can be made into several logic-algebraic structures. In [7], for instance, the attention was focused on semi-simple Nelson algebras, Heyting algebras, double Stone algebras, three-valued Łukasiewicz algebras and Chain Based Lattices. In the present paper, Rough Set Systems are analyzed from the point of view of co-Heyting algebras. This new chapter in the algebraic analysis of Rough Sets does not follow from esthetic or completeness issues, but it is a pretty immediate consequence of interpreting the basic features of co-Heyting algebras (originally introduced by C. Rauszer [8] and investigated by W. Lawvere in the context of Continuum Physics) through the lenses of incomplete information analysis. Indeed in [3] and in [4], Lawvere pointed out the role that the co-intuitionistic negation "non" (dual to the intuitionistic negation "not") plays in grasping the geometrical notion of "boundary" as well as the physical concepts of "sub-body" and "essential core of a body" and we aim at providing an outline of how and to what extent they are mirrored by the basic features of incomplete information analysis.

1.1 Indiscernibility Spaces and Rough Sets

Given a (finite) universe U, when we take into account an equivalence relation $E \subseteq U \times U$ we assume that any equivalence class collects together the elements of U that are considered *indiscernible* from some point of view: in Rough Set

L. Polkowski and A. Skowron (Eds.): RSCTC'98, LNAI 1424, pp. 123–130, 1998.

Analysis, typically $a, b \in U$ are indiscernible if they share all the possible (observational) properties we are provided by some Information System (see [6]). In this perspective, any equivalence class modulo E will be called a *basic category* and the ordered pair (U, E) will be called an *indiscernibility space*. Let $\mathbf{AS}(U)$ be the atomic Boolean algebra having U/E as set of atoms. Then $(U, \mathbf{AS}(U))$ is a 0-dimensional topological space. This space is called an *Approximation Space*, but with the same term we will also refer to the topology $\mathbf{AS}(U)$ itself. As a topology, $\mathbf{AS}(U)$ will induce a closure operator $\mathcal{C}$ and an interior operator $\mathcal{I}$ from $\wp(U)$ to $\mathbf{AS}(U)$.

In Rough Set Analysis, $\forall G \in U, \mathcal{C}(G)$ is called the *upper approximation* of G (with respect to E) and it is usually denoted by $(uE)(G)$, while $\mathcal{I}(G)$ is called the *lower approximation* of G and it is denoted by $(lE)(G)$. If $(lE)(G) = \emptyset$ (if $(uE)(G) = U$, then G is said to be *internally undefinable* (*externally undefinable*).

If two sets $G, G' \subseteq U$ are such that $(uE)(G) = (uE)(G')$ and $(lE)(G) = (lE)(G')$, then G and G' are said to be *rough equal*, $G \approx G'$. Thus *rough equality* is an equivalence relation on $\wp(U)$. Any equivalence class of subsets of $\wp(U)$ modulo the rough equality relation is called a *rough set*.

1.2   Basic algebraic concepts

Let $\mathbf{L} = (L, \vee, \wedge, 0, 1)$ be a bounded distributive lattice.

Definition 1. *The* RIGHT ADJOINT *to* MEET*: given $a, b \in L$, set $\forall x \in L$,*

$$(i)\ a \wedge x \leq b \text{ if and only if } x \leq a \supset b; \quad (ii) \div a = a \supset 0.$$

Then $a \supset b$ is called the *pseudo-complement of a relative to b* and by definition it is the largest element x of L such that $a \wedge x$ is less than or equal to b, while the element $\div a$ is called the *pseudo-complement* (or *intuitionistic negation*) of a and it is the largest element x of L such that $a \wedge x = 0$.

Definition 2. HEYTING ALGEBRAS*: if for any $a, b \in L$, the operation $a \supset b$ is defined, then $\mathbf{H} = (L, \vee, \wedge, \supset, \div, 0, 1)$ is called a Heyting algebra.*

Definition 3. *The* LEFT ADJOINT *to* JOIN*: given $a, b \in L$, set $\forall x \in L$,*

$$(i)\ a \vee x \geq b \text{ if and only if } x \geq a \subset b; \quad (ii) \neg a = a \subset 1.$$

Then $a \subset b$ is called the *dual pseudo-complement of a relative to b*, and it is the smallest element x of L such that $a \vee x$ is greater than or equal to b, while $\neg a$ is called the *dual pseudo-complement* (or *co-intuitionistic negation*) of a and it is the smallest element x of L such that $a \vee x = 1$.

Definition 4. CO-HEYTING ALGEBRAS*: if for any $a, b \in L$, $a \subset b$ is defined, then $\mathbf{CH} = (L, \vee, \wedge, \subset, \neg, 0, 1)$ is called a co-Heyting algebra.*

Definition 5. *A* BI-HEYTING ALGEBRA *is a bounded distributive lattice that is both a Heyting and a co-Heyting algebra.*

EXAMPLES

1. The system of all closed subsets of a topological space is a co-Heyting algebra. In fact in this case given two closed subsets X, Y, $X \subset Y = \mathcal{C}(Y \cap -X)$. It follows that $\neg X = \mathcal{C}(-X) = -\mathcal{I}(X)$.
2. Any finite Heyting algebra is a bi-Heyting algebra.

We must now remark that being a bi-Heyting algebra is an interesting condition *in se* but not *per se*: indeed if we consider a finite Heyting algebra $\mathbf{H}$ such that the top element, 1, is co-prime (that is, for any $X \subseteq \mathbf{H}$ such that $1 = \bigvee X$, $1 \in X$) then $\mathbf{H}$ is surely a bi-Heyting algebra, but for any $a \in \mathbf{H}$, $\neg a = 1$ if $a \neq 1$, $\neg a = 0$ otherwise: the co-intuitionistic negation carries a poor information, in this case. Therefore, though one can "a priori" say that any Rough Set System is bi-Heyting algebra (since it is a finite distributive lattice, thus a finite Heyting algebra), nevertheless this framework becomes interesting if we are able to identify the bi-Heyting algebra operations and to relate them to the features of Rough Set Analysis. So let us start with the operations that we can define on rough sets.

We have seen that any rough set Z is an equivalence class of subsets of U modulo the rough equality relation $\approx$. Thus $Z \in \wp(\wp(U))$. But since Z is uniquely determined by the lower and the upper approximation of any of its element, we can denote it by $< \mathcal{C}(Z_i), \mathcal{I}(Z_i) >$ for $Z_i \in Z$. Therefore we introduce the following map:

Definition 6. $rs : \wp(U) \longmapsto \mathbf{AS}(U) \times \mathbf{AS}(U); rs(X) = < \mathcal{C}(X), \mathcal{I}(X) >$.

Then, we have: $[X]_{\approx} = rs^{-1}(rs(X)), \forall X \subseteq U$. If the pre-image of rs is a singleton $\{X\}$, then we will denote it directly by X.
We denote the image of $\wp(U)$ along rs, by $RS(U)$ and we call it the *Rough Set System* induced by $\mathbf{AS}(U)$. Clearly, if $< X, Y > \in RS(U)$, then $X \supseteq Y$. Particularly, if $G \in \mathbf{AS}(U)$ then $\mathcal{C}(G) = \mathcal{I}(G)$; hence any element of $rs(\mathbf{AS}(U))$ has the form $< X, X >$. We will call any rough set of this kind an *exact rough set*.
Then, let us consider the following families, for any Boolean algebra $\mathbf{B}$:
$\mathcal{P}(\mathbf{B}) = \{< a_1, a_2 > \in \mathbf{B}^2 : a_1 \geq a_2\}$; $\mathcal{B}(\mathbf{B}) = \{< a_1, a_2 > \in \mathbf{B}^2 : a_1 = a_2\}$.
Clearly $rs(\mathbf{AS}(U)) = \mathcal{B}(\mathbf{AS}(U)) \subseteq RS(U) \subseteq \mathcal{P}(\mathbf{AS}(U))$.
Before short, we will identify precisely $RS(U)$.
From now on by $a = < a_1, a_2 >, x = < x_1, x_2 >$ and so on, we will denote elements of $\mathcal{P}(\mathbf{B})$.

BASIC OPERATIONS ON $\mathcal{P}(\mathbf{B})$
1. $1 = < 1, 1 >$;
2. $a \vee b = < a_1 \vee b_1, a_2 \vee b_2 >$;
3. $a \longrightarrow b = < a_2 \longrightarrow b_1, a_2 \longrightarrow b_2 >$ (weak implication);
4. $\sim a = < \neg a_2, \neg a_1 >$ (strong negation);
where $1, \vee, \longrightarrow$ and $\neg$ applied inside the ordered pairs are the operations of the underlying Boolean algebra $\mathbf{B}$.

From the above basic set, we can define other derived operations. Some of these are the operations that make it possible to made $\mathcal{P}(\mathbf{B})$ into a Heyting algebra and a co-Heyting algebra.

DERIVED OPERATIONS ON $\mathcal{P}(\mathbf{B})$

5. $0 =\sim 1 =< 0, 0 >$;
6. $a \wedge b =\sim (\sim a \vee \sim b) =< a_1 \wedge b_1, a_2 \wedge b_2 >$;
7. $a \supset b =\sim \neg \sim a \vee b \vee (\neg a \wedge \neg \sim b)$ (intuitionistic implication);
8. $\div a =\sim \neg \sim a$ (intuitionistic negation);
9. $a \subset b =\sim (\sim a \supset \sim b) = \neg a \wedge b \wedge (\sim \neg \sim a \vee \sim \neg b)$;
10. $\neg a = a \longrightarrow 0 =< \neg a_2, \neg a_2 >$ (weak negation),

where $0, \wedge$ and $\neg$ applied inside the ordered pairs are the operations of $\mathbf{B}$.

Lemma 7. *For any $a \in \mathcal{P}(\mathbf{B})$, $\div a = a \supset 0$.*

Proposition 8. FACTS ABOUT NEGATIONS: *For any $a \in \mathcal{P}(\mathbf{B})$*

1. a) $\sim\sim a = a$; b) $\neg\neg\neg a = \neg a$; c) $\div \div \div a = \div a$.
2. a) $\div \div a = \neg \div a =\sim \div a = \neg \sim a$; b) $\neg\neg a = \div\neg a = \div \sim a =\sim \neg a$;
3. If $a =< x, x >$, then $\sim a = \neg a = \div a$.

Proposition 9. *For any Boolean algebra $\mathbf{B}$,*

1. $\mathbf{N}(\mathbf{B}) = (\mathcal{P}(\mathbf{B}), \wedge, \vee, \longrightarrow, \sim, \neg, 0, 1)$ *is a semi-simple Nelson algebra.*
2. $\mathbf{H}(\mathbf{B}) = (\mathcal{P}(\mathbf{B}), \wedge, \vee, \supset, \div, 0, 1)$ *is a Heyting algebra.*

(The proofs for the above Lemma, Facts and Propositions are in [7]).

Proposition 10. *For any Boolean algebra $\mathbf{B}$, $\mathbf{CH}(\mathbf{B}) = (\mathcal{P}(\mathbf{B}), \wedge, \vee, \subset, \neg, 0, 1)$ is a co-Heyting algebra.*

Proof For any $a, b \in \mathcal{P}(\mathbf{B})$, $a \leq b$ iff $\sim b \leq\sim a$ and $\sim (a \wedge b) =\sim a \vee \sim b$: these properties of $\sim$ are inherited from the same properties of the negation of $\mathbf{B}$. Thus, since from Proposition 9, $a \supset b = \bigvee\{z \in \mathcal{P}(\mathbf{B}) : z \wedge a \leq b\}$, it follows that $\sim (\sim a \supset \sim b) =\sim \bigvee\{z \in \mathcal{P}(\mathbf{B}) : z\wedge \sim a \leq\sim b\} =\sim \bigvee\{z \in \mathcal{P}(\mathbf{B}) :\sim\sim b \leq\sim (z\wedge \sim a)\} =\sim \bigvee\{z \in \mathcal{P}(\mathbf{B}) : b \leq\sim z \vee a\} = \bigwedge\{\sim z \in \mathcal{P}(\mathbf{B}) : b \leq\sim z \vee a\}$. Since $\sim$ is an involutive anti-isomorphism we obtain that $\subset$ is a dual relative pseudocomplementation.
Moreover, $a \subset 1 = \neg a \wedge 1 \wedge (\sim \neg \sim a \vee \sim \neg 1) = \neg a \wedge (\sim \neg \sim a \vee 1) = \neg a$. QED

Corollary 11. *For any Boolean algebra $\mathbf{B}$, $\mathbf{BH}(\mathbf{B}) = (\mathcal{P}(\mathbf{B}), \wedge, \vee, \subset, \neg, \supset, \div, 0, 1)$ is a bi-Heyting algebra.*

2   Rough Sets and bi-intuitionistic Operations

Proposition 12. $\forall X \subseteq U$,
1) $\sim (rs(X)) = rs(-X)$; 2) $\div(rs(X)) = rs(-\mathcal{C}(X))$; 3) $\neg(rs(X)) = rs(-\mathcal{I}(X))$.

Proof. 1) $\sim (rs(X)) = \sim< \mathcal{C}(X), \mathcal{I}(X) >=< -\mathcal{I}(X), -\mathcal{C}(X) >$
$=< \mathcal{C}(-X), \mathcal{I}(-X) >= rs(-X);$
2) $\div (rs(X)) = \div < \mathcal{C}(X), \mathcal{I}(X) >=< -\mathcal{C}(X), -\mathcal{C}(X) >= rs(-\mathcal{C}(X));$
3) $\neg (rs(X)) = \div < \mathcal{C}(X), \mathcal{I}(X) >=< -\mathcal{I}(X), -\mathcal{I}(X) >= rs(-\mathcal{I}(X)).$ QED
Combining 1 with 2 and 1 with 3 we obtain:

Corollary 13. *For any Rough Set System $RS(U)$, for any $X \subseteq U$,*
1) $\neg \sim (rs(X)) = rs(\mathcal{C}(X));$ 2) $\sim \neg(rs(X)) = rs(\mathcal{I}(X)).$
3) $\sim \neg(RS(U)) = \neg \sim (RS(U)) = rs(\mathbf{AS}(U)) = \neg(RS(U)) = \div(RS(U)).$

Proof. 1) $\neg \sim (rs(X)) = rs(- - \mathcal{C}(X));$ 2) $\sim \neg(rs(X)) = rs(- - \mathcal{I}(X)).$
3) Since both $\mathcal{C}$ and $\mathcal{I}$ map $\wp(U)$ onto $\mathbf{AS}(U)$. QED
It follows that $\neg \sim$ (that is $\div\div$) and $\sim \neg$ (that is $\neg\neg$) can be considered as
modal operators that parallel with (uE) and, respectively, (lE).

3   Bodies and Boundaries via Co-Heyting Algebras

Given an element a of a co-Heyting algebra **CH**, Lawvere calls $\neg\neg a$ the *regular
core* of a. Generally $\neg\neg a \leq a$. In order to appreciate this term, consider the
above results: if $a = rs(X)$ for $X \subseteq U$, then $rs^{-1}(\neg\neg a) = (lE)(X)$, that is
the *necessary* part of X, (in a literal sense when $\mathbf{AS}(U)$ is interpreted as an S5
modal space -see for instance [5]). Moreover in [3], it is claimed that a part a may
be considered a *sub-body* (or shortly a *body*) if and only if $\neg\neg a = a$. Thus the
notion of "sub-body" coincides in Rough Set Systems with that of *exact rough
set*, that is rough sets "(deductively) closed" and "perfect": and not by chance
we are following the terminology used by Leibniz for describing the notion of
"individual substance" (*Discourse on Metaphysics*).
But everything is centered on the fact that in co-Heyting algebras we can recap-
ture the geometrical notion of "boundary". Indeed Lawvere points out that this
notion is definable by means of the co-intuitionistic negation, $\neg$, in the following
manner (for a belonging to any co-Heyting algebra):

$$\partial(a) = a \wedge \neg a.$$

First of all, $\partial(a)$ is the boundary of a in a topological sense: if the given co-
Heyting algebra is that of Example 1, then a is a closed set of some topological
space. Thus $a \cap \neg(a) = \mathcal{C}(a) \cap -\mathcal{I}(a)$, that is exactly the topological boundary,
$\mathcal{F}(a)$, of a.
More generally, $\partial(a)$ is a boundary since for any a, b of any co-Heyting algebra,
it formally fulfills the rules:
1) $\partial(a \wedge b) = (\partial(a) \wedge b) \vee (a \wedge \partial(b));$ 2) $\partial(a \wedge b) \vee \partial(a \vee b) = \partial(a) \vee \partial(b).$
The first formula is called "Leibniz formula" by W. Lawvere who underlines
that though its validity for boundaries of closed sets is supported by our space
intuition (think of two partially overlapping ovals), nevertheless it is virtually
unknown in general topology literature.
Indeed we can notice that it is essentially the usual Leibniz rule for differentiation
of a product (but see also the Grassmann rule).

Moreover, Lawvere notices that any element a in a co-Heyting algebra is the join of its core and its boundary: $a = \neg\neg a \vee \partial(a)$.

All these relations apply to any rough set. Particularly, if $a = rs(X)$, then $\partial(a) = a \wedge \neg a = < a_1 \cap -a_2, a_2 \cap -a_2 > = < \mathcal{C}(X) \cap -\mathcal{I}(X), \emptyset > = < \mathcal{F}(X), \emptyset >$ (since the interior of $\partial(a)$ is always empty, one can deduce the two-dimensionality of the rough sets with non empty interior, that is rough sets corresponding to internally definable subsets of the Approximation Space). Moreover, the boundary of X in the topological space $(U, \mathbf{AS}(U))$ is given by: $\mathcal{F}(X) = rs^{-1}(\neg \sim (\partial(a)))$. In fact if $Z \in rs^{-1}(\partial(a))$, then $\mathcal{C}(Z) = \mathcal{F}(X)$ and the preceding equation follows from Corollary 13.

Now let us notice, that no boundary can contain *isolated* points of the given topology. This means that if a basic category $A \in U/E$ is a singleton, say $\{y\}$, then for no $X \subseteq U$, $A \subseteq \mathcal{F}(X)$. In fact, either $y \in X$ or not. But this means $A \subseteq \mathcal{I}(X)$ or $A \subseteq -\mathcal{C}(X)$: *tertium non datur*, whereas any non-singleton basic category M, say $\{y, z\}$, a third possibility is allowed. In fact, we can find at least two sets Y, Z such that $y \in Y, z \notin Y$ and $z \in Z, y \notin Z$ (for instance we can trivially consider $Y = \{y\}$ and $Z = \{z\}$). In this case neither $M \subseteq \mathcal{I}(Y)$ nor $M \subseteq -\mathcal{C}(Y)$, (the same for Z).

But in Rough Set Analysis x is an isolated point if and only if we have a *complete information* about it: the basic category including x reduces to a singleton and this means that x can be discerned from any other object. Let us denote by B the union of all the isolated points of $(U, \mathbf{AS}(U))$, and let us set $P = U \cap -B$. Since $a \vee \div a = \sim \partial(a)$ and $\sim < P, \emptyset > = < U, B >$, we have just established that:

Proposition 14. *For any Approximation Space* $\mathbf{AS}(U)$, *for any* $a \in \mathcal{P}(\mathbf{AS}(U))$, *the following are equivalent:*

1. $\exists X \subseteq U$ *such that* $a = rs(X)$;
2. $a_1 \cap B = B \cap a_2$;
3. $a \vee \div a \geq < U, B >$;
4. $\partial(a) \leq < P, \emptyset >$.

It is worth noticing that condition 14.3 claims that the *tertium non datur* $a \vee \div a$ must hold with respect to the *local top* $< U, B >$. Dually, 14.4 says that the *contradiction* $a \wedge \neg a$ must be invalid with respect to the *local bottom* $< P, \emptyset >$. Now the algebraic task is to find an appropriate filter in order to distinguish $RS(U)$ within $\mathcal{P}(\mathbf{AS}(U))$. In and [7] a suitable filter for condition 14.2 has been introduced. Now we show that it makes the other new constraints work as well. Indeed in the quoted papers, it is proved that $RS(U)$ can be recovered from $\mathcal{P}(\mathbf{AS}(U))$ via the filter generated by $< U, P >$. Shortly, let us set for $a, b \in \mathcal{P}(\mathbf{AS}(U))$: a) $a \equiv b$ iff $\exists x \geq < U, P >$ s. t. $a \wedge x = x \wedge b$; b) $J(a) = \bigvee [a]_\equiv$. Then we have:

Lemma 15. *For any* $a \in \mathcal{P}(\mathbf{AS}(U))$, *if* $a = J(a)$, *then* $a_1 \cap B = B \cap a_2$.

Corollary 16. $J^{<U,P>}(\mathcal{P}(\mathbf{AS}(U))) = RS(U)$.

Now we will prove:

Proposition 17. *For any $a \in \mathcal{P}(\mathbf{AS}(U))$, if $a = J(a)$ then*
1) $a \vee \div a \geq < U, B >$;   2) $a \wedge \neg a \leq < P, \emptyset >$.

Proof: $a \vee \div a = < a_1, a_2 > \vee < \neg a_1, \neg a_1 > = < U, a_2 \cup \neg a_1 >$. But since a is a fixed point of the operator J, in view of the previous Lemma we have $a_2 = a_2 \cup (B \cap a_1)$. Hence $a_2 \geq (B \cap a_1)$. It follows that $a_2 \cup \neg a_1 \geq (a_1 \cap B) \cup \neg a_1 = \neg a_1 \cup B \geq B$. The proof of the second statement is obtained by duality. QED

4   Bi-Heyting Algebras, Modalities and Rough Sets

We have seen that the two double negation sequences $\neg \sim$ and $\sim \neg$ can be considered a couple of modality operators. Indeed, according to the physical interpretation, in view of Corollary 13(3), given any $a \in RS(U)$, $\neg \sim a$ sends a to the smallest "sub-body" that contains a, whereas $\sim \neg a$ sends a to the largest sub-body contained in a. Since $\forall a \in \mathcal{P}(\mathbf{B})$, $\neg \sim a = \neg \div a$ and $\sim \neg a = \div \neg a$, we can connect our results to some interesting general properties of bi-Heyting algebras. Namely, let us define two operators $\Box_n$ and $\Diamond_n$ as follows:

Definition 18. *Given a $\sigma-$complete bi-Heyting algebra $\mathbf{BH}$, $\forall a \in \mathbf{BH}$,*
(i) $\Box_0 = \Diamond_0 = Id$; (ii): $\Box_{n+1} = \div \neg \Box_n$, $\Diamond_{n+1} = \neg \div \Diamond_n$;
(iii) $\Box(a) = \bigwedge_{i=1}^{n} \Box_i(a)$; (iv) $\Diamond(a) = \bigvee_{i=1}^{n} \Diamond_i(a)$.

Then in [9] it is shown that for any a, $\Box(a)$ is the largest complemented element of $\mathbf{BH}$ below a, while $\Diamond(a)$ is the smallest complemented element above a (the interested reader must take great care of the different notations: in the quoted paper the intuitionistic negation $\div$ is denoted by $\neg$ and the co-intuitionistic negation $\neg$ is denoted by $\sim$. So in that paper $\Box_1 = \neg \sim$ and $\Diamond_1 = \sim \neg$); it follows that in Rough Set Systems $\Box = \Box_1$ and $\Diamond = \Diamond_1$. In other words both the sequences $\Box_1, \Box_2, \dots$ and $\Diamond_1, \Diamond_2, \dots$ stabilize at step 1. These facts, as pointed out by in [9], are related to the De Morgan laws that, generally, fail in Heyting and, respectively, in co-Heyting algebras:

Definition 19. *Let $\mathbf{H}$ and $\mathbf{CH}$ be a Heyting and, respectively, a co-Heyting algebra. Then: 1) $\mathbf{H}$ satisfies the De Morgan's law for $\div$, if $\forall x, y$, $\div(x \wedge y) = \div x \vee \div y$; 2) $\mathbf{CH}$ satisfies the De Morgan's law for $\neg$, if $\neg(x \vee y) = \neg x \wedge \neg y$.*

One can show that in bi-Heyting algebras the law for $\div$ implies $\Box(a) = \div \neg a$ and that the law for $\neg$ implies $\Diamond(a) = \neg \div a$. The reverse of the implications does not hold, but one can prove that both the laws actually hold in Rough Set Systems:

Proposition 20. *In any rough set system $RS(U)$ the two laws of Definition 19 hold.*

Proof: In a Heyting algebra $\mathbf{H}$ the De Morgan law for $\div$ is equivalent to the fact that $\div \div (\mathbf{H})$ is a sublattice of $\mathbf{H}$ (see [2]). Dually for $\neg$ in co-Heyting algebras. But this is precisely the case for $RS(U)$: from Proposition 8, $\div \neg a = \neg \neg a$ and $\neg \div a = \div \div a$, any a, but from Corollary 13 we have immediately that $\neg \neg$ and

$\div\div$ are both multiplicative and additive ($\mathcal{C}$ is additive in any topological space and $\div\div$ is multiplicative in any Heyting algebra. Dually for $\neg\neg$ and $\mathcal{I}$. -cf. also [1]-). QED

Thus in Rough Set Systems the De Morgan rules for $\div$ and $\neg$ hold. Since this is also equivalent to the fact that $\forall a$, $\neg a$ and $\div a$ are complemented, from Corollary 13(3) we obtain that for any exact rough set a, $\partial a = 0$. Particularly, since $\neg\neg(RS(U)) = rs(\mathbf{AS}(U)) = \mathcal{B}(\mathbf{AS}(U))$, it follows that Rough Set Systems are examples of those "lucky" situations in which sub-bodies form a Boolean algebra.

References

1. M. Banerjee & M. Chakraborthy, Rough Sets Through Algebraic Logic. *Fundamenta Informaticae*, 28, 1996, pp. 211–221.

2. P.T. Johnstone, Conditions related to De Morgan's law. In M.P. Fourman, C.J. Mulvey & D. S. Scott (eds.), **Applications of Sheaves** (Durham 1977), LNM 753, Springer-Verlag, 1979, pp. 479–491.

3. F.W. Lawvere, Introduction to F. W. Lawvere & S. Schanuel (eds.), **Categories in Continuum Physics** (Buffalo 1982), LNM 1174, Springer-Verlag, 1986.

4. F. W. Lawvere, Intrinsic co-Heyting boundaries and the Leibniz rule in certain toposes. In A. Carboni, M.C. Pedicchio & G. Rosolini (eds.), **Category Theory** (Como 1990), LNM 1488, Springer-Verlag 1991, pp. 279–297.

5. E: Orłowska, Logic for reasoning about knowledge. In W.P. Ziarko (ed.) **Rough Sets, Fuzzy Sets and Knowledge Discovery**, Springer-Verlag, 1994, pp. 227–236.

6. Z. Pawlak, **Rough Sets: A Theoretical Approach to Reasoning about Data**. Kluwer, 1991.

7. P. Pagliani, Rough Set System and Logic-algebraic Structures. In E. Orłowska (ed.): **Incomplete Information: Rough Set Analysis**, Physica Verlag, 1997, pp. 109–190.

8. C. Rauszer, Semi-Boolean algebras and their application to intuitionistic logic with dual operations. *Fundamenta Mathematicae* LXXIII, 1974, pp. 219–249.

9. G.E. Reyes & N. Zolfaghari, Bi-Heyting Algebras, Toposes and Modalities. *Journ. of Philosophical Logic*, 25, 1996, pp. 25–43.

Multifunctions as Approximation Operations in Generalized Approximation Spaces

P. Maritz

Stellenbosch, South Africa
pm@maties.sun.ac.za

Abstract. An approximation space can be defined as a quintuple $\mathcal{A} = (T, U, F, \Phi, \Gamma)$, where $F : T \to U$ is a multifunction and Φ and Γ are unary operations on the power set of U.

1 Introduction

The use of multifunctions for approximations has been studied before under names such as *compatibility relations*, *binary relations* and *multi-valued mappings*, see [9, p.63] for further references. Also, in [8], Yao and Lin use various binary relations for defining approximation operators.

2 Preliminaries

Denote the class of all subsets of U by $\mathcal{P}(U)$, and the *diagonal* of U by $\triangle_U = \{(x, x) : x \in U\}$. If $R \subset U \times U$ is a binary relation on U, then its *inverse* is denoted by $R^{-1} = \{(y, x) : (x, y) \in R\}$. The relation R on U is *reflexive* if $\triangle_U \subset R$, *symmetric* if $R = R^{-1}$ and *transitive* if $R \circ R \subset R$.

2.1 Definition.
(1) A binary relation R on U is called:
    (a) a *tolerance* relation if it is reflexive and symmetric;
    (b) a *preordering* (or *quasi-ordering*, see [7]) of U if it is reflexive and transitive;
    (c) an *equivalence* relation if it is reflexive, symmetric and transitive.
(2) ([7, p.54]). An operation $\Psi : \mathcal{D} \to \mathcal{E}$ is said to be a *unit operation* on $\mathcal{D}$ if $\Psi(D) = \cup\{\Psi(\{x\}) : x \in D\}$ for every set $D \in \mathcal{D}$.
(3) An operation $\Psi : \mathcal{P}(U) \to \mathcal{P}(U)$ is
    (a) *reflexive* if it satisfies $x \in \Psi(\{x\})$ for every $x \in U$;
    (b) *symmetric* if it satisfies $x \in \Psi(\{y\}) \Rightarrow y \in \Psi(\{x\})$ for every $x, y \in U$;
    (c) *transitive* if it satisfies $(C1)$ $\Psi(\Psi(\{x\})) \subset \Psi(\{x\})$ for every $x \in U$.
(4) An operation $\Psi : \mathcal{P}(U) \to \mathcal{P}(U)$ is *tolerance* (respectively, a *preordering*; *equivalence*) on $\mathcal{P}(U)$ if it is *reflexive and symmetric* (respectively, *reflexive and transitive*; *reflexive, symmetric and transitive*) on $\mathcal{P}(U)$.

2.2 Lemma. If $\Psi : \mathcal{P}(U) \to \mathcal{P}(U)$ is a unit operation, then the conditions $(C1), (C2)$ and $(C3)$ are equivalent, where
    $(C2)$ $y \in \Psi(\{x\}) \Rightarrow \Psi(\{y\}) \subset \Psi(\{x\})$ for every $x, y \in U$;
    $(C3)$ $(z \in \Psi(\{y\})$ and $y \in \Psi(\{x\})) \Rightarrow z \in \Psi(\{x\})$ for every $x, y, z \in U$.

L. Polkowski and A. Skowron (Eds.): RSCTC'98, LNAI 1424, pp. 131–138, 1998.
© Springer-Verlag Berlin Heidelberg 1998

3   The approach that uses relations

Pawlak [5] defines an approximation space to be an ordered pair $\mathcal{A} = (U, R)$, where $R \subset U \times U$ an *indiscernibility* relation on U.

3.1 Tolerance relations

If R is a tolerance relation on U, if $x \in U$ and $|x|_R$ is the tolerance class determined by x, then the *upper approximation* operation $\widetilde{R} : \mathcal{P}(U) \rightarrow \mathcal{P}(U)$ and the *lower approximation* operation $\underset{\sim}{R} : \mathcal{P}(U) \rightarrow \mathcal{P}(U)$ with respect to R are defined, for every set $A \in \mathcal{P}(U)$, by

$\widetilde{R}(A) = \{x \in U : |x|_R \cap A \neq \emptyset\}$ and $\underset{\sim}{R}(A) = \{x \in U : |x|_R \subset A\}$ respectively.

3.1.1 Example. Define the tolerance relation R on $\mathbb{R}$ by $xRy \Leftrightarrow (x, y \in (t - 1, t + 1)$ for some $t \in \mathbb{R})$. Then $\widetilde{R}([0, 1]) = \{x \in \mathbb{R} : |x|_R \cap [0, 1] \neq \emptyset\} = (-2, 3)$ and $\underset{\sim}{R}([0, 1]) = \emptyset$. Also, $\widetilde{R}([0, 4]) = (-2, 6)$ and $\underset{\sim}{R}([0, 4]) = \{2\}$.

3.2 Equivalence relations

If R is an equivalence relation on U, if $x \in U$ and $[x]_R$ is the equivalence class determined by x, then $\overline{R}(A) = \{x \in U : [x]_R \cap A \neq \emptyset\}$ defines the *upper approximation* and $\underline{R}(A) = \{x \in U : [x]_R \subset A\}$ the *lower approximation* operation respectively. It is clear that the operations $\overline{R}$ and $\underline{R}$ are equivalence on $\mathcal{P}(U)$ by definition 2.1(4)

3.2.1 Examples.

(1) Define the equivalence relation R on $\mathbb{R}$ by
$xRy \Leftrightarrow (x, y \in [\lfloor x \rfloor, \lfloor x \rfloor + 1)$ for some $n \in N)$, where $\lfloor x \rfloor$ is the largest integer less than or equal to x. If $x \in \mathbb{R}$, then $[x]_R = [\lfloor x \rfloor, \lfloor x \rfloor + 1)$. $\overline{R}([0, 1]) = \{x \in \mathbb{R} : [x]_R \cap [0, 1] \neq \emptyset\} = [0, 2)$ and $\underline{R}([0, 1]) = \{x \in \mathbb{R} : [x]_R \subset [0, 1]\} = [0, 1)$.

(2) Let $a < b$ and $U = (a, b)$. Let $xRy \Leftrightarrow x - y$ is a rational number. If $A = (c, d)$, where $a < c < d < b$, then $\underline{R}((c, d)) = \emptyset$ and $\overline{R}((c, d)) = U$. If $x \in U$ is a rational number, then $\underline{R}([x]_R) = [x]_R = \overline{R}([x]_R)$.

A Pawlak approximation space $\mathcal{A} = (U, R)$ induces a triple $(U, \widetilde{R}, \underset{\sim}{R})$ or $(U, \overline{R}, \underline{R})$, depending on whether R is a tolerance or and equivalence relation on U.

3.3 Definition. Let R be an indiscernibility relation on U and let $\mathcal{A} = (U, R)$.

(1) If R is a tolerance relation, then $(U, \widetilde{R}, \underset{\sim}{R})$ is called the *generalized approximation space of Pawlak* induced by $\mathcal{A}$.

(2) If R is an equivalence relation, then $(U, \overline{R}, \underline{R})$ is called the *approximation space of Pawlak* induced by $\mathcal{A}$.

4   The approach that uses covers

The approach to approximation spaces through the means of a cover for the universe U is due to Żakowski [10]. For a binary relation R on U, let the operation $\boldsymbol{R} : \mathcal{P}(U) \to \mathcal{P}(U)$ be defined, for every set $A \in \mathcal{P}(U)$, by

(α) $\boldsymbol{R}(A) = \{y \in U : \exists x \in A \text{ such that } xRy\}$.

If R is reflexive, then $\boldsymbol{R}$ is a reflexive operation. Consequently, every reflexive relation R on U uniquely determines a cover for U:

(β) $\mathcal{C}_R = \{\boldsymbol{R}(\{x\}) : x \in U\}$.

If R is a tolerance (an equivalence) relation on U, then $\mathcal{C}_R$ is the collection of all tolerance (equivalence) classes of R, determined by the elements of U. If R is an equivalence relation on U, then $\mathcal{C}_R$ is a partition of U. In fact, it is common knowledge that if R is an equivalence relation on U, then the collection of equivalence classes is a partition $\mathcal{C}_R$ of U that induces the relation R, and if $\mathcal{C}$ is a partition of U, then the induced relation $R_{\mathcal{C}}$ on U is an equivalence relation whose collection of equivalence classes is exactly $\mathcal{C}$, see for example, Halmos [2, p.28]. Every cover (partition) $\mathcal{C}$ of U induces a tolerance (an equivalence) relation $R_{\mathcal{C}}$ on U:

(γ) $xR_{\mathcal{C}}y \Leftrightarrow (x, y \in U \text{ and } \exists C \in \mathcal{C} \quad (x, y \in C))$.

Note further, from the definition of $R_{\mathcal{C}}$ in (γ) above, that

(δ) $\boldsymbol{R}_{\mathcal{C}}(\{x\}) = \cup\{C \in \mathcal{C} : x \in C\} = |x|_{R_{\mathcal{C}}}$ for every $x \in U$.

Consequently, the sets of $\mathcal{C}$ need not be the tolerance classes of $R_{\mathcal{C}}$, whereas, from (β),

($\in$) $\mathcal{C}_{R_{\mathcal{C}}} = \{\boldsymbol{R}_{\mathcal{C}}(\{x\}) : x \in U\}$ is the collection of tolerance classes of $R_{\mathcal{C}}$.

If $\mathcal{C}$ is a partition then the sets of $\mathcal{C}$ are the equivalence classes of $R_{\mathcal{C}}$.

4.1 Definition. If U is a nonempty set and $\mathcal{C}$ a cover for U, then the ordered pair $(U, \mathcal{C})$ is called an *approximation space* of Żakowski [10].

Given an approximation space $\mathcal{A} = (U, \mathcal{C})$ as in definition 4.1. Following Żakowski, we define two approximation operations on $\mathcal{P}(U)$. For every set $A \in \mathcal{P}(U)$, the *upper approximation operation* $\overline{\mathcal{C}} : \mathcal{P}(U) \to \mathcal{P}(U)$ and the *lower approximation operation* $\underline{\mathcal{C}} : \mathcal{P}(U) \to \mathcal{P}(U)$ are defined by

(ζ) $\overline{\mathcal{C}}(A) = \cup\{C \in \mathcal{C} : C \cap A \neq \emptyset\}$ and (η)$\underline{\mathcal{C}}(A) = \cup\{C \in \mathcal{C} : C \subset A\}$

respectively.

We also define a *weak lower approximation* operation $\underline{\mathcal{C}}' : \mathcal{P}(U) \to \mathcal{P}(U)$, see [6, p.656]:

(θ) $\underline{\mathcal{C}}'(A) = \{y \in U : \overline{\mathcal{C}}(\{y\}) \subset A\}$ for every $A \in \mathcal{P}(U)$.

In general, $\underline{\mathcal{C}}'(A) \subset \underline{\mathcal{C}}(A)$ for every set $A \in \mathcal{P}(U)$, and $\underline{\mathcal{C}}' = \underline{\mathcal{C}}$ if $\mathcal{C}$ is a partition of U.

An approximation space of Żakowski can induce at least two triples, $(U, \overline{\mathcal{C}}, \underline{\mathcal{C}})$ or $(U, \overline{\mathcal{C}}, \underline{\mathcal{C}}')$.

If we restrict both concepts of approximation spaces to the spaces determined by equivalence relations and partitions respectively, then both approaches to them are equivalent. The notion of approximation spaces determined by a cover (the approach by Żakowski) is in some sense a generalization of the classical one defined by Pawlak, as given in definition 3.3(2).

5   The equivalence of the two approaches

5.1 Question Under what conditions are the two presented approaches to approximation spaces equivalent? Wybraniec–Skardowska [7], established, among others, the following.

5.2 Results

(1) If C is a cover for U, then $(U, \overline{C}, \underline{C}') = (U, \widetilde{R}_C, \underset{\sim C}{R})$.

(2) If C is a partition of U, then $(U, \overline{C}, \underline{C}) = (U, \overline{C}, \underline{C}') = (U, \overline{R}_C, \underline{R}_C)$.

(3) If R is a tolerance relation on U, then, for any set $A \subset U$,

    (a) $\widetilde{R}(A) = \overline{C}_R(A)$

    (b) $\underset{\sim}{R}(A) = \underline{C}_R(A)$.

(4) If R is an equivalence relation on U, then for any set $A \subset U$,

    (a) $\overline{R}(A) = \overline{C}_R(A)$

    (b) $\underline{R}(A) = \underline{C}_R(A)$.

    and then $(U, \overline{R}, \underline{R}) = (U, \overline{C}_R, \underline{C}_R)$ on U.

Results 5.2 (1),(2) and (4) yield the statement that if we restrict both concepts of approximation spaces to the spaces determined by partitions and equivalence relations respectively, then both approaches to them are equivalent.

6   The approach by multifunctions

Henceforth, we assume that T and U are nonempty sets, with U not necessarily finite, and that $F : T \to U$ is a strict multifunction. If $A \subset U$, then $F^+(A) = \{t \in T : F(t) \subset A\}$ and $F^-(A) = \{t \in T : F(t) \cap A \neq \emptyset\}$, where F^+ is the *strong* and F^- the *weak* inverse of F, respectively. Furthermore,

    (ι) $F(B) = \cup\{F(t) : t \in B\}$ for every set $B \subset T$

If $F : T \to U$ is a multifunction, then F can also be regarded as a unary operation $F : \mathcal{P}(T) \to \mathcal{P}(U)$, and (ι) now becomes, for every set $B \in \mathcal{P}(T)$,

    (κ) $F(B) = \cup\{F(\{t\}) : t \in B\}$ so that F is a unit operation on $\mathcal{P}(T)$.

The set $F(T)$ is called the *range* of F. If $F(T) = U$, then F is a *surjection*. *We assume throughout this section that F is a surjection.* The class $\mathcal{F}(T) = \{F(t) : t \in T\}$ is then a cover for U. Let R be a tolerance relation on U. By section 4, R uniquely determines a cover C_R for U, and C_R is the collection of all tolerance classes of R, determined by the elements of U. Write

    (λ) $C_R = \{E_t : t \in T\}$

where $x R y$ if and only if $x, y \in E_t$ for some $t \in T$, where T is the index set for the collection of tolerance classes of R. This clearly defines a surjective multifunction $F_R : T \to U$ by $F_R(t) = E_t$ for every $t \in T$. Consequently, a tolerance relation R on U uniquely determines a surjective multifunction $F_R : T \to U$, and hence a cover $C_R = \mathcal{F}_R(T) = \{F_R(t) : t \in T\}$ for U. Similarly, if R is an equivalence relation on U, then it uniquely determines a partition C_R of U, where the elements of C_R are exactly the equivalence classes of R : the class

$\mathcal{C}_R = \mathcal{F}_R(T) = \{F_R(t) : t \in T\}$ is then a partition of U by equivalence classes of R. Let $F : T \to U$ be a multifunction. By definition 4.1, the ordered pair $\mathcal{A} = (U, \mathcal{F}(T))$ is an approximation space of Żakowski. The cover $\mathcal{F}(T)$ for U induces a tolerance relation $R_{\mathcal{F}(T)}$ on U, so that by (γ) in section 4,

(μ) $xR_{\mathcal{F}(T)}y \Leftrightarrow (x, y \in U$ and $\exists t \in T$ such that $x, y \in F(t))$.

In the notation of [3, Definition 7.2.1], let

(ν) $G(x) = \cup\{F(t) \in \mathcal{F}(T) : x \in F(t)\}$ for every $x \in U$.

The set $G(x)$ is said to be an *indiscernibility neighborhood* of x. By (δ),

(ξ) $R_{\mathcal{F}(T)}(\{x\}) = \{y \in U : xR_{\mathcal{F}(T)}y\} = \cup\{F(t) \in \mathcal{F}(T) : x \in F(t)\} =$ $|x|_{R_{\mathcal{F}(T)}} = G(x)$ for every $x \in U$, and for every set $A \subset U$, (α) in section 4 becomes

(o) $R_{\mathcal{F}(T)}(A) = \{y \in U : \exists x \in A$ such that $x \in R_{\mathcal{F}(T)}y\} = \cup\{R_{\mathcal{F}(T)}(\{x\}) : x \in A\} = \cup\{G(x) : x \in A\} = G(A)$ by (κ).

Henceforth, we shall use G instead of $R_{\mathcal{F}(T)}$; G can be regarded as a multifunction, $G : U \to U$, or as an operation $G : \mathcal{P}(U) \to \mathcal{P}(U)$, where $G(x) = G(\{x\})$ for every $x \in U$. The operation $G : \mathcal{P}(U) \to \mathcal{P}(U)$ is clearly tolerance and, if $x \in F(t)$, then $F(t) \subset G(x)$. The sets in $\mathcal{F}(T)$ are not necessarily the tolerance classes of $R_{\mathcal{F}(T)}$. By $(\in)$ in section 4, the collection

(π) $\mathcal{C}_{R_{\mathcal{F}(T)}} = \{G(x) : x \in U\}$ is the collection of tolerance classes of $R_{\mathcal{F}(T)}$.

For the approximation space $\mathcal{A} = (U, \mathcal{F}(T))$, we define, in the notation of [3], the following four operations on $\mathcal{P}(U)$. For $A \subset U$, let

(ϱ) $FF^-(A) = F(F^-(A)) = \cup\{F(t) \in \mathcal{F}(T) : F(t) \cap A \neq \emptyset\}$,

the *upper approximation* of A in $\mathcal{A}$ (corresponding to (ζ) in section 4);

(σ) $FF^+(A) = F(F^+(A)) = \cup\{F(t) \in \mathcal{F}(T) : F(t) \subset A\}$,

the *lower approximation* of A in $\mathcal{A}$ (corresponding to (η) in section 4);

(τ) $G_1(A) = \{x \in A : \forall F(t) \in \mathcal{F}(T)$ $(x \in F(t) \Rightarrow F(t) \subset A)\}$,

the *weak lower approximation* of A in $\mathcal{A}$ (it will follow from (ψ) below that this corresponds to (θ) in section 4);

(υ) $G_2(A) = U \backslash FF^+(U \backslash A)$

the *strong upper approximation* of A in $\mathcal{A}$. For every set $A \subset U$ and every $x \in U$,

(ϕ) $FF^-(A) = \{x \in U : \exists t \in T$ such that $x \in F(t)$ and $F(t) \cap A \neq \emptyset\}$

(χ) $FF^-(\{x\}) = \cup\{F(t) \in \mathcal{F}(T) : x \in F(t)\} = G(x) = |x|_{R_{\mathcal{F}(T)}}$

(ψ) $G_1(A) = \{x \in A : FF^-(\{x\}) \subset A\} = G^+(A)$

(ω) $G_2(A) = \{x \in U : \forall F(t) \in \mathcal{F}(T)$ $(x \in F(t) \Rightarrow F(t) \cap A \neq \emptyset)\}$

Also $G_1(A) \subset FF^+(A)$, $G_2(A) \subset FF^-(A)$. Thus, if $F : T \to U$ is a (surjective) multifunction, then $\mathcal{F}(T)$ is a cover for U, and the approximation space $\mathcal{A} = (U, \mathcal{F}(T))$ induces the triples (U, FF^-, FF^+), (U, FF^-, G_1), (U, G_2, FF^+) and (U, G_2, G_1).

6.1 Example. Let $F : \mathbb{R} \to \mathbb{R}$ be defined by $F(t) = (t - 1, t + 1)$ for every $t \in \mathbb{R}$. Then $\mathcal{A} = (\mathbb{R}, \mathcal{F}(\mathbb{R}))$ is an approximation space in the sense of Żakowski, where $\mathcal{F}(\mathbb{R}) = \{(t - 1, t + 1) : t \in \mathbb{R}\}$ is a cover for $\mathbb{R}$. Define the tolerance relation $R_{\mathcal{F}(\mathbb{R})}$ on $\mathbb{R}$ by $xR_{\mathcal{F}(\mathbb{R})}y \Leftrightarrow (x, y \in (t - 1, t + 1)$ for some $t \in \mathbb{R})$. If $x \in \mathbb{R}$, then, by (ξ) $R_{\mathcal{F}(\mathbb{R})}(\{x\}) = |x|_{R_{\mathcal{F}(\mathbb{R})}} = G(x) = (x - 2, x + 2)$ so that, for example, $|3|_{R_{\mathcal{F}(\mathbb{R})}} = (1, 5)$, $R_{\mathcal{F}(\mathbb{R})}(\{4\}) = (2, 6)$ and $R_{\mathcal{F}(\mathbb{R})}(\{1, 4, 6\}) = (-1, 8)$.

Clearly, the sets in $\mathcal{F}(\mathbb{R})$, namely the intervals $(t-1, t+2)$, are not the tolerance classes of $R_{\mathcal{F}(\mathbb{R})}$. By (π) and example 3.1.1, $\mathcal{C}_{R_{\mathcal{F}(\mathbb{R})}} = \{(t-2, t+2) : t \in \mathbb{R}\}$ is the collection of tolerance classes of $R_{\mathcal{F}(\mathbb{R})}$. Let $A = [-6, 8]$. Then, by $(\rho), (\sigma), (\tau)$ and (υ), we have that $FF^-(A) = (-8, 10)$, $FF^+(A) = (-6, 8)$, $G_1(A) = (-4, 6)$ and $G_2(A) = [-6, 8]$. It is clear that the approximation spaces (U, FF^-, FF^+), (U, FF^-, G_1), (U, G_2, FF^+) and (U, G_2, G_1) are mutually distinct.

By (o) and (ϕ), $\mathbf{R}_{\mathcal{F}(\mathbb{R})}([0,1]) = \cup\{R_{\mathcal{F}(T)}(\{x\}) : x \in [0,1]\} = G([0,1]) = \cup\{G(x) : x \in [0,1]\} = FF^-([0,1]) = (-2, 3)$. If $\mathcal{F}(T)$ is a partition of U, then $R_{\mathcal{F}(T)}$ is an equivalence relation on U, $G(x) = [x]_{R_{\mathcal{F}(T)}} = F(t)$ if $x \in F(t)$, $FF^+ = G_1$ on $\mathcal{P}(U)$, $FF^- = G_2$ on $\mathcal{P}(U)$ because $FF^-(A) \cap G_1(U \setminus A) = \emptyset$.

6.2 Lemma. ([4, p.167]). A multifunction $F : T \to U$ is semi-single-valued if and only if there exists an equivalence relation R on U and a function $f : T \to U$ such that $F = R \circ f$.

If $F : T \to U$ is semi-single-valued, then $\mathcal{F}(T)$ is a partition of U.

6.3 Lemma.

(1) If $F : T \to U$ is a multifunction, then for $A \subset U$ and $x \in U$:
   (a) $FF^-(A) = \cup\{FF^-(\{x\}) : x \in A\}$
   (b) $R_{\mathcal{F}(T)}(A) = G(A) = \cup\{R_{\mathcal{F}(T)}(\{x\}) : x \in A\} = \cup\{G(x) : x \in A\} = G(A)$
   (c) $FF^-(\{x\}) = G(x) = |x|_{R_{\mathcal{F}(T)}}$
   (d) $R_{\mathcal{F}(T)}(A) = \cup\{G(x) : x \in A\} = \cup\{FF^-(\{x\}) : x \in A\} = FF^-(A)$
   (e) $R_{\mathcal{F}(T)}(A) = \{y \in U : |y|_{R_{\mathcal{F}(T)}} \cap A \neq \emptyset\} = \widetilde{R}_{\mathcal{F}(T)}(A)$.
(2) If $F : T \to U$ is semi-single-valued, then for $A \subset U$ and $x \in U$:
   (f) $R_{\mathcal{F}(T)}(\{x\}) = [x]_{R_{\mathcal{F}(T)}} = G(x) = F(t)$ if $x \in F(t)$
   (g) $R_{\mathcal{F}(T)}(A) = \{y \in U : [y]_{R_{\mathcal{F}(T)}} \cap A \neq \emptyset\} = \overline{R}_{\mathcal{F}(T)}(A)$.

Lemmas 6.4 and 6.5 below stipulate correspondences between some of the ordered triples of definition 3.3 and those introduced in this section.

6.4 Lemma.

(1) If $F : T \to U$ is a multifunction, then $(U, FF^-, G_1) = (U, \widetilde{R}_{\mathcal{F}(T)}, R_{\underset{\sim}{\mathcal{F}(T)}})$.
(2) If $F : T \to U$ is a semi-single-valued multifunction, then
$(U, FF^-, FF^+) = (U, FF^-, G_1) = (U, G_2, FF^+) = (U, G_2, G_1) = (U, \overline{R}_{\mathcal{F}(T)}, \underline{R}_{\mathcal{F}(T)})$.

6.5 Lemma.

(1) If R is a tolerance relation on U, $F_R : T \to U$ the corresponding multifunction and A any subset of U, then:
   (a) $\widetilde{R}(A) = F_R F_R^-(A)$
   (b) $\underset{\sim}{R}(A) = F_R F_R^+(A)$.

(2) If R is an equivalence relation on U, $F_R : T \to U$ the corresponding multi-function and A any subset of U, then:

(c) $\overline{R}(A) = F_R F_R^-(A)$

(d) $\underline{R}(A) = F_R F_R^+(A)$ and $(U, \overline{R}, \underline{R}) = (U, F_R F_R^-, F_R F_R^+)$.

7   An approximation space

Wybraniec–Skardowska [7, p.54] defines an approximation space by $\mathcal{A} = (U, \Phi, \Gamma)$ where U is the universe of $\mathcal{A}$ and $\Phi : \mathcal{P}(U) \to \mathcal{P}(U)$ and $\Gamma : \mathcal{P}(U) \to \mathcal{P}(U)$ such that, for $A \subseteq U$

(a1) $\Phi(A) \subseteq U$ and $\Gamma(A) \subseteq A$

(a2) $\Phi(A) = \cup \{\Phi(\{x\}) : x \in A\}$

(a3) $\Gamma(A) = \{x \in U : \emptyset \neq \Phi(\{x\}) \subseteq A\}$.

The operations Φ and Γ are respectively called the *upper* and *lower* approximation operations associated with $\mathcal{A}$. We take this one step further.

7.1 Definition. An approximation space is an ordered quintuple $\mathcal{A} = (T, U, F, \Phi, \Gamma)$, where T and U (the universe) are nonempty sets, $F : T \to U$ is a multifunction,$\Phi : \mathcal{P}(U) \to \mathcal{P}(U)$ and $\Gamma : \mathcal{P}(U) \to \mathcal{P}(U)$ are unary operations satisfying the properties (a1)-(a3).

We now employ the work on multifunctions from section 6. Note that the operation $FF^- : \mathcal{P}(U) \to \mathcal{P}(U)$ satisfies $FF^-(A) = \cup \{FF^-(\{x\}) : x \in A\}$ for every $A \in \mathcal{P}(U)$, so that FF^- is a unit operator on $\mathcal{P}(U)$. If we let $T = U$ and $F = \Phi$ on the class $\mathcal{S}$ of singleton subsets of U, then because $F(A) = \cup\{F(t) : t \in A\} = \cup\{\Phi(\{t\}) : t \in A\} = \Phi(A)$, and $F^+(A) = \{x \in U : F(x) \subset A\} = \{x \in U : \emptyset \neq F(x) \subset A\} = \{x \in U : \emptyset \neq \Phi(\{x\}) \subset A\} = \Gamma(A)$ (F is strict by assumption), we can reduce the quintuple $\mathcal{A} = (T, U, F, \Phi, \Gamma)$ in definition 7.1 to the triple $\mathcal{A} = (U, \Phi, \Gamma) = (U, F, F^+)$ in the definition of Wybraniec–Skardowska.

7.2 Remarks.

(1) If $F : T \to U$ is a multifunction, then the cover $\mathcal{F}(T)$ induces the tolerance relation $R_{\mathcal{F}(T)}$ on U. By lemma 6.4, $FF^-(A) = \tilde{R}_{\mathcal{F}(T)}(A)$ for every $A \in \mathcal{P}(U)$ and $G_1(A) = R_{\sim \mathcal{F}(T)}(A)$ where, by (ψ) in section 6, $G_1(A) = \{x \in A : FF^-(\{x\}) \subset A\}$ and $\mathcal{A} = (T, U, F, FF^-, G_1) = (T, U, F, \tilde{R}_{\mathcal{F}(T)}, R_{\sim \mathcal{F}(T)})$ is an approximation space determined by F, with $\tilde{R}_{\mathcal{F}(T)}$ and $R_{\sim \mathcal{F}(T)}$ the *upper* and *lower* approximations, respectively. Example 3.1.1 illustrates this.

(2) If the multifunction $F : T \to U$ is semi-single-valued, then the class $\mathcal{F}(T)$ is a partition of U and $R_{\mathcal{F}(T)}$ is the induced equivalence relation. Then, lemma 6.4, and [3, Corollary 7.4.3(2)],
$$\mathcal{A} = (T, U, F, FF^-, FF^+) = (T, U, F, FF^-, G_1) = (T, U, F, G_2, FF^+) = (T, U, F, G_2, G_1) = (T, U, F, \overline{R}_{\mathcal{F}(T)}, \underline{R}_{\mathcal{F}(T)}).$$

(3) If R is a tolerance relation on U, then R uniquely determines a cover $\mathcal{C}_R$ (section 4) which in turn defines a surjective multifunction $F_R : T \to U$ (see (λ) in section 6), where by (β) and (δ) in section 4, $\mathcal{C}_R = \{R(\{x\}) : x \in U\}$, and $R(\{x\}) = |x|_R$ is the R-tolerance class determined by x. Define $\underset{\to}{R} : \mathcal{P}(U) \to \mathcal{P}(U)$ by $\underset{\to}{R}(A) = \{y \in U : \emptyset \neq \overrightarrow{R}(\{y\}) \subseteq A\}$. Then $\underset{\to}{R}(A) = R_{\underset{\sim}{\mathcal{F}_R(T)}}(A)$ and we have the approximation space $\mathcal{A} = (T, U, F_R, FF^-, G_1) =$

$$(T, U, F_R, \overrightarrow{R}, \underset{\to}{R}) = (T, U, F_R, \overrightarrow{R}, R_{\underset{\sim}{\mathcal{F}_R(T)}}) = (T, U, F_R, \overrightarrow{R}, G^+),$$

by lemma 6.3(d) and (ψ) in section 6. If R is an equivalence relation, then $\overrightarrow{R} = \overline{R}$ and $\underset{\to}{R} = \underline{R}$. Then the class $\mathcal{F}_R(T) = \{F_R(t) : t \in T\}$ is a partition of U, and as noticed just above Lemma 6.2, $FF^+ = G_1$ and $FF^- = G_2$. In this case we also have that $\mathcal{A} = (T, U, F_R, G_2, FF^+) = (T, U, F_R, \overline{R}, \underline{R})$.

(4) If $\mathcal{C}$ is a cover for U, then by 5.2(1), $(U, \overline{\mathcal{C}}, \underline{\mathcal{C}}') = (U, \widetilde{R}_{\mathcal{C}}, R_{\underset{\sim \mathcal{C}}{}})$ and again we can arrive at $\mathcal{A} = (T, U, F, \widetilde{R}_{\mathcal{C}}, R_{\underset{\sim \mathcal{C}}{}}) = (T, U, F, \overline{\mathcal{C}}, \underline{\mathcal{C}}')$ where $F : T \to U$ is defined as indicated by (λ) in section 6. If $\mathcal{C}$ is a partition of U, then by 5.2(2), $(U, \overline{\mathcal{C}}, \underline{\mathcal{C}}) = (U, \overline{\mathcal{C}}, \underline{\mathcal{C}}') = (U, \overline{R}_{\mathcal{C}}, \underline{R}_{\mathcal{C}})$ and we have that $\mathcal{A} = (T, U, F, \overline{R}_{\mathcal{C}}, \underline{R}_{\mathcal{C}}) = (T, U, F, \overline{\mathcal{C}}, \underline{\mathcal{C}})$.

(5) For other properties of the operations $FF^- : \mathcal{P}(U) \to \mathcal{P}(U)$ and $FF^+ : \mathcal{P}(U) \to \mathcal{P}(U)$ and for the topologies associated by FF^- and FF^+, see [3].

References

1. P.M. Cohn. *Universal algebra*. D. Reidel, Dordrecht, 1981.
2. P.R. Halmos. *Naive set theory*. Springer, New York, 1974.
3. P. Maritz. *Glasnik Matematički* **31(51)** (1996), 159–178.
4. A. Münnich and A. Száz. *Publ. Inst. Math. (Beograd) (N.S.)* **33(47)** (1983), 163–168.
5. Z. Pawlak. *Internat. J. Comput. Inform. Sci.* **11**(1982), 341–356.
6. J.A. Pomykała. *Bull. Polish Acad. Sci. Math.* **35** (1987), 653–662.
7. U. Wybraniec–Skardowska. *Bull. Polish Acad. Sci. Math.* **37** (1989),51–62.
8. Y.Y. Yao and T.Y. Lin. *Intelligent Automation and Soft Computing, An International Journal* **2** (1996), 103–120.
9. Y.Y. Yao, S.K.M. Wong and T.Y. Lin. In: *Rought sets and data mining, analysis for imprecise data*. Ed. T.Y. Lin and N. Cercone. Kluwer, Boston, 1997, 47–75.
10. W. Żakowski. *Demon. Math.* **16** (1983), 761–769.

Preimage Relations and Their Matrices

Jouni Järvinen

University of Turku, Department of Mathematics, FIN-20014 TURKU

1 Introduction

Information systems (in the sense of Pawlak) are used for representing properties of objects by the means of attributes and their values. However, sometimes there are situations in which we cannot give an exact value of an attribute for an object; we can only approximate the incompletely known value by a subset of values, in which the actual value is expected to be.

In these kind of information systems we can define several information relations on the object set. It seems that these relations have many common properties. Here we introduce strong and weak preimage relations, which are suitable for the investigation of those common features.

Dependence spaces are general settings for the study of reducts of attribute sets, for example. In here we consider especially dense families of dependence spaces, and we give a characterization of reducts by the means of dense families. Dependence spaces defined by strong and weak preimage relations are also studied. In addition to this, we introduce matrices of preimage relations and show how we can determine dense families of dependence spaces defined by strong and weak preimage relations by using these matrices.

This paper is structured as follows. In the next section we define information systems and some information relations. Section 3 contains the definition of strong and weak preimage relations. In Section 4 we give some basic concepts concerning dependence spaces and show how strong and weak preimage relations define dependence spaces. Finally, in Section 5 dense families of dependence spaces and matrices of preimage relations are studied.

2 Information Systems and Information Relations

An *information system* is a triple $\mathcal{S} = (U, A, \{V_a\}_{a \in A})$, where U is a set of *objects*, A is a set of *attributes*, and $\{V_a\}_{a \in A}$ is an indexed set of *value sets of attributes*. All these sets are assumed to be finite and nonempty. Each attribute is a function $a \colon U \to \wp(V_a)$ which assigns subsets of values to objects such that $a(x) \neq \emptyset$, for all $a \in A$ and $x \in U$ (see e.g. [5,6]).

If $|a(x)| = 1$, then the information of the attribute a for the object x is *complete* (or *deterministic*), and we usually write $a(x) = \{v\}$ simply by $a(x) = v$. If $|a(x)| > 1$, then the information of the attribute a for the object x is *incomplete* (or *nondeterministic*). For example, if a is "age", and x is 25 years old, then $a(x) = 25$. It is also possible

L. Polkowski and A. Skowron (Eds.): RSCTC' 98, LNAI 1424, pp. 139–146, 1998.
© Springer-Verlag Berlin Heidelberg 1998

that we know the age of x only approximately, say between 20 and 28. In this case $a(x) = \{20, \ldots, 28\}$.

In the following we shall present 16 information relations between objects, which are based on their values of attributes. These relations can be found in [5,6], for example. Suppose $\mathcal{S} = (U, A, \{V_a\}_{a \in A})$ is an information system and let $B(\subseteq A)$ be a subset of attributes. The first eight relations reflect **indistinguishability** between objects:

- *strong (weak) indiscernibility:*

  $(x, y) \in ind(B)$ $(wind(B))$ iff $a(x) = a(y)$ for all (some) $a \in B$;
- *strong (weak) similarity:*

  $(x, y) \in sim(B)$ $(wsim(B))$ iff $a(x) \cap a(y) \neq \emptyset$ for all (some) $a \in B$;
- *strong (weak) forward inclusion:*

  $(x, y) \in fin(B)$ $(wfin(B))$ iff $a(x) \subseteq a(y)$ for all (some) $a \in B$;
- *strong (weak) backward inclusion:*

  $(x, y) \in bin(B)$ $(wbin(B))$ iff $a(y) \subseteq a(x)$ for all (some) $a \in B$,

where $-a(x)$ denotes the complement of $a(x)$ in V_a.

For example, two objects are strongly B-indiscernible if they have the same values for all attributes in B, and they are weakly B-similar if there is an attribute in B such that these objects have at least one common value for this attribute.

The following eight relations reflect **distinguishability** between objects:

- *strong (weak) diversity:*

  $(x, y) \in div(B)$ $(wdiv(B))$ iff $a(x) \neq a(y)$ for all (some) $a \in B$;
- *strong (weak) right orthogonality:*

  $(x, y) \in rort(B)$ $(wrort(B))$ iff $a(x) \subseteq -a(y)$ for all (some) $a \in B$;
- *strong (weak) right negative similarity:*

  $(x, y) \in rnim(B)$ $(wrnim(B))$ iff $a(x) \cap -a(y) \neq \emptyset$ for all (some) $a \in B$;
- *strong (weak) left negative similarity:*

  $(x, y) \in lnim(B)$ $(wlnim(B))$ iff $-a(x) \cap a(y) \neq \emptyset$ for all (some) $a \in B$.

For example, two objects are weakly B-diverse if their values for all attributes in B are not the same, and two objects are strongly right B-orthogonal if they have no common value for any attribute in B.

3   Preimage Relations

All information relations defined in the end of the previous section are similar in the following sense. Two objects belong to a certain strong (resp. weak) information relation with respect to an attribute set B if and only if their all (resp. some) value sets of B-attributes are in a specified relation. For example, objects x and y are in the relation $sim(B)$ if and only if $a(x) \cap a(y) \neq \emptyset$ for all attributes a in B.

In this section we introduce preimage relations. This notion allows us to study properties of information relations in a more abstract setting. We denote by $\mathrm{Rel}(A)$ the set of all binary relations on a set A. The *complement* of any relation $R(\in \mathrm{Rel}(A))$ is $-R = \{(x, y) \in A^2 \mid (x, y) \notin R\}$. The set of all maps from A to B is denoted by B^A. Moreover, we assume that U and Y are nonempty sets, and $R \in \mathrm{Rel}(Y)$.

For any map $f \in Y^U$, the *preimage relation* of R is

$$f^{-1}(R) = \{(x, y) \in U^2 \mid f(x)Rf(y)\}.$$

Thus, two elements x and y are in the relation $f^{-1}(R)$ if and only if their images $f(x)$ and $f(y)$ are in the relation R. For example, if R is the equality relation, $=$, then $f^{-1}(R)$ is the kernel of the map f, $\ker f = \{(x, y) \mid f(x) = f(y)\}$.

The following obvious lemma shows that $f^{-1}(R)$ inherits many properties from R.

[**Lemma 1.**]*If R is reflexive, then $f^{-1}(R)$ is reflexive, and similar conditions hold when R is irreflexive, symmetric, or transitive*   □

It is also true that

$$f^{-1}(-R) = -f^{-1}(R).$$

[*Example 1.*]Suppose $S = (U, A, \{V_a\}_{a \in A})$ is an information system, which is described in the following table.

	Age (years)	Weight (kg)	Height (cm)
P1	$\{22, \ldots, 26\}$	$\{48, \ldots, 54\}$	$\{154, \ldots, 157\}$
P2	$\{26, \ldots, 33\}$	$\{73, \ldots, 78\}$	$\{170, \ldots, 175\}$
P3	$\{24, \ldots, 29\}$	$\{51, \ldots, 58\}$	$\{159, \ldots, 162\}$
P4	$\{31, \ldots, 37\}$	$\{75, \ldots, 82\}$	$\{157, \ldots, 165\}$

We denote $V = \bigcup_{a \in A} V_a$ and $Y = \wp(V) - \{\emptyset\}$. Let us define the binary relation SIM on Y by setting

$$(W_1, W_2) \in SIM \iff W_1 \cap W_2 \neq \emptyset.$$

The preimage relations of SIM with respect to the attributes Age, Weight, and Height are the following.

$$\begin{aligned}
\text{Age}^{-1}(SIM) &= \{(x, y) \in U^2 \mid \text{Age}(x) \cap \text{Age}(y) \neq \emptyset\} \\
&= \Delta \cup \{(P1, P2), (P2, P1), (P1, P3), (P3, P1), \\
&\qquad (P2, P3), (P3, P2), (P2, P4), (P4, P2)\} \\
\text{Weight}^{-1}(SIM) &= \{(x, y) \in U^2 \mid \text{Weight}(x) \cap \text{Weight}(y) \neq \emptyset\} \\
&= \Delta \cup \{(P1, P3), (P3, P1), (P2, P4), (P4, P2)\} \\
\text{Height}^{-1}(SIM) &= \{(x, y) \in U^2 \mid \text{Height}(x) \cap \text{Height}(y) \neq \emptyset\} \\
&= \Delta \cup \{(P1, P4), (P4, P1), (P3, P4), (P4, P3)\}
\end{aligned}$$

Here Δ is the *diagonal* relation of U; that is, $\Delta = \{(x, x) \mid x \in U\}$. Two objects are in the relation $\text{Age}^{-1}(SIM)$ if and only if their ages are possibly the same, for example.

Next we shall extent the notion of preimage relations in a natural way. Let $A(\subseteq Y^U)$ be a nonempty set of functions. The *strong* and *weak preimage relations of* a subset $B(\subseteq A)$ are defined by

$$\begin{aligned}
S_R(B) &= \{(x, y) \in U^2 \mid (\forall f \in B)f(x)Rf(y)\}; \\
W_R(B) &= \{(x, y) \in U^2 \mid (\exists f \in B)f(x)Rf(y)\},
\end{aligned}$$

respectively. The following properties are clear by the definition of strong and weak preimage relations.

[**Lemma 2.**]*If* $B, C \subseteq A$ *and* $f \in A$, *then*
  (a) $S_R(\{f\}) = W_R(\{f\}) = f^{-1}(R)$;
  (b) $S_R(B) = \bigcap\{f^{-1}(R) \mid f \in B\}$ *and* $W_R(B) = \bigcup\{f^{-1}(R) \mid f \in B\}$;
  (c) $S_R(\emptyset) = U \times U$ *and* $W_R(\emptyset) = \emptyset$;
  (d) $S_R(B \cup C) = S_R(B) \cap S_R(C)$ *and* $W_R(B \cup C) = W_R(B) \cup W_R(C)$;
  (e) *If* $B \subseteq C$, *then* $S_R(C) \subseteq S_R(B)$ *and* $W_R(B) \subseteq W_R(C)$;
  (f) *If* $B \neq \emptyset$, *then* $S_R(B) \subseteq W_R(B)$;
  (g) $-S_R(B) = W_{(-R)}(B)$ *and* $-W_R(B) = S_{(-R)}(B)$.   □

The following obvious proposition shows that also strong and weak preimage relations inherit many properties from the original relation.

[**Proposition 1.**]*Let* $\emptyset \neq B \subseteq A$. *If* R *is reflexive, then* $S_R(B)$ *and* $W_R(B)$ *are reflexive, and similar conditions apply when* R *is irreflexive or symmetric. Moreover, if* R *is transitive, then* $S_R(B)$ *is transitive.*   □

Information relations are preimage relations, as we see in the following example.

[*Example 2.*]Assume $\mathcal{S} = (U, A, \{V_a\}_{a \in A})$ is an information system. As before, we set $V = \bigcup_{a \in A} V_a$ and $Y = \wp(V) - \{\emptyset\}$.
  Now we can define the following relations on the set Y.

$$(W_1, W_2) \in IND \iff W_1 = W_2;$$
$$(W_1, W_2) \in SIM \iff W_1 \cap W_2 \neq \emptyset;$$
$$(W_1, W_2) \in FIN \iff W_1 \subseteq W_2;$$
$$(W_1, W_2) \in BIN \iff W_1 \supseteq W_2;$$
$$(W_1, W_2) \in DIV \iff W_1 \neq W_2;$$
$$(W_1, W_2) \in RORT \iff W_1 \subseteq -W_2;$$
$$(W_1, W_2) \in RNIM \iff W_1 \cap -W_2 \neq \emptyset;$$
$$(W_1, W_2) \in LNIM \iff -W_1 \cap W_2 \neq \emptyset.$$

It can be easily seen that $-IND = DIV$, $-SIM = RORT$, $-FIN = RNIM$, and $-BIN = LNIM$.
  For any subset $B(\subseteq A)$ of attributes,

$ind(B) = S_{IND}(B)$	and	$wind(B) = W_{IND}(B)$;
$sim(B) = S_{SIM}(B)$	and	$wsim(B) = W_{SIM}(B)$;
$fin(B) = S_{FIN}(B)$	and	$wfin(B) = W_{FIN}(B)$;
$bin(B) = S_{BIN}(B)$	and	$wbin(B) = W_{BIN}(B)$;
$div(B) = S_{DIV}(B)$	and	$wdiv(B) = W_{DIV}(B)$;
$rort(B) = S_{RORT}(B)$	and	$wrort(B) = W_{RORT}(B)$;
$rnim(B) = S_{RNIM}(B)$	and	$wrnim(B) = W_{RNIM}(B)$;
$lnim(B) = S_{LNIM}(B)$	and	$wlnim(B) = W_{LNIM}(B)$.

Because the relation IND is trivially reflexive, symmetric, and transitive, then by Proposition 1, the relation $ind(B) = S_{IND}(B)$ is reflexive, symmetric, and transitive. Moreover, by Lemma 2(g), $-IND = DIV$ implies that $-ind(B) = -S_{IND}(B) = W_{(-IND)}(B) = W_{DIV}(B) = wdiv(B)$, for example.

Preimage relations allow us to define also different kind of information relations, as we see in the following example.

[*Example 3.*]Suppose $\mathcal{S} = (U, A, \{V_a\}_{a \in A})$ is an information system, which is given by the following table.

	Height (cm)	Weight (kg)
P1	186	80
P2	157	59
P3	172	64
P4	166	52

Let us now consider the usual order relation $>$ on $\mathbb{N}$. The preimage relations of $>$ with respect to the attributes Height and Weight are

$$\text{Height}^{-1}(>) = \{(P1, P2), (P1, P3), (P1, P4), (P3, P2), (P3, P4), (P4, P2)\}$$
$$\text{Weight}^{-1}(>) = \{(P1, P2), (P1, P3), (P1, P4), (P3, P2), (P3, P4), (P2, P4)\}.$$

For all $B \subseteq A$,

$$S_>(B) = \{(x, y) \in U^2 \mid (\forall a \in B) a(x) > a(y)\};$$
$$W_>(B) = \{(x, y) \in U^2 \mid (\exists a \in B) a(x) > a(y)\}.$$

For example,

$$S_>(A) = \{(P1, P2), (P1, P3), (P1, P4), (P3, P2), (P3, P4)\} \text{ and}$$
$$W_>(A) = S_>(A) \cup \{(P2, P4), (P4, P2)\}.$$

4 Dependence Spaces of Preimage Relations

Dependence spaces are algebraic structures which are suitable for the study of reducts of attribute sets, for example.

An equivalence relation Θ on $\wp(A)$ is a *congruence* on the semilattice $(\wp(A), \cup)$ if for all $X_1, X_2, Y_1, Y_2 \subseteq A$, $X_1 \Theta X_2$ and $Y_1 \Theta Y_2$ implies $X_1 \cup Y_1 \Theta X_2 \cup Y_2$. The *congruence class* of a subset $B(\subseteq A)$ is $B/\Theta = \{C \subseteq A \mid B\Theta C\}$.

If A is a finite nonempty set and Θ is a congruence on the semilattice, $(\wp(A), \cup)$ then, by the definition of Novotný and Pawlak, the pair $\mathcal{D} = (A, \Theta)$ is called a *dependence space* (see e.g. [1,2,3,4]).

Assume U and Y are nonempty sets, let $R \in \text{Rel}(Y)$, and let $A(\subseteq Y^U)$ be a finite subset of functions. Let us now define the following two binary relations Θ_R^S and Θ_R^W on the set $\wp(A)$:

$$(B, C) \in \Theta_R^S \iff S_R(B) = S_R(C);$$
$$(B, C) \in \Theta_R^W \iff W_R(B) = W_R(C).$$

So, two subsets of functions are in the relation Θ_R^S (resp. Θ_R^W) iff they define the same strong (resp. weak) preimage relation.

By Lemma 2(d) it is easy to see that the following proposition holds.

[**Proposition 2.**]*The pairs (A, Θ_R^S) and (A, Θ_R^W) are dependence spaces.*   □

By the previous proposition we can now define in an information system $\mathcal{S} = (U, A, \{V_a\}_{a \in A})$ a dependence space with respect to any information relation presented in Section 2. For example, the dependence space defined by strong diversity is $\mathcal{D}_{div} = (A, \Theta_{div})$, where $\Theta_{div} = \{(B, C) \in \wp(A)^2 \mid div(B) = div(C)\}$.

In the theory of information systems the notion of reducts is defined usually only with respect to the strong indiscernibility relations. In such cases a reduct of a set B of attributes is a minimal subset C of B, which defines the same strong indiscernibility relation as B. In here we define reducts in a more abstract setting of dependence spaces.

Let $\mathcal{D} = (A, \Theta)$ be a dependence space and $B \subseteq A$. We say that a subset C ($\subseteq A$) is a *reduct* of B, if $C \subseteq B$ and C is minimal in B/Θ.

In the framework of dependence spaces we can study reducts of subsets of attributes in information systems with respect to any type information relation, as we see in the next example.

[*Example 4.*]Let $\mathcal{S} = (U, A, \{V_a\}_{a \in A})$ be an information system. Let us consider the dependence space $\mathcal{D}_{sim} = (A, \Theta_{SIM}^S)$. Two subsets B and C of attributes are now in the relation Θ_{SIM}^S if and only if they define the same strong similarity relation; that is, $sim(B) = sim(C)$. A reduct of a subset $B(\subseteq A)$ of attributes is a minimal subset C of B, which defines the same strong similarity relation as B.

5   Dense Families and Matrices of Preimage Relations

Dense families of dependence spaces are families of subsets which contain enough information about the structure of dependence spaces. In this section we shall show how we can find reducts of subsets by applying dense families. Moreover, we will study how in dependence spaces defined by preimage relations, dense families can be determined by using matrices of preimage relations.

Suppose A is a set. Then each family $\mathcal{H}$ ($\subseteq \wp(A)$) defines a binary relation $\Gamma(\mathcal{H})$ on $\wp(A)$ as follows.

$$(B, C) \in \Gamma(\mathcal{H}) \text{ iff for all } X \in \mathcal{H}, B \subseteq X \iff C \subseteq X.$$

It is easy to see that $\Gamma(\mathcal{H})$ is a congruence on the semilattice $(\wp(A), \cup)$.

Let $\mathcal{D} = (A, \Theta)$ be a dependence space. We say that a family $\mathcal{H}$ ($\subseteq \wp(A)$) is *dense* in $\mathcal{D}$ if $\Gamma(\mathcal{H}) = \Theta$ [4].

Next we shall show how we can find dense families of dependence spaces which are defined by strong and weak preimage relations.

Assume Y is a nonempty set and let $U = \{x_1, \ldots, x_n\}$ be finite. If R is a binary relation on Y and A ($\subseteq Y^U$) is a finite subset of functions, then the *matrix of preimage relations* of R is an $n \times n$-matrix $\mathbf{M}(R) = (c_{ij})_{n \times n}$ such that

$$c_{ij} = \{f \in A \mid f(x_i) R f(x_j)\}$$

for all $1 \leq i, j \leq n$. Thus, the entry c_{ij} consists of functions $f \in A$ such that x_i and x_j are in the preimage relation $f^{-1}(R)$ (cf., discernibility matrices defined in [7]).

The following lemma is trivial.

[**Lemma 3.**]*If A $(\subseteq Y^U)$ is a finite nonempty set of functions and $\mathbf{M}(R) = (c_{ij})_{n \times n}$ is the matrix of preimage relations of R ($\in \text{Rel}(Y)$), then for all $B \subseteq A$,*
  (a) *$(x_i, x_j) \in S_R(B)$ iff $B \subseteq c_{ij}$;*
  (b) *$(x_i, x_j) \in W_R(B)$ iff $B \cap c_{ij} \neq \emptyset$.*   □

Our next proposition shows how matrices of preimage relations define dense families of dependence spaces.

[**Proposition 3.**]*Assume A $(\subseteq Y^U)$ is a finite nonempty set of functions and $\mathbf{M}(R) = (c_{ij})_{n \times n}$ is the matrix of preimage relations of R ($\in \text{Rel}(Y)$). Then*
  (a) *$\{c_{ij} \mid 1 \leq i, j \leq n\}$ is dense in the dependence space (A, Θ_R^S);*
  (b) *$\{-c_{ij} \mid 1 \leq i, j \leq n\}$ is dense in the dependence space (A, Θ_R^W).*

[*Proof.*](a) Let us denote $\mathcal{H} = \{c_{ij} \mid 1 \leq i, j \leq n\}$. We have to show that $\Gamma(\mathcal{H}) = \Theta_R^S$. If $(B, C) \in \Theta_R^S$, then for all $1 \leq i, j \leq n$, $B \subseteq c_{ij}$ iff $(x_i, x_j) \in S_R(B)$ iff $(x_i, x_j) \in S_R(C)$ iff $C \subseteq c_{ij}$, which implies $(B, C) \in \Gamma(\mathcal{H})$. Hence, $\Theta_R^S \subseteq \Gamma(\mathcal{H})$.

If $(B, C) \in \Gamma(\mathcal{H})$, then for all $1 \leq i, j \leq n$, $(x_i, x_j) \in S_R(B)$ iff $B \subseteq c_{ij}$ iff $C \subseteq c_{ij}$ iff $(x_i, x_j) \in S_R(C)$, which implies $S_R(B) = S_R(C)$. Thus, also $\Gamma(\mathcal{H}) \subseteq \Theta_R^S$ and so $\Gamma(\mathcal{H}) = \Theta_R^S$.

(b) Let us denote $\mathcal{K} = \{-c_{ij} \mid 1 \leq i, j \leq n\}$. If $(B, C) \in \Theta_R^W$, then for all $1 \leq i, j \leq n$, $B \subseteq -c_{ij}$ iff $B \cap c_{ij} = \emptyset$ iff $(x_i, x_j) \notin W_R(B)$ iff $(x_i, x_j) \notin W_R(C)$ iff $C \cap c_{ij} = \emptyset$ iff $C \subseteq -c_{ij}$, which implies $(B, C) \in \Gamma(\mathcal{K})$. Hence, $\Theta_R^W \subseteq \Gamma(\mathcal{K})$.

If $(B, C) \in \Gamma(\mathcal{K})$, then for all $1 \leq i, j \leq n$, $(x_i, x_j) \in W_R(B)$ iff $B \cap c_{ij} \neq \emptyset$ iff $B \not\subseteq -c_{ij}$ iff $C \not\subseteq -c_{ij}$ iff $C \cap c_{ij} \neq \emptyset$ iff $(x_i, x_j) \in W_R(C)$, which implies $W_R(B) = W_R(C)$. So, also $\Gamma(\mathcal{K}) \subseteq \Theta_R^W$ and hence $\Gamma(\mathcal{K}) = \Theta_R^W$.

The next proposition, which can be found in [2], characterizes the reducts of given subset of attributes by the means of dense sets.

[**Proposition 4.**]*Let $\mathcal{D} = (A, \Theta)$ be a dependence space and let $\mathcal{H}$ $(\subseteq \wp(A))$ be dense in $\mathcal{D}$. If $B \subseteq A$, then C is a reduct of B iff C is a minimal set which contains an element from each nonempty differences $B - X$, where $X \in \mathcal{H}$.*   □

[*Example 5.*]Suppose $\mathcal{S} = (U, A, \{V_a\}_{a \in A})$ is an information system, which is described in the following table.

	Age (years)	Weight (kg)	Height (cm)
P1	$\{22, \ldots, 26\}$	$\{48, \ldots, 54\}$	$\{154, \ldots, 157\}$
P2	$\{26, \ldots, 33\}$	$\{73, \ldots, 78\}$	$\{170, \ldots, 175\}$
P3	$\{24, \ldots, 29\}$	$\{51, \ldots, 58\}$	$\{159, \ldots, 162\}$
P4	$\{31, \ldots, 37\}$	$\{75, \ldots, 82\}$	$\{157, \ldots, 165\}$

Let us denote $a = $ Age, $b = $ Weight, and $c = $ Height. If $R = SIM$, then the preimage matrix of R is the following.

	P1	P2	P3	P4
P1	A	$\{a\}$	$\{a,b\}$	$\{c\}$
P2	$\{a\}$	A	$\{a\}$	$\{a,b\}$
P3	$\{a,b\}$	$\{a\}$	A	$\{c\}$
P4	$\{c\}$	$\{a,b\}$	$\{c\}$	A

By Proposition 3, the family $\mathcal{H} = \{\{a\}, \{c\}, \{a,b\}, A\}$ is dense in the dependence space $\mathcal{D}_{sim} = (A, \Theta^S_{SIM})$. Next we determine the reducts of the set A in this dependence space.

The differences $A - X$, where $X \in \mathcal{H}$, are

$$A - \{a\} = \{b,c\}, A - \{c\} = \{a,b\}, A - \{a,b\} = \{c\}, \text{ and } A - A = \emptyset.$$

The first three of them are nonempty, and clearly $\{a,c\}$ and $\{b,c\}$ are minimal sets which contain an element from these differences. Then by Proposition 4, the set A has the reducts $\{a,c\}$ and $\{b,c\}$ in the dependence space $\mathcal{D}_{sim}$. Thus, $\{a,c\}$ and $\{b,c\}$ are minimal sets which define the same strong similarity relation as A.

Similarly, the family $\mathcal{K} = \{-\{a\}, -\{c\}, -\{a,b\}, -A\} = \{\emptyset, \{c\}, \{a,b\}, \{b,c\}\}$ is dense in the dependence space $\mathcal{D}_{wsim} = (A, \Theta^W_{SIM})$. The differences $A - X$, where $X \in \mathcal{K}$, are

$$A - \emptyset = A, A - \{c\} = \{a,b\}, A - \{a,b\} = \{c\}, \text{ and } A - \{b,c\} = \{a\}.$$

They all are nonempty and obviously $\{a,c\}$ is the only minimal set which contains an element from these differences. This means that $\{a,c\}$ is the only reduct of A in the dependence space $\mathcal{D}_{wsim}$. So, $\{a,c\}$ is the unique minimal set which defines the same weak similarity relation as A.

References

1. J. JÄRVINEN, *A Representation of Dependence Spaces and Some Basic Algorithms*, Fundamenta Informaticae 29 (1997), 369–382.
2. J. JÄRVINEN, *Representations of Information Systems and Dependence Spaces, and Some Basic Algorithms*, Licentiate's Thesis, Department of Mathematics, University of Turku, April 1997. Available at http://www.utu.fi/~jjarvine/licence.ps.
3. M. NOVOTNÝ, *Applications of Dependence Spaces*, In E. ORŁOWSKA (ed.), *Incomplete Information: Rough Set Analysis*, Physiga Verlag, 1997.
4. M. NOVOTNÝ, *Dependence Spaces of Information Systems*, In E. ORŁOWSKA (ed.), *Incomplete Information: Rough Set Analysis*, Physiga Verlag, 1997.
5. E. ORŁOWSKA, *Introduction: What You Always Wanted to Know About Rough Sets*, In E. ORŁOWSKA (ed.), *Incomplete Information: Rough Set Analysis*, Physiga Verlag, 1997.
6. E. ORŁOWSKA, *Studying Incompleteness of Information: A Class of Information Logics*, In K. KIJANIA–PLACEK, J. WOLEŃSKI (eds.), *The Lvov–Warsaw School and Contemporary Philosophy*, Kluwer Academic Publishers, 1997.
7. A. SKOWRON, C. RAUSZER, *The discernibility matrices and functions in information systems*. In R. SLOWINSKI (ed.), *Intelligent decision support, Handbook of applications and advances of the rough set theory*, Kluwer Academic Publisher, 1991.

Cellular Neural Networks for Navigation of a Mobile Robot

Barbara Siemiątkowska and Artur Dubrawski

Institute of Fundamental Technological Research, Polish Academy of Sciences
21 Świętokrzyska Str., 00-049 Warszawa, Poland

Abstract. This paper summarizes applications of cellular neural networks to autonomous mobile robot navigation tasks, which have been developed at the Institute of Fundamental Technological Research. They include map building, path planning and self-positioning of an indoor robot equipped with a laser range sensor. Efficiency of navigation based on cellular neural networks has been experimentally verified in a variety of natural, partially structured environments.

1   Introduction

Autonomous navigation of a mobile robot deals with the following fundamental tasks: identifying a free space and the objects which compose the robot's surroundings, determining the robot's location within its operating environment, planning a suitable path that leads from the robot's current position to some desired location, and monitoring a safe travel along the planned trajectory. During the past decade or so there has been a vast amount of research conducted in the field of mobile robotics. As a result, an according portion of published material illustrates a variety of approaches taken to solve the fundamental problems of autonomous navigation. In this paper we present our attempts to provide world modeling, path planning and self-localization capabilities to an autonomous indoor mobile robot, by using a methodology of cellular neural networks.

In our experiments we use the RWI Pioneer-1 vehicle equipped with the HelpMate LightRanger scanning laser range finder (see Fig. 1). The laser device provides planar scans covering 330^0 of angular width at a rate of 660 distance measurements per second. The power level of the laser beam enables measuring distances up to 10m with an accuracy of 3cm.

Subsequent laser scans are being aggregated into a grid-based map of the robot's environment. The map may be built from scratch, using the series of immediate range measurements only, or it may be an effect of combining the current readouts with an a priori model of the workspace. In any case, the map is represented as a two-dimensional, rectangular grid of square cells. Each cell may be *free*, *occupied* by an object (perhaps an obstacle) or its status may be *unknown*. The actual status of a given cell depends on the set of current readings provided by the range sensor and on the prior state. The status update rules are based on Bayesian approach to data aggregation.

Path planning module is realized in a form of a two layer cellular network, which implements a diffusion (also known as a wave-front) algorithm. The first

L. Polkowski and A. Skowron (Eds.): RSCTC'98, LNAI 1424, pp. 147–154, 1998.
© Springer-Verlag Berlin Heidelberg 1998

layer neurons are first exposed to the external signals provided by the map cells, which carry the occupancy information. Also, the neuron which corresponds to the goal location is purposively biased. Then the neuron attached to the current robot's position is activated and the signal wave propagates through the network from there until it reaches the goal neuron. The steady state activation levels are then propagated to the second layer of the network, which performs a downhill search for the shortest path linking the goal and the robot. The planned path is then represented as a sequence of the map cells to follow.

Fig. 1. Mobile robot RWI Pioneer-1 equipped with HelpMate LightRanger scanning laser range finder.

Relevance of the planned path depends very much on the accuracy and relevance of the map. All the process can be distorted very seriously, if the actual location of the robot at the scan collection points were substantially different than their estimates. Some way of providing accurate fixes of the current position, and most importantly the current orientation of the robot, is a neccesary prerequisite of efficient autonomous navigation. Our approach to orientation and position tracking is well suited to the range-based navigation, and it is implemented in a form of a layered, circular cellular neural network.

2   Cellular neural networks

The concept of cellular neural networks was introduced by O. Chua and L. Yang in 1988 [1], and since then these structures have found many applications, especially in signal and image processing [2,3]. A generic cellular neural network is an n-dimensional array of identical processing units. All interactions between the cells are local within a finite radius, and interconnection effect is represented by the equations (a 2-D case):

$$\frac{dx_{ij}}{dt} = -x_{ij} + \sum_{k=-r}^{r} \sum_{l=-r}^{r} a_{ij}^{i+k,j+l} y_{i+k,j+l} + \sum_{k=-r}^{r} \sum_{l=-r}^{r} b_{ij}^{i+k,j+l} u_{i+k,j+l} + I \quad (1)$$

$$y_{ij} = f(x_{ij}) \quad (2)$$

where x_{ij} is a state, y_{ij} is an output and u_{ij} is an input of a cell (ij), r is a neighborhood range, I is a scalar constant, $a_{ij}^{i+k,j+l}$ and $b_{ij}^{i+k,j+l}$ represent weights of the interconnection between the cell (ij) and its (kl)-th neighbor, and $f(\cdot)$ is some nonlinear, bounded function. Note, that because of the propagation effect the cells which are not directly connected can still affect each other.

Usually values of $a_{ij}^{i+k,j+l}$ and $b_{ij}^{i+k,j+l}$ are memorized in matrices A and B, and they are symmetrical (ie. $a_{ij}^{i+k,j+l} = a_{i+k,j+l}^{ij}$ and $b_{ij}^{i+k,j+l} = b_{i+k,j+l}^{ij}$. The triple (A,B,I) is called a cloning template. Most networks are spatially invariant, ie. all cells in one network have the same cloning template.

3   Map building and path planning

A grid-based map of the robot's workspace aggregates a prior knowledge about the location of the objects and geometry of the traversable area, with a posterior information in the form of range measurements provided by the on-board sensors. The aggregation formula follows Bayes rule:

$$p_{ij} = P(o \mid ij) = \frac{P(ij \mid o) \cdot P_{ij}(o)}{P(ij \mid o) \cdot P_{ij}(o) + [1 - P(ij \mid o)] \cdot [1 - P_{ij}(o)]} \quad (3)$$

where p_{ij} is the aggregate probability that the map cell (ij) is occupied, $P(o \mid ij)$ is the posterior probability of the occupancy given the location (ij), $P(ij \mid o)$ is the prior (possibly a result of a previous aggregation), and $P_{ij}(o)$ is the immediate information provided by the sensor. We assume that all the cells located on the way of a laser beam measuring some distance r are free for almost sure $(P_{ij}(o) = 0.1)$, the cell corresponding to r is almost surely occupied $(P_{ij}(o) = 0.9)$, and the cells located behind that one have their immediate status unknown and are not subject to update.

For path planning purposes we use a two layer cellular neural network. The geometry of both layers is identical and it corresponds to the geometry of the map. Each cell of the map (or, more practically, of a cut-off of the map) has its counterpart in the first layer of the planning structure, and there is one corresponding unit in the second layer. The interlayer connections of the planning network have one-to-one topology. In both layers the neighborhood range equals one, ie. each neuron is interconnected only with its immediate neighbors of the same layer.

Each unit of the map sends a binary signal s_{ij} to the corresponding neuron of the first layer of the planning network. A basic scheme of computing the values of s_{ij} would set it to one if the cell (ij) is *free* or *unknown* and to zero otherwise. But, the map raster dimension is taken smaller than the robot's size, so the sensed objects (the obstacles) dimensions are expanded a little bit to secure a margin for a safe travel. The following obstacle expansion function is then introduced:

$$s_{ij} = \begin{cases} 0 \ if \ \sum_{kl \in N_{ij}} h(p_{kl}) < d \\ 1 \ if \ \sum_{kl \in N_{ij}} h(p_{kl}) \geq d \end{cases} \qquad (4)$$

where $h(\cdot)$ is a threshold function (in our case $h(x) = 0$ if $x \leq 0.4$ and $h(x) = 1$ otherwise), d is a half of the largest dimension of the robot (in pixels) and N_{ij} is a set of neurons belonging to the neighborhood of the cell (ij). In this way, the input to the planning network unit (ij) equals $s_{ij} = 1$ if the map cell (ij) and all of its neighbors are unoccupied, and $s_{ij} = 0$ otherwise.

$$A = \begin{array}{|c|c|c|} \hline \beta & \alpha & \beta \\ \hline \alpha & 1 & \alpha \\ \hline \beta & \alpha & \beta \\ \hline \end{array} \quad B = \begin{array}{|c|c|c|} \hline 1 & 1 & 1 \\ \hline 1 & 1 & 1 \\ \hline 1 & 1 & 1 \\ \hline \end{array} \quad I = 0 \quad A = \begin{array}{|c|c|c|} \hline 1 & 1 & 1 \\ \hline 1 & 1 & 1 \\ \hline 1 & 1 & 1 \\ \hline \end{array} \quad B = \begin{array}{|c|c|c|} \hline 0 & 0 & 0 \\ \hline 0 & 1 & 0 \\ \hline 0 & 0 & 0 \\ \hline \end{array} \quad I = 0$$

Fig. 2. Planner's cloning templates: the first (left) and the second layer (right).

The cloning templates of the first planning layer are depicted in the left part of Fig. 2. The values of α and β are smaller than one and $\alpha > \beta$. The initial states of the cells are all set to $x_{ij}(0) = 0$, except for the cell corresponding to the goal location, for which $x_{ij}^{(goal)}(0) = F$, where $F >> 1$. The interaction between the neighboring cells is described by the following discrete-time equations:

$$x_{ij}(t+1) = S_{ij} \cdot \max_{kl \in N_r^{ij}} (a_{ij}^{kl} \cdot y_{kl}(t)) \sum_{kl \in N_r^{ij}} b_{ij}^{kl} \cdot u_{kl}(t) \qquad (5)$$

$$y_{ij}(t+1) = x_{ij}(t) \qquad (6)$$

The above formulas induce the following properties:

1. $\forall_{(ij \ and \ t)} \ x_{ij}(t) \leq F$
2. If the cell (ij) is occupied by an obstacle, then $x_{ij}(t) = 0$.
3. If $u_{ij}(t)$=const, then $x_{ij}(t) \leq x_{ij}(t+1)$
4. The network is stable, ie. $\forall_{t \geq \bar{t}} \ |x_{ij}(t) - x_{ij}(t+1)| \leq \epsilon$

The signals propagate until the first layer of the planning network reaches a stable state:

$$\forall_{ij} \ x_{ij}(t) = x_{ij}(t+1) \qquad (7)$$

The topology of the second layer is analogical to the first layer. Respective neurons of the first and the second layer are connected one-to-one. Values of the respective interlayer convection weights are set to one, so the initial activations are equal to the values of the units' input signals: $x_{ij}^{(2)}(t) = u_{ij}^{(2)}(t)$, where $x_{ij}^{(2)}(t)$ is the second layer neuron activation at the moment t, and $u_{ij}^{(2)}(t)$ the value of its input signal. Cells of the second layer of the planner have cloning templates shown in the right part of Fig. 2.

When the second layer neuron which corresponds the robot's location on the grid-based map receives a signal form its counterpart in the first layer the path generation process begins. The path starts at the robot's position, the

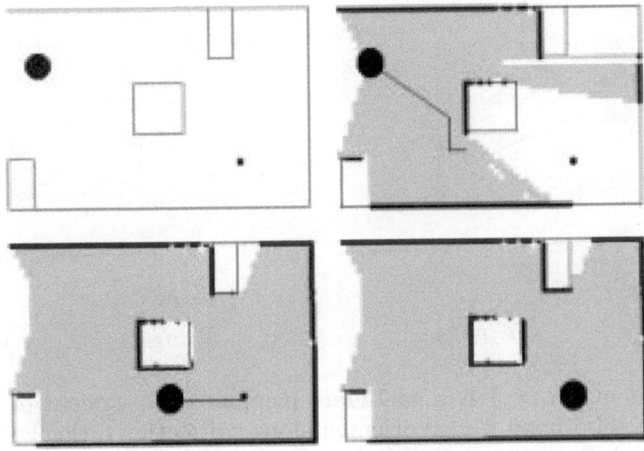

Fig. 3. Stages of path planning.

next position is indicated by the neighboring neuron which has the smallest activation, and so on. The process of path generation continues until the goal position is reached or until none of the neurons change their activation. The second condition means that there is no free path to the goal.

The experimental verification of the feasibility of the presented map building and path planning method revealed a few of its practical advantages. Unlike many other techniques of planning trajectories for indoor mobile robots it does not suffer from local minima problems. In particular dead-end corridors are easily recognized and avoided. Also the "no way" case, in which the goal is completely surrounded by the obstacles does not lead to an oscillatory behavior of the robot. Moreover, the presented method may be mapped into a parallel hardware, which would provide processing frequencies sufficient for reactive navigation. Figure 3 presents the subsequent stages of map building and path planning. The robot is depicted as a big black dot, and the goal location is shown as a small dot. First, the map is built (or updated) using a recent set of laser range finder indications. The gray pixels represent the map cells corresponding to the free space, black ones represent parts of the obstacles, the status of the white cells is *unknown*. First picture presents the initial situation, then the map of the environment is built, the path is planned, the robot is moving along the path, while it is continuously updating the map. When a new obstacle is detected at a colliding location, the path may be replanned on the fly.

4   Self-localization

A scanning range sensor collects *scans*, ie. sets of m readouts $\{R_i, \varphi_i\}$, where R_i represents an observed distance to an object placed in the way of the laser beam in the direction determined by a scanning angle φ_i. The scanning angle takes m discrete values ranging from 0° to 360° (in general; in our case the working range of the scanning angle is limited by the hardware to 330°), so that $i = 1, \ldots, m$ and $\Delta\varphi = \varphi_{i+1} - \varphi_i = const$, and the indices i are additive modulo m.

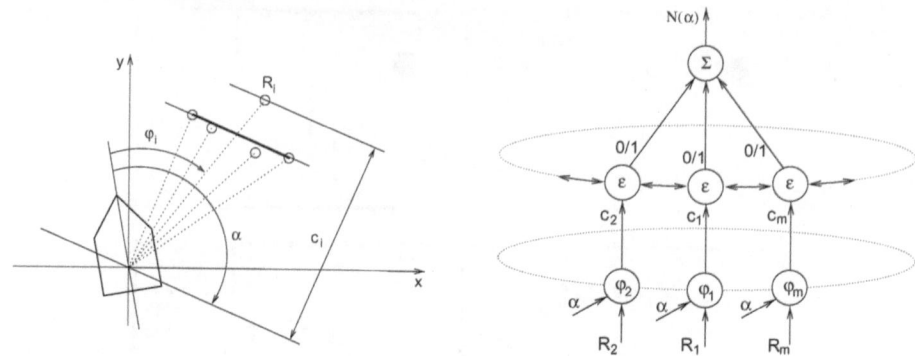

Fig. 4. Readout $\{R_i, \varphi_i\}$ is a candidate member of a segment of five collinear readouts (neighborhood $n = 2$) oriented along a direction α (left). Cyclic cellular neural network in a setup suitable for processing laser range scans in order to obtain segment orientation histograms (right).

We may write a normal equation of a line in the plane (x,y) oriented along the direction α as follows: $x \cdot sin\alpha - y \cdot cos\alpha + c = 0$, where $|c|$ determines the distance between the line and the origin. By taking into account that $x_i = R_i \cdot \cos\varphi_i$ and $y_i = R_i \cdot \sin\varphi_i$ we may determine the distance c_i for the line which crosses a given point (x_i, y_i):

$$c_i = R_i \cdot sin(\alpha - \varphi_i) \tag{8}$$

Two points $\{R_i, \varphi_i\}$ and $\{R_j, \varphi_j\}$ belong to the same line oriented along α if $c_i = c_j$. This is the essential idea for testing colinearity of the neighboring range readouts in the presented method. We may say that the readout $\{R_i, \varphi_i\}$ belongs to a line segment oriented along a specified direction α (see left part in Fig. 4) if it and its $2n$ neighbors fulfill the following condition:

$$\mathop{\forall}_{i-n \leq k \leq i+n} \; |c_k - \frac{1}{2n+1} \sum_{t=i-n}^{i+n} c_t| \leq \epsilon \tag{9}$$

So, for the colinearity check purpose, it is sufficient to compare the neighboring c-values computed for a particular α using the set of current readouts $\{R_i, \varphi_i\}$.

In the neural implementation of the described stage of processing (right part of Fig. 4), the first ring-shaped input layer is composed of processing units which model the equation (8). Each of the m units receives a separate input signal R_i and a common input α. The φ_i's are hardware dependent and may be assigned constant values, one for each of the input layer's units. The second layer is composed of m laterally interconnected neurons which receive signals proportional to c_i's computed by the units of the input layer. The neurons take into account signals received from their neighbors (of course in general the range of lateral connections may be larger than $n = 1$) and check the condition (9). A second layer neuron returns 1 if its colinearity condition is met and 0 otherwise. Consecutive 1's in the ring-shaped output vector from the second layer of processing units represent wall segments oriented along the specified α with the accuracy specified by ϵ. Zeros represent the components of the given scan, which do not seem to be the members of line segments parallel to the direction α.

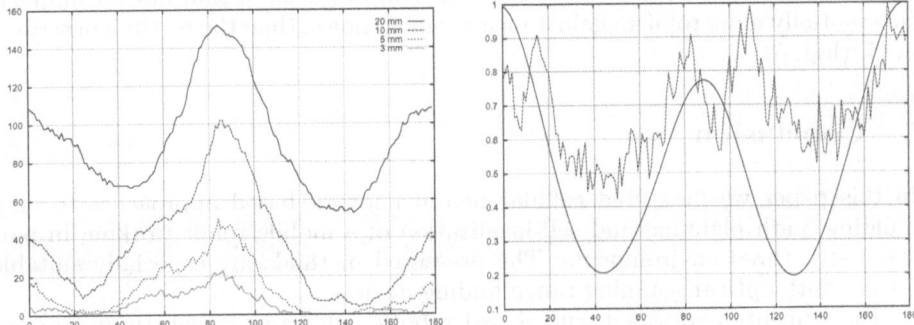

Fig. 5. Left: histograms $N(\alpha)$ with $n = 1$ and ϵ set to subsequently 20, 10, 5 and 3 millimeters. Right: normalized crosscorrelations of a pair of histograms computed for the two consecutive scans taken enroute; solid line is a result obtained using the method described in this paper with $n = 2$ and $\epsilon = 10mm$, dashed line was obtained with *angle histograms* implemented after [5] with $n = 7$.

We compute sets of c values for all the range of possible line segments orientations, ie. from 0 to +180 degrees in the robot's coordinates. For each α from that range we sum up 1's present on the outputs of the second layer of neurons, and then we plot these sums, denoted $N(\alpha)$, against α's. In such a way we obtain a sort of angle histogram, which actually represents the cummulative dimensions of scene segments oriented in particular directions as seen by the sensor. The histograms, shown in Fig. 5 for $n = 1$ and several values of ϵ, position their extrema at the locations of the most and the least predominant directions of the scene segments. In a typical structured or partially structured indoor environment there are usually two maxima, which correspond to (most often) roughly perpendicular directions.

If we compared two histograms computed for the scans taken at nearby locations, we could expect them to reveal changes in position and orientation of the viewpoint in respect to the elements of the scene. The difference in α's which correspond to the maxima of the histograms would be roughly equal to the robot's orientation change if the scans were being collected sufficiently often. But apparently, computing a maximum of a normalized cross-correlation of the two histograms is a much more robust way of tracking orientation changes (location of a maximum of the crosscorrelation curve indicates the orientation change of -2° for a sample pair of histograms, as shown in the right part of Fig. 5). It is possible to track changes of the robot's position in the same framework. Due to the space limits we have to skip a discussion on that topic here, but an interested reader will find more in [4].

In [5] there is presented a technique of tracking orientation and position called *angle histograms*. In that method the histograms reflect local gradients of orientation of the lines crossing the subsequent components of the scans. Our method reflects the cumulative lengths of colinear segments, which are present in the scan, instead. The experiments conducted in a realistic indoor surroundings, composed of the wall segments made of various materials, painted in various colors and of various surface textures and geometries, show that our technique is substantially more robust against measurement noise, than the method described in [5] (Fig. 5).

5   Conclusion

In this paper we presented cellular neural network based approaches to map building, path planning and self-localization of a mobile robot working in partially structured environments. The presented methods are especially suitable for use with optical scanning range finding devices.

Experiments performed with a real robotic vehicle in a realistic indoor environment revealed computational efficiency of the discussed methods. The presented self-localization technique may serve as an add-on to the existing autonomous navigation systems. It can provide high accuracy and efficiency of on-line map building and navigation in known or unknown, partially structured environments. Path planning method has some important advantages too. In case of parallel realization it would provide processing frequencies in a range typical to reactive the control schemes. Moreover it does not suffer from local minima, and it does not fail when the goal is unreachable.

References

1. Chua O., Yang L. Cellular Neural Networks: Theory, *IEEE Transactions on Circuits and Systems*, 35, 1257–1272, 1988.
2. Chua O., Yang L. Cellular Neural Networks: Applications, *IEEE Transactions on Circuits and Systems*, 35, 1273–1290, 1988.
3. Chua O., Roska T. The CNN Paradigm, *IEEE Transactions on Circuits and Systems*, 40, 147–156, 1993.
4. Dubrawski A., Siemiątkowska B. A Neural Method for Self-Localization of a Mobile Robot Equipped with a 2-D Scanning Laser Range Finder, *5th International Symposium on Intelligent Robotic Systems SIRS'97*, 159–168, Stockholm, Sweden, July, 1997.
5. Weiss G., Wetzler C., von Puttkamer E. Keeping Track of Position and Orientation of Moving Indoor Systems by Correlation of Range-Finder Scans, *IEEE/RSJ International Conference on Intelligent Robots and Systems IROS'94*, 595–601, Munich, Germany, September, 1994.

The Takagi-Sugeno Fuzzy Model Identification Method of Parameter Varying Systems*

Xie Keming**   T. Y. Lin***   and   Zhang Jianwei**

** Department of Automation, Taiyuan University of Technology,
Taiyuan, 030024, P. R. China
*** Department of Mathematics and Computer Science, San Jose State University,
San Jose, CA 95192, U. S. A.

Abstract - **This paper presents the TS model identification method by which a great number of systems whose parameters vary dramatically with working states can be identified via Fuzzy Neural Networks (FNN). The suggested method could overcome the drawbacks of traditional linear system identification methods which are only effective under certain narrow working states and provide global dynamic description based on which further control of such systems may be carried out. Simulation results of a second-order parameter varying system demonstrate the effectiveness of the method.**

Keywords - **Parameter Varying Systems, TS Fuzzy Model, Fuzzy Neural Networks (FNN), Identification.**

1 Introduction

Controlled systems whose parameters vary dramatically with working states, namely parameter varying systems, are widely encountered in practical industrial situations. Although traditional linear system identification methods have been well established in the last twenty years, it can only be used under a certain narrow range of working conditions. Moreover, traditional controllers based on such models can not cope with the changes on process dynamic effectively. Therefore, developing global dynamic model and establishing the corresponding control schemes for the parameter varying systems are desrirable.

Takagi and Sugeno [2,3,4] proposed a new type of fuzzy model (TS model) which has been widely used. The model provides succinct description of complex systems, and is convenient for designing controllers. Recently, the authors [5] suggested an identification method of TS fuzzy model for nonlinear systems via Fuzzy Neural Networks (FNN). It is very effective in describing systems. In this paper, the TS fuzzy model is generalized to the parameter varying systems, and an identification method based on FNN is presented. Simulation results of a second-order system illustrate and verify the effectiveness of the method.

L. Polkowski and A. Skowron (Eds.): RSCTC'98, LNAI 1424, pp. 155-162, 1998

2. TS Fuzzy Model

Parameter varying systems which possess m working state characteristic variables, q inputs and single output can be described by the TS fuzzy model consisting of R rules where the i-th rule can be represented as:

Rule i: if z_1 is A_1^{i,k_1}, z_2 is A_2^{i,k_2}, $\cdots$, and z_m is A_m^{i,k_m}

$$\text{then } y^i = a_1^i x_1 + a_2^i x_2 + \cdots + a_q^i x_q \tag{1}$$

$$i = 1,2,\cdots,R. \ k_j = 1,2,\cdots,r_j.$$

where R is the number of rules in the TS fuzzy model. $z_j \ (j=1,2,\cdots,m)$ is the j-th characteristic variable on the working state of the systems and can be selected as input, output or other parameters of the system. $x_l \ (l=1,2,\cdots,q)$ is the l-th model input. y^i is the output of the i-th rule. For the i-th rule, A_j^{i,k_j} is the k_j-th fuzzy subset of the domain of z_j. a_l^i is the coefficient of the consequent. r_j is the fuzzy partition number of the domain of z_j. This rules says if the variables z_j of the working state stay in the domains A_j^{i,k_j} as described, then the output y^i and input $x_l \ (l=1,2,\cdots,q)$ are related by (1). For simplicity, we write $r_j = r$. r is determined by the complexity and accuracy of the model. Once a set of a set of working state variables $(z_{10},z_{20},\cdots,z_{m0})$ and model input variables $(x_{10},x_{20},\cdots,x_{q0})$ are available, then the output of TS model can be calculated by the weighted-average of each y^i:

$$y = \sum_{i=1}^R \mu^i y^i \Big/ \sum_{i=1}^R \mu^i \tag{2}$$

where y^i is determined by consequent equation of the i-th rule. The truth-value μ^i of the i-th rule can be calculated as [1, pp. 382]:

$$\mu^i = \min_{j=1}^m A_j^{i,k_j}\left(z_{j0}\right) \tag{3}$$

Equation (2) can be rewritten as:

$$y = \left(\sum_{i=1}^R \mu^i a_1^i x_1 + \cdots + \sum_{i=1}^R \mu^i a_q^i x_q\right) \Big/ \sum_{i=1}^R \mu^i \tag{4}$$

From (4), one can see that the TS fuzzy model can be expressed as an ordinary linear equation. As μ^i varies with working state, TS fuzzy model is a coefficient-varying linear equation. For all possible varying ranges of working states, the TS fuzzy model reflects the relationships between model parameters and working states. The global dynamic characteristics of the parameter varying systems is represented.

3. Fuzzy Neural Networks TS Fuzzy Model Identification Method

A. Structure of the FNN

According to (1~3), the structure of FNN presented here consists of premise, consequent and fuzzy inference. For systems which posses m working state characteristic variables, q inputs and single output, the FNN used for the TS model identification is shown in Fig. 1. The circles and the squares in the figure represent the units of the networks. The notations between the units denote the connection weights. The units without any notation just deliver the signals from input to output.

1) Normalization of the working state variables

Layers (A)~(B) of the FNN are used to normalize the working state variables in case of saturation of the premise nodes. Assuming P samples $\left(z_1^p, z_2^p, \cdots, z_m^p\right)$ $(p = 1,2,\cdots,P)$ are available for training the networks, the j-th working state variable of the p-th sample can be normalized as:

$$\bar{z}_j^p = \left(w_s\right)_j \left(z_j^p - \left(w_t\right)_j\right) \tag{5}$$

where $\bar{z}_j^p$ is the normalized working state variable of $\bar{z}_j^p$; $\left(w_s\right)_j$ and $\left(w_t\right)_j$ are the coefficients and biases of normalization respectively:

$$\left(w_s\right)_j = \frac{2}{\max\left(z_j^p\right) - \min\left(z_j^p\right)}$$

$$\left(w_t\right)_j = \frac{\max\left(z_j^p\right) + \min\left(z_j^p\right)}{2}$$

$$j = 1,2,\cdots,m. \ p = 1,2,\cdots,P$$

2) Premise

The premise parts of the FNN include Layers (C)~(F) which are used for fuzzy partition and truth-value calculations. Signature 'Σ' in layer (D) ,which is the sum node, realizes the following operations for the k-th fuzzy subset of $\bar{z}_j$:

$$\Sigma: \begin{cases} I_{j,k}^{(D)} = \left(w_g\right)_{j,k} \bar{z}_j - \left(w_c\right)_{j,k} \\ O_{j,k}^{(D)} = I_{j,k}^{(D)} \end{cases} \tag{6}$$

Signature 'Λ' in layer (F) is the fuzzy minimum node and the input-output relationships for the i-th rule can be written as:

$$\Lambda: \begin{cases} I_i^{(F)} = \min_{j=1 \text{ and } k=\phi(i,j)}^{m} O_{j,k}^{(E)} \\ O_i^{(F)} = I_i^{(F)} \end{cases} \tag{7}$$

$$\phi(i,j) = \mathrm{int}\left(\frac{i - \sum_{l=0}^{j-1}\left[(\phi(i,l)-1)r^{(m-l)}\right] - 1}{r^{(m-j)}} \right) + 1 \tag{8}$$

$\phi(i,0) = 1.$ $i = 1,2,\cdots,R.$ $j = 1,2,\cdots,m.$ $k = 1,2,\cdots,r$

where $I_{j,k}^{(\cdot)}$ and $O_{j,k}^{(\cdot)}$ are input and output of the nodes which correspond to the k-th fuzzy subset of $\bar{z}_j$ in layer (·) respectively; $I_i^{(\cdot)}$ and $O_i^{(\cdot)}$ are input and output of the nodes which correspond to the i-th rule in layer (·) respectively; the central point and gradient of the k-th fuzzy subset for $\bar{z}_j$ are determined by both $\left(w_g\right)_{j,k}$ and $\left(w_c\right)_{j,k}$; $\phi(i,j)$ represents the connective relationship between the i-th rule and the k-th fuzzy subset of $\bar{z}_j$. The membership functions of the working state variables are determined by activation functions of the nodes in layer (E). In this paper, the following activation functions are taken:

$$f_k(x) = \begin{cases} e^{-x}/\left(1+e^{-x}\right) & , k = 1 \\ e^{-x^2} & , k = 2,3,\cdots,r-1 \\ 1/\left(1+e^{-x}\right) & , k = r \end{cases} \tag{9}$$

which realize fuzzy partition as shown in Fig. 2.

3) Consequent and fuzzy inference

Layers (G)~(J), which are used to implement the linear equations of the TS fuzzy model, are consequent parts of the FNN. As for the i-th rule of the consequent, input-output relation realized can be written as:

$$O_i^{(J)} = \sum_{j=1}^{q}\left(w_a\right)_{j,i} x_j \tag{10}$$

where $\left(w_a\right)_{j,i}$ is the coefficient of x_i in rule i .

Layers (K)~(M) realize the fuzzy inference as shown in (2).

B. Learning algorithm

Parameters $\left(w_g\right)_{j,k}$ and $\left(w_a\right)_{j,k}$, which determine central points and gradients of the membership functions in the premise, and $\left(w_a\right)_{j,i}$, which determines local linear relationships of the consequent are need to be learnt by the FNN. Assuming P samples $\left(z_1^p, z_2^p, \cdots, z_m^p, x_1^p, x_2^p, \cdots, x_q^p\right)$ $(p = 1, 2, \cdots, P)$ are available for training the FNN and the corresponding teacher signal is t^p. Once the p-th sample is put on the networks, the actual output y^p of the networks can be obtained. Thus, the learning error function of the sample can be defined as:

$$E^p = \frac{1}{2}\left(t^p - y^p\right)^2 \tag{11}$$

Under this definition, the total error function of all the samples can be written as:

$$E = \sum_{p=1}^{P} E^p = \frac{1}{2}\sum_{p=1}^{P}\left(t^p - y^p\right)^2 \tag{12}$$

According to the Gradient-Descent learning algorithm, one can obtain:

$$\frac{\partial E}{\partial\left(w_a\right)_{l,i}} = \sum_{P=1}^{P}\left\{\frac{\partial E^p}{\partial y^p}\cdot\frac{\partial y^p}{\partial O_l^{(J)p}}\cdot\frac{\partial O_l^{(J)p}}{\partial\left(w_a\right)_{l,i}}\right\}$$

$$= \sum_{p=1}^{P}\left\{\left(t^p - y^p\right)\cdot\left(O_l^{(J)p}\middle/\sum_{k=1}^{R} O_k^{(F)p}\right)\cdot x_i^p\right\} \tag{13}$$

$$\frac{\partial E}{\partial\left(w_g\right)_{j,k}} = \sum_{p=1}^{P}\left\{\frac{\partial E}{\partial y^p}\cdot\frac{\partial y^p}{\partial\left(w_g\right)_{j,k}}\right\}$$

$$= \sum_{p=1}^{P}\left\{\frac{\partial E}{\partial y^p}\cdot\sum_{i=1 \text{ and } k=\phi(i,j)}^{R}\left[\frac{\partial y^p}{\partial O_i^{(F)p}}\cdot\frac{\partial O_i^{(F)p}}{\partial O_{j,k}^{(E)p}}\right]\cdot\frac{\partial O_{j,k}^{(E)p}}{\partial I_{j,k}^{(E)p}}\cdot\frac{\partial I_{j,k}^{(E)p}}{\partial\left(w_g\right)_{j,k}}\right\} \tag{14}$$

$$= -\sum_{p=1}^{P}\left\{\left(t^p - y^p\right)\cdot\sum_{i=1 \text{ and } k=\phi(i,j)}^{R}\left[\left(O_i^{(F)p} - y^p\right)\middle/\sum_{l=1}^{R} O_l^{(F)p}\cdot\frac{\partial O_i^{(F)p}}{\partial O_{j,k}^{(E)p}}\right]\right.$$

$$\left.\cdot\frac{\partial O_{j,k}^{(E)p}}{\partial I_{j,k}^{(E)p}}\cdot \bar{z}_j^p\right\}$$

In order to solve $\partial O_i^{(F)p}/\partial O_{j,k}^{(E)p}$ in (14), the following equivalent transition for (7) is needed:

$$O_i^{(F)} = \min_{j=1 \text{ and } k=\phi(i,j)}^{m} O_{j,k}^{(E)} = \sum_{j=1 \text{ and } k=\phi(i,j)}^{m} \left[O_{j,k}^{(E)} \cdot \prod_{l=1 \text{ and } l \neq j}^{m} \mathrm{I}\left(O_{j,k}^{(E)} - O_{l,k}^{(E)}\right) \right] \quad (15)$$

where:

$$\mathrm{I}\left(O_{j,k}^{(E)} - O_{l,k}^{(E)}\right) = \begin{cases} 1 & , \; O_{j,k}^{(E)} \leq O_{l,k}^{(E)} \\ 0 & , \; O_{j,k}^{(E)} > O_{l,k}^{(E)} \end{cases} \quad (16)$$

Therefore, $\partial O_i^{(F)p}/\partial O_{j,k}^{(E)p}$ can be calculated by

$$\frac{\partial O_i^{(F)p}}{\partial O_{j,k}^{(E)p}} = \prod_{l=1 \text{ and } l \neq j}^{m} \mathrm{I}\left(\partial O_{j,k}^{(E)p} - \partial O_{l,k}^{(E)p}\right) = \begin{cases} 1 & , \partial O_{j,k}^{(E)p} \leq \partial O_{l,k}^{(E)p} \\ 0 & , \partial O_{j,k}^{(E)p} > \partial O_{l,k}^{(E)p} \end{cases} \quad (17)$$

Moreover, $\partial O_{j,k}^{(E)p}/\partial I_{j,k}^{(E)p}$ can be obtained from (10) as follows:

$$\frac{\partial O_{j,k}^{(E)p}}{\partial I_{j,k}^{(E)p}} = \begin{cases} O_{j,1}^{(E)p} \cdot \left(O_{j,1}^{(E)p} - 1\right) & , \; k=1 \\ -2 I_{j,k}^{(E)p} \cdot O_{j,k}^{(E)p} & , \; k=2,3,\cdots,r-1 \\ O_{j,r}^{(E)p} \cdot \left(1 - O_{j,r}^{(E)p}\right) & , \; k=r \end{cases} \quad (18)$$

From (14), (17) and (18), $\partial E / \partial \left(w_g\right)_{j,k}$ can be obtained.

Using the same method mentioned above, $\partial E / \partial \left(w_c\right)_{j,k}$ can also be represented by

$$\begin{aligned}
\frac{\partial E}{\partial (w_c)_{j,k}} &= \sum_{p=1}^{P} \left\{ \frac{\partial E^p}{\partial y^p} \cdot \frac{\partial y^p}{\partial (w_c)_{j,k}} \right\} \\
&= \sum_{p=1}^{P} \left\{ \frac{\partial E^p}{\partial y^p} \cdot \sum_{i=1 \text{ and } k=\phi(i,j)}^{R} \left[\frac{\partial y^p}{\partial O_i^{(F)p}} \cdot \frac{\partial O_i^{(F)p}}{\partial O_{j,k}^{(E)p}} \right] \cdot \frac{\partial O_{j,k}^{(E)p}}{\partial I_{j,k}^{(E)p}} \cdot \frac{\partial I_{j,k}^{(E)p}}{\partial (w_c)_{j,k}} \right\} \\
&= -\sum_{p=1}^{P} \left\{ \left(t^p - y^p\right) \cdot \sum_{i=1 \text{ and } k=\phi(i,j)}^{R} \left[\left(O_i^{(F)p} - y^p\right) \middle/ \sum_{l=1}^{R} O_i^{(F)p} \cdot \frac{\partial O_i^{(F)p}}{\partial O_{j,k}^{(E)p}} \right] \right. \\
&\quad \left. \cdot \frac{\partial O_{j,k}^{(E)p}}{\partial I_{j,k}^{(E)p}} \cdot (-1) \right\}
\end{aligned} \quad (19)$$

Therefore, the final tuning equations of the premise and consequent parameters of the FNN can be written as:

$$\left(w_a\right)_{j,i}(n+1) = \left(w_a\right)_{j,i}(n) - \zeta \cdot \partial E / \partial \left(w_a\right)_{j,i} \quad (20)$$

$$\left(w_g\right)_{j,k}(n+1) = \left(w_g\right)_{j,k}(n) - \xi \cdot \partial E / \partial\left(w_g\right)_{j,k} \qquad (21)$$

$$\left(w_c\right)_{j,k}(n+1) = \left(w_c\right)_{j,k}(n) - \xi \cdot \partial E / \partial\left(w_c\right)_{j,k} \qquad (22)$$

where n is the training times; ζ and ξ are learning rates. In this paper, we use the adaptive back-propagation algorithm suggested by the authors [6].

4. Simulation Example

Considering the following second-order parameter varying system:

$$\frac{y(s)}{u(s)} = \frac{1}{(1+Ts)^2} \qquad (23)$$

where the time constant T is affected by a working state variable z $(z \in [0.3, 0.9])$. Suppose the relationship between them is:

$$T = 20 + 20 \cdot (z - 0.3) \qquad (24)$$

Once the sample time T_0 is given, the discrete time description of the system could be obtained:

$$y(k) = \left[2 \cdot \left(T^2 + T \cdot T_0\right) \cdot y(k-1) - T^2 \cdot y(k-2) + T_0^2 \cdot u(k-1)\right] / (T + T_0)^2 \qquad (25)$$

In this paper, the sample time T_0 is taken as 5 seconds. Curves 1, 2 and 3 in Fig. 3 show the unit step response of the system at $z = 0.3$, $z = 0.6$ and $z = 0.9$ respectively and one can see that variations between the different z's are very large.

Using the suggested FNN TS model identification method, we select z as a working state variable for the input of premise in the FNN and take $u(k-1)$, $y(k-1)$ and $y(k-2)$ as input variables for the TS model. The aim of the identification is to obtain the global model which is suitable for all the possible working states of the system. First, ten states are selected randomly, and 310 groups of training data are obtained by exerting 5-order M sequels which have the range of 1 on the system. All of the weights in the consequent of the FNN are selected between -0.1 and 0.1 randomly, and the fuzzy partition number r is selected to 7 as shown in Fig. 2. In order to fasten the convergence rate of the networks, the following parameters are used as the initial value of the adaptive BP algorithm shown in [6]:

$$\zeta(1) = 0.9, \ \xi(1) = 0.4, \ \alpha_0 = 1.4, \ \alpha_1 = 0.6, \ E_s = 0.5$$

The final convergence conditions are taken as:

1) the number of the samples which have satisfied $\left(t^p - y^p\right) / t^p \leq 0.05$ has exceeded 95 percent of the total samples.

2) Training times has exceeded the maximum times specified as 10000.

After training the FNN 868 times, the networks converged by satisfying condition 1) and the final simulation results are shown in Fig 4 and Fig. 5. As shown in Fig. 4, where the solid line and the dotted line denote the expected output of the system and the actual output of the networks respectively, most of the samples have good performance to describe the actual outputs of the system. Finally, we use another ten groups of z to verify the performance of the resulted FNN TS fuzzy model and the results are shown in Fig. 6. The same conclusion can be drawn from it. Therefore, the suggested TS model identification method is strongly effective in obtaining the global dynamic model of parameter varying systems.

5. Conclusions

This paper generalizes the TS model to the parameter varying systems and presents the corresponding identification method via FNN. The proposed method can effectively realize the identification of parameter varying systems whereas the traditional linear system identification methods can not. Furthermore, control of such systems based on the well-established TS fuzzy model can be carried out and this further research fields creates for us. Simulation results of a second-order parameter varying system have fully verified the effectiveness of this method. It should be noted that a more effective way is provided to establish fuzzy control rules for the multi-working-states situations. Based on the model, performance of fuzzy controller will be greatly improved under such situations.

References

[1] B. Kosko, Neural Networks and Fuzzy Systems, Prentice Hall, 1992
[2] T. Takagi and M. Sugeno, "Fuzzy identification of system and its applications to modeling and control", *IEEE Trans. on System Man and Cybernetics*, vol. SMC-15, no. 1, 1985, pp. 116-132.
[3] M. Sugeno and G. T. Kang, "Structure identification of fuzzy model", *Fuzzy Sets and Systems*, vol. 28, 1988, pp. 15-33.
[4] M. Sugeno and K. Tanaka, "Successive identification of a fuzzy model and its applications to prediction of a complex system", *Fuzzy Sets and Systems*, vol. 42, 1992, pp. 315-334.
[5] Xie Keming and Zhang Jianwei. "A linear fuzzy model identification method based on fuzzy neural networks", *in Proceedings of the 2nd Worldwide Chinese Intelligence Control and Intelligence Automation Conference*, 1997.
[6] Xie Keming and Zhang Jianwei, "An Adaptive Backpropagation Algorithm Based on Error Rate of Change", *submitted to Journal of Taiyuan University of Technology*.

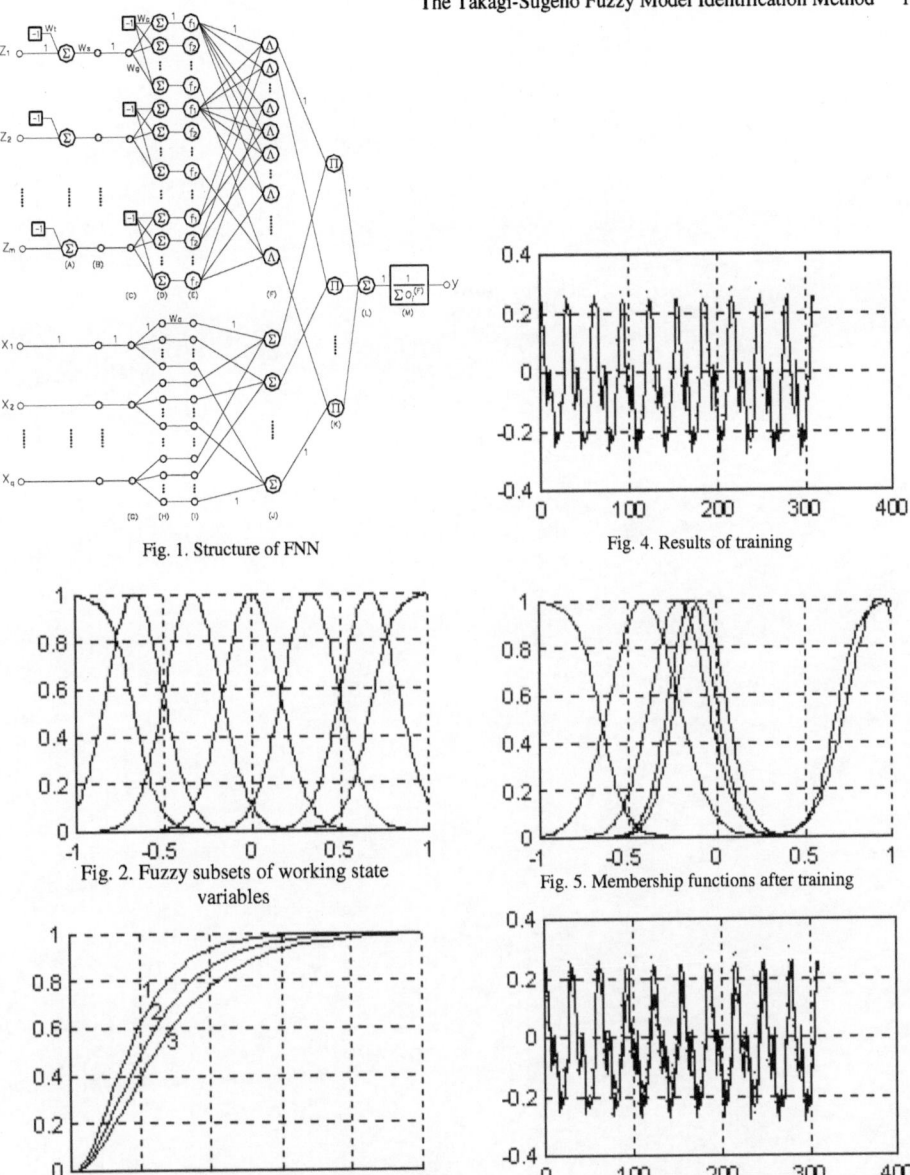

Fig. 1. Structure of FNN

Fig. 4. Results of training

Fig. 2. Fuzzy subsets of working state variables

Fig. 5. Membership functions after training

Fig. 3. Unit step response under different working state

Fig. 6. Results of verification

Sensing of Arc Length and Wire Extension Using Neural Network in Robotic Welding

Kazuhiko Eguchi[1], Satoshi Yamane[2], Hideo Sugi[2] and Kenji Oshima[2]

[1]Semiconductor & IC Division Hitachi, Ltd. 5-20-1 Josuihon-cho,
Kodairashi Tokyo 187-8858, Japan

[2]Department of Environmental Science & Human Engineering,
Saitama University, Urawa 338-8570,JAPAN
yamane, sugi, oshima} @d-kiki.ees.saitama-u.ac.jp

Abstract. It is important to design intelligent welding robots to obtain a good quality of the welding. It is required to detect bead height and deviation from center of gap and to control arc length and back bead by adjusting welding conditions such as welding speed, power source voltage and wire feed rate. Authors propose arc sensor using neural networks to detect arc length and wire extension. By using them and by means of geometric method, the bead height is detected. Moreover, authors propose switch back welding method to get stable back bead. That is, welding torch is not only woven in the groove, but also moved backward and forward.

1 Introduction

Arc welding process plays one of important modern technologies to join metal plates[1]. It is important to design intelligent welding robots so as to obtain a good quality of the welding[2]. It is also important to control arc length regardless of disturbances such as variations of the torch height, feed rate and so on. The method is discussed to control arc length by adjusting the power source characteristic. If groove gap is narrow, the welding speed is faster and the current is bigger than the groove gap is wider. The power source characteristic of the constant voltage is used. If the gap becomes wide, the pulsed current is applied to the welding and it is made by using the power source with rising characteristic.

Authors propose the switch back welding method to obtain a stable back bead. The welding torch is not woven with 10Hz on the groove, but also moved backward and forward. Therefore it is required to detect groove gap. During forward process, edges of the groove gap are melted and the droplet metal is deposited to the edges, so that the weaving width of the torch is equal to the gap. During backward process, the deposited metals bridge base metals.

If the weaving center of the torch is just at the center of the groove gap, the waveforms of the voltage and the current on right side and left side are symmetric about

L. Polkowski and A. Skowron (Eds.): RSCTC'98, LNAI 1424, pp. 163-170, 1998.
© Springer-Verlag Berlin Heidelberg 1998

the center of the gap. Sums of voltage or current on the right side of the weaving are equal to that on the left side. If there is difference of the sums, the weaving center is not at the center of the gap. In conventional arc sensor, the difference of the sums between right and left sides is used to detect the deviation from center of the gap. It is difficult to detect the gap by the conventional arc sensor.

In order to find the gap, authors propose neural network (N.N.) arc sensor which uses the sampled data of both current and voltage at the current pick-up point. That is, the extension of an electrode wire and the arc length are found by the N.N. The groove shape is estimated by outputs of the N.N. The training data are constructed from the experimental results. The performance of the N.N. is examined by using the testing data.

2 A proposal of switch back method to get a stable back bead

By using the knowledge of weld pool phenomena, authors have developed a new welding method of controlling the torch motion. Figure 1 shows the switch back welding method. The weaving width depends on the gap. If root gap is 4mm, the torch is moved forward 18mm with 90 cm/min and backward 9.5mm with 22 cm/min in one cycle. This motion is repeated and in this example the average welding speed is 13.4 cm/min. During the forward process, the amplitude of torch weaving of 10Hz is controlled to be equal to the root gap, such that the heat input and droplet are given to each root edge of the base metals. The pulse peak is made after passing the weaving center to transfer the droplet to each root edge of both base metals in the case of relatively wide root gap. In Fig.2, top of the electrode wire is near the root edge of the groove, where the current is synchronized with pulse duration. During the backward process the back weld pool tends to become big enough to get a good back bead. Before the molten metal burns through, the torch should begin to move away forward from the weld pool. Welding speed in forward process is faster than that in backward process.

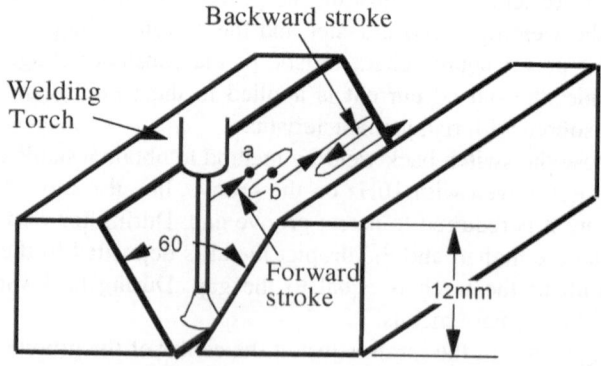

Fig. 1. Switch-back welding method.

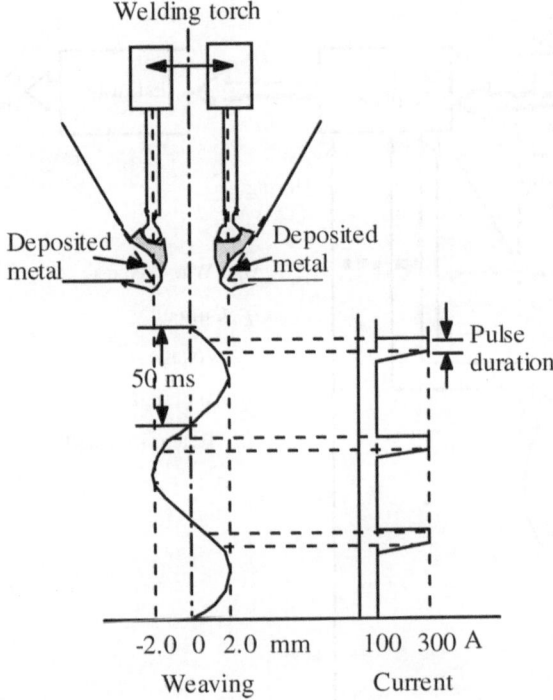

Welding torch

Deposited metal

Deposited metal

50 ms

Pulse duration

-2.0 0 2.0 mm   100 300 A

Weaving   Current

Fig. 2. Relationship between voltage,torch position and pulse duration.

3 Estimation of the arc length and the wire extension length by the neural network

The voltage of the current pick-up point (the torch voltage) consists of the arc volt-age and the voltage drop in wire extension. The arc voltage may be experimentally decided by the arc length and current. The voltage drop in wire extension may be determined by the resistance and inductance at each part of the extension and its length which depends on the wire feed rate, the melting rate at the top of the wire and the temperature or current given till that time. In discrete time system, the phenomena (the current and the voltage) at the present sampling instance may be described by the difference equations concerning the current, the voltage, the arc length, the wire extension and wire feed rate at previous sampling instances. We tried to construct the N.N. for obtaining the arc length $l(k)$ and the wire extension length $L(k)$ at the present sampling instance kT (T: sampling period 1ms) from the current $i(k) \sim i(k\text{-}49)$ and the voltage $v(k) \sim v(k\text{-}49)$, the wire feed rate $v_f(k) \sim v_f(k\text{-}49)$ at the present and previous 50 sampling instances in the past time 50 ms as illustrated in Fig.3. Number of units in input layer is 150. Number of units in the hidden layer is 15. We tried to train the N.N. as shown in Fig.4 by means of the back propagation method and by using the training data obtained in experiments, of which one example is shown in Fig. 5.

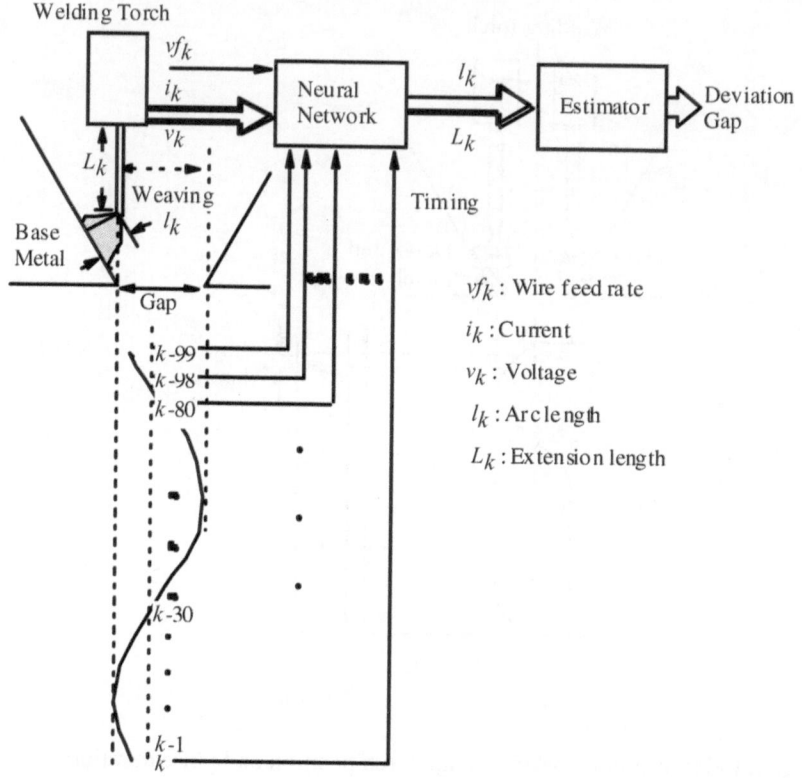

Fig. 3. Relationship between sampling instant of current and voltage, and torch position.

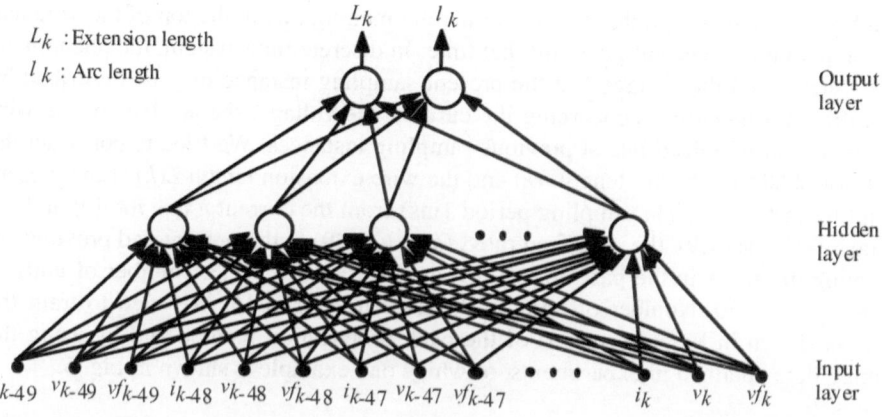

Fig. 4. Neural network to estimate arc length and wire extension length

In Fig.5, the arc length and the wire extension length are investigated by the motion picture taken by high speed video CCD camera. Figure 5 shows the experimental results in the cases where the root gap is 4mm and the center of torch weaving agrees with the center of the groove. Figure (a) and (b) show the experimental results in the forward process at (a) in Fig.1 and in the backward process at (b) in Fig.1, respectively.

While the torch weaves in the groove, the arc discharge between the top of the electrode wire and the base metal or weld pool surface through the shortest path. However, when the temperature of the base metals is low and the gap between both base metals is wider than 3mm, the arc continues to form in one side of the base metal until the top of the wire come close to the opposite base metal as illustrated in Fig. 6.

In the forward process , the neural network (arc sensor) detects the torch position in the groove to perform the seam tracking and the N.N. also detects the groove gap to determine the weaving width of torch.

In the backward process, the N.N. measures the bead height in the groove to investigate whether both edges of the base metals bridge or not with the deposited metal and to control the welding conditions such as the current waveform, wire feed rate, power source voltage and welding speed.

The outputs of the trained N.N. are also shown in Fig.5. Good results of the estimation of the arc length l_k and extension L_k under the training data and testing data are obtained.

4 Estimation of the gap and deviation from the groove gap center

During the forward process, the arc discharges with a right angle from the base metal surface. The angle of the groove is 60 degrees. The point on the surface of the base metal, from which the arc discharges, is estimated from the arc length and the extension.

Let the origin of coordinate x be the center of the torch weaving, and origin of the coordinate y be the end of the wire. First, coordinate of the top of the wire is found from the torch position calculated from encoder pulses and the extension found by N.N. Secondly, the circle is drawn with radius l_k corresponded to the arc length. The surface of the groove corresponds to the tangent line of the circle. If the arc discharges to the left side of the groove, the coordinate (x, y) of the tangent point of the arc on the left side is calculated by using the following equation.

$$x = tp - l_k \cos 30°$$
$$y = L_k + l_k \sin 30° \tag{1}$$

where tp is torch position.

On the other hand, if the arc discharges to the right side of the groove, the coordinate (x, y) of the tangent point of the arc on the right side is calculated by using the following equation.

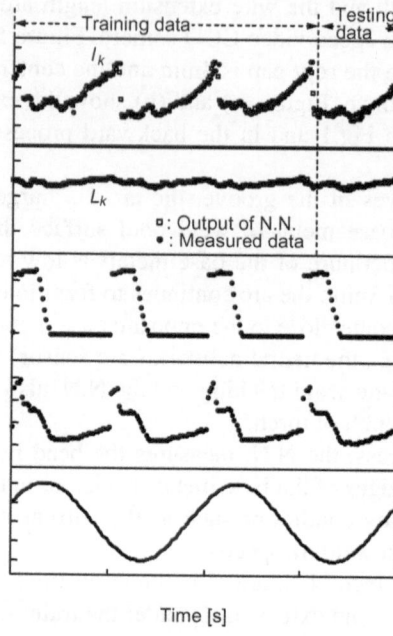

(a) Forward process.

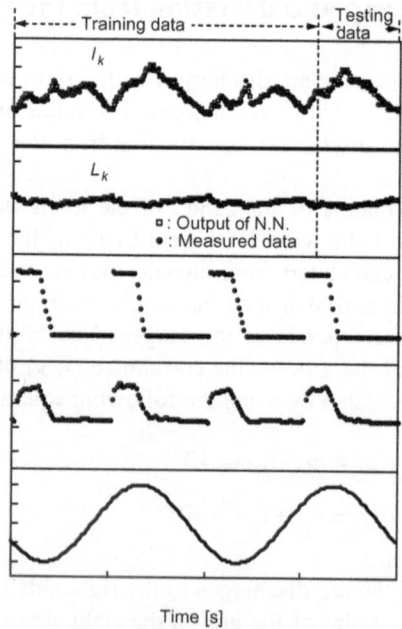

(b) Backward process.

Fig. 5. Estimation result when gap is 4mm and deviation is 0mm.

$$x = tp + l_k \cos 30°$$

$$(2)$$

$$y = L_k + l_k \sin 30°$$

The groove shape is found by using estimated points. Since the length from the bottom of the base metals to the torch is 20mm, the coordinate of root edge is calculated. The deviation from the weaving center to that of the root edges is found. The surface of the groove is calculated by using estimation result in Fig.5 (a) and is shown in Fig.6. The dot corresponds to the estimated surface and is close to the groove surface. The coordinate of the root edges is found. The deviation D is calculated and is almost equal to the value found by high speed video image.

During the backward process, the surface of the weld pool is found. First, the coordinate of the top of the electrode wire is calculated by using the extension and torch position. Secondly, let the center be the top of the wire. The circle is drawn with the radius corresponded to the arc length lk. Thirdly, the envelope of circles is drawn and corresponds to the surface of the weld pool. By using estimation results, the surface of the weld pool is calculated and is shown in Fig.7. The bead height can be found from the coordinate of the root edges and the weld pool surface.

5 Robotic welding system for back bead control

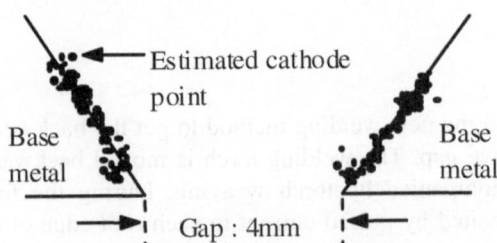

Fig.6. Estimation result of groove.

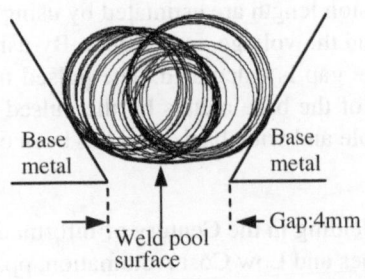

Fig.7. Estimation result of weld pool surface
when gap is 4mm.

It is important to give the pulsed droplet to each root edge of base metals alternately in one cycle of torch weaving. For this purpose, the center of torch weaving should be adjusted to the center of groove, and the arc length should also be controlled. The control block diagram is shown in Fig.8. The control of torch position and arc length are performed by giving the pulse number u_k to manipulator and by changing the voltage u_k of the power source, respectively.

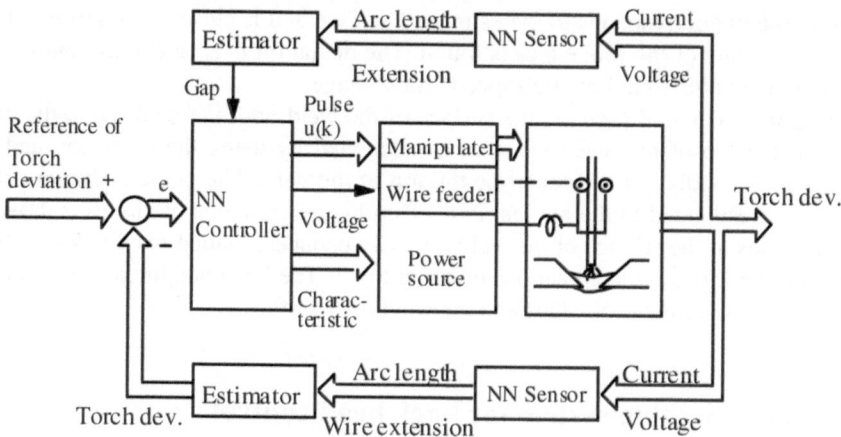

Fig.8. Control system using neural network.

6 Conclusions

The authors proposed the new welding method to get the back bead regardless of the variation of the groove gap. The welding torch is moved backward and forward like the switch back accompanied by torch weaving. During the forward process, the droplet metal is deposited by pulsed current to each root edge of both base metals in the case of relatively wide root gap. During the backward process, the base metals bridge and the back weld pool tends to become big enough to get a good back bead.

The amplitude of the torch weaving is controlled according to the gap. The arc length and the wire extension length are estimated by using the neural network sensor, which takes the current and the voltage as the input. By using the estimated value and the geometric method, the gap is calculated. The pulsed transfer metal has been deposited to the root edges of the base metals by the pulsed current synchronized with the torch motion. The stable and wide back bead has been obtained.

References
1. P.Drews, G.Starke, "Welding in the Century of Information Technology", Int.Conf. on Advanced Techniques and Low Cost Automation, pp.3-22(1994)
2. K.Ohshima, et. al., "Digital Control of Torch Position and Weld Pool in MIG-Welding Using Image Processing Device", Trans. of IEEE IAS, Vol.28,No.3, pp.607- 612(1992)

Traffic Signal Control Using
Multi-layered Fuzzy Control

Satoshi Yamane, Kazuhiro Okada, Kenji Shinoda

and Kenji Oshima

Department of Environmental Science & Human Engineering,
Saitama University, Urawa 338-8570, Japan
{yamane, sinoda, okada, oshima}@d-kiki.ees.saitama-u.ac.jp

Abstract. This paper deals with problems concerning traffic signal control. It is important to reduce waiting time of each car on the intersection, in order to avoid traffic congestion. Authors apply multi-layered fuzzy inference to the traffic signal control. First, number of cars is detected by using sensors set on road. Elapsed time is measured after change of signal from blue (go) to red (stop). Secondly, the degree to change the traffic signal is inferred by using multi-layered fuzzy inference of which the inputs are the elapsed time and number of cars. A performance of a fuzzy controller depends on the fuzzy variables and the control rules. In this paper, the control rules are constructed from expert's knowledge . Numerical simulations are carried out. A good performance is obtained by using multi-layered fuzzy control.

1 Introduction

It is important to improve traffic congestion in urban traffic. The degree of the traffic congestion depends on performance of the signal control and urban planning. Moreover, traffic volume changes by time and date. If the method in which cycle of changing the signal is fixed is applied to the signal control, a good performance is obtained in case where there is no variation of the traffic volume. If there is unexpected variation of the traffic volume, the waiting time of the car becomes long and there is traffic congestion. In order to improve the situation of the traffic, the traffic volume is detected. According to it, the traffic signal is controlled.

L. Polkowski and A. Skowron (Eds.): RSCTC'98, LNAI 1424, pp. 171-177, 1998

For this purpose, authors apply fuzzy control using expert's knowledge to the traffic signal control. The performance of the fuzzy inference depends on the fuzzy variables and the inference rules. When the conventional fuzzy inference is used, many rules are required. The authors propose the multi-layered fuzzy inference to easily present the skills and the experiences of the expert. The tuning of the fuzzy variables is performed by using the steepest decent method. In order to evaluate the control performance, the performance function J defined as the sum of the waiting time is introduced. The optimum fuzzy variables are calculated by the steepest descent method, so as to minimize the performance function J. The validity of the proposed control system is verified by carrying out the numerical simulation.

2 Modeling of Intersection

The intersection is shown in Fig. 1. There are two roads from east to west and from north to south. The road is composed of 3 lanes. In Japan, cars run on the left lane. The traffic signal of each lane has 5 kinds of sign as follows: red, blue, yellow, blue for right turn, red for right turn. Two detectors are set on each lane to count the cars. The distance between the detectors is 150m. Cars enter into the intersection according to random numbers in simulation. In the intersection, there is no passing.

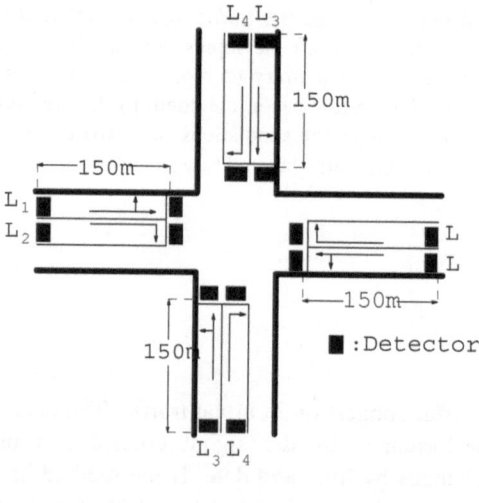

Fig. 1. Intersection.

3 Multi-layered fuzzy control for traffic signal control

The controller is shown in Fig. 2. In the fuzzy inference, the degree to change the signal is inferred. In the decision section, if the degree is over a threshold, the signal is changed from red to blue.

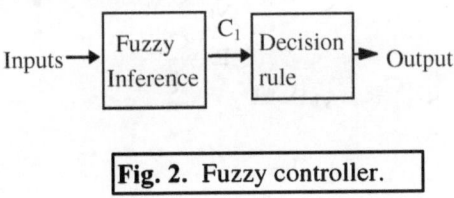

Fig. 2. Fuzzy controller.

In order to make the control rules, the laws to change the signal are considered as follows:
1. If there is no car on the lane of the blue sign, there is a few cars on the lane of the red sign and the elapsed time from the change to the red sign is not short, then the signal is changed.
2. If cars on the lane of the blue sign are few and the elapsed time is short, then the sign is holding. That is, because this case corresponds to the beginning of moving of the cars.
3. If the elapsed time is long, the number of the cars on the lane of the red sign is greater than that on the lane of the blue sign, then the signal is changed.

By considering the above mentioned, the following input variables are selected.

1. T : Elapsed time from the sign change to red.
2. G : Number of cars on the lane of the blue sign.
3. W_1 : Number of cars on the lane of next blue sign. For example, when the signal of lane L_1 is blue, the signal of the lane L_2 becomes blue after the signal of L_1 changes. Number of cars on the lane L_2 is W_1.
4. W_2 : Number of cars waiting the change of the signal except for the lanes of blue sign and next blue sign.

The relationship between the degree C to change the signal and T, G, W_1 and W_2 are represented by using the linguistic rules shown in Fig.3. Fuzzy rule is composed of 4 layers. The authors represent T, G, W_1 and W_2 by using three kinds of linguistic representation (membership function of Large, Medium, Small) illustrated in Fig.4. In fourth layer of Fig.3, the relationship between W_1 and C is described. For example, the rules are as follows, in case where W_2 is small:

1. If W_1 is Small, then C is Small.
2. If W_1 is Medium, then C is Medium.
3. If W_1 is Large, then C is Large.

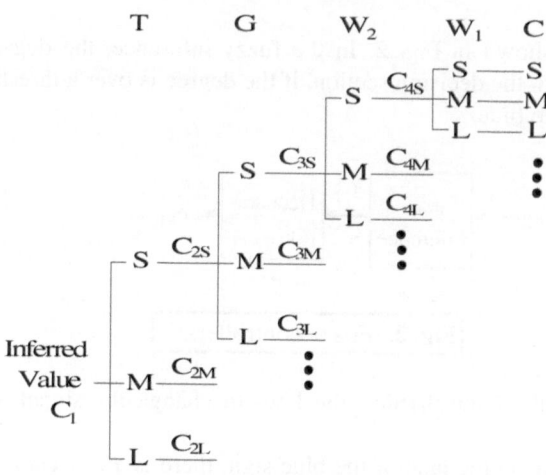

Fig. 1. Control rules of multi-layered fuzzy inference.

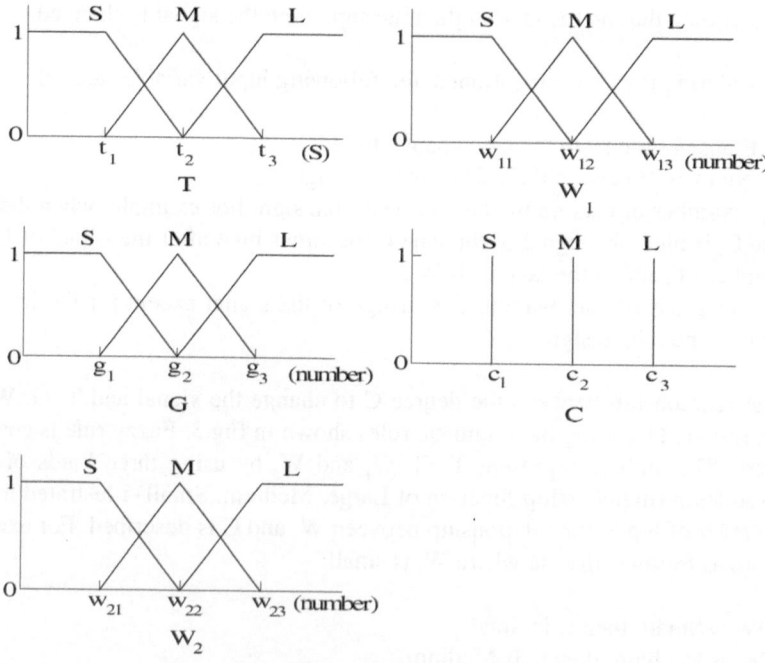

Fig. 2. Fuzzy membership function.

If W_1 is given, the fitness w_{1s}, w_{1m} and w_{1l} for small, medium, and large in C is calculated from the membership functions in Fig.4. Then-part in case where W_2 is small is the inference result C_{4s} of the fourth layer given by

$$C_{4s} = w_{1s}S + w_{1m}M + w_{1l}L \qquad (1)$$

First of all, the inference result of the fourth layer is calculated. Finally, the inference C_1 is obtained.

4 Tuning up fuzzy variables

Since the performance of the estimator depends on the fuzzy variables, its parameter is adjusted to obtain the good performance. The optimum parameters are found by steepest decent method so as to minimized the performance function J, which is the sum of the waiting time f of the cars

$$J = \frac{\sum_{i=0}^{N-1} f}{N} \qquad (2)$$

where N is total number of cars.
The parameter w_{11} is given by

$$w_{11}^{i+1} = w_{11}^{i} - \alpha \frac{\partial J}{\partial w_{11}} \qquad (3)$$

where α is small positive number and corresponds to the training rate.

The partial differential values are calculated by using membership functions illustrated in Fig.4. Other parameters are calculated in the same manner as the above equations. In the tuning, the parameter of the fuzzy variables is adjusted from the first layer to the fourth layer like the back propagation method.

5 Simulation of traffic

Sequence of signal change is as follows: red, blue, yellow, blue for right turn, yellow for right turn, and red. Holding time of yellow is 4s. First, simulation is carried out in case where traffic volume is 400 cars /h for east-west direction and north-south direction. Two kinds method of signal control are applied to the simulation. Secondly, after 1 hour, the traffic volume changes from 400 cars /h to 800 cars/h for north-south direction only. Simulation result shows in Fig. 5. After the change of the traffic volume, the waiting time increases in case where the method of fixed timing is

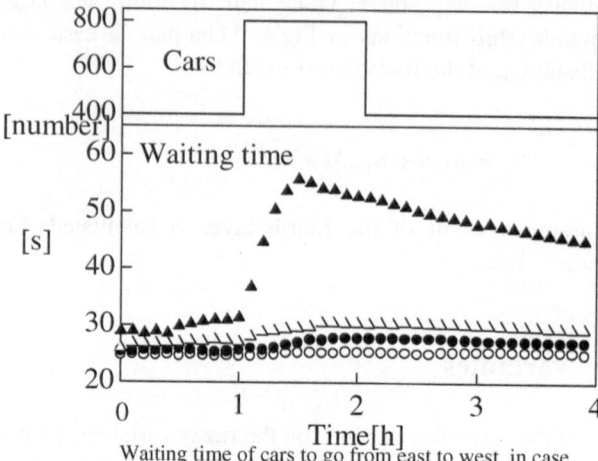

Fig. 3. Simulation result.

applied. Moreover, after the traffic volume is return to the normal situation (400 cars/h), the waiting time is not recovered. On the other hand, the waiting time is almost kept constant without the variation of the traffic volume in case where the traffic signal is controlled by the fuzzy controller. A good performance is obtained by applying the multi-layered fuzzy controller.

6 Conclusions

This paper dealt with the problem concerning traffic control. A new method based on the multi-layered fuzzy inference was proposed. The fuzzy variables and the control rules play the important role in the performance of the controller. The validity of the proposed method was verified by the simulation.

References

1. T.Fukuda and T.Shibata, "Theory and Application of Neural Networks for industrial Control Systems", IEEE Trans. Industrial Electron., vol.39, no.6, 472-489(1992).
2. K.Ohshima, Y.Mori, Y.Kaneko, A.Mimura, T.Kumazawa, T.Kubota, and S.Yamane, "Groove Gap Detection and Weld Pool Control Using CCD Camera", The 5th Int. Sym. of the Japan Welding Society: 489-494(1990).
3. K.Ohshima, Y.Mori, P.Ma, S.Yamane, "Fuzzy Control in Tracking of Welding line", Int. Conf. on New Advances in Welding and Allied Process, Beijing, China: 138-141(1991).
4. P.Ma, K.Ohshima, S.Yamane, " Application of Neural Network and Fuzzy Control to 2 Dimensional Orbit Tracking", Int. Conf. on Automation, Robotics and Computer Vision, Singapore: co.7.5.1-co.7.5.5(1992).

Approximation Region-Based Decision Tables

Wojciech Ziarko

Computer Science Department
University of Regina
Regina, Saskatchewan, S4S 0A2
Canada

Abstract. Approximation region-based decision tables are tabular specifications of three, in general uncertain , decision rules corresponding to rough approximation regions: positive, boundary and negative regions. The focus of the paper is on the extraction of such decision tables from data, their relationship to conjunctive rules and probabilistic assessment of decision confidence. The theoretical framework of the paper is a variable precision model of rough sets.

1   Introduction

Rough sets theory has been an active area of research for almost twenty years since the first pioneering works on the topic were published by Pawlak [3]. This novel set representation model, capturing the relationship between the ability to discern observations, and the ability to define a subset of observations, turned out to be both fundamental and general enough to stimulate significant developments in logic, machine learning, data mining, control theory and a number of application areas (see, for example [2,5,7,10]). Despite its relative popularity and the interest in this research paradigm, some fundamental questions still remain to be answered. One of these questions concerns the major difference between rough set-based approaches to machine learning and data mining versus similar methodologies developed outside the rough sets community(e.g. [13]). Other questions relate to whether we are utilizing the capabilities of the rough sets approach properly in the context of machine learning and data mining applications, and whether we are taking full advantage of the power of the methodology.

These questions are motivated by the fact that rough set-based approaches to machine learning and data mining are primarily concerned with acquisition (so-called "induction") of conjunctive rules from data, contributing new, or improved rule extraction algorithms to already existing ones (e.g. [4,8]). In this sense, the results of rough sets research cannot be distinguished from results of other fields and, consequently, lead to questions regarding the existence of essentially new unique contribution of this methodology in comparison to other approaches.

It has been my opinion for some time however, that the rough sets methodology has a unique "product of its own" to offer, a "product" which has numerous advantages over conjunctive rules, and despite of this seems to have been

L. Polkowski and A. Skowron (Eds.): RSCTC'98, LNAI 1424, pp. 178–185, 1998.

neglected in much of recent research and applications dominated by rule computation techniques. This "product" is a whole methodology of decision tables analysis and acquisition from observation data.

Therefore, the main goal of this paper is a presentation of the methodology of extended decision tables extracted from data, referred to as *approximation region-based* decision tables. The approximation region-based decision tables encapsulate uncertain, in general, decision rules extracted from data. The decision rules correspond to approximation regions of the target set. In the following sections, the capabilities of approximation region-based decision tables as tools for predictive decision making are investigated and compared to conjunctive rules in the context of the variable precision model (VPRS) of rough sets [9,12]. The VPRS model has been chosen here rather than the original model of rough sets since in most of the practical problems occurring in machine learning, pattern recognition and data mining the inter-data relationships are probabilistic in nature, leading to non-deterministic decision tables.

2   Approximation Region Rules

Decision tables are fundamental to decision making, and hardware/software design methodologies [14]. In their original formulation they represent functional relationship between sets of observations and decisions. The relationship is expressed in the form of a collection of conjunctive deterministic rules:

$\langle condition_1 \rangle \wedge \langle condition_2 \rangle \wedge ... \wedge \langle condition_m \rangle \rightarrow \langle decision \rangle.$

In this approach the decision tables are designed manually, reflecting processing requirements of the designer. However, in problems related to machine learning, pattern recognition and data mining, deterministic decision tables cannot be constructed due to lack of proper decision knowledge or the inherent non-determinism (i.e. partial functionality, or lack of functionality) of the relationship between conditions (observations) and decisions (recognitions, predictions, etc.). Consequently, in many problems there is a need to use observation data, rather than the designer's knowledge, to derive the decision table (non-deterministic in general) .

In the rough sets theory, the decision table is a central notion. Specifically, the data-extracted decision table represents partial functional dependency [1,2] occurring in the data. For the purpose of this article, it can be defined as follows:

Given the subset $X \subseteq U$ of the universe U, the decision table with respect to deciding (predicting) whether an object $e \in U$ also belongs to the set X, corresponds to rough characterization [2] of the set X in the approximation space (U,R), where R is the equivalence relation (indiscernibilty relation) on U, typically defined in terms of combinations of values of attributes $a \in A$ (we assume that the set of attributes A is finite and each attribute has finite domain). In other words, the decision table for the target set X (the *concept*) , in the simplest case, is a set of disjoint conjunctive rules, $r^*_{E,X}$ referred to as *elementary rules*, conforming to either of the following three possible formats:

- $r^+_{E,X} : \langle a_1, v_{a_1} \rangle \wedge \langle a_2, v_{a_2} \rangle \wedge ... \langle a_m, v_{a_m} \rangle \rightarrow +$ with $supp(r^+_{E,X}) \subseteq POS(X)$

$$- r^?_{E,X} : \langle a_1, v_{a_1} \rangle \wedge \langle a_2, v_{a_2} \rangle \wedge ... \langle a_m, v_{a_m} \rangle \to ? \text{ with } supp(r^?_{E,X}) \subseteq BND(X)$$
$$- r^-_{E,X} : \langle a_1, v_{a_1} \rangle \wedge \langle a_2, v_{a_2} \rangle \wedge ... \langle a_m, v_{a_m} \rangle \to - \text{ with } supp(r^-_{E,X}) \subseteq NEG(X)$$

where E is an elementary set of the relation R, $POS(X)$, $BND(X)$ and $NEG(X)$ are respectively the positive, boundary and negative regions of the set X [1] and $supp(r^*_{E,X})$ is a rule support set, that is a set of objects belonging to U and matching the conditions of the rule.

The above set of rules can be re-expressed in the form of a maximum three rules in disjunctive normal form (DNF), each of which corresponding to exactly one rough approximation region $POS(X)$, $BND(X)$ and $NEG(X)$:

$$- r^+_X : \vee(\langle a_1, v_{a_1} \rangle \wedge \langle a_2, v_{a_2} \rangle \wedge ... \langle a_m, v_{a_m} \rangle) \to + \quad , \text{ for positive region}$$
$$- r^?_X : \vee(\langle a_1, v_{a_1} \rangle \wedge \langle a_2, v_{a_2} \rangle \wedge ... \langle a_m, v_{a_m} \rangle) \to ? \quad , \text{ for boundary region}$$
$$- r^-_X : \vee(\langle a_1, v_{a_1} \rangle \wedge \langle a_2, v_{a_2} \rangle \wedge ... \langle a_m, v_{a_m} \rangle) \to - \quad , \text{ for negative region.}$$

Thus, when deriving decision tables from data in the rough sets approach we are essentially facing the problem of acquisition of a maximum of three rules in the very uniform DNF format. We will refer to these rules as an *approximation region* rules, or simply as positive rule, negative rule or boundary rule. The decision table with approximation region rules will be referred to as an *approximation region-based* decision table.

3   Probabilistic Decision Tables

In most practical problems involving processing large amounts of data, all equivalence classes of the relation R, and the relationship between classes and the target set (concept) X are unknown. Usually, the data in our possession is a proper subset, a sample of the universe U. We will assume that this is a uniformly distributed random sample, although in practice this is normally not the case. Also, in typical problems the space U is infinite comprising all past and future occurrences of some observations (called elementary events in probability theory). Our objective, in case of probabilistic approximation region-based decision tables (probabilistic decision tables in short), will be to use the data sample to identify, with sufficient confidence, the following:

- All or almost all equivalence classes of the relation R by their discriminating descriptions;
- The discriminating descriptions of the generalized VPRS approximation regions of the target set X , that is of $POS_u(X)$, $BND_{l,u}(X)$ and $NEG_l(X)$ where $0 \leq l < u \leq 1$ are lower and upper limits parameters respectively of VPRS model as defined below.

The main ideas of VPRS model [12] approximation regions are summarized as follows:

Let R^* be the set of equivalence classes of the indiscernibilty relation R and let $E \in R^*$ be an equivalence class (elementary set) of the relation R. With each class E we can associate the estimate of the conditional probability $P(X|E)$ by the formula: $P(X|E) = card(X \cap E)/card(E)$ if sets X and E are finite.

For any subset $X \subseteq U$ we define the *u-positive* region of X, $POS_u(X)$ as a union of those elementary sets whose conditional probability $P(X|E)$ is higher than the upper limit , that is

$$POS_u(X) = \bigcup \{E \in R^* : P(X|E) \geq u\}$$

The *(l,u)-boundary* region $BNR_{\ell,u}(X)$ of the set X with respect to the lower and upper limits ℓ and u is a union of those elementary sets E for which the conditional probability $P(X|E)$ is higher than the lower limit ℓ and lower than the upper limit u. Formally,

$$BNR_{\ell,u}(X) = \bigcup \{E \in R^* : \ell < P(X|E) < u\}$$

In practical decision situations, the boundary area represents objects which cannot be classified with sufficiently high confidence (represented by u) into set X and which also cannot be excluded from X with the sufficiently high confidence (represented by $1 - l$).

The *l-negative* region $NEG_l(X)$ of the subset X, is a collection of objects which can be excluded from X with the confidence higher than $1 - l$, that is,

$$NEG_l(X) = \bigcup \{E \in R^* : P(U - X|E) \geq 1 - l\}$$

Based on the above definitions of approximation regions, the approximation region rules defined in the context of the original model of rough sets in the previous section can now be generalized to non-deterministic approximation region rules with attached estimates of conditional probabilities of their respective consequents (confidence factors):

- $r_X^+ : \vee(\langle a_1, v_{a_1}\rangle \wedge \langle a_2, v_{a_2}\rangle \wedge ...\langle a_m, v_{a_m}\rangle) \to +$ with conditional probability $P(+| \langle a_1, v_{a_1}\rangle \wedge \langle a_2, v_{a_2}\rangle \wedge ...\langle a_m, v_{a_m}\rangle) \geq u$ for positive region;
- $r_X^? : \vee(\langle a_1, v_{a_1}\rangle \wedge \langle a_2, v_{a_2}\rangle \wedge ...\langle a_m, v_{a_m}\rangle) \to ?$ with "inconclusive" conditional probability $l < P(+| \langle a_1, v_{a_1}\rangle \wedge \langle a_2, v_{a_2}\rangle \wedge ...\langle a_m, v_{a_m}\rangle) < u$ for boundary region; the boundary area rule is relatively inconclusive in the sense that the probability of positive outcome is not high enough (i.e. at least u) and not low enough (i.e. at most l) .
- $r_X^- = \vee(\langle a_1, v_{a_1}\rangle \wedge \langle a_2, v_{a_2}\rangle \wedge ...\langle a_m, v_{a_m}\rangle) \to -$ with conditional probability $P(-| \langle a_1, v_{a_1}\rangle \wedge \langle a_2, v_{a_2}\rangle \wedge ...\langle a_m, v_{a_m}\rangle) \geq 1-l$ for negative region.

These three rules can be summarized in the form of a probabilistic approximation region-based decision table as illustrated in Table 1.

REGION	AGE	SEX	SKILL	INCOME	$P(+|conditions)$
	young	female	yes		
POS	young	male	no	+	0.85
	medium	male	yes		
BND	young	female	no	?	0.6
	old	male	no		
NEG	old	female	yes	-	0.9
	medium	male	no		

Table 1. Example probabilistic decision table with $l = 0.1$ and $u = 0.85$

4 Estimating Probability Distributions

The problem of estimating probability distributions based on a random sample has been investigated in probability theory. In brief, the main theorems of probability theory, known as laws of large numbers, assert that if the sample is collected at random and without bias, and if its size is sufficiently large, then the frequency distribution-based estimator of the probability of an event will converge to the actual probability with growth of the sample size. Consequently, if the sample size is greater than the assumed threshold level n then the computed estimates are very likely to be close approximations of the actual probabilities. The choice of the threshold n depends on the required confidence level and the actual value of the estimated probability (which is unknown) meaning that in practice somewhat arbitrary value of n usually have to be selected. The implications of these well-known facts in the process of decision table construction from data are summarized below.

Given approximation space (U,R), let $U' \subset U$ be the random sample and let n be the threshold sample size. Also, let X', R' be the restrictions of the target concept X and of the relation R to the sample, that is $X' = X \cap U'$ and $R' = R \cap (X' \times X')$. To construct an approximation region-based decision table with credible assessments of probability distributions the following requirements need to be satisfied:

- $\{[y]_{R'} : y \in U'\} = \{[y]_R : y \in U\}$ that is, all equivalence classes of the relation R must be represented among the classes of the relation R'.
- $\text{card}(POS_u(X))) > n$, $\text{card}(BND_{l,u}(X')) > n$, $\text{card}(NEG_l(X') > n$ that is, credible confidence factors are to be associated with the computed approximation region rules.

The satisfaction of at least one condition in the second requirement guarantees that $\text{card}(U') > n$ which means that a credible estimate of the probability $P(X)$ of the target concept $X \subseteq U$ can also be calculated by taking the ratio $\text{card}(X')/\text{card}(U')$. The probabilities $P(X)$, $1 - P(X)$ can be associated with two "empty antecedent" rules:

-      $r_X^+ : \phi \rightarrow +$ with the probability $P(+) = P(X)$

$-$ $\qquad$ $r_X^- : \phi \to -$ with the probability $P(\text{-}) = P(U\text{-}X)$.

The above two rules reflect the situation of extreme lack of knowledge, when there is no other information about objects $e \in U'$, with the exception of frequency distribution of positive and negative cases. Therefore, these two rules are not very useful if the set of attributes is non-empty and if approximation region rules of sufficiently high confidence can be computed. To take full advantage of this extra knowledge (i.e. attributes and their values) the parameters lower limit l and upper limit u for the approximation regions should be set in such a way as to provide for some gain of the predictive capability over the "empty antecedent rules", that is, it is required that: (1) $u > P(+)$ and (2) $l < P(+)$.

The requirement (1) guarantees that the positive region rule will predict $+$ with higher confidence and smaller error than random choice and the requirement (2) guarantees that the negative region rule will predict - with higher confidence and smaller error than random selection.

5   Conjunctive Rules v.s. Approximation Region Rules

The number of data cases matching the rule condition part is referred to in data mining literature as the strength of the rule (alternative naming conventions used are scope of the rule or rule coverage). For acceptable assessments of the conditional probabilities (confidence factors) associated with the rules the strength should be higher than the threshold n. In practical applications, however, particularly when the number of attributes is large, the elementary rules tend to be weak, often supported by just a few cases. This means that their confidence factors are likely to be unreliable. On the other hand, even with relatively many weak elementary rules contained in an approximation region rule, the support level for the approximation region rule may exceed the threshold n resulting in a credible estimate of the rule's conditional probability. In fact, all that is required is that the sum of support levels over all elementary rules contained in a given approximation region rule be greater than the threshold n, that is, for the rule
$r_X^* = \vee r_{E,X}^*$ it is required that $\sum supp(r_{E,X}^*) > n$.

The above condition indicates that even in the absence of sufficient support for the elementary rules, the approximation region rules would still be acceptably supported and, consequently, could be used for predictive decision making. This leads to the conclusion that elementary rules should not be used individually for decision making but in sets corresponding to sufficiently strong approximation region rules.

Most of the interest in rule "induction" from data using rough sets methodology is focused on computation of minimal length conjunctive rules corresponding to prime implicants of the associated discernibility function [6]. Outside the rough sets approach, similar rules are computed by other methods [14]. The common feature of all these approaches is that they attempt to minimize the number of conditions in the antecedent parts of the rules. We will refer to such rules as minimal rules. An important advantage of minimal rules over other kinds of rules, such as those corresponding to root-leaf paths in decision trees or elementary rules, is the relatively high strength of the minimal rules. In what follows we compare the minimal rules to the approximation region-based rules. The comparison is conducted with respect to such criteria as completeness of the rule set, its incrementality and support level. To ensure fairness of the comparison we will assume that we are comparing an approximation region rule versus a set of minimal rules whose support sets cover the same approximation region computed from the sample $U' \subseteq U$.

- **Completeness:** Since the relationships between minimal rules, in terms of the overlaps between their support sets, are not known, it is impossible to determine how complete the set of minimal rules is. On the other hand, the elementary rules have disjoint support sets and their number is combinatorially bounded which means that definite completeness criteria can be established.
- **Incrementality:** A new minimal rule added to the existing collection of rules may have its support set covered by support sets of existing rules which essentially means that the new rule is redundant. An extra elementary rule for an approximation region will always contribute, in terms of covering previously uncovered cases, to the set of existing elementary rules. This means that the approximation region rule would grow incrementally as new sample cases are added to the database.
- **Support:** Support level of an approximation region rule is typically higher, and never lower than support of a minimal rule for the same approximation region, for obvious reasons. On the other hand, one could argue that a disjunction of all minimal rules would have the support level as good as the corresponding approximation region rule. Such a rule, however, would be ill-defined in the context of our requirements, as there is no guarantee that its confidence factor would not be lower than the threshold level u, even if all component minimal rules would satisfy that requirement. The other aspect is that it would not be possible to determine the support level of such a rule in terms of support levels of the component minimal rules leading to serious implementation difficulties when dealing with large databases.

6   Conclusions

The article introduced approximation region-based decision tables, both in probabilistic and non-probabilistic setting. It appears that decision tables of this kind are superior data-extracted knowledge representation technique to conjunctive rules as it has been argued in the last section of the paper. The necessary software for generation of such tables is currently being incorporated into KDD-R system for data mining [11]. Significant potential for applications of this technique seems also to exist in pattern recognition, machine learning and trainable control.

Acknowledgment

The research reported in this article was partially supported by a research grant awarded by Natural Sciences and Engineering Research Council of Canada.

References

1. Pawlak, Z., Grzymała–Busse, J., Słowiński, R., Ziarko, W. (1995). Rough sets. *Communications of the ACM*, 38, 88–95.
2. Pawlak, Z. (1991). *Rough Sets - Theoretical Aspects of Reasoning about Data.* Kluwer Academic Publishers, Boston, London, Dordrecht.
3. Pawlak, Z. (1982). Rough sets. *International Journal of Computer and Information Sciences*, 11, 341–356.
4. Son, N. (1997). Rule induction from continuous data. in: P.P. Wang (ed.), *Joint Conference of Information Sciences,*March 1-5, Duke University, Vol. 3, 81–84.
5. Słowiński, R. (ed.) (1992). *Intelligent Decision Support. Handbook of Applications and Advances of the Rough Set Theory*, Kluwer Academic Publishers, Boston, London, Dordrecht.
6. Skowron A., Rauszer, C. (1992). The discernibility matrices and functions in information systems. In: R. Słowiński (ed.) *Intelligent Decision Support. Handbook of Applications and Advances of the Rough Set Theory.* Kluwer Academic Publishers, Boston, London, Dordrecht, 311–362.
7. Ziarko, W. (ed.) (1994) Rough Sets, Fuzzy Sets and Knowledge Discovery, Springer Verlag, 326–334.
8. Yang, A., Grzymała–Busse J. (1997). Modified algorithms LEM1 and LEM2 for rule induction form data with missing attribute values., In: P.P. Wang (ed.), *Joint Conference of Information Sciences*, March 1-5, Duke University, Vol. 3, 69–72.
9. Ziarko, W. (1993). Variable precision rough sets model.*Journal of Computer and Systems Sciences*, vol. 46, no. 1, 39–59.
10. Ziarko, W., Katzberg, J.(1993). Rough sets approach to system modelling and control algorithm acquisition. *Proceedings of IEEE WASCANEX 93 Conference*, Saskatoon, 154–163.
11. Ziarko, W., Shan, N. (1994). KDD-R: a comprehensive system for knowledge discovery using rough sets. *Proceedings of the International Workshop on Rough Sets and Soft Computing*, RSSC'94, San Jose 1994, 164–173.
12. Katzberg, J., Ziarko, W. (1996). Variable precision extension of rough sets. *Fundamenta Informaticae*, vol. 27, no. 2-3, 155–168.
13. Michalski, R., Tecuci, G. (eds.) (1994). *Machine Learning.* Morgan Kaufman Publishers, San Francisco.
14. Hurley, R. (1983) *Decision Tables in Software Engineering.* Van Nostrand Reinhold Data Processing Series.

A Model of RSDM Implementation

María C. Fernández–Baizán * Ernestina Menasalvas Ruiz
Anita Wasilewska
{cfbaizan, emenasalvas}@fi.upm.es, anita@sunysb.edu

Departamento de Lenguajes y Sistemas Informáticos e Ingeniería del Software,
Facultad de Informática, Campus de Montegancedo, Madrid

Department of Computer Science, State University of New York
Stony Brook NY 11794 New York

Abstract. Today's Data Base Management Systems do not provide
functionality to extract potentially hidden knowledge in data. This problem gave rise in the 80's to a new research area called Knowledge Discovery in Data Bases (KDD). In spite the great amount of research that has
been done in the past 10 years, there is no uniform mathematical model
to describe various techniques of KDD. The main goal of this paper is to
describe such a model. The Model integrates in an uniform framework
various Rough Sets Techniques with standard, non *Rough Sets* based
techniques of KDD.
The Model has been already partially implemented in RSDM (Rough Set
Data Miner) and we plan to complete the implementation by integrating
all the operations in the code of database management systems.
Operations that are defined in the paper have successfully been implemented as part of RSDM.

1 Introduction

KDD process was first defined as the '*non trivial process of extracting valid potentially useful and ultimately understable patterns in data*' (see [8]). Since the
appearance of the KDD term a lot of research has been developed around the
efficient implementation of algorithms to extract knowledge from data [12,10,5].
Some of the algorithms that have been studied come from different research fields
such as statistics, machine learning, artificial intelligence. They have just been
adapted to tackle extralarge databases.
Some implementations deal with the integration of the algorithms with Relational Databases (see [4], [9], [2], [13]). In such systems, the algorithms are run
over the relational database making use of SQL to select and order the objective data. For some of these algorithms a tightly coupled version has been
programmed [1] has been shown to improve the efficiency but still improvements
could be obtained if such algorithms would be programmed as operations of the

---

* This work is supported by the Spanish Ministry of Education under project PB95-
0301

L. Polkowski and A. Skowron (Eds.): RSCTC'98, LNAI 1424, pp. 186–193, 1998.
© Springer-Verlag Berlin Heidelberg 1998

RDBMS because the execution of the algorithms will take advantages of the optimizer of the system.

In spite of the great amount of research that has been around KDD there is no mathematical model to describe in an uniform way the whole process of discovering patterns, although a first approach of a model of generalization is presented in [3].

In this paper we present a first approach to such a model. Basic functions have been modelized and some properties of such functions have been extracted.

The rest of the paper is as follows: in section 2 the universe of the functions as well as the preliminaries of the model are established. In section 3 basic functions are defined in mathematical terms what allows us to extract the similarities among such functions. To conclude, in section 5 the future lines of research are outlined.

2   DMM: Data Base Mining Model

Database Mining, as we understand it, is an extraction of knowledge from a certain collection of information. We extract the knowledge by identifying similar information. We assume that information is given by means of attributes and its values. In order to describe formally the information similarity and knowledge extraction, we also assume that every piece of information in the collection is uniquely identified. We will refer to these unique identifiers as **objects**.

Remark: In relational databases identifiers are represented by the key attribute and the information is determined by the values of the rest of the attributes in a certain table.

Similarity of information is a nice concept, but difficult to implement. So we make an stronger assumption: We deal only with some forms of **equivalent** information.

Given a set of attributes we can define many equivalence relations on the set of objects. We say that two objects in a database are equivalent with respect to certain attributes if they both have the same values associated with these attributes.

2.1   Preliminaries

In any database mining situation, we have to assume that we have at least objects and attributes.

So, in our investigation we assume that we have a finite set O not empty of objects and a finite not empty set AT of attributes. We also have to assume that attributes have associated values with them. On the other hand, as we are trying to mine databases looking for information, we can assume that we deal with non empty databases.

For the sake of simplicity, we assume that objects and attributes are disjoint sets.

We see elements of the universe through the information about them. So, we assume in our investigation that with each element x of the universe O ($x \in O$) we associate a certain information $\mathbf{i}(x)$. Moreover, the information $\mathbf{i}(x)$ is the set of all elements of the universe that are the same as x with respect to the information $\mathbf{i}$.

In the case of databases, the function $\mathbf{i}$ refers to an attribute or combination of attributes (descriptor). If we impose every attribute to be a monovalued function (this means one object for an attribute can only take a value), any $\mathbf{i}$, induces a partition of O into equivalence classes. So, it happens that $\mathbf{i}(x)$ can be seen as the name of the equivalence class of x regarding information $\mathbf{i}$. That's why, we can effectively define $\mathbf{i}$ as a function that maps O in $\mathcal{P}(O)$. Formally, we define the information function $\mathbf{i}$ about the universe O as any function:

$$\mathbf{i}: \quad O \quad \longrightarrow \quad \mathcal{P}(O)$$

such that $y \in \mathbf{i}(x)$ means that y and x are the same with respect to information $\mathbf{i}$.

$$\forall x \in O, \mathbf{i}(x) = \{y \in O : \mathbf{i}(x) = \mathbf{i}(y)\} = [x]_i$$

Obviously

$$O/\approx_i = \{[x]_i, x \in O\}$$

Definition: We define a function r in the following way:

$$r: \quad \mathcal{P}(\mathcal{AT}) \times \mathcal{P}(\mathcal{O}) \quad \longrightarrow \quad \mathcal{P}(\mathcal{P}(\mathcal{O}))$$

such that for all $A_x \subset AT$, $O = \{x_1, x_2, \ldots, x_n\}$, $O_x \subset O$:

$$r(AT, O) = \{\{x_1\}, \{x_2\}, \ldots \{x_n\}\}$$

$$r(\emptyset, O) = \{\emptyset\}$$

$$r(A_x, \emptyset) = \{\emptyset\}$$

$$r(A_x, O_x) = \{O_x/\approx_{A_x}\}$$

2.2   Data Mining Universe

Let $O \neq \emptyset$, O finite set of **objects** and let $AT \neq \emptyset$, AT finite set of **attributes** Let $\mathcal{R}$ be the set every equivalence relation defined on O using the descriptors from AT.

Definition : The set

$$U = \mathcal{P}(\bigcup\{O/\approx: \approx \in \mathcal{R}\})$$

will be called Data Mining Universe from now on.

Observe that the universe can also be defined as:

$$U = \mathcal{P}(\bigcup\{r(A_X, O), A_X \subseteq AT\})$$

Remark: Observe that the universe generation complexity is exponential but it is important noting that it is a theoretical construction only needed for the model to be establish. In any case it will be needed to generate it while running any algorithm.

In what follows we will try to express rough set operations as functions over this universe. In order to express all the operations in a uniform way we define a generic function that will be called data mining operator.

Definition: We define *Data Mining* operator as any partial function of the form:

$$g: \quad U \times \mathcal{P}(AT) \quad \longrightarrow \quad U \times \mathcal{P}(AT)$$

such that if

$$g(X, A_X) = (Y, A_Y)$$

then:

$$X \subseteq r(A_X, O)$$

$$Y \subseteq r(A_Y, O)$$

$$A_Y \subseteq A_X$$

Once the basic function has been defined we will try to defined basic *Data Mining* operations in this terms.

3 Modelization of Operations of *Data Mining* Using *Rough Sets*

3.1 Projection operation

Projection operation selects certain attributes (columns in a relational table), of the objects taken into account and eliminates the rest of the attributes.

Fact: If $A_P \subseteq A_X$

$$O/\approx_{A_P} \ \ni O/\approx_{A_X}$$

meaning that :

$$\forall x \in O/\approx_{A_P} \ \exists y \in O/\approx_{A_X}$$

such $x \supseteq y$

Having $A_P \subseteq A_X$, the family of partial functions P_{A_P} is defined in the following way:

$$P_{A_P} : U \times \mathcal{P}(AT) \longrightarrow U \times \mathcal{P}(AT)$$

$$P_{A_P}(X, A_X) = (Y, A_P)$$

where, $Y = \{[x]_{A_P}\}$ for all X such that, $A_P \subseteq A_X$ and $X \subseteq r(A_X, O)$ Observe that,

$$\forall y \in Y, \quad \exists x \in X : y \supseteq x$$

where: $\bigcup_{y \in Y} y = \bigcup_{x \in X} x = O$

3.2   Selection operation

This operation is used to select the subset of elements of the universe of objects that match a condition. The condition is a boolean expression specified on the attributes and their values.

The condition of selection is expressed as a well formed formula of first order logic:

$$\{x \in O : f(x) is\ true\}$$

Definition: The condition f is expressed as:

$$\bigwedge_{i=1}^{n} A_i(x) \alpha v_i$$

where:

$v_i \in Dom(A_i) \forall i$

α is a comparison operator that is, $\alpha \in \{=,] \geq, >, \leq, <\}$

We get now two sets of objects:

1. $O_f = \{x \in O \ \ such \ \ that \ \ f(x) = True\}$
2. $A_f = \{A \in AT \ \ belonging \ \ to \ \ the \ \ alphabet \ \ defined \ \ by \ \ f\}$

Property: Given a set of attributes A_f, and a set of objects O, The set of attributes A_f defines a equivalence relation on O $O/ \approx_{A_f}$

Definition: Selection operation: Given f, O_f, A_f the family of functions S_f is defined in the following way:

$$S_f : U \times \mathcal{P}(AT) \longrightarrow U \times \mathcal{P}(AT)$$

$$S_f(X, A_X) = (\{x \in X : x \cap O_f \neq \emptyset\}, A_X)$$

for all X and for all A_f such that $X \subseteq r(A_X, O)$, $A_f \subseteq A_X$

3.3   Lower operator

Remark: For a further explanation of *Rough Sets* theory see [6,7,11,14].
A family of functions l_C, is defined where C represents the concept which lower
approximation is going to be defined with respect to the descriptor A_X:

$$l_C : \quad U \times \mathcal{P}(\mathcal{AT}) \quad \longrightarrow \quad \mathcal{U} \times \mathcal{P}(\mathcal{AT})$$

$$l_C(X, A_X) = (\{z \in X : x \subseteq C\}, A_X)$$

$$X \subseteq r(A_X)$$

Property
If $l_C(X, A_X) = (Y, A_Y)$, then $Y \subset X$

3.4   Upper operator [6,7,11,14]

The family of functions u_C, is defined as the upper approximation of objects
belonging to C with respect to to the descriptor A_X :

$$u_C : \quad U \times \mathcal{P}(AT) \quad \longrightarrow \quad U \times \mathcal{P}(AT)$$

$$u_C(X, A_X) = (\{z \in X : z \cap C \neq \emptyset\}, A_X)$$

$$X \subseteq r(A_X)$$

Let's see now how all the operations that have been studied can be expressed by
means of this basic operators so we can modelize the process by using them.

3.5   Key Elimination

Let $T(K, C_1, \dots, C_n, D)$ be the relational decision table containing the target
data. Key elimination can be expressed in terms of the operators of the model
as: $\Pi_{C_1, \dots, C_n, D}(T)$, where K represents the key attribute.

3.6   Eliminating those Equivalence Classes which Cardinality is not Higher than a Predetermined Value

Users will be interested in rules or descriptions of the concepts that are supported
by a sensible number of objects. So those rows corresponding to descriptions not
supported by enough number of elements can be eliminated. Observe that this
operation be expressed by means of selection operation. Let's called the function
$E_{threshold}$. Then f can be expressed as $f = counter \geq threshold$ supposing that
the counter attribute exists in the table. So we have:

$$E_{threshold} = S_f(X, AT)$$

3.7   Reduction of Attributes

This step deals with the finding of a small set of attributes to represent the data, hence it eliminates all the superfluous information.

Dealing with the basic operations that have already been described, the result of applying reduct calculation to the table can be expressed as a projection over the selected attributes.

3.8   Generalization

Should you have a concept hierarchy available for some attributes, in preprocessing phase a generalized attriould be obtained. Thus before applying the mining algorithm if a particular generalization is desired in this point all that rest is to project over the generalized attribute. So Let A_g be the generalized attribute of attribute A that is wanted for a particular query. Let G_A a notation for the operation of generalization. It can be expressed:

$$G_{A_{1_g}}(T(A_1, A_2, \ldots, A_n, A_{1_g})) = \overline{\varPi}_{A_{1_g}, A_2, \ldots, A_n}$$

3.9   Discretization of Values of Attributes

Discretization of attributes is a particular case of generalization in which values of the attribute are generalized, so it can be expressed by means of projection operation.

4   Conclusions

We have presented here a Model, DMM,which identifies stages of Data Mining process and describes them in terms of various Data Base mining operations. Some of the operations that have been presented are *Rough Sets* basic functions such as the lower and upper approximation of a concept.
These operations have been successfully defined as operations of the model. Also basic KDD operations not belonging to *Rough Sets* theory have been studied and described as part of the model such as discretization, generalization and preprocessing operations. As a consequence we can establish that the model allows to define in a uniform way KDD operations.

5   Future Developments

The model is a result of the detailed study of the implementation of RSDM system. The functions that have been described as part of the model have successfully been implemented. On the other hand, the model has allowed to design a graphical interface in which the process of data analysis is displayed to the final user.

Implementation of algorithms to obtain association rules that have been shown to be able to be described as operations of the model is under development.

Acknowledgments

Thanks are due to Dr. Ziarko and Dr. Skowron. for several helpful comments.

References

1. R. Agrawal, K. Shim *Developing Tightly-Coupled Data Mining Applications on a Relational Database System* Proceedings of the second International Conference on Knowledge Discovery and Data Mining, ed. Simoudis, Han, Fayyad, 1996, pp. 287–291

2. G. H. John, *Enhancements to the Data Mining Process* PHD Dissertation Stanford University, July 1997.

3. M. Hadjimichael *A Model for Generalization*, Proceedings of JCIS'97, pp. 113–120 vol 3.

4. J. Han et al., *DBMiner: A System for mining knowledge in Large Relational Databases* In Proceedings The Second International Conference on Knowledge discovery and Data Mining, pp. 250–255. August 1996

5. X. Hu, N. Cercone, *Learning in Relational Databases: A Rough Set Approach* Computational Intelligence: An International Journal, vol. 11, issue2, May 1995, pp. 323–339

6. Z. Pawlak, *Rough Sets - Theoretical foundations*, Information systems, 6, No.4, 1993, pp. 299–297.

7. Z. Pawlak, *Rough Sets - Theoretical Aspects of Reasoning about Data*, Kluwer, 1991.

8. G. Piatesky–Shaphiro, *An Overview of Knowledge Discovery in Databases: Recent Progress and Challenges*, Rough Sets, Fuzzy Sets and Knowledge Discovery, 1994, pp. 1–11.

9. J. Komorowski, A. Ohrn *Rosetta: A Rough Set Toolkit for Analysis of Data* In Proceedings, JCIS'97, North Carolina, March-97, pp. 403–407

10. N. Shan, W. Ziarko, J. Hamilton, N. Cercone, *Using Rough Sets as Tools for Knowledge Discovery* Proccedings of the first International Conference on Knowledge Discovery and Data Mining. Ed. U. Fayyad and R. Uthurusamy, pp. 263–269

11. A. Skowron, *The Discernibility Matrices and Functions in Information Systems*, Decision Support by Experience, R. Slowinski(ed.) Kluwer Academic Publishers, 1992.

12. A. Wasileska and M. Hadjimichael *Rules Reduction for Knowledge Representation Systems* Bulletin of Polish Academy of Science. Vol. 38, No. 1-12, (1990), 113–120.

13. W. Ziako, N. Shan *KDD-R: A Comprehensive System for Knowledge Discovery in Databases using rough sets* RSSC'94 the 3rd International Workshop on Rough Sets and Soft Computing. Conference Proceedings, Nov. 10-12, pp. 164–173

14. W. Ziarko, *Variable Precision Rough Sets Model*, Journal of Computer and System Sciences, vol. 46. 1993, 39–59.

Handling Queries in Incomplete CKBS through Knowledge Discovery

Zbigniew W. Raś

University of North Carolina, Dept. of Comp. Science, Charlotte, N.C. 28223, USA
Polish Academy of Sciences, Dept. of Comp. Science, 01-237 Warsaw, Poland
email: ras@uncc.edu or ras@wars.ipipan.waw.pl

[Abstract.] In this paper, we propose a new query answering system for an incomplete Cooperative Knowledge-Based System (CKBS). CKBS is a collection of autonomous knowledge-based systems called agents which are capable of interacting with each other. In the first step of the query processing strategy, the contacted site of CKBS will identify all locally incomplete attributes used in a query. An attribute is locally incomplete if there is an object in a local information system with an incomplete information on this attribute. The values of all locally incomplete attributes are treated as concepts to be learned at other sites of CKBS (see [6]). Rules discovered at all these sites are sent to the site contacted by the user and used locally by the query answering system to replace an incomplete information by values provided by the rules.
In the second step of the query processing strategy, an incomplete information is removed from the local information system in a maximal number of places. Next, the query answering system finds the answer to a user query in a usual way (similar to CKBS query answering system).

Key Words: incomplete information system, cooperative query answering, rough sets, multi-agent system, knowledge discovery.

1   Introduction

By a cooperative knowledge-based system ($CKBS$) we mean a collection of autonomous knowledge-based systems called agents (sites) which are capable of interacting with each other. Each agent is represented by an information system (either complete or incomplete) and a collection of rules called a knowledge base. Any site of $CKBS$ can be a source of a local or a global query. By a local query for a site i (or i-reachable query) we mean a query entirely built from attributes which are complete and local at site i. Local queries need only to access an information system of the site where they were issued and they are completely processed on the system associated with that site. In order to resolve a global

L. Polkowski and A. Skowron (Eds.): RSCTC'98, LNAI 1424, pp. 194–201, 1998.
© Springer-Verlag Berlin Heidelberg 1998

query for a site i (built from attributes not necessarily complete or local at site i) successfully, we have to access an information system at more than one site of $CKBS$ and discover rules describing attributes (used in a query) which are either not complete or not local at the site i. Rules discovered by neighbors of i are sent to the site i and used locally by the query answering system to replace some of the incomplete vales in a local information system by values provided by the rules. After the process of removing as many incomplete vales as possible in the information system of site i, the query answering system finds the answer to a user query in a usual way (similarly to CKBS query answering system).

There is a number of strategies which allow us to find rules describing decision attributes in terms of classification attributes. We should mention here such systems like $LERS$ (developed by J. Grzymala–Busse), $DQuest$ (developed by W. Ziarko), $AQ15$ (developed by R. Michalski) or rules discovery system based on discriminant functions proposed by A. Skowron (see [8]). Most of these strategies have been developed under the assumption that the database part of KBS is complete. Problem of inducing rules from attributes with incomplete values was discussed in ([2], [3], [4]). Our strategy shows how to compute such rules with certainty factors not necessarily equal to 1 and next how to use them to make local information system more complete. The Chase algorithm presented for instance in [1] is using dependencies to make a database more complete. We use rules learned at remote sites to achieve a similar goal.

2   Basic definitions

In this section, we introduce the notion of an information system, distributed information system, a knowledge base, and $s(i)$-queries which can be processed locally at site i.

By an information system ([5], [4]) we mean a structure $S = (X, A, V, f)$, where X is a finite set of objects, A is a finite set of attributes (or properties), V is the set-theoretical union of domains of attributes from A, and f is a classification function which describes objects in terms of their attribute values. We assume that:

- $V = \bigcup \{V_a : a \in A\}$ is finite,
- $V_a \cap V_b = \emptyset$ for any $a, b \in A$ such that $a \neq b$,
- $f : X \times A \longrightarrow 2^V$ where $f(x, a) \in 2^{V_a} - \{\emptyset\}$ for any $x \in X$, $a \in A$.

If $f(x, a) = V_a$, then the value of the attribute a for the object x is unknown. We will call system S incomplete if there is $a \in A$, $x \in X$ such that $card(f(x, a)) \geq 2$. Also, if $card(f(x, a)) \geq 2$, then the attribute a is called incomplete. Otherwise system S as well as the attribute a are called complete. The set of all incomplete attributes in S we denote by $In(A)$ and the set $\bigcup \{V_a : a \in In(A)\}$ by $In(V)$. For simplicity

reason any complete or incomplete information system will be called, in this paper, an information system.

Let $S_1 = (X_1, A_1, V_1, f_1)$, $S_2 = (X_2, A_2, V_2, f_2)$ be information systems.

- S_2, S_1 are consistent if $f_1(x,a) \subseteq f_2(x,a)$ or $f_2(x,a) \subseteq f_1(x,a)$ for any $a \in A_1 \cap A_2$, $x \in X_1 \cap X_2$.

By a distributed information system [8] we mean a pair $DS = (\{S_i\}_{i \in I}, L)$ where:

- $S_i = (X_i, A_i, V_i, f_i)$ is an information system for any $i \in I$,
- L is a symmetric, binary relation on the set I,
- I is a set of sites.

System DS is called incomplete, if $(\exists i \in I)[S_i$ is *incomplete*].

Systems S_i, S_j (sites i, j) are called neighbors in DS if $(i,j) \in L$. The transitive closure of L in I is denoted by L^*.

A distributed information system $DS = (\{S_i\}_{i \in I}, L)$ is consistent if:
$(\forall i)(\forall j)(\forall x \in X_i \cap X_j)(\forall a \in A_i \cap A_j)$
$[(x,a) \in Dom(f_i) \cap Dom(f_j) \longrightarrow f_i(x,a) \subseteq f_j(x,a)$ or $f_j(x,a) \subseteq f_i(x,a)]$.

By a set of $s(i)$-terms we mean a least set T_i such that:

- $\mathbf{0}, \mathbf{1} \in T_i$,
- $(a,w) \in T_i$ for any $a \in A_i$ and $w \in V_{ia}$,
- if $t_1, t_2 \in T_i$, then $(t_1 + t_2), (t_1 \star t_2), \sim t_1 \in T_i$.

We say that:

- $s(i)$-term t is *atomic* if it is of the form (a,w) or $\sim (a,w)$ where $a \in B_i \subseteq A_i$ and $w \in V_{ia}$
- $s(i)$-term t is *positive* if it is of the form $\prod\{(a,w) : a \in B_i \subseteq A_i$ and $w \in V_{ia}\}$
- $s(i)$-term t is *primitive* if it is of the form $\prod\{t_j : t_j$ is atomic $\}$
- $s(i)$-term is in *disjunctive normal form* (DNF) if $t = \sum\{t_j : j \in J\}$ where each t_j is primitive.

By a query for a site i ($s(i)$-query) we mean any element in T_i which is in DNF.

Before we give the interpretation of $s(i)$-queries, we introduce the notion of X-algebra. So, let us assume that X is a set of objects. By an X-algebra we mean a sequence $(\mathbf{P}, \bigoplus, \bigotimes, \neg)$ where:

- $\mathbf{P} = \{P_i : i \in J\}$ where $P_i = \{(x, p_{<x,i>}) : p_{<x,i>} \in [0,1] \ \& \ x \in X\}$,
- $P_i \bigotimes P_j = \{(x, p_{<x,i>} \cdot p_{<x,j>}) : x \in X\}$,
- $P_i \bigoplus P_j = \{(x, max(p_{<x,i>}, p_{<x,j>})) : x \in X\}$,
- $\neg P_i = \{(x, 1 - p_{<x,i>}) : x \in X\}$,
- $\mathbf{P}$ is closed under the above three operations.

[**Theorem 1.**] Let $P_i, P_j, P_k \in \mathbf{P}$. Then:

- $(P_i \bigotimes P_j) \bigotimes P_k = P_i \bigotimes (P_j \bigotimes P_k)$,
- $(P_i \bigoplus P_j) \bigoplus P_k = P_i \bigoplus (P_j \bigoplus P_k)$,
- $(P_i \bigoplus P_j) \bigotimes P_k = (P_i \bigotimes P_k) \bigoplus (P_j \bigotimes P_k)$,
- $P_i \bigotimes P_j = P_j \bigotimes P_i$,

- $P_i \bigoplus P_j = P_j \bigoplus P_i$,
- $P_i \bigoplus P_i = P_i$.

Let $DS = (\{S_j\}_{j \in I}, L)$ be a distributed information system where $S_j = (X_j, A_j, V_j, f_j)$ and $V_j = \bigcup\{V_{ja} : a \in A_j\}$, for any $j \in I$. By a standard interpretation of $s(i)$-queries in DS we mean a partial function M_i, from the set of $s(i)$-queries into X_i-algebra, defined as follows:

- $Dom(M_i) \subseteq \mathbf{T}_i$,
- $M_i((a, w)) = \{(x, p) : x \in X_i$ & $w \in f_i(x, a)$ & $p = 1/card(f_i(x, a))\}$ for any $w \in V_i$,
- $M_i(\sim (a, w)) = \neg M_i((a, w))$
- for any atomic term $t_1(a) \in \{(a, w), \sim (a, w)\}$ and any primitive term $t = \prod\{s(b) : (s(b) = (b, w_b)$ or $s(b) =\sim (b, w_b))$ & $(b \in B_i \subset A_i)$ & $(w_b \in V_{ib})\}$ we have

$$M_i(t \star t_1(a)) = M_i(t) \bigotimes M_i(t_1) \text{ if } a \notin B_i$$
$$M_i(t \star t_1(a)) = \emptyset \text{ if } a \in B_i \text{ and } t_1(a) \neq s(a),$$
$$M_i(t \star t_1(a)) = M_i(t) \text{ if } a \in B_i \text{ and } t_1(a) = s(a).$$

- for any $s(i)$-terms t_1, t_2

$$M_i(t_1 + t_2) = M_i(t_1) \bigoplus M_i(t_2).$$

By (k, i)-rule in $DS = (\{S_j\}_{j \in I}, L)$, $k, i \in I$, we mean a pair (t, c) such that:

- either $c \in In(V_i) \cap V_k$ or $c \in V_k - V_i$,
- t is a positive $s(k)$-term which belongs to $\mathbf{T}_k \cap \mathbf{T}_i$,
- if $(x, p1) \in M_k(t)$ then $(\exists p2)[(x, p2) \in M_k(c)]$.

An object x satisfies a rule $r = (t, c)$ with a certainty p at site k, if $p = p1 \cdot p2$, $(x, p1) \in M_k(t)$, and $(x, p2) \in M_k(c)$.

We say that (k, i)-rule (t, c) is in k-optimal form if there is no other subterm $t1 \in \mathbf{T}_k \cap \mathbf{T}_i$ of $s(k)$-term t, such that: if x satisfies rule (t, c) with certainty p, then x satisfies rule $(t1, c)$ with the same or higher certainty.

Let $X = \{x_i : 1 \leq i \leq n\}$ and x_i satisfies the rule $r = (t, c)$ with a certainty p_i at site k for any $i \in \{1, 2, ..., n\}$. We say that r has certainty p, if $p = [\Sigma\{p_i : p_i \neq 0$ & $1 \leq i \leq n\}]/[card\{i : p_i \neq 0$ & $1 \leq i \leq n\}]$.

By a knowledge base D_{ki} we mean any set of (k, i)-rules satisfying the condition below:

$$\text{if } (t, c) \in D_{ki} \text{ then } (\exists t_1)(t_1, \sim c) \in D_{ki}.$$

We say that a knowledge base D_{ki} is in k-optimal form if all its rules are in k-optimal form.

In [6] we proposed an algorithm to construct a knowledge base D_{ki} in k-optimal form. Let us assume that $L(D_{ki}) = \{(t, c) \in D_{ki} : c \in In(V_i)\}$. The algorithm, given below, converts system S_i in DS to a new more complete information system $Chase(S_i)$.

> **Algorithm** $Chase(S_i, In(A_i), L(D_{ki}))$;
> Input system $S_i = (X_i, A_i, V_i, f_i)$, set of incomplete attributes
> $In(A_i) = \{a_1, a_2, ..., a_k\}$, and a set of rules $L(D_{ki})$
> Output a system $Chase(S_i)$.
> **begin**
> $j := 1$;
> while $j \leq k$ do
> > for all $c \in V_{a_j}$ do
> > > while there is $x \in X_i$ and a rule $(t, c) \in L(D_{ki})$
> > > such that $x \in M_i(t)$ and $card(f_i(x, a_j)) \neq 1$ do
> > > $f_i(x, a_j) := c$;
> > $j := j + 1$
> $Chase(S_i) := S_i$
> **end**

By a standard chase-interpretation $\hat{M}_i$ of $s(i)$-queries in a distributed system DS, we mean the standard interpretation M_i of $s(i)$-queries in a distributed information system $Chase_i(DS) = (\{\hat{S}_j\}_{j \in I}, L)$, where:

- $\hat{S}_j = S_j$ if $j \neq i$,
- $\hat{S}_j = Chase(S_j)$ if $j = i$.

3   Cooperative knowledge-based system

In this section, we define a Cooperative Knowledge Based System ($CKBS$) and introduce the notion of its consistency. We also give an example of CKBS.

Let $\{D_{ki}\}_{k \in K_i}$, $K_i \subseteq I$, be a collection of knowledge bases where D_{ki} was created at site $k \in I$ for any $k \in K_i$ and $D_i = \bigcup\{D_{ki} : k \in K_i\} \cup R_i$. By R_i we mean a set of rules (t, c) created by an expert and stored at site i. Additionally, we assume here that t is an $s(i)$-term. System $(\{(S_i, D_i)\}_{i \in I}, L)$, introduced in ([6], [7]), is called a cooperative knowledge-based system ($CKBS$).

Rules $(t1, w1) \in D_{ki}$, $(t2, w2) \in D_{ni}$ are consistent at Site i if $At(w1) \neq At(w2)$ or $w1 = w2$ or $M_i(t1 \star t2) = \emptyset$. Otherwise, we call them possibly inconsistent. We say that the knowledge base D_i is consistent at Site i if any two rules in D_i are consistent at Site i. Similarly, we say that the cooperative knowledge based system $DS = (\{(S_i, D_i)\}_{i \in I}, L)$ is consistent if D_i is consistent at Site i for any $i \in I$.

Figure 1 gives an example of $CKBS$. Rules in the knowledge base of Site 2 have been computed at another site of $CKBS$. It can be easily checked that these rules are consistent at Site 2.

X2	F	C	D	E	G
a1	f1	c1	d1	e1	g1
a6	f2	c1	d2	{e2, e3}	g2
a8		c2	d1	e3	g1
a9	f2	c1		e3	g1
a10	f2	c2	d1	{e1, e3}	g1
a11	f1	c2	d1	e3	g2
a12	f1	c1	d2	{e3, e2}	g1

Links

to other sites

Rules	Certainty Factor
(e1, b1)	23/48
(c1*e2, b1)	1/4
(c2, b1)	7/16
(c1*e1, b2)	5/24
(e3, b2)	11/18
(c3, b2)	8/18
(c2*e2, b3)	5/24

SITE 2

Fig. 1. Site 2 of CKBS

4 Query Language and Its Interpretation.

In this section we introduce a query language and propose its optimistic interpretation in a $Site(i)$ of $CKBS$. A formal system for handling queries in $CKBS$ will be presented in a separate paper.

Standard chase-interpretation $\hat{M}_i$, introduced in Section 2, shows how to interpret $s(i)$-queries in a $Site(i)$ of $CKBS$. The question of interpreting DNF queries built from values of attributes belonging to a superset of V_i in $Site(i)$ remains open. Such queries are called global for a Site i. Their standard interpretation at Site i of a cooperative knowledge based system $(\{(S_j, D_j)\}_{j \in I}, L)$, where $D_j = \bigcup\{D_{nj} : n \in K_j\}$, $S_i = (X_i, A_i, V_i, f_i)$ is proposed. To simplify our notation, we write S instead of S_i ,we write w instead of and atomic term (a, w) and assume that $V = V_i = \bigcup\{V_{ia} : a \in A_i\}$ and $C_S = \bigcup\{V_j : j \in I\} - V$. Elements in C_S are called concepts at site i.

By a query language $L(S, C_S)$ we mean a sequence (A, T, F), where A is an alphabet, T is a set of DNF terms (queries), and F is a set of atomic formulas.

The alphabet A of $L(S, C_S)$ contains:

- constants: w where $w \in V_i \cup C_S$
- constants: $\mathbf{0}, \mathbf{1}$
- functors: $+, \star, \sim$
- predicate: $=$
- auxiliary symbols: $($, $)$.

The set of terms T is a least set such that:

- constants $\mathbf{0}, \mathbf{1}$ are terms,
- if w is a constant, then $w, \sim w$ are terms,
- if t_1, t_2 are terms, then $t_1 \star t_2$ is a term.

The set of DNF terms is a least set such that:

- if t is a term, then t is a DNF term,
- if t_1, t_2 are DNF terms, then $t_1 + t_2$ is a DNF term.

Parentheses are used, if necessary, in the obvious way. As will turn out later, the order of a sum or product is immaterial. So, we will abbreviate finite sums and products as $\sum\{t_j : j \in J\}$ and $\prod\{t_j : j \in J\}$, respectively.

The set of atomic formulas F is a least set such that:

- if t_1, t_2 are DNF terms, then $(t_1 = t_2)$ is an atomic formula.

Let $\hat{M}_i$ be a standard chase-interpretation of local $s(i)$-queries in $DS = (\{S_j\}_{j \in I}, L)$. By a standard interpretation of DNF queries and atomic formulas from $L(S, C_S)$ in S-consistent cooperative knowledge based system $(\{S_j, \{D_{kj}\}_{k \in K_j}\}_{j \in I}, L)$, where $S = (X_i, A_i, V_i, f_i)$ and $V_i = \bigcup\{V_{ia} : a \in A_i\}$, we mean a partial function N_i from the set of DNF queries into X_i-algebra $(\mathbf{P}, \bigoplus, \bigotimes, \neg)$ such that:

(1)   for any $w \in V_{ia}$,
$$N_i(w) = \hat{M}_i(a, w), \quad N_i(\sim w) = \neg N_i(w)$$

(2)   if $w \in C_S$,
$$N_i(w) = max(\{(x, p) : x \in X_i \ \& \ (\exists n \in K_i)(\exists p > 0)(\exists t)[(t, w) \in D_{ni} \ \& \ (x, p) \in \hat{M}_i(t)]\}),$$
$$N_i(\sim w) = max(\{(x, p) : x \in X_i \ \& \ (\exists n \in K_i)(\exists p > 0)(\exists t)[(t, \sim w) \in D_{ni} \ \& \ (x, p) \in \hat{M}_i(t)]\}) \text{ where } (x, p) \in max(D) \text{ iff}$$
$$\sim (\exists q > p)((x, p) \in D \ \& \ (x, q) \in D)$$

(3) $N_i(\mathbf{0}) = N_i(\sim \mathbf{1}) = \emptyset, \quad N_i(\mathbf{1}) = N_i(\sim \mathbf{0}) = X_i$

(4)   for any terms t, w
$$N_i(t \star w) = N_i(t) \bigotimes N_i(w), \quad N_i(t \star (\sim w)) = N_i(t) \bigotimes N_i(\sim w)$$

(5)   for any DNF terms t_1, t_2
$$N_i(t_1 + t_2) = N_i(t_1) \cup N_i(t_2),$$

(6)   for any DNF terms t_1, t_2
$N_i((t_1 = t_2)) = ($if $N_i(t_1) = N_i(t2)$ then T else $F)$
(T stands for True and F for False)

From the point of view of site i, the interpretation N_i represents a pessimistic approach to query evaluation. If (x, p) belongs to the response of a query t, it means that x satisfies the query t with a confidence not less than p.

5 Conclusion

We have proposed a new query answering system (QAS) for an incomplete cooperative knowledge based system (CKBS). The Chase algorithm based on rules discovered at remote sites helps to make the data at the local site more complete and the same improve the previous QAS (see [6]) for CKBS.

References

1. Atzeni, P., DeAntonellis, V., "Relational database theory", The Benjamin Cummings Publishing Company, 1992
2. Grzymala–Busse, J., *On the unknown attribute values in learning from examples*, Proceedings of ISMIS'91, LNCS/LNAI, Springer-Verlag, Vol. 542, 1991, 368–377
3. Kodratoff, Y., Manago, M.V., Blythe, J."Generalization and noise", in *Int. Journal Man-Machine Studies*, Vol. 27, 1987, 181–204
4. Kryszkiewicz, M., Rybinski, H., *Reducing information systems with uncertain attributes*, Proceedings of ISMIS'96, LNCS/LNAI, Springer-Verlag, Vol. 1079, 1996, 285–294
5. Pawlak, Z., "Rough sets and decision tables", in *Proceedings of the Fifth Symposium on Computation Theory*, Springer Verlag, Lecture Notes in Computer Science, Vol. 208, 1985, 118–127
6. Ras, Z.W., Joshi, S., "Query approximate answering system for an incomplete DKBS", in *Fundamenta Informaticae Journal*, IOS Press, Vol. XXX, No. 3/4, 1997, 313–324
7. Ras, Z.W., "Cooperative knowledge-based systems", in *Intelligent Automation and Soft Computing*, AutoSoft Press, Vol. 2, No. 2, 1996, 193–202
8. Skowron, A., "Boolean reasoning for decision rules generation", in *Methodologies for Intelligent Systems, Proceedings of the 7th International Symposium on Methodologies for Intelligent Systems*, (eds. J. Komorowski, Z. Ras), Lecture Notes in Artificial Intelligence, Springer Verlag, No. 689, 1993, 295–305

Learning Logical Descriptions for Document Understanding: A Rough Sets-Based Approach

Emmanuelle Martienne & Mohamed Quafafou

Institut de Recherche en Informatique de Nantes (IRIN)
2, rue de la Houssinire - B.P. 92208 - 44322 Nantes cedex 3
E-mail: {martien,quafafou}@irin.univ-nantes.fr

Abstract. Inductive learning systems in a logical framework are prone to difficulties when dealing with huge amount of information. In particular, the learning cost is greatly increased, and it becomes difficult to find descriptions of concepts in a reasonable time. In this paper, we present a learning approach based on Rough Set Theory, and more especially on its basic notion of concept approximation. In accordance with RST, a learning process is splitted into three steps, namely (1) partitioning of knowledge, (2) approximation of the target concept, and finally (3) induction of a logical description of this concept. The second step of approximation reduces the volume of the learning data, by computing well-chosen portions of the background knowledge which represent approximations of the concept to learn. Then, only one of these portions is used during the induction of the description, which allows for reducing the learning cost. In the first part of this paper, we report how RST's basic notions namely indiscernibility, as well as lower and upper approximations of a concept have been adapted in order to cope with a logical framework. In the remainder of the paper, some empirical results obtained with a concrete implementation of the approach, i.e., the EAGLE system, are given. These results show the relevance of the approach, in terms of learning cost gain, on a learning problem related to the document understanding.

1 Introduction

Inductive Logic Programming [6] is a symbolic approach to machine learning within a logical framework. The aim of ILP is to learn logical descriptions of concepts from their ground examples and counter-examples, as well as an initial model of the domain called a background knowledge. Learning systems such as FOIL [13], FOCL [10] and PROGOL [7] are based on the paradigm of ILP. Given a concept to learn, called the *target concept*, the learning task of ILP is formalized as a search problem inside the space of the whole candidate descriptions [5], also called *hypotheses*. Some operators, such as generalization and specialization, allow for exploring the space from an hypothesis to better ones. The relevance of an hypothesis in comparison with others is estimated with different criteria, which are most often related to the number of examples and counter-examples

L. Polkowski and A. Skowron (Eds.): RSCTC'98, LNAI 1424, pp. 202–209, 1998.
© Springer-Verlag Berlin Heidelberg 1998

it characterizes, according to the given background knowledge. The search stops when an hypothesis which characterizes all the examples but no counter-example of the target concept is found. The accuracy of the learned description is then estimated through the classification of unseen examples. The cost of an exhaustive exploration of the search space relies on several parameters, such as the language used to represent hypotheses, which determines the size of the space, but also the number of examples and the size of the background knowledge, the search strategy, and so forth. In practice, this cost is prohibitive, and a lot of research in the ILP field is devoted to develop *learning biases* [2, 14, 8], i.e., techniques in order to prune the search space.

In this paper, we propose a learning approach which aims at reducing the learning cost, by limitating the volume of the background knowledge to be used for the search of the description. To achieve this goal, this approach is based on Rough Set Theory [9], and more especially on its notion of concept approximation. A learning process comprises three steps, namely (1) partitioning of the knowledge, (2) approximation of the target concept, and finally (3) induction of a suitable description. In order to decrease the learning cost, the second step computes well-chosen portions of the background knowledge, which represent approximations of the concept to learn. Then, only one of these portions is used during the last induction step. This approach has been implemented through the EAGLE system, and evaluated on a real-world dataset related to the problem of document understanding [3]. The empirical results obtained by EAGLE show the relevance of the approach on this problem, since the reduction of the background knowledge leads to a reduction of the learning time, without any loss of accuracy.

In section 2, a more detailed presentation of the ILP framework is given, together with the principles of the RST-based learning approach. This latter is then described more precisely in sections 3, 4 and 5. In particular, it is shown how RST's basic notions, namely indiscernibility as well as the lower and upper approximations of a concept, have been defined in the context of ILP. Finally, experimentations conducted with the EAGLE system are reported and discussed in section 6.

2   Introduction of RST within an ILP framework

In ILP, the language used to represent the knowledge is the First Order Logic. Each concept C is basically defined by a boolean predicate $C(x_1, \ldots, x_n)$, where each variable x_i can be assigned either a *nominal* value (e.g., an identifier standing for a domain entity), or a *qualitative* value (e.g., a constant). Thus, ground examples of a concept C consists of a collection of n-ary relational tuples of the form $C(t_1, \ldots, t_n)$, where each function-free term t_i is a possible assignment for variable x_i (nominal or qualitative). For instance, Height(Owen,Small) and Height(Garp,High) are examples of the Height(x,y) concept, x and y being respectively nominal and qualitative variables. Each example is labeled as positive ($\oplus$) or negative ($\ominus$), depending on its membership to the concept.

Given a target concept, a *training set* composed of its labeled examples, as well as a *background knowledge* including examples of other concepts, the goal of ILP can be stated as the induction of a logical description of the target concept, which defines it in terms of other concepts of the background knowledge. A description is a set of definite Horn clauses, which come in the following form:

$$\underbrace{L_{Target}}_{Head} \longleftarrow \underbrace{L_1 \wedge \ldots \wedge L_m}_{Body}$$

Each L_i is a literal corresponding to a concept C_i which includes variables or constants. For instance, Height(x,High) $\longleftarrow$ Parent(x,y) $\wedge$ Height(y,High) is a possible description of the Height(x,y) concept. The learned description must characterize most positive but least negative training examples of the target concept in the given background knowledge.

In accordance with RST, the approach proposes three different steps to achieve the learning goal. The first one performs a *partitioning* of the training set and the background knowledge, according to an *indiscernibility relation*. The resulting knowledge consists of a collection of equivalence classes, also called *granules* of concepts, grouping together indistinct examples. Then, a second step computes the *lower* and *upper approximations* of the target concept. Both approximations are defined as portions of the background knowledge, which include respectively the minimal (lower approximation) and maximal (upper approximation) amounts of information about the target concept. They are computed from granules, according to criteria related to the *nominal information* of the target concept. Thus, the underlying goal of this approximation step is the reduction of the data which will be used for inducing the description. Finally, the third step consists in inducing the searched description from a chosen approximation, either lower or upper one. Thus, this last step is the one which explores the search space. The use of the resulting description through the classification of test examples, i.e., examples which did not take part to the learning process, allows for evaluating its accuracy.

3   Partitioning of knowledge and granularity of concepts

In the ILP framework of the approach, the RST's basic notion of indiscernibility is an equivalence relation I, which is defined on (1) the representation of the same concept, (2) the equality of the label ($\oplus$ or $\ominus$), and (3) the equality of constant terms.

Definition 1. *Let $T_1 : P(t_1, \ldots, t_p)$ and $T_2 : Q(t_1, \ldots, t_q)$ be two examples, T_1 and T_2 are indiscernible according to I, denoted $(T_1 \ I \ T_2)$, if and only if:*
(1) $P = Q$
(2) $Label(T_1) = Label(T_2)$
(3) $\forall t_i \in T_1, \forall t_j \in T_2$ such as $Constant(t_i)$ and $Constant(t_j)$, $i = j \Longrightarrow t_i = t_j$

Partitioning the knowledge consists in grouping together into *granules* examples which are indiscernible according to I, i.e., the information they describe is insufficient to distinguish them. Granules including positive (resp. negative) examples are called positive (resp. negative) granules. After this partitioning of ground examples, the representation of a concept consists of the collection of positive and negative granules including its examples. The granularity of a concept, i.e., its number of granules, varies from a concept to another. Indeed, it results from the above definition of the indiscernibility relation that the examples of concepts including constant terms are distributed among several granules, depending on the different possible values for the constant terms. On the contrary, concepts whose examples contain exclusively nominal terms are represented by only two granules, including respectively their positive and negative examples.

4 Approximation of concepts

In the context of ILP, the lower and upper approximations of a concept are well-chosen portions of the background knowledge, which are defined according to some criteria based on the nominal information of this concept.

Definition 2. *The nominal information of a concept C, denoted $Nom(C)$, is the set of nominal terms occuring in its positive or negative examples: $Nom(C) = \{t_i \ / \ Nominal(t_i) \ and \ \exists C(\ldots, t_i, \ldots) \in [C]^\oplus \cup [C]^\ominus\}$ where $[C]^\oplus$ (resp. $[C]^\ominus$) stands for the set of positive (resp. negative) granules representing C.*

Since nominal terms designate domain entities, the nominal information of a concept thus consists of the set of entities which appear in its representation. Given this information, the lower and upper approximations are defined as follows:

Definition 3. *The lower approximation of a concept C, denoted $\underline{R}C$, is the set of positive and negative examples of the background knowledge, whose nominal terms are included in or equal to the nominal information of C: $\underline{R}C = \{T_i \ / \ NT(T_i) \subseteq Nom(C)\}$, where $NT(T_i)$ stands for the set of nominal terms occurring in example T_i.*

Definition 4. *The upper approximation of a concept C, denoted $\overline{R}C$, is the set of positive and negative examples of the background knowledge, whose nominal terms intersect with the nominal information of C: $\overline{R}C = \{T_i \ / \ NT(T_i) \cap Nom(C) \neq \emptyset\}$.*

Actually, approximating a concept consists in selecting, into each granule of the background knowledge, examples which share common information with this concept. Indeed, the lower (resp. upper) approximation is the portion of the background knowledge whose nominal information is identical (resp. close) to the one of the concept. Both represent respectively the minimal and the maximal amounts of data available about the target concept. Indeed, it follows from the previous definitions that both approximations verify the following property: $\underline{R}C \subseteq \overline{R}C$, i.e., the lower approximation of a concept is a subset of its upper approximation. Thus, the approximation step reduces the volume of the background knowledge which will be used for inducing the searched description. This reduction does not modify the granularity of the data, but the size of the granules.

5   Induction of logical descriptions of concepts

The problem of inducing a description of a target concept during the last induction step can be formalized as follows. Given

- the set of target concept's positive granules: $[Target]^{\oplus}$,
- the set of target concept's negative granules: $[Target]^{\ominus}$,
- the target concept's chosen approximation: either lower one $\underline{R}Target$ or upper one $\overline{R}Target$,

Find a description, i.e., a set of clauses, which characterizes most positive examples (in $[Target]^{\oplus}$) but least negative ones (in $[Target]^{\ominus}$) of the target concept. To address this goal, an inductive learning algorithm which overlaps the one of the FOIL system [13] is performed. Since positive training examples may be distributed among several granules, consequently to the partitioning of the knowledge, several stages are required in order to find a suitable description. Indeed, positive examples coming from distinct granules can not be handled at the same time, since they are not indiscernible. Thus, in order to respect the granularity of the target concept, each stage deals with a single positive granule of $[Target]^{\oplus}$ and consists in finding a subset of clauses which characterize all its included examples but no negative one. In its initial state, each clause h which is searched has an empty body: $L_{Target} \longleftarrow ?$. The head literal L_{Target} is a generalized form of the examples included in the current positive granule, obtained by replacing nominal terms by distinct variables and constant terms by the corresponding value. Then, the construction of h consists in adding literals one by one to its body. Before any addition, the space of candidate literals comprises all the possible generalized forms of granules which can be obtained from the chosen approximation, by replacing nominal terms with bound variables, i.e., variables which already appear in the clause, but also new ones. In order to prevent from a large space, each candidate literal must contain at least one bound variable. Then, the literal that is chosen among all possible candidates to be added to the body of h is the one which has the maximal *gain*, as it is defined in the FOIL system [13]. This heuristic allows for choosing the best literal, according to the number of positive and negative training examples which are characterized by the resulting clause. The addition of literals to the clause's body stops when one of the three following stopping criteria is verified: (1) the clause in progress is complete, i.e., it characterizes no more negative example in the training set, (2) all the candidate literals have a negative gain or (3) the number of literals in the body is greater than or equal to a maximal number that is fixed by the user.

6   Experimental results

This section reports an experimentation, performed with the EAGLE system, which is a concrete implementation of the approach. The chosen real-world dataset is related to the problem of document understanding, and can be found at the MLnet Archive [1]. Through this experimentation, our purpose is to analyze the effects of the background knowledge reduction, i.e., the approximation advocated by the approach, on the learning cost, i.e., the time spent to induce the descriptions, and the accuracies of the learned descriptions. The learning problem consists in discovering rules for identifying the logical components of a document (for instance sender, receiver, logo, reference or date), according to their layout information. This learning problem has already been the object of several studies, such as for example the learning of contextual rules, which can be found in [3]. The complete dataset contains the descriptions of 30 single page documents, which are letters sent by Olivetti. For each document, a description consists of the list of its logical components, together with their respective layout features, which are expressed by means of various concepts. The labels of the logical components are indicated by the five following target concepts, namely sender(x), receiver(x), logo(x), ref(x) and date(x). Other concepts of the background knowledge, such as width_small(x), height_large(x), type_picture(x), position_center(x) and above(x,y), allow for expressing the components layout features as well as their relationships.

The global experimentation on this dataset has been performed under a 6-cross validation protocol, which means that 6 experiments have been conducted. At each experiment, a sample of labeled components coming from a random subset of 20 documents is used as training examples. Other components coming from the remaining 10 documents are used as test examples. In both samples, only positive examples of each target concept are given. Negative examples are derived from positive ones, by considering that for each target concept, the components which have a different label are its negative examples. For instance, negative examples of senders are components which are labeled as receiver, logo, reference or date.

Each experiment consists in inducing definitions for each target concept, from the complete background knowledge, but also from the lower and upper approximations, which represent respectively (on average on the six experiments and for each target concept) 60% and 66.3% of the whole background knowledge. For each target concept, the classification accuracy of its induced description is computed from the test examples, as the number of examples which match the description (i.e., the number of $\oplus$ examples which are characterized by the description, added to the number of $\ominus$ examples which are not characterized by the description), divided by the total number of test examples. The results reported on table 1 are averages of the results obtained during the six experiments. These latter comprise the classification accuracies of the learned descriptions, as well as the learning times (in seconds), according to the portion of background knowledge used (complete, lower or upper). In case of a learning from the complete background knowledge, the learning time corresponds only to the time spent by the induction step, since no approximation is computed. In case of a learning from lower or upper approximation, it additionally comprises the time spent by

---

[1] http://www.gmd.de/ml-archive/frames/datasets/datasets-frames.html

the approximation step. By comparison, the accuracies obtained by the FOCL [10] system for each target concept are also given.

Target	Lower	Upper	Complete	Accuracy
Sender	1.19s	1.35s	1.59s	**99.03%**
Receiver	1.99s	3.29s	4.56s	**97.78%**
Logo	0.72s	0.75s	0.81s	**100%**
Ref	4.4s	5s	7.17s	**97.65%**
Date	5s	5.67s	8.27s	**98.48%**

	Complete
Sender	85.9%
Receiver	59.9%
Logo	100%
Ref	93.1%
Date	74.6%

(a) EAGLE   (b) FOCL

Fig. 1. Experimental results

A first observation is that the accuracies obtained by EAGLE are identical, whatever the portion of the background knowledge (lower, upper, complete) used for learning. These latter are better by comparison with the ones obtained by FOCL. It appears from the table that learning from the lower approximation is more interesting than using the whole background knowledge, since it reduces the learning time without any loss of accuracy. Actually, since an approximation is only a portion of the background knowledge, the time spent by the system for inducing descriptions is reduced. In particular, the evaluation of each candidate literal's gain is faster because there are less possible bindings for the variables. The time used for computing an approximation, which is on average equal to 0.38 seconds, appears to be insignificant by comparison with the time which is further saved during the inductive step. This phenomenon is observable on each target concept, however it is less obvious for Logo(x). Indeed, for this concept the learning goal is quite easy to achieve, i.e., EAGLE finds a description in a short time, at each experiment, from any portion of the background knowledge. This latter is the following: Logo(x) ⟵ Type_picture(x). As a consequence, the approximation step consumes time without significantly decreasing the time of induction. In that case, the search can straightforwardly be performed on the complete background knowledge. On the contrary, in case of concepts such as Sender(x), Receiver(x), Ref(x) and Date(x), the learning goal is more difficult to achieve, since a description with several clauses (5 on average) including more literals in their bodies (4 on average) are searched. In that case, computing an approximation is interesting since it reduces significantly the learning time. For example, on concept Receiver, the gain of time reaches approximatively 56%.

7   Conclusion

In this paper, we have pointed out the problem of handling large amount of data faced by inductive learning systems, especially those which fit into a logical framework. In order to cope with this problem, we have proposed a learning approach, which introduces and adapts the basic notions of Rough Set Theory within the paradigm of Inductive Logic Programming. In particular, it argues in favor of performing induction from only well-chosen portions of the background knowledge, which represent approximations of the concept to learn. Empirical results obtained with a first concrete implementation of the proposed approach, show that the approximation allows for reducing the learning cost without losing the learning results accuracy. The current direction of research aims at improving the flexibility and the adaptability of the approach to various learning problems, and especially various forms of data. In particular, an extension of this approach, which provides a suitable framework for handling uncertain data and inducing flexible concepts, has already been proposed and evaluated in [11,12,4].

References

1. Botta M. 1994, *Learning First Order Theories*, Lecture Notes in Artificial Intelligence, **869**, pp. 356–365.
2. Champesme M. 1995, *Using Empirical Subsumption to Reduce the Search Space in Learning*, In Proceedings of the International Conference on Conceptual Structures (ICCS'95), Santa Cruz, California, USA.
3. Esposito F., Malerba D. & Semeraro G. 1993, *Automated Acquisition of Rules for Document Understanding*, In proceedings of the 2nd International Conference on Document Analysis and Recognition, Tsukuba Science City, Japan, pp. 650–654.
4. Martienne E. & Quafafou M. 1998, *Learning Fuzzy Relational Descriptions Using the Logical Framework and Rough Set Theory*, In proceedings of the International Conferenc on Fuzzy Systems (FUZZ-IEEE'98), to appear.
5. Mitchell T.M. 1982, *Generalization as search*, Artificial Intelligence, **18**, 203–226.
6. Muggleton S. 1991, *Inductive Logic Programming*, New Generation Computing, **8(4)**, 295–318.
7. Muggleton S. 1995, *Inverse entailment and PROGOL*, New Generation Computing, **13**; 245–286.
8. Ndellec C., Ad H., Bergadano F. & Tausend B. 1996, *Declarative Bias in ILP*, Advances in Inductive Logic Programming, De Raedt L. Ed., IOS Press, pp. 82–103.
9. Pawlak Z. 1991, *Rough Sets: Theorical Aspects of Reasoning About Data*, Kluwer Academic Publishers, Dordrecht, Netherlands.
10. Pazzani M. & Kibler D. 1992, *The utility of knowledge in inductive learning*, Machine Learning, **9(1)**.
11. Quafafou M. 1997, *Learning Flexible Concepts from Uncertain Data*, In Proceedings of the tenth International Symposium on Methodologies for Intelligent Systems (ISMIS'97), Charlotte, USA.
12. Quafafou M. 1997, *α-RST: A Generalization of Rough Sets Theory*, In Proceedings of the tenth International Conference on Rough Sets and Soft Computing (RSSC'97).
13. Quinlan J.R. 1990, *Learning Logical Definitions from Relations*, Machine Learning, **5**, J. Mostow (Ed.), 239–266.
14. Tausend B. 1995, *A guided tour through hypothesis spaces in ILP*, Lectures notes in Artificial Intelligence: ECML 95, Springer-Verlag Ed., **912**, pp. 245–259.

Integrating KDD Algorithms and RDBMS Code

María C. Fernandez–Baizán*, Ernestina Menasalvas Ruiz,
José M. Peña Sánchez, Borja Pardo Pastrana
{cfbaizan, emenasalvas}@fi.upm.es, {chema, borja}@orion.ls.fi.upm.es

Departamento de Lenguajes y Sistemas Informáticos e Ingeniería del Software,
Facultad de Informática, Campus de Montegancedo, Madrid

Abstract. In this paper we outline the design of a RDBMS that will provide the user with traditional query capabilities as well as KDD queries.
Our approach is not just another system which adds KDD capabilities,
this design is aimed to integrate these KDD capabilities into RDBMS
core. The approach also defines a generic engine of Data Mining algorithms that allows easy enhancement of system capabilities as a new
algorithm is implemented.

1 Introduction

Most of the KDD systems that have been implemented up to the present moment apply just one particular methodology or implement a particular algorithm
(rough sets[9], attribute-induction[5], a priori[1,2]). When designing this architecture we wanted a system that integrates data mining capabilities within the
RDBMS. We wanted the system to be extensible, that is, we wanted to build
a system in which adding new algorithms would be easy. This goal is achieved
dividing KDD algorithms into basic operations that will be implemented as particular instance of a structure that has been called *operators*.
The paper is organized as follows: The division of KDD algorithms into basic
operations is explained in section 2 as well as the main structure that operators
must have in order to be included in the system. Extension of main modules of
traditional database systems to handle new operations is discussed in section 3.

2 Algorithms

It is easy to observe that many KDD algorithms have similar behavior during,
at least, an important part of their execution. This fact has led us to consider
the division of the algorithms into several parts, in order to achieve reusability
of code. Moreover, as one of the goals of the design is the integration with a
RDBMS, we have tried to make each of those parts as similar to RDBMS basic
operations as possible. In our desing the basic operations will be performed by

---

* This work is supported by the Spanish Ministry of Education under project PB95-
0301

L. Polkowski and A. Skowron (Eds.): RSCTC'98, LNAI 1424, pp. 210–213, 1998.
© Springer-Verlag Berlin Heidelberg 1998

the **operator** structure. We will call operator any operation that is made up of: Relational tables both as **Input** and **Output**, **Auxiliary Structures** that will be used to keep input/output information of the operation and **Parameters** than guide their behavior. The main result of this process could be represented as **Extracted Information**. In figure 1 the basic structure of an operator is depicted. A data mining query will then be defined as the sequence of operators

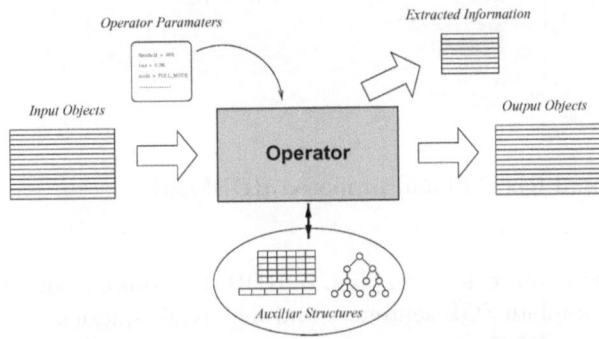

Fig. 1. Operator interface

that, given a particular table containing certain objects, gets as a result the set of patterns to describe the knowledge asked by such a query.

3   Architecture

The decomposition of algorithms into operators guided the design of the system we plan to build. The identification of these simple components could be used to split most KDD queries into atomic elements that could be managed directly by a new RDBMS. This system may be named as **RDBMAS** (**R**elational **D**ata**B**ase **M**anagement and **A**nalysis **S**ystem) and would be designed (see figure 2) as an extension of traditional RDBMS with new capabilities.

In figure 2 we show how a generic traditional RDBMS would have to be modified. As a result next generation of database systems would provide the user with traditional queries as well as the possibility to analyze data.

- **Query Analyzer**: This module parses each submitted query to the system and translates it into a internal representation. In traditional RDBMS this module analyzes only SQL commands, for RDBMAS it will have to analyze also new KDD sentences.
- **Optimizer**: This component gets internal sentence representation supplied by the previous module and optimize the order of execution of its clauses. As a result returns a specific execution plan for the user query. If new operations provide the optimizer with the same measures traditional operations do (weights, restrictions, ...) this module would not have to be changed significantly.

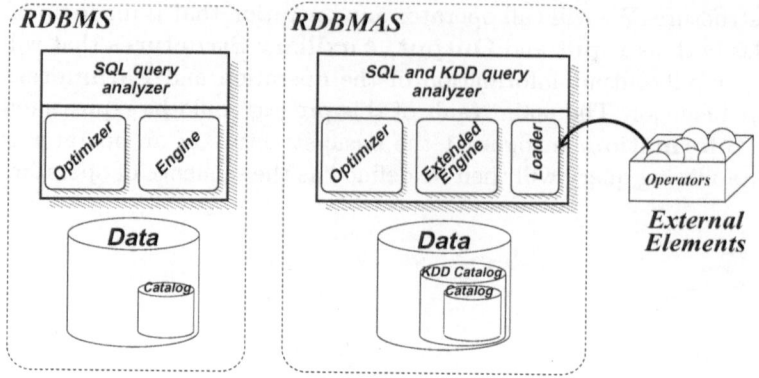

Fig. 2. Traditional RDBMS and proposed RDBMAS

- **Engine**: This module carries out execution plan. In RDBMAS this execution engine must be able to complete SQL sentences and new KDD queries.
- **Loader**: In order to achieve KDD queries some new code must be executed in the RDBMAS. Almost all KDD systems provide their functionalities with static defined code. In this system all algorithms must be split into elemental operations and may be implemented in external dynamically-loaded components (**operators**).
- **Catalog**: Information about data (metadata) is stored as additional tables for RDBMS. This information is required for management functions, but a new information about data may also be necessary in order to support analysis (KDD capabilities) functions. So, the system catalog must be extended.

3.1 Comparison with other systems

Approaches like Data Surveyor [6] propose similar system architectures that modify RDBMS in order to achieve better performance, but in most cases KDD algorithms run outside of these enhanced RDBMS. With our design all queries (SQL queries and KDD queries) are managed by the same RDBMAS.
RSDM [4] has been conceived as an engine of KDD algorithms instead of a system that adds some particular capabilities. This approach has its advantages as well as disadvantages. On the one hand, the idea of building an engine of algorithms in contrast to all the existing Data Mining systems, will allow to add new capabilities with the only task of building the module that will execute such capability. This avoids the complex process of codifying programs for an integrated system in which you have to care not only of the coding of the algorithm but of the communication, storing of intermediate results and so on. On the other hand, the process of construction of the architecture is more complex. However, we must emphasize once again that adding any capability will be a straightforward task once the architecture has been finished.

3.2   Conclusion and future research

The design of a new generation of database systems that will provide the user with query and analysis of data has been outlined in this paper. We will called these system RDMAS.

At the present moment the design is being further studied to tackle the problems that are arising as a result of addition the new capabilities. Also implementation of a first prototype has just began. We hope to have in the near future a prototype available.

We have to remark once again that the implementation of the operation inside the core of the RDBMS is twofold on the one hand efficiency gain due to maximization of optimizer functions on the other next generation of RDBMS will provide users with data analysis capabilities.

Acknowledgments

We are very much indebted for inspiration to Dr. Ziarko and Dr. Pawlak.

References

1. Agrawal R., Imielinski T., Swami A. *Mining association rules between sets of Item in large Databases*, Proceedings of ACM SIGMOD, pp. 207–216, May 1993.
2. R. Agrawal, *Mining Association Rules Between Sets of Items in Large Databases*, In Prooceedings of ACM SIGMOD Int. Conf. on Management of data, Washington DC, pp. 207–216, 1993.
3. R. Agrawal et al., *The Quest Data Mining System* In Proceedings The Second Int. Conf. on Knowledge discovery and Data Mining, pp. 244–249. August 1996
4. M. Fernandez–Baizan, E. Menasalvas, J.M. Peña. *Integrating RDBMS and Rough Set Theory* To Appear in Fuzzy Databases in August 1998
5. J. Han et al., *DBMiner: A System for mining knowledge in Large Relational Databases* In Proceedings The Second Int. Conf. on Knowledge discovery and Data Mining, pp. 250–255, August 1996
6. M.L. Kersten, A.P.J.M. Siebes, *Data Surveyor: Searching the nuggets in parallel* Advances in Knowledge Discovery and Data Mining, AAAI Press, pp. 447–467
7. J. Komorowski, A. Ohrn, *ROSETTA: A Rough Set Toolkit for Analysis of Data* In Proceedings JCIS 97, pp. 403–407. March 1997
8. Z. Pawlak, *Rough Sets - Theoretical Aspects of Reasoning about Data*, Kluwer, 1991.
9. Z. Pawlak, *Information Systems-Theoretical foundations*, Information systems, 6, No.4, 1993, pp. 299–297.
10. G. Piatesky–Shaphiro, *An Overview of Knowledge Discovery in Databases: Recent Progress and Challenges*, Rough Sets, Fuzzy Sets and Knowledge Discovery, 1994, pp. 1–11
11. A. Skowron, C. Rauszer, *The Discernibility matrices and Functions in Information System*, ICS PAS Report 1/91, Technical University of Warsaw 1991, pp. 1–44
12. W. Ziarko, *Variable Precision Rough Sets Model* Journal of Computer and System Sciences, vol. 46. 1993, 39-59.
13. W. Ziarko, N. Shan *On Discovery of Attribute Interactions and Domain Classifications*, CSC'95 23 Annual Computer Science Conf. on Rough Sets and Data Mining

Fast Discovery of Representative Association Rules

Marzena Kryszkiewicz

Institute of Computer Science, Warsaw University of Technology
Nowowiejska 15/19, 00-665 Warsaw, Poland
mkr@ii.pw.edu.pl

Abstract. Discovering association rules among items in a large database is an important database mining problem. The number of association rules may be huge. To alleviate this problem, we introduced in [1] a notion of representative association rules. Representative association rules are a least set of rules that covers all association rules satisfying certain user specified constraints. The association rules, which are not representative ones, may be generated by means of a cover operator without accessing a database. In this paper, we investigate properties of representative association rules and offer a new efficient algorithm computing such rules.

1 Introduction

Discovering association rules among items in large databases is recognized as an important database mining problem. The problem was introduced in [2] for sales transaction database. The association rules identify sets of items that are purchased together with other sets of items. For example, an association rule may state that 90% of customers who buy butter and bread buy also milk. Several extensions of the notion of an association rule were offered in the literature (see e.g. [3-4]). One of such extensions is a generalized rule that can be discovered from a taxonomic database [3]. Applications for association rules range from decision support to telecommunications alarm diagnosis and prediction [5-6].

The number of association rules is usually huge. A user should not be presented with all of them, but rather with these which are original, novel, interesting. There were proposed several definitions of what is an interesting association rule (see e.g. [3,7]). In particular, pruning out uninteresting rules which exploits the information in taxonomies seems to be quite useful (resulting in the rule number reduction amounting to 60% [3]). The interestingness of a rule is usually expressed by some quantitative measure. In [1] we offered a different approach. We did not introduce any measure defining interestingness of a rule, but we showed how to derive the set of association rules from a given association rule by means of a cover operator without accessing a database. A least set of association rules that allows to deduce all other rules satisfying user specified constraints is called a set of representative association rules. In [1], it was offered the *GenAllRepresentatives* algorithm computing representative

L. Polkowski and A. Skowron (Eds.): RSCTC'98, LNAI 1424, pp. 214-222, 1998.

association rules. To check whether a candidate rule is representative the algorithm required comparing the rule with longer representative rules, which was quite time-consuming operation. In this paper, we investigate some properties of representative association rules that allow us to propose a new efficient algorithm for representative association rules mining. The new algorithm generates representative rules independently from other representative rules.

2 Association Rules

The definition of a class of regularities called *association rules* and the problem of their discovering were introduced in [2]. Here, we describe this problem after [2,8]. Let $I = \{i_1, i_2, ..., i_m\}$ be a set of distinct literals, called *items*. In general, any set of items is called an *itemset*. Let D be a set of transactions, where each transaction T is a set of items such that $T \subseteq I$. An *association rule* is an expression of the form $X \Rightarrow Y$, where $\varnothing \neq X, Y \subset I$ and $X \cap Y = \varnothing$. X is called the antecedent and Y is called the consequent of the rule.

Statistical significance of an itemset X is called *support* and is denoted by $sup(X)$. $Sup(X)$ is defined as the number of transactions in D that contain X. Statistical significance (*support*) of a rule $X \Rightarrow Y$ is denoted by $sup(X \Rightarrow Y)$ and defined as $sup(X \cup Y)$. Additionally, an association rule is characterized by *confidence*, which expresses its strength. The confidence of an association rule $X \Rightarrow Y$ is denoted by $conf(X \Rightarrow Y)$ and defined as the ratio $sup(X \cup Y) / sup(X)$.

The problem of mining association rules is to generate all rules that have support greater than some user specified minimum support $s \geq 0$ and confidence not less than a user specified minimum confidence $c > 0$. In the sequel, the set of all association rules whose support is greater than s and confidence is not less than c will be denoted by $AR(s,c)$. If s and c are understood then $AR(s,c)$ will be denoted by AR.

In the paper, we apply also the following simple notions:

The number of items in an itemset will be called the *length of the itemset*. An itemset of the length k will be referred to as a *k-itemset*. Similarly, the *length of an association rule* $X \Rightarrow Y$ will be defined as the total number of items in the rule's antecedent and consequent ($|X \cup Y|$). An association rule of the length k will be referred to as a *k-rule*. An association *k*-rule will be called *shorter* than, *longer* than or *of the same length* as an association *m*-rule if $k < m$, $k > m$, or $k = m$, respectively.

3 Cover Operator

A notion of a *cover operator* was introduced in [1] for deriving a set of association rules from a given association rule without accessing a database.

The *cover* C of the rule $X \Rightarrow Y$, $Y \neq \varnothing$, is defined as follows:

$$C(X \Rightarrow Y) = \{X \cup Z \Rightarrow V | Z, V \subseteq Y \text{ and } Z \cap V = \varnothing \text{ and } V \neq \varnothing\}.$$

Each rule in $C(X \Rightarrow Y)$ consists of a subset of items occurring in the rule $X \Rightarrow Y$. The antecedent of any rule r covered by $X \Rightarrow Y$ contains X and perhaps some items from Y, whereas r's consequent is a non-empty subset of the remaining items in Y. It was proved in [1] that each rule r in the cover $C(r')$, where r' is an association rule having support s and confidence c, belongs in $AR(s,c)$. Hence, if r belongs in $AR(s,c)$ then every rule r' in $C(r)$ also belongs in $AR(s,c)$. The number of different rules in the cover of the association rule $X \Rightarrow Y$ is equal to $3^m - 2^m$, where $m = |Y|$ (see [1]).

Example 3.1

Let $T_1 = \{A,B,C,D,E\}$, $T_2 = \{A,B,C,D,E,F\}$, $T_3 = \{A,B,C,D,E,H,I\}$, $T_4 = \{A,B,E\}$ and $T_5 = \{B,C,D,E,H,I\}$ are the only transactions in the database D. Let r: $(AB \Rightarrow CDE)$. Fig. 1 contains all rules belonging in the cover $C(r)$ along with their support and confidence in D. The support of r is equal to 3 and its confidence is equal to 75%. The support and confidence of all other rules in $C(r)$ are not less than the support and confidence of r.

#	Rule r' in $C(r)$	Support of r'	Confidence of r'
1.	$AB \Rightarrow CDE$	3	75%
2.	$AB \Rightarrow CD$	3	75%
3.	$AB \Rightarrow CE$	3	75%
4.	$AB \Rightarrow DE$	3	75%
5.	$AB \Rightarrow C$	3	75%
6.	$AB \Rightarrow D$	3	75%
7.	$AB \Rightarrow E$	4	100%
8.	$ABC \Rightarrow DE$	3	100%
9.	$ABC \Rightarrow D$	3	100%
19.	$ABC \Rightarrow E$	3	100%
11.	$ABD \Rightarrow CE$	3	100%
12.	$ABD \Rightarrow C$	3	100%
13.	$ABD \Rightarrow E$	3	100%
14.	$ABE \Rightarrow CD$	3	75%
15.	$ABE \Rightarrow C$	3	75%
16.	$ABE \Rightarrow D$	3	75%
17.	$ABCD \Rightarrow E$	3	100%
18.	$ABCE \Rightarrow D$	3	100%
19.	$ABDE \Rightarrow C$	3	100%

Fig. 1. The cover of the association rule r: $(AB \Rightarrow CDE)$

Below, we present two simple properties, which will be used further in the paper.

Property 3.1

Let r: $(X \Rightarrow Y)$ and r': $(X' \Rightarrow Y')$ be association rules. Then:

$$r \in C(r') \text{ iff } X \cup Y \subseteq X' \cup Y' \text{ and } X \supseteq X'.$$

Property 3.2

(i) If an association rule r is longer than an association rule r' then $r \notin C(r')$.

(ii) If an association rule r: $(X \Rightarrow Y)$ is shorter than an association rule r': $(X' \Rightarrow Y')$ then $r \in C(r')$ iff $X \cup Y \subset X' \cup Y'$ and $X \supseteq X'$.

(iii) If $r: (X \Rightarrow Y)$ and $r': (X' \Rightarrow Y')$ are different association rules of the same length then $r \in C(r')$ iff $X \cup Y = X' \cup Y'$ and $X \supset X'$.

4 Representative Association Rules

In this section we describe a notion of representative association rules which was introduced in [1]. Informally speaking, a set of all representative association rules is a least set of rules that covers all association rules by means of the cover operator.

A set of *representative association rules* wrt. minimum support s and minimum confidence c will be denoted by $RR(s,c)$ and defined as follows:

$$RR(s,c) = \{r \in AR(s,c) | \neg \exists r' \in AR(s,c), r' \neq r \text{ and } r \in C(r')\}.$$

If s and c are understood than $RR(s,c)$ will be denoted by RR. Each rule in RR is called a *representative association rule*. By the definition of RR no representative association rule may belong in the cover of another association rule.

Example 4.1
Given minimum support $s = 3$ and minimum confidence $c = 75\%$, the following representative rules $RR(s,c)$ would be found for the database D from Example 3.1:

$$\{A \Rightarrow BCDE, C \Rightarrow ABDE, D \Rightarrow ABCE, B \Rightarrow CDE, E \Rightarrow BCD, B \Rightarrow AE, E \Rightarrow AB\}.$$

There are 7 representative association rules in $RR(s,c)$, whereas the number of all association rules in $AR(s,c)$ is 165. Hence, the representative association rules constitute 4.24% of all association rules.

We may expect that a user will often request the set of representative association rules RR rather than the set of all association rules AR. If RR is provided then the user may formulate queries about the association rules represented by RR. Clearly, $AR(s,c) = \bigcup\{C(r) | r \in RR(s,c)\}$. However, we expect the user to ask rather about the covers of specific representative rules. The queries might contain not only the cover operator, but also the set-theoretical operators of union, difference and intersection.

5 The Algorithm

The problem of generating association rules is usually decomposed into two subproblems:
1. Generate all itemsets whose support exceeds the minimum support s. The itemsets of this property are called *frequent (large)*.
2. From each frequent itemset generate association rules whose confidence is not less than the minimum confidence c. Let Z be a frequent itemset and $\emptyset \neq X \subset Z$. Then any rule $X \Rightarrow Z \backslash X$ holds if $sup(Z)/sup(X) \geq c$.

In the paper we restrict the second subproblem to generation of all representative association rules whose confidence is not less than the minimum confidence c.

Several efficient solutions were proposed to solve the first subproblem (see [3,8-9]). We will remind briefly the main idea of the *Apriori* algorithm [8] computing frequent itemsets. Then, we will propose a new efficient algorithm computing representative association rules from the found frequent itemsets.

5.1 Computing Frequent Itemsets

The *Apriori* algorithm exploits the following properties of frequent and non-frequent itemsets: All subsets of a frequent itemset are frequent and all supersets of a non-frequent itemset are non-frequent. The following notation is used in the *Apriori* algorithm:

- C_k - set of candidate k-itemsets;
- F_k - set of frequent k-itemsets;

The items in itemsets are assumed to be ordered lexicographically. Associated with each itemset is a *count* field to store the support for this itemset.

```
Algorithm Apriori
  F₁ = {frequent 1-itemsets};
  for (k = 2; F_{k-1} ≠ ∅; k++) do begin
  C_k = AprioriGen(F_{k-1});
  forall transactions T ∈ D do
    forall candidates Z ∈ C_k do
      if Z ⊆ T then
        Z.count++;
  F_k = {Z ∈ C_k | Z.count > s};
  endfor;
return ∪_k F_k;
```

First, the support of all 1-itemsets is determined during one pass over the database D. All non-frequent 1-itemsets are discarded. Then the loop "for" starts. In general, some k-th iteration of the loop consists of the following operations:
1. *AprioriGen* is called to generate the candidate k-itemsets C_k from the frequent $(k-1)$-itemsets F_{k-1}.
2. Supports for the candidate k-itemsets are determined by a pass over the database.
3. The candidate k-itemsets that do not exceed the minimum support are discarded; the remaining k-itemsets F_k are found frequent.

```
function AprioriGen(frequent (k-1)-itemsets F_{k-1});
  insert into C_k
    select (Z[1], Z[2], ... , Z[k-1], Y[k-1]) from F_{k-1} Z, F_{k-1} Y
    where Z[1] = Y[1] ∧ ... ∧ Z[k-2] = Y[k-2] ∧ Z[k-1] < Y[k-1];
  delete all itemsets Z ∈ C_k such that some (k-1)-subset of Z
    is not in F_{k-1};
  return C_k;
```

The *AprioriGen* function constructs candidate k-itemsets as supersets of frequent $(k-1)$-itemsets. This restriction of extending only frequent $(k-1)$-itemsets is justified since any k-itemset, which would be created as a result of extending a non-frequent $(k$-1)-itemset, would not be frequent either. The last operation in the *AprioriGen* function prunes the candidates from C_k that do not have all their $(k$-1)-subsets in the frequent $(k$-1)-itemsets F_{k-1}. If k-itemset Z does not have all its $(k$-1)-subsets in F_{k-1} then there is some non-frequent $(k$-1)-itemset $Y \notin F_{k-1}$ which is a subset of Z. This means that Z is non-frequent as a superset of a non-frequent itemset.

5.2 Computing Representative Association Rules

In this subsection we offer an efficient algorithm for computing representative association rules. Unlike the *GenAllRepresentatives* algorithm proposed in [1], the new *FastGenAllRepresentatives* algorithm exploits solely the information about the supports of frequent itemsets. The *Apriori* algorithm may be run to calculate all frequent itemsets and their supports. *FastGenAllRepresentatives* is based on Properties 5.2.1-5.2.2, which we present and prove below.

Lemma 5.2.1
Let $\varnothing \neq X \subset Z \subseteq I$ and r be an expression of the form $(X \Rightarrow Z \backslash X)$.

$$\exists Z' \subseteq I, Z' \supset Z \text{ and } sup(Z') > s \text{ and } sup(Z')/sup(X) \geq c \text{ iff}$$

$$\exists r' \in AR(s,c), r' \text{ is longer than } r \text{ and } r \in C(r').$$

Proof:
    $(\Rightarrow)$ Let r' be an expression of the form: $X \Rightarrow Z' \backslash X$. Clearly, r' is longer than r since $Z' \supset Z$. The rule $r' \in AR(s,c)$ because $X \neq \varnothing$, the support $sup(r') = sup(Z') > s$ and the confidence $conf(r') = sup(Z')/sup(X) \geq c$. Additionally, Property 3.1 allow us to conclude that $r \in C(r')$.

    $(\Leftarrow)$ Let $Z' \subseteq I$ and r' be an expression of the form: $X \Rightarrow Z' \backslash X$. By the assumption, $r' \in AR(s,c)$. Hence, $sup(r') = sup(Z') > s$ and $conf(r') = sup(Z')/sup(X) \geq c$. Additionally, r' is longer than r and $r \in C(r')$. Therefore, we can conclude from Property 3.2.ii that $Z' \supset Z$.

Lemma 5.2.2
Let $\varnothing \neq X \subset Z \subseteq I$ and r be an expression of the form $(X \Rightarrow Z \backslash X)$. Let $maxSup = max(\{sup(Z') | Z \subset Z' \subseteq I\} \cup \{0\})$.

    $maxSup > s$ and $maxSup/sup(X) \geq c$ iff $\exists r' \in AR(s,c), r'$ is longer than r and $r \in C(r')$.

Proof: Lemma 5.2.2 follows immediately from Lemma 5.2.1.

Property 5.2.1
Let $\varnothing \neq X \subset Z \subseteq I$ and r be a rule: $(X \Rightarrow Z \backslash X) \in AR(s,c)$. The rule r belongs in $RR(s,c)$ if the two following conditions are satisfied:
(i) $maxSup \leq s$ or $maxSup/sup(X) < c$, where
      $maxSup = max(\{sup(Z') | Z \subset Z' \subseteq I\} \cup \{0\})$,

(ii) $\neg \exists X',\ \varnothing \neq X' \subset X$, such that $(X' \Rightarrow Z\backslash X') \in AR(s,c)$.

Proof: Property 3.2.i tells us that an association rule does not belong in the cover of any shorter rule. So, the association rule r is representative if it does not belong in the cover of any association rule different from r which is longer than r or which is of the same length as r. The first condition (i) guarantees that the rule r does not belong in the cover of any association rule longer than r (see Lemma 5.2.2). The second condition (ii) ensures that the rule r does not belong in the cover of any association rule of the same length as r (see Property 3.2.iii).

Property 5.2.2

Let $\varnothing \neq Z \subset Z' \subseteq I$. If $sup(Z) = sup(Z')$ then no rule $(X \Rightarrow Z\backslash X) \in AR(s,c)$, where $\varnothing \neq X \subset Z$, belongs in $RR(s,c)$.

Proof: Let $(X \Rightarrow Z\backslash X) \in AR(s,c)$. Then, $\varnothing \neq X$, $sup(Z) > s$ and $conf(X \Rightarrow Z\backslash X) \geq c$. Now, let us consider a rule: $X \Rightarrow Z'\backslash X$. $(X \Rightarrow Z'\backslash X) \in AR(s,c)$ because $\varnothing \neq X$, $sup(Z') = sup(Z) > s$ and $conf(X \Rightarrow Z'\backslash X) = sup(Z')/sup(X) = conf(X \Rightarrow Z\backslash X) \geq c$. Additionally, $(X \Rightarrow Z\backslash X) \in C(X \Rightarrow Z'\backslash X)$. Hence, $X \Rightarrow Z\backslash X$ is not representative.

```
procedure FastGenAllRepresentatives(all frequent itemsets F);
 forall Z ∈ F do begin
  k = |Z|; maxSup = max({sup(Z') | Z⊂Z'∈F_{k+1}} ∪ {0});
  if Z.sup ≠ maxSup then begin            // see Property 5.2.2
   A_1 = {{Z[1]}, {Z[2]}, ... , {Z[k]}}; // create 1-antecedents
   /* Loop1 */
   for (i = 1; (A_i ≠ ∅) and (i < k); i++) do begin
    forall X ∈ A_i do begin
     find Y∈F_i such that Y = X;
     XCount = Y.count;
     /* Is X ⇒ Z\X an association rule? */
     if (Z.count/XCount ≥ c) then begin
      /* Aren't there representatives longer than X ⇒ Z\X? */
      if (maxSup/XCount < c) then      // see Property 5.2.1.i
       print(X, "⇒", Z\X, " with support: ", Z.count,
             " and confidence: ", Z.count / XCount);
      /* Antecedents of association rules are not extended */
      A_i = A_i \ {X};                 // see Property 5.2.1.ii
      endif;
     endfor;
     A_{i+1} = AprioriGen(A_i);         // compute i+1-antecedents
    endfor;
   endif;
 endfor;
endproc;
```

The *FastGenAllRepresentatives* algorithm computes representative association rules from each itemset in F. Let Z be a considered itemset in F. Only k-rules, $k = |Z|$, are generated from Z. First, *maxSup* is determined as a maximum from the supports of these itemsets in F_{k+1} which are supersets of Z. If there is no superset of Z in F_{k+1} then *maxSup*=0. Let us note that the supports of other proper supersets of Z, which do not belong in F_{k+1}, are not greater than *maxSup*. Clearly, *maxSup*>s or *maxSup*=0. If $sup(Z)$ is the same as *maxSup* then no representative rule can be generated from Z (see

Property 5.2.2). Otherwise, single-item antecedents of candidate k-rules are created. Loop1 starts. In general, the i-th iteration of Loop1 looks as follows:

Each candidate $X \Rightarrow Z\backslash X$, where $X \subset Z$ belongs in i-itemsets A_i, is considered. Z is frequent, so X, which is a subset of Z, is also frequent. In order to check if $X \Rightarrow Z\backslash X$ is an association rule its confidence: $sup(Z)/sup(X)$ has to be determined. $sup(Z)=Z.count$, while $sup(X)$ is computed as $sup(Y)$ of a frequent itemset Y in F_i such that $Y=X$. Only association rules that satisfy both conditions of Property 5.2.1 are representative. Condition (ii) is satisfied for any antecedent X which is 1-itemset. Proper generating of antecedents makes this condition true also for consequent sets A_i. So, in order to state whether an association rule is representative it is enough to check if condition (i) of Property 5.2.1 holds, i.e. whether $maxSup \leq s$ or $maxSup/sup(X)<c$. If $maxSup=0$ then both subconditions are satisfied, so $X \Rightarrow Z\backslash X$ is representative. Otherwise $maxSup>s$, which means that the letter subcondition will decide if $X \Rightarrow Z\backslash X$ is representative. The antecedent X of each association rule $X \Rightarrow Z\backslash X$ is removed from A_i. Having found all representative k-rules with i-antecedents from Z, $(i+1)$-itemset antecedents A_{i+1} are built from A_i by the *AprioriGen* function. In the result A_{i+1}, does not contain any itemset X such that $X \Rightarrow Z\backslash X$ would belong in the cover of another association rule $X' \Rightarrow Z\backslash X'$ such that $X' \subset X$. Therefore statement (ii) of Property 5.2.1 is an invariant of the algorithm.

6 Conclusions

In the paper we investigated properties of association rules that allowed us to construct an efficient algorithm computing representative association rules. Unlike the algorithm proposed in [1], the new algorithm exploits solely the information about the supports of frequent itemsets.

References

1. Kryszkiewicz, M.: Representative Association Rules. In: Proc. of PAKDD '98. Melbourne, Australia. Lecture Notes in Artificial Intelligence. Springer-Verlag (1998)
2. Agraval, R., Imielinski, T., Swami, A.: Mining Associations Rules between Sets of Items in Large Databases. In: Proc. of the ACM SIGMOD Conference on Management of Data. Washington, D.C. (1993) 207-216
3. Srikant, R., Agraval, R.: Mining Generalized Association Rules. In: Proc. of the 21st VLDB Conference. Zurich, Swizerland (1995) 407-419
4. Meo, R., Psaila, G., Ceri, S.: A New SQL-like Operator for Mining Association Rules. In: Proc. of the 22nd VLDB Conference. Mumbai (Bombay), India (1996)
5. Communications of the ACM, November 1996, Vol. 39. No 11. (1996)
6. Fayyad, U.M., Piatetsky-Shapiro, G., Smyth, P., Uthurusamy, R. (eds.): Advances in Knowledge Discovery and Data Mining. AAAI, Menlo Park, California (1996)

7. Piatetsky-Shapiro, G.: Discovery, Analysis and Presentation of Strong Rules. In: Piatetsky-Shapiro, G., Frawley, W. (eds.): Knowledge Discovery in Databases. AAAI/MIT Press, Menlo Park, CA (1991) 229-248

8. Agraval, R., Mannila, H., Srikant, R., Toivonen, H., Verkamo, A.I.: Fast Discovery of Association Rules. In: [6] (1996) 307-328

9. Savasere, A, Omiecinski, E., Navathe, S.: An Efficient Algorithm for Mining Association Rules in Large Databases. In: Proc. of the 21st VLDB Conference. Zurich, Swizerland (1995) 432-444

Rough Classifiers Sensitive to Costs Varying from Object to Object

Aleksandra Lenarcik and Zdzisław Piasta

Kielce University of Technology
Mathematics Department
Al. 1000-lecia P.P. 6, 25-314 Kielce, Poland
e-mail: lapiasta.lenarcik@tu.kielce.pl@tu.kielce.pl

Abstract. We present modification of the Probrough algorithm for inducing decision rules from data. The generated rough classifiers are now sensitive to cost splitting from object to object in the training data. The individual costs are reflected by new cost attributes defined for every single decision. In this approach the decision attribute is discretionary. Grouping of objects and defining prior probabilities are made on the basis of the group attributes. Most of this attributes may have no relations with the decision. The proposed approach is a generalisation of the methodology incorporating the cost issue. Behaviour of the algorithm is illustrated on the data concerning the credit-evaluation task.

1. Introduction

In supervised machine learning tasks the task-domain is constituted individually by examination of a number of objects representing several classes, and generalisation from these objects. The search process in a space of models (hypotheses or rulesets) is usually directed by identifying the classification accuracy manifested by the percentage of new cost objects correctly classified. The use of this classification accuracy criterion tacitly assumes that the distribution of classes in the real world is the same as in the training data, and the costs of the classification are equal. This is hardly a universal assumption in real life classification tasks (Provost and Kohavi, 1997). In such cases it is often more appropriate to take into the cost of mistake-then-the-less (Pazzani et al., 1994). The reason is that it frequently costs more to make one kind of classification error than the other. For example, in a credit-evaluation task it rather costs more in the long-run that a credit applicant will be given a line-up when in actuality is a defaulter than to establish that a client with a good credit record will default on payments. They are some less clear situations that offer opportunities for the allocation tasks with unequal misclassification costs. However, in situations in which problems misclassification costs vary from object to object. Unfortunately by now this new challenge has no satisfactory solution in (Brazdil et al., 1994).

¹ Feature to set Aleksander (Sokal 1997) P. 101-121, no ancestors pole.
² Systematical value Rough Heuristics Pole.

Rough Classifiers Sensitive to Costs Varying from Object to Object

Andrzej Lenarcik and Zdzisław Piasta

Kielce University of Technology
Mathematics Department
Al. 1000-lecia P.P. 5, 25-314 Kielce, Poland
e-mail: {zpiasta, lenarcik}@sabat.tu.kielce.pl

[Abstract.] We present modification of the ProbRough algorithm for inducing decision rules from data. The generated rough classifiers are now sensitive to costs varying from object to object in the training data. The individual costs are represented by new cost attributes defined for every single decision. In this approach the decision attribute is dispensable. Grouping of objects and defining prior probabilities are made on the basis of the group attribute. Values of this attribute may have no relations with the decisions. The proposed approach is a generalization of the methodology incorporating the cost matrix. Behavior of the algorithm is illustrated on the data concerning the credit evaluation task.

1   Introduction

In supervised machine learning a classification model is constructed inductively by exploration of a number of objects representing several classes and generalization from these objects. The search process in a space of models (e.g. decision trees or rulesets) is usually directed by maximizing the classification accuracy, measured by the percentage of new test objects correctly classified. The use of the classification accuracy criterion tacitly assumes that the distribution of classes in the real world is the same as in the training data, and the costs of misclassification are equal. This is rather a rare situation in real-life classification tasks (Provost and Fawcett, 1997). In such tasks it is often more appropriate to reduce the cost of misclassified objects (Pazzani et al., 1994). The reason is that it frequently costs more to make one kind of classification error than the other one. For example, in a credit evaluation task, it rather costs more to determine that a credit applicant will be paying debts back, when in reality he is a defaulter; than to establish that a client with a good credit record will default on payments. There are some learning algorithms that offer supports for classification tasks with unequal misclassification costs. However, in some real-life problems misclassification costs vary from object to object. Unfortunately, by now this new challenge has no satisfactory solution (Ezawa et al., 1996).

L. Polkowski and A. Skowron (Eds.): RSCTC'98, LNAI 1424, pp. 222–230, 1998.
© Springer-Verlag Berlin Heidelberg 1998

In this paper we present a modified version of the ProbRough algorithm (Piasta and Lenarcik; 1996, 1998). A source of inspiration for a construction of ProbRough was the rough set theory (Pawlak, 1991). Rough classifiers generated by the modified ProbRough algorithm are sensitive to costs varying from object to object. The individual costs are represented by new cost attributes defined for every single decision. The decision attribute is not necessary in this approach. Grouping of objects and defining prior probabilities are made on the basis of the group attribute. The values of this attribute may have no relations with the decisions.

Modification of the algorithm is primarily concerned with the way of computing values of the cost criterion. Also a method of determining a set of intermediate values for continuous attributes is modified.

The proposed approach is more general than the methodology incorporating the cost matrix. Behavior of the algorithm is illustrated on the data concerning the credit evaluation task.

2   Learning task

In our paper we discuss the problem of inducing the decision rules of the condition-decision type from a given training sample of objects. The training sample is a subset of a given universe of objects.

Our goal is to minimize costs of making a decision about every new object by using the induced ruleset. The condition parts of the rules contain information about the values of condition attributes, characterizing objects. These attributes are denoted by $X_1, X_2, \ldots, X_m$. We consider a finite set of l possible decisions. For each single decision, we define the cost attribute. The value of this attribute for an object is equal to the cost of relating the decision with the object. The cost attributes are denoted by $Y_1, Y_2, \ldots, Y_l$.

Every condition attribute has its own specific set of values that determines the type of the attribute: discrete unordered, discrete ordered, or continuous one. The cost attributes are number-valued. We denote elements of the value sets for condition and cost attributes by small letters: $x_1, \ldots, x_m$ and $y_1, y_2, \ldots, y_l$, respectively. A condition attribute space is a collection of all possible m-tuples $(x_1, \ldots, x_m)$ of condition attribute values. Any object characterized by given values of condition attributes has its unique position in the condition attribute space.

The training data is a finite set of objects with known values of attributes. The information about these objects has a form of a table. There are two general approaches to collecting the training data. In the first approach the training data is a representative sample of the whole universe, obtained, for example, by random sampling. In the second approach we assume that the whole universe is partitioned into disjoint groups. The true proportions of objects representing these groups have to be known and given as the *prior probabilities*. In this case, the training data is composed of learning objects that are taken at random only within the groups. In this approach it is convenient to use a group attribute.

We denote this attribute by G, and its values by g. The second approach to collecting training data is more general and therefore in the sequel we take into consideration only this case. All formulae for the first case can be easily obtained by substituting the prior probability of the group g by the ratio

$$\frac{\text{number of learning objects related to } g}{\text{number of all learning objects}}.$$

Let us notice that we can assign a set of decisions to every learning object in a natural way by taking these decisions that correspond to the minimum cost. In this way we can obtain a multivalued substitute of the decision attribute which is typical for classification tasks. It is worth stressing that generally, in the proposed approach, the group attribute and the decision attribute can play quite different roles.

The cost criterion presented in the next section refers to the estimates of the probability that a new object from the universe belongs to different subsets of the condition attribute space. For any subset Δ of the condition attribute space, we denote the event that an object taken at random from the universe belongs to Δ by $\mathbf{X} \in \Delta$. Similarly, for any group g, $G = g$ is the event that an object taken at random from the universe comes from the group g. We denote the probabilities of the above events by $P(\mathbf{X} \in \Delta)$ and $P(G = g)$, respectively. We use the symbol $P(\mathbf{X} \in \Delta, G = g)$ to represent the probability that the both events will occur simultaneously. To get the conditional probability $P(\mathbf{X} \in \Delta | G = g)$ that event $\mathbf{X} \in \Delta$ occurs given that event $G = g$ occurs, we divide the probability that both events occur by the probability that the second event occurs, i.e.

$$P(\mathbf{X} \in \Delta | G = g) = P(\mathbf{X} \in \Delta, G = g)/P(G = g). \tag{1}$$

Let us notice that the probability $P(G = g)$ is equal to the prior probability corresponding to the group g. Since the learning objects are taken at random within the groups, the conditional probability $P(\mathbf{X} \in \Delta | G = g)$ can be estimated by the ratio

$$\frac{\text{number of learning objects being in } \Delta \text{ that come from the group } g}{\text{number of all learning objects from the group } g}. \tag{2}$$

The probability $P(\mathbf{X} \in \Delta, G = g)$ is estimated by a product of the ratio (2) and $P(G = g)$.

The learning algorithm tries to find an optimum partition of the condition attribute space by minimizing the criterion value. The resulting partition can be described in a compact way by a set of simple decisions rules, if we allow partition elements of the special form only. We need some additional notions to define the form of partition elements. By a segment Δ_q in the value set of the attribute X_q we mean an interval when the values of the attribute X_q are ordered, or an arbitrary set otherwise. By a feasible subset Δ in the condition attribute space, determined by segments $\Delta_1, \Delta_2, \ldots, \Delta_m$, corresponding to the condition attributes, we mean a collection of all m-tuples $(x_1, x_2, \ldots, x_m)$, $x_1 \in \Delta_1$, $x_2 \in \Delta_2$, $\ldots$, $x_m \in \Delta_m$.

An assignment of the set $\kappa(\Delta)$ of decisions to a feasible subset Δ can be written in the form of the decision rule:

$$\text{if } x_1 \in \Delta_1 \text{ and } \ldots \text{ and } x_m \in \Delta_m \text{ then } d \in \kappa(\Delta). \tag{3}$$

The subset Δ is the domain of the rule, while $\kappa(\Delta)$ is the set of decisions that the rule assigns to the objects from Δ. When the set Δ_q is the whole value set of the condition attribute X_q, then this attribute may be omitted in the rule. In practice, the set of decisions $\kappa(\Delta)$ assigned to the element Δ of the partition usually contains a single decision. We assign several equivalent decisions to the same Δ, when we are not able to distinguish them with respect to the given criterion.

By a rough classifier we mean every set of the decision rules (3) with the domains that form a partition of the condition attribute space into disjoint subsets. Each rough classifier assigns the unique set of decisions to every element of the condition attribute space. When the assignments obtained with different rough classifiers are identical then these classifiers are equivalent.

Finally, we describe the convention of the decision-making by using the ruleset of the rough classifier. In order to classify an object $(x_1, x_2, \ldots, x_m)$ with an unknown value of the decision attribute it is sufficient to find the domain Δ which contains $(x_1, x_2, \ldots, x_m)$ and assign the set of decisions $\kappa(\Delta)$ to the object. When $\kappa(\Delta)$ contains a single decision, then the assignment is unique and the corresponding rule is decisive. Otherwise, the rule is not decisive. In this latter case any decision from $\kappa(\Delta)$, chosen at random with the probability card $\kappa(\Delta)^{-1}$, can be assigned to the object.

3   Cost criterion

The procedure of searching for the optimal rough classifier has to be guided by the precise criterion which unables a comparison of classifiers. Below, we present the criterion based on the estimation of the expected value of the cost of decision-making concerning new objects. This expected value is denoted by $E(cost)$ in the sequel. This value, for a fixed classifier, can be interpreted as the average cost when we make a great number of decisions about new objects by using the classifier.

Suppose that a rough classifier is given. Domains of the decision rules of this classifier determine a partition of the condition attribute space into disjoint feasible subsets. A set of decisions $\kappa(\Delta)$ is associated with each domain Δ. Let us fix a domain Δ and a group g. We start computing the average cost of decision-making with the assumption that an object taken at random from the universe belongs to Δ, and comes from the group g. Assume first that a decision d is assigned to every object being in Δ. Then the average cost $E(cost\ of\ assigning\ d | \mathbf{X} \in \Delta, G = g)$ can be estimated by the ratio

$$\frac{\text{sum of costs of assigning } d \text{ to objects in } \Delta \text{ that come from } g}{\text{number of objects in } \Delta \text{ that come from } g}. \tag{4}$$

By using the well-known property of the conditional expected value we obtain

$E(cost\ of\ assigning\ d|\mathbf{X} \in \Delta) =$

$= \sum_{g} E(cost\ of\ assigning\ d|\mathbf{X} \in \Delta, G = g)P(G = g|\mathbf{X} \in \Delta) =$

$= \dfrac{1}{P(\mathbf{X} \in \Delta)} \sum_{g} E(cost\ of\ assigning\ d|\mathbf{X} \in \Delta, G = g)P(\mathbf{X} \in \Delta, G = g)\ ,$

where $\sum_{g}$ stands for summing across all the groups. Denoting the sum in the last formula by $C(\Delta \to d)$, we obtain

$$E(cost\ of\ assigning\ d|\mathbf{X} \in \Delta) = P(\mathbf{X} \in \Delta)^{-1}C(\Delta \to d).$$

Now, let us take all the decisions in $\kappa(\Delta)$. According to our convention of the decision-making by use of the rough classifier, the conditional average cost $E(cost|\mathbf{X} \in \Delta)$ is equal to the mean value of the average costs $E(cost\ of\ assigning\ d|\mathbf{X} \in \Delta)$ corresponding to all d from $\kappa(\Delta)$. Thus,

$$E(cost|\mathbf{X} \in \Delta) = P(\mathbf{X} \in \Delta)^{-1}\dfrac{1}{card\ \kappa(\Delta)} \sum_{d' \in \kappa(\Delta)} C(\Delta \to d')\ ,$$

Finally, the average cost of decision-making concerning a new object by using the rule-set of the rough classifier is

$$E(cost) = \sum_{\Delta} E(cost|\mathbf{X} \in \Delta)P(\mathbf{X} \in \Delta) = \sum_{\Delta} \dfrac{1}{card\ \kappa(\Delta)} \sum_{d' \in \kappa(\Delta)} C(\Delta \to d')\ ,$$

(5)

where the symbol $\sum_{\Delta}$ stands for summing across all the rule domains of the rough classifier.

While generating a rough classifier, we try to find not only the optimum partition of the condition attribute space, but also to assign the optimum set of decisions to each partition element. For a fixed partition, if the probabilities $P(\mathbf{X} \in \Delta, G = g)$ were known then this assignment could be done in an optimal way. To explain this, we introduce a notion of a set of admissible decisions for Δ. We define it as the set of these decisions that lead to the minimum value of the cost $C(\Delta \to d)$, where d runs over the set of all decisions. Clearly, for a particular Δ, the mean cost

$$\dfrac{1}{card\ \kappa(\Delta)} \sum_{d' \in \kappa(\Delta)} C(\Delta \to d') \tag{6}$$

in formula (5) takes its minimum value, when $\kappa(\Delta)$ is an arbitrary and non-empty subset of the set of admissible decisions for Δ. In order to avoid the ambiguity, we take the whole set of admissible decisions as $\kappa(\Delta)$. In this case the mean (6) is equal to the minimum value of $C(\Delta \to d)$, where d runs over the set of all decisions.

Now, it is clear that the value of the cost criterion can be treated as a function a partition of the condition attribute space. The value of (5) can be rewritten in the form

$$E(cost) = \sum_{\Delta} \min_{d} \ C(\Delta \to d) \,, \tag{7}$$

where the sum is taken across all the partition elements while the minimum is taken across all the decisions. The value of the criterion is determined by the partition of the condition attribute space, the probabilities $P(\mathbf{X} \in \Delta, G = g)$, and the costs $E(cost\ of\ assigning\ d | \mathbf{X} \in \Delta, G = g)$. The value of the criterion can be estimated by using the training data, prior probabilities, and estimators (2) and (4).

4   General idea of the ProbRough algorithm

The modification of the ProbRough algorithm, presented in this paper, concerns mainly the representation of data (inclusion of new cost attributes), and the way of estimating values of the criterion that directs the searching process in the space of models. The main ideas and structure of the algorithm described in (Piasta and Lenarcik; 1996, 1998) are preserved.

ProbRough consists of two main phases. In the first phase, the condition attribute space is partitioned into feasible subsets in the iteration process that is guided by the cost criterion (7). Every partition in this phase is determined by a division of the value set of a particular condition attribute. During this process the value of the cost criterion can either decrease or remain unchanged. In the basic version of ProbRough the number of iterations in the first phase of the algorithm is given in advance. The number of iterations can be optimized by using the k-fold cross validation method (Piasta and Lenarcik, 1996). The resulting classifiers are generated in the second phase of the algorithm by joining the elements of partitions obtained in the first phase. Joining the feasible subsets is permitted when they have a common admissible decision. The value of the cost criterion is preserved in this phase of the algorithm.

Illustrative example.
Now, we illustrate the ProbRough algorithm incorporating costs varying from object to object on synthetic data concerning a simplified problem from the area of evaluating credit applications. Each object is characterized by two condition attributes. The first one, *duration* of credit in months, is treated as a continuous one. The second attribute, *account*, is a binary one, and takes on the value "yes" when the credit applicant has a bank account, or value "no" otherwise. We also consider two cost attributes associated with two decisions: granting a credit (*decision1*), or refusal of an application (*decision2*). Table 1 includes the hypothetical data. We assume that credits were granted to all clients. Thus, it is possible to estimate the costs associated with the *decision1* (see, Table 1). Negative costs mean profits for the bank. We also assume that the refusal of a credit application does not bring any cost for the bank (attribute *decision2* in Table 1).

Table 1. The *Illustrative* data set.

object	duration	account	decision1	decision2
1	10	YES	750	0
2	12	NO	-250	0
3	16	NO	-200	0
4	18	NO	250	0
5	24	YES	-300	0
6	24	YES	2250	0
7	36	NO	500	0
8	42	NO	-350	0
9	48	YES	2000	0
10	48	NO	-200	0

We use the data from Table 1 as the learning set to explain the process of rough classifier generation with the given in advance number of iterations equal to 3. We assume prior probabilities equal to the frequencies of groups in the data. In such a case the value of the cost criterion (7) for a partition of the condition attribute space has a simple interpretation. For a fixed element of the partition and a given decision, we obtain the costs related with that decision as the sum of individual costs across all the objects belonging to the partition element. Next, we repeat this step with all the decisions and choose the minimum value of the cost. The above procedure is performed with every partition element. Finally, we compute the value of the cost criterion as the ratio of the sum of minimum values of the costs across all the partitions, to the number of all learning objects.

We start the first phase of the algorithm with a trivial partition of the condition attribute space. The only element of this partition consists of the whole space. In order to find the criterion value for this partition, we obtain the sums of costs related to making the *decision1* and the *decision2* as equal to 4450 and 0, respectively. Hence, the cost value, as $\min(4450, 0)/10$, is 0. In the first iteration of the algorithm, we choose the partition of the condition attribute space that yields the minimum value of the criterion. This is a partition induced by the segmentation $(NO) - (YES)$ of the value set of the condition attribute *account*. The cost value is equal to -25. In the next iteration, we choose a segmentation, which combined with the previous one, yields the minimum value of the cost. This segmentation is determined by the intermediate value 39 of the condition attribute *duration*. The resulting partition of the condition attribute space consists of four elements, and the criterion value corresponding to this partition is equal to -55. In the third iteration we choose, in the same way, the segmentation determined by the intermediate value 17 of *duration*. The outcome partition consists of six elements and the corresponding cost -100. In the second phase of the algorithm we join three elements of the outcome partition with a common admissible decision *decision2* into one feasible subset. Finally, we obtain a unique rough classifier illustrated in Figure 1. The induced classifier can be presented as the set of four decision rules:

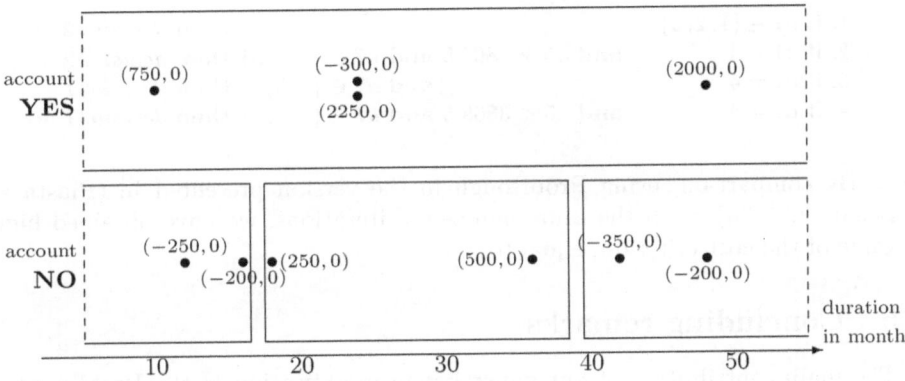

Fig. 1. Partition of the condition attribute space induced by the unique rough classifier induced from the *Illustrative* data set.

1. **if** $\quad\quad\quad\quad\quad\quad\quad\quad\quad account = YES$ **then** *decision2*,
2. **if** $\quad 17 \leq duration < 39$ **and** $account = NO$ **then** *decision2*,
3. **if** $\quad\quad duration \geq 39$ **and** $account = NO$ **then** *decision1*,
4. **if** $\quad\quad duration < 17$ **and** $account = NO$ **then** *decision1*.

5   Behavior of the algorithm on real–life data

In this section we illustrate the application of our algorithm on the *German credit* data set that was provided to the repository of machine learning databases at the University of California by H. Hofmann from the University of Hamburg. Credit applicants (objects) are described by the condition attributes of mixed types: from unordered qualitative ones, e.g., marital status, job, reason for loan request, to continuous quantitative ones, e.g., age, credit amount, duration of current account. Each object belongs to one of the two groups: *good* or *bad* credit applicants. We consider two decisions: granting a credit (*decision1*) and refusal of a credit application (*decision2*).

The learning set consists of 1000 objects. We assumed the priors: $P(G = good) = 0.7$ and $P(G = bad) = 0.3$, that reflect the structure of the learning set. Since the representation of data in the *German credit* set does not fully fit the representation used in our approach, values of the cost attributes were based on the *amount of credit* attribute. We assumed that while granting a credit, the bank profit is 5% of the credit value in the case of a *good* client, and the bank loss is 25% of the credit value in the case of a *bad* client.

Using ProbRough with the number of iterations equal to 3 we have obtained a number of classifiers corresponding to the same value of the cost criterion $-102, 7$. One of these classifiers is of the form:

1. **if** $a1 \in \{1, 2, 3\}$ **then** *decision2*
2. **if** $a1 = 4$ **and** $a5 \geq 3865.5$ **and** $a7 \in \{1, 2, 3\}$ **then** *decision2*
3. **if** $a1 = 4$ **and** $a7 \in \{4, 5\}$ **then** *decision1*
4. **if** $a1 = 4$ **and** $a5 < 3865.5$ **and** $a7 \in \{1, 2, 3\}$ **then** *decision1*

By comparison, using ProbRough in the version presented in (Piasta and Lenarcik, 1996), with the same number of iterations, we have obtained higher value of the cost criterion, equal to -65.0.

6   Concluding remarks

The main contribution of our paper lies in modification of the ProbRough algorithm that enables us to incorporate individual misclassification costs varying from object to object. The modification concerns mainly the inclusion of new cost attributes to the representation of data, and the way of estimating the cost criterion that directs the search process in the space of models. The prior probabilities related to the groups of objects can also be incorporated. The algorithm makes room for non-determinism in assignment of decisions to partition elements of the underlying attribute space, which is important when several decisions with very close values of the cost compete for assignment.

Acknowledgments

This work was supported by grant no 8 T11C 010 12 from State Committee for Scientific Research (KBN).

References

1. Ezawa, K. J., Singh, M., Norton, S. W. (1996). Learning goal oriented networks for telecommunications risk management. *Proceedings of the 13th International Conference on Machine Learning*, Bari, Italy, July 3-6, 1996. Avaliable at `ftp://ftp.cis.upenn.edu/pub/msingh/ml96_alt.ps.Z`
2. Pawlak, Z. (1991). *Rough Sets: Theoretical Aspects of Reasoning About Data*. Dordrecht: Kluwer Academic Publishers.
3. Pazzani, M., Merz, C., Murphy, P., Ali, K., Hume, T., Brunk, C. (1994). Reducing misclassification costs. *Proceedings of the 11th International Conference on Machine Learning* (pp. 217–225). Morgan Kaufmann.
4. Piasta, Z., Lenarcik, A. (1996), Rule induction with probabilistic rough classifiers. ICS Research Report 24/96, Warsaw University of Technology, to appear in *Machine Learning*.
5. Piasta, Z., Lenarcik, A. (1998), Learning rough classifiers from large databases with missing values. In: L. Polkowski, A. Skowron (eds): *Rough Sets in Knowledge Discovery*. Physica-Verlag (Springer), forthcoming.
6. Provost, F., Fawcett, T. (1997). Analysis and visualization of classifier performance: comparison under imprecise class and cost distributions. *Proceedings of the Third International Conference on Knowledge Discovery and Data Mining (KDD-97)*, Huntington Beach, CA.

Soft Techniques to Data Mining

Ning Zhong[1], Juzhen Dong[1], and Setsuo Ohsuga[2]

[1] Dept. of Computer Science and Sys. Eng., Yamaguchi University
[2] Dept. of Information and Computer Science, Waseda University

Abstract. This paper describes two soft techniques, GDT-NN and GDT-RS, for mining *if-then* rules in databases with uncertainty and incompleteness. The techniques are based on a Generalization Distribution Table (GDT), in which the probabilistic relationships between concepts and instances over discrete domains are represented. The GDT provides a probabilistic basis for evaluating the strength of a rule. We describe that a GDT can be represented by connectionist networks (GDT-NN for short), and *if-then* rules can be discovered by learning on the GDT-NN. Furthermore, we combine the GDT with the *rough set* methodology (GDT-RS for short). Thus, we can first find the rules with larger strengths from possible rules, and then find minimal relative reducts from the set of rules with larger strengths. The strength of a rule represents the uncertainty of the rule, which is influenced by both unseen instances and noises. We compare GDT-NN with GDT-RS, and describe GDT-RS is a better way than GDT-NN for large, complex databases.

1 Introduction

Over the last two decades, several inductive methods for learning *if-then* rules and concepts from instances have been proposed. Based on the viewpoint of the style of information processing, the inductive methods can be divided into two styles: *top-down* and *bottom-up*. Usually, the methods belonging to top-down style such as ID3 [6] can learn rules very fast, but it is difficult to handle data change, to use background knowledge in the learning process, and to perform in a parallel-distributed cooperative mode. On the other hand, the methods belonging to bottom-up style such as version-space [1] and back-propagation [2] are incremental ones, in which learning a concept is possible not only when instances are input simultaneously but also when they are given one by one. Although the methods belonging to bottom-up style have no the problems that the methods belonging to top-down style have, some issues on real-world applications such as

- How can rules be learned in the environment with noise and incompleteness?
- How can unseen instances be predicted, and how can the uncertainty of a rule including the prediction be represented explicitly?
- How can biases be selected and altered dynamically for constraint and search control?
- How can the use of background knowledge be selected according to whether background knowledge exists or not?

L. Polkowski and A. Skowron (Eds.): RSCTC'98, LNAI 1424, pp. 231–238, 1998.
© Springer-Verlag Berlin Heidelberg 1998

are still the ones to which no satisfactory solution has been found.

In this paper, we describe two soft techniques: GDT-NN and GDT-RS, for mining *if-then* rules in databases with uncertainty and incompleteness. The techniques are based on a Generalization Distribution Table (GDT), in which the probabilistic relationships between concepts and instances over discrete domains are represented. The GDT provides a probabilistic basis for evaluating the strength of a rule. We describe that a GDT can be represented by connectionist networks (GDT-NN for short), and *if-then* rules can be discovered by learning on the GDT-NN. Furthermore, we combine the GDT with the *rough set* methodology (GDT-RS for short). Thus, we can first find the rules with larger strengths from possible rules, and then find minimal relative reducts from the set of rules with larger strengths. The strength of a rule represents the uncertainty of the rule, which is influenced by both unseen instances and noises. We compare GDT-NN with GDT-RS, and describe GDT-RS is a better way than GDT-NN for large, complex databases.

2   GDT-NN

The central idea of our methodology is to use *Generalization Distribution Table (GDT)*, as a hypothesis search space for generalization, in which the probabilistic relationships between concepts and instances over discrete domains are represented [7,8]. We define that a GDT consists of three components: The *possible instances*, which are denoted in columns in a GDT, are all possible combinations of attribute values in a database; The *possible generalizations* for instances, which are denoted in rows in a GDT, are all possible generalization for all possible instances; The *probabilistic relationships* between the possible instances and the possible cases of generalization, which are denoted in the elements G_{ij} in a GDT, are the probabilistic distribution for describing the strength of the relationship between every possible instance and every possible generalization. The default prior probability distribution is equiprobable, that is,

$$
p(PI_j|PG_i) = \begin{cases} \dfrac{1}{N_{PG_i}} & \text{if } PI_j \supset PG_i \\[2mm] 0 & \text{otherwise} \end{cases} \tag{1}
$$

where PI_j is the jth possible instance, PG_i is the ith possible generalization. and N_{PG_i} is the number of the possible instances satisfying the ith possible generalization, that is,

$$
N_{PG_i} = \prod_{j}^{m} n_j, \tag{2}
$$

where $j = 1, \ldots, m$, and $j \neq$ the attribute that is contained by the ith possible generalization (i.e., j just contains the attributes expressed by the wild card as shown in Table 1).

Furthermore, background knowledge can be used as a bias to constrain the possible instances and the prior probabilistic distributions. For example, if we use a background knowledge,

"when the air temperature is very high, it is not possible there exists some frost at ground level",

then we do not consider the possible instances that are contradictory with this background knowledge in all possible combination of different attribute values in a *earthquake* database for creating a GDT. Thus, we can get the more refined rules by using background knowledge.

Table 1. Generalizations Distribution Table for a sample database

	a0b0c0	a0b0c1	a0b1c0	a0b1c1	...	a1b2c1
*b0c0	1/2				...	
*b0c1		1/2			...	
*b1c0			1/2		...	
*b1c1				1/2	...	
*b2c0					...	
*b2c1					...	1/2
a0*c0	1/3		1/3		...	
a0*c1		1/3		1/3	...	
a1*c0					...	
a1*c1					...	1/3
a0b0*	1/2	1/2			...	
a0b1*			1/2	1/2	...	
a0b2*					...	
a1b0*					...	
a1b1*					...	
a1b2*					...	1/2
**c0	1/6		1/6		...	
**c1		1/6		1/6	...	1/6
b0	1/4	1/4			...	
b1			1/4	1/4	...	
b2					...	1/4
a0**	1/6	1/6	1/6	1/6	...	
a1**					...	1/6

Table 2. A sample database

No	a	b	c	d
u1	a0	b0	c1	y
u2	a0	b1	c1	y
u3	a0	b0	c1	y
u4	a1	b1	c0	n
u5	a0	b0	c1	n
u6	a0	b2	c1	n
u7	a1	b1	c1	y

In our approach, the basic process of hypothesis generation is that of generalizing the instances observed in a database by searching and revising the GDT. Table 1 shows a GDT that is generated by using three attributes, *a, b, c,* in a sample database shown in Table 2. In Table 1, "*" is a wild card that means the attribute can be any value, and the elements that are not displayed are also all zero.

A GDT can be represented by connectionist networks for rule discovery in an evolutionary, parallel-distributed cooperative mode (GDT-NN for short) [8]. The connectionist networks consist of three layers: the *input unit* layer, the *hidden unit* layer, and the *output unit* layer.

A unit that receives instances from a database is called an *input unit*. A unit that receives a result of learning in a hidden unit, which is used as one of the rule candidates discovered, is called an *output unit*. A unit that is neither input nor output unit is called a *hidden unit*. Let the hidden unit layer be further divided into *stimulus units* and *association units*. The stimulus units are used to represent the possible instances like the columns in a GDT, and the association units are used to represent the possible generalizations for instances like the rows in a GDT. Furthermore, there is a link between the stimulus units and an association unit if the association unit represents a possible generalization for some possible instances in the stimulus units. Moreover, the probabilistic relationships between the possible instances and the possible generalizations are denoted in the weights of the links, and the initial weights are equiprobable like the G_{ij} of an initial GDT if we do not use any prior background knowledge for creating the initial weights. Furthermore, there are two kinds of links: excitatory link and inhibitory link that can be changed dynamically.

We have developed an algorithm (called GDT-NN) for learning *if-then* rules based on the connectionist representation [8]. One good feature of the GDT-NN is that every instance in a database is only searched once, and if the data in a database are changed (added, deleted, or updated), then we only need to modify the connectionist networks and the discovered rules related to the changed data, but the database is not searched again. Here we would like to stress that the connectionist networks do not need to be explicitly created in advance. They can be embodied in the learning algorithm, and we only need to record the weights and the units stimulated by instances in a database. However, the recording number still is quite large when processing large, complex databases. We need to find much better way to solve the problem.

3   GDT-RS

In order to solve the problem stated above, we combine the GDT with the rough sets methodology (GDT-RS for short). Using the *rough set* theory as a methodology of rule discovery is effective in practice [5,4]. The discovery process based on the rough set methodology is that of knowledge reduction in such a way that the decision specified could be made by using minimal set of conditions. The process of knowledge reduction is similar to the process of generalization in a hypothesis search space. By combining the GDT with rough sets, we can first find the rules with larger strengths from possible rules, and then find minimal relative reducts from the set of rules with larger strengths. Thus, a minimal set of rules with larger strengths can be acquired from databases with noisy, incomplete data by using GDT-RS.

3.1   Rule Representation and Condition/Decision Attributes

Let $T = (U, A, C, D)$ be a decision table, U a universe of discourse, A a family of equivalence relations over U, and $C, D \subset A$ two subsets of attributes that are called condition and decision attributes, respectively. The learned rules are typically expressed in

$$X \rightarrow Y \text{ with } S. \qquad (X \in C, Y \in D)$$

That is, "a rule $X \rightarrow Y$ has a strength S in a given decision table T". Where X denotes the conjunction of the conditions that a concept must satisfy, Y denotes a concept that the rule describes, and S is a "measure of strength" of which the rule holds.

3.2   Rule Strength

We define the strength S of a rule $X \rightarrow Y$ in a given decision table T as follows:

$$S(X \rightarrow Y) = s(X) \times (1 - r(X \rightarrow Y)). \tag{3}$$

From Eq. (3) we can see that the strength S of a rule is affected by the following two factors:

1. The strength of the generalization X (i.e., the condition of the rule), s. It is given by Eq. (4).

$$s(PG_i) = \sum_j p(PI_j | PG_i) = \frac{N_{ins-rel,i}}{N_{PG_i}} \tag{4}$$

   where $N_{ins-rel,i}$ is the number of the observed instances satisfying the ith generalization. The initial value of $s(PG_i)$ is 0. The value will be dynamically updated according to giving an input one by one. If all of the instances satisfying ith generalization appear, the strength will be the maximal value, 1. The larger the value of $s(PG_i)$, the stronger the ith generalization.

2. The rate of noises, r. It shows the quality of classification, that is, how many instances as the conditions that a rule must satisfy can be classified into some class.

$$r(X \rightarrow Y) = \frac{N_{ins-rel}(X) - N_{ins-class}(X, Y)}{N_{ins-rel}(X)}, \tag{5}$$

   where $N_{ins-rel}(X)$ is the number of the observed instances satisfying the generalization X, $N_{ins-class}(X, Y)$ is the number of the instances belonging to the class Y within the instances satisfying the generalization X.

From the GDT, we can see that a generalization is 100% true if and only if all of instances belonging to this generalization appear. Let us again use the example shown in Table 2. Considering the generalization $\{a_0 b_1\}$, if instances both $\{a_0 b_1 c_0\}$ and $\{a_0 b_1 c_1\}$ appear, the strength $s(\{a_0 b_1\})$ is 1; if only one of $\{a_0 b_1 c_0\}$ and $\{a_0 b_1 c_1\}$ appears, the strength $s(\{a_0 b_1\})$ is 0.5, as shown in Figure 1. We can see that both $\{a_0 b_1\}$ and $\{b_1 c_1\}$ are generalizations for the instance $\{a_0 b_1 c_1\}$. But the strengths of them are $s(\{a_0 b_1\}) = 0.5$ and $s(\{b_1 c_1\}) =$

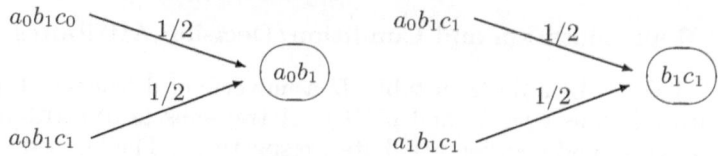

Fig. 1. Probability of a generalization rule

1, respectively. No matter what value of noise rate r may be, the strength of the rule $\{b_1 c_1\} \rightarrow y$ is greater than the strength of the rule $\{a_0 b_1\} \rightarrow y$.

If a generalization contains the instances belonging to different decision classes, the rule acquired from the generalization is noisy. Furthermore, when the value of noise is over the threshold, the rule is contradictory. As the example in Table 2, the generalization $\{a_1 b_1\}$ is such a contradictory generalization. Since the strength $s(\{a_1 b_1\}) = 1$ and the noise rates for decision classes y and n are $\frac{1}{2}$, $S(a_1 \wedge b_1 \rightarrow y) = S(a_1 \wedge b_1 \rightarrow n) = 0.5$. Furthermore, a user can specify an allowed noise rate as the threshold value. Thus, the rules with the larger rates than the threshold value will be deleted.

3.3   Simplifying a Decision Table by Using the GDT

By using the GDT, it is obvious that one instance can be expressed by several possible generalizations, and several instances can be also expressed by one possible generalization. Simplifying a decision table is to find such a set of generalizations, which cover all of the instances in a decision table and the number of generalizations is minimal.

The method of computing the reducts of condition attributes in our approach, in principle, is equivalent to the discernibility matrix method [3,5], but we do not remove dispensable attributes. This is because

- The greater the number of dispensable attributes, the more difficult it is to acquire the best solution;
- Some values of a dispensable attribute may be indispensable for some values of a decision attribute.

Figure 2.(1) gives the relationship among generalizations. We can see that every generalization in upper levels contains all generalizations related to it in lower levels. That is, $\{a_0\} \supset \{a_0 b_1\}, \{a_0 c_1\} \supset \{a_0 b_1 c_1\}$. In other words, $\{a_0\}$ can be specialized into $\{a_0 b_1\}$ and $\{a_0 c_1\}$ only. In contrast, $\{a_0 b_1\}$ and $\{a_0 c_1\}$ can be generalized into $\{a_0\}$. If the rule $\{a_0\} \rightarrow y$ is true, the rules $\{a_0 b_1\} \rightarrow y$ and $\{a_0 c_1\} \rightarrow y$ are also true.

It is clear that if a generalization for some instances is contradictory, the related generalizations in upper levels than this generalization are also contradictory. As shown in Figure 2.(2), $\{a_0 c_1\}$ is a contradictory generalization for the instance $\{a_0 b_1 c_1\}$, so that the generalizations $\{a_0\}$ and $\{c_1\}$ are also contradictory. Hence, for the instance $\{a_0 b_1 c_1\}$, the generalizations $\{a_0 c_1\}, \{b_1\}, \{a_0\}$,

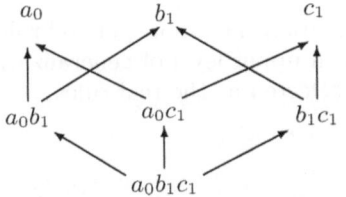

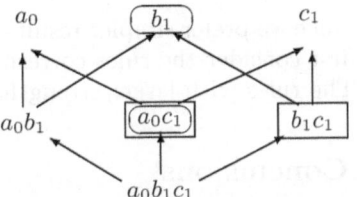

(1) The relationship among generalizations   *(2) The generalizations of $\{a_0 b_1 c_1\}$*

☐ *the generalization with instances in the same class*

⬭ *the generalization with instances in different classes*

Fig. 2. The relationships among generalizations

and $\{c_1\}$ are contradictory. Thus, only the generalizations $\{a_0 b_1\}$ and $\{b_1 c_1\}$ can be used.

This result is the same as the one of the discernibility matrix method when no noise exists in the database [3,5]. Let G_- be contradictory generalizations, G_T be all possible consistent generalizations obtained from a discernibility matrix. Clearly, $G_T = \overline{G_-}$. That is,

$$G_T = \{b_1\} \cap (\{a_0\} \cup \{c_1\}) = \{b_1 a_0\} \cup \{b_1 c_1\}$$

$$\overline{G_-} = \overline{\{a_0 c_1\} \cup \{b_1\}} = \overline{\{a_0 c_1\}} \cap \overline{\{b_1\}} = \{b_1\} \cap (\{a_0\} \cup \{c_1\})$$
$$= \{b_1 a_0\} \cup \{b_1 c_1\}.$$

For the database with noises, the generalization that contains instances with different classes should be checked. If a generalization contains more instances belonging to a class than those belonging to other classes, and the noise rate is smaller than a threshold value, the generalization is regarded as a consistent generalization of that class. Otherwise, the generalization is contradictory. Furthermore, if two generalizations in the same level have different strengths, the one with larger strength will be selected first.

3.4   Rule Selection

There are several possible ways for rule selection. For example,

- Select the rules that contain as many instances as possible;
- Select the rules in the levels as high as possible according to the first type of biases stated above;
- Select the rules with larger strengths.

Here we would like to describe a method of rule selection for our purpose as follows:

- Since our purpose is to simplify the decision table, the rules that contain less instances will be deleted if a rule that contains more instances exists.

- Since we prefer simpler results of generalization (i.e., more general rules), we first consider the rules corresponding to an upper level of generalization.
- The rules with larger strengths are first selected as the real rules.

4 Conclusions

In this paper, we presented two soft techniques, GDT-NN and GDT-RS, for mining *if-then* rules in databases with uncertainty and incompleteness. We described basic concepts and principles of our methodology. Some of databases such as postoperative patient, earthquack, weather, mushroom, and cancer have been tested or are being tested for our approaches. Although both GDT-NN and GDT-RS are very soft techniques for data mining, GDT-RS is a better one than GDT-NN for large, complex databases. By using the GDT-RS, we can first find the rules with larger strengths from possible rules, and then find minimal relative reducts from the set of rules with larger strengths. Thus, a minimal set of rules with larger strengths can be acquired from databases with noisy, incomplete data.

References

1. T.M. Mitchell. "Generalization as Search", *Artif. Intell.*, Vol.18 (1982) 203–226.
2. D.E. Rumelhart, G.E. Hinton, and R.J. Williams, "Learning Internal Representations by Back-Propagation Errors", *Nature* Vol.323 (1986) 533–536.
3. A. Skowron and C. Rauszer. "The discernibility matrics and functions in information systems", R. Slowinski (ed.) *Intelligent Decision Support* (1992) 331–362.
4. T.Y. Lin and N. Cercone (ed.) *Rough Sets and Data Mining: Analysis of Imprecise Data*, Kluwer Academic Publishers (1997)
5. Z. Pawlak. *ROUGH SETS, Theoretical Aspects of Reasoning about Data*, Kluwer Academic Publishers (1991).
6. J.R. Quinlan, "Induction of Decision Trees", *Machine Learning*, Vol.1, (1986).
7. N. Zhong and S. Ohsuga, "Using Generalization Distribution Tables as a Hypotheses Search Space for Generalization", *Proc. 4th Inter. Workshop on Rough Sets, Fuzzy Sets, and Machine Discovery (RSFD-96)* (1996) 396–403.
8. N. Zhong, S. Fujitsu, and S. Ohsuga, "Generalization Based on the Connectionist Networks Representation of a Generalization Distribution Table", *Proc. First Pacific-Asia Conference on Knowledge Discovery and Data Mining (PAKDD-97)*, World Scientific (1997) 183–197.
9. N. Zhong, J.Z. Dong, and S. Ohsuga, "Discovering Rules in the Environment with Noise and Incompleteness", *Proc. 10th Inter. Florida AI Reaserch Symposium (FLAIRS-97)* edited in the *Special Track on Uncertainty in AI* (1997) 186–191.

Business Process Understanding: Mining Many Datasets

Jan M. Żytkow[1], Arun P. Sanjeev[2]

[1] Computer Science Department, UNC Charlotte, Charlotte, N.C. 28223, U.S.A.
also Institute of Computer Science, Polish Academy of Sciences
[2] Office of Institutional Research, Wichita State Univ., Wichita, KS 67260, U.S.A.
zytkow@uncc.edu, sanjeev@cs.twsu.edu

[**Abstract.**] **Institutional databases can be instrumental in understanding a business process, but additional data may broaden the empirical perspective on the investigated process. We present a few data mining principles by which a business process can be analyzed and the results represented. Sequential and parallel process decomposition can apply in a data driven way, guided by a combination of automated discovery and human judgment. Repeatedly, human operators formulate open questions, use queries to prepare the data, issue quests to invoke automated search, and interpret the discovered knowledge. As an example we use mining for knowledge about student enrollment, which is an essential part of the university educational process. The target of discovery has been the understanding of the university enrollment. Many discoveries have been made. The particularly surprising findings have been presented to the university administrators and affected the institutional policies.**

1 Business process analysis

Many databases have been developed to store detailed information about business processes. By design, the data capture the key information about events that add up to the entire process. For instance, university databases keep track of student enrollment, grades, financial aid and other key information recorded each semester or each year.

KDD can facilitate process understanding. In addition to the known elements of the process, which have been used in database design, further knowledge can be discovered by empirical analysis. We can use a discovery mechanism to mine a database in search of knowledge useful in postulating a particularly justified hidden structure within the process. For instance, it may turn out that different groups of students finish their degrees in different proportion or take drastically different numbers of credit hours.

In this paper we focus on knowledge derived from data, in distinction to expert knowledge. We present a discovery process that results in knowledge which aids the business process understanding. The discovery process is driven by two factors. The first is the basic structure of the business process and the way it is represented by database schemas and attributes. It is used to plan data preparation and search problems. The second factor is the data and empirical knowledge they can provide. The knowledge discovered from data can lead to further data preparation and search for knowledge.

Quests generate knowledge while queries prepare data. Automated search for knowledge can use discovery systems such as EXPLORA Klösgen, 1992; KDW: Piatetsky–Shapiro and Matheus, 1991; 49er: Żytkow & Zembowicz, 1993; KDD-R: Ziarko & Shan, 1994; Rosetta: Ohrn, Komorowski, Skowron, Synak, 1998. A discovery system requires a well defined search problem, which we call a quest. It also requires data, which are defined by a query. Queries can be supported by a DBMS, but when data come from

L. Polkowski and A. Skowron (Eds.): RSCTC'98, LNAI 1424, pp. 239–246, 1998.
© Springer-Verlag Berlin Heidelberg 1998

several databases, data miners must use their own application programs. An extended KDD process can be described by a sequence of quests and queries.

Database design knowledge includes temporal relations between attributes. When data describe a business process, the temporal order of events captured by different attributes is clear most of the time. Since an effect cannot precede the cause, questions about possible causal relations are constrained within the temporal relation between attributes. Much of the sophisticated search for causes, performed by systems such as TETRAD (Spirtes, Glymour & Scheines, 1993) is not needed. A typical question about causes that still can be asked is "which among the temporarily prior attributes influences a given set of target attributes?"

Vital characteristics of a process include throughput, output and duration. We do not need to argue that the output and/or throughput are the most important effects of each business process. But it is also important to know how long is the process active. This is needed to develop process effectiveness metrics and also to decide for how long should the data be kept in active records. The quest for knowledge about process output, throughput and duration will guide our data mining effort. In practical applications, a business process consists of many elementary processes added together. For instance, the university "production of credit hours" is the sum of the enrollment histories of individual students. For a given cohort of students, the total throughput can be described by the histogram of the attribute "credit hours".

Process can be split into sequential and parallel components. Each of the parallel subprocesses uses a part of the input and contributes a part of the output or throughput. The inputs and outputs of parallel subprocesses add up to the input and output of the entire process.

Process P can be decomposed sequentially into subprocess P_1 followed by P_2, when the output of process P_1 is the input to process P_2.

Subprocesses can be further decomposed in sequence or in parallel. We can also seek explanation of the input to the business process P by processes that are prior to P and supply parts of P's input. The data come from various sources, external to the business process database.

Parallel decomposition can be guided by regularities between input and output/throughput attributes. Let C be an attribute which describes the throughput of process P. Let V_C be the range of values of C. The histogram of C is the mapping $h : V_C \longrightarrow N$, where for each $c \in V_C, h(c) = n$ is the number of occurrences of c in the data. We can use the histogram of C to measure the efficiency of the process P.

Consider a regularity in the form of a contingency table for an attribute A, that describes the input of P, and the throughput C. For each $a \in V_A$ and each $c \in V_C$, $P(a, c)$ is the probability of the value combination (a, c) derived from data. Using the distribution of inputs given by the histogram $h(a), a \in V_A$ we can compute the throughput histogram, by converting probabilities $p(a, c)$ into $p(c|a)$ and using:

$$h(c) = \sum_{a \in A} p(c|a), \forall c \in V_C. \tag{1}$$

When the histograms of A and C and the contingency table have been derived from the same data, the equation (1) is an identity. But the probability distributions $p(c|a)$ for each a and equation (1) can provide valuable predictions for a new similar process, when only the histogram of the input A is known.

When probability profiles (vectors) $p(x|a), a \in V_A$, differ for different values of A, it makes sense to think about process P as a combination of parallel processes P_a for each value $a \in V_A$. Parallel decomposition is particularly useful when:

1. differences between probability profiles are big;
2. attribute A can be controlled by the business process manager;
3. the histograms of A undergo big variations for inputs made at different time;
4. we expect that process P goes on differently when P_a's differ.

Sequential analysis is useful when some attributes describe stages of a process. Often, in addition to the input and output attributes, other attributes describe intermediate results. Suppose that an attribute B allows us decompose process P sequentially into P_1 and P_2, and can be used as a measure of effectiveness of P_1. This is particularly useful, when the subprocess P_1 applies only to some inputs while some other inputs can be used as a control group, so that the effectiveness of P_1 can be compared with the control group. We will discuss remedial education and financial aid and their effectiveness as examples.

Sequential analysis is also important when it leads to knowledge useful in predicting process input. For instance, the number of new students depends on the number of high school graduates, on the cost of study per credit hour, and so on. Regularities derived from past data lead to predictions about student enrollment in the future.

Since the attributes that can provide predictions of new input to the business process are typically not available in the business process database (**B**), other databases must be searched for relevant information. A relevant database **D** must include at least one attribute A_1 that provides information temporarily prior to the input attribute A_2 in **B** and at least one attribute J that can be used to join **B** and **D**. Further, the join'ed data table **B+D** must yield a regularity between A_1 and A_2. For instance, a regularity between high school graduation and freshmen enrollment has been detected from tables (1), (3), and (5), listed below.

2   A walk-through example: university enrollment

Understanding the factors in enrollment decline and increase is critical for universities, as often the resources available to the university depend on the number of credit hours the students enroll. Many specific steps to increase enrollment may not be productive because enrollment is a complex phenomenon, especially in metropolitan institutions where the student population is diverse in age, ethnic origin and socio-economic status.

Student databases kept at every university can be instrumental in understanding the enrollment. We have applied the process analysis methodology to a university database exploration and step after step expanded our understanding of the enrollment process. The initial discovery goals have been simple but their subsequent refinement led to sophisticated knowledge that surprised us and influenced university administrators. Within the limits of this paper we only describe a few steps and a few results. Our previous research on enrollment has been reported by Sanjeev & Zytkow (1996).

Our data came from several sources. Consider a student database that consists of the following files (tables) 1-4 and an additional database (5):
   (1) Grade Tape for each academic term (Mainframe, Sequential File)
   (2) Student History File (Mainframe, VSAM file)
   (3) Student Transcript file (Mainframe, VSAM file)
   (4) Student Financial Assistance (Mainframe, IMS/DL1 database)
   (5) High School Graduates; Kansas State Board of Education

We used temporal precedence to group the attributes into three categories. **Category 1** describes students prior to their university enrollment. It includes *demographics:* age at first term, ethnicity, sex, and so forth, as well as *high school information:* the graduation year, high school name, high school grade point average (HSGPA), rank in the graduating class, the results on standardized tests (COMPACT) and so on. All these attributes come from the tables (1) and (5).

The attributes in **Category 2** describe events in the course of study: hours of remedial education in the first term (REMHR), performance in basic skills classes during the first term, cumulative grade point average (CUMGPA), number of academic terms skipped, maximum number of academic terms skipped in a row, number of times changed major, number of times placed on probation, and academic dismissal. All these attributes come from the tables (2), (3), and (4).

Category 3 includes the goal attributes. They capture the global characteristics of a business process: the output, throughput and duration. In our example we use academic degrees received (DEGREE) as a direct measure of desired output. Bachelor degrees are awarded after completing approximately 120 credit hours. But all credit hours taken by a student contribute to the total credit hours, which determines the university's budget. Thus the total number of credit hours taken (CURRHRS) measures the process throughput. Process duration can be measured by the number of academic terms enrolled (NTERM) by a student. All these attributes come from table (3).

Query-1 prepared the initial data table. In our walk-through example we use attributes in all three categories. We analyze a homogeneous yet large group of students, containing first-time, full-time freshmen with no previous college experience, from the Fall 1986. The choice of the year provides sufficient time for the students to receive a bachelor degree by the time we conducted our study, even after a number of stop-outs.

Query-1 prepares data in a number of steps: (a) Select from the table (1) for the Fall 1986 all freshmen (**class=1**) without previous college experience **Prev=0** and **sex=1** or **sex=2** for new males and/or new females; (b) join the result with the transcript file (3) by SSN; (c) create the attributes that total the credit hours, the remedial classes, the number of semesters enrolled, average the grades, etc. (d) project the attributes in Categories 1, 2, and 3, including all the attributes created in step (c).

We used the 49er KDD system to search for knowledge. 49er (Żytkow & Zembowicz, 1993) discovers knowledge in the form of statements "Pattern P holds for data in range R". A range of data is the whole dataset or a data subset distinguished by conditions imposed on one or more attributes. Examples of patterns include contingency tables, equations, and logical equivalence. Contingency tables are very useful as a general tool for expressing statistical knowledge which cannot be summarized into specialized patterns such as equations or logical expressions. Since enrollment data lead to fuzzy knowledge, in this paper we will only consider contingency tables, although some enrollment knowledge has been approximated by equations.

49er can be used on any relational table (data matrix). It systematically searches patterns for different combinations of attributes and data subsets. Initially, 49er looks for contingency tables, but if the data follow a more specific pattern, it can turn on a specialized discovery mechanism, such as a search in the space of equations.

If the statistical test of significance exceeds the acceptance thresholds, a hypothesis is qualified as a regularity. The significance indicates sufficient evidence. It is measured by the (low) probability Q that a given sample is a statistical fluctuation of random distribution. While in typical "manual" applications of statistics researchers accept regularities with $Q < 0.05$, 49er typically uses much lower thresholds, on the order of $Q < 10^{-5}$, because in a single run it can examine many thousands of hypotheses, so many random patterns look significant at the level 0.05.

49er's principal, if crude, measurement of predictive strength of contingency tables is based on Cramer's V coefficient

$$V = \sqrt{\chi^2/(N \min(M_{row} - 1, M_{col} - 1))},$$

for a given $M_{row} \times M_{col}$ contingency table, and a given number N of records. Both Q and V are derived from the χ^2 statistics which measures the distance between tables of actual and expected counts. We have used V to detect tables which can be used for parallel process decomposition. To the same end we also use correspondence analysis to capture large differences between different probability profiles, but the details go beyond the scope of this paper.

Quest-1, the initial discovery tasks, has been a broad search request. We requested all regularities between attributes in Category 1 and 2 as independent variables against attributes in Category 3.

In response to quest-1, the 49er's discovery process resulted in many regularities. In this paper, we focus on a selected few. Let us mention a few other examples (Sanjeev & Żytkow, 1996): big differences exist in college persistence among races and among students of different age; students who changed their majors several times received degrees at the highest percentage.

Table 1. Count tables for HSGPA vs{CURRHRS, NTERM, DEGREE}; COMPACT vs CURRHRS

(a)

CURRHRS					
120 +	0	11	102	92	73
90-119	0	13	67	26	32
60-89	0	6	54	25	25
30-59	0	34	100	32	22
1-29	4	164	243	60	29
0	0	14	17	5	3
HSGPA	F	D	C	B	A

$\chi^2 = 229, Q = 1.7 \cdot 10^{-32}, V = 0.19$

(b)

CURRHRS					
120 +	40	78	59	108	5
90-119	40	44	24	38	3
60-89	26	32	28	29	0
30-59	57	68	43	36	0
1-29	262	196	65	56	1
0	38	8	4	2	0
COMPACT	missing	< 19	≤ 22	≤ 29	> 29

$\chi^2 = 221, Q = 6 \cdot 10^{-34}, V = 0.2$

(c)

NTERM					
12 +	0	10	41	26	8
9-11	0	16	107	70	42
6-8	0	17	98	47	67
3-5	1	42	110	31	31
1-2	3	158	228	69	36
HSGPA	F	D	C	B	A

$\chi^2 = 168, Q = 3 \cdot 10^{-23}, V = 0.2$

(d)

DEGREE					
Bachelor's	0	15	128	97	91
Associate	0	2	14	8	13
No-degree	4	226	443	139	81
HSGPA	F	D	C	B	A

$\chi^2 = 157, Q = 1.5 \cdot 10^{-28}, V = 0.25$

Academic results in high school turned out to be the best predictor of persistence and superior performance in college. Similar conclusions have been reached by Druzdzel and Glymour (1994) through application of TETRAD (Spirtes, Glymour & Scheines, 1993). They used summary data for many universities, in which every university has been represented by one record of many attributes that represent various totals and averages. Since we considered records for individual students we have been able to derive further interesting conclusions.

Among the measures of high school performance and academic ability, our results indicate that composite ACT score is a better predictor than either high school grade point average (HSGPA) or the ranking in the graduating class. This can be seen by comparing Tables 1-a and b. Table 1-b shows a regularity which is slightly stronger (V:0.20 vs 0.19).

Analogous patterns of approximately the same strength and significance relate COMPACT and HSGPA with all three goal variables. The corresponding tables cannot be reproduced due to the space limit.

Parallel decomposition of the process by the ACT scores has been very useful, and led to further findings, when different subprocesses have been analyzed in detail. We will see that in the case of remedial instruction. Since the values of ACT and HSGPA are closely related, it does not make sense to create separate parallel processes for HSGPA.

A sequential analysis problem: does financial aid help retention? Financial aid attributes belong to Category 2, the events that occur during the study process. Aid is available in the form of grants, loans and scholarships. The task of **Query-2** has been to utilize the financial aid data. By joining the source table (4) with the table obtained as a result of query-1 we augmented that table with 64 attributes: eight types of financial aid awarded to students in each of the 8 fiscal years 1987-94.

Quest-2 confronted the new aid attributes with with our goal attributes (Category 3). The results were surprising. No evidence has been found that financial aid causes students to enroll in more terms, take more credit hours and receive degrees. For instance, the patterns for financial aid received in the first fiscal year indicated a random relation with very high probability $Q = 0.88$ (for terms enrolled), $Q = 0.24$ (for credit hours taken) and $Q = 0.36$ (for degrees received). None would pass even the least demanding threshold of significance.

These negative results stimulated us to use **query-3** and **query-4** to select the subgroups of students at two extremes of the spectrum: those needing remedial instruction and those who had received high school grade 'A'/'B'. In each of two subgroups we tried **quest-3** and **quest-4** analogous to quest-2.

244   J.M. Żytkow, A.P. Sanjeev

Table 2. Actual Tables for DEGREE vs REMHR (a) all students (b) remedial needing

DEGREE (a)

Bachelor's	302	0	27	10	1	7
Associate	32	0	3	3	1	0
No-degree	735	2	119	82	10	47
REMHR	0	2	3	5	6	

$\chi^2 = 31.2, Q = 0.0, V = 0.106$

DEGREE (b)

Bachelor's	19	4	1	0	4
Associate	2	1	1	0	0
No-degree	174	39	36	4	21
REMHR	0	3	5	6	8

$\chi^2 = 5.06, Q = 0.89, V = 0.091$

In the quest-3, for students needing remedial instruction, we sought a possible impact on the Category 3 variables of financial aid received in the first fiscal year, when the remedial instruction has been provided. The results were negative: the patterns among the amount of financial aids received and the goal variables had the following high probabilities of randomness: $Q = 0.11$ (for terms enrolled), $Q = 0.22$ (for credit hours taken) and $Q = 0.86$ (for degrees received). In response to quest-4, in the group of students receiving high school grade 'A'/'B', the corresponding probabilities were $Q = 0.99$ (for terms enrolled), $Q = 0.99$ (for credit hours taken) and $Q = 0.94$ (for degrees received). These findings indicate that financial aid received by students in the first year was not helpful in their retention.

Using **query-5** we created an attribute which provides the total dollar amount of aid received in fiscal years 1987 – 1994. Now, with **quest-5** we sought the impact of this variable on the goal attributes in Category 3. Finally, a positive influence of financial aid has been detected (Sanjeev & Żytkow, 1996), but the results are due to the fact that in order to receive financial aid the student must be enrolled. We could not demonstrate that financial aid plays the role of seed money by increasing the enrollment in the years when it hasn't been received.

An example of sequential analysis: remedial instruction. One of the independent variables in Category 2, used in quest-1 has been REMHR (total number of remedial hours taken in the first term). An intriguing regularity has been returned by the search (Table 2-a): *"Students who took remedial hours in their first term are less likely to receive a degree"*. The percentage of students receiving a degree decreased from 31% for REMHR=0 to 13% for REMHR=8. This is a disturbing result, since the purpose of remedial classes is to prepare students for the regular classes.

Query-6: select students needing remedial instruction. After a brief analysis we realized that Table 2-a is misleading. Remedial instruction is intended only for the academically under-prepared students, while students who do not need remedial instruction are not the right control group to be compared with. In order to obtain relevant data we had to identify students for whom remedial education had been intended and analyze the success only for those students. After discussing with several administrators, the need for remedial instruction was defined as query-6: select a composite ACT score of less than 20 and either having high school GPA of 'C'/'D'/'F' or graduating in the bottom 30% of the class. Those students for whom the remedial instruction was intended but did not take it, played the role of the control group.

Quest-6: for data selected by query-6 search for regularities between REMHR and process performance attributes. Use attributes in Category 1 to make subsets of data. 49er's results were again surprising because no evidence has been found that remedial instruction helps the academically under-prepared students to enroll in more terms, take more credit hours and receive degrees.

Table 2-b indicates that taking remedial classes does not improve the chances for a student to reach a degree. For instance, those students who did not take remedial classes, but needed them according to our criteria, received bachelor and associate degrees at about the same percentage (10.8% vs 9.9%) when compared to those who took from 3 to 8 hours of remedial class. A similar table indicates no relationship ($Q = 0.98$) between hours of remedial classes taken and number of terms enrolled.

Finally, let us briefly discuss how **external data can be used to predict part of the input.** Table (5) provides the number of high school graduates by year and county. Knowing the fraction of that number who enroll at WSU, we can predict in June, a part of university enrollment in August. **Query-7** joined tables (1) and (5) by the year, but to make a join operation possible it aggregated each table (1) for the corresponding years by the county, counting the numbers of students per county. Now, **quest-7** has determined the percentage of students who transfer from high school to WSU. That number, recently at about 20% can be used to predict new enrollment.

3   Impact of findings on the business process

In 1995-97 the results of our enrollment research have been presented to senior university administrators including *Vice President of Academic Affairs, Associate Vice Presidents, Director of Budget and academic college Deans.* Many of them chaired or were part of executive purpose committees like *Strategic Plan Task Force, Academic Affairs Management Group, and University Retention Management Committee.*

Starting in the Spring 1997, for the fourth consecutive enrollment period, WSU's enrollment has increased. It is the most consistent enrollment increase in the 1990s. While business decisions are not always based entirely on empirical evidence, such evidence helps to make well-informed decisions. We discuss here some of the strategic decisions, and outline how our findings have formed their underlying empirical foundation.

"WSU will recruit and retain high quality students from a variety of ethnic and socioeconomic backgrounds" is the second of the five *Goals and Objectives* stated in the draft *Strategic Plan for Wichita State University.* This strategic plan, outlined in 1997, is currently being presented to the various university constituencies, such as the faculty senate, for review and acceptance.

In 1995 and 1996, our research uncovered that academic results in high school are the best predictor of persistence and superior performance in college. Tables 1 show regularities between composite ACT and average grade in high school (HSGPA) as predictors and the college performance attributes: cumulative credit hours taken (CURRHRS), total academic terms enrolled (NTERM), and degrees received (DEGREE).

The eight year graduation rate measure has been included as WSU's performance indicator. A strategic planning process called VISION 2020, initiated by the State of Kansas, requires the universities to formulate a set of performance indicators and report the results. The first of the core indicators concerns *undergraduate student retention and graduation rates.* The report mandated by the Regents asked for graduation rate measures after four, five, and six years. But the students at WSU take often longer than six years to graduate: they tend to stop-out for several academic terms during their college careers and enroll in less than 15 hrs in one semester. This phenomenon has been a conclusion from our studies in 1995 and 1996. It can be partially observed in Table 1-c. It shows that a significant percentage of students enroll above 11 terms. In addition many students stop out for few semesters. As a result, among the six Regents universities in Kansas, WSU is the only institution in which the graduation rate is also measured at the end of the eight year.

Our negative results increased the awareness of the cost of remedial education. Although the upcoming replacement of open admission by entry requirements is pushing aside the question of reforming the remedial education, university administrators are increasingly questioning the effectiveness of remedial classes. In the Fall 1997 a cost study has been conducted on remedial education programs. *Can those costs be justified in the absence of empirically provable success?* is currently being discussed.

4   Process networks vs. Bayesian networks

While Bayesian networks (Heckerman, 1996; Spirtes Glymour & Scheines, 1993) emphasize the relationships between attributes, the business process networks capture the relations between states of affairs at different time. They resemble the physical approach to causality: the state of a system at time T_1 along with the domain regularities causes the state at time T_2. This is a more basic understanding of causes.

In distinction to Bayesian networks that relate entire data through probabilistic relations between attributes, a subprocess often involves only a slice of data. For instance, remedial classes are taken by only some students. One part of a process can be decomposed differently than another part and the corresponding subsets of records can hold different regularities. For instance, a causal relation between attributes may differ or not exist in a subset of data characterized by a subset of attribute values.

Conclusions. We have introduced a number of knowledge discovery techniques useful in analyzing business processes. A process can be divided into parallel components when it improves the predictions. Processes can be analyzed sequentially, to find out how the preceding process influences the next in sequence. We also demonstrated how queries that seek data can be combined with quests that seek knowledge. In a KDD process, queries are instrumental to quests. The data that we used as a walk-through example have been obtained from a large student database, augmented with statistics kept by the State of Kansas. Both have been explored by 49er, leading to many discoveries. In this paper we used a few examples of practically important findings that have influenced the University policies.

References

Druzdzel, M. & Glymour, C. 1994. Application of the *TETRAD II* Program to the Study of Student Retention in U.S. Colleges. *Proc. AAAI-94 KDD Workshop*, 419–430.

Heckerman, D. 1996. Bayesian Networks for Knowledge Discovery, in Fayyad, Piatetsky-Shapiro, Smyth & Uthurusamy eds. *Advances in Knowledge Discovery and Data Mining*, AAAI Press, pp. 59–82.

Klösgen, W. 1992. Patterns for Knowledge Discovery in Databases. In Proc. of the ML-92 Workshop on Machine Discovery, 1-10. National Institute for Aviation Research, Wichita, KS: Zytkow, J. ed.

Ohrn, A., Komorowski, J., Skowron, A. & Synak, P. 1998. The Design and Implementation of a Knowledge Discovery Toolkit Based on Rough Sets - The ROSETTA System, *Rough Sets in Knowledge Discovery*, L. Polkowski & A. Skowron eds. Physica Verlag.

Piatetsky–Shapiro, G. & Matheus, C. 1991. Knowledge Discovery Workbench. In Proc. of AAAI-91 KDD Workshop, 11–24. Piatetsky–Shapiro, G. ed.

Sanjeev, A., & Zytkow, J. 1996. A Study of Enrollment and Retention in a University Database, *Journal of the Mid-America Association of Educational Opportunity Program Personnel*, Volume VIII, pp. 24–41.

Spirtes, P., Glymour, C. & Scheines, R. 1993. *Causation, Prediction and Search*, Springer-Verlag.

Ziarko, W. & Shan, N. 1994. KDD-R: A Comprehensive System for Knowledge Discovery in Databases Using Rough Sets, in T.Y.Lin & A.M. Wildberger eds. Proc. of Intern. Workshop on Rough Sets and Soft Computing, pp. 164–73.

Żytkow, J. & Zembowicz, R. 1993. Database Exploration in Search of Regularities. Journal of Intelligent Information Systems 2:39–81.

A Genetic Algorithm for Switchbox Routing Problem

Akinori KANASUGI, Takashi SHIMAYAMA, Naoshi NAKAYA
and Takeshi IIZUKA

Faculty of Engineering, Saitama University,
255, Shimo-okubo, Urawa 338, Japan

Abstract. A genetic routing method for switchbox problem with novel coding technique is presented. The principle of proposed method and results of computer experiments are described in detail.

1 Introduction

Genetic algorithm (GA) is a method to search optimum solution by avoiding to fall down into local minima which is based on mechanics of natural selection and genetics[1]. In VLSI physical design, wiring region is composed of channels and switchboxes. Channels limit their interconnection terminals to one pair of parallel sides. Switchboxes are generalizations of channels and allow terminals on all four sides of the region[2]. Several channel routers using GA have been reported[3][4], while switchbox router by GA haven't been proposed. Switchbox routing problem is more difficult than channel routing problem. Several attempts have been reported on switchbox routing problem. "WEAVER" which is based on knowledge[5] and "BEAVER" which is based on computational geometry [6] are famous switchbox routers. The quality of solution by them are better than or comparable to previously reported solutions. However these systems are complicated and required long calculation time.

In this paper, a genetic routing method for switchbox problem is presented. The method is not complicated and effective for practical switchbox problem.

2 Routing method

The order of routing is very important in switchbox routing problems.

Figure 1(a) shows the example that the net a-a' was connected first, while figure 1(b) shows the same example that net b-b' was connected first. It is impossible to connect the net a-a' in figure 1(b).

The presented method isn't influenced by order of routing, because all nets are routed in the same time. That is, the shapes of all wires are supposed first, then all wires are improved by using genetic algorithm. However, if the coding technique and the crossover operation are not appropriate, the result will be the same one by random search.

L. Polkowski and A. Skowron (Eds.): RSCTC'98, LNAI 1424, pp. 247–254, 1998.
© Springer-Verlag Berlin Heidelberg 1998

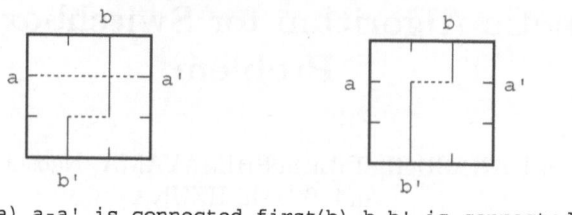

(a) a-a' is connected first(b) b-b' is connected first.

Fig. 1. Influence of routing order

2.1   Model

In this paper, the switchbox model has two layers. The lower layer has only x-direction wires, while the upper layer has only y-direction wires.

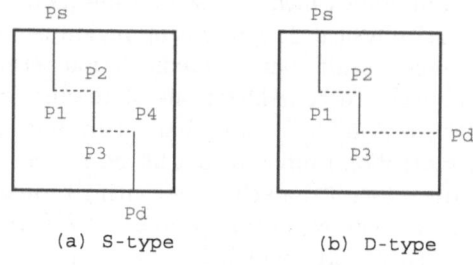

Fig. 2. Types of wire shape

In this paper, two types of wire shape are defined. One of them is "S-type" where the source terminal and the destination terminal are *same* layer. The other one is "D-type" where the source terminal and the destination terminal are *different* layer. Figure 2 (a) and (b) show the corresponding examples. The S-type wire is defined by four vias[1] $P_1 \sim P_4$, while the D-type wire is defined by three vias $P_1 \sim P_3$.

Figure 3 shows some special cases. These cases can be regarded as cases that vias P_2 and P_3 are overlapped. However, in some cases, some additional operations are required. Figure 4 show the example which has wasteful wire. The operations $P_2 \rightarrow P_4$ and $P_3 \rightarrow P_4$ have to be applied in this case.

---

[1] "via" is a hole to contact wires in different layers.

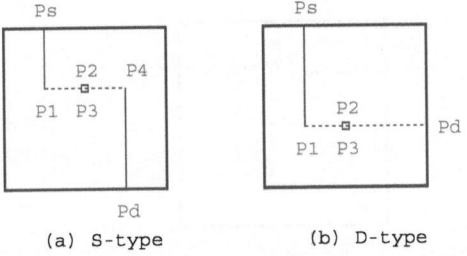

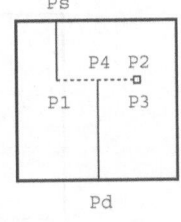

Fig. 4.   Example   with
wasteful wire

Fig. 3. Examples with overlapped vias

2.2 Coding

In the coding technique, it is important to generate no lethal gene and to keep schemata. As a preparation of coding, the coordinates of vias are summarized in figure 5. It is clear that following parameters are required to define wire shape.

(Case 1) S-type; P_s is on the x-axis. $\cdots \{x_a, y_a, y_b\}$
(Case 2) S-type; P_s is on the y-axis. $\cdots \{x_a, y_a, x_b\}$
(Case 3) D-type; P_s is on the x-axis. $\cdots \{x_a, y_a\}$
(Case 4) D-type; P_s is on the y-axis. $\cdots \{x_a, y_a\}$

Therefore, the following code can be obtained as an example.

$$\{x_a, y_a, z\} \tag{1}$$

where, z is y_b in case 1, x_b in case 2, and ignored in case 3 and 4. However, it is more effective to change x-coordinate and y-coordinate at the same time in the search of optimum vias. Therefore, we use grid number instead of x,y coordinates. Figure 6 shows grid numbers. In this paper, eq.(2) is used as a code for one wire,

$$\{G_a, G_b\} \tag{2}$$

where, G_a and G_b mean grid numbers. In the case 1, only y-coordinate of grid G_b is used, while only x-coordinate is used in case 2. In the case 3 and 4, grid G_b is ignored. Figure 7 shows the format of chromosome for all wires.

2.3 Crossover

The uniform crossover technique is used in the presented method.

2.4 Mutation

In the mutation operation of the presented method, sometimes the wire shape doesn't change, because the grid G_b is partially used or not used.

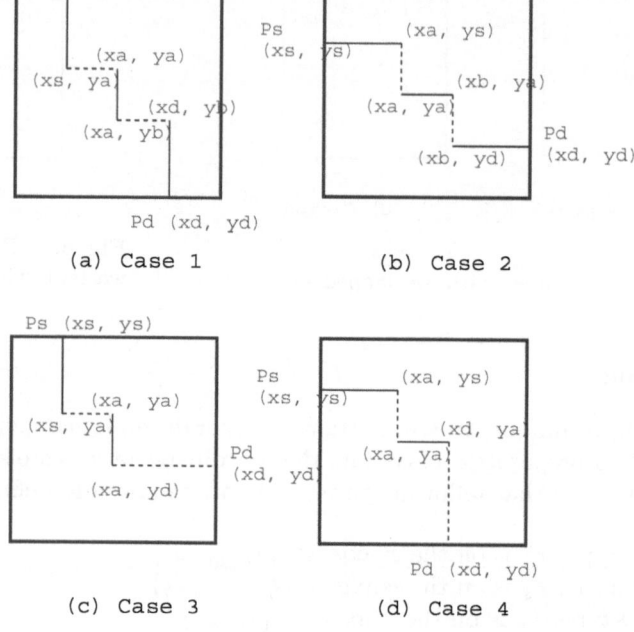

(a) Case 1   (b) Case 2

(c) Case 3   (d) Case 4

Fig. 5. Coordinates of vias in wires

2.5 Evaluation function

The evaluation function is defined as follows,

$$f = \exp\left\{-\left(\alpha\frac{L_d}{L_m} + \beta\frac{V_d}{V_s} + \gamma\frac{L_t - L_m}{L_m}\right)\right\} \tag{3}$$

where L_d is the overlapped wire length, L_t is the obtained total wire length, L_m is the minimum total wire length[2], V_d is the number of overlapped vias, V_s is the standard via numbers, $\alpha C \beta C \gamma$ are weight constants. In the presented method, V_s is calculated as follows,

$$V_s = 4N_S + 3N_D \tag{4}$$

where N_S and N_D are the numbers of S-type and D-type wires, respectively.

2.6 Generation of initial population

To improve the convergence of GA, the following initial population are used.

---

[2] L_m is calculated by the summation of Manhattan lengths between source and destination terminals.

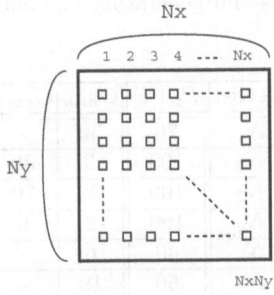

Fig. 6. Grid numbers

Net 1		Net 2			Net N	
Ga	Gb	Ga	Gb	----------	Ga	Gb

Fig. 7. Format of chromosome

(**Case 1**) S-type; P_s is on the x axis.$\cdots G_a = (x_s, m), G_b = (x_d, m)$
(**Case 2**) S-type; P_s is on the y axis.$\cdots G_a = (n, y_s), G_b = (n, y_d)$
(**Case 3**) D-type; P_s is on the x axis.$\cdots G_a = (x_s, y_d)$
(**Case 4**) D-type; P_s is on the y axis.$\cdots G_a = (x_d, y_s)$

where, m and n are random numbers ($1 \leq m \leq N_y$, $1 \leq n \leq N_x$). These value represent simple wire shapes which are illustrated in figure 3. However, 50% of initial populations are made from random value to keep large initial solution space.

3   Computer experiments

In order to examine the proposed method, we made a routing program using *Visual BASIC* and executed computer experiments on a PC with Pentium MPU (133 MHz).

Table 1 summarizes the results of 10 examples with 10×10 size and 10 nets. The routing program executed 10 times for each netlist. In the table 1, P_s means the probability for 100% connection. G_{min} and G_{max} mean the minimum and maximum generation numbers to obtain 100% connection, respectively. The symbol "-" means that 100% connection was failed.

In these experiments, the total wire length is not considered ($\gamma = 0$). The parameters of GA are as follows; population number is 100, mutation ratio is 2%, elite number is 20, maximum generation number is 200, and $\alpha = \beta = 10$. The execution time for one generation is 50 [msec]. The results show that 100% connections are obtained in almost all cases.

Table 1. Probabilities and generation numbers for 100% connection

Netlist	P_s [%]	G_{min}	G_{max}
N_1	70	30	-
N_2	100	5	30
N_3	100	22	70
N_4	100	2	5
N_5	40	6	-
N_6	50	19	-
N_7	100	3	53
N_8	0	-	-
N_9	10	135	-
N_{10}	40	20	-

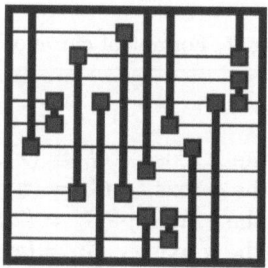

Fig. 8. An example of routing (10 × 10 size, 10 nets)

Figure 8 shows an example of routing result (netlist N_1).

Figure 9 and figure 10 show the results of example with 20 × 20 size and 20 nets. In figure 9, the total wire length is not considered ($\gamma = 0$), while in figure 10, it is considered ($\gamma = 1$). The minimum generation numbers for 100% connection are 42 and 100, respectively. The calculation time for one generation is 130 [msec].

Finally, a result for famous Berstein's switchbox problem[7] is shown in figure 11 (23 × 15 size, 24 nets). In this case, two nets couldn't be connected, while WEAVER and BEAVER performed 100% connection. The reason must be low freedom of wire shapes. That is, via number and wire direction in each layer are restricted in the presented method. There are 500 population in this case. The improvement of solutions saturated at the 13-th generation. The nets which includes more than two terminals were connected by divided into two terminal nets.

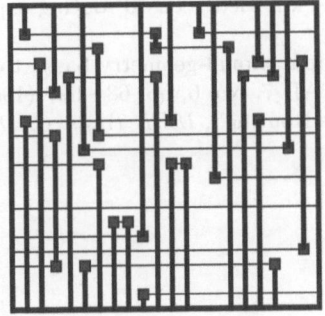
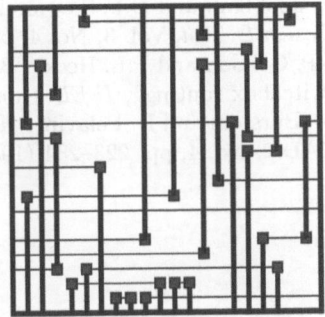

Fig. 9. An example of routing (20 × 20 size, 20 nets, $\gamma = 0$)

Fig. 10. An example of routing (20 × 20 size, 20 nets, $\gamma = 1$)

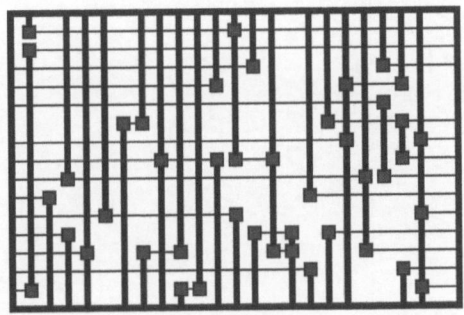

Fig. 11. An example of Berstein's switchbox problem

4 Conclusions

A novel routing method for switchbox problem using genetic algorithm was presented. The computer experiments showed that the proposed method is effective for practical routing problems.

The future works are to improve the connection ability and to reduce the calculation time.

References

1. L. Davis ed.: "Handbook of Genetic Algorithms", Van Nostrand Reinhold (1991).
2. S.M. Sait and H. Youssef: "VLSI Physical Design Automation", IEEE Press (1995).
3. A.T. Rahmani and N. Ono : "A Genetic Algorithm for Channel Routing Problem", *Proc. of the 5th Int. Conf. on Genetic Algorithms*, pp. 494–498 (1993)
4. V.N. Davidenko, V.M. Kureichik and V.V. Miagkikh: "Genetic Algorithm for restrictive Channel Routing Problem", *Proc. of the 7th Int. Conf. on Genetic Algorithms*, pp. 636–642 (1997)

5. R. Joobbani and D.P. Siewiorek: "WEAVER: A knowledge based routing expert", *Design & Test*, Vol. 3, No. 4, pp. 12–23 (1983)
6. J.P. Cohoon and P.L. Heck: "BEAVER: A computational-geometry-based tool for switchbox routing", *IEEE Trans. CAD*, Vol. CAD-7, No. 6, pp. 684–697 (1988)
7. M. Burstein and R. Pelavin: "Hierarchical Wire Routing", *IEEE Trans. CAD*, Vol. CAD-2, No. 4, pp. 223–234 (1983)

On the Benefits of Random Memorizing in Local Evolutionary Search

Hans-Michael Voigt and Jan Matti Lange

GFaI – Center for Applied Computer Science
Rudower Chaussee 5, Geb. 13.7, D-12484 Berlin, Germany
voigt@gfai.de
URL: http://www.amspr.gfai.de

Abstract. For the calibration of laser induced plasma spectrometers robust and efficient local search methods are required. Therefore, several local optimizers from nonlinear optimization, random search and evolutionary computation are compared. It is shown that evolutionary algorithms are superior with respect to reliability and efficiency. To enhance the local search of an evolutionary algorithm a new method of random memorizing is introduced. It leads to a substantial gain in efficiency for a reliable local search.

1 Introduction and Motivation

Laser induced plasma spectrometry is one of the latest developments in measurement technologies. It is very precise, can be applied to multi-element analysis without any preparation, and can be used especially in process measurement such as glass and steel processing.

Since laser spectrometry is very precise the calibration of such devices has to be also very precise. This is a difficult task. The calibration accuracy depends on the time dependent modeling of the atomic light emission of a mircoplasma. The calibration model is usually nonlinear because of different physical effects like re-absorption and light scattering. It equally well depends on the precise adaptation of the calibration model parameters to the calibration samples. This problem is ill-posed. Regularization has to be done due to facts from laser physics. Furthermore, from experience, it is ill-conditioned with condition indices $\lambda_{max}/\lambda_{min} > 10^6$. It is also high dimensional. For the calibration of 60 elements 120 to 180 parameters have to be adapted.

For the calibration we designed a hierarchical adaptation procedure. At higher levels the search space will be constrained by evidences from plasma physics. Coarse search procedures are applied. At the lowest level a fine tuning of the model parameters is done. It is outside the scope of this paper to describe the whole procedure. Here we will explain to some extent the lowest level of calibration parameter fine tuning. For this task the goal was to find an efficient and robust local search algorithm which does not use analytic gradient information.

There was made an extensive exploration of the available algorithms. A comparison is given in section II.

For the fine tuning we apply the simplest evolutionary algorithm, the $(1+1)$-Evolution Strategy – but with a memory. This memory is accessed at random to find search direction based on former good solutions. By a beam search this memorized directions will be exploited. In section III we present a conceptual algorithm for the enhancement of evolutionary algorithms with global step size adaptation by random memorizing. For an implementation with an $(1+1)$-Evolution Strategy results for difficult unimodal test functions are presented and compared.

Finally, in section IV the application to the calibration of laser spectrometers is described in more detail.

2   Searching for a Robust and Efficient Local Optimizer

The assessment of the problem solving capacity of different local optimizers[1] was done using the well known Rosenbrock function f_R (see next section) for $n = 2, ..., 100$ variables. The initial point was set to $x_i = 0, i = 1, ..., n$. The number of function evaluations was counted until the function value $f_{stop} \leq 10^{-9}$ was reached or a maximum number of generations had passed. The results are summarized in the following table where MEAN is the average number of function evaluations to reach $f_{stop} \leq 10^{-9}$, SD the corresponding standard deviation, SUCC the number of successful runs related to the total number of runs (succ/total), and BF gives the best function value reached for non-converging runs.

A natural first step in looking for good local optimizers is to consult available program packages including procedures for mathematical programming or nonlinear optimization (e.g. [14]). Among the recommended procedures are the simplex method [10] and different variants of second order optimization methods based on conjugate directions [12] or quasi Newton procedures [11].

The results of the application of these methods to the minimization of function f_R are given in Table 1. As it can be seen at once non of these methods is reliable enough to locate the optimum with the desired accuracy for dimensions up to 100 variables. Similar observations have been made already for other functions in [3]. In a next step a random search technique [13] and the dynamic hill-climber [7] were analyzed (viz. Table 1). The effort to reach the optimum with the random search technique for a very moderate dimension of $n = 10$ was much higher than with the conjugate direction method.

And even the dynamic hill-climber, though very reliable for dimensions up to $n = 30$, could not locate the optimum for $n = 100$.

Therefore we looked for alternatives which may be offered by evolutionary algorithms.

As has been shown [8] [9] the Breeder Genetic Algorithm with fuzzy gene pool recombination, the utilization of covariances and generalized elitist selection is a very reliable search method. The same is true for the $(1, \lambda)$-Evolution

[1] The results of Table 1, except for the method of Hansen et.al., were kindly provided by Dirk Schlierkamp-Voosen. Many helpful discussions with Heinz Mühlenbein and Dirk Schlierkamp-Voosen from the GMD are gratefully acknowledged.

Table 1. Experimental results for the different analyzed methods

n	MEAN	SD	SUCC	BF	MEAN	SD	SUCC	BF
	Nelder and Mead				Powell			
2	1.690E+02	0.000E+00	0/1	4.384E-01	4.350E+02	0.000E+00	1/1	
3	2.030E+02	0.000E+00	0/1	1.603E+00	6.380E+02	0.000E+00	1/1	
10	3.382E+03	0.000E+00	1/1		5.975E+03	0.000E+00	1/1	
20	1.009E+05	0.000E+00	1/1		1.907E+04	0.000E+00	1/1	
30	1.705E+05	0.000E+00	0/1	1.535E+01	4.278E+04	0.000E+00	1/1	
50					6.248E+07	0.000E+00	1/1	
60					2.098E+05	0.000E+00	0/1	4.766E-09
70					3.748E+05	0.000E+00	0/1	3.892E-09
80					3.569E+05	0.000E+00	0/1	4.551E-09
100	2.000E+05	0.000E+00	0/1	8.635E+01	5.507E+05	0.000E+00	0/1	
n	Stewart				Solis and Wets			
2	7.350E+02	0.000E+00	0/1	2.553E-02	3.780E+03	0.000E+00	1/1	
3	2.985E+03	0.000E+00	0/1	7.549E-01	3.780E+03	0.000E+00	1/1	
10	4.820E+02	0.000E+00	0/1	8.713E+00	2.770E+05	0.000E+00	1/1	
20	1.580E+02	0.000E+00	0/1	1.880E+01				
30	2.170E+02	0.000E+00	0/1	2.870E+01				
100	6.370E+02	0.000E+00	0/1	9.799E+01				
n	Yuret and de la Maza				Hansen, Ostermeier, Gawelczyk			
2	3.910E+02	0.000E+00	30/30		5.043E+02	1.237E+02	5/5	
3	1.998E+03	0.000E+00	30/30		1.025E+03	1.966E+02	5/5	
10	5.866E+04	0.000E+00	30/30		1.027E+04	7.875E+02	5/5	
20	1.418E+05	0.000E+00	30/30		4.398E+04	3.627E+03	5/5	
30	2.345E+05	0.000E+00	30/30		1.129E+05	6.576E+02	5/5	
100	5.827E+03	0.000E+00	0/30	9.698E+00				

Strategy with covariance matrix adaptation [5]. Unfortunately, both methods require to solve complete eigenvalue problems. This operation is of order $\mathcal{O}(n^3)$ thus increasing the computational overhead considerably for high-dimensional problems. The best results came up with the $(1, \lambda)$-Evolution Strategy with the generating set adaptation [6] (Table 1). This procedure was very robust with in general less computational effort compared to the other methods. But the effort was still too high for problems having more than 100 variables.

3   Adding Random Memorizing to Evolutionary Algorithms

What are the lessons to be learned from the previous experiments? First of all, good local search techniques do have a memory – either a sequential one like the conjugate direction method [12], the quasi Newton method [11], the Evolution Strategies with generating set adaptation [6] or with covariance matrix adaptation [5]– or a parallel one like the Breeder Genetic Algorithm with fuzzy gene pool recombination and covariance utilization [9]. Having a closer look at all methods which use a sequential memory reveals that it is used in general in a rather deterministic or derandomized way for the construction of new search directions. On the other hand there are already preliminary investigations into direction mutations [2] without using a memory. But the results are unsatisfactory. Therefore it was decided to analyze the impact of randomly accessing a sequential memory for the determination of new search directions.

3.1 Conceptual Local Evolutionary Algorithm with Random Memorizing

The basic idea is the following: Use any Evolutionary Algorithm (EA) which has a global step size control. Add a sequential memory to the Evolutionary Algorithm were previous best solutions in search space are stored up to a certain depth. Determine by means of the EA a better solution. Take a solution from memory at random and compute from the solution of the EA and the randomly memorized solution a search direction. Follow this search direction by increasing multiples of the global step size as long as better solutions will be found. Then go ahead with the EA, etc.

This procedure is formalized on the following page. Obviously, there are two parameters concerning the memory access. That is the memory depth d_M and the beam search factor b. Furthermore the type of the probability density distribution for the random access of the sequential memory is important. For simplicity we used a uniform distribution. Other distributions may be more appropriate.

3.2 The Enhanced (1 + 1)-Evolution Strategy

To verify the outlined algorithm we instantiated it with the most simple EA, the (1 + 1)-Evolution Strategy. A thorough theoretical analysis of the (1 + 1)-Evolution Strategy is given in [1] [4]. The parameters of the EA were set as recommended there. The memory depth was set rather high to $d_m = 2 \cdot n...n^3$ and the beam search factor to $b = 2$.

A first experiment was made for Rosenbrock's function with $n = 20$ and the initial point $x_i = 0, i = 1, ..., n$ resulting in a real surprise. This experiment is shown in Figure 1. The (1 + 1)-Evolution Strategy with random memorizing is about four times faster then the $(1, \lambda)$-Evolution Strategy with generating set adaptation. To add a further surprise it is about twice as fast then the $(1, \lambda)$-Evolution Strategy with covariance matrix adaptation [5]. There is almost no computational overhead for the extraction of information from the memory. We then computed the minimum of f_R for dimensions $n = 5, ..., 200$ and $\epsilon = 10^{-10}$ as shown in Table 2. For this function this method was superior to all other considered approaches with respect to efficiency and robustness.

Finally, we made a benchmarking with the following functions:

- $f_R = \sum_{i=1}^{n-1} \left(100(x_i^2 - x_{i+1})^2 + (x_i - 1)^2\right)$ Rosenbrock function

- $f_S = \sum_{i=1}^{n} x_i^2$ Sphere function
- $f_E = \sum_{i=1}^{n} \left(1000^{(i-1)/(n-1)} x_i\right)^2$ Hyper ellipsoid with condition index 10^6 [6]
- $f_P = -x_1 + \sum_{i=2}^{n} x_i^2$ Parabolic ridge [1]

Outline of the Local Evolutionary Search with Random Memorizing

---

set generation counter $g := 0$;
set memory counter $c := 0$;
set beam search factor b;
set maximal number of generations g_{max};
set memory depth d_m;
set initial values of variables x_g;
set initial global step size d_g;
set termination criterion ϵ;
set initial memory content m_i, $i = 0, ..., d_m - 1$;
compute fitness $f := f(x_g)$;
do
    do
        $g := g + 1$;
        apply the EA for this generation $f_g, x_g, d_g := EA(x_{g-1}, d_{g-1})$;
    until $f_g < f_{g-1}$;
    update memory $m_{(c \bmod d_m)} := x_g$;
    select a memory content m_r **with** $r \in \{0, ..., d_m\}, r \neq c \bmod d_m$
    randomly drawn with uniform probability $1/(d_m - 1)$;
    compute the direction from $s := (x_g - m_r)/\|(x_g - m_r)\|$;
    set beam search counter $c_b = 0$;
    set actual beam search factor $b_a := 1$;
    set $x_{c_b} := x_g$;
    do
        $c_b := c_b + 1$;
        $b_a := b_a \cdot b$;
        compute new beam point $x_{c_b} := x_{c_b-1} + s \cdot b_a \cdot d_g$;
    while $f(x_{c_b}) < f(x_{c_b-1})$;
    update memory $m_{(c \bmod d_m)} := x_{c_b}$;
    $c := c + 1$;
    $x_g := x_{c_b}$;
until $(f(x_{c_b}) < \epsilon$ **or** $(g \geq g_{max}))$;

The functions f_E and f_P were transformed for each run to a randomly generated basis system [6]. The initial points were set in the corresponding basis system to $x_i = 0$ for f_R and f_P and $x_i = 1$ for f_S and f_E, $i = 1, ..., n$. The runs were terminated at $f_{stop} < \epsilon = 10^{-10}$ for f_R, f_S and f_E. For f_P it was set to $f_{stop} < \epsilon = -10^5$. For dimensions $n = 5, 10, 20, 40, 100$ we made 30 runs each. The average number of function evaluation to reach the minimum with the given

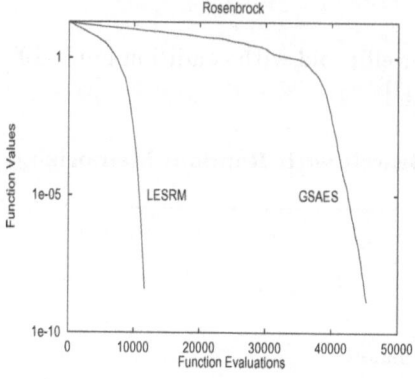 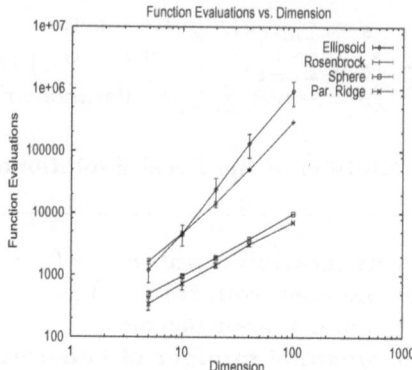

Fig. 1. Results for f_R for the $(1, \lambda)$-Evolution Strategy with generating set adaptation (GSAES) and the $(1 + 1)$-Evolution Strategy with random memorizing (LESRM), $n = 20$ (left), Results for the minimization of the functions given in the text, the parameter settings are explained in the text, average values are shown together with the standard deviations (right)

Table 2. Results for the $(1 + 1)$-Evolution Strategy with Local Memorizing for Rosenbrock's Function

n	MEAN	SD	SUCC
5	1.643533e+03	1.600629e+02	30/30
10	4.631000e+03	4.114435e+02	30/30
20	1.449407e+04	9.077249e+02	30/30
30	2.887270e+04	2.795075e+03	30/30
40	5.057945e+04	1.719543e+03	30/30
100	2.948448e+05	3.209187e+03	30/30
200	1.107836e+06	9.806003e+04	30/30

precision together with the standard deviation is depicted in Figure 1. It is notable that all runs for all functions converged with the given precision. Comparing the results with data from [6] and [5] the local evolutionary search algorithm with random memorizing needs about half the number of function evaluations than the $(1, \lambda)$-Evolution Strategy with covariance matrix adaptation and only a quarter of the number of function evaluations of the $(1, \lambda)$-Evolution Strategy with generating set adaptation.

4   Calibration of a Laser Induced Atomic Emission Plasma Spectrometer

Laser induced atomic emission plasma spectrometry is the latest development in a new generation of innovative measuring technologies. Such devices are employed in various industrial applications for analytical control of raw materials, products and processes. The advanced measuring method is based on the time-resolved spectral analysis of a light emitting laser-induced micro-plasma. The spectral range is 180 - 750 nm with a resolution of a few pm. An image intensifier and a camera are installed on the focal plane of the spectrograph. The measurement of spectra in two dimensions allows the simultaneous analysis of all relevant spectral lines. 60 elements can be determined simultaneously. The measurement device works in wavelength ranges higher 190 nm without protective gas. Due to the small diameter of the laserbeam and to the fact that no protective gas is necessary, a sample preparation is not required. For the calibration of such devices it is assumed that there is a relation between the intensity I_{ik} of emitted light at certain spectral lines and the concentration c_{ik} for element i in calibration sample k, i.e. $c_{ik} = \frac{f_i(I_{ik}, p_i)}{\sum_{j=1}^{n} f_j(I_{jk}, p_j)}$.

The normalization has to be done to cope with concentrations. The vectors p_i are the calibration model parameters which has to be adapted such that a calibration error measure based on the known calibration sample concentrations $\hat{c}_{ik}$ is minimized. Because of the normalization term all parameters are mutually dependent. This makes the problem difficult. Furthermore, it cannot be assumed that the model functions are continuous differentiable. Figure 2 shows calibration results for real data from five samples and five elements with a rather small calibration error.

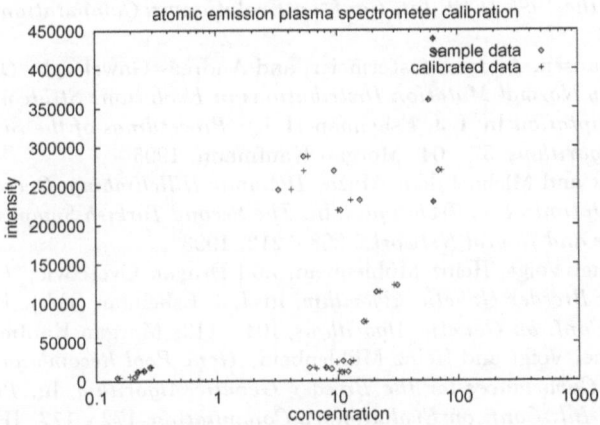

Fig. 2. Intensities vs. known calibration sample concentrations ◊ and estimated concentrations + based on the calibration model adaptation

5   Conclusions

Evolutionary Algorithms can be upgraded by a sequential memory which is accessed randomly to generate promising new search directions. The computational overhead for exploiting the memory is neglectible. With a beam search along these new directions it has been shown that even for the simplest $(1 + 1)$-Evolution Strategy a robust and efficient algorithm can be designed which needs less function evaluations than more sophisticated approaches.

One prerequisite that this approach will work is that the used Evolutionary Algorithm must be able to generate better solutions. For the considered $(1 + 1)$-Evolution Strategy this need not necessarily happen. It cannot work for points where the level set is such that increasing or decreasing the global step size leaves the success probability for finding a better solution at a constant very low value.

References

1. Ingo Rechenberg, *Evolutionsstrategie '94*, Frommann–Holzboog, 1994.
2. Adam Ghozeil and David B. Fogel, *A Preliminary Investigation into Directed Mutations in Evolutionary Algorithms*, In: H.-M. Voigt, W. Ebeling, I. Rechenberg, and H.-P. Schwefel (Eds.). *Parallel Problem Solving from Nature – PPSN IV, 329 - 335*. Lecture Notes in Computer Science 1141. Springer, 1996
3. Hans-Paul Schwefel, *Evolution and Optimum Seeking*, Wiley, 1995.
4. Hans-Georg Beyer, *Towards a Theory of Evolution Strategies: Some Asymptotical Results from the $(1,^+ \lambda)$ - Theory*, Evolutionary Computation 1 (2) 165 - 188, 1993.
5. Nikolaus Hansen and Andreas Ostermeier, *Adapting Arbitrary Normal Mutation Distributions in Evolution Strategies: The Covariance Matrix Adaptation* In: *Proceedings of the 1996 IEEE Int. Conf. on Evolutionary Computation*, 312 - 317. IEEE Press, 1996
6. Nikolaus Hansen, Andreas Ostermeier, and Andreas Gawelczyk, *On the Adaptation of Arbitrary Normal Mutation Distributions in Evolutions Strategies: The Generating Set Adaptation* In: L.J. Eshelman (Ed.). *Proceedings of the Sixth Int. Conf. on Genetic Algorithms*, 57 - 64, Morgan Kaufmann, 1995
7. Deniz Yuret and Michael de la Maza, *Dynamic Hillclimbing: Overcoming the Limitations of Optimization Techniques.* In: *The Second Turkish Symposium on Artificial Intelligence and Neural Networks*, 208 - 212, 1993
8. Hans-Michael Voigt, Heinz Mühlenbein, and Dragan Cvetcovic, *Fuzzy Recombination for the Breeder Genetic Algorithm.* In: L.J. Eshelman (Ed.). *Proceedings of the Sixth Int. Conf. on Genetic Algorithms*, 104 - 113, Morgan Kaufmann, 1995
9. Hans-Michael Voigt and Heinz Mühlenbein, *Gene Pool Recombination and the Utilization of Covariances for the Breeder Genetic Algorithm* In: *Proceedings of the 1995 IEEE Int. Conf. on Evolutionary Computation*, 172 - 177. IEEE Press, 1995
10. J.A. Nelder and R. Mead, *A Simplex Method for Function Minimization*, Comp. J. 7, 308 - 313, 1965
11. G.W. Stewart, *A Modification of Davidon's Minimization Method to Accept Difference Approximations of Derivatives* JACM 14, 72 - 83, 1967
12. M.J.D. Powell, *An Efficient Method for Finding the Minimum of a Function of Several Variables without Calculating Derivatives* Comp. J. 7, 155 - 162, 1964
13. Francisco J. Solis and Roger J.-B. Wets, *Minimization by Random Search Techniques* Operations Research, 19 - 30, 1981
14. William H. Press, Saul A. Teukolsky, William T. Vetterling, and Brian P. Flannery, *Numerical Recipes in C, 2nd Edition* Cambridge University Press, 1992

An Application of Genetic Algorithms to Floorplanning of VLSI

Kazuhiko Eguchi[1] , Junya Suzuki[2] , Satoshi Yamane[2], Kenji Oshima[2]

[1] Semiconductor & IC division, Hitachi, Ltd.,
5-20-1 Josuihon-cho, Kodaira-shi, Tokyo 187-8858 Japan

[2] Dept. of Environmental Science & Human Engineering
Graduate School of Science & Engineering
Saitama University
225 Shimo-okubo, Urawa, Saitama 338-8570 JAPAN

Abstract. Floorplanning of VLSI design is one of the key design flows which decides chip size, electrical characteristics, timing constrains, etc., of final silicon chip. Many useful floorplan tools are available in the industry. Those tools provide very user-friendly interactive environment and also provide useful information to proceed with chip design. However, construction and decision-making of floorplan design itself relies on the insight of human being. Therefore, the result varies depending on "who did it" and what initial condition was given at first. In this paper, authors propose an application of Genetic Algorithms to floorplanning for the purpose of providing better initial conditions as a starting point of design work to novice designers. A floorplan placement model suitable to Genetic Algorithm is discussed. Computational experiment is also carried out and results suggests practical possibility.

1. Introduction

According to the rapid increase of complexity & design difficulty of VLSI, EDA(Electronic Design Automation) tools have been widely adopted and have automated the many design flows. Such tools have been fulfilling the so called design gap between the increasing speed of silicon complexity and the improvement speed of design efficiency.

In the case of layout design, the placement of primitive cells and routing problems are well automated by the long time effort [1][2][3]. Many useful layout tools are available in the marketplace or proprietary within companies already.

Most of advance digital VLSIs include not only sets of primitive cells but also memory blocks, large macros, sometimes analog blocks, and so on. Therefore, floor planning has become the key design flow which decides the final chip size and performance of silicon. Several floor plan tools are available in the marketplace, and usually advanced semiconductor companies have their own floorplan tools.

L. Polkowski and A. Skowron (Eds.): RSCTC'98, LNAI 1424, pp. 263-270, 1998.
© Springer-Verlag Berlin Heidelberg 1998

Almost all such tools provide very user-friendly interactive environment of floorplan manipulation and provide useful information to improve the chip design such as estimated chip size, congestion of routing, connectivity among blocks etc. for chip designers.

Several trials to automate floorplan placement have been reported [4] [5] [6]. However, in a usual industry field, construction and decision-making of floor plan design itself completely relies on the insight of designer, i.e. human being. Therefore, the result of floorplanning is heavily depend on "who did it". In general, an expert designer gives much better result than a novice or not well trained one.

Authors are proposing an application of Genetic Algorithms to floorplan design of VLSI to provide a better initial condition as a starting point of complicated design for a novice chip designer.

Genetic Algorithms (refereed as GAs hereafter) are popular methods in the area of Soft Computing. Several applications of GAs have been reported in the area of Design Automation [5] [7] [8].

In this paper, we will discuss how to apply GAs to floorplan problems and computational experiments are also presented.

2. Genetic Algorithms(GAs)

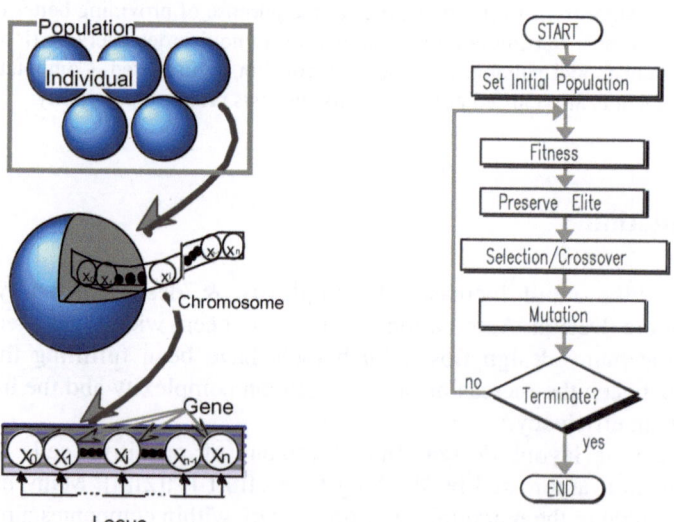

Fig. 1. Virtual Life   **Fig. 2.** Flow Chart of GAs Operation

2.1 Virtual Life

Let a set of virtual lives be introduced as illustrated in Fig. 1. Only chromosomes are considered and they correspond to the array on the computer memory which will be a set of solution of the problem. Parameters which are the objects of optimization occupy the locus as gene. The relation or correspondence between those parameters and the nature of optimization problem is called as the problem of coding. It is a very important portion of the application of GAs. GAs solve optimization problems by manipulating those chromosomes.

2.2 Basic Operations of GAs

Fig. 2 shows the flow chart of GAs operation. Basic operations are following.
  (1) Generation of an initial population.
     Generate an initial population by using a randomization method.
  (2) Evaluation of fitness
     Evaluate and compare the virtual lives by the numerical value called fitness.
  (3) Preservation of elite
     Preserve superior or outstanding individuals not to be destroyed by the next operation, crossover or mutation.
  (4) Selection
     Select two individuals for the crossover operation. Basically it is preferable to select superior individuals if possible. However, if it is too much inclined to one side, there is some danger to fall into a local optimization. There are many kind of methods for the selection. We employed roulette wheel selection because higher probability of selection is expected for superior individuals but also there remains the possibility to choose individuals with lower fitness.
  (5) Crossover
     This operation means the generation of children's chromosomes by recombining parents' chromosomes. Two children are generated from one pair of parents. One point or multi-point crossover operation is possible. In the case of one point crossover, the first child's chromosome is identical to the first parent until the crossing point. And after the crossing point, it is identical to the second parent.
     Separated from the above, uniform crossover operation can be considered, which is, each chromosome position is crossed with some probability. In this paper, the uniform crossover operation is employed as illustrated in Fig. 3.
  (6) Mutation
     Mutation is an operation to change gene(s) randomly with some probability as shown in Fig. 4. Suitable mutation will give variety to the population and make possible to obtain a solution which cannot be derived from only the set of initial population.

One generation of GAs is one cycle of evaluation of fitness, selection, crossover, and mutation. A solution will be found by increasing superior individuals in the population by repeating alternation of generation.

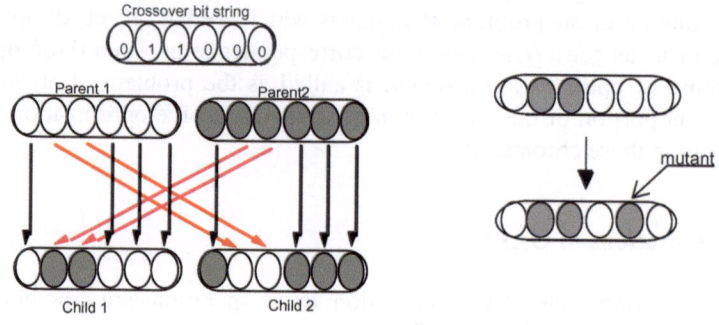

Fig. 3. Crossover   **Fig. 4.** Mutation

3. Floorplanning

The purpose of floorplan is to decide a macroscopic relative placement of blocks on a chip in advance to detail placement and routing with careful evaluation of final chip size, wire length, total timing constraints, electrical characteristic and so on. Well experienced designers proceed with those operations comprehensively, however, time required to complete the design is affected by the quality of initial condition. In the case of not well trained designers the design efficiency of both quality of resulted chip and design time needed varies heavily depending on what kind of initial condition was given.

If a tool could give better initial condition of floorplan placement, it would be beneficial for both expert and not well experienced designers.

4. Application of Genetic Algorithms to Floorplan Placement

A model of objective for placement is shown in Fig. 5. Rectangular blocks are connected by wires and an aspect ratio and size of each rectangular is given. A designated chip area(both size and aspect ratio) is also given. The following is basic operations:

(1)Randomly generate an initial population
(2)Evaluate fitness by calculating performance function
(3)Generate children: Preserve Elite, Select, Crossover, Mutate
(4)Repeat (2) (3)
(5)After certain generation, choose the best individual

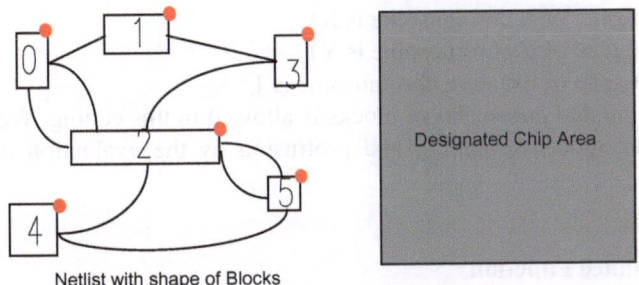

Netlist with shape of Blocks

Fig. 5. Model of Objective for Placement

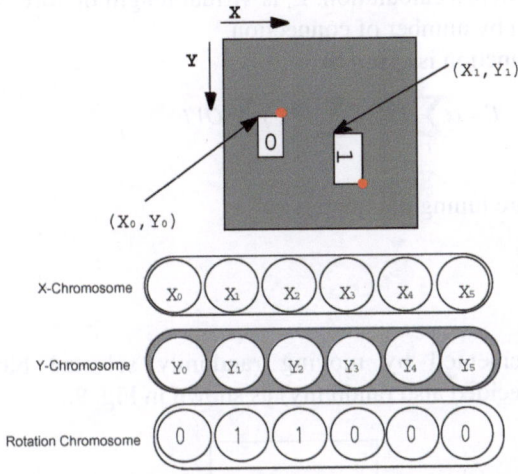

Fig. 6. Coding

4.1 Coding

Correspondence between the chromosome and the target optimization problem is called coding.

First of all, define the X-Y coordinate as shown in Fig. 6. Rotation of block is also taken into consideration. Prepare three(3) chromosomes and let them be X chromosome, Y chromosome, and Rotation chromosome, respectively. In Figure 6, #0 block of coordinate (X0, Y0) and rotation 0 is expressed as follows:

- 0th gene of X chromosome is X0
- 0th gene of Y chromosome is Y0
- 0th gene of Rotation chromosome is 0

#1 block of (X1, Y1) and 90 degree rotation is expressed:

- 1st gene of X chromosome is X1
- 1st gene of Y chromosome is Y1
- 1st gene of Rotation chromosome is 1

Overlapping and protrusion of blocks is allowed in this coding. We employed the method to exclude overlapping and protrusion by the evaluation of performance function.

4.2 Performance Function

Based on the behavior of expert designer, we chose minimum items for the component of performance function. Virtual wire length L_i, area of overlapping S_i, and area out of designated boundary OV_i are taken into consideration for evaluation. Fig. 7 shows those item's calculation. L_i is virtual length of wires between centers of each block weighed by number of connection.
The performance function is given by

$$E = \alpha \sum_i Li + \beta \sum_i Si + \gamma \sum_i OVi \qquad (4.1)$$

where α, β, and γ are tuning parameters.

4.3 Mutation

Mutation is implemented by moving randomly selected blocks with certain distance(distance decided also randomly) as shown in Fig. 8.

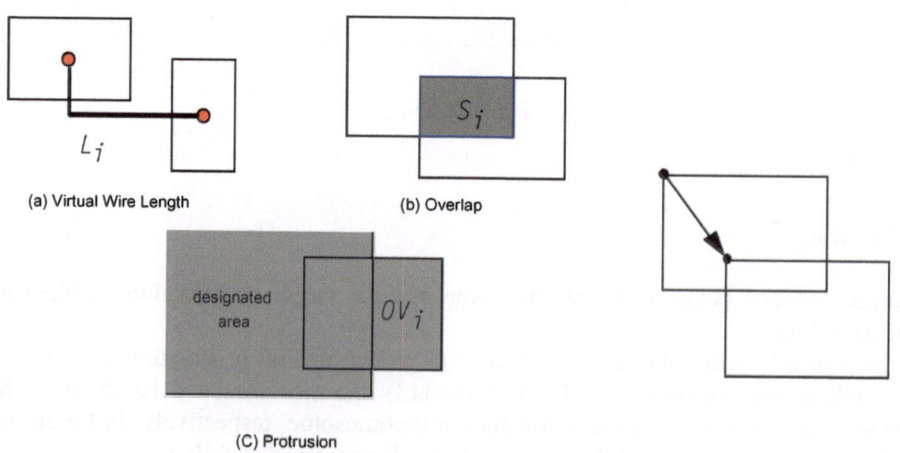

(a) Virtual Wire Length   (b) Overlap

(C) Protrusion

Fig. 7. Evaluation Items of Performance Function   **Fig. 8.** Mutation Method

5. Computational Experiment

Computational experiment has been carried out. We have applied GAs to a test circuit consist of 16 blocks with 1000, 3000, and 10000 individuals as an initial population respectively. Repetition of generation evolution was set 300. The mutation probability was employed as 0.5% based on experimental result.

Fig. 9 shows the value E of the performance function defined by (4.1) against the number of generation. In the case of 1000 individuals the performance function was not improved after 80 generations. It is considered that a local optimization occurred. In the case of 3000 individuals and 10000 individuals, E converges at about 160th generation and both cases gave similar results. The case of 3000 individuals as an initial population seems enough for this circuit.

Some results of placement obtained by this trial are shown in Fig. 10, 11, 12, and 13. Fig. 10, Fig. 11 and Fig. 12 show the best individual at the initial population, at the 50th generation, and 100th generation respectively. Fig. 13 shows the final result after the repetition of 300 generations.

Impression obtained by interviewing to experienced designers showed sensuously positive opinions as initial conditions for a starting point of floorplanning.

6. Conclusion

An application of Genetic Algorithms to VLSI floor planning was proposed.

A computational experiment on a test case consist of 16 blocks shows a practical result as an initial condition for novice designers.

Also for the size of above stated case, the result showed that 3000 initial individuals was enough but 1000 was not enough for the application of GAs.

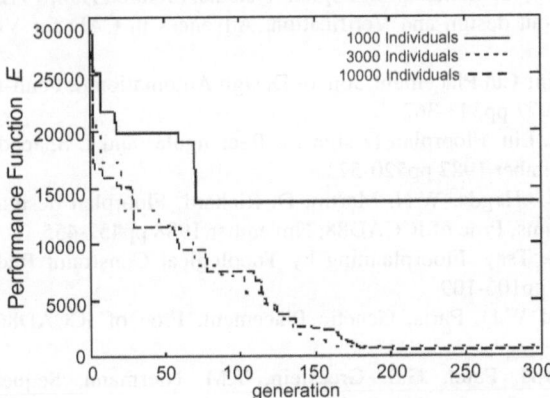

Fig. 9. Performance Function of the Best Individual in Each Generation

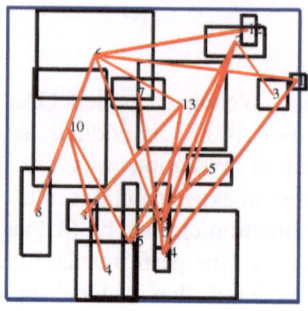

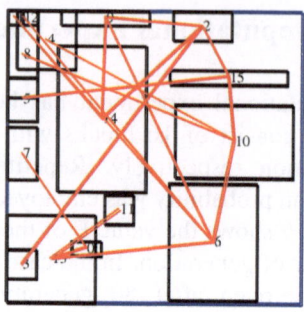

Fig. 10. 1st Generation   **Fig. 11.** 50th Generation

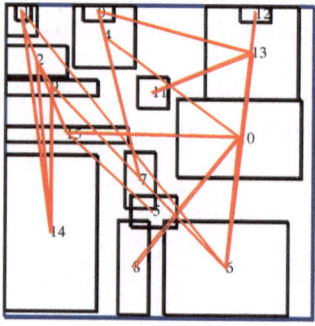

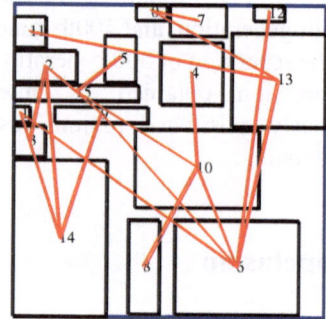

Fig. 12. 100th Generation   **Fig. 13.** 300th Generation

References

[1]M.A. Breuer, Design automation of Digital Systems, Prentice-Hall, 1972   Hall, 1972

[2]T. Ohtsuki, Layout design and Verification: Advances in CAD for VLSI Vol.14, North-Holland 1986

[3]M.A. Breuer, Mini-Cut Placement, Jou. of Design Automation & Fault-tolerant Computing Vol.1 October 1977 pp343-362

[4]D.F. Wong, C.L. Liu, Floorplan Design for Rectangular and L-Shaped Moddule, Proc of ICCAD87, November 1987 pp520-523

[5]J.P. Cohoon, S.U. Hegde, W.N. Martin, D. Richard, Floorplan design Using distributed Genetic Algorithms, Proc of ICCAD88, November 1988 pp452-455

[6]G. Vijayan, R.S. Tsay, Floorplanning by Topological Constraint Reduction, ICCAD90, November 1990 pp106-109

[7]J.P. Cohoon and W.D. Paris, Genetic Placement, Proc of ICCAD86, November 1986 pp422-425

[8]E.M. Rudnik, J.H. Patel, G.S. Greestein, T.M. Niermann, Sequential Circuit Test Generation in a Genetic Algorithm Framework, Proc. oc 31st design Automation Conference, June 1994 pp698-704

Learning with Delayed Rewards in Ant Systems for the Job–Shop Scheduling Problem

Urszula Boryczka[1]

University of Silesia, Institute of Computer Science, ul. Zeromskiego 3, 41-200
Sosnowiec, Poland

Abstract. We apply the idea of learning with delayed rewards to improve performance of the Ant System. We will mention different mechanisms of delayed rewards in the Ant Algorithm (AA). The AA for JSP was first applied in classical form by A. Colorni and M. Dorigo. We adapt an idea of an evolution of the algorithm itself using the methods of the learning process. We accentuate the co-operation and stigmergy effect in this algorithm. We propose the optimal values of the parameters used in this version of the AA, derived as a result of our experiments.

1 Introduction

The optimization algorithm that we propose in this paper was inspired by previous works on ant systems, and in general, by the term - stigmergy. This phenomenon, called stigmergy, was first introduced by P.P.Grasse [GRAS59] to present the indirect communication taking place among individuals through modifications induced in their environment.

The first application of the AA to the JSP was described in [COLO94]. In this paper authors showed how a new heuristic called ant system can be successfully applied to find good solutions of this problem. The effectiveness of the algorithm is due to the co-operation among the agents (a positive feedback or autocatalytic process). It was shown in the experiments with MT10, ORB1, ORB4 and LA21. Our intention is to present a new concept of learning among the ants. We propose different versions of reward functions.

2 Background

This paper introduces a family of ant algorithms where distributed agents (artificial ants) with some kinds of knowledge solve the JSP problem. The ant system we are going to apply simulates the real world and features of this system are presented in [BORY97].

The Ant System utilizes a weighted graph (N, E) with the set of nodes (N) and the set of edges between nodes weighted by an Utility measure (E). Each edge (i, j) has also a Reward measure $\Delta\tau(i, j)$, called pheromone, which is updated by artificial ants. Different choices of increasing the Reward measure (pheromone) cause different versions of the AA: AntDensity, AntQuantity and AntCycle [COLO92].

L. Polkowski and A. Skowron (Eds.): RSCTC'98, LNAI 1424, pp. 271–??, 1998.
© Springer-Verlag Berlin Heidelberg 1998

The mentioned AntCycle algorithm (the best of the AA) may be presented in the notation of Evolution Computing [MICH96] in the way described in [BORY97]. Below we discuss the node transition rule and the pheromone reinforcement rules. The node transition rule used in this version of Ant System, called random proportional rule, is given by Eq.(1), which describes the probability with which agent k in node i choose the next node j to move to:

$$
p_t(i,j) = \begin{cases} \dfrac{[Util_t(i,j)]^\delta \cdot [He_t(i,j)]^\beta}{\sum\limits_{j \in allowed} [Util_t(i,j)]^\delta \cdot [He_t(i,j)]^\beta} & \text{if } j \in \text{allowed} \\ \\ 0 & \text{otherwise} \end{cases}
\tag{1}
$$

where: *allowed* δ, β are described in [COLO92], $Util(i,j)$ is a utility measure, and $He(i,j)$ is some kinds of heuristics helping to solve the JSP (in our experiments the rule LRT).

The pheromone reinforcement rule is intended to allocate a greater amount of pheromone to a shorter schedule. This rule is similar to the reinforcement learning scheme [SING97], [GAMB95], in which better solutions get a higher reinforcement.

Pheromone trail laid on the edges plays the rule of a delayed reward and memory for the each agents will probably chose this edge. This allows an indirect form of communication and co-operation between these agents.

The Utility measure is a sum of present rewards and long–term rewards received during the iteration time, according to the formula:

$$
Util_t(i,j) = Rew_t(i,j) + \gamma \cdot \max \sum_{n=1}^{N-1} Rew_{t+n}(i,j)
\tag{2}
$$

where $Rew_t(i,j) = \Delta\tau_t(i,j)$ — delayed reward, $\Delta\tau_{t=0}(i,j) = \tau_0$, n is the number of steps of the algorithm and γ is a discount factor (like in the Reinforcement Learning [SING97]) ($0 \le \gamma \le 1$). In our experiments we tested $\gamma = 1$. l is the next node and k is the number of ants. We must emphasize that we lay the pheromone trail on appropriate edges after the cycle algorithm is complete.

The Reward measure, connected with increase of the pheromone trail (equal to the global reinforcement rule) is computed according to the formula:

$$
\Delta\tau(i,j) = \begin{cases} \frac{Q_L}{L_{Gl}} & \text{if } (i,j) \in \text{global–best allocation} \\ 0 & \text{otherwise} \end{cases}
\tag{3}
$$

where L_{Gl} is a length of the globally best solution going from the beginning of the trail and Q is a parameter (like in Q–learning [DORI94]).

Finally, in the ant system, the pheromone level is updated by the formula:

$$
Util_{t+1}(i,j) = (1-\alpha) * Util_t(i,j) + \alpha \Delta Util_{t+1}^k(i,l)
\tag{4}
$$

where l is the next node and k is the number of ants and α is the learning factor, equal to a pheromone evaporation parameter.

We also tested another type of updating rule, called local–best, which used L_{loc} (the length of the best allocation in the current iteration of the trail) instead of previous one rule. As a matter of the local–best rule, the edges which receive reinforcement are those belonging to to the best allocation received in the current iteration. Experiments have shown that the difference between those two schemes is minimal, with a slight preference aimed to the global–best, which is therefore applied in following experiments.

An application of the AA (in the classical form) to the JSP was described (with many details) in [COLO94].

3   Examining the efficiency of the AA for the JSP. Conclusions

First, the classical AA — AntCycle was tested. The problem 8×10 was tested using the different sets of parameters [COLO92]. Analyzing the results, we can conclude that the classical AA was finding the best results (the shortest processing time) for the greatest value of β and for the small values of δ. The influence of parameter ρ (in analyzed range) was unimportant.

Now we presents the results of tests for the 10×10 JSP with the sets of parameters described in previous section. Parameters Q and Q_L had values 700 and 1400, respectively. All the results were averaging after 10 experiments of 1000 iterations. The investigations were carried on for the following delayed rewards [BORY97]:

name	denotation
global leader	global I
reinforced global leader	global II
élite	global IV
local leader	local I
reinforced local leader	local II

For the set of parameters (Fig.1.), the advantage of the rules based on the global reinforcement may be observed. The results were about 4% better than for the classic algorithm. It was a result of decrease the relative importance of heuristics β (from 20 to 5). Using this set of parameters the shortest processing time for the 10×10 JSP was obtained (646 time units for the global I).

From the described results we can derive the following conclusions: the reinforcement rules we introduced increased the efficiency of the AA regardless of the size of the analyzed JSP; suitably chosen values of the parameters (δ, β, ρ, Q and Q_L) ensure the AA with any delayed rewards based on the global information (global reinforcement rule) to give the results about 5% better than for the classic algorithm; the global IV seems to be the best rule; the values assigned to the parameters Q and Q_L are important just as the value of the parameter δ.

We have also examined, how the number of agents (ants) influences the behavior of the generative policies for the 10×10 JSP. The results for 100 agents

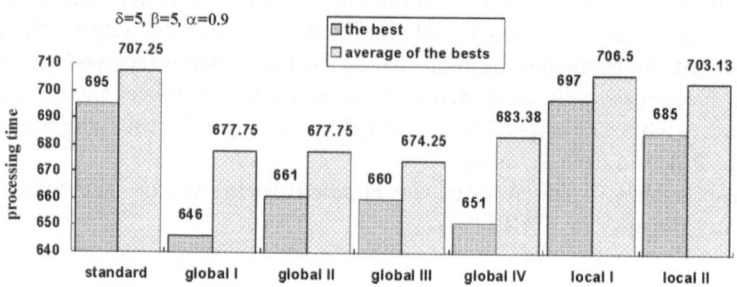

Fig. 1. Results for $\delta = 5, \beta = 5$ and $\rho = 0.9$.

were: 625 time units for the shortest processing time and 645 time units for the average of the bests results.

Artificial ants co–operate by updating information via the pheromone trail which help them to find the optimal solution very quickly. The experiments show very important role of heuristics that agents use, and of how this heuristics influences the transition rule. Thus, in further experiments with the JSP, the different types of heuristics will be tested.

References

[BORY97] Boryczka M., Boryczka U.: Generative policies in ant system. In: Proceedings of the EUFIT'97 Conference, Aachen, September, 1997, 857–861.

[COLO92] Colorni A., Dorigo M., Maniezzo U.: An Investigation of same Properties of an Ant Algorithm. In: Proceedings of the Parallel Problem Solving from Nature Conference (PPSN 92), Brussels, Belgium, Elsevier Publishing, 1992.

[COLO94] Colorni A., Dorigo M., Maniezzo U., Trubian M.: Ant system for Job–shop Scheduling, *Belgian Journal of Operations Research, Statistic and Computer Science*, 1994.

[DORI94] Dorigo M., Bersini H.: A comparison of Q–learning and classifier systems. In: Proceedings of From Animats to Animals Third International Conference on Simulation of Adaptive Behavior (SAB 94), Brighton UK, August 8–12, 1994.

[GAMB95] Gambarella L.M., Dorigo M.: AntQ: A Reinforcement Learning approach to the travelling salesman problem. Proceedings of ML–95, Twelfth International Conference On Machine Learning, Morgan Kaufmann Publishers, 1995, 252–260.

[GRAH79] Graham R.L., Lawler E.L., Lenstra J.K., Rinnooy Kan A.H.G.: Optimization and approximation in deterministic sequencing and scheduling: a survey. *Annals of Discrete Mathematics*, 5(1979), 287–326.

[MICH96] Michalewicz, Z.: Genetic Algorithms + Data Structures = Evolution Programs. Berlin: Springer Verlag, 1996.

[SING97] Singh S., Norving P., Cohn D.: Agents and Reinforcement Learning. *Dr. Dobb's Journal*, March, 1997.

FUZZY EXTENSION OF ROUGH SETS THEORY
(EXTENDED ABSTRACT)

GIANPIERO CATTANEO

1. ABSTRACT ROUGH APPROXIMATION SPACES

In this paper we consider many approximation spaces for rough theories, all of which are concrete realizations of an abstract structure stronger than the one introduced in [2] and defined in the following way:

$$\mathfrak{A} := \langle \Sigma, \mathcal{D}(\Sigma), \leq, 0, 1 \rangle$$

where:

1. $\langle \Sigma, \wedge. \vee, 0, 1 \rangle$ is a complete lattice with respect to the partial order relation $\leq$, bounded by the least element 0 ($\forall x \in \Sigma$, $0 \leq x$) and the greatest element 1 ($\forall x \in \Sigma$, $x \leq 1$); elements from Σ are interpreted as concepts, data, etc., and are said to be *approximable* elements;
2. $\mathcal{D}(\Sigma)$ is a sublattice of Σ whose elements are called *definable*;

and satisfying the following axioms:

(Ax1): For any approximable element $x \in \Sigma$, there exists (at least) one element $i(x)$ such that:

(1.1a) $$i(x) \in \mathcal{D}(\Sigma)$$
(1.1b) $$i(x) \leq x$$
(1.1c) $$\forall \alpha \in \mathcal{D}(\Sigma), \ (\alpha \leq x \ \Rightarrow \ \alpha \leq i(x))$$

(Ax2): For any approximable element $x \in \Sigma$, there exists (at least) one element $o(x)$ such that:

(1.2a) $$o(x) \in \mathcal{D}(\Sigma)$$
(1.2b) $$x \leq o(x)$$
(1.2c) $$\forall \gamma \in \mathcal{D}(\Sigma), \ (x \leq \gamma \ \Rightarrow \ o(x) \leq \gamma)$$

i.e., $i(x)$ [resp., $o(x)$] is the best approximation of the "vague" element x from the bottom [resp., top] by definable elements.

For any approximable element $x \in \Sigma$, the inner and the outer definable elements $i(x), o(x) \in \mathcal{D}(\Sigma)$, whose existence is assured by (Ax1) and (Ax2), are unique. Thus it is possible to introduce the *inner approximation mapping*, $i : \Sigma \mapsto \mathcal{D}(\Sigma)$ and the *outer approximation mapping* $o : \Sigma \mapsto \mathcal{D}(\Sigma)$, defined for every $x \in \Sigma$ respectively as

(1.3) $$i(x) := \max\{\alpha \in \mathcal{D}(\Sigma) : \alpha \leq x\} \qquad o(x) := \min\{\gamma \in \mathcal{D}(\Sigma) : x \leq \gamma\}$$

The *rough approximation* of any approximable element $x \in \Sigma$ is then the ordered inner–outer pair

(1.4) $$r(x) := (i(x), o(x)) \quad [\text{with } i(x) \leq x \leq o(x)]$$

which is the image of the element x under the *rough approximation mapping* $r : \Sigma \mapsto \mathcal{D}(\Sigma) \times \mathcal{D}(\Sigma)$ pictured by the following diagram:

L. Polkowski and A. Skowron (Eds.): RSCTC'98, LNAI 1424, pp. 275--282, 1998.

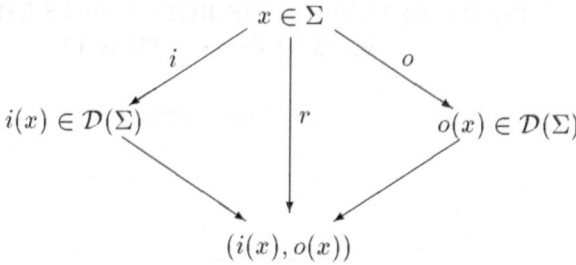

Following [2], an element x of X is said to be *crisp* (also *exact*) if and only if its inner and outer approximations coincide: $i(x) = o(x)$, equivalently, iff its rough approximation is the trivial one $r(x) = (x, x)$. Owing to (1.3) this happens iff x is definable; therefore, $\mathcal{D}(\Sigma)$ is the collection of all crisp elements.

2. THE ORTHODOX PAWLAK APPROACH TO ROUGH SET THEORY

The usual approach to rough set theory as introduced by Pawlak [10, 11] is formally based on a pair $(X, \pi(X))$ consisting of a nonempty set X, the *universe* [with corresponding power set $\mathcal{P}(X)$, the collection of *approximable sets*], and a partition $\pi(X) := \{M_i \in \mathcal{P}(X) : i \in I\}$ of X whose elements are the *elementary sets*. The partition $\pi(X)$ can be characterized by the induced equivalence relation $\mathcal{R} \subseteq X \times X$, defined as $(x, y) \in \mathcal{R}$ iff $\exists M_j \in \pi(X) : x, y \in M_j$; in this case we say that x, y are *indistinguishable* with respect to $\mathcal{R}$ and the equivalence relation $\mathcal{R}$ is called an *indistinguishability* relation.

A *definable set* (or *simple proposition*) is any subset of X obtained as the set theoretic union of elementary subsets: $M_J = \cup\{M_j \in \pi(X) : j \in J \subseteq I\}$; the collection of all such definable sets plus the empty set will be denoted by $\Pi(X)$ and it turns out to be a Boolean algebra (orthocomplemented atomic distributive complete lattice) $\langle \Pi(X), \cap, \cup, {}^c, \emptyset, X \rangle$ with respect to set theoretic intersection, union, and complementation.

To any subset of the universe $H \in \mathcal{P}(X)$ one can associate the *lower* (also *inner*) *approximation*

$$(2.1) \qquad i(H) := \cup\{M_J \in \Pi(X) : M_J \subseteq H\}$$

and the *upper* (also *outer*) *approximation*

$$(2.2a) \qquad o(H) := \cap\{M_J \in \Pi(X) : H \subseteq M_J\}$$
$$(2.2b) \qquad = \cup\{M_J \in \Pi(X) : H \cap M_J \neq \emptyset\}$$

In this way, according to Section 1, we can construct the concrete *approximation space*

$$(2.3) \qquad \mathfrak{A} = \langle \mathcal{P}(X), \Pi(X), i, o \rangle$$

consisting of: (1) the boolean (complete) lattice $\mathcal{P}(X)$ of all *approximable* subsets of the universe X; (2) the boolean (complete) lattice $\Pi(X)$ of all *definable* subsets of X; (3) the *inner approximation map* $i : \mathcal{P}(X) \mapsto \Pi(X)$ associating with any approximable set H its inner approximation $i(H)$ defined by (2.1); (4) the *outer approximation map* $o : \mathcal{P}(X) \mapsto \Pi(X)$ associating to any approximable set H its outer approximation $o(H)$ defined by (2.2).

The *rough approximation* of an approximable set H is the pair

$$(2.4) \qquad r(H) := (i(H), o(H)), \quad \text{with} \quad i(H) \subseteq H \subseteq o(H).$$

3. APPROXIMATION SPACES INDUCED FROM A KNOWLEDGE REPRESENTATION SYSTEM

Knowledge representation systems are generally introduced as models for *concepts* and *relations* among them; a number of such models have been proposed in the literature, mainly in connection with Artificial Intelligence. In this paper we are concerned with knowledge representation systems as they are understood in Rough Set Theory: "The main theoretical concept in this [...] approach is the notion of *knowledge representation system* (KR–system) introduced by Pawlak in [9] under the name of *information system*. A KR–system is a formalism for representing knowledge about some objects in terms of *attributes* (e.g., color) and *values* of attributes (e.g., green)." [13]. To be precise, a KR–system is a structure

$$\mathcal{KR} := \{X, Att(X), val(X), F\}$$

where X is a nonempty set of *objects* (*situations, entities, states*); $Att(X)$ is a nonempty set of *attributes* valuable on objects of the set X; $val(X)$ is the set of *possible values* which can be assumed in any observation on objects from X; F is a mapping $F : X \times Att(X) \to val(X)$ called *information mapping*.

For any fixed attribute $\alpha \in Att(X)$ we denote by $val(\alpha) := \{\lambda \in Att(X) \,|\, \exists x \in X : F(x, \alpha) = \lambda\}$ the set of all *possible values* of attribute α; the pair (α, λ) is an *elementary question* describing the *sentence*: "A test of the attribute α yields the value λ.". Moreover, the chosen attribute determines a map $f_\alpha : X \to val(\alpha)$, $f_\alpha(x) := F(x, \alpha)$. which, in analogy with physics, can be viewed as an *observable magnitude* defined on the *phase space* X and assigning to every state $x \in X$ the value $f_\alpha(x)$ assumed by the observable α when the system is in the state x.

With any elementary question $\langle \alpha, \lambda \rangle$ we can associate the subset of states

$$(3.1) \qquad M_\alpha(\lambda) := f_\alpha^{-1}(\lambda) = \{x \in X : f_\alpha(x) = \lambda\}$$

consisting of all states in which a measure of the observable α produces the result λ. Trivially, the collection of all such subsets of X,

$$(3.2) \qquad \pi_\alpha(X) := \{M_\alpha(\lambda) \in \mathcal{P}(X) : \lambda \in val(\alpha)\}.$$

is a partition of X, called the α–*partition*. The indistinguishability relation induced from this α–partition is $(x, y) \in \mathcal{R}_\alpha$ iff $f_\alpha(x) = f_\alpha(y)$.

Example 3.1. Consider the KR–system:

$$X = \{1, 2, 3, 4\}, \quad Att(X) = \{\alpha_0, \alpha_1, \alpha_2\}, \quad val(X) = \{Y, R, G, M, L, S, A, T\}$$

The observables describing the attributes under consideration are given in the following table:

$x \in X$	$f_{\alpha_0}(x)$	$f_{\alpha_1}(x)$	$f_{\alpha_2}(x)$
1	A	G	S
2	A	R	S
3	T	Y	M
4	T	Y	L

We have three α–partitions, one for each attribute:

$$\pi_{\alpha_0}(X) = \{\{1, 2\}, \{3, 4\}\}, \quad \pi_{\alpha_1}(X) = \{\{1\}, \{2\}, \{3, 4\}\}, \quad \pi_{\alpha_2}(X) = \{\{1, 2\}, \{3\}, \{4\}\}.$$

The following are rough approximations of the same subset of states:

$$r_{\alpha_0}(\{2, 3, 4\}) = (\{3, 4\}, \{1, 2, 3, 4\}) \qquad r_{\alpha_1}(\{2, 3, 4\}) = (\{2, 3, 4\}, \{2, 3, 4\})$$

$$r_{\alpha_2}(\{2, 3, 4\}) = (\{3, 4\}, \{1, 2, 3, 4\}).$$

Notice that the α_1–approximation of $\{2, 3, 4\}$ is crisp.

The meaning of the attributes described in this example could be the following: α_0 is the *shape* of the object (A as *arched* and T as *thin*), α_1 is the *color* (Y as *yellow*, G as *green* and R as *red*), α_2 is the *dimension* (S as *small*, M as *medium* and L as *large*).

Coming back to the general case, consider the *measurable space* of (micro) states $(X, \mathcal{P}(X))$; moreover, for any attribute α, consider the pair $(val(\alpha), \mathcal{P}(val(\alpha))$, consisting of the α–value set and of its power set, as the *measurable space* induced by α. Then, the α–definable sets are of the form:

$$(3.3a) \qquad \forall \Delta \in \mathcal{P}(val(\alpha)), \; M_\alpha(\Delta) := \cup \{M_\alpha(\lambda) \in \pi_\alpha(X) : \lambda \in \Delta\}$$

$$(3.3b) \qquad\qquad\qquad = \{x \in X : f_\alpha(x) \in \Delta\} = f_\alpha^{-1}(\Delta)$$

In this way we can also describe the attribute α as the *random variable* $f_\alpha^{-1} : \mathcal{P}(val(\alpha)) \mapsto \mathcal{P}(X)$ satisfying the following conditions:

(RV_1) $f_\alpha^{-1}(val(\alpha)) = X$;

(RV_2) $\Delta_1 \cap \Delta_2 = \emptyset$ implies $f_\alpha^{-1}(\Delta_1) \cap f_\alpha^{-1}(\Delta_2) = \emptyset$;

(RV_3) for any family $\{\Delta_n\}$ of pairwise disjoint subsets of $val(\alpha)$ we have that

$$f_\alpha^{-1}(\cup\Delta_n) = \cup f_\alpha^{-1}(\Delta_n)$$

Note that on the basis of conditions $(RV_1) - (RV_3)$ it is possible to infer the following further properties of a random variable: (a) $f_\alpha^{-1}(\emptyset) = \emptyset$; (b) $f_\alpha^{-1}(\Delta^c) = X \setminus f_\alpha^{-1}(\Delta)$; (c) for any family $\{\Delta_n\}$ of pairwise disjoint subsets of $val(\alpha)$: $f_\alpha^{-1}(\cap\Delta_n) = \cap f_\alpha^{-1}(\Delta_n)$.

4. REALIZATION OF PAWLAK ROUGH APPROXIMATION SPACES BY SHARP IDENTITY RESOLUTIONS

The power set $\mathcal{P}(X)$ of the *universe* X and the family of all *characteristic functionals* ($\{0,1\}$-valued functions on X) are in a one-to-one correspondence with respect to the map associating to any subset A of X the functional $\chi_A : X \mapsto \{0,1\}$ defined as $\chi_A(x) := 1$ if $x \in A$, 0 otherwise.

In the case of an universe X which is a *compact* metric space with respect to some distance function $d : X \times X \mapsto \mathbb{R}_+$ defined on it, the set $\{0,1\}^X$ of all $\{0,1\}$-valued functions on X can be considered as embedded in the C^* algebra $\langle \mathcal{B}(\mathbb{C}^X), +, \cdot, *, \|\cdot\|, \underline{0}, \underline{1} \rangle$ of all complex-valued bounded functions defined on X, equipped with the usual pointwise sum and product operations, the adjoint involution operation $*$ which yields the complex conjugation of functions, and the norm of uniform convergence $\|f\| := sup\{|f(x)| : x \in X\}$. In particular $\{0,1\}$-valued functions χ_A satisfy the property of being *projections* of the C^* algebra $\mathcal{B}(\mathbb{C}^X)$: that is, they are (1) self-adjoint (i.e., real valued); (2) bounded; (3) idempotent $[(\chi_A)^2 = \chi_A]$.

In what follows, extending this terminology to the case of a general universe X, $\{0,1\}$-valued functions on X are said to be *projections*.

The set $\{0,1\}^X$ of all characteristic functionals on X determines an atomic Boolean (complete) lattice $\langle \{0,1\}^X, \wedge, \vee, \neg, \underline{0}, \underline{1} \rangle$, where $\underline{0}$ and $\underline{1}$ are the characteristic functionals of the empty set $[\forall x \in X, \underline{0}(x) = 0]$ and of the whole universe $[\forall x \in X, \underline{1}(x) = 1]$, respectively. The operations $\wedge$, $\vee$ and $\neg$ are defined $\forall x \in X$ by the laws:

(4.1a)
$$(\chi_A \wedge \chi_B)(x) = \min\{\chi_A(x), \chi_B(x)\}$$

(4.1b)
$$= \max\{0, \chi_A(x) + \chi_B(x) - 1\}$$

(4.1c)
$$= \chi_A(x) \cdot \chi_B(x)$$

(4.2a)
$$(\chi_A \vee \chi_B)(x) = \max\{\chi_A(x), \chi_B(x)\}$$

(4.2b)
$$= \min\{1, \chi_A(x) + \chi_B(x)\}$$

(4.2c)
$$= \chi_A(x) + \chi_B(x) - \chi_A(x) \cdot \chi_B(x)$$

(4.3a)
$$(\neg\chi_A)(x) = (1 - \chi_A)(x)$$

(4.3b)
$$= \begin{cases} 1, & \text{iff } \chi_A(x) = 0 \\ 0, & \text{iff } \chi_A(x) = 1 \end{cases}$$

The mapping $\chi : \mathcal{P}(X) \mapsto \{0,1\}^X$, $A \mapsto \chi_A$ is clearly a boolean lattice isomorphism identifying $\mathcal{P}(X)$ and $\{0,1\}^X$, since

(4.4a)
$$\chi_{A \cap B} = \chi_A \wedge \chi_B$$

(4.4b)
$$\chi_{A \cup B} = \chi_A \vee \chi_B$$

(4.4c)
$$\chi_{A^c} = \neg\chi_A$$

This isomorphism preserves also the partial ordering relations since $A \subseteq B$ if and only if $\forall x \in X$, $\chi_A(x) \leq \chi_B(x)$.

Consider now a rough approximation space $\mathfrak{A}$ determined on an universe X of *finite* cardinality by the (finite) partition $\pi(X) = \{M_i : i \in I\}$ of *elementary* set (the extension to a generic universe with a countable partition is straightforward). The corresponding collection of characteristic functionals $\{P(i) := \chi_{M_i} : i \in I\}$, borrowing the terminology from Functional Analysis, is a *sharp identity resolution* $P : I \mapsto \{0,1\}^X$, $i \to P(i) = \chi_{M_i}$, in the sense that it satisfies the following properties:

(sir–1) all $P(i)$ are nonzero projections [real valued, bounded, and idempotent functions];
(sir–2) $\forall i \neq j$, $P(i) \leq \neg P(j)$ equivalently $P(i) \cdot P(j) = \underline{0}$ [orthogonality condition];
(sir–3) $\sum_{i \in I} P(i) = \underline{1}$ [identity resolution].

The corresponding Boolean lattice (σ–algebra) $\Pi(X) = \{M_J : J \in \mathcal{P}(I)\}$ of *definable* sets can be represented by the family of characteristic functions (projections) $\{P(J) := \chi_{M_J} : J \in \mathcal{P}(I)\}$ which generate a *Projection Valued* (PV) measure $P : \mathcal{P}(I) \mapsto \{0,1\}^X$, $J \to P(J) = \chi_{M_J}$, in the sense that it satisfies the following properties:

(PV_1) $P(I) = \underline{1}$;
(PV_2) $J_1 \cap J_2 = \emptyset$ implies $P(J_1) \leq \neg P(J_2)$;
(PV_3) for any family $\{J_n\}$ of pairwise disjoint subsets of I we have that

$$P(\cup J_n) = \sum P(J_n).$$

In the case of a KR–system of *finite* cardinality and with *real* set of values $val(X) \subseteq \mathbb{R}$, the α-partition $\pi_\alpha(X) = \{M_\alpha(\lambda_1), M_\alpha(\lambda_2), \ldots, M_\alpha(\lambda_{n(\alpha)})\}$, determined by an attribute α with finite set of possible real values $val(\alpha) = \{\lambda_1, \lambda_2, \ldots, \lambda_{n(\alpha)}\}$, introduced for simplicity the notation

(4.5) $$P_\alpha(\lambda) := \chi_{M_\alpha(\lambda)},$$

gives rise not only to the identity resolution $\{P_\alpha(\lambda_1), P_\alpha(\lambda_2), \ldots, P_\alpha(\lambda_{n(\alpha)})\}$ but also to a *spectral* identity resolution of the real observable f_α. This means that the following condition is also satisfied:

(sir–4)

$$f_\alpha = \sum_{i=1}^{n(\alpha)} \lambda_i \cdot P_\alpha(\lambda_i)$$

Example 4.1. If in Example 3.1 we represent the values Y, R, G, M, L, S, A, T by distinct real numbers, then obviously

$$f_{\alpha_0} = A \cdot \chi_{\{2,3\}} + B \cdot \chi_{\{3,4\}}, \qquad f_{\alpha_1} = G \cdot \chi_{\{1\}} + R \cdot \chi_{\{2\}} + Y \cdot \chi_{\{3,4\}},$$
$$f_{\alpha_2} = S \cdot \chi_{\{1,2\}} + M \cdot \chi_{\{3\}} + L \cdot \chi_{\{4\}}.$$

Note that in the present context of a crisp representation of $\mathcal{P}(X)$, if for the sake of simplicity we introduce the notation

(4.6) $$P_\alpha(\Delta) := \chi_{f_\alpha^{-1}(\Delta)},$$

in analogy with the sharp approach to axiomatic foundations of quantum mechanics, the conditions $(RV_1) - (RV_3)$ above assume the form of a *Projection Valued* (PV) measure:

(PV_1) $P_\alpha(val(\alpha)) = \underline{1}$;
(PV_2) $\Delta_1 \cap \Delta_2 = \emptyset$ implies $P_\alpha(\Delta_1) \leq \neg P_\alpha(\Delta_2)$;
(PV_3) for any family $\{\Delta_n\}$ of pairwise disjoint subsets of $val(\alpha)$ we have that

$$P_\alpha(\cup \Delta_n) = \sum P_\alpha(\Delta_n).$$

5. Fuzzy Rough Approximation Spaces

Recall that the notion of characteristic functional on the universe X can be generalized to the notion of *fuzzy set* defined as a $[0,1]$–valued function on X, $f : X \mapsto [0,1]$. The most interesting algebraic structure involving fuzzy sets is the $BZMV$ algebra of De Morgan type (see [4, 3]):

(5.1) $$\langle [0,1]^X, \oplus, \odot, \neg, \sim, \underline{0}, \underline{1} \rangle$$

where the operations are defined as follows:

(5.2a) $$(f \oplus g)(x) := \min\{1, f(x) + g(x)\}$$

(5.2b) $$(f \odot g)(x) := \max\{0, f(x) + g(x) - 1\}$$

(5.2c) $$\neg f(x) := (1 - f)(x)$$

(5.2d) $$\sim f(x) := \begin{cases} 1, & f(x) = 0 \\ 0, & f(x) \neq 0 \end{cases}$$

Following Chang [7, 8], the algebraic substructure $\langle [0,1]^X, \oplus, \odot, \neg, \underline{0}, \underline{1} \rangle$ is a standard MV algebra. Recall that in any MV algebra the following operations can also be introduced:

(5.3a) $$(f \wedge g)(x) := [(f \oplus \neg g) \odot g](x)$$

(5.3b) $$(f \vee g)(x) := [(f \odot \neg g) \oplus g](x)$$

(5.3c) $$(f \rightarrow_L g)(x) := (\neg f \oplus g)(x)$$

The first two new operations are binary lattice operation of meet and join generating the partial ordering

(5.4) $$f \leq g \quad \text{iff} \quad f \wedge g = f \quad \text{iff} \quad f \rightarrow_L g = \underline{1}$$

Trivially, on fuzzy sets the above operations assume the forms: (1) $(f \wedge g)(x) = \min\{f(x), g(x)\}$ and (2) $(f \vee g)(x) = \max\{f(x), g(x)\}$; moreover, the partial ordering (5.4) turns out to be the pointwise ordering on real valued functions: $f \leq g$ iff $\forall x \in X$, $f(x) \leq g(x)$. The third binary operation corresponds the Lukasiewicz implication connective for many-valued logics [12]: $(f \rightarrow_L g)(x) = \min\{1, 1 - f(x) + g(x)\}$.

With respect to the partial ordering the substructure $\langle [0,1]^X, \wedge, \vee, \neg, \sim, \underline{0}, \underline{1} \rangle$ is a Brouwer–Zadeh (BZ) distributive (complete) lattice equipped with two nonclassical negations: the Kleene negation $\neg$ (possibly violating the noncontradiction principle, $f \wedge \neg f \neq \underline{0}$, and the excluded-middle principle, $f \vee \neg f \neq \underline{1}$) and the Brouwer negation $\sim$ (possibly violating the strong double negation law, $f \leq \sim\sim f$, and the excluded middle principle, $f \vee \sim f \neq \underline{1}$) [5, 6].

To any fuzzy set $f \in [0,1]^X$ we can associate the two subsets of the universe X

(5.5) $$I(f) := \{x \in X : f(x) = 1\}, \qquad O(f) := \{x \in X : f(x) \in (0,1]\},$$

called the *inner support* and the *outer support* respectively.

In the context of fuzzy set theory, one can construct the *fuzzy rough approximation space*

(5.6) $$\mathfrak{A}_f = \langle [0,1]^X, \{0,1\}^X, \Box, \Diamond \rangle$$

consisting of: (1) the BZ distributive (complete) lattice $[0,1]^X$ of all fuzzy sets as *approximable* elements; (2) the boolean (complete) lattice $\{0,1\}^X$ of all crisp sets as *definable* elements; (3) the *inner approximation* map $\Box : [0,1]^X \mapsto \{0,1\}^X$ associating with any fuzzy set f its (crisp) *necessity* $\Box(f) = \chi_{I(f)}$, i.e., the best approximation of f from the bottom by crisp sets. (4) the *outer approximation* map $\Diamond : [0,1]^X \mapsto \{0,1\}^X$ associating with any fuzzy set f its (crisp) *possibility* $\Diamond(f) = \chi_{O(f)}$, i.e., the best approximation of f from the top by crisp sets.

Recall that the mappings $\Box$ and $\Diamond$ are S_5 modal operators on the Kleene distributive lattice of all fuzzy sets, i.e., they constitute an algebraic K–S_5 model of modal logic, where K means that the basic lattice structure is not a Boolean algebra, but a Kleene algebra [3].

The corresponding rough approximation of a fuzzy set f is then the pair:

(5.7) $$r(f) := \left(\chi_{I(f)}, \chi_{O(f)} \right), \quad \text{with} \quad \chi_{I(f)} \leq f \leq \chi_{O(f)},$$

which can be identified with the pair of subsets of X consisting of the inner and the outer supports:

(5.8) $$r(f) \equiv (I(f), O(f)), \quad \text{with} \quad I(f) \subseteq O(f),$$

6. Realization of Rough Approximation Spaces by Unsharp Identity Resolutions

In analogy with the approach to *unsharp quantum mechanics*, we can construct rough approximation spaces generalizing the notion of identity resolution to the case of fuzzy identity resolution.

A *fuzzy identity resolution* (also *fuzzy partition* of the universe, see [1]) is any collection $\pi_f(X) = \{f_1, f_2, \ldots, f_n\}$ of fuzzy sets satisfying the following conditions:

(fir–1) all f_i are real valued, non zero $[f_i \neq \underline{0}]$, positive and absorbing $[\underline{0} \leq f_i \leq \underline{1}]$ (in other words $f_i \in [0,1]^X$);

(fir–2) $\forall i \neq j$, $f_i \leq \neg f_j$ (which in general does not imply $f_i \cdot f_j = \underline{0}$) [orthogonality condition];

(fir–3) $\sum_{i=1}^{n} f_i = \underline{1}$ [identity resolution].

To any fuzzy identity resolution $\pi_f(X)$ of X we can associate two families of subsets:

$$(6.1) \quad \mathcal{I}(\pi_f(X)) := \{I(f_1), I(f_2), \ldots, I(f_n)\}, \quad \mathcal{O}(\pi_f(X)) := \{O(f_1), O(f_2), \ldots, O(f_n)\}$$

called the *inner (partial) covering* and the *outer covering* of X induced from $\pi_f(X)$ respectively.

The *inner granule* and the *outer granule* determined by the fuzzy partition $\pi_f(X)$ on the point $x \in X$ are defined as:

$$(6.2) \quad g_i(x) := \cap \{I(f_i) : x \in I(f_i)\}, \quad g_o(x) := \cap \{O(f_i) : x \in O(f_i)\} .$$

The inner granule may of course be empty, but the following chain of inclusions holds:

$$g_i(x) \subseteq \{x\} \subseteq g_o(x) .$$

We now want to show how an unsharp (fuzzy) realization of any sharp identity resolution can be obtained in a canonical way. Consider a random variable $f_\alpha : X \mapsto val(X)$ associated to an attribute α from a KR–system; and let $\{P_\alpha(\lambda) = \chi_{f_\alpha^{-1}(\lambda)} : \lambda \in val(\alpha)\}$ be the corresponding spectral identity resolution of f, satisfying in particular the spectral condition (sir–4).

Then, for any function $u : val(\alpha) \mapsto [0,1]$ (whose range is necessarily finite) we can introduce

$$(6.3) \quad F_\alpha^{(u)} := \sum_{\lambda \in val(\alpha)} u(\lambda) \cdot P_\alpha(\lambda)$$

which is a fuzzy set $F_\alpha^{(u)} : X \mapsto u(val(\alpha))$ defined for any $x \in X$ by the law:

$$(6.4) \quad F_\alpha^{(u)}(x) = \sum_{\lambda \in val(\alpha)} u(\lambda) \cdot \chi_{f_\alpha^{-1}(\lambda)}(x)$$

This fuzzy set is realized on the same partition $\pi_\alpha(X) = \{f_\alpha^{-1}(\lambda) : \lambda \in val(\alpha)\}$ of the attribute α; the set of possible values is however changed from $\{\lambda_1, \lambda_2, \ldots, \lambda_n\} \in \mathbb{R}$ to the new values $\{u(\lambda_1), u(\lambda_2), \ldots, u(\lambda_n)\} \subseteq [0,1]$.

We can state the following result.

Theorem 6.1. *Let* $\mathcal{G} = \{u_k : val(\alpha) \mapsto [0,1] \mid k \in \mathbb{K} \subseteq \mathbb{R}\}$, *with* $|\mathbb{K}| < \infty$, *be a finite family of* *[0,1]–valued maps all defined on the value set* $val(\alpha)$. *For every* k *construct the corresponding fuzzy set*

$$(6.5) \quad F_\alpha^{(u)}(k) := \sum_{\lambda \in val(\alpha)} u_k(\lambda) \cdot P_\alpha(\lambda) \in [0,1]^X .$$

Then, the induced map

$$(6.6) \quad F_\alpha^{(u)} : \mathbb{K} \mapsto [0,1]^X, \quad k \to F_\alpha^{(u)}(k)$$

is an fuzzy identity resolution $[\forall k \in \mathbb{K}, \ F_\alpha^{(u)}(k) \neq \underline{0}, \ and \ \sum_{k \in \mathbb{K}} F_\alpha^{(u)}(k) = \underline{1}]$ *iff the [0,1]–valued* *"matrix"* $(u_k(\lambda) : k \in \mathbb{K}, \ \lambda \in val(\alpha))$ *is stochastic, i.e.,*

$$(6.7) \quad \forall \lambda \in val(\alpha), \quad \sum_{k \in \mathbb{K}} u_k(\lambda) = 1 .$$

REFERENCES

[1] J. BEZDEK AND J. HARRIS, *Fuzzy partitions and relations: an axiomatic basis for clustering*, Fuzzy Sets and Systems, 1 (1978), pp. 112–127.

[2] G. CATTANEO, *Abstract approximation spaces for rough theories*, in Rough Sets in Data Mining and Knowledge Discovery, L. Polkowski and A. Skowron, eds., Soft Computing Series, Physica Verlag, Springer, Berlin, 1998, ch. 4, pp. 1–40. in print.

[3] G. CATTANEO, M. L. D. CHIARA, AND R. GIUNTINI, *Some algebraic structures for many-valued logics*. accepted in *Tatra Mountains Mathematical Publication*, 1998.

[4] G. CATTANEO, R. GIUNTINI, AND R. PILLA, *BZMV^{dM} and Stonian MV algebras (applications to fuzzy sets and rough approximations)*. in print in *Fuzzy Sets Syst.*, 1998.

[5] G. CATTANEO AND G. MARINO, *Brouwer-Zadeh posets and fuzzy set theory*, in Proceedings of the 1st Napoli Meeting on Fuzzy Systems, A. D. Nola and A. Ventre, eds., Napoli, June 1984, pp. 34–58.

[6] G. CATTANEO AND G. NISTICÒ, *Brouwer-Zadeh posets and three valued Lukasiewicz posets*, Fuzzy Sets Syst., 33 (1989), pp. 165–190.

[7] C. C. CHANG, *Algebraic analysis of many valued logics*, Trans. Amer. Math. Soc., 88 (1958), pp. 467–490.

[8] ———, *A new proof of the completeness of the Lukasiewicz axioms*, Trans. Amer. Math. Soc., 95 (1959), pp. 74–80.

[9] Z. PAWLAK, *Information systems - theoretical foundations*, Information Systems, 6 (1981), pp. 205–218.

[10] ———, *Rough sets*, Internat. J. Inform. Comput. Sci., 11 (1982), pp. 341–356.

[11] ———, *Rough sets: A new approach to vagueness*, in Fuzzy Logic for the Management of Uncertainty, L. A. Zadeh and J. Kacprzyc, eds., J. Wiley and Sons, New York, 1992, pp. 105–118.

[12] N. RESCHER, *Many-valued Logic*, Mc Graw-Hill, New York, 1969.

[13] D. VAKARELOV, *A modal logic for similarity relations in Pawlak knowledge representation systems*, Fundamenta Informaticae, XV (1991), pp. 61–79.

DIPARTIMENTO DI SCIENZE DELL'INFORMAZIONE, UNIVERSITÀ DI MILANO, VIA COMELICO 39, I-20135 MILANO (ITALIA)

E-mail address: `cattang@dsi.unimi.it`

Fuzzy Similarity Relation as a Basis for Rough Approximations

Salvatore Greco[1], Benedetto Matarazzo[1], and Roman Slowinski[2]

[1] Faculty of Economics, University of Catania,
55, Corso Italia, I-95129 Catania, Italy
[2] Institute of Computing Science, Poznan University of Technology,
3a, Piotrowo 60-965 Poznan, Poland

1 Introduction

The rough sets theory proposed by Pawlak [8,9] was originally founded on the idea of approximating a given set by means of indiscernibility binary relation, which was assumed to be an equivalence relation (reflexive, symmetric and transitive). With respect to this basic idea, two main theoretical developments have been proposed: some extensions to a fuzzy context (e.g. Dubois and Prade, [1,2], Slowinski and Stefanowski, [13,14,15], Yao, [19]) and some extensions of the indiscernibility relation by means of more general binary relations (e.g. Nieminen, [7], Lin, [5], Marcus, [6], Polkowski, Skowron and Zytkow, [10], Skowron and Stepaniuk, [11], Slowinski, [12], Slowinski and Vanderpooten, [16,17,18], Yao and Wong, [20]). In the latter extensions, we wish to point out the proposal of Slowinski and Vanderpooten([16,17,18]) who introduced and characterized a general definition of rough approximations using a similarity relation which is a reflexive binary relation, relaxing the assumption of symmetry and transitivity.

In this paper, based on Greco, Matarazzo and Slowinski [4], we put together these two extensions, considering within a fuzzy context the approach proposed by Slowinski and Vanderpooten. More specifically we propose to approximate a given fuzzy set by means of reflexive fuzzy binary relations.

The paper is structured as follows. In section 2 we introduce the rough approximation of a given fuzzy set by a fuzzy similarity relation. In section 3 we present an application of the proposed approach to an exemplary problem. Section 4 groups conclusions.

2 Rough approximation by reflexive fuzzy binary relation

In the following the negation and the classic connectives of fuzzy logic are used in a suitable way, in particular those of the t-norm T as conjunction, of the t-conorm S as disjunction, of the fuzzy negation N and of fuzzy implication $I^{\rightarrow}$ (for a brief but thorough introduction to fuzzy logic see the first chapter of Fodor and Roubens [3]). Let U be a (non empty) finite set of objects called universe and R a reflexive *fuzzy* binary relation defined on U, i.e. $R : U \times U \rightarrow [0,1]$ such that $R(x,x) = 1$ for each $x \in U$. R represents a certain form of *similarity*.

L. Polkowski and A. Skowron (Eds.): RSCTC'98, LNAI 1424, pp. 283–289, 1998.

On this basis we can extend the concepts of positive and negative objects, well known within classical rough sets theory (e.g. see Pawlak [8]), to a fuzzy context.

Let X be a fuzzy set in U and let also $\mu_X : U \to [0,1]$ be the membership function of X. Given $x \in U$, we say that:

1. the membership function of x to the set of *positive objects* with respect to X, denoted by $\text{Pos}(x, X)$, is the credibility that "for each $y \in U$ x is not similar to y or y belongs to X", i.e.

$$\text{Pos}(x, X) = T_{y \in U}(S(N(R(x,y)), \mu_X(y)));$$

2. the membership function of x to the set of *negative objects* with respect to X, denoted by $\text{Neg}(x, X)$, is the credibility that "for each $y \in U$ x is not similar to y or y does not belong to X", i.e.

$$\text{Neg}(x, X) = T_{y \in U}(S(N(R(x,y)), N(\mu_X(y)))).$$

Let us remark that, remembering the definition of S-implication $I_{S,N}^{\to}$ (see e.g. Fodor and Roubens [3]), we can write

$$\text{Pos}(x, X) = T_{y \in U}(I_{S,N}^{\to}(R(x,y), \mu_X(y))), \tag{1}$$

$$\text{Neg}(x, X) = T_{y \in U}(I_{S,N}^{\to}(R(x,y), N(\mu_X(y)))). \tag{2}$$

On the basis of Equation 1, $\text{Pos}(x, X)$ can be seen as the credibility that "for each $y \in U$ the similarity of x with respect to y implies that y belongs to X". Analogously from Equation 2, $\text{Neg}(x, X)$ can be seen as the credibility that "for each $y \in U$ the similarity of x with respect to y implies that y does not belong to X".

Considering a fuzzy set X in U and a reflexive fuzzy binary relation R defined on U, the lower approximation of X, denoted by $\underline{R}(X)$, and the upper approximation of X, denoted by $\overline{R}(X)$, are fuzzy sets of U whose membership functions are respectively defined as

$$\mu(x, \underline{R}(X)) = T_{y \in U}(S(N(R(x,y)), \mu_X(y))),$$

$$\mu(x, \overline{R}(X)) = S_{y \in U}(T(R(x,y), \mu_X(y))).$$

Let us observe that $\mu(x, \underline{R}(X)) = \text{Pos}(x, X)$, i.e. the lower approximation of X is the set of positive objects with respect to X as in the classical rough sets theory. $\mu(x, \overline{R}(X))$ represents the credibility of the proposition "there is at least one $y \in U$ such that x is similar to y and y belongs to X".

Considering a fuzzy set X in U and a reflexive fuzzy binary relation R defined on U, we obtain the following results (Greco, Matarazzo and Slowinski [4]).

Theorem 1.

$$\mu(x, \underline{R}(X)) \leq \mu_X(x) \leq \mu(x, \overline{R}(X)), \forall x \in U.$$

Theorem 2. *If N is involutory and (S, T, N) is a De Morgan triple (see Fodor and Roubens [3]), then $\overline{R}(X)$ is the complement of the set of negative objects with respect to X, i.e.*

$$\mu(x, \overline{R}(X)) = N(\mathrm{Neg}(x, X)), \forall x \in U.$$

Theorem 3. *If (S, T, N) is a De Morgan triple and $\mu_{U-X}(x) = N(\mu_X(x))$, then*

$$\mu(x, \underline{R}(X)) = N(\mu(x, \overline{R}(U - X))), \forall x \in U.$$

Theorems 1 to 3 can be read as the fuzzy counterparts of the following results well known within classical rough set approach: theorem 1 says that set X includes its lower approximation and is included in its upper approximation; theorem 2 says that the upper approximation is the complement of the set of negative objects; theorem 3 says that lower approximation of X is the complement of the upper approximation of the complement of X.

3   An illustrative example

In order to illustrate the use of the rough approximation by means of a fuzzy similarity relation, we propose a simple example already considered in Slowinski [12] and Slowinski and Vanderpooten [16]. This example is based on a decision table describing 12 firms which have got an approximately equal credit in a bank. The firms are characterized by three condition attributes: A1: value of fixed capital, A2: value of sales in the year preceding the application, A3: kind of activity. Attributes A1 and A2 are quantitative, while attribute A3 is a qualitative one with three possible values. Decision attribute d makes a dichotomic partition of the firms: $d = 1$ if the firm paid back its credit and $d = 2$ otherwise. The decision table is shown in Table 1. In this application we considered the following fuzzy logical operators: $\forall a, b \in [0, 1]$,

$$T(a, b) = min(a, b),$$

$$S(a, b) = max(a, b),$$

$$N(a) = 1 - a.$$

For each attribute q we consider a valued binary relation R_q, i.e. a function $R_q : U \times U \rightarrow [0, 1]$ where, $\forall x, y \in U$, $R_q(x, y)$ represents the credibility of similarity between x and y with respect to the attribute q. Practically $R_q(x, y)$ is computed on the basis of the evaluations of x and y by means of attribute q, denoted by $f(x, q)$ and $f(y, q)$. More precisely, for each attribute q and $\forall x, y \in U$,

$R_q(x, y) = 0$ means absence of similarity between x and y,
$R_q(x, y) = 1$ means that x is absolutely similar to y, $(R_q(x, x) = 1)$,
$R_q(x, y) \geq R_q(w, z)$ means that the similarity between x and y is at least as credible as the similarity between w and z.

Table 1. Sample of firms

Firm	A1	A2	A3	d
F1	43	78	0	1
F2	54	75	1	2
F3	124	50	1	1
F4	102	65	1	2
F5	98	80	2	2
F6	88	102	2	2
F7	130	57	0	1
F8	128	92	1	2
F9	82	59	1	1
F10	134	103	2	2
F11	58	55	0	1
F12	126	71	1	2

The similarity relation with respect to $A1$ and $A2$ was defined, $\forall x, y \in U$, as

$$
R_q(x,y) = \begin{cases} 1 & \text{if } |f(x,q) - f(y,q)| \leq 0.3f(y,q) \\ \frac{0.4f(y,q) - |f(x,q) - f(y,q)|}{0.4f(y,q) - 0.3f(y,q)} & \text{if } 0.3f(y,q) < |f(x,q) - f(y,q)| \leq 0.4f(y,q) \\ 0 & \text{if } |f(x,q) - f(y,q)| > 0.4f(y,q), \end{cases}
$$

where $q = 1, 2$. With respect to $A3$ the similarity relation was defined, $\forall x, y \in U$, as

$$
R_3 = \begin{cases} 1 & \text{if } f(x,3) = f(y,3) \\ 0 & \text{if } f(x,3) \neq f(y,3). \end{cases}
$$

For attribute $A3$ we shall write therefore "x has the same evaluation of y with respect to $A3$", rather then "x is similar to y with respect to $A3$", $\forall x, y \in U$.

To model the comprehensive similarity between x and $y \in U$ with respect to $P \subseteq \{A1, A2, A3\}$, denoted by $R_P(x,y)$, we consider the credibility of the proposition "x is similar to y with respect to all the attributes of P ". Thus we obtain

$$
R_P(x,y) = T_{q \in P}(R_q(x,y)).
$$

We approximate the set of the firms which paid back, i.e.

$X_1 = \{F1, F3, F7, F9, F11\}$, and the set of firms which did not pay back, i.e. $X_2 = \{F2, F4, F5, F6, F8, F10, F12\}$.

Let us observe that X_1 and X_2 are crisp sets. Of course, we can also in this case apply our approach by stating $\mu_X(x) = 1$ if $x \in X$ and $\mu_X(x) = 0$ if $x \notin X$, $X = X_1, X_2$.

We calculated the lower and upper approximation of X_1 and X_2 obtaining the results presented in Table 2.

The knowledge contained in the considered decision table (Table 1) can be expressed in terms of *certain* or *possible fuzzy* decision rules, which are specific "if. . ., then . . ." propositions with an associated credibility (see Greco, Matarazzo, Slowinski [4]).

Table 2. Lower and upper approximations with respect to $\{A_1, A_2, A_3\}$

Firm	$\mu(x, \underline{R}(X_1))$	$\mu(x, \underline{R}(X_2))$	$\mu(x, \overline{R}(X_1))$	$\mu(x, \overline{R}(X_2))$
$F1$	1	0	1	0
$F2$	0	0.41	0.59	1
$F3$	0	0	1	1
$F4$	0	0	1	1
$F5$	0	1	0	1
$F6$	0	1	0	1
$F7$	1	0	1	0
$F8$	0	1	0	1
$F9$	0	0	1	1
$F10$	0	1	0	1
$F11$	1	0	1	0
$F12$	0	1	0	1

The following set of minimal certain rules was obtained from the considered decision table (each pair (y, c) in parenthesis shows the object $y \in U$ which supports the considered rule with a credibility equal to $c \in [0, 1]$):

1. if $f(x, 1)$ is similar to 88 and $f(x, 2)$ is similar to 102, then $x \in X_2$ with a credibility equal to 1 $((F2, 0.14), (F4, 0.37), (F5, 1), (F6, 1))$;
2. if $f(x, 1)$ is similar to 128 and $f(x, 2)$ is similar to 92, then $x \in X_2$ with a credibility equal to 0.59 $((F4, 1), (F5, 1), (F6, 0.875), (F8, 1), (F10, 1), (F12, 1))$;
3. if $f(x, 1)$ is similar to 134 and $f(x, 2)$ is similar to 103, then $x \in X_2$ with a credibility equal to 1 $((F4, 0.31), (F5, 1), (F6, 0.57), (F8, 1), (F10, 1), (F12, 0.89)$;
4. if $f(x, 3) = 0$, then $x \in X_1$ with a credibility equal to 1 $((F1, 1), (F7, 1), (F11, 1))$;
5. if $f(x, 3) = 2$, then $x \in X_2$ with a credibility equal to 1 $((F5, 1), (F6, 1), (F10, 1))$;
6. if $f(x, 1)$ is similar to 54 and $f(x, 3) = 1$, then $x \in X_2$ with a credibility equal to 0.41 $((F2, 1))$;
7. if $f(x, 2)$ is similar to 92 and $f(x, 3) = 1$, then $x \in X_2$ with a credibility equal to 0.59 $((F2, 1), (F4, 1), (F8, 1), (F12, 1))$.

Also some possible decision rules can be obtained from the considered decision table, e.g.:

8. if $f(x, 2)$ is similar to 50, then $x \in X_1$ with a credibility equal to 1 $((F3, 1), (F7, 1), (F9, 1), (F11, 1))$.
9. if $f(x, 2)$ is similar to 98, then $x \in X_2$ with a credibility equal to $1((F2, 1), (F4, 1), (F5, 1), (F6, 1), (F8, 1), (F10, 1), (F12, 1))$.

Let us remark that since rule 8. is a possible decision rule it must be read "if $f(x, 2)$ is similar to 50, then x could belong to X_1". An analogous interpretation should be given to rule 9..

4 Conclusions

We introduced rough approximations of fuzzy sets by means of similarity relation defined as a reflexive fuzzy binary relation. The proposed framework represents a theoretical development with respect to the extension of rough set approach to a fuzzy context and also with respect to the approximations by means of binary relations more general than classical indiscernibility. Some fuzzy decision rules can be extracted from the approximations obtained by the fuzzy similarity binary relations.As shown by a simple example, the new rough set approach to decision table analysis gives quite comprehensible results.

5 Acknowledgments

The research of the first two authors has been supported by grant no. 96.01658. $ct10$ from Italian National Research Council (CNR). The third author wishes to acknowledge financial support from State Committee for Scientific Research, KBN research grant no. 8 T11C 013 13, and from CRIT 2 - Esprit Project no. 20288. For the task of typing this paper, we are indebted to the high qualification of Ms Silvia Angilella.

References

1. Dubois, D., Prade, H.: Rough Fuzzy Sets and Fuzzy Rough Sets, Int. J. of General Systems, **17**, (1990), 191–200.
2. Dubois, D., Prade, H.: Putting Rough Sets and Fuzzy Sets Together, in R. Slowinski (ed.), Intelligent Decision Support, Handbook of Applications and Advances of the Rough Sets Theory, Kluwer, Dordrecht, (1992), 203–233.
3. Fodor, J., Roubens, M.: Fuzzy Preference Modelling and Multicriteria Decision Support, Kluwer, Dordrecht, (1994).
4. Greco, S., Matarazzo, B. Slowinski, R.: Rough approximations by fuzzy similarity relations, Research Report No.1, Facolta di Economia, Universita di Catania, Catania, (1997).
5. Lin, T.: Neighborhood systems and approximation in database and knowledge base systems, in Proceedings of the 4th International Symposium on Methodologies for Intelligent Systems, (1989).
6. Marcus, S.:Tolerance rough sets, Cech topologies, learning processes, Bull. of the Polish Academy of Sciences, Technical Sciences, **42 (3)**, (1994), 471–487.
7. Nieminen, J.: Rough tolerance equality, Fundamenta Informaticae, **11 (3)**, (1988), 289–296.
8. Pawlak, Z.: Rough sets, International Journal of Information & Computer Sciences, **11**, (1982), 341–356.

9. Pawlak, Z.: Rough Sets, Theoretical Aspects of Reasoning about Data, Kluwer Academic Publishers, Dordrecht, (1991).

10. Polkowski, L., Skowron, A., Zytkow, J.: Tolerance based rough sets, in T. Lin and A. Wildberger, eds., Soft Computing: Rough Sets, Fuzzy Logic, Neural Networks, Uncertainty Management, Simulation Councils, Inc., San Diego, (1995), 55–58.

11. Skowron, A., Stepaniuk, J.: Generalized approximation spaces, in T. Lin and A. Wildberger, eds., Soft Computing: Rough Sets, Fuzzy Logic, Neural Networks, Uncertainty Management, Simulation Councils, Inc., San Diego, (1995), 18–21.

12. Slowinski, R.: A generalization of the indiscernibility relation for rough set analysis of quantitative information, Rivista di matematica per le scienze economiche e sociali,**15**, (1993), 65–78.

13. Slowinski, R., Stefanowski, J.: Rough Classification in Incomplete System, Math. Comput. Modelling,**12 (10/11)**, (1989), 1347–1357.

14. Slowinski, R., Stefanowski, J.: Handling Various Types of Uncertainty in the Rough Set Approach, in W. P. Ziarko (ed.) Rough Sets, Fuzzy Sets and Knowledge Discovery, Springer-Verlag, London, (1994), 366–376.

15. Slowinski, R., Stefanowski, J.: Rough set reasoning about uncertain data, Fundamenta Informaticae, **27**, (1996), 229–243.

16. Slowinski, R., Vanderpooten, D. : Similarity relation as a basis for rough approximations, ICS Research Report **53**, Warsaw University of Technology,(1995) Warsaw.

17. Slowinski, R., Vanderpooten, D.: A generalized definition of rough approximation, ICS Research Report **4**, Warsaw University of Technology, Warsaw (1996).

18. Slowinski, R., Vanderpooten, D.: A generalized definition of rough approximations based on similarity, IEEE Transactions on Data and Knowledge Engineering, (1998) (to appear).

19. Yao, Y.: Combination of rough sets and fuzzy sets based on (α-level sets, in T.Y. Lin, N. Cercone, eds., Rough Sets and Data Mining, Kluwer Academic Publishers, Boston, (1996), 301–321.

20. Yao, Y., Wong, S.:Generalization of rough sets using relationships between attribute values, in Proceedings of the 2nd Annual Joint Conference on Information Sciences, (1995) 30–33.

Approximation Spaces in Extensions of Rough Set Theory

Jaroslaw Stepaniuk

Institute of Computer Science
Bialystok University of Technology
Wiejska 45A, 15-351 Bialystok, Poland
e-mail: jstepan@ii.pb.bialystok.pl

Abstract. In this paper we present a generalization of the approximation space notion. We also present different notions of rough relations. We point out the role of searching for proper approximation space.

1 Introduction

Rough set theory was proposed [7] as a new approach for processing of incomplete data.

Investigations on relation approximation are well motivated both from theoretical and practical points of view. Let us mention two examples. The equality approximation is fundamental for a generalization of the rough set approach based on a similarity relation approximating the equality relation in the value sets of attributes. Rough set methods in control processes require function approximation.

One can distinguish several directions in research on relation approximations. Below we list some examples of them. In [6,15] properties of the rough relations are presented. The relationships of rough relations and modal logics have been investigated by many authors (see e.g. [20,10]). We will refer to [10], where the upper approximation of the input-output relation $R(P)$ of a given program P with respect to indiscernibility relation IND is treated as the composition $IND \circ R(P) \circ IND$ and where a special symbol for the lower approximation of $R(P)$ is introduced. Properties of relation approximations in generalized approximation spaces are presented in [11,18]. The relationships of rough sets with algebras of relations are investigated for example in [1].

Relationships between rough relations and a problem of objects ranking are presented for example in [2]. It is shown that the classical rough set approximations based on indiscernibility relation do not take into account the ordinal properties of the considered criteria. This drawback is removed by considering rough approximations of the preference relations by graded dominance relations [2].

One of the problems we are interested in is the following: given a subset $X \subseteq U$ or a relation $R \subseteq U \times U$, define X or R in terms of the available information. In this paper we discuss an approach based on generalized approximation spaces

L. Polkowski and A. Skowron (Eds.): RSCTC'98, LNAI 1424, pp. 290–297, 1998.
© Springer-Verlag Berlin Heidelberg 1998

introduced and investigated in [11,13]. There are several modifications of the original approximation space definition [7].

The first one concerns the so called uncertainty function. Information about an object, say x is represented for example by its attribute value vector. The set of all objects with similar (to attribute value vector of x) value vectors creates the set $I(x)$. In [7] all objects with the same value vector create the indiscernibility class. The relation $y \in I(x)$ is in this case an equivalence relation. We consider a more general case when it can be any relation.

The second modification of approximation space definition introduces a generalization of a rough membership function. We assume that to answer a question whether an object x belongs to an object set X we have to answer a question whether $I(x)$ is in some sense included in X. Hence we take as a primitive notion a rough inclusion function rather than rough membership function. Our approach allows to unify different cases considered in [7,22].

2   Approximations in Approximation Spaces

In this section we present general definition of approximation space [11,13] which can be used for example for introducing the tolerance based rough set model and the variable precision rough set model.

2.1   Approximation Spaces

An approximation space is a system $AS = (U, I, \nu)$, where

- U is a non-empty set of objects,
- $I : U \longrightarrow P(U)$ is an uncertainty function ($P(U)$ denotes the set of all subsets of U),
- $\nu : P(U) \times P(U) \longrightarrow [0, 1]$ is a rough inclusion function.

An uncertainty function defines a neighborhood of every object x.

The rough inclusion function defines the value of inclusion between two subsets of U.

Definitions of the lower and the upper approximations can be written as follows:

$$L(AS, X) = \{x \in U : \nu(I(x), X) = 1\},$$

$$U(AS, X) = \{x \in U : \nu(I(x), X) > 0\}.$$

For examples of approximation spaces for the variable precision rough set model [22] and generalized approximation spaces in information retrieval problem see [17].

We recall the notion of positive region in the case of generalized approximation spaces. Let $AS = (U, I, \nu)$ be an approximation space and let $\{X_1, \ldots, X_r\}$ be a classification of objects (i.e. $X_1, \ldots, X_r \subseteq U$, $\bigcup_{i=1}^{r} X_i = U$ and $X_i \cap X_j = \emptyset$ for $i \neq j$, where $i, j = 1, \ldots, r$).

The positive region of the classification $\{X_1, \ldots, X_r\}$ with respect to the approximation space AS is defined by

$$POS(AS, \{X_1, \ldots, X_r\}) = \bigcup_{i=1}^{r} L(AS, X_i).$$

2.2   Properties of Approximations and Rough Definability

In this subsection we first list properties of approximations in generalized approximation spaces. Next we present algorithms for checking rough definability, internal undefinability etc..

Let $AS = (U, I, \nu)$ be an approximation space. We define for two sets $X, Y \subseteq U$ the equality with respect to the rough inclusion ν in the following way: $X =_\nu Y$ if and only if $\nu(X, Y) = 1 = \nu(Y, X)$

Assume that the following conditions are satisfied:

- $x \in I(x)$, for every $x \in U$,

- ν is a standard rough inclusion i.e. $\nu(X, Y) = \begin{cases} \frac{card(X \cap Y)}{card(X)} & \text{if } X \neq \emptyset \\ 1 & \text{if } X = \emptyset \end{cases}$ for any $X, Y \subseteq U$.

Then one can show the following properties of approximations:

1. $\nu(L(AS, X), X) = 1$ and $\nu(X, U(AS, X)) = 1$.
2. $L(AS, \emptyset) =_\nu U(AS, \emptyset) =_\nu \emptyset, L(AS, U) =_\nu U(AS, U) =_\nu U$.
3. $U(AS, X \cup Y) =_\nu U(AS, X) \cup U(AS, Y)$.
4. $L(AS, X \cap Y) =_\nu L(AS, X) \cap L(AS, Y)$.
5. $\nu(X, Y) = 1$ implies $\nu(L(AS, X), L(AS, Y)) = \nu(U(AS, X), U(AS, Y)) = 1$.
6. $\nu(L(AS, X) \cup L(AS, Y), L(AS, X \cup Y)) = 1$.
7. $\nu(U(AS, X \cap Y), U(AS, X) \cap U(AS, Y)) = 1$.
8. $L(AS, U - X) =_\nu U - U(AS, X)$.
9. $U(AS, U - X) =_\nu U - L(AS, X)$.
10. $\nu(L(AS, L(AS, X)), L(AS, X)) = \nu(L(AS, X), U(AS, L(AS, X))) = 1$.
11. $\nu(L(AS, U(AS, X)), U(AS, X)) = \nu(U(AS, X), U(AS, U(AS, X))) = 1$.

By analogy with standard rough set theory we define the following four types of sets:

1. X is *roughly AS-definable* iff $L(AS, X) \neq_\nu \emptyset$ and $U(AS, X) \neq_\nu U$.
2. X is *internally AS-undefinable* iff $L(AS, X) =_\nu \emptyset$ and $U(AS, X) \neq_\nu U$.
3. X is *externally AS-undefinable* iff $L(AS, X) \neq_\nu \emptyset$ and $U(AS, X) =_\nu U$.
4. X is *totally AS-undefinable* iff $L(AS, X) =_\nu \emptyset$ and $U(AS, X) =_\nu U$.

The algorithms for checking corresponding properties of sets have $O\left(n^2\right)$ time complexity, where $n = card\left(U\right)$. For example we sketch algorithm for *rough AS-definability*.

```
function rough_definability(I, ν, X): boolean;
var temp1, temp2: boolean;
begin
temp1:=FALSE; temp2:=FALSE;
for x ∈ U do
begin
if ν (I (x), X) = 1 then temp1:=TRUE;
if ν (I (x), X) = 0 then temp2:=TRUE;
if temp1 AND temp2 then return(TRUE)
end;
return(FALSE)
end;
```

Let us also observe that using properties of approximations:

$$L\left(AS, U - X\right) =_{\nu} U - U\left(AS, X\right) \text{ and } U\left(AS, U - X\right) =_{\nu} U - L\left(AS, X\right)$$

one can obtain that X is *internally AS-undefinable* if and only if $U - X$ is *externally AS-undefinable*. Thus we can use the same algorithm in both cases.

2.3 Approximation Spaces and Rough Relations

In this subsection we discuss approximations of relations with respect to different rough inclusions.

Let $AS = (U, I, \nu)$ be an approximation space, where $U = U_1 \times U_2$ and I defines a partition of U. The rough inclusion function is defined for any relations $S, R \subseteq U_1 \times U_2$ as follows: $\nu\left(S, R\right) = \begin{cases} \frac{card(S \cap R)}{card(S)} & \text{if } S \neq \emptyset \\ 1 & \text{if } S = \emptyset \end{cases}$.

For any relation $R \subseteq U_1 \times U_2$, we define two relations $L\left(AS, R\right)$ and $U\left(AS, R\right)$ called the lower and the upper approximation of R, respectively, and defined as follows:

$$L\left(AS, R\right) = \{(x_1, x_2) \in U : \nu\left(I\left((x_1, x_2)\right), R\right) = 1\},$$

$$U\left(AS, R\right) = \{(x_1, x_2) \in U : \nu\left(I\left((x_1, x_2)\right), R\right) > 0\}.$$

By $\pi_i\left(R\right)$ we denote the projection of the relation R onto the $i - th$ axis i.e. for example for $i = 1$

$$\pi_1\left(R\right) = \{x_1 \in U_1 : \exists_{x_2 \in U_2} (x_1, x_2) \in R\}.$$

The definition of the rough inclusion function for relations can be based on the cardinality of the projections. Let us assume

$$\nu_{\pi_i}\left(S, R\right) = \begin{cases} \frac{card(\pi_i(S \cap R))}{card(\pi_i(S))} & \text{if } S \neq \emptyset \\ 1 & \text{if } S = \emptyset \end{cases},$$

where $i = 1, 2$.

The standard lower approximation of a relation $R \subseteq U_1 \times U_2$ has the following property: any objects $x_1 \in U_1, x_2 \in U_2$ are connected by the lower approximation of R if and only if any objects (y_1, y_2) from $I((x_1, x_2))$ are in the relation R.

One can propose some less restrictive definitions of the lower approximation using the rough inclusions ν_{π_1} and ν_{π_2}. Let $AS_{\pi_i} = (U_1 \times U_2, I, \nu_{\pi_i})$ and

$$L(AS_{\pi_i}, R) = \{(x_1, x_2) \in U_1 \times U_2 : \nu_{\pi_i}(I((x_1, x_2)), R) = 1\}$$

for $i = 1, 2$. Assuming $i = 1$ we have that the pair (x_1, x_2) is in the lower approximation $L(AS_{\pi_i}, R)$ if and only if for every y_1 there is y_2 such that the pair (y_1, y_2) is from $I((x_1, x_2))$ and in the relation R. Similar interpretation we obtain for $i = 2$.

One can also propose some more restrictive definition of the lower approximation using the approximation space $AS_{res} = (U_1 \times U_2, I, \nu_{res})$ with rough inclusion ν_{res} defined by:

$$\nu_{res}(S, R) = \nu(S, R) * \nu(\pi_1(R_1), \pi_1(S)) * \nu(\pi_2(R_2), \pi_2(S))$$

where $R_1 = R \cap (U_1 \times \pi_2(S))$ and $R_2 = R \cap (\pi_1(S) \times U_2)$.

Proposition 1. *For the lower and the upper approximations the following conditions are satisfied:*

1. $L(AS_{res}, R) \subseteq L(AS, R) \subseteq R$.
2. $L(AS, R) \subseteq L(AS_{\pi_1}, R)$ and $L(AS, R) \subseteq L(AS_{\pi_2}, R)$.
3. $U(AS, R) = U(AS_{\pi_1}, R) = U(AS_{\pi_2}, R) = U(AS_{res}, R)$.

Proposition 2. *The computational complexity of algorithms for computing approximations of relations is equal to $O\left((card(U))^2\right)$.*

3   Searching for Approximation Spaces

In this section we consider problem of searching for adequate uncertainty function in approximation space. The searching for proper uncertainty function is the crucial and most difficult task related to decision algorithm synthesis based on uncertainty functions.

One approach to searching for an uncertainty function is based on the assumption that there are given some metrics (distances) on attribute values. Distance and similarity are closely related. Relations obtained on attribute values by using metrics are reflexive and symmetrical i.e. are tolerance relations. For review of different metrics defined on attribute values see [21]. Here we only present two examples of such metrics.

The Value Difference Metric (VDM) provides an appropriate distance function for nominal attributes. A simplified version of VDM (without the weighting schemes) defines the distance between two values v and v' of an attribute a as:

$$vdm_a(v, v') = \sum_{i=1}^{r(d)} (\nu(X_v, X_i) - \nu(X_{v'}, X_i))^2,$$

where $r(d)$ is a number of decision classes, ν is the standard rough inclusion, $X_i = \{x \in U : d(x) = i\}$ and $X_v = \{x \in U : a(x) = v\}$.

Using the distance measure VDM, two values are considered to be closer if they have more similar classifications. For example, if an attribute color has three values red, green and blue, and the application is to identify whether or not an object is an apple, then red and green would be considered closer than red and blue because the former two both have correlations with decision apple.

If this distance function is used directly on continuous attributes, all values can potentially be unique. Some approaches to the problem of using VDM on continuous attributes are presented in [21].

One can also use some other distance function for continuous attributes, for example

$$diff_a(v, v') = \frac{|v - v'|}{\max_a - \min_a},$$

where $\max_a$ and $\min_a$ are the maximum and minimum values, respectively, for attribute $a \in A$.

Let $\delta_a : V_a \times V_a \longrightarrow [0, \infty)$ be a given distance function on attribute values, where V_a is the set of all values of attribute $a \in A$.

One can define the following uncertainty function

$$y \in I_a^{\varepsilon_a}(x) \text{ if and only if } \delta_a(a(x), a(y)) \leq \varepsilon_a,$$

where $\varepsilon_a \geq 0$ is a given real number.

Some further examples of uncertainty functions we can also derive from the literature. In [14] strict and weak indiscernibility relations were considered which can define some kind of uncertainty functions. In some cases it is natural to consider relations defined by ε-indiscernibility [3]. For more details on corresponding uncertainty functions see [17].

Different methods of searching for parameters of proper uncertainty functions are discussed for example in [3,8,13,16,5]. In [16,4] genetic algorithm was applied for searching for adequate uncertainty functions of the type $I_a^{\varepsilon_a}$.

Now we discuss the following problem:

How large is a set of possible uncertainty functions for a given decision table? In other words how large is a search space for searching for optimal uncertainty function?

Let $DT = (U, A \cup \{d\})$ be a decision table, where U is a non-empty set such that $card(U) = n$ and $A = \{a_1, \ldots, a_m\}$ is a set of condition attributes, where $n, m > 0$ are given natural numbers.

In the further analysis we assume that for every attribute $a \in A$ there is some metric δ_a on the set of values and we consider uncertainty functions $I_a : U \to P(U) - \{\emptyset\}$ such that the following conditions are satisfied:

- $x \in I_a(x)$,
- for every $x, y, z \in U$ if $\delta_a(a(x), a(y)) \leq \delta_a(a(x), a(z))$ and $z \in I_a(x)$, then $y \in I_a(x)$ (i.e. if a distance between x and y is not greater than a distance between x and z, and z is related to x, then y is related to x).

For example the above conditions are satisfied by uncertainty function $I_a^{\varepsilon_a}$.

The global uncertainty function is defined as the intersection i.e. $I_A(x) = \bigcap_{a \in A} I_a(x)$.

We introduce some notions which are used in the next theorem. For every object $x \in U$ we define an equivalence relation $E(\delta_a, x)$ as follows:

$(y, z) \in E(\delta_a, x)$ if and only if $\delta_a(a(x), a(y)) = \delta_a(a(x), a(z))$.

We number equivalence classes of the relation $E(\delta_a, x)$ starting from the closest to the object x. Let $p_{(\delta_a, x)} : U \to \{1, \ldots, card(U/E(\delta_a, x))\}$ be a numbering such that for every $y, z \in U p_{(\delta_a, x)}(y) \leq p_{(\delta_a, x)}(z)$ if and only if $\delta_a(a(x), a(y)) \leq \delta_a(a(x), a(z))$.

Let $Y_{(\delta_a, x)} \in U/E(\delta_a, x)$ be a set of objects $y \in U$ with different decision than x and as close as possible to x i.e. $y \in Y_{(\delta_a, x)}$ if and only if $d(x) \neq d(y)$ and $\forall_{z \in U}((\delta_a(a(x), a(z)) < \delta_a(a(x), a(y))) \to d(x) = d(z))$.

Theorem 1.   *1. For every attribute $a \in A$ the number of possible different uncertainty functions is equal to $\prod_{x \in U} card(U/E(\delta_a, x))$.*

   2. Let y be a given object. The number of the functions I_a, where $a \in A$ such that $y \notin I_A(x)$ is equal to $n^{m-1} * \left(\sum_{a \in A} \left(p_{(\delta_a, x)}(y) - 1\right)\right)$.

   3. The number of the functions I_a, where $a \in A$ such that for every $y \in Y_{(\delta_a, x)}$ $y \notin \bigcap_{a \in A} I_a(x)$ is not greater than n^m.

Conclusions. We have presented different notions of rough relations. We hope that they can be considered as good starting point for investigations of important problems related to relation approximation for example the optimization problem for controllers design. We point out the role of searching for proper uncertainty functions. Adequate approximation spaces are important for extracting laws from decision tables.

Acknowledgments. The author would like to thank Andrzej Skowron and Dominik Slezak for valuable discussions. This work has been supported by the grant 8T11C01011 from the State Committee for Scientific Research (Komitet Badan Naukowych) and by Technical University of Bialystok Rector's grant.

References

1. Duentsch, I.: Rough Sets and Algebras of Relations, in: (ed.) E. Orlowska, Incomplete Information: Rough Set Analysis, Physica-Verlag, 1998, pp. 95–108.
2. Greco, S., Matarazzo, B., Slowinski, R.: Rough Approximation of a Preference Relation in a Pairwise Comparison Table, (eds.) L. Polkowski, A. Skowron, Rough Sets in Knowledge Discovery Part I and II, Physica-Verlag, 1998.
3. Krawiec, K., Slowinski, R., Vanderpooten, D.: Learning of Decision Rules from Similarity Based Rough Approximations, (eds.) L. Polkowski, A. Skowron, Rough Sets in Knowledge Discovery, Part I and II, Physica-Verlag, 1998.
4. Kretowski, M., Stepaniuk, J.: Selection of Objects and Attributes, a Tolerance Rough Set Approach, Proceedings of the Poster Session of Ninth International Symposium on Methodologies for Intelligent Systems, June 10-13, 1996, Zakopane, Poland, pp. 169–180.

5. Nguyen, S.H., Skowron, A., Synak, P.: Discovery of Data Patterns with Applications to Decomposition and Classification Problems, (eds.) L. Polkowski, A. Skowron, Rough Sets in Knowledge Discovery, Part I and II, Physica-Verlag, 1998.
6. Pawlak, Z.: Rough Relations, Bulletin of the Polish Academy of Sciences, Technical Sciences vol. 34 (9-10), 1986, pp. 587-590.
7. Pawlak, Z.: Rough Sets. Theoretical Aspects of Reasoning about Data, Kluwer Academic Publishers, 1991.
8. Skowron, A., Polkowski, L.: Synthesis of Decision Systems from Data Tables, (eds.) T.Y. Lin, N. Cercone, Rough Sets and Data Mining Analysis of Imprecise Data, Kluwer Academic Publishers, 1997, pp. 259-299.
9. Skowron, A., Polkowski, L., Komorowski J.: Learning Tolerance Relations by Boolean Descriptors: Automatic Feature Extraction from Data Tables, Proceedings of the Fourth International Workshop on Rough Sets, Fuzzy Sets, and Machine Discovery, November 6-8, 1996, Tokyo, Japan, pp. 11-17.
10. Skowron, A., Stepaniuk, J.: Towards an Approximation Theory of Discrete Problems, Fundamenta Informaticae 15(2), 1991, pp. 187-208.
11. Skowron, A., Stepaniuk, J.: Approximations of Relations, In: W. Ziarko, ed., Rough Sets, Fuzzy Sets and Knowledge Discovery, Springer Verlag 1994, pp. 161-166.
12. Skowron, A., Stepaniuk, J.: Generalized Approximation Spaces, Proceedings of the Third International Workshop on Rough Sets and Soft Computing, San Jose, November 10-12, 1994, pp. 156-163.
13. Skowron A., Stepaniuk, J.: Tolerance Approximation Spaces, Fundamenta Informaticae, 27 (1996) pp. 245-253.
14. Slowinski, R.: Strict and Weak Indiscernibility of Objects Described by Quantitative Attributes with Overlapping Norms, Foundations of Computing and Decision Sciences, Vol. 18, 1993, pp. 361-369.
15. Stepaniuk, J.: Properties and Applications of Rough Relations, ICS WUT Research Report 26/96, 1996 and In: Proceedings of the Fifth International Workshop on Intelligent Information Systems, Deblin, Poland, June 2-5, 1996, pp. 136-141.
16. Stepaniuk, J., Kretowski, M.: Decision System Based on Tolerance Rough Sets, Proceedings of the Fourth International Workshop on Intelligent Information Systems, Augustow, Poland, June 5-9, 1995, pp. 62-73.
17. Stepaniuk, J.: Approximation Spaces, Reducts and Representatives, (eds.) L. Polkowski, A. Skowron, Rough Sets in Knowledge Discovery, Part I and II, Physica-Verlag, 1998.
18. Stepaniuk, J.: Rough Relations and Logics, (eds.) L. Polkowski, A. Skowron, Rough Sets in Knowledge Discovery, Part I and II, Physica-Verlag, 1998.
19. Yao, Y.Y., Wong, S.K.M., Lin, T.Y.: A Review of Rough Set Models, (eds.) T. Y. Lin, N. Cercone, Rough Sets and Data Mining Analysis of Imprecise Data, Kluwer Academic Publishers, 1997, pp. 47-75.
20. Vakarelov, D.: Rough Polyadic Modal Logics, Journal of Applied Non-Classical Logics, vol. 1(1), 1991, pp. 9-36.
21. Wilson, D.A., Martinez, T.R.: Improved Heterogeneous Distance Functions, Journal of Artificial Intelligence Research, Vol. 6, 1997, pp. 1-34.
22. Ziarko, W.: Variable Precision Rough Sets Model, Journal of Computer and Systems Sciences, Vol. 46, No. 1, 1993, pp. 39-59.

On Generalizing
Pawlak Approximation Operators

Y.Y. Yao

Department of Computer Science, Lakehead University
Thunder Bay, Ontario, Canada P7B 5E1
E-mail: yyao@flash.lakeheadu.ca

[Abstract.] **This paper reviews and discusses generalizations of Pawlak rough set approximation operators in mathematical systems, such as topological spaces, closure systems, lattices, and posets. The structures of generalized approximation spaces and the properties of approximation operators are analyzed.**

1 Introduction

In the development of the theory of rough sets, approximation operators are typically defined by using equivalence relations [10,11]. Researchers have proposed many generalized notions of approximation operators [2,12,15,17,19,20]. Based on the results of these studies, we review and discuss generalizations of Pawlak rough set approximation operators, and show their connections with other mathematical systems.

We interpret the rough set theory as an extension of set theory with two additional unary set-theoretic operators referred to as approximation operators [16,18]. Such an interpretation is consistent with interpreting modal logic as an extension of classical two-valued logic with two added unary operators [5]. With respect to an equivalence relation on a finite and nonempty universe, one can construct a subsystem of the power set of the universe, namely, the σ-algebra or the topology generated by the equivalence classes. Every subset of the universe is approximated by two sets of the subsystem. By generalizing this formulation, one may obtain generalized approximation operators. The resulting systems are related to topological spaces and closure systems. These systems are well-known in logic and algebraic literature [3,7,9,13,14]. We will first review and apply the relevant results for the present study of approximation operators. Further generalizations of approximation operators are studied using posets based on the recent work of Cattaneo [2].

2 Set-theoretic Approximation Operators

In this section, we apply results from the algebraic study of logic [3,7,13,9,14] to Pawlak type approximation operators. A common formulation is adopted from a recent paper by Cattaneo [2]. Two subsystems of the power set of a universe are considered. They are the family of inner definable subsets of the universe

L. Polkowski and A. Skowron (Eds.): RSCTC'98, LNAI 1424, pp. 298–307, 1998.

and the family of outer definable subsets. An arbitrary subset of the universe is approximated by an inner definable subset and an outer definable subset.

Let $E \subseteq U \times U$ be an equivalence relation on a finite and nonempty universe U. That is, the relation E is reflexive, symmetric, and transitive. The pair $apr = (U, E)$ is called a Pawlak approximation space. The equivalence relation E partitions the universe U into disjoint subsets called equivalence classes. Elements in the same equivalence class are said to be indistinguishable. Equivalence classes of E are called elementary sets. A union of elementary sets is called a definable (composed) set [10,11]. The empty set is considered to be a definable set [16]. The family of all definable sets is denoted by $\mathrm{Def}(U)$. It is an σ-algebras of subsets of U. A Pawlak approximation space defines uniquely a topological space $(U, \mathrm{Def}(U))$, in which $\mathrm{Def}(U) \subseteq 2^U$ is the family of all open and closed sets [10].

For a subset $X \subseteq U$, one can approximate X by a pair of subsets of U. The lower approximation $i(X)$ is the greatest definable set contained in X, and the upper approximation $c(X)$ is the least definable set containing X. They correspond to the interior and closure of X in the topological space $(U, \mathrm{Def}(U))$. Thus we have the definition:

$$\textbf{(P)} \qquad i(X) = \bigcup \{Y \mid Y \in \mathrm{Def}(U), Y \subseteq X\},$$

$$c(X) = \bigcap \{Y \mid Y \in \mathrm{Def}(U), X \subseteq Y\}.$$

One may interpret $i, c : 2^U \longrightarrow 2^U$ as unary set-theoretic operators [8,15]. They are dual operators in the sense that $i(X) = \neg c(\neg X)$ and $c(X) = \neg i(\neg X)$. The system $(2^U, \neg, i, c, \cap, \cup)$ is called a Pawlak rough set algebra. It is an extension of the classical set algebra $(2^U, \neg, \cap, \cup)$.

Pawlak approximation operators have the following properties:

(i1) $i(X \cap Y) = i(X) \cap i(Y)$,

(i2) $i(X) \subseteq X$,

(i3) $i(i(X)) = i(X)$,

(i4) $i(U) = U$,

(i5) $c(X) = i(c(X))$,

and

(c1) $c(X \cup Y) = c(X) \cup c(Y)$,

(c2) $X \subseteq c(X)$,

(c3) $c(c(X)) = c(X)$,

(c4) $c(\emptyset) = \emptyset$,

(c5) $i(X) = c(i(X))$.

The above sets of properties are not independent. In fact, (i2) and (i5) imply (i3), and (c2) and (c5) imply (c3). The first four properties are the Kuratowski axioms for topological interior and closure operators. Axioms (i5) and (c5) show that an inner definable subset is also outer definable, and vice versa. Conversely, given a pair of approximation operators $i, c : 2^U \longrightarrow 2^U$ satisfying axioms (i1)-(i5) and axioms (c1)-(c5), respectively, their fixed elements:

$$Def(U) = \{X \mid i(X) = X\} = \{X \mid c(X) = X\}, \tag{1}$$

are the open, and the closed, sets of an 0-dimensional topological space.

In the formulation of Pawlak approximation operators, a special type of topological space is used, in which the set of inner definable subsets (open sets) is the same as the set of outer definable subsets (closed sets). For an arbitrary topological space, the family of open sets is different from the family of closed sets. This immediately leads to a generalization of Pawlak approximation operators. Let $(U, O(U))$ be a topological space, where $O(U) \subseteq 2^U$ is a family of subsets of U called open sets. The family of open sets contains $\emptyset$ and U, and is closed under union and finite intersection. The family of all closed sets $C(U) = \{\neg X \mid X \in O(U)\}$ contains $\emptyset$ and U, and is closed under intersection and finite union. Following Cattaneo [2], a pair of approximation operators is defined by:

$$\textbf{(T)} \qquad i(X) = \bigcup \{Y \mid Y \in O(U), Y \subseteq X\},$$
$$c(X) = \bigcap \{Y \mid Y \in C(U), X \subseteq Y\}.$$

They satisfy axioms (i1)-(i4), and axioms (c1)-(c4), respectively. Conversely, given approximation operators, $i, c : 2^U \longrightarrow 2^U$, satisfying axioms (i1)-(i4) and axioms (c1)-(c4), the sets of their fixed points:

$$O(U) = \{X \mid i(X) = X\},$$
$$C(U) = \{X \mid c(X) = X\}, \tag{2}$$

are families of open and, respectively closed, sets of a topological space.

The notion of closed sets in a topological space may be further generalized. A family $\mathcal{C}(U)$ of subsets of U is called a closure system if it contains U and is closed under intersection [3]. By collecting the complements of members of $\mathcal{C}(U)$, we obtain another system $\mathcal{O}(U) = \{\neg X \mid X \in \mathcal{C}(U)\}$. According to properties of $\mathcal{C}(U)$, the system $\mathcal{O}(U)$ contains the empty set $\emptyset$ and is closed under union. In this case, we define two approximation operators in a closure system:

$$\textbf{(C)} \qquad i(X) = \bigcup \{Y \mid Y \in \mathcal{O}(U), Y \subseteq X\},$$
$$c(X) = \bigcap \{Y \mid Y \in \mathcal{C}(U), X \subseteq Y\}.$$

They satisfy axioms (i2) and (i3), and axioms (c2) and (c3), as well as the following weaker version of (i1) and (c1):

(i0)   If $X \subseteq Y$, then $i(X) \subseteq i(Y)$,
(c0)   If $X \subseteq Y$, then $c(X) \subseteq c(Y)$.

Conversely, for a closure operator $c : 2^U \longrightarrow 2^U$ satisfying axioms (c0), (c2), and (c3), the set of its fixed points:

$$\mathcal{C}(U) = \{X \mid c(X) = X\}, \tag{3}$$

is a closure system. Similar results can be stated between the system $\mathcal{O}(U)$:

$$\mathcal{O}(U) = \{X \mid i(X) = X\}, \tag{4}$$

and the dual operator $i(X) = \neg(c(\neg(X)))$.

In defining set-theoretic approximation operators, three definitions (**P**), (**T**), and (**C**), in the order of generality, are used. For this formulation, inner definable sets must be closed under union and outer definable sets must be closed under intersection. A closure system is therefore the most generalized structure, and one cannot generalize set-theoretic approximation operators further under the same formulation.

3 Approximation Operators in Lattices

The power set of the universe is a special lattice. The results of the last section can be generalized as follows.

Suppose $(\mathcal{B}, \neg, \wedge, \vee, 0, 1)$ is a finite Boolean algebra and $(\mathcal{B}_0, \neg, \wedge, \vee, 0, 1)$ is a sub-Boolean algebra. By using (**P**), one may approximate an element of $\mathcal{B}$ using elements of $\mathcal{B}_0$:

$$\textbf{(LP)} \quad i(x) = \bigvee \{y \mid y \in \mathcal{B}_0, y \leq x\},$$

$$c(x) = \bigwedge \{y \mid y \in \mathcal{B}_0, x \leq y\}.$$

Any finite Boolean algebra is a complete Boolean algebra, and hence the above definition is well defined. Operators i and c satisfy the axioms:

$$\begin{aligned}
&\text{(i1)} && i(x \wedge y) = i(x) \wedge i(y), \\
&\text{(i2)} && i(x) \leq x, \\
&\text{(i3)} && i(i(x)) = i(x), \\
&\text{(i4)} && i(1) = 1, \\
&\text{(i5)} && c(x) = i(c(x)),
\end{aligned}$$

and

$$\begin{aligned}
&\text{(c1)} && c(x \vee y) = c(x) \vee c(y), \\
&\text{(c2)} && x \leq c(x), \\
&\text{(c3)} && c(c(x)) = c(x), \\
&\text{(c4)} && c(0) = 0, \\
&\text{(c5)} && i(x) = c(i(x)).
\end{aligned}$$

Conversely, one may define a pair of approximation operators directly, and use their fixed points as inner and outer definable elements. Gehrke and Walker [4] considered a more generalized definition in which the Boolean algebra $\mathcal{B}$ is replaced by a completely distributive lattice. Like the Pawlak rough set model, one subsystem is used.

Consider a subsystem $O(\mathcal{B})$ of $\mathcal{B}$ satisfying the following axioms:

(O1) $0 \in O(\mathcal{B}), 1 \in O(\mathcal{B})$;

(O2)   For any subsystem $\mathcal{D} \subseteq O(\mathcal{B})$, if there exists a least upper bound $LUB(\mathcal{D}) = \bigvee \mathcal{D}$, then it belongs to $O(\mathcal{B})$;

(O3) $O(\mathcal{B})$ is closed under finite meet, i.e.,
for any $x, y \in O(\mathcal{B})$, we have $x \wedge y \in O(\mathcal{B})$.

For a finite Boolean algebra, axiom (O2) in fact states that the system $O(\mathcal{B})$ is closed under join. Elements of $O(\mathcal{B})$ are referred to as inner definable elements. The complement of an inner definable element is called an outer definable element. The set of outer definable elements $C(\mathcal{B}) = \{\neg x \mid x \in O(\mathcal{B})\}$ is characterized by the axioms:

(C1) $0 \in C(\mathcal{B}), 1 \in C(\mathcal{B})$;

(C2)   For any subsystem $\mathcal{D} \subseteq C(\mathcal{B})$, if there exists a greatest lower bound $GLB(\mathcal{D}) = \bigwedge \mathcal{D}$, then it belongs to $C(\mathcal{B})$;

(C3) $C(\mathcal{B})$ is closed under finite join, i.e.,
for any $x, y \in C(\mathcal{B})$, we have $x \vee y \in C(\mathcal{B})$.

¿From the sets of inner and outer definable elements, we define the following approximation operators:

$$\textbf{(LT)} \quad i(x) = \bigvee \{y \mid y \in O(\mathcal{B}), y \leq x\},$$
$$c(x) = \bigwedge \{y \mid y \in C(\mathcal{B}), x \leq y\}.$$

They satisfy axioms (i1)-(i4), and axioms (c1)-(c4), respectively, and are the topological interior and closure operators. The sets of fixed points of i and c are inner and outer definable elements, respectively. The system $(\mathcal{B}, \neg, i, c, \wedge, \vee, 0, 1)$ is a topological Boolean algebra [14], which is an extension of Boolean algebra with added operators.

Let $(L, \leq, 0, 1)$ be a bounded lattice. Suppose $O(L)$ is a subset of L such that it contains 0 and is closed under join, and $C(L)$ a subset of L such that it contains 1 and is closed under meet. They are complete lattices, although the meet of $O(L)$ and the join of $C(L)$ may be different from that of L. Based on these two systems, we can define two approximation operators as follows:

$$\textbf{(LC)} \quad i(x) = \bigvee \{y \mid y \in O(L), y \leq x\},$$
$$c(x) = \bigwedge \{y \mid y \in C(L), x \leq y\}.$$

The approximation operators satisfy axioms (i2) and (i3), axioms (c2) and (c3), as well as the following weaker version of (i1) and (c1), respectively:

(i0)   If $x \leq y$, then $i(x) \leq i(y)$,

(c0)   If $x \leq y$, then $c(x) \leq c(y)$.

The sets $O(L)$ and $C(L)$ are the fixed points of i and c, respectively. The operator c is a closure operator [1]. $C(L)$ corresponds to the closure system in the set-theoretic framework. But since a lattice may not be complemented, we must explicitly consider both $O(L)$ and $C(L)$. That is, the system $(L, O(L), C(L))$, or equivalently the system (L, i, c), is used for the generalization of Pawlak approximation operators.

4   Approximation Operators in Posets

Instead of using a Boolean algebra or a lattice, one may use a poset. Such generalizations of Pawlak rough set model were considered by Iwinski [6], and were systematically studied by Cattaneo recently [2]. Let $(\Sigma, \leq, 0, 1)$ be a poset with respect to a partial order relation $\leq$ bounded by the least element 0 and the greatest element 1. If the sets of inner, and respectively outer, definable elements of Σ are chosen to be complete lattices, one may immediately use (**LC**) to define approximation operators. A similar idea is in fact used by Iwinski [6], although the definition is different from (**LC**). The formulation suggested by Cattaneo [2] is consistent with (**LC**).

For an arbitrary subset $X \subseteq \Sigma$, a least upper bound or a greatest lower bound of X may not exist. If a least upper bound of X exists, it is unique and is denoted by $LUB(X)$. Similarly, if a greatest lower bound of X exists, it is unique and is denoted by $GLB(X)$. Any subset Σ_0 of a poset Σ is itself a poset under the same order relation, and is called a subposet. In the subposet Σ_0, if the least upper bound of a subset $X \subseteq \Sigma_0$ exists, we denote it by $LUB_{\Sigma_0}(X)$ or simply $LUB(X)$ when Σ_0 is clear from the context. If the greatest lower bound of X exists, we denote it by $GLB_{\Sigma_0}(X)$ or simply $GLB(X)$. Both Boolean algebras and lattice can be understood as posets with additional properties. For the three definitions in the last section, approximation operators are defined through the LUB of a subsystem of inner definable elements, and the GLB of a subsystem of outer definable elements. Following the same argument, we will use two subposets $O(\Sigma)$ and $C(\Sigma)$ to represent inner and outer definable elements. But we must require that in some sense the subposet $O(\Sigma)$ be closed with respect to LUB, and $C(\Sigma)$ be closed with respect to GLB.

Given $x \in \Sigma$ and $Y \subseteq \Sigma$, the order ideal relative to Y generated by x is defined by:

$$\downarrow x | Y = \{ y \in Y \mid y \leq x \}. \tag{5}$$

Dually, the order filter relative to Y generated by x is defined by:

$$\uparrow x | Y = \{ y \in Y \mid x \leq y \}. \tag{6}$$

With these notions, we can construct two families of subsets of $O(\Sigma)$ and $C(\Sigma)$:

$$SO(\Sigma) = \{ \downarrow x | O(\Sigma) \mid x \in \Sigma \},$$
$$SC(\Sigma) = \{ \uparrow x | C(\Sigma) \mid x \in \Sigma \}. \tag{7}$$

Each element of $SO(\Sigma)$ is a subsystem of $O(\Sigma)$, and each element of $SC(\Sigma)$ is a subsystem of $C(\Sigma)$.

For the definition of abstract approximation operators in posets, we adopt and generalize the proposal of Cattaneo [2]. However, the formulation is slightly different. In particular, we explicitly specify the structure of the set of inner definable and the structure of the set of outer definable elements. Consider a triple $(\Sigma, O(\Sigma), C(\Sigma))$ called an abstract approximation space, where Σ is a bounded poset, $O(\Sigma)$ is assumed to be the set of inner definable elements, and $C(\Sigma)$ is assumed to be the set of outer definable elements. The set of inner definable elements is characterized by the axioms:

(O1) $0 \in O(\Sigma), 1 \in O(\Sigma)$;

(O2*)   With respect to $O(\Sigma)$, the least upper bound of $\downarrow x | O(\Sigma)$

          exists and satisfies the condition :

$$LUB_{O(\Sigma)}(\downarrow x | O(\Sigma)) \leq x, \quad \text{for every } x \in \Sigma.$$

Axiom (O2*) suggests that $LUB_{O(\Sigma)}(\downarrow x | O(\Sigma))$ is also an inner definable element. That is, $O(\Sigma)$ is closed under LUB at least for any subsystem of $SO(\Sigma)$. For an arbitrary subset of $O(\Sigma)$, the least upper bound may not exist. Hence the system $O(\Sigma)$ may not be a lattice. The set of outer definable elements is defined by the axioms:

(C1) $0 \in C(\Sigma), 1 \in C(\Sigma)$;

(C2*)   With respect to $C(\Sigma)$, the greatest lower bound of $\uparrow x | C(\Sigma)$

          exists and satisfies the condition :

$$x \leq GLB_{C(\Sigma)}(\uparrow x | C(\Sigma)), \quad \text{for every } x \in \Sigma.$$

The subposet $C(\Sigma)$ is closed under GLB for at least any subsystem of $SC(\Sigma)$. It may not be a lattice. The two systems $O(\Sigma)$ and $C(\Sigma)$ are usually different subsets of Σ. The set $O(\Sigma) \cap C(\Sigma)$ consists of those elements which are both inner and outer definable.

An element of Σ may be approximated by a pair of inner and outer definable elements from $O(\Sigma)$ and $C(\Sigma)$. We define a pair of inner and outer approximation operators, $i : \Sigma \longrightarrow O(\Sigma)$ and $c : \Sigma \longrightarrow C(\Sigma)$, as follows:

$$
\begin{aligned}
\textbf{(PC)} \quad i(x) &= LUB(\downarrow x | O(\Sigma)) \\
&= LUB(\{y \in O(\Sigma) \mid y \leq x\}), \\
c(x) &= GLB(\uparrow x | C(\Sigma)) \\
&= GLB(\{y \in C(\Sigma) \mid x \leq y\}),
\end{aligned}
$$

where LUB and GLB are defined with respect to $O(\Sigma)$ and $C(\Sigma)$. By definition, $i(x)$ is the best approximation of x from below using the inner definable elements $O(\Sigma)$, while $c(x)$ is the best approximation of x from above using outer definable elements $C(\Sigma)$. More specifically, i satisfies axioms (i0) and (i2)-(i4), and c satisfies axioms (c0) and (c2)-(c4). Assume $x \leq y$. We have $\downarrow x | O(\Sigma) \subseteq \downarrow y | O(\Sigma)$. Hence, $i(x) = LUB(\downarrow x | O(\Sigma)) \leq LUB(\downarrow y | O(\Sigma)) = i(y)$, namely, (i0) holds. By (O2*), i satisfies (i2). For any $y \in O(\Sigma)$, we have $y \in \downarrow x | O(\Sigma)$. This implies $y \leq i(y)$. By combining it with (i2), we have $i(y) = y$ for any $y \in O(\Sigma)$. For any $x \in \Sigma$, $i(x) \in O(\Sigma)$. Thus, $i(i(x)) = i(x)$, namely, (i3) holds. By the assumption

$1 \in O(\Sigma)$ and the definition of i, it follows that i satisfies (i4). Similarly, we can show that c obeys (c0) and (c2)-(c4).

Alternatively, we may define an abstract approximation space by a triple (Σ, i, c), where i and c are mappings characterized by axioms (i0) and (i2)-(i4), and axioms (c0) and (c2)-(c4), respectively. The sets of i-fixed and c-fixed elements:

$$O(\Sigma) = \{x \in \Sigma \mid i(x) = x\},$$
$$C(\Sigma) = \{x \in \Sigma \mid c(x) = x\}, \tag{8}$$

are the sets of inner and outer definable elements, respectively. By axioms (i2) and (i4), it can be easily verified that $0, 1 \in O(\Sigma)$. Consider an element $x \in \Sigma$. Suppose $y \in \downarrow x | O(\Sigma)$. We have $y \leq x$ and $y = i(y)$. By axiom (i0), it follows $y = i(y) \leq i(x)$. Thus, $i(x)$ is an upper bound of $\downarrow x | O(\Sigma)$. Now, we want to show that it is in fact the least upper bound. Suppose z is an upper bound of $\downarrow x | O(\Sigma)$. We have $y \leq z$ for all $y \in \downarrow x | O(\Sigma)$. By axiom (i2), $i(x) \leq x$. By axiom (i3), $i(x) \in O(\Sigma)$. Therefore, $i(x) \in \downarrow x | O(\Sigma)$. It immediately follows that $i(x) \leq z$. This implies that (O2*) holds. Similarly, one can show that C is the family of outer definable elements satisfying axioms (C1) and (C2*).

In the previous formulation of approximation operators, two methods have been used. One starts from the system $(\Sigma, O(\Sigma), C(\Sigma))$ where the two subposets $O(\Sigma)$ and $C(\Sigma))$ are given specific structures. From this system one can define two approximation operators, i and c, enjoying certain properties. Dually, the second method starts from a pair of approximation operators, namely, the system (Σ, i, c) satisfying some axioms, and a system $(\Sigma, O(\Sigma), C(\Sigma))$ is recovered by the sets of fixed points of i and c. In the formulation of Cattaneo [2], the structures of two subposets are stated implicitly by using the approximation operators. Our formulation avoided such a problem. This makes our discussion of approximation operators to be conform to the commonly used approaches for the study of approximation operators in systems such as topological spaces and closure systems.

Approximation operators in posets can be further generalized. Suppose that the set of inner definable elements $O(\Sigma)$ satisfies the axiom:

$$(\text{O1}^*) \qquad 0 \in O(\Sigma),$$

and axiom (O2*). The approximation operator i defined by (**PC**) only satisfies axioms (i0), (i2), and (i3). Similarly, if the set of outer definable elements $C(\Sigma)$ satisfies the axiom:

$$(\text{C1}^*) \qquad 1 \in C(\Sigma),$$

and axiom (C2*), the approximation operator c defined by (**PC**) only satisfies axioms (c0), (c2), and (c3). They correspond to approximation operators in closure systems. A subposet $C(\Sigma)$ is said to be a closure system if it satisfies axioms (C1*) and (C2*). In the set-theoretic setting, a closure system must be closed under intersection for any of its subsystem. A closure system on a poset is closed under GLB for only certain subsystems.

Suppose the set of inner definable elements $O(\Sigma)$ satisfies the axioms (O1*) and (O2**): for $x \in \Sigma$,

(O2**)   With respect to $O(\Sigma)$, the least upper bound of $\downarrow x | O(\Sigma)$ exists,

and the set of outer definable elements $C(\Sigma)$ satisfies the axioms (C1*) and (C2**): for $x \in \Sigma$,

(C2**)   With respect to $C(\Sigma)$, the greatest lower bound of $\uparrow x | C(\Sigma)$ exists.

In this case, approximation operator i defined by (**PC**) satisfies axioms (i0), (i3), and $i(0) = 0$. Approximation operator c satisfies axioms (c0), (c3), and $c(1) = 1$.

A more detailed and systematic study of approximation operators in special types of lattice and posets, as well as examples, can be found in a recent paper by Cattaneo [2]. A different formulation of approximation operators in poset can be found in an earlier paper by Iwinski [6].

5   Conclusion

In generalizing Pawlak approximation operators, we have considered four systems. From more particular instantiations to more general cases, they are Pawlak approximation spaces (0-dimensional topological spaces), topological Boolean algebras (topological spaces), closure systems, and abstract approximation spaces. For the definition of approximation operators, two subsystems, corresponding to the set of inner definable elements and the set of outer definable elements, are used. An arbitrary element is approximated from blow by using inner definable elements through LUB, and from the above by using outer definable elements through GLB. In Pawlak approximation spaces, the set of inner definable elements is the same as the set of outer definable elements. It contains 0 and 1, and is closed under both LUB and GLB. For topological Boolean algebras, the set of inner definable elements is closed under LUB and finite GLB, while the set of outer definable elements is closed under GLB and finite LUB. For closure system, the set of inner definable elements is only closed under LUB, and the set of outer definable elements is only closed under GLB. For abstract approximation spaces, the set of inner definable elements is partially closed under LUB (i.e., for only some subsystems), while the set of outer definable elements is partially closed under GLB. The structure of the family of inner definable elements determines the properties of inner approximation operator. Likewise, the structure of the family of outer definable elements determines the properties of outer approximation operator. Dually, one may start from a pair of approximation operators. The set of inner definable elements can be obtained by the fixed points of the inner approximation operator, while the set of outer definable elements can be obtained by the fixed points of the outer approximation operator.

References

1. G. Birkhoff, *Lattice Theory*, American Mathematical Society, Providence, 1967.
2. G. Cattaneo, Abstract approximation spaces for rough theories, in: *Rough Sets in Data Mining and Knowledge Discovery*, edited by L. Polkowski and A. Skowron, Physica Verlag, Berlin, to appear.
3. P.M. Cohn, *Universal Algebra*, Harper and Row Publishers, New York, 1965.
4. Gehrke, M., and Walker, E., On the structure of rough sets, *Bulletin of the Polish Academy of Sciences, Mathematics*, 40: 235–245, 1992.
5. S. Haack, *Philosophy of Logics*, Cambridge University Press, Cambridge, 1978.
6. T.B. Iwinski, Rough orders and rough concepts, *Bulletin of the Polish Academy of Sciences, Mathematics*, 36: 187–192, 1988.
7. G.E. Hughes, and M.J. Cresswell, *An Introduction to Modal Logic*, Methuen, London, 1968.
8. T.Y. Lin and Q. Liu, Rough approximate operators: axiomatic rough set theory, in: *Rough Sets, Fuzzy Sets and Knowledge Discovery*, edited by W.P. Ziarko, Springer-Verlag, London, 256-260, 1994.
9. C.C. McKinsey and A. Tarski, On closed elements in closure algebras, *Annals of Mathematics*, 47: 122–162, 1946.
10. Z. Pawlak, Rough sets, *International Journal of Computer and Information Science*, 11:341–356 (1982).
11. Z. Pawlak, Rough classification, *International Journal of Man-Machine Studies*, 20:469-483 (1984).
12. J.A. Pomykala, Approximation operations in approximation space, *Bulletin of the Polish Academy of Sciences, Mathematics*, 35:653–662 (1987).
13. A.N. Prior, Tense logic and the continuum of time, *Studia Logica*, 13: 133–148, 1962.
14. H. Rasiowa, *An Algebraic Approach to Non-classical Logics*, North-Holland, Amsterdam, 1974.
15. U. Wybraniec–Skardowska, On a generalization of approximation space, *Bulletin of the Polish Academy of Sciences: Mathematics*, 37: 51–61 (1989).
16. Y.Y. Yao, Two views of the theory of rough sets in finite universes. *International Journal of Approximate Reasoning*, 15:291–317 (1996).
17. Y.Y. Yao, Relational interpretations of neighborhood operators and rough set approximation operators, *Information Sciences*, to appear.
18. Y.Y. Yao, A comparative study of fuzzy sets and rough sets, *Information Sciences*, to appear.
19. Y.Y. Yao and T.Y. Lin, Generalization of Rough Sets using Modal Logic, *Intelligent Automation and Soft Computing, an International Journal*, 2:103-120 (1996).
20. W. Zakowski, Approximations in the space (U, Π), *Demonstratio Mathematica*, XVI: 761–769 (1983).

Real-Time Real-World Visual Classification –
Making Computational Intelligence Fly

Jan Matti Lange[1], Hans-Michael Voigt[1], Steffen Burkhardt[2], and Ralph Göbel[2]

[1] GFaI - Center for Applied Computer Science, Berlin, Germany
voigt@gfai.de
URL: http://www.amspr.gfai.de
[2] Parsytec GmbH, Aachen/Chemnitz, Germany
sb@parsytec.de

Abstract. An Intelligent Inspection Engine (IIE) for classification of non-regular shaped objects from images is described and evaluated using real-world data from a waste package sorting application. The entire system is self-organizing. Principal component analysis and additional a priori knowledge on color properties are used for feature extraction. As classifiers growing neural networks provide robustness and minimize the number of runs for parameter tuning. We propose a method to encompass feature extraction and classification within a bootstrap procedure. These method reduces the immense memory requirement for the computation of principal components if number and size of training images are huge without to much loss of recognition quality.

1 Introduction

Visual classifier systems have been intensely studied in the past three decades. A lot of algorithms, theoretical results and systems have resulted from this research. Especially systems for object recognition have been proposed, e.g. [10]. Most of these systems are designed to recognize a large number of unique objects from different appearances while the objects itself do not vary much. That means the appearance of an object changes mainly because of changing illumination, angle of view and background but less because of changes of the object itself. Application areas are face recognition, scene recognition and search in image databases.

In this paper we face a more unstructured recognition task. We present a visual classifier designed to recognize the membership of a large number of different objects to classes. Visual object properties like shape and color within each class vary in a wide range. There are no objects with unique appearance in a single class. In terms of statistics we can say the density distribution of the classes have much more modes in comparison to the tasks explained above. Applications of this type of visual classifiers are e.g. the sorting of non-regular shaped objects like agricultural products, waste packages, food or also the search in image databases using concept queries. For real-world applications as industrial automation these systems have to be robust, fast and flexible to use. Especially flexibility is often

L. Polkowski and A. Skowron (Eds.): RSCTC'98, LNAI 1424, pp. 308–315, 1998.

very important. The user must be able to use the same system for different visual classification tasks without much cost-intensive modifications. The system must provide full self-organization including the preprocessing and should offer multi-sensor fusion ability. Changes of the classification task must be manageable by non-pattern-recognition-experts. The task described is not solvable by pattern matching or by classification from descriptive shape representations. We will show in this paper how different classifiers compare for the real-world visual classification task of waste package sorting. The system has been implemented on a Parsytec MIMD parallel computer and recognizes (depending on the number of processors) for this special application up to 10 objects per second.

2   System architecture and algorithms

2.1   Image capture and preprocessing

The goal of preprocessing is to reduce the noise and disturbations. That is to reduce the variance of images of same classes. The images are taken from objects at a black conveyor belt illuminated by a flash light because of the speed of belt. The objects orientation at the belt is not defined. To reduce the resulting high inner class variance a translation and rotation invariant transformation is done: First the background is removed by thresholding. Than the object is centered by shifting the mass point (center of gravity) of the relating grey image to the geometric center point. Thereafter the image is rotated with respect to the first and second inertial moment of the grey image. It is clear that this preprocessing method reduces the inner class variance because objects of same shape which might have different position and orientation at the conveyor belt get a more similar representation for preprocessed images.

2.2   Feature Extraction

The system should provide full self-organization. Therefore a feature extraction method is desired which can be parameterized by *Learning from Examples*. The discrete Karhunen–Lóeve–Transformation (KLT) is such a method under assumption of a normal distribution. It became popular for image retrieval and object recognition within the last few years because of the availability of high-performance computers. The KLT transforms an image linearly to the orthogonal space spanned by the eigenvectors of the estimated covariance matrix of all sample images. Because of high correlation of data in image regions it is only necessary to use a space spanned by eigenvectors of the n largest eigenvalues.

The problem with the KLT is the computation of the eigenvectors of the very large estimated covariance matrix. The matrix contains about $3.5 \cdot 10^{11}$ real elements for example if the training set contains RGB images of size 256×256. That is 500 G-Bytes if we consider the symmetry of the matrix and assume 4 Bytes used to code an element of the matrix! To overcome that problem a method introduced in [11] has been used: If the number N of sample images is smaller than the image size it is only necessary to compute the implicit eigenvectors of the implicit covariance matrix which has the size $N \times N$. Thereafter the $N - 1$ eigenvectors are computed as linear combination of the sample images with the implicit eigenvectors as coefficients and normalized. The idea is that there exist only $N - 1$ eigenvectors with non-zero eigenvalues. The implicit eigenvectors are computed using a conjugated gradient procedure.

Still one limitation remains: To compute the implicit covariance matrix in an acceptable time all the images must be in the computer main memory. To overcome this a bootstrap method explained later is used.

2.3   The TACOMA classifier

To reduce the engineering efforts for the design of neural network architectures a data driven algorithm is desirable which constructs a network during the training process automatically. Especially if the neural network is part of a system subject to be managed by a non-neural-network-expert a very robust algorithm is necessary. Within the IIE the TACOMA growing neural network architecture is used [6].

The idea TACOMA is based on is to build a feed-forward neural network bottom-up by cyclically inserting cascaded hidden layers. The activation function of a hidden layer unit combines the local characteristics of radial basis function units (or other window functions) with sigmoid units. With each growth step a hidden layer consisting of such local attention neurons will be inserted. The number of units to be inserted and their attention regions, means and ratios are given by an approximation of the mapping of the residual error from output to input space. The attention region of a unit becomes restricted to a region in the input space where the residual error is still high. The attention regions of the hidden units of the same hidden layer do not overlap (no lateral influence) and can be considered as units of different subnetworks. Contrary to the Cascade-Correlation [3] Learning Architecture different correlation measures are used to train the units. The TACOMA algorithm provides good generalization properties as shown with different experiments [6], [7], [8], [9] .

3   Experiments

3.1   Data and error measures

The prototype of the Intelligent Inspection Engine has been trained and evaluated with 3402 24 bit RGB images from real-world waste packages. The image size is 376 x 280. A training and test set has been generated each containing 1701 images, both reflecting the number of representation in each class. For the bootstrap method described later four training sets each with 423 images have been used. Table 1 shows the descriptions of the seven classes. The classes are very different represented in the set which makes the classification even harder.

Table 1. Description of object classes and frequencies

label	description	freq.	%
C1	plastic bottles white	72	2.12
C2	plastic bottles transparent	158	4.64
C3	plastic bottles colored w. stickers, printed	346	10.17
C4	plastic or metal cans, cups white	400	11.76
C5	plastic or metal cans, cups transparent	254	7.47
C6	plastic or metal cans, cups colored w. stickers, printed	1420	41.74
C7	different tetra packs	752	22.10
sum		3402	100.00

To evaluate the classifier results we define the overall error rate, typed as *wrong* in the tables, the class specific cleanness typed as *C1* to *C7* and a rejection rate, typed as *rejected*. The overall error rate is the percentage of wrong classified objects within the test set (without the number of rejected objects). An image will be rejected if both of the biggest class probabilities of an outcome of the TACOMA classifier are very similar. This threshold and all tunable parameters of the TACOMA algorithm were the same for all experiments. The class specific cleanness is typical for real-world sorting applications. It is defined as the percentage of wrong classified objects within the objects classified to a class. Imagine a sorting application where the objects fall in class-boxes depending on the classification. Than this describes the percentage of wrong classified objects within each of the boxes - the cleanness of objects within a box. The rejection rate is the percentage of rejected objects as explained above (only for the TACOMA classifier). In all runs the TACOMA algorithm stops automatically after 20 layer have been inserted or earlier if the training set is classified 100 % correct. Parameters have been the same for all runs. All measures within the tables or text are with respect to the test set.

3.2   Results and Comparison

As it is well known from the *Curse of dimensionality* the dimension of features should be well adapted to the density of data available for training and to the classifier [5], [4]. The more the dimension increases with constant number of feature vectors, the less dense are the data in the input space and the greater is the risk for bad generalization on unseen pattern.

The eigenvalues represent the variance of the pattern projected to the directions given by the eigenvectors. That means the sum over all eigenvalues represent all information available for classification after the KLT. The plot of the eigenvalues sorted decreasingly gives an imagination of how many eigenvectors should be used for transformation. Figure 1 shows the plot of up to the 30'th eigenvalue. As one can see the eigenvalues decrease rapidly and quasi-linear (because of log)

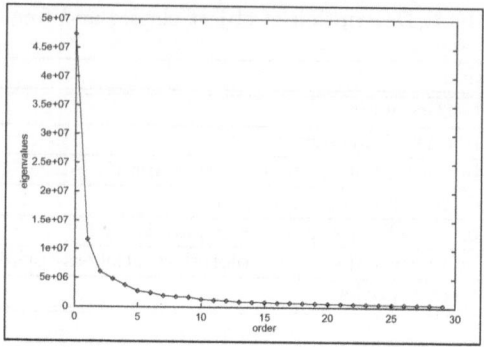

Fig. 1. Plot of the eigenvalues sorted decreasingly.

is entered early. Nearly 90% of variance are in the first 20 eigenvalues. With this visualization we are able to chose roughly the number of eigenvectors to be used. A more detailed view is given in Figure 2 (left). Here the percentage of wrong classified images for different classifier (Cascade Correlation, K-nearest neighbor with $k = 1, 3, 5$ and TACOMA with and without rejection) are shown. The TACOMA classifier shows the best performance with 18 to 22 eigenvectors. Table 3 shows the averaged TACOMA results for 20 eigenvectors with rejection (the column headed with T) over five runs. Especially the percentages of wrong classified objects of low represented classes do not satisfy. Theoretically the best would be to add more images from those classes. But remember the general demands to a real-world system. Within the waste package sorting problem the classes are unequal represented in reality and creating training sets where all classes are equal represented would add cost-intensive overhead while adapting the system to a new task.

4   Enhancements Using A Priori Knowledge

Additionally to the features resulting from the KLT a color feature vector should be computed because it is known a priori that color plays an important role to distinguish between some classes, compare table 1. The color feature vector should give a compressed description of the pixel distribution in the RGB space of an image. An objects illumination varies depending on its orientation at the conveyor belt. To reduce the resulting inner class variance only the direction of the RGB vectors (given for each pixel by the values of the red, green and blue channel) are used. The color feature vector is computed as a histogram of all RGB vector directions of an image. The histogram intervals are given by prototype vectors computed by vector quantization. The vector quantization moves the prototype vectors iteratively to approximate the density distribution of the RGB vector directions of all images from the training set. The averaged TACOMA results with 20 eigenvectors and rejection (0.05 and 0.1) over five

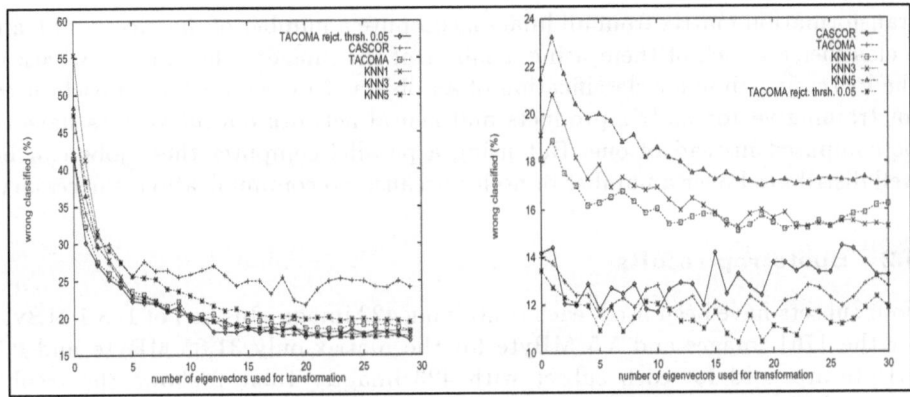

Fig. 2. Percentage of wrong classified images for different classifier using different numbers of eigenvectors for the KLT (left). Percentage of wrong classified images for different classifier using different numbers of eigenvectors for the KLT and and additional color feature vector of size 18 (right).

runs are much better using these a priori knowledge as one can see from Table 3 (column headed with TC). The number of wrong classified images decrease from 18.59 to 11.13%.

Figure 2 (right) shows the results for different classifiers using one to 30 eigenvectors and a color feature vector with 18 dimensions. Again the best results are with the TACOMA classifier with 18 - 22 eigenvectors.

5   Using bootstrapping to reduce the memory requirements

5.1   The bootstrap algorithm

Bootstrapping is well known as a method to manage the bias-variance dilemma [5], [1], [2] and offers a simple framework to improve the generalization ability of classifier systems. The idea is to use a number of density estimators instead of one and to average the class probabilities of the estimators as input to the decision function. For a theoretical description of bootstrapping see [5].

Here we show the use of bootstrapping to reduce the immense memory requirements to compute the principal components of huge sets of images. As explained a the smart method [11] computes the eigenvectors and -values using the implicit covariance matrix. To compute this matrix within an acceptable time all images of the training set have to be held within the computer memory. Additionally, the size of the implicit covariance matrix increases quadratically with the number of images. Assuming 3000 24 bit RGB images of size 256 x 256 562.5 MByte for the images and 17.2 MByte for the matrix are required. The simple concept to reduce the memory requirements is to encompass the computation of the KLT within the bootstrap procedure. Instead to compute one transformation matrix from all images, compute a number of matrices and train a classifier for each of them using a subset of the images. The only drawback is the increasing time for classification of an object. Now as much as subsets used for training vector matrix products and neural network output vectors have to be computed instead of one. But using a parallel computer these jobs can be well distributed over a number of nodes because no communication is necessary.

5.2 Bootstrap results

Four subsets have been used, each containing 423 images. Instead of 128.1 MByte for the 1701 images and 5.5 MByte for the matrix only 31.85 MByte and 0.7 MByte are used for each subset with 423 images. Table 2 shows the results achieved by the single networks (column headed with s1 - s4) and by the bootstrapped networks (column headed with boot) for a typical run. The bootstrap results are much better for all error rates than the results of the single classifier.

Table 2. Bootstrapping using multiple matrices, results of a typical run, rejection threshold = 0.1 (left part). Summary of results for 20 eigenvectors. The columns contain the average rates over 5 runs (right part)

	Boot	s1	s2	s3	s4	T 0.05	TC 0.05	TC 0.1	TCB 0.05	TCB 0.1
wrong (%)	10.47	17.35	15.72	17.07	15.84	18.59	11.13	10.10	12.49	10.99
rejected (%)	8.47	8.88	9.52	9.41	8.70	3.88	2.83	5.26	4.08	7.4
C1 wrong (%)	8.33	57.89	50.00	40.00	37.93	41.86	34.00	31.43	21.11	13.36
C2 wrong (%)	10.20	35.29	22.45	21.88	31.67	26.92	18.56	18.44	11.64	10.35
C3 wrong (%)	15.38	26.53	21.68	31.94	13.67	20.12	13.40	11.44	13.56	11.47
C4 wrong (%)	10.49	27.69	27.57	10.14	18.18	19.53	15.33	13.71	13.68	12.58
C5 wrong (%)	22.64	29.73	23.81	34.48	28.71	29.70	22.01	21.18	23.49	21.97
C6 wrong (%)	9.46	12.37	10.93	12.88	14.84	17.04	8.23	7.42	11.62	10.24
C7 wrong (%)	6.87	8.71	10.17	11.27	9.36	13.77	6.07	5.40	9.5	8.44

The averaged results of five runs are shown in table 3 (column headed with TCB) for rejection thresholds of 0.05 and 0.1. The overall error rates are about one percent above the rates of the TC runs. The differences are not large and might be smaller if the classes would be more equal distributed. In that case one should use more subsets. It seems to be a good compromise to use the bootstrap method.

6 Summary and Conclusions

Classification of non-regular shaped objects from images becomes more and more interesting for real-world industrial applications. We proposed an Intelligent Inspection Engine able to solve the very hard task of waste package sorting with good results. The system fulfills the demands for self-organization, flexibility, robustness and manageability by non-pattern-recognition-experts. The results demonstrate that the techniques underlying KLT, TACOMA and bootstrapping are general. Especially the extension of bootstrapping over the feature extraction by principal component analysis is a smart method to decrease memory requirements important for practical application (system costs). The scalable parallel implementation of the system guarantees the fulfillment of real-time demands for a wide range of applications. This has led to the development of a software package called Intelligent Inspection Engine for classification of non-regular shaped objects from images. The image data set is available via FTP on request. Send an E-Mail to voigt@gfai.de.

References

1. Leo Breiman. *Bias, variance, and arcing classifiers.* Technical Report, Department of Statistics, University of California, Berkely, 1996.
2. Leo Breiman. *Bagging predictors*, Technical Report, Department of Statistics, University of California, Berkely, 1995.
3. S. E. Fahlman and C. Lebiere. The Cascade–Correlation Learning Architecture. Tech. Report CMU-CS-90-100, School of Computer Science, Carnegie Mellon University, Pittsburgh, PA, Februar 1990.
4. Jerome H. Friedman.   *On Bias, Variance, 0/1 - loss, and the Curse-of-Dimensionality*, Technical Report, Department of Statistics and Stanford Linear Accelerator Center, Stanford University, April 1996.
5. Stuart Geman, Elie Bienenstock, and René Doursat.   Neural networks and the bias/variance dilemma. *Neural Computation*, 4:1–58, 1992.
6. J. M. Lange, H.-M. Voigt, and D. Wolf. Growing Artificial Neural Networks Based on Correlation Measures, Task Decomposition and Local Attention Neurons. In *Proceedings of the IEEE International Conference on Neural Networks 1994 as Part of the IEEE World Congress on Computational Intelligence*, pages 1355–1358, Orlando, 1994. Vol. 2.
7. J. M. Lange, H.-M. Voigt, and D. Wolf. The tacoma algorithm for reflective growing of neural networks. In *Proceedings of the IEE Colloquium on Advances in Neural Networks for Control and Systems, Berlin, IEE-Digest 94/136*, pages 2/1 – 2/2, London, 1994. The Institution of Electrical Engineers.
8. J. M. Lange, H.-M. Voigt, and D. Wolf. Task Decomposition and Correlations in Growing Artificial Neural Networks. In P. G. M. M. Mariano, editor, *Proceedings of the International Conference on Artificial Neural Networks*, pages 735–738, Sorrento, Mai 1994. Springer.
9. J. M. Lange. *Cyclical Local Structural Risk Minimization with Growing Neural Networks*, Technical Report TR–96–015, International Computer Science Institute, Berkeley, CA, April 1996.
10. S. K. Nayar, S. A. Nene, and H. Murase. Real-Time 100 Object Recognition System. In *Proceedings of ARPA Image Understanding Workshop*, San Fransisco, February 1996
11. S. K. Nayar, H. Murase, and S. A. Nene. Learning, Positioning, and Tracking Visual Appearance. In *IEEE International Conference on Robotics and Automation*, San Diego, May 1994.

Fractal Operator Convergence
by Analysis of Influence Graph

Władysław Skarbek*

Department of Electronics and Information Technology
Warsaw University of Technology

Abstract. The convergence of fractal operator F used in image compression is investigated. A sufficient condition for eventual contractivity is derived by using the adjacency matrix of an influence graph which is determined by the fractal encoder.

1 Introduction

In fractal image compression scheme, the original image f is decoded as the fixed point $\tilde{f}$ of a fractal operator F which has been built for f in the encoding stage (see Jacquin [5]). This operator works in a space $\mathcal{I}$ of images defined on a fixed discrete domain D and real valued:

$$\mathcal{I} \doteq \{g \mid g : D \to R\} \ .$$

Using the classical contractive mapping theorem of Banach (see Banach [1], Dugundi and Granas [3]) the fixed point $\tilde{f}$ is approximated by successive iterations $F^{\circ i}(x_0)$ of the contractive operator F (see Barnsley and Hurd [2], Fisher [4]). However, practical use of this scheme encountered several difficulties as for most existing encoding schemes the essential condition of contractivity for F cannot be guaranteed.

It was observed that if F is not contractive then a certain iteration $F^{\circ k}$ may be contractive. This situation is recognized as eventual contractivity. In this paper, there is given a new sufficient condition for eventual contractivity using algebraic operations on adjacency matrix of an influence graph defined between image domains, i.e. between parts of the given image.

2 Fractal operators

Let $f \in \mathcal{I}$. We assume that its domain $D = dom(f)$ is a rectangular array of $N = |D|$ pixels. Two kinds of subdomains are considered in the definition of a fractal operator F:

---

* The work was sponsored in part by ECU grant CRIT2.

L. Polkowski and A. Skowron (Eds.): RSCTC'98, LNAI 1424, pp. 316–321, 1998.
© Springer-Verlag Berlin Heidelberg 1998

— *target domains* (shortly t-Domains) which create a partition Π of D, i.e.:

$$\Pi = \{T_1, \ldots, T_a\}, \ D = \bigcup_{i=1}^{a} T_i \ ,$$

$$T_i \cap T_j = \emptyset, \ i \neq j \ ;$$

— *source domains* (shortly s-Domains) which create a cover Γ of D, i.e.:

$$\Gamma = \{S_1, \ldots, S_b\}, \ D = \bigcup_{i=1}^{b} S_i \ .$$

We say that the pair (Π, Γ) is a *regular pair of domain sets* if the following conditions are satisfied:

1. each t-Domain $T_i \in \Pi$ is a discrete square of size $|T_i| = 2^{t_i} \times 2^{t_i}$ pixels, where typically $t_i = 2, 3, 4$;
2. each s-Domain $S_i \in \Gamma$ is a discrete square of size $|S_i| = 2^{s_i} \times 2^{s_i}$ pixels, where typically $s_i = 3, 4, 5$;
3. each s-Domain $S_i \in \Gamma$ is a sum of certain t-Domains:

$$S_i = \bigcup T_{j_k}$$

Let U be domain from $\Pi \cup \Gamma$. Then by χ_U we mean the characteristic function of U, i.e. for any pixel $p \in U$:

$$\chi_U(p) \doteq \begin{cases} 1 \text{ if } p \in U \ , \\ 0 \text{ otherwise} \ . \end{cases}$$

We are modeling the subimage of the image g restricted to a domain $U \in \Pi \cup \Gamma$ by $g_U = g\chi_U$. Note that

$$g = \sum_{T \in \Pi} g_T \quad \text{for any } g \in \mathcal{I} \tag{1}$$

The main idea in the construction of the fractal operator F for the given image f, is the elaboration of a *matching function* $\mu : \Pi \to \Gamma$ such that for any $T \in \Pi$, the subimage $f_{\mu(T)}$ is the most *similar* to the subimage f_T and the size of $\mu(T)$ is greater than size of T.

In simple, but practical case the similarity is defined by *affine mapping* acting separately between domains of subimages and between ranges of subimages, i.e. between gray scale intervals.

The degree of similarity between subimages is measured by a p-norm $\| \cdot \|_p$, $1 \leq p \leq \infty$.

Having Π and μ, the *fractal operator* is defined for any $g \in \mathcal{I}$ as follows:

$$F(g) \doteq \sum_{T \in \Pi} [c_T \cdot \mathbf{R}_T(g) + o_T \cdot \chi_T] \tag{2}$$

where

- c_T is the *contrast* between subimages $g_{\mu(T)}$ and g_T;
- o_T is the *offset* between subimages $g_{\mu(T)}$ and g_T;
- $\mathbf{R}_T \doteq \mathbf{P}_T \circ \mathbf{A}_T$ is the *reducing mapping*;
- $\mathbf{A}_T$ is the *averaging mapping* effectively reducing subimage $g_{\mu(T)}$ by averaging pixels from a subsquare in the bigger s-Domain $\mu(T)$ and putting the result into the corresponding pixel of the smaller t-Domain T. For the regular (Π, Γ) the mapping $\mathbf{A}_T = (1/4)^k \mathbf{A}'_T$ where:
  - T is of size $2^t \times 2^t$, $\mu(T)$ is of size $2^s \times 2^s$, and $k = s - t$;
  - at certain numbering of pixels in D the mapping $\mathbf{A}'_T$ can be written in matrix notation which has:
    * $N \doteq |D|$ rows and N columns;
    * exactly 4^k ones in rows corresponding to elements from T;
    * only zeros in rows corresponding to elements outside of T;
    * only single one in columns corresponding to elements from $\mu(T)$;
    * only zeros in columns corresponding to elements outside of $\mu(T)$;
- $\mathbf{P}_T$ is the *affine permutation mapping* on T, i.e. with single ones in rows and columns corresponding to elements from T.

3   Standard results on convergence of fractal operators

If we take the matrix form of the mappings $\mathbf{R}_T$ and of the characteristic functions χ_T $(T \in \Pi)$ then any fractal operator F can be written in a vector form:

$$\mathbf{Fg} = \mathbf{Lg} + \mathbf{o} \tag{3}$$

where $\mathbf{g}, \mathbf{o}$ are vectors and $\mathbf{L}$ is $N \times N$ matrix:

$$\mathbf{L} \doteq \sum_{T \in \Pi} c_T \cdot \mathbf{R}_T, \quad \mathbf{o} = \sum_{T \in \Pi} o_T \cdot \chi_T \tag{4}$$

Using matrix notation we can also represent the iterations of F by matrix powers of $\mathbf{L}$:

Lemma 1.

For any natural $k > 0$:

$$\mathbf{F}^{\circ k}(\mathbf{g}) = \mathbf{L}^k \mathbf{g} + \sum_{i=0}^{k-1} \mathbf{L}^{i-1} \mathbf{o} \tag{5}$$

Obviously F $(F^{\circ k})$ is Lipshitzian with factor $\|F\|_p$ $(\|F^{\circ k}\|_p)$ equal to the p-norm of the matrix $\mathbf{L}$ $(\mathbf{L}^k)$. Hence we can easily prove that

Lemma 2.
1. *F is contractive in p-norm if and only if $\|\mathbf{L}\|_p < 1$;*
2. *F is eventually contractive in p-norm, i.e. there exists k such that the operator $F^{\circ k}$ is contractive if and only if there exists k such that, $\|\mathbf{L}^k\|_p < 1$.*

From properties of mapping $\mathbf{A}_T$ it follows immediately that for the regular pair (Π, Γ), in the supremum norm $\|\cdot\|_\infty$, its operator norm $\|\mathbf{A}_T\| = 1$. Applying of the permutation $\mathbf{P}_T$ from the left side to $\mathbf{A}_T$ results in permutation of its rows in matrix representation. The interchange of rows does not change the supremum norm. Therefore the supremum norm of $\mathbf{R}_T$ is equal to one too. Hence the supremum norm of $\mathbf{L}$ can be easily derived:

Lemma 3.

If the pair of domain sets (Π, Γ) is regular then

$$\|\mathbf{L}\|_\infty = c^* \doteq \max_{T \in \Pi} |c_T| \tag{6}$$

From the lemmas 3 and 2, we get:

Corollary 1.

If $c^* < 1$ then fractal operator F is contractive in supremum norm.

For finite dimensional vector spaces all p-norms are equivalent. Therefore the condition $c^* < 1$ is also sufficient for the convergence of the fractal operator iterations with any p-norm:

Theorem 1.

Let the fractal operator F be given by the equation 2. If $c^* < 1$ then

1. *there exists a unique fixed point $\tilde{f}$ of the operator F;*
2. *for any initial image $g_0 \in \mathcal{I}$ and for any p-norm $(1 \le p \le \infty)$:*

$$\lim_{i \to \infty} \|F^{\circ i}(g_0) - \tilde{f}\|_p = 0 \ .$$

4   Influence graph for fractal operator

Assuming that (Π, Γ) is regular, we are going to generalize the above results to eventual contractive fractal operators.

We say that a t-Domain T_i is *influenced* by a t-Domain T_j if T_j is included in the most similar s-Domain $\mu(T_i)$, i.e.:

$$T_i \leftarrow T_j \quad \Longleftrightarrow \quad T_j \subset \mu(T_i) \tag{7}$$

The *influence graph* $G = (V, E)$ can be easily defined using the relation $\leftarrow$. Namely, the set of vertices V is identified with integer labels of t-Domains, i.e. $V \doteq \{1, \ldots, a\}$, while the set of directed edges E is specified as follows:

$$E \doteq \{(j, i)|\ T_i \leftarrow T_j\} \ .$$

Note, that the influence graph is completely determined by the matching function μ and therefore it can be built in fractal image encoding stage.

The following lemma explains the significance of the influence graph G :

Lemma 4.

1.
$$\text{if } \mathbf{R}_{T_i}\mathbf{R}_{T_j} \neq 0 \text{ then } (j,i) \in E \; ;$$

2.
$$L^2 = \sum_{i=1}^{a} \sum_{j:(j,i)\in E} c_{T_i} c_{T_j} \mathbf{R}_{T_i}\mathbf{R}_{T_j} \; ;$$

3.
$$L^k = \sum_{j_1=1}^{a} \sum_{j_2:(j_2,j_1)\in E} \cdots \sum_{j_k:(j_k,j_{k-1})\in E} c_{T_{j_1}} c_{T_{j_2}} \cdots c_{T_{j_k}} \mathbf{R}_{T_{j_1}} \mathbf{R}_{T_{j_2}} \cdots \mathbf{R}_{T_{j_k}} \; .$$

For the influence graph G, a weighted adjacency matrix $\mathbf{C} = [c_{ij}]$ is defined as follows:

$$c_{ij} \doteq \begin{cases} |c_{T_i}| & \text{if } T_i \leftarrow T_j \; , \\ 0 & \text{otherwise.} \end{cases} \tag{8}$$

By $\max(\mathbf{M})$ we denote a maximum element of the matrix $\mathbf{M}$. Let $\otimes$ stands for the composition of matrices in which the operation $+$ is replaced by max operation. $\mathbf{M}^{\bullet k}$ denotes $k-1$ compositions of type $\otimes$:

$$\mathbf{M}^{\bullet 2} \doteq \mathbf{M} \otimes \mathbf{M}, \; \mathbf{M}^{\bullet(k+1)} \doteq \mathbf{M} \otimes \mathbf{M}^{\bullet k} \; .$$

We can prove by the mathematical induction the following lemma which gives an upper bound for the supremum operator norm of powers of $\mathbf{L}$:

Lemma 5. *For any $k > 1$:* $\|\mathbf{L}^k\|_\infty \leq \max\left(\mathbf{C}^{\bullet k}\right)$.

Using the lemmas 5 and 2 we give a sufficient condition for the convergence which is based on eventual contractivity:

Theorem 2.

Let the fractal operator F be given by the equation 2. If there exist $k > 0$ such that $\max\left(\mathbf{C}^{\bullet k}\right) < 1$ then

1. *there exists a unique fixed point $\tilde{f}$ of the operator F;*
2. *for any initial image $g_0 \in \mathcal{I}$ and for any p-norm ($1 \leq p \leq \infty$):*

$$\lim_{i\to\infty} \|F^{\circ i}(g_0) - \tilde{f}\|_p = 0 \; .$$

References

1. S. Banach, "Sur les operations dans les ensembles abstraits et leur applications aux equations integrales", Fundamenta Mathematica, vol.3, pp. 133–181, 1922.
2. M. F. Barnsley and L. P. Hurd, *Fractal image compression*, AK Peters. Ltd, Wellesley, MA, 1993.
3. J. Dugundi and A. Granas, *Fixed point theory*, Polish Scientific Publishers, Warszawa, 1982.
4. Y. Fisher ed., *Fractal image compression, Theory and Application*, Springer Verlag, New York, 1995.
5. A. E. Jacquin, "Image coding based on a fractal theory of iterated contractive image transformations", *IEEE Trans. on Image Processing*, vol. 1, no. 1, pp. 18–30, 1992.

Pattern Recognition by Invariant Reference Points

Krystian Ignasiak, Władysław Skarbek *

Department of Electronics and Information Technology
Warsaw University of Technology

Abstract. New methodology for pattern recognition is presented which is based on design of invariant reference points. It is shown that the k-NN distance classifier is a special case of this methodology. New classifiers within this framework are also described.

1 Introduction

Pattern recognition is an area in artificial intelligence relating to synthesis and analysis of procedures for classifying objects on the basis of their physical measurements (for instance visual images). In case of images, the object recognition process consists of three main stages:

1. object localization in the image (*segmentation stage*);
2. feature vector extraction (*measurement stage*);
3. object classifying (*decision stage*).

For instance in zip code recognition: the zip area is detected, digits are separated, feature vector extracted for each digit, and finally the feature vectors are classified.

The recognition rate in real life systems never attains 100%. The basic reason is in the measurement process, which can give for two objects from different classes two equal or very close measurement vectors. A good recognition system reduces the probability of such events.

This work concerns a new classifying methodology, called here the *IRP method (Invariant Reference Points)*.

The IRP method as a methodology which is capable to generate a number of classifiers, gives no restrictions on segmentation methods and feature extraction methods. It is based on a construction for the given class a pair of type $(x_i, F_i) = $ (the reference point from the space of measurements, the operator in the space of measurements) such that x_i is an invariant point of F_i, i.e. $F_i(x_i) = x_i$.

The classification of the vector y is performed on the basis of all distance values $\|F_i(y) - y\|$ computed for all reference points x_i, which are designed for all classes for the given recognition problem.

It is interesting that the IRP method not only defines several known classifiers but it leads to new classification schemes with very simple feature extraction step.

---

* The work was sponsored by the grant of Institute of Radioelectronics

L. Polkowski and A. Skowron (Eds.): RSCTC'98, LNAI 1424, pp. 322–329, 1998.

2   Problem of classifying measurement vectors

In any classification problem, we deal with a distinguished object class Ω, which is subdivided into nonintersecting subclasses Ω_i, $i = 1, \ldots, n$:

$$\Omega = \Omega_1 \cup \ldots \cup \Omega_n, \quad \Omega_i \cap \Omega_j = \emptyset \ \ i, j = 1, \ldots, n, \quad i \neq j.$$

For each object $\omega \in \Omega$ we can measure its N–dimensional feature vector $x = X(\omega)$. The classifier using a particular vector $X(\omega) \in R^N$ elaborates a decision $\delta(x) = i$ of the membership for the object ω from the class Ω_i :

$$\delta : R^N \to \{1, \ldots, n\}.$$

If $\omega \in \Omega_i$, but $\delta(X(\omega)) \neq i$, then the decision is wrong. An optimal decision procedure δ gives the minimum for the recognition error.

Assuming that the probability of each class $P(\Omega_i) = \text{Prob}(\omega \in \Omega_i)$ and probability distribution density in each class $p_i(x) = \text{Prob}(X(\omega)) = x|\omega \in \Omega_i)$ are known. Then it can be proved that the following decision function:

$$\delta^{\text{opt}}(x) = \arg \max_{1 \leq i \leq n} p_i(x) P(\omega_i),$$

is optimal, i.e. δ^{opt} has the minimum of wrong decisions. It is equivalent to so called *maximum likelihood rule:* given the measurement $x = X(\omega)$, choose the class index i for which the likelihood defined by the formula:

$$\log \text{Prob}(x = X(\omega) \wedge \omega \in \Omega_i),$$

is maximal. The classifier δ^{opt} is rather of theoretical significance as the practical estimates of probability distributions are not known for large N and relatively sparse training set with cardinality L. The only exception is a *flat* Gaussian distribution when the data is concentrated around K–dimensional subspace with $K \ll N$. It means that a search for new effective classifiers for a particular application is still important.

Certain classification problems require more than one feature vector for the given object (e.g.: face or fingerprint recognition). In such cases we can say that a generalized classifier operating on k measurement vectors $x_1 = X(\omega_1), \ldots, x_k = X(\omega_k)$ for certain objects $\omega_1, \ldots, \omega_k$ which belong to the same class Ω_j elaborates a decision $\nu(x_1, \ldots, x_k) = i$ about a membership of this sequence to the class Ω_i :

$$\nu : \underbrace{R^N \times \ldots \times R^N}_{k} \to \{0, 1, \ldots, n\},$$

where the symbol 0 corresponds to *no decision* category. Using a *majority* rule for results of a single vector classifier δ, we can build a generalized classifier for $k > 1$ feature vectors:

$$\nu(x_1, \ldots, x_k) = majority(\delta(x_1), \ldots, \delta(x_k)).$$

The *majority* function can be defined in many ways. We consider here two definitions:

$$majority_1(i_1, \ldots, i_k) = \arg \max_{1 \leq j \leq n} |\{a : i_a = j\}| \tag{1}$$

At ties, the ambiguity is dissolved by random choice of index j.

$$majority_2(i_1, \ldots, i_k) = j \text{ if } |\{a : i_a = j\}| > \frac{k}{2} \qquad (2)$$

If such j is missing – the result is zero.

3   Invariant reference points

Let us assume that the learning set $X^{(i)} \subset R^N$, $i = 1, \ldots, n$, consists of measurement vectors obtained for objects $\omega \in \Omega_i$.

The IRP method is based on a construction of an *IRP collection* $\mathcal{Z}_i$, $i = 1, \ldots, n$ which is built using information included in the learning set X_i. $\mathcal{Z}_i$ consists of certain number k_i of pairs $(x_j^{(i)}, F_j^{(i)})$, where $x_j^{(i)}$ is the j-th reference point in the collection $\mathcal{Z}_i$, while $F_j^{(i)}$ is the operator in measurement space ($F_j^{(i)} : R^N \to R^N$) (we call it also as *eigenoperator*) such that $x_j^{(i)}$ is its invariant point:

$$F_j^{(i)}(x_j^{(i)}) = x_j^{(i)}, \quad j = 1, \ldots, k_i, \quad i = 1, \ldots, n,$$

$$\mathcal{Z}_i = \left(k_i, (x_1^{(i)}, F_1^{(i)}), \ldots, (x_{k_i}^{(i)}, F_{k_i}^{(i)}) \right).$$

Let $\mathcal{R}_i = \{x_1^{(i)}, \ldots, x_{k_i}^{(i)}\}$ be the set of reference points for the class Ω_i and $\mathcal{F}_i = \{F_1^{(i)}, \ldots, F_{k_i}^{(i)}\}$ be the set of eigenoperators for the class Ω_i.

Roughly, the IRP collection $\mathcal{Z}_i$ approximates the learning set X_i with reference points. Eigenoperators in proximity of their invariant points have a low dynamics. This property is used in the classifier. We define here three general [8,9] schemes for constructing reference sets $\mathcal{R}_i$:

1. **Centroid method:** we define only one reference point in the class ($k_i = 1$) which is defined as the centroid of the learning set X_i :

$$x_1^{(i)} \doteq \frac{1}{|X_i|} \sum_{y \in X_i} y, \qquad \mathcal{R}_i = \{x_1^{(i)}\};$$

2. **Clustering method:** the algorithm finds a set of reference points $\mathcal{R}_i = \{x_1^{(i)}, \ldots, x_{k_i}^{(i)}\}$ optimizing a certain cost function, e.g.:

$$\text{cost}(\mathcal{R}_i) = \alpha \text{MSE}(\mathcal{R}_i, X_i) + \beta k_i \qquad (3)$$

where α, β are weights,

$$\text{MSE}(\mathcal{R}) \doteq \frac{1}{|X_i|} \sum_{y \in X_i} (y - x_{j(y)}^{(i)})^2,$$

$$j(y) = \arg \min_{1 \le j \le k_i} \|y - x_j^{(i)}\|.$$

Let us notice that the cost function 3 makes possible adequate choice of the number of reference points;

3. **Learning set technique:** here the set of reference points is equal to the learning set, i.e. $\mathcal{R}_i = X^{(i)}$. This technique has a special meaning when the class is represented by several measurement vectors (for instance in face recognition we usually have only few face poses of the given person).

4 Design of eigenoperators

The design of the eigenoperator $F_j^{(i)}$ such that $F_j^{(i)} = x_j^{(i)}$ is generally hard problem if the only criterion is the recognition error for the resulting classifier. We present here three universal design techniques:

1. **Constant operator technique:** for each reference point $x_j^{(i)}$, the operator $F_j^{(i)}$ is defined as follows:

$$F_j^{(i)}(x) \doteq x_j^{(i)} \text{ for each } x \in R^N.$$

It will be shown how this technique combined with a clustering method for reference point design, leads to k–NN method.

2. **Technique of projection onto principal subspace:** let $X_j^{(i)} \doteq \{y \in X^{(i)} : i(y) = j\}$ be the set of learning vectors which are closer to $x_j^{(i)}$ than to any other reference point in $\mathcal{R}_i$. If the reference points are found by the centroid algorithm then $x_j^{(i)}$ is the centroid of the set $X_j^{(i)}$.

Assuming an adequate cardinality of this set we can implement the principal component analysis (PCA – [2,6]) with K components. PCA gives K principal vectors which are placed into columns of the matrix $W_j^{(i)}$. Then the eigenoperator can be defined by the following formula:

$$F_j^{(i)}(x) \doteq x_j^{(i)} + W_j^{(i)} W_j^{(i)^T} (x - x_j^{(i)}) \text{ for any } x \in R^N.$$

It is obvious that $F_j^{(i)}$ is invariant in $x_j^{(i)}$, i.e. $F_j^{(i)}(x_j^{(i)}) = x_j^{(i)}$.

We should emphasize that the above design has a practical sense for large learning sets.

3. **Fractal operator technique:** the algorithm performs an extensive search of local mappings of fragments of the vector $x = x_j^{(i)}$ of cardinality pk, $p > 1$ (so called source domains) to fragments of cardinality k (so called target domains) [4,7].

Let us assume that k divides N, $(N = kl)$, and let $\Pi = \{T_1, \ldots, T_l\}$ be the set of target domains, i.e. a partition of index set $\{1, \ldots, N\}$ into l subsets of cardinality k. Let $\gamma = \{S_1, \ldots, S_l\}$ be the set of the source domains which are best matched to corresponding target domains, $|S_i| = pk$. Matching here denotes an optimal local affine mapping $L_a^{(x)}$ which reduces the fragment $x|_{S_a}$ to $x|_{T_a}$, where $\cdot|_{T_a}$ denotes a restriction to the domain T_a. Then the fractal operator, i.e. the eigenoperator for the reference point $x = x_j^{(i)}$ is defined as follows:

$$F_j^{(i)}(y) = \sum_{a=1}^{l} L_a^{(x)}(y).$$

5   Class selection techniques

In the IRP method, for the classification of a measurement vector x we consider a sequence of distance values $D_i(x) = (\|x - F_1^{(i)}(x)\|, \ldots, \|x - F_{k_i}^{(i)}(x)\|)$, $i = 1, \ldots, n$. The classifier uses a selection technique S_a which for the given x on the basis of $D_i(x)$ returns a class id, x probably belongs to.

Let $D(x) = (D_1(x), \ldots, D_{k_i}(x))$. There exists many possibilities to aggregate information about distances, and next the class selection. Few of them are defined below:

1. **Minimum distortion technique:**

$$S_1 \doteq \arg \min_{1 \leq i \leq n} \min(D_i(x)).$$

The class for which the minimum distortion $\|x - F_j^{(i)}(x)\|$ is achieved;

2. **Minimum average distortion technique:**

$$S_2 \doteq \arg \min_{1 \leq i \leq n} \text{avg}(D_i(x)),$$

where $\text{avg}(D_i)$ is the average value of the sequence D_i. Here the average distortion introduced by eigenoperators in the given class decides about the choice of the given class.

3. **Basis functions technique:**

$$S_3 \doteq \left\lfloor 0.5 + \sum_{i=1}^{n} \sum_{j=1}^{k_i} c_{ij} \phi_{ij}(x) \right\rfloor,$$

where

$$\phi_{ij}(x) \doteq e^{-\|x - F_j^{(i)}(x)\|}, \quad x \in R^N$$

is a basis function concentrated in the point $x_j^{(i)}$ [5]. The coefficients c_{ij} can be found using recursive least square method for the learning data sequence $X = \bigcup_{i=1}^{n} X^{(i)}$.

4. **k least distortions technique k–NZ:** Let $I_k(x) \doteq (i_1, \ldots, i_k)$ be the sequence of class ids for which we have found k least distortions from the sequence $\|x - F_j^{(i)}(x)\|$. Then:

$$S_4 \doteq \arg \max_{1 \leq j \leq n} |\{a : i_a = j\}|,$$

where i_a is a-th coordinate of the sequence $I_k(x)$. In this approach we choose the class id which occurs in k least distortions most frequently.

6   The IRP method definition

Let $X^{(i)} \subset R^N$ be the set of learning sequence for the class Ω_i, $i = 1, \ldots, n$. Suppose that for these learning sets we can build IRP ensembles $Z_i = (k_i; (x_1^{(i)}, F_1^{(i)}), \ldots, (x_{k_i}^{(i)}, F_{k_i}^{(i)}))$.

Let $x \in R^N$ be a testing measurement vector. Then the classification procedure in the IRP method is of the form:

1. Compute sequences $D_i(x) = (\|x - F_i^{(j)}(x)\|),\ 1 \le j \le k_i,\ i = 1, \ldots, n$;
2. Let $D(x) = (D_1(x), \ldots, D_n(x))$;
3. Assign x to the class with id $\delta(x) \doteq S_a(D(x))$.

Let us notice that using one of majority rules (1,2) we can generalize the IRP method to sequences of measurement vectors.

The above algorithm is of generic type. A specific algorithm is obtained by specifying:

1. a method for constructing the IRP ensemble $\mathcal{Z}_i,\ i = 1, \ldots, n$;
2. a norm $\|x - y\|$;
3. a selection technique S_a;
4. majority rule (for sequence recognition only).

In the figure 1 we give a graphical intuition for an IRP classifier. Concentric circles stand for N-dimensional spheres. Left family of concentric spheres illustrate the eigenoperator of the first class. The larger sphere is mapped onto a smaller sphere by the eigenoperator. Now if x belongs to the smaller sphere with the center in the reference point of the first class and at the same time it belong to the larger sphere with the center in the reference point of the second class than as it is shown on the picture the distance $\|x - F_1^{(1)}(x)\|$ is smaller than $\|x - F_1^{(2)}(x)\|$. Therefore x is assigned to the first class.

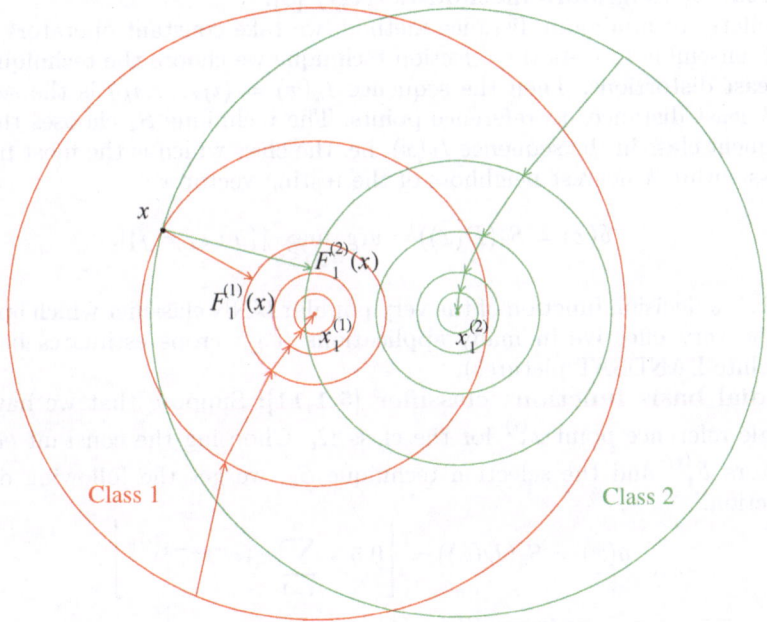

Fig. 1. Illustration of the IRP method.

7 Specifying some known classifiers by Invariant Reference Points method

In this section, we show that several known classifiers which are based on distance from a pattern can be specified as special cases of the IRP method.

1. **The minimum distance from reference points method [8,9,3]:** Let us assume that in the i-th class there are reference points $\mathcal{R}_i = \{x_1^{(i)}, \ldots, x_{k_i}^{(i)}\}$. Building an IRP ensemble we choose constant operator technique, i.e. we choose $F_j^{(i)}(x) = x_j^{(i)}$ for each $x \in R^N$. Let us notice that the distortion introduced by the operator $F_j^{(i)}$ equals to the distance of x to the reference point $x_j^{(i)}$:

$$\|x - F_j^{(i)}(x)\| = \|x - x_j^{(i)}\| \qquad (4)$$

Therefore by choosing the selection technique S_1 which is based on the minimum of the distortions we get a classifier which chooses the class to which the closest reference point belongs:

$$\delta(x) \doteq S_1(D(x)) = \arg \min_{1 \leq i \leq n} \min_{1 \leq j \leq k_i} \|x - x_j^{(i)}\|;$$

2. **k nearest neighbors method (k–NN) [3]:**
   Similarly to minimum distance method, we take constant operators for the IRP ensemble too. As the selection technique we choose the technique S_4 of k least distortions. Then the sequence $I_k(x) = (i_1, \ldots, i_k)$ is the sequence of k least distances to reference points. The technique S_4 chooses the most frequent class in the sequence $I_k(x)$, i.e. the class which is the most frequent class within k nearest neighbors of the testing vector x :

$$\delta(x) \doteq S_4(D(x)) = \arg \max_{1 \leq j \leq n} |\{a : i_a = j\}|.$$

This is a decision function of the very popular k-NN classifier which appeared to be very effective in many applications (e.g.: crops estimates based on satellite LANDSAT pictures);

3. **Radial basis functions classifier [5,1,11]:** Suppose that we have only single reference point $x_1^{(i)}$ for the class Ω_i. Choosing the constant eigenoperators $F_1^{(i)}$ and the selection technique S_3, we get the following decision function:

$$\delta(x) \doteq S_3(D(x)) = \left\lfloor 0.5 + \sum_{i=1}^{n} c_{i1} e^{-\|x - x_1^{(i)}\|} \right\rfloor.$$

Denoting $w_i = c_{i1}$, $x^{(i)} = x_1^{(i)}$ let us consider the function

$$g(x) \doteq \sum_{i=1}^{n} w_i e^{-\|x - x^{(i)}\|}.$$

The function $g(x)$ is a linear combination of Gaussian basis functions. There are well known techniques searching parameters $x^{(i)}$ and w_i such that $g(x)$ estimates the given function f which is known only by specifying of learning pairs (y, i), i.e. measurement vectors y for learning objects and class ids i, $(i = 1, \ldots, n)$, the object belongs to. The obtained classifier is the known radial basis functions neural network.

8 Conclusion

New methodology for pattern recognition was elaborated. It is based on design of invariant reference points. The methodology is capable to define new classifiers such as fractal operator classifying system. It is shown that many prominent classical recognition schemes, for instance the k-NN distance classifier, are special cases of the IRP method.

References

1. C. Bishop (1995) *Neural Networks for Pattern Recognition*, Clarendon Press, Oxford.
2. K.I. Diamantaras, S.Y. Kung (1996) *Principal Component Neural Networks*, John Wiley & Sons, New York.
3. R.O. Duda, P.E. Hart (1973) *Pattern Classification and Scene Analysis*, Wiley, New York.
4. Y. Fisher, ed. (1995) *Fractal Image Compression – Theory and Application*, Springer Verlag.
5. S. Haykin (1994) *Neural networks – A Comprehensive Foundation*, Maxwell Macmillan International.
6. H. Hotelling (1933) *Analysis of a complex of statistical variables into principal components*, Journal of Educational Psychology 24, 417–441.
7. A. Jacquin (1992) Image coding based on a fractal theory of iterated contractive image transformations, *IEEE Transactions on Image Processing*, 1, 18–30.
8. T. Kohonen (1995) *Self-Organizing Maps*, Springer, Berlin.
9. Y. Linde, A. Buzo, R.M. Gray (1980) An algorithm for vector quantizer design, *IEEE Trans. Comm.*, COM-**28** 1980 28–45.
10. E. Oja (1983) *Subspace methods of pattern recognition*, Research Studies Press, England.
11. B.D. Ripley (1996) *Pattern Recognition and Neural Networks*, Cambridge University Press, Cambridge.

An Analysis of Context Selection in Embedded Wavelet Coders

Juan Miguel del Rosario[+] and Czesław Jędrzejek[+,*]

[+,*] Institute of Telecommunications and Information Technologies
ul. Rakoniewicka 20, 60-111, Poznań, Poland;
[*] Institute of Telecommunication, ATR Bydgoszcz, Poland
email: jedrzeje@itti.com.pl

Abstract

In image compression using wavelet transforms the final stage of process-ing often involves entropy encoding, out of which arithmetic coding is most essential. A significant contributor to the effectiveness of the arithmetic encod-ing is the selection of coding contexts. We show for various context selection schemes, that the interbit correlations in the multi-symbol alphabet is a pri-mary source of compression gain in the entropy coding of the image. Further, we analyze the use of more conventional context selection schemes and show that full image histograms contain information not yet available to the decoder in embedded algorithms. The use of predictors in the embedded algorithm can be quite ineffective.

1 Introduction

Methods of image compression involving the use of orthonormal wavelet trans-forms have proven to be effective [1–3]. Shapiro [4], noticed that, since wavelet transforms retain pixel spatial positions, an insignificant coefficient in the lower frequency subbands likely entails the existence of insignificant coefficients in the same spatial locations in the higher frequency subbands. He introduced a method of encoding entire trees of zeroes with a single symbol, called embedded zerotree wavelet (EZW) coding. The essence of the EZW method is scalar quantisation based on notion of significance:

Given n, if $|c_{ij}| \geq 2^n$, a coefficient c_{ij} is *significant* with respect to a given threshold n; otherwise it is called *insignificant*. The similar notion of significance holds for sets (at least one of its elements has to be significant).

The basic structure for wavelet transformed coding of images is a three-directional pyramidal subband structure [2,4] coming from the decomposition. Pixels of the lowest LL (low-low subbband) are roots of trees. The parent-child relationship and dependency is central to the idea of zerotrees. In an embedded coder trees are processed piecewise by scanning coefficients in the transformed image through subbands in an assumed order. In the Shapiro algorithm each coefficient was examined independently of any others (no context model). Fol-lowing Shapiro, Said and Pearlman (S&P) proposed an algorithm using "spatial-orientation trees" which was an extension of the idea of zerotrees [5], but with some significant implementational differences.

In all of the schemes the final stage of processing involves entropy (arithmetic) coding [6], based on a selection of coding contexts. The theoretical underpinning

L. Polkowski and A. Skowron (Eds.): RSCTC'98, LNAI 1424, pp. 330--337, 1998.
© Springer-Verlag Berlin Heidelberg 1998

of gain due to a context use is mutual information I(X,Y). This quantity is non-negative and equal

$$I(X,Y) = H(X) - H(X|Y),$$

where H(X) is entropy of distribution X and $H(X|Y)$ the conditional entropy of X due to the knowledge of Y, which can be treated as a context.

With S&P coder (one of the best existing) as a basis, we examine the subject of context selection for embedded wavelet image codecs. In section two we discuss the use of a multi-symbol alphabet versus a binary alphabet within the scope of the Said and Pearlman scheme. In section three we discuss the use of popular context selection strategies in embedded coders and present sample results from extensive experimental data.

2   Analysis of Said and Pearlman Context Selection

The first two stages of the S&P algorithm, the significants selection and the descendants selection, correspond to what Shapiro called the dominant pass in his original paper. In the significants selection section, each coefficient is determined to be significant or insignificant relative to the current threshold level; in the descendants selection section, the coefficients are determined to have or not to have significant descendants, thus informing the decoder as to whether or not to proceed down the tree or whether the rest of the tree is a zerotree relative to the current threshold level. The resulting bitstream from these two sections are entropy coded in the third stage in sets of a quartet of coefficients (in a 2x2 block) rather than on a coefficient by coefficient basis. A state is stored for each of the quartets as shown in Figure 1, where A_x, B_x, C_x i D_x represent state information for the $(0,0)$, $(0,1)$, $(1,0)$ and $(1,1)$ coefficients of the quartet respectively, and x with subscripts denotes $\{s, d\}$ *significant* (s) and *descendant significant* (d) states for each of the quartet coefficients. For instance, if A_s and B_d have the value 1, then this indicates that the coefficient in the $(0,0)$ position of the quartet has already been found to be significant, furthermore, that the coefficient in the $(0,1)$ position has been found to have a significant descendant.

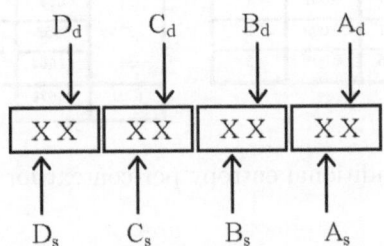

Figure 1: State structure for the coefficient quartet.

2.1   Said and Pearlman Multi-Symbol Alphabet

Each quartet of coefficients is classified into a set of equivalence classes and then assigned numbers which act as the index into context table used by the arithmetic coder. Thus, contexts will have a variable number of symbols (ie, bit length) based on the degrees of freedom of the equivalence class. There are 34 such equivalence classes for the significance selection section, and 34 equivalence classes plus 5 special case classes for the descendants section. As an example of the context formation, the context indexed as number 2 has all states with 2 coefficients still not found to be significant while all the others have been found significant. This means that there are 2 degrees of freedom implying a 4 symbol alphabet for this context. In Table 1, we show the entropy breakdown produced from the bitstream of the significance selection section of the S&P. The average bit compression for the conditional entropy is 0.889.

Model (S = n)	Total	H(X|S)	Comp	p(X=n)	bits per symbol	Model (S = n)	Total	H(X|S)	Comp	p(X=n)	bits per symbol
0	2463	1.000	1.000	0.054	1	17	94	3.888	0.972	0.002	4
1	2586	1.983	0.991	0.057	2	18	525	1.000	1.000	0.011	1
2	3112	2.839	0.946	0.068	3	19	450	1.991	0.996	0.010	2
3	**11440**	**3.141**	**0.785**	**0.250**	**4**	20	150	2.961	0.987	0.003	3
4	0			0.000	NA	21	664	1.000	1.000	0.015	1
5	466	1.000	1.000	0.010	1	22	300	1.982	0.991	0.007	2
6	200	1.997	0.998	0.004	2	23	809	0.996	0.996	0.018	1
7	48	2.765	0.922	0.001	3	24	378	1.969	0.984	0.008	2
8	0			0.000	NA	25	337	2.915	0.972	0.007	3
9	234	1.000	1.000	0.005	1	26	156	3.852	0.963	0.003	4
10	62	1.987	0.993	0.001	2	27	775	1.945	0.972	0.017	2
11	0			0.000	NA	28	454	2.921	0.974	0.010	3
12	100	0.977	0.977	0.002	1	29	2133	1.938	0.969	0.047	2
13	0			0.000	NA	30	507	2.831	0.944	0.011	3
14	331	0.997	0.997	0.007	1	31	665	3.605	0.901	0.015	4
15	413	1.981	0.991	0.009	2	32	3952	2.760	0.920	0.087	3
16	211	2.929	0.976	0.005	3	**33**	**11663**	**3.358**	**0.839**	**0.255**	**4**
						total	45678				

Table 1: mth order conditional entropy per context for Lenna at 1.0 bpp rate.

Two rows (most high frequency and contributing the most towards the compression ratios contexts) are highlighted in the table. These contexts are: context 3 - all nodes with significant descendants found, but no significant members in the quartet; and context 33 - no found significants in the quartet, neither descendants nor quartet members. In the second case there are 4 degrees of freedom implying 16 symbols for this context.

2.2   Binary Alphabet Performance

The multi-symbol alphabet employs the mth order entropy of the symbol. A binary alphabet would employ a zeroth order entropy and is more efficiently handled by arithmetic coders. In Table 2, we show the results of recomputing the entropy of the bitstream into the zeroth order.

Model (S = n)	Input symbol 0	1	Total	H(X|S)	p(X=n)	Model (S = n)	Input symbol 0	1	Total	H(X|S)	p(X=n)
0	1222	1241	2463	1.000	0.017	17	218	158	376	0.982	0.003
1	2841	2331	5172	0.993	0.036	18	261	264	525	1.000	0.004
2	5925	3411	9336	0.947	0.065	19	463	437	900	0.999	0.006
3	30745	15015	45760	0.913	0.321	20	251	199	450	0.990	0.003
4	0	0	0	0.000	0.000	21	338	326	664	1.000	0.005
5	235	231	466	1.000	0.003	22	331	269	600	0.992	0.004
6	204	196	400	1.000	0.003	23	373	436	809	0.996	0.006
7	75	69	144	0.999	0.001	24	422	334	756	0.990	0.005
8	0	0	0	0.000	0.000	25	590	421	1011	0.980	0.007
9	117	117	234	1.000	0.002	26	370	254	624	0.975	0.004
10	58	66	124	0.997	0.001	27	925	625	1550	0.973	0.011
11	0	0	0	0.000	0.000	28	787	575	1362	0.982	0.010
12	41	59	100	0.977	0.001	29	2516	1750	4266	0.977	0.030
13	0	0	0	0.000	0.000	30	967	554	1521	0.946	0.011
14	176	155	331	0.997	0.002	31	1785	875	2660	0.914	0.019
15	449	377	826	0.995	0.006	32	7870	3986	11856	0.921	0.083
16	363	270	633	0.984	0.004	33	33650	13002	46652	0.854	0.327

Table 2: Zeroth entropy from contexts using binary alphabet at 1.0 bpp.

For the descendants selection section in the tree most of the gain originates from the use of "escape" routines which involve special contexts; these encode the predominant zero condition in the bitstream. For the lack of space we do not give results. The interbit correlations are higher in descendant selection than for significance selection - mth order entropy is much lower than the binary entropy and significantly lower than for the significance selection portion of the algorithm.

3   Conventional context selection schemes

Conventional context selection schemes typically involve the use of linear predictors which are formed by employing a weighted linear combination of these variables - neighboring coefficients (and/or parent coefficient). In contrast, the Said and Pearlman context does not use any additional information external to the quartet. In particular, it does not employ the parent-child correlation. In order to study context formation, we use an extended version of the technique used by Buccigrossi and Simoncelli [7]. They showed the correlation between magnitudes of predictors and the coefficients to be predicted by constructing a joint histogram of the coefficients. In addition, they compute the mutual information as a measure of gain in coding obtained from knowledge of the conditional information. In this paper, we compute the mutual information from the full image histograms. Then, the equivalent values from data that is obtained from an embedded implementation of the image oder. As a simplified case we have verified that the histogram results (not shown here) for the full Lenna image, and a single predictor (the parent coefficient) reveal a strong magnitude correlation.

3.1   Full Image Joint Statistics

The context variables used in this study are the parent, the upper neighbor, the left neighbor and the cousin (ie, in the same spatial position as that of the coefficient to be predicted but located on a different subband) as shown in Figure 2.

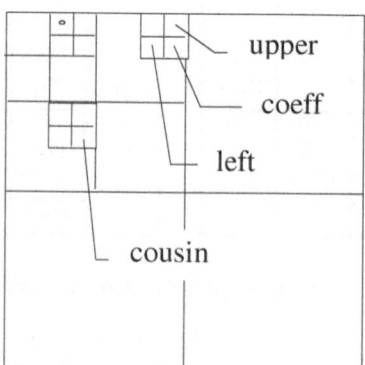

Figure 2: Neighboring coefficients to be used in linear predictor construction.

In Table 3, on the left we show the mutual information computed from such histograms for each of the four predictor coefficients for each of the decomposition directions. In order to use the full predictor, we compute the optimal weights to form the weighted linear combination of the predictor variables (the neighboring node values). By minimizing the MSE of the predictor

$$w = E(P \odot P^T)^{-1} E(C \odot P),$$

where w is the vector of parameter weights, P is the set of parameter magnitudes, C is the coefficient magnitude being predicted, and $E(\cdot)$ is the expected value.

Computing the mutual information of the full predictor, we get the values on the right side of Table 3. We see that the compression percentages are very competitive with those of Said and Pearlman's. However, it is important to note that the gain that can be expected from the use of the predictor variable, as indicated by the individual mutual information results, is not additive towards the construction of the full predictor as can be seen when comparing the %comp values of both tables in Table 3. This, we believe, is caused by some correlation between the predictor variables as well, and that there exists an overlap in the conditional knowledge provided by each predictor variable.

		H(C)	H(C|P)	I(C;P)	%comp
Parent	horizontal	2.354	2.096	0.258	10.96
	diagonal	1.997	1.823	0.173	8.663
	vertical	2.133	1.957	0.176	8.251
Left	horizontal	2.355	2.129	0.225	9.554
	diagonal	1.997	1.847	0.15	7.511
	vertical	2.129	1.968	0.162	7.609
Up	horizontal	2.352	2.078	0.274	11.65
	diagonal	1.997	1.841	0.156	7.812
	vertical	2.133	1.973	0.16	7.501
Cousin	horizontal	2.355	2.218	0.136	5.775
	diagonal	1.997	1.861	0.136	6.81
	vertical	2.133	1.987	0.146	6.845

	H(C)	H(C|P)	I(C;P)	%comp
Horizontal	2.355	1.989	0.366	15.541
Diagonal	1.997	1.768	0.229	11.467
Vertical	2.133	1.881	0.252	11.814

Table 3: Mutual information of coefficient histograms for Lenna.

3.2 Conventional Context in Embedded Algorithms

The problem with the mutual information values in the table above is that it assumes full knowledge of the image statistics. However, as previously mentioned, this is not true of embedded coders in the style of Said and Pearlman, and significant side information would be required.

We apply the procedure above to the statistics available to the decoder as it actually executes in the Said and Pearlman code in order to compare the possible gains from using a different type of context. The data presented in Table 4 applies only to the significance selection portion of the codec. The predictor index indicates which of the predictors is available for use at the time the current coefficient is being evaluated (ie, its value predicted), the zero case is not included in the table. The predictor weights were computed for the data per pass of the algorithm. Table 4 contains entropies of the horizontal direction coefficients (results for diagonal and the vertical direction are similar). The overall entropies are in general worst than those of the simple Said and Pearlman contexts. More importantly, the results provide much less compression gain than is expected by inspection of the results of a computation of the mutual entropy from the knowledge of the full image histogram.

pass#	1	2	3	4	5	6	7	8	9	10	11	12	13	14	15
					Predictor index		$P = cousin*8 + up*4 + left*2 + parent$								
1	0.000	0.000	0.000	0.000	0.000	0.000	0.000	0.000	0.000	0.000	0.000	0.000	0.000	0.000	1.000
2	0.918	0.000	0.811	0.000	0.000	0.000	0.000	0.000	0.000	0.000	0.000	0.000	0.000	0.000	1.000
3	0.944	0.000	0.918	0.000	0.000	0.000	1.000	0.000	0.000	0.000	0.000	0.000	0.000	0.000	0.000
4	0.947	0.000	0.991	0.954	0.963	0.000	0.786	0.000	0.000	0.000	0.000	0.000	0.000	0.000	0.000
5	0.882	0.811	0.891	0.000	0.910	0.406	0.976	0.650	0.918	0.000	0.918	0.000	0.000	0.000	0.000
6	0.885	0.840	0.912	0.938	0.908	0.971	0.951	0.974	0.541	0.918	0.985	0.000	0.845	0.811	0.785
7	0.922	0.932	0.931	0.934	0.977	0.992	0.977	0.870	0.841	1.000	0.952	0.985	0.966	0.918	0.970
8	0.923	0.891	0.932	0.993	0.974	0.948	0.976	0.800	0.936	0.863	0.985	0.863	0.990	0.902	0.989
9	0.929	0.921	0.947	0.940	0.973	0.957	0.974	0.873	0.929	0.974	0.938	0.911	0.974	0.920	0.984
10	0.893	0.869	0.944	0.912	0.964	0.956	0.982	0.936	0.954	0.968	0.989	0.910	0.983	0.923	0.988
11	0.951	0.986	0.996	0.888	0.998	0.998	0.998	0.998	0.995	0.967	0.985	0.911	0.992	0.965	0.999
Ave.	0.913	0.890	0.955	0.936	0.975	0.968	0.985	0.920	0.957	0.968	0.980	0.946	0.986	0.973	0.993

Table 4: Conditional entropy of the coefficient given the predictor (horizontal direction).

Several factors contribute to the worse than expected results. First, the context is binary; we have seen that for both the significance selection and descendants determinations portions of the embedded codec, the interbit correlations are considerable and in this case is not capitalized upon. Second, there is significant decrease in performance from not having all the predictor variables available at the time of evaluation of each coefficient. Third, there is a problem in computing the weights on the per pass basis which would also affect any other less detailed version, and this is that the inverse of the autocorrelation function is not always possible to obtain. We have to average over possibly a large number of passes worth of bits. This averaging will result in errors in prediction which will increase the further down the number of passes the averaging has to take place.

4   Summary

We have evaluated various context selection schemes. We have shown that the interbit correlations are the primary source of compression gain in the entropy coding of the image. Further, we have studied the use of more conventional context selection schemes and shown that, in this case, the gain expected from a study of the full image histograms can be deceiving and that one has to analyze the predictor in the light of the information currently available to the decoder; under these constraints using predictors in the embedded algorithm can be quite ineffective. We have found that all patterns of results applied not only to the Lenna image case, but held over all the images we have tested; among these were the commonly used images of: baboon, boats, goldhill, peppers, and cheyenne, and over a broad range of image types and at a wide range of bit rates.

Recently, Marpe and Cycon [8] improved somewhat upon the Said and Pearlman code. They use run-length coding prior/instead of arithmetic coding with no bitrate gain but 30% complexity reduction. This is probably where most of the (small) gain comes from in the non-context/predicted version. They, however, got up to 0.2 dB gain using Pearlman type maps but then employed the non-embedded fashion, which allows them to use more information for context formation.

C. Jędrzejek acknowledges the partial support by the Polish Scientific Committee (KBN) grant 8T11E035 10, and ECU grant CRIT2.

References

1. J. W. Woods, S. D. O'Neil (1986) Subband coding of images, *IEEE Trans. Acoust. Speech Signal Proc.*, ASSP-34, no. 5, pp. 1278-1288, October 1986.
2. E. H. Adelson, E. P. Simoncelli, R. Hingorani (1987) Orthogonal pyramid transforms of image coding *Proceedings of SPIE*, vol. 845, pp. 50-58, Cambridge, MA, October 1987.
3. M. Vetterli (1984) Multidimensional subband coding: some theory and algorithms, *Signal Processing*, vol. 6, no. 2, pp. 97-112.
4. J. M. Shapiro (1993) Embedded image coding using zerotrees of wavelets coefficients, *IEEE Trans. Signal Proc.*, vol. 41, pp. 3445-3462, December 1993.
5. A. Said, W. A. Pearlman (1996) A new fast and efficient image codec based on set partitioning in hierarchical trees, *IEEE Trans. Circuits and Systems for Video Technology*, June 1996.
6. I. H. Whitten, R. M. Neal, J. G. Cleary (1987) Arithmetic Coding for Data Compression, *Communications of the ACM*, vol. 30, no. 6, pp. 520-540, June 1987.
7. R. W. Buccigrossi, E. P. Simoncelli (1997) Image Compression via Joint Statistical Characterization in the Wavelet Domain, *GRASP Laboratory Technical Report # 414*, University of Pennsylvania, May 1997.
8. D. Marpe, H. L. Cycon (1997) Efficient Pre-Coding techniques for Wavelet-Based Image Compression, *Proceedings of the 1997 Picture Coding Symposium*, pp. 45-50.

Equivalent Characterization of a Class of Conditional Probabilistic Independencies

S.K.M. Wong and C.J. Butz

University of Regina, Regina, SK, Canada, S4S 0A2

Abstract. Markov networks utilize nonembedded probabilistic conditional independencies in order to provide an economical representation of a joint distribution in uncertainty management. In this paper we study several properties of nonembedded conditional independencies and show that they are in fact equivalent. The results presented here not only show the useful characteristics of an important subclass of probabilistic conditional independencies, but further demonstrate the relationship between relational theory and probabilistic reasoning.

1 Introduction

Belief networks [5, 6] utilize probabilistic conditional independencies to provide an economical representation of a joint distribution for managing uncertainty. A Bayesian network is a directed acyclic graph, explicitly specifying the *embedded* and *nonembedded* independency information, coupled with a corresponding set of conditional probability distributions. As performing inference in such a network may easily become intractable [2, 3], it is useful to transform a Bayesian network into a Markov network [5] albeit sacrificing the embedded independencies. That is, a Markov network is defined with respect to a hypergraph which explicitly specifies nonembedded conditional independencies only. This subclass of nonembedded independencies, called *generalized multivalued dependencies* (GMVDs) [13], is defined in a similar fashion to multivalued dependencies in relational database theory.

In this paper we study several properties of GMVDs and show that they are in fact equivalent. The results presented here not only show the useful characteristics of this important subclass of conditional independencies, but further demonstrate the intriguing relationship between relational database theory and probabilistic reasoning. It is perhaps worth mentioning that probabilistic conditional independencies do *not* have a complete axiomatization [8, 9] contrary to Pearl's conjecture [6]. However, it has been shown that GMVDs *do* in fact have a finite complete axiomatization [12].

Although the discussion in this paper draws heavily from [1], the exposition presented here is more general. For instance, in relational databases multivalued dependency [4] is a necessary and sufficient condition for a relation to be losslessly decomposed into two projections. However, multivalued dependency is a necessary but *not* a sufficient condition for the defining the notion of probabilistic conditional independence [13].

L. Polkowski and A. Skowron (Eds.): RSCTC'98, LNAI 1424, pp. 338–345, 1998.
© Springer-Verlag Berlin Heidelberg 1998

This paper is organized as follows. Section 2 contains background knowledge of our extended relational model [10–13] for probabilistic reasoning. In Section 3, we analyze several properties of GMVDs and show that they are in fact equivalent.

2   Background

2.1   Hypergraphs and Hypertrees

A *hypergraph* [1, 7] is a pair $(\mathcal{N}, \mathbf{R})$, where $\mathcal{N}$ is a finite set of nodes and $\mathbf{R} = \{R_1, R_2, \ldots, R_n\}$ is a finite set of hyperedges which are arbitrary subsets of $\mathcal{N}$. An ordinary undirected graph without self loops is a hypergraph whose every hyperedge consists of two nodes. In this paper we assume $\mathcal{N} = \cup_{i=1}^{n} R_i$ and henceforth will often refer to the hypergraph $\mathbf{R}$ without mentioning the set $\mathcal{N}$ of nodes.

We call an element $R_i \in \mathbf{R}$, a *twig*, if there exists another distinct element $R_j \in \mathbf{R}$, such that $R_i \cap (\cup(\mathbf{R} - \{R_i\})) = R_i \cap R_j$. (By this definition, the hyperedge in a hypergraph consisting of a single hyperedge is not a twig). This means that the intersection of R_i and the hypergraph is contained in one hyperedge of the hypergraph. We call any such R_j a *branch* for the twig R_i, and note that a twig R_i may have many possible branches. A hypergraph is called a *hypertree* (an acyclic hypergraph [1,7]) if its elements can be ordered, $R_1, R_2, \ldots, R_i$, such that R_i is a twig in the sub-hypergraph $R_1, R_2 \ldots, R_i$ for $i = 1, \ldots, n$. We call any ordering satisfying this condition a hypertree *construction ordering* for $\mathbf{R}$. (A hypertree construction ordering can also be represented as a *join tree* [1].) The first hyperedge R_1 in the hypertree construction ordering is called the root. Given a particular hypertree construction ordering, we can choose an integer $b(i)$, for $i = 2, \ldots, n$, such that $1 \leq b(i) \leq i - 1$ and $R_{b(i)}$ is a branch for R_i in $R_1, R_2, \ldots, R_i$. We call such a function $b(i)$ satisfying this condition a *branching function* for $\mathbf{R}$. Note that a particular construction ordering may have many branching functions.

For example, the hypergraph $\mathbf{R} = \{R_1 = \{A_1, A_2, A_3\}, R_2 = \{A_1, A_2, A_4\}, R_3 = \{A_1, A_2, A_5\}, R_4 = \{A_5, A_6\}\}$, depicted in Figure 1, is in fact a hypertree; for instance, the ordering R_1, R_2, R_3, R_4 is a hypertree construction ordering and $b(2) = 1, b(3) = 1, b(4) = 3$ defines one possible branching function.

A *path* from a node A_i to a node A_j is a sequence of hyperedges $R_1, R_2, \ldots, R_k$ ($k \geq 1$) such that $A_i \in R_1$, $A_j \in R_k$, and $R_l \cap R_{l+1} \neq \emptyset$ if $1 \leq l < k$. We also say that the above sequence of hyperedges is a *hyperedge path* (or simply path when no confusion arises) from R_1 to R_k.

Two nodes (attributes) are *connected* if there is a path from one to the other. Similarly, two hyperedges are connected if there is a hyperedge path from one to the other. A set of nodes or hyperedges is connected if every pair is connected. The *connected components* are the maximal connected set of edges or hyperedges.

The following undirected graph terminology will also be useful in describing characteristics of GMVDs. A *clique* in a graph is a set of nodes such that every

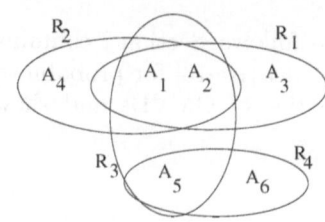

Fig. 1. A graphical representation of the hypergraph $\mathbf{R} = \{R_1, R_2, R_3, R_4\}$.

pair forms an edge of the graph. A *cycle* in a graph is a sequence $(A_1, A_2, \ldots, A_m)$ of nodes ($m \geq 3$) such that each A_i is distinct, except that $A_1 = A_m$, and (A_i, A_{i+1}) is an edge for $1 \leq i \leq m$.

Let $\mathbf{R}$ be a hypergraph. The *graph* of $\mathbf{R}$, denoted $G(\mathbf{R})$, has the same nodes as $\mathbf{R}$ and an edge between every pair of nodes that are in the same hyperedge of $\mathbf{R}$. Thus, the edges of $G(\mathbf{R})$ are precisely the set of all pairs (A_i, A_j) for which there is a hyperedge $R \in \mathbf{R}$ with $A_i, A_j \in R$.

2.2   Extended Relational Definitions

Let $\mathcal{N}$ be a finite set of distinct variables and let $X \subseteq \mathcal{N}$. Following [1], we define a X-tuple (or simply a *tuple* if X is understood) to be a function with domain X. Thus a tuple is a mapping that associates a value with each attribute in X. If $Y \subseteq X$ and t is a X-tuple, then $t[Y]$ denotes the Y-tuple obtained by restricting the mapping to Y. A X-relation (or a *relation over X*, or more simply a *relation* if X is understood) is a finite set of X-tuples. If r is a X-relation and $Y \subseteq X$, then the *projection* of r onto Y, denoted $r[Y]$, is the set of all tuples $t[Y]$ where t is in r.

We now extend the traditional relational concepts in order to express corresponding probabilistic concepts used in managing uncertain knowledge. Let r be a fixed relation representing the domain of a finite set of variables $\mathcal{N}$. A *joint probability distribution* [5, 6] over r, is a function ϕ on r assigning to each tuple $t \in r$ a real number $0 \leq \phi(t) \leq 1$ such that $\sum_{t \in r} \phi(t) = 1$. (We say the distribution is over $\mathcal{N}$ when the domain r is understood, and write ϕ as $\phi_{\mathcal{N}}$.) Suppose $\phi_{\mathcal{N}}$ is a joint probability distribution over $\mathcal{N} = \{A_1, A_2, \ldots, A_l\}$. We can succinctly express the probability distribution $\phi_{\mathcal{N}}$ as an *extended relation* $\Phi_{\mathcal{N}}$ with attributes $\{A_1, A_2, \ldots, A_l, f_{\phi_{\mathcal{N}}}\}$ as shown in Figure 2. Each row in $\Phi_{\mathcal{N}}$ corresponds to a tuple $t_i \in r$, and s is the cardinality of r.

Let ϕ_X and ϕ_Y be two distributions over X and Y respectively. We can express the product of these distributions $\phi_X \cdot \phi_Y$ as the *product join* of the extended relations Φ_X and Φ_Y, denoted by $\Phi_X \times \Phi_Y$, as follows:

$\Phi_X \times \Phi_Y$ is the *extended* relation obtained by adding a new column with attribute $f_{\phi_X \cdot \phi_Y}$ to the relation $r = \Phi_X[X] \bowtie \Phi_Y[Y]$, where $\bowtie$ is the *natural join* operator [1]. (r is referred to as the domain of $\Phi_X \times \Phi_Y$.) For each tuple $t \in \Phi_X \times \Phi_Y$, $t[XY] = r[XY]$, and $t[f_{\phi_X \cdot \phi_Y}] = \phi_X(t[X]) \cdot \phi_Y(t[Y])$.

$$\Phi_{\mathcal{N}} = \begin{array}{|ccccc|}
\hline
A_1 & A_2 & \ldots & A_l & f_{\phi_{\mathcal{N}}} \\
\hline
t_1[A_1] & t_1[A_2] & \ldots & t_1[A_l] & t_1[f_{\phi_{\mathcal{N}}}] = \phi_{\mathcal{N}}(t_1[\mathcal{N}]) \\
t_2[A_1] & t_2[A_2] & \ldots & t_2[A_l] & t_2[f_{\phi_{\mathcal{N}}}] = \phi_{\mathcal{N}}(t_2[\mathcal{N}]) \\
\vdots & \vdots & \vdots & \vdots & \vdots \\
t_s[A_1] & t_s[A_2] & \ldots & t_s[A_l] & t_s[f_{\phi_{\mathcal{N}}}] = \phi_{\mathcal{N}}(t_s[\mathcal{N}]) \\
\hline
\end{array}$$

Fig. 2. A joint distribution $\phi_{\mathcal{N}}$ expressed as an *extended relation* $\Phi_{\mathcal{N}}$.

If $\Phi_{\mathcal{N}}$ is an extended relation and $X \subseteq \mathcal{N}$, then the marginalization of $\Phi_{\mathcal{N}}$ onto X is the *marginal* distribution, denoted $\Phi_{\mathcal{N}}^{\downarrow X}$, with attributes $X \cup \{f_{\phi_{\mathcal{N}}^{\downarrow X}}\}$, defined by:

$$\Phi_{\mathcal{N}}^{\downarrow X} = \{t \mid t[X] \in \Phi_{\mathcal{N}}[X], \text{ and } t[f_{\phi_{\mathcal{N}}^{\downarrow X}}] = \phi_{\mathcal{N}}^{\downarrow X}(t[X]) = \sum_{t' \in \Phi_{\mathcal{N}}} \phi_{\mathcal{N}}(t')$$

where $t[X] = t'[X]\}$.

Due to limited space, we refer the reader to [10–13] for examples illustrating the product join $\times$ and marginalization $\downarrow$ operators.

Let $\Phi_{\mathcal{N}}$ be an extended relation and $X \subseteq \mathcal{N}$. The *inverse* extended relation $(\Phi_{\mathcal{N}}^{\downarrow X})^{-1}$ is defined from $\Phi_{\mathcal{N}}^{\downarrow X}$ by renaming attribute $f_{\phi_{\mathcal{N}}^{\downarrow X}}$ as $f_{(\phi_{\mathcal{N}}^{\downarrow X})^{-1}}$, and

$$t[f_{(\phi_{\mathcal{N}}^{\downarrow X})^{-1}}] = \begin{cases} 1 / t[f_{\phi_{\mathcal{N}}^{\downarrow X}}] & \text{if } t[f_{\phi_{\mathcal{N}}^{\downarrow X}}] > 0 \\ 0 & \text{otherwise.} \end{cases}$$

If Φ is an extended relation over $\mathcal{N}$ and $X, Y \subseteq \mathcal{N}$, then we say that the *generalized multivalued dependency* (GMVD) [12], denoted $X \multimap Y$ holds for Φ if

$$\Phi = \Phi^{\downarrow XY} \otimes \Phi^{\downarrow XZ} = \Phi^{\downarrow X} \times \Phi^{\downarrow Y} \times (\Phi^{\downarrow X \cap Y})^{-1},$$

where $Z = \mathcal{N} - XY$, and $\otimes$ is called the *generalized join* operator.

For example, consider the joint distribution Φ_R on $R = \{A_1, A_2, A_3, A_4, A_5\}$ shown in Figure 3. Φ_R satisfies the GMVD $\{A_1\} \multimap \{A_2, A_3, A_4\}$ since $\Phi_R = \Phi_R^{\downarrow\{A_1, A_2, A_3, A_4\}} \otimes \Phi_R^{\downarrow\{A_1, A_5\}}$ as shown in Figure 4.

We say that an extended relation Φ_R over attributes $R = R_1 \cup R_2 \cup \ldots \cup R_n$ obeys the *generalized acyclic join dependency* (GAJD), denoted $\otimes\{R_1, \ldots, R_n\}$ or $\otimes \mathbf{R}$, if Φ_R can be expressed as

$$\Phi_R = (\ldots((\Phi^{\downarrow R_1} \otimes \Phi^{\downarrow R_2}) \otimes \Phi^{\downarrow R_3}) \ldots \otimes \Phi^{\downarrow R_n}),$$

where $R_1, \ldots, R_n$ is a hypertree construction ordering for $\mathbf{R} = \{R_1, R_2, \ldots, R_n\}$.

Let $\sum$ be a set of dependencies and σ be a single dependency. We say that $\sum$ *logically implies* σ if whenever every dependency in $\sum$ holds for an extended relation Φ, then σ also holds for Φ. That is, there is no "counterexample extended

$$\Phi_R = \begin{array}{|ccccc|c|} A_1 & A_2 & A_3 & A_4 & A_5 & f_{\phi_R} \\ \hline 0 & 0 & 0 & 0 & 1 & 0.2 \\ 0 & 0 & 1 & 1 & 1 & 0.2 \\ 0 & 1 & 0 & 0 & 1 & 0.2 \\ 0 & 1 & 1 & 0 & 1 & 0.2 \\ 1 & 2 & 1 & 1 & 0 & 0.2 \end{array}$$

Fig. 3. A joint distribution ϕ_R on $R = \{A_1, A_2, A_3, A_4, A_5\}$ expressed as the *extended relation* Φ_R.

A_1 A_2 A_3 A_4 D_1		A_1 A_5 D_2		A_1 A_2 A_3 A_4 A_5	D_3
0 0 0 0 0.2	$\otimes$	0 1 0.8	$=$	0 0 0 0 1	$((0.2)(0.8)/0.8) = 0.2$
0 1 1 1 0.2		1 0 0.2		0 0 1 1 1	$((0.2)(0.8)/0.8) = 0.2$
0 1 0 0 0.2				0 1 0 0 1	$((0.2)(0.8)/0.8) = 0.2$
0 1 1 0 0.2				0 1 1 0 1	$((0.2)(0.8)/0.8) = 0.2$
1 2 1 1 0.2				1 2 1 1 0	$((0.2)(0.2)/0.2) = 0.2$

Fig. 4. The distribution Φ_R in Figure 3 satisfies the GMVD $\{A_1\} \multimap\!\!\longrightarrow \{A_2, A_3, A_4\}$ since $\Phi_R = \Phi_R^{\downarrow\{A_1,A_2,A_3,A_4\}} \otimes \Phi_R^{\downarrow\{A_1,A_5\}}$. $D_1 = f_{\phi_R^{\downarrow\{A_1,A_2,A_3,A_4\}}}$; $D_2 = f_{\phi_R^{\downarrow\{A_1,A_5\}}}$; $D_3 = f_{\phi_R^{\downarrow\{A_1,A_2,A_3,A_4\}} \cdot \phi_R^{\downarrow\{A_1,A_5\}} / \phi_R^{\downarrow\{A_1\}}}$

relation" Φ such that every dependency in $\sum$ holds for Φ but σ does not hold for Φ.

Since the concept of nonembedded conditional independence is the focus of our discussion, we conclude this section by distinguishing it from an embedded conditional independence. Consider the joint probability distribution ϕ_R on $R = \{A_1, A_2, A_3, A_4, A_5\}$ as shown in Figure 3. It can be verified that the conditional independence of $\{A_2\}$ and $\{A_3\}$ given $\{A_1\}$ does not hold with respect to ϕ_R. That is,

$$\phi_R \neq \frac{\phi_R^{\downarrow\{A_1,A_2\}} \cdot \phi_R^{\downarrow\{A_1,A_3,A_4,A_5\}}}{\phi_R^{\downarrow\{A_1\}}}.$$

However, the conditional independence of $\{A_2\}$ and $\{A_3\}$ given $\{A_1\}$ holds in the marginal distribution $\phi_R^{\downarrow\{A_1,A_2,A_3\}}$, namely:

$$\phi_R^{\downarrow\{A_1,A_2,A_3\}} = \frac{(\phi_R^{\downarrow\{A_1,A_2,A_3\}})^{\downarrow\{A_1,A_2\}} \cdot (\phi_R^{\downarrow\{A_1,A_2,A_3\}})^{\downarrow\{A_1,A_3\}}}{(\phi_R^{\downarrow\{A_1,A_2,A_3\}})^{\downarrow\{A_1\}}}.$$

We call such an independency an embedded conditional independency with respect to the distribution ϕ_R.

3   Equivalent Characterizations of GMVDs

In this section we derive several characterizations the set M of generalized multivalued dependencies (GMVDs) such that M is the consequence of a given generalized acyclic join dependency (GAJD). Since it has been shown [12] that every GMVD $X -\!\circ\!\!\to Y$ is equivalent to the GMVD $X -\!\circ\!\!\to Y - X$, we will only consider those GMVDs with $X \cap Y = \emptyset$ to simplify the notation. Thus M^+ will denote the set of all GMVDs $X -\!\circ\!\!\to Y$, with $X \cap Y = \emptyset$, that are logically implied by the set M.

If $\mathcal{H}$ is a hypergraph, then the set of GMVDs *generated by* $\mathcal{H}$ is the set of GMVDs $X -\!\circ\!\!\to Y$, where Y is the union of some connected components of the hypergraph $\mathcal{H} - X$ obtained from $\mathcal{H}$ by deleting the set X of nodes. That is, $\mathcal{H} - X = \{E - X \mid E \text{ is an edge of } \mathcal{H}\} - \{\emptyset\}$. We then say that X *separates off* Y from the rest of the nodes. A set M of GMVDs is *hypergraph generated* if there is a hypergraph that generates M. Similarly, M is *graph generated* if there is a graph (treated as a hypergraph) that generates M.

A GMVD $X -\!\circ\!\!\to Y$ ($X \cap Y = \emptyset$) *splits* two attributes A_i and A_j if one of them is in Y and the other is in $\mathcal{N} - XY$, where $\mathcal{N}$ is the set of all the attributes. A set M of GMVDs splits A_i and A_j if some GMVD in M splits them.

Lemma 1. *Two attributes A_i and A_j are split by a set M of generalized multivalued dependencies if and only if they are split by its closure.*

Lemma 1 indicates that two logically equivalent sets of GMVDs split exactly the same pairs of attributes.

Given a set M of GMVDs, we can construct a graph $G(M)$ with the attributes as nodes and an edge (A_i, A_j) between two attributes A_i and A_j if A_i and A_j are not split by M. For example, let $\mathcal{N} = \{A_1, A_2, A_3, A_4\}$ and $M = \{A_1 -\!\circ\!\!\to A_3, A_3 -\!\circ\!\!\to A_4\}$. The first GMVD splits A_2 and A_3, and A_3 and A_4, while the second GMVD splits A_1 and A_4, and A_2 and A_4. The set of edges in the graph $G(M)$ is $\{(A_1, A_2), (A_1, A_3)\}$.

Lemma 2. *Let M be a set of GMVDs, $G(M)$ its graph, and N the set of GMVDs generated by $G(M)$. Then $M^+ \subseteq N$.*

The converse to Lemma 2 does not necessarily hold. In the last example, the GMVD $\emptyset -\!\circ\!\!\to A_4$ is graph generated by $G(M)$ but is not logically implied by M. It will be shown that the converse holds exactly for those sets of GMVDs that form a cover of the set of GMVDs implied by a given generalized acyclic join dependency, where M_1 is a *cover* of M_2 if $M_1^+ = M_2^+$.

We say that M has the *intersection property* if whenever the GMVDs $X -\!\circ\!\!\to Z$ and $Y -\!\circ\!\!\to Z$ are implied by M (with Z disjoint from both X and Y), then $X \cap Y -\!\circ\!\!\to Z$ is also implied by M.

Let M be a set of GMVDs. Two disjoint sets X and Y are called *orthogonal* if the GMVD $\mathcal{N} - XY -\!\circ\!\!\to X$ (or equivalently, by the complementation rule for GMVDs [12], $\mathcal{N} - XY -\!\circ\!\!\to Y$) is implied by M. It follows from Lemma 1 and the rules for manipulating GMVDs [12] that two attributes A_i and A_j are orthogonal (i.e., the singleton sets $\{A_i\}$ and $\{A_j\}$ are orthogonal) if and only if they are split by M. It follows from the rules for manipulating GMVDs [12] that if X and Y are orthogonal, then for every pair A_i and A_j of attributes where $A_i \in X$ and $A_j \in Y$, necessarily A_i and A_j are orthogonal. We say that M has the orthogonal property if the converse also holds. M has the *orthogonal property* if every two sets X and Y are orthogonal whenever every attribute of X is orthogonal to every attribute of Y.

Theorem 1. *Let M be a set of generalized multivalued dependencies. The following are equivalent:*

(1) M is a cover of the set of GMVDs implied by some GAJD.
(2) M^+ is hypergraph generated.
(3) M^+ is graph generated.
(4) There is exactly one graph that generates M^+.
(5) M^+ is the set of GMVDs generated by $G(M)$.
(6) M^+ has the intersection property.
(7) M^+ has the orthogonal property.

Proof. We will show that (1) and (2), and (2) and (3) are equivalent. We then show (3) $\Rightarrow$ (6) $\Rightarrow$ (7) $\Rightarrow$ (5) $\Rightarrow$ (3). Hence, conditions (1), (2), (3), (5), (6), (7) are equivalent. Finally, we show (5) $\Rightarrow$ (4) $\Rightarrow$ (3) which shows (4) is equivalent to the others.

(1) $\Leftrightarrow$ (2): The set of GMVDs implied by a generalized acyclic join dependency $\otimes \mathbf{R}$ is exactly the set of GMVDs generated by the hypergraph $\mathbf{R}$ [12]. Thus, if M is the set of GMVDs implied by the generalized acyclic join dependency $\otimes \mathbf{R}$, then we know that M^+ is the set of GMVDs generated by the hypergraph $\mathbf{R}$.

(2) $\Rightarrow$ (3): Let $\mathcal{H}$ be a hypergraph, and let $G = G(\mathcal{H})$ be the graph of $\mathcal{H}$. It is easy to see that a set X of nodes separates off another set Y in $\mathcal{H}$ if and only if X separates off Y in G. Therefore, the set of GMVDs generated by $\mathcal{H}$ is the same set of GMVDs generated by G.

(3) $\Rightarrow$ (2): Obvious, since every graph is a hypergraph.

(3) $\Rightarrow$ (6): Let $\mathcal{H}$ be the hypergraph that generates M^+. Suppose $X \multimap Z$ and $Y \multimap Z$ are in M^+. Since $X \multimap Z$ is in M^+, X separates off Z from $\mathcal{N} - XZ$. Thus, no node in Z is connected to a node in $\mathcal{N} - XZ$. Since $Y \multimap Z$ is in M^+, Y separates off Z from $\mathcal{N} - YZ$. Thus, no node in Z is connected to a node in $\mathcal{N} - YZ$. Therefore, no node in Z is connected to a node in $\mathcal{N} - [(X \cap Y) \cup Z]$. Therefore, $X \cap Y$ separates off Z from $\mathcal{N} - [(X \cap Y) \cup Z]$. By definition, $X \cap Y \multimap Z$ is in M^+.

(6) $\Rightarrow$ (7): Let $X = \{A_1, A_2, \ldots, A_k\}$ and $Y = \{B_1, B_2, \ldots, B_m\}$ be two disjoint sets with every A_i orthogonal to every B_j. Let $Z = \mathcal{N} - XY$, $X_i = X - A_i (i = 1, 2, \ldots, k)$, and $Y_j = Y - B_j (j = 1, 2, \ldots, m)$. Since every A_i is orthogonal to every B_j, we have $ZX_iY_j \multimap A_i$ for each i and j. Since M has the intersection property, we have $\cap \{ZX_iY_j \mid j = 1, 2, \ldots, m\} \multimap A_i$. But $\cap \{ZX_iY_j \mid j = 1, 2, \ldots, m\} = ZX_i$. Hence, $ZX_i \multimap A_i$, or equivalently by the complementation rule for GMVDs [12], $ZX_i \multimap Y$ for each i. Again from the intersection property,

$\cap\{ZX_i \mid i=1,2,\ldots,k\} \dashrightarrow Y$. Since $\cap\{ZX_i \mid i=1,2,\ldots,k\} = Z$, we have $Z \dashrightarrow Y$, or equivalently, $Z \dashrightarrow X$.

(7) $\Rightarrow$ (5): Suppose that M has the orthogonal property, and let N be the set of GMVDs graph generated from $G(M)$. By Lemma 2, $M^+ \subseteq N$. For the other inclusion, let $X \dashrightarrow Y$ be a GMVD in N, and let $Z = \mathcal{N} - XY$. By definition of N, there is no edge in $G(M)$ connecting a node in Y to a node in Z. Thus every attribute of Y is orthogonal to every attribute of Z. Then by the orthogonality property, Y is orthogonal to Z, and $X \dashrightarrow Y$ is in M^+.

(5) $\Rightarrow$ (3): Obvious.

(5) $\Rightarrow$ (4): Let G be the graph that generates M^+. Attributes A_i and A_j are split by M if and only if the edge (A_i, A_j) is not in G. But also, A_i and A_j are split by M if and only if the edge (A_i, A_j) is not in $G(M)$. Therefore, $G = G(M)$.

(4) $\Rightarrow$ (3): Obvious.

References

1. Beeri, C. and Fagin, R. and Maier, D. and Yannakakis, M.: On the Desirability of Acyclic Database Schemes. J. ACM **30**(3) (1983) 479–513
2. Cooper, G.F.: The Computational Complexity of Probabilistic Inference Using Bayesian Belief Networks. Artificial Intelligence. **42** (1990) 393–402
3. Dagum, P., Luby, M.: Approximating Probabilistic Inference in Bayesian Belief Networks is NP-hard. Artificial Intelligence. **60**(1) (1993) 141–153
4. Fagin, R.: Multivalued Dependencies and a New Normal Form for Relational Databases. ACM Transactions on Database Systems. **2**(3) (1977) 262–278
5. Hajek, P., Havranek, T., Jirousek, R.: Uncertain Information Processing in Expert Systems. CRC Press. (1992)
6. Pearl, J.: Probabilistic Reasoning in Intelligent Systems: Networks of Plausible Inference. Morgan Kaufmann. San Francisco, California. (1988)
7. Shafer, G.: An Axiomatic Study of Computation in Hypertrees. University of Kansas. School of Business Working Papers (232). (1991)
8. Studeny, M.: Conditional Independence Relations Have No Finite Complete Characterization. Eleventh Prague Conference on Information Theory, Statistical Decision Foundation and Random Processes. (1990)
9. Wong, S.K.M., Wang, Z.W.: On Axiomatization of Probabilistic Conditional Independence. Tenth Conference on Uncertainty in Artificial Intelligence. (1994) 591–597
10. Wong, S.K.M., Butz, C.J., Xiang, Y.: A Method for Implementing a Probabilistic Model as a Relational Database. Eleventh Conference on Uncertainty in Artificial Intelligence. (1995) 556–564
11. Wong, S.K.M.: Testing Implication of Probabilistic Dependencies. Twelfth Conference on Uncertainty in Artificial Intelligence. (1996) 545–553
12. Wong, S.K.M.: The Relational Structure of Belief Networks. (submitted for publication) (1997)
13. Wong, S.K.M.: An Extended Relational Data Model for Probabilistic Reasoning. Journal of Intelligent Information Systems. **9** (1997) 181–202

A New Qualitative Rough-Set Approach to Modeling Belief Functions

Mieczysław A. Kłopotek, Sławomir T. Wierzchoń

Institute of Computer Science, Polish Academy of Sciences
Warszawa, Poland
e-mail: klopotek,stw@ipipan.waw.pl

Abstract. The paper presents a novel view of the Dempster–Shafer belief function as a measure of diversity in relational data bases. The Dempster rule of evidence combination corresponds to the join operator of the relational database theory. This rough-set based interpretation is qualitative in nature and can represent a number of belief function operators.

1 Introduction

A case-based interpretation of Mathematical Theory of Evidence (MTE) has been a hot issue for a long time (compare discussions in [1]). Several models have been proposed, including rough set theory based ones (see an overview in [8]). However, none seems to be both complete and intuitively simple [6]. Failure to achieve this goal is attributed to relying on frequencies [6], but so far a non-frequency interpretation appears to be missing. Below we present a modification of a decision-table based rough-set interpretation [5], in which we abandon object identities. The new measure of support of a decision does not rely on the number of records containing the decision, but rather the number of records with *distinct* information part. This means that instead of frequencies we use *diversity* of support. Such an approach seems to be reasonable in cases where the decision table does not reflect the actual frequencies of decision situations, but is meant to present their diversity. The new measure of support will be *higher* for *more universal* types of decisions [1]. The new non-frequency interpretation fulfills the requirement to be qualitative in nature and still to be case-based. Its appealing nature is illustrated by some examples.

We assume familiarity with basic concepts of mathematical theory of evidence [3], rough sets [8], decision tables [5], SQL language, in which the new interpretation has been implemented by the authors, and relational databases in general [7].

Denotation: Let a tuple μ mean a function $\mu : A \to DOM(A)$, with A being a set of attributes A_j, $DOM(A_j)$ being the domain of the attribute A_j, $DOM(A) = \bigcup_{A_j \in A} DOM(A_j)$. A be called the scheme of μ, $A = S(\mu)$.

---

[1] Shafer [1] recalls the fact that MTE was a generalization of legal reasoning rules. In that case the new interpretation would mean "discarding" witnesses that give suspiciously similar testimonies and would count those ones that differ on non-essential details reflecting the usual subjective impressions of individuals.

L. Polkowski and A. Skowron (Eds.): RSCTC'98, LNAI 1424, pp. 346–354, 1998.
© Springer-Verlag Berlin Heidelberg 1998

A relational table TAB be any set of tuples with identical scheme. This common scheme be denoted by $S(TAB)$. Let $\mu[R]$ with $R \subseteq S(\mu)$ denote the restriction of the tuple μ to the scheme R: $\mu[R] = \{(A_j, a_{jk}) | A_j \in R \wedge a_{jk} = \mu(A_j)\}$. The restriction of a relational table TAB to R, denoted $TAB[R]$, be defined $TAB[R] = \{\mu[R] | \mu \in TAB\}$. A relational join of two relational tables TAB_1, TAB_2 be defined as: $TAB_1 \otimes TAB_2 = \{\mu_1 \cup \mu_2 | \mu_1 \in TAB_1, \mu_2 \in TAB_2 \wedge \forall_{A_j \in S(\mu_1) \cap S(\mu_2)} \mu_1(A_j) = \mu_2(A_j)\}$.

A decision table is a relational table in which we split the scheme into two distinct parts: the information part and the decision part.

Let $card(SET)$ denote the cardinality of the set SET.

2 New Interpretation

Let us define the plausibility $Pl_{TAB}(SET)$ derived from a decision table TAB with decision variable D and the set $\mathbf{I}$ of information variables as: $Pl_{TAB}(SET) = card(\{\mu[\mathbf{I}] | \mu \in TAB \wedge \mu(D) \in SET\})/card(TAB[\mathbf{I}])$, implemented as

create view tmpTAB(No) as select count(distinct $\mathbf{I}$) from TAB;
create view plTAB as select count(distinct $\mathbf{I}$)/No from TAB,tmpTAB where TAB.D in SET;

Example 1 explains the detailed numerical procedure for calculation of Pl from the above SQL expression.

Table 1. Decision table: BUILD. I - the firm; D - the object to be erected.

	I	D
1.	ABD A.G.	center
2.	LQR Inc.	school
3.	PTS Ltd.	center
	PTS Ltd.	restaurant
4.	XYZ Inc.	center
5.	ZZZ Ltd.	restaurant
	ZZZ Ltd.	school

Example 1. Assume a public offering for erection of buildings of a school, a restaurant and a shopping center where the offers presented by various firms have been summarized in the decision table BUILD (tab. 1) with "information part" (I) - the firm and "decision part" (D) - the object to be erected. The domain of the decision variable D is {center, restaurant, school}. What is the share of firms that would build either the school or the restaurant ? To answer this question we need to calculate the plausibility Pl({school, restaurant}) from this table. There are 7 cases (rows) in the dataset. But there are only 5 cases with distinct information part (firms) **I**. And there are only 3 cases with decision either *school* or *restaurant* with distinct information part **I** (LQR Inc., PTS

Ltd., ZZZ Ltd).So the plausibility is equal[2] Pl({school, restaurant})=3/5. One can check that Pl({school})=2/5 and Pl({restaurant})=2/5.

Theorem 1. *The function $Pl_{TAB}(SET)$ derived from a decision table TAB with decision variable D and the set **I** of information variables is a plausibility function $Pl(SET)$ in the sense of Dempster–Shafer theory.*

For the proof see [2].

Relational views for other MTE measures may be also derived (numerical illustration of these SQL-based definitions is given in example 2):

Belief $Bel_{TAB}(SET) = 1 - card(\{\mu[\mathbf{I}] | \mu \in TAB \wedge \mu(D) \notin SET\})/card(TAB[\mathbf{I}])$, implemented as

*create view belTAB as select 1-count(distinct **I**)/No from TAB,tmpTAB where not (TAB.D in SET);*

Commonality $Q_{TAB}(SET) = card(\{\mu[\mathbf{I}] | \forall_{d \in SET} \ \mu[\mathbf{I}] \cup \{(D,d)\} \in TAB\})$ $/card(TAB[\mathbf{I}])$, implemented as

create view tmp1TAB(CN) as select count(distinct D) from TAB
*where TAB.D in SET group by **I** ;*
create view qTAB as select count(*)/No from tmpTAB,tmp1TAB where CN=card(set); (card() is a function counting the elements of the set passed as its argument)

Basic belief assignment $m_{TAB}(SET) = card(\{\mu[\mathbf{I}] | \forall_{d \in SET} \ \mu[\mathbf{I}] \cup \{(D,d)\} \in TAB\} \wedge \forall_{d \notin SET} \ \mu[\mathbf{I}] \cup \{(D,d)\} \notin TAB\})/card(TAB[\mathbf{I}])$, implemented as

*create view tmp11TAB(**I**,D) as select TAB.**I**, TAB.D+XX.D from TAB, TAB XX where TAB.D in SET and XX.**I**=TAB.**I**;*
*create view tmp12TAB(**I**,CN) as select **I**,count(distinct D) from tmp11TAB group by **I**;*
create view m as count(*)/No from tmpTAB, tmp12TAB
where CN=card(SET)*card(SET);

Example 2. From tab. 1 we easily calculate that:
Commonality Q({school, restaurant}=1/5 (Number of firms ready to build either the school and the restaurant: ZZZ Ltd).
Belief Q({school, restaurant}=2/5 Number of firms ready to build nothing but the school or the restaurant (LQR Inc., ZZZ Ltd)
bpa - No of firms exactly offering erecting of: m({school, restaurant}=1/5 (ZZZ Ltd), m({restaurant}=0 (none), m({school}=1/5 (LQR Inc.)

Theorem 2. *The functions $Bel_{TAB}(SET), Q_{TAB}(SET), m_{TAB}(SET)$ derived from a decision table TAB with decision variable D and the set **I** of information variables are belief, commonality, basic probability/belief assignment functions resp. $Bel(SET), Q(SET), m(SET)$ in the sense of Dempster–Shafer theory.*

---

[2] Under Skowron/Busse interpretation [5] we get a different *Pl* value: $Pl = 4/7$. The difference stems from the fundamental difference between frequency [5] and relational (ours) view.

3   Reasoning as Selection of a Subtable

A natural way to understand a belief function Bel_{TAB} conditioned on evidence B (denoted $Bel_{TAB}(.||B)$) is to select from TAB records fitting the condition B $TAB_B = \{\mu|\mu \in TAB \wedge \mu(D) \in B\}$, implemented as

create view TAB_B(I,D) as select I,D from TAB where TAB.D in B;

and to calculate $Bel_{TAB}(.||B)$ as the belief function $Bel_{TAB\_B}(.)$ over TAB_B. In our example, $Bel_{BUILD}(.||\{school, restaurant\})$ is just Bel calculated from tab. 2. Our conditional belief function matches perfectly the Shafer's definition of $Bel(.||B) = Bel \oplus Bel_B$.

Table   2.   Relational   modeling   of   reasoning   in   MTE: $Bel(.||\{school, restaurant\})$.

Prescription *select I,D from BUILD where D=school or D= restaurant*

	I	D
1.	LQR Inc.	school
2.	PTS Ltd.	restaurant
5.	ZZZ Ltd.	restaurant
	ZZZ Ltd.	school

$Pl(\{restaurant\}||\{school, restaurant\}) = 2/3(PTS\ Ltd., ZZZ\ Ltd.)$
$m(\{restaurant\}||\{school, restaurant\}) = 1/3(PTS\ Ltd.)$

4   Combination as Relational Join

Let us calculate the table FINISH as a join of BUILD and EQUIP (tab. 3) over the common column D so that the new decision table has as its decision column D and as its information part I,I2 (tab.3).

create table FINISH (I,I2,D);
insert into table FINISH from
select I,I2,D from BUILD, EQUIP where BUILD.D=EQUIP.D;

Notice that in BUILD (tab.1), there were 5 cases with distinct information part, in EQUIP (tab.3) - 3, and in BUEQ (tab.3) there are only 10. You can easily check that $Bel_{FINISH} = Bel_{BUILD} \oplus Bel_{EQUIP}$.

Theorem 3. *If the decision table DT1(I1,D) and DT2(I2,D) with non-overlapping information parts I1,I2 are combined by relational join operation DT1 $\otimes$ DT2, implemented as*

select I1,I2,DT1.D from DT1,DT2 where DT1.D=DT2.D;

yielding table DT12(I,D) with I=I1$\cup$I2, then $Bel_{DT12} = Bel_{DT1} \oplus Bel_{DT2}$.

For the proof see [2].

5   Relational Marginalization and Decombination

Notice that BUILD and EQUIP in our example are both in first normal form and the domain of the attribute D is identical in both tables. Therefore marginalization of FINISH over I,D (*select distinct I,D from FINISH;*) is exactly identical with BUILD. In general:

Table 3. Combination of Independent Evidence. Example: Public offering: equipment for buildings of a school, a restaurant and a shopping center. Decision table: EQUIP "Information part" - the firm, "Decision part" - the object to be equipped

	I2	D
1.	AAA GmbH	school
2.	BBB Ltd.	center
	BBB Ltd.	restaurant
3.	CCC Inc.	center
	CCC Inc.	restaurant

Independence of evidence means that no pair of firms (one from BUILD, one from EQUIP) refuse to cooperate on erecting and equipping an object. How many pairs of firms do we have to finish a set of objects mentioned in the offerings? Answer: *create view BUEQ(I,I2,D) as select I,I2,D from BUILD, EQUIP where BUILD.D=EQUIP.D,* yielding the table below.

	I	I2	D
1.	ABD A.G.	BBB Ltd.	center
2.	ABD A.G.	CCC Inc.	center
3.	LQR Inc.	AAA GmbH	school
4.	PTS Ltd.	BBB Ltd.	center
	PTS Ltd.	BBB Ltd.	restaurant
5.	PTS Ltd.	CCC Inc.	center
	PTS Ltd.	CCC Inc.	restaurant
6.	XYZ Inc.	BBB Ltd.	center
7.	XYZ Inc.	CCC Inc.	center
8.	ZZZ Ltd.	BBB Ltd.	restaurant
9.	ZZZ Ltd.	CCC Inc.	restaurant
10.	ZZZ Ltd.	AAA GmbH	school

Pl({school, restaurant})=8/10, Bel({school, restaurant})=3/10

Theorem 4. *If the information part* **I** *of the decision table DT(**I**,D) can be split into two such parts* **I1,I2** *that* **I1** ∪ **I2** = **I** *and* **I1** ∩ **I2** = ∅ *and the relation DT is identical with DT1⊗DT2, implemented*

select **I1,I2**,*DT1.D from DT1,DT2 where DT1.D=DT2.D;*

*where DT1 and DT2 are DT1=DT[**I1**,D], DT2=DT[**I2**,D], implemented*

create view DT1 as select distinct **I1**,*D from DT;*
create view DT2 as select distinct **I2**,*D from DT;*

that is there is a multivariate dependency between **I1** *and* **I2** *given D, then* $Bel_{DT} = Bel_{DT^{\downarrow I1,D}} \oplus Bel_{DT^{\downarrow I2,D}}$.

Proof. Follows directly from theorem 3.

Let consider the unnormalized MTE measures of decision tables m'_{TAB}, Bel'_{TAB}, Pl'_{TAB}, Q'_{TAB}, such that $f'_{TAB} = f_{TAB} \cdot card(TAB^{\downarrow I})$ (card - number of distinct rows, f - m or Bel or Pl or Q) and the unnormalized combination operator $\oplus'$ such that $Bel'_{E_1,E_2} = Bel'_{E_1} \oplus' Bel'_{E_2}$ is defined as follows: $m'_{E_1,E_2}(A) = \sum_{B,C:A=B \cap C} m'_{E_1}(B) \cdot m'_{E_2}(C)$.

What may be more surprising, a kind of a reverse theorem holds:

Theorem 5. *The information part I of the decision table DT(I,D) can be split into two such parts* **I1,I2** *that* **I1** $\cup$ **I2** $= I$ *and* **I1** $\cap$ **I2** $= \emptyset$ *and* $Bel'_{DT} = Bel'_{DT^{\downarrow I1,D}} \oplus' Bel'_{DT^{\downarrow I2,D}}$ *if and only if the relation DT is identical with DT1* $\otimes$ *DT2, implemented*

select **I1,I2,**DT1.D *from DT1,DT2 where DT1.D=DT2.D;*

where DT1 and DT2 are DT1=DT[I1,D], DT2=DT[I2,D], implemented

create view DT1 as select distinct **I1,**D *from DT;*
create view DT2 as select distinct **I2,**D *from DT;*

that is there is a multivariate dependency between **I1** *and* **I2** *given D,*

For the proof see [2].

6   Multivariate Beliefs

Our definition of MTE measures is easily extended to tables with multiple decision variables which may model multivariate belief distributions (in all the decision variables). It is trivial to see that dropping a decision variable does not diminish the diversity of the information part. Hence dropping a decision variable Di from the set of decision variables **D** has the same effect as dropping a variable (so-called marginalization or projection operation $\downarrow$) in the belief function. That is for any set B of decision vectors in variables **D**-$\{Di\}$: $m_{TAB^{\downarrow D-\{Di\}}}(B) = m_{TAB}^{\downarrow D-\{Di\}}(B) = \sum_{A;B=A^{\downarrow D-\{Di\}}} m(A)_{TAB}$. See tab. 4 for an example.

The operator of projection $\downarrow$ should be understood as the MTE projection operator applied to a belief function. Let DTM be a decision table with decision variables D1 and D2. Let the information part consist of two disjoint parts **I1** and **I2**. Let us consider the following views:

create view DTM1 (I1,D1) as select distinct **I1,**D1 *from DTM;*
create view DTM2 (I2,D2) as select distinct **I2,**D2 *from DTM;*
create view DTM12 as select distinct **I1,I2,**D1,D2 *from DTM1,DTM2;*

If now the table DTM12 is relationally identical with DTM, then we shall say that the decision variables D1 and D2 are independent in the decision table DTM. It is not surprising that: $Bel_{DTM} = Bel_{DTM}^{\downarrow D1} \oplus Bel_{DTM}^{\downarrow D2}$. This means that independence of decision variables in a decision table implies independence of variables in the corresponding belief function. See tab. 5 for an example.

Table 4. Multivariate MTE. A table and its projection

Decision table MADEOF

	I	D	D1
1.	ABD A.G.	center	wooden
2.	LQR Inc.	school	stone
	LQR Inc.	school	wooden
3.	PTS Ltd.	center	stone
	PTS Ltd.	center	wooden
	PTS Ltd.	restaurant	stone
4.	XYZ Inc.	center	stone
5.	ZZZ Ltd.	restaurant	stone
	ZZZ Ltd.	restaurant	wooden
	ZZZ Ltd.	school	wooden

Its projection onto I,D:
select distinct I,D from MADEOF

	I	D
1.	ABD A.G.	center
2.	LQR Inc.	school
3.	PTS Ltd.	center
	PTS Ltd.	restaurant
4.	XYZ Inc.	center
5.	ZZZ Ltd.	restaurant
	ZZZ Ltd.	school

$m(\{(school,wooden)\})=1/5$, $\quad m(\{(school,stone)\})=2/5$
$m^{\downarrow D}(\{(school)\})=m(\{(school,stone)\})+m(\{(school,wooden)\})=1/5$

Table 5. Variable Independence

D2 - heating

	I2	I3	D	D2
1.	AAA GmbH	EC	school	electric
2.	AAA GmbH	GC	school	gas
3.	BBB Ltd.	EC	center	electric
	BBB Ltd.	EC	restaurant	electric
4.	BBB Ltd.	GC	center	gas
	BBB Ltd.	GC	restaurant	gas
5.	CCC Inc.	EC	center	electric
	CCC Inc.	EC	restaurant	electric
6.	CCC Inc.	GC	center	gas
	CCC Inc.	GC	restaurant	gas

Variables D and $D2$ are independent ($Bel = Bel^{\downarrow D} \oplus Bel^{\downarrow D2}$) because the above table represents a cross product of the tables (without common columns)

	I2	D
1.	AAA GmbH	school
2.	BBB Ltd.	center
	BBB Ltd.	restaurant
3.	CCC Inc.	center
	CCC Inc.	restaurant

and

	I3	D2
1.	EC	electric
2.	GC	gas

Let DTX be a decision table with decision variables D1, D2 and D3. Let the information part consist of two disjoint parts **I1** and **I2**. Let us consider the following views:

*create view DTX1 (**I1**,D1,D3) as select distinct **I1**,D1,D3 from DTX;*
*create view DTX2 (**I2**,D2,D3) as select distinct **I2**,D2,D3 from DTX;*
*create view DTX12 as select distinct **I1**,**I2**,D1,D2,D3 from DTX1, DTX2 where DTX1.D3=DTX2.D3;*

If now the table DTX12 is relationally identical with DTX, then we shall say that the decision variables D1 and D2 are independent given D3 in the decision table DTX. It is not surprising that: $Bel_{DTX} = Bel_{DTX}^{\downarrow D1,D3} \oplus Bel_{DTX}^{\downarrow D2,D3}$. But this means that the variables D1 and D2 are independent given D3 in the belief function Bel_{DTX} in the sense of Shenoy's VBS. See tab. 6 for an example.

Table 6. Conditional Variable Independence

I4 - painting company, $D4$ - color, $D5$ - finish.

	I2	I4	D	D5	D4
1.	AAA GmbH	Messer	school	wood	green
	AAA GmbH	Messer	school	wood	red
	AAA GmbH	Messer	school	plastic	green
	AAA GmbH	Messer	school	plastic	red
2.	BBB Ltd.	Messer	center	metallic	white
	BBB Ltd.	Messer	center	metallic	yellow
	BBB Ltd.	Messer	center	marble	white
	BBB Ltd.	Messer	center	marble	yellow
3.	BBB Ltd.	Gabel	restaurant	wood	red
4.	CCC Inc.	Messer	center	metallic	white
	CCC Inc.	Messer	center	metallic	yellow
	CCC Inc.	Messer	center	laminated	white
	CCC Inc.	Messer	center	laminated	yellow
5.	CCC Inc.	Gabel	restaurant	plastic	red

In Bel of the above table variables $D4$ and $D5$ are conditionally independent given D because the above table is a relational join of the tables below (with D as a common column)

	I2	D	D5
1.	AAA GmbH	school	wood
	AAA GmbH	school	plastic
2.	BBB Ltd.	center	metallic
	BBB Ltd.	center	marble
	BBB Ltd.	restaurant	wood
3.	CCC Inc.	center	metallic
	CCC Inc.	center	laminated
	CCC Inc.	restaurant	plastic

&

	I4	D	D4
1.	Gabel	restaurant	red
2.	Messer	school	green
	Messer	school	red
	Messer	center	white
	Messer	center	yellow

References

1. *International Journal of Approximate Reasoning*. Special issues on MTE, 1990:4 and 1992:6.
2. Kłopotek M.A., Wierzchoń S.T.: Basic Formal Properties of A Relational Model of The Mathematical Theory of Evidence. Submitted to the Journal *Demonstratio Mathematica*.
3. Shafer G.: *A Mathematical Theory of Evidence*. Princeton University Press, Princeton, 1976
4. Shafer G.: Allocation of probability,*Ann. Probab.* 7 (5) (1979), 827–839
5. Skowron A., Grzymała-Busse J.W.: From rough set theory to evidence theory. [in:]Yager R.R., Kasprzyk J. and Fedrizzi M., eds, *Advances in the Dempster-Shafer Theory of Evidence*. J. Wiley, New York (1994), 193–236.
6. Smets Ph.: Resolving misunderstandings about belief functions, *International Journal of Approximate Reasoning* 1992:6:321–344.
7. Vang A.: *SQL and Relational Databases*. Microtrend Books, Slawson Communications Inc., 1991.
8. Yao Y.Y., Lingras P.J.: Interpretations of belief functions in the Theory of Rough Sets. *Information Sciences 104*(1998)1–2, 81–106.

On Stability of Oja Algorithm

Władysław Skarbek¹, Mirosław Skarbek²

¹ Institute of Mathematics, Polish Academy of Sciences
² Department of Electronics and Information Technology
Warsaw University of Technology

Abstract. By elementary tools of matrix analysis we show that the discrete dynamic system defined by Oja algorithm is stable in the ball $B(0, r)$ if its polynomial are bounded by $(2r)^{-1}$ where $r \geq e^{-1}$ and e is the limit for the learning sequence. We also define a neural based Oja's system (with point attractors, stochastic convergence conditions) which leads to the infinity with exponential rate if only their initial states start sufficiently far from the zero point.

1. Introduction

Oja algorithm [6] is a neural type modifier scheme

$$W_{t+1} = W_t + \eta_t(x_t y_t - y_t^2 W_t) \tag{1}$$

used for stochastic approximation of a principal vector w that from random variables $X \in R^n$. Within mean value $E[X] = 0$.

The sample $x \in R^n$ is a value of X chosen at the t-dimension of the same time, randomly and independently of other samples. The real positive number η_t is called the gain in the learning rate coefficient at time t. The function $y(x, w)$ has the following form:

$$y(x, w) = y(x, w_t). \tag{2}$$

When vectors for the system are written in the column form and w denotes the input transposed. Components of the vector are on above that vector, telling us that can be interpreted as weights of a significant neuron, with output $y = w^T x$. Thus neuron is taught to minimize the variance of the neuron variable $y = w^T x$. Contrary to a number of algorithms or eigenvector computation, the Oja algorithm does not require an estimate of the covariance matrix C. They simply input w_t of t modified incrementally after receiving a new data vector x. The sensitivity to current input data is the basic reason for the wide-spread use of this approach.

The algorithm is tuned to be implemented and results are enough, such as for such applications as image compression and pattern recognition (cf. instance.

---

¹ The work was supported in part by University of Technology grant for a real task.

L. Polkowski and A. Skowron (eds.): RSCTC'98, LNAI 1424, pp. 354–360, 1998.
© Springer-Verlag Berlin Heidelberg 1998

On Stability of Oja Algorithm

Radosław Sikora[1], Władysław Skarbek[2] *

[1] Institute of Mathematics, Polish Academy of Sciences
[2] Department of Electronics and Information Technology
Warsaw University of Technology

Abstract. By elementary tools of matrix analysis, we show that the discrete dynamical system defined by Oja algorithm is stable in the ball $K(0, 81/64)$ if only gains β_n are bounded by $(2B)^{-1}$, where $B = b^2$ and b is the bound for the learning sequence. We also define a general class of Oja's systems (with gains satisfying stochastic convergence conditions) which tend to the infinity with exponential rate if only their initial states are chosen too far from the zero point.

1 Introduction

Oja algorithm ([6]) is a neural type iterative scheme

$$w_{n+1} = w_n + \beta_n f(x_n, w_n) \tag{1}$$

used for stochastic approximation of a principal vector $w \in R^N$ for the given random variable $X : \Omega \to R^N$ with zero mean value ($E[X] = 0$).

The sample $x_n \in R^N$ is a value of X chosen at the n-th moment of discrete time, randomly and independently of other samples. The real positive number β_n is called the gain or the learning rate coefficient at time n. The function $f(x, w)$ has the following form:

$$f(x, w) \doteq y(x - yw) \tag{2}$$

where $y = x^T w$. The vectors are written in the column form and ^{T} denotes the matrix transposition. Components of the vector w are also called weights as they can be interpreted as weights of one linear neuron with output $y = x^T w$. This neuron is taught to maximize the variance of the random variable $Y = X^T w$.

Contrary to numerical algorithms for eigenvector computation, the Oja algorithm does not require an estimate of the covariance matrix C – the principal vector is modified incrementally after receiving a new data vector x. This adaptivity to current input data is the basic reason for the widespread use of this approach.

The algorithm is trivial to be implemented and results are enough accurate for such applications as image compression and pattern recognition (for instance

---

* The work was sponsored in part by University (Rector) priority grant *New materials*.

L. Polkowski and A. Skowron (Eds.): RSCTC'98, LNAI 1424, pp. 354–360, 1998.
© Springer-Verlag Berlin Heidelberg 1998

the classification of materials based on textures extracted from digital images of material surfaces).

The only drawback of this approach is the need for large training sequences. For instance to find useful approximation of the principal subspace in handwritten digit recognition we had to use about 500 training examples per one digit. Using only 50 examples had reduced the recognition rate from 99.4% to 96.5% ([7]). In other words we need not only representative sequence for the random variable but this sequence must be delivered several times to reach adequate accuracy.

The iterative scheme (1) with the function f defined by (2) is also known as the Oja's learning rule. Its convergence was shown by reducing the difference equation to the corresponding ordinary differential equation. Namely, using the famous theorem of Kushner and Clark [4] many authors (e.g.:[1]) prove the stochastic convergence with probability one ($a.s.$ – $almost\ sure$) to the principal vector of X under the following conditions for gains:

$$\sum_n^\infty \beta_n = \infty, \quad \sum_n^\infty \beta_n^2 < \infty \tag{3}$$

Additionally it is required that the covariance matrix $C = E[XX^T]$ for X has the principal vector e_1 with the eigenvalue λ_1 of multiplicity one (i.e. $\lambda_1 > \lambda_2$) and it is not orthogonal to the initial weight vector w_0 (i.e. $w_0^T e_1 \neq 0$).

It is interesting that in their proofs, all authors using the stochastic approximation ignored the basic condition of this theory: the boundeness of the stochastic sequence w_n. Therefore the proofs were incomplete.

Recently there were some efforts to find a proof independent of the Kushner and Clark stochastic approximation theory using only general facts from martingale theory. The work of Duflo [2] includes such a proof. She has shown a.s. convergence of the Oja algorithm but it seems that the overhead of theorems it is based on, is comparable with the previous approach. She has also proved the stability of the algorithm by showing that any trajectory of Oja's dynamical system beginning from the ball $K(0, 5.0)$ (with center in zero, with radius 5) will stay within this ball provided the sequence β_n is bounded by $(2B)^{-1}$, where $B = b^2$ and b is the bound on the learning sequence:

$$b \doteq \max_n \|x_n\|, \quad B \doteq \max_n \|x_n\|^2 \tag{4}$$

In this work we improve the above result to the ball of radius about four times less. Namely we show in the section 2 that the Oja algorithm is stable in the ball $K(0, 81/64)$.

2   Stability of Oja algorithm

Let us consider the Oja algorithm in the following classical form which follows from (1) and (2) by eliminating variable y:

$$w_{n+1} = w_n + \beta_n x_n^T w_n \left(x_n - (x_n^T w_n) w_n \right) \tag{5}$$

The equivalent form we obtain by multiplying first term in the parenthesis by $x_n^T w_n$ from the right and the second one by $w_n^T x_n$ from the left:

$$w_{n+1} = w_n + \beta_n(x_n x_n^T w_n - w_n^T x_n x_n^T w_n w_n) \tag{6}$$

Extracting w_n, by introducing identity matrix $I \in R^{N \times N}$ we get:

$$w_{n+1} = w_n + \beta_n(x_n x_n^T - w_n^T x_n x_n^T w_n I)w_n \tag{7}$$

Denoting the matrix $x_n x_n^T$ by C_n we have:

$$w_{n+1} = w_n + \beta_n(C_n - w_n^T C_n w_n I)w_n \tag{8}$$

Using equality $\|w\|^2 = w^T w$ we obtain: $\|w_{n+1}\|^2 =$

$$\|w_n\|^2 + 2\beta_n(1 - \|w_n\|^2)w_n^T C_n w_n + \beta_n^2 w_n^T C_n^2 w_n + \beta_n^2(\|w_n\|^2 - 2)(w_n^T C_n w_n)^2 \tag{9}$$

Assuming the bounds defined in (4), by simple manipulations we get the following useful facts:

$$(w_n^T C_n w_n)^2 = (w_n^T C_n^2 w_n)\|w_n\|^2, \tag{10}$$

$$w_n^T C_n^2 w_n = \|x_n\|^2(w_n^T C_n w_n) \leq B \cdot w_n^T C_n w_n, \tag{11}$$

$$w_n^T C_n w_n = \|x_n\|^2\|w_n\|^2 \leq B \cdot \|w_n\|^2 \tag{12}$$

The following theorem gives us a tight bound on the stability region of Oja's dynamical system.

Theorem 1.
If for each $n \geq 0$ $\|x_n\|^2 \leq B$ a.s. and $\beta_n \leq \frac{1}{2B}$ then for each $n > 0$ $\|w_n\|^2 \leq 81/64$ a.s. if only $\|w_0\|^2 \leq 81/64$.

Proof: We show the thesis by induction on n. In the inductive step we consider three cases:

1. $\|w_n\|^2 \leq 1$:
   In the formula (9) we ignore the last term which is negative, next we apply the inequalities (11,12) to the third term, the inequality (12) to the second one and finally the one bound the first term:

   $$\|w_{n+1}\|^2 \leq \|w_n\|^2 + 2\beta_n(1 - \|w_n\|^2)w_n^T C_n w_n + \beta_n^2 w_n^T C_n^2 w_n$$

   $$\leq \|w_n\|^2 + 2\beta_n(1 - \|w_n\|^2)\|w_n\|^2 B + \beta_n^2 B^2\|w_n\|^2$$

   $$\leq \|w_n\|^2 + (1 - \|w_n\|^2)\|w_n\|^2 + \tfrac{1}{4}\|w_n\|^2 \leq 81/64$$

   The last line above was obtained by using the assumption $B\beta_n \leq 1/2$ and computing the maximum value of the polynomial $t + (1 - t)t + 0.25t$ in the interval $[0, 1]$.

2. $1 < \|w_n\|^2 \le 1 + \frac{\beta_n B}{2}$:

This time in the formula (9) we ignore the last term which is negative and next we apply the inequality (11) to the third term only:

$$\|w_{n+1}\|^2 \le \|w_n\|^2 + 2\beta_n(1 - \|w_n\|^2)w_n^T C_n w_n + \beta_n^2 w_n^T C_n^2 w_n$$

$$\le \|w_n\|^2 + 2\beta_n(1 - \|w_n\|^2)w_n^T C_n w_n + \beta_n^2 B w_n^T C_n w_n \qquad (13)$$

$$= \|w_n\|^2 + 2\beta_n w_n^T C_n w_n \left(1 - \|w_n\|^2 + \frac{\beta_n B}{2}\right)$$

Applying the bound for $\beta_n B$ we get:

$$\|w_{n+1}\|^2 \le \|w_n\|^2 + \|w_n\|^2 \left(\frac{5}{4} - \|w_n\|^2\right) = \|w_n\|^2 \left(\frac{9}{4} - \|w_n\|^2\right) \le \frac{9^2}{8^2} = \frac{81}{64}$$

The bound $81/64$ is the maximum value of the polynomial $t(9/4 - t)$. Hence the upper bound $81/64$ is held.

3. $1 + \frac{\beta_n B}{2} < \|w_n\|^2 \le 81/64$:

As $81/64$ is less than 2 we can use the last inequality in (13) in which we can drop the negative second term:

$$\|w_{n+1}\|^2 \le \|w_n\|^2 + 2\beta_n w_n^T C_n w_n \left(1 - \|w_n\|^2 + \frac{\beta_n^2 B}{2}\right) \le \|w_n\|^2 \le \frac{81}{64}$$
$$(14)$$

□

As a byproduct of the above proof we get the corollary about monotonic behavior of the norm $\|w_n\|$ in certain regions:

Corollary 1.

1. If $\|w_n\| \le 81/64$ and $\|w_n\|^2 \ge 1 + \frac{\beta_n B}{2}$ then $\|w_n\| \ge \|w_{n+1}\|$;
2. If $\|w_n\| \le 1$ then $\|w_n\| \le \|w_{n+1}\|$;

Proof: The first fact is included in the inequality (14). The second requires more careful bounding the right side of the equation (9). Firstly we break the factor $\|w_n\|^2 - 2$ into $\|w_n\|^2 - 1$ and -1. Multiplying, substituting the equality (10) and reordering the terms produces:

$$\|w_{n+1}\|^2 = \|w_n\|^2 + 2\beta_n(1 - \|w_n\|^2)w_n^T C_n w_n +$$

$$+ \beta_n^2(1 - \|w_n\|^2)w_n^T C_n^2 w_n - \beta_n^2(1 - \|w_n\|^2)(w_n^T C_n^2 w_n)\|w_n\|^2$$

$$\ge \|w_n\|^2 + 2\beta_n(1 - \|w_n\|^2)w_n^T C_n w_n + \beta_n^2(1 - \|w_n\|^2)w_n^T C_n^2 w_n -$$

$$- \beta_n^2(1 - \|w_n\|^2)B w_n^T C_n w_n \|w_n\|^2$$

$$= \|w_n\|^2 + \beta_n^2(1 - \|w_n\|^2)w_n^T C_n^2 w_n +$$

$$2\beta_n(1 - \|w_n\|^2)w_n^T C_n w_n \left(1 - \frac{B\beta_n \|w_n\|^2}{2}\right)$$

$$\ge \|w_n\|^2 .$$

The last inequality is valid as all remaining terms are nonnegative. In particular $B\beta_n\|w_n\|^2 \leq 1/2$ what implies positivity of the factor $1 - B\beta_n\|w_n\|^2/2$.
□

3 Unbounded trajectories in Oja scheme

In this section we construct a general class of Oja's dynamical systems defined by the recurrent equation (5) which exhibits very fast convergence to infinity despite the training sequence is bounded.

Let $e \in R^N$ be a unit length vector and two real constants $0 < A \leq B, A \leq 1$. Assume that for each $n \geq 0$, $x_n \in \text{span}(e)$ and $A \leq \|x_n\| \leq B$. Assume also that $w_0 \in \text{span}(e)$. Then obviously for each $n > 0$, $w_n \in \text{span}(e)$, too. As x_n and w_n are collinear then we have immediately:

$$w_n^T C_n w_n = (x_n^T w_n)^2 = \|x_n\|^2 \|w_n\|^2 \tag{15}$$

Let us assume the gain coefficients of the form

$$\beta_n \doteq \frac{1}{2B + n} \leq \frac{1}{2B} \tag{16}$$

We see that β_n satisfies the assumption of stability Theorem 1.

Denote the difference $\|w_{n+1}\|^2 - \|w_n\|^2$ by $\Delta\|w_n\|^2$. Then from the equation (9) we have:

$$\Delta\|w_n\|^2 = 2\beta_n w_n^T C_n w_n - 2\beta_n \|w_n\|^2 w_n^T C_n w_n$$
$$+ \beta_n^2 w_n^T C_n^2 w_n + \beta_n^2 (\|w_n\|^2 - 2)(w_n^T C_n w_n)^2 \tag{17}$$

By dropping the first and the third nonnegative terms and next using the equation (15) and bounds for the sequence x_n, we get the lower bound for $\|w_n\|^2 \geq 2$:

$$\Delta\|w_n\|^2 \geq -2\beta_n \|w_n\|^4 \|x_n\|^2 + \beta_n^2 (\|w_n\|^2 - 2)\|w_n\|^4 \|x_n\|^4$$
$$\geq -2\beta_n \|w_n\|^4 B + \beta_n^2 (\|w_n\|^2 - 2)\|w_n\|^4 A^2 \tag{18}$$
$$\geq \beta_n \|w_n\|^4 (-2B + \beta_n A^2 \|w_n\|^2 - 2\beta_n A^2)$$

Hence we get the lower bound $\Delta\|w_n\|^2 \geq \|w_n\|^2$ provided

$$\beta_n \|w_n\|^2 \geq 1 \quad \text{and} \quad -2B + \beta_n A^2 \|w_n\|^2 - 2\beta_n A^2 \geq 1 .$$

The second condition is equivalent to

$$\|w_n\|^2 \geq \frac{1 + 2B + 2\beta_n A^2}{\beta_n A^2} \tag{19}$$

Note that assumed condition $A \leq 1$ implies that the right side of the inequality (19) is greater than $1/\beta_n$. It implies the first condition $\beta_n\|w_n\|^2 \geq 1$. Therefore to have $\|w_{n+1}\|^2 \geq 2\|w_n\|^2$ it is enough to satisfy the condition (19).

We have to show that this condition is true for all $n > 0$ if only it is true for $n = 0$. The inductive step assumes that

$$\|w_{n+1}\|^2 \geq 2\|w_n\|^2 \quad \text{and} \quad \|w_n\|^2 \geq \frac{1 + 2B + 2\beta_n A^2}{\beta_n A^2} .$$

Hence

$$\|w_{n+1}\|^2 \geq 2(2B + n)\frac{1 + 2B + 2\beta_n A^2}{A^2} > (2B + n + 1)\frac{1 + 2B + 2\beta_n A^2}{A^2} >$$

$$> \frac{1 + 2B + 2\beta_{n+1} A^2}{\beta_{n+1} A^2} .$$

Therefore both relations

$$\|w_{n+1}\|^2 \geq 2\|w_n\|^2 \quad \text{and} \quad \|w_n\|^2 \geq \frac{1 + 2B + 2\beta_n A^2}{\beta_n A^2}$$

are true for all $n \geq 0$.

In conclusion, choosing as $w_0 = \alpha e$ where

$$\alpha \geq \sqrt{\frac{2B + 4B^2 + 2A^2}{A^2}}$$

we get

$$\|w_n\| \geq 2^{n/2}\alpha .$$

This proves unboundness of the weight sequence in Oja algorithm.

4 Conclusions

By elementary tools of matrix analysis, we have shown that the discrete dynamical system associated with Oja algorithm is stable in the ball $K(0, 81/64)$ if only gains β_n are bounded by $(2B)^{-1}$, where $B = b^2$ and b is the bound for the learning sequence. We have also defined a general class of Oja's systems (with gains satisfying stochastic convergence conditions) which tend to the infinity with exponential rate if only their initial states are chosen too far from the zero point.

Assuming that zero vectors are skipped in the training random sequence, the above results imply the following heuristic rule for the initial choice of the weight in Oja algorithm:

$$w_0 \doteq \frac{x_0}{\|x_0\|} .$$

References

1. Diamantaras, K.I., Kung, S.Y. (1995) *Principal component neural networks – Theory and applications*, John Wiley & Sons, Inc.
2. Duflo, M. (1997) *Random iterative models*, Appl. Math., 34 Springer.
3. Karhunen, J. (1994) *Stability of Oja's PCA subspace rule*, Neural Computation.
4. Kushner, H.J., Clark, D.S. (1978) *Stochastic approximation for constrained and unconstrained systems*, Appl. Math. Sci., 26. Springer.
5. Kushner, H.J., Yin, G. (1997) *Stochastic approximation and applications*, Springer.
6. Oja, E. (1982) *A simplified neuron model as a principal component analyzer*, J. Math. Biology, 15, 267–273.
7. Skarbek, W., Ignasiak, K. (1997) *Handwritten digit recognition by local principal components analysis*, ISMIS'97, International Symposium for Methodology of Intelligent Systems, 217–226, Charlotte, USA, October 1997.

Transform Vector Quantization of Images in One Dimension

Remigiusz J. Rak

Warsaw University of Technology, Pl. Politechniki 1,
PL00-660, Warsaw, Poland.

Abstract. In this paper there is enclosed a description of the hybrid system for black and white (256 by 256 pixels) images compression. The system includes the following procedures: image decomposition (8x8, 16x16 blocks), DCT transformation, "zig-zag" scanning, product code vector quantization (one- dimensional block) and a bit allocation. The standard LBG algorithm for codebook design has been enriched with the simulated annealing procedure for avoiding the local minima. Standard vector quantization in two-dimensional transform space has been investigated for comparison.

1 Introduction

Compression of 2D signals establishes the most popular application of vector quantization. A natural way to apply basic VQ to images directly is to decompose a sampled image into rectangular 2D blocks of fixed size (M by N points - pixels) and then use these blocks as vectors. It seems that much simpler analysis could be done by moving along rows (horizontally), reading pixel values and then using ready, verified algorithms in 1D space. But in that case, 2D intercorrelations between pixels would be lost irreversibly. After moving into transform (frequency) domain, where the transform coefficients are much less correlated, such a procedure seems to be quite reasonable. However because of a very characterized amplitude distribution of the coefficients (at the plane), using of "zig-zag" scanning is preferred. As a good example there could be shown JPEG standards (DCT, "zig-zag", SQ), where the property of energy compactness was first successfully exploited in practice. Investigations realized by the author has proved that the efficient hybrid method for image compression should consist of:image decomposition, 2D-transformation, "zig-zag" scanning of the coefficients, product code vector quantization and a bit allocation.

2 System Description

Independent coding of amplitude and phase of the spectrum in a case of DFT, effectively implemented in speech coding (phase degradation) [12], does not bring good results in the area of picture coding. Quite opposite, there is needed a higher precision in the quantization of phase coefficients. So that, the investigations

L. Polkowski and A. Skowron (Eds.): RSCTC'98, LNAI 1424, pp. 361–368, 1998.
© Springer-Verlag Berlin Heidelberg 1998

has been reduced to the DCT domain. There has been taken into consideration two versions of the coding system: first in two-dimensional and second in one dimensional space of the frequency domain. Simplified block diagrams of two versions of the system are presented in figures 2 and 3 properly. The symbol "Z" denotes "zig-zag" scanning.

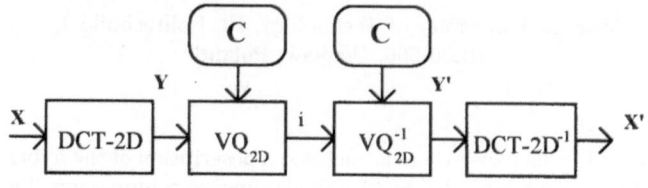

Fig. 1. The block diagram of the system: DCTVQ-2D

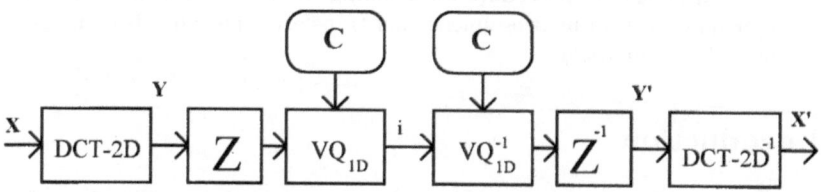

Fig. 2. The block diagram of the system: DCTVQ-1D

2.1   Analytical Description

Let us assume that $\mathbf{x}(m, n)$ denotes a two-dimensional data array (real numbers) with size MxN. The two-dimensional pair of orthogonal transforms (forward and inverse) is defined as:

$$y(k, l) = \sum_{m=0}^{M-1} \sum_{n=0}^{N-1} x(m, n) kern_f(k, l, m, n)$$

$$x(k, l) = \sum_{k=0}^{M-1} \sum_{l=0}^{N-1} y(m, n) kern_f(k, l, m, n)$$

(1)

for $k, m = 0, 1, ..., M - 1$ and $n, l = 0, 1, ..., N - 1$.

In place of kernels it is necessary to introduce proper trigonometric expressions. For the Discrete Cosine Transform:

$$kern_f(k,l,m,n) = \frac{2\alpha(k)\alpha(l)}{\sqrt{MN}} \cos\left(\frac{(2m+1)k\pi}{2M}\right) \cos\left(\frac{(2n+1)l\pi}{2N}\right) \quad (2)$$

where: $\alpha(i) = \frac{1}{\sqrt{2}}$ for $i = 0$, $\alpha(i) = 1$ for the others and $kern_f = kern_i$.

Procedures implemented for the transform computing were based on the modification of the well known FFT algorithm. Standard 256x256 pixel images encoded with the resolution r=8bpp: bird, cat, couple, face, hat, jet, and peppers were used for system design (training sequence) and baboon, lena for checking system parameters. Before the transformation entire image was decomposed into different data blocks: 2x2, 4x4, 8x8 and 16x16. In a case of the small blocks (2x2, 4x4 and 8x8) there was implemented a direct vector quantization in 2D space. In a case of the larger block sizes (8x8 and 16x16) the total values of pixels were 64 and 256 properly. So that there was used a coding procedure with divided data blocks after "zig-zag" scanning, in 1D space. The last system (16x16) was quite compatible with a one implemented to speech signal [12]. The change of the achieved data format to the form adequate for implemented VQ algorithms has been done by using a logarithmic compandor (with an experimentally chosen curve).

2.2 Codebook Design

For the codebooks design there have been used the LBG algorithm (based on GLA) equipped with simulated annealing procedure. The necessary conditions for optimality of the quantizer (iterative Lloyd algorithm) provide the basis for iteratively improving a given codebook. The Lloyd iteration begins with a vector quantizer having already its codebook, and the corresponding optimal nearest neighbor partition. So that, the first problem to be solved is to obtain an initial codebook. There are the variety of techniques for generation of an initial codebook. The simplest conceptual approach towards filling a codebook of N code words is to randomly select the code words according to the source distribution, which can be viewed as a Monte Carlo codebook design. In a case when there is an access to the training sequence, it is simply possible to select N first or randomly chosen training vectors. The Shannon source coding theorems imply that such a random selection will on the average yield a good code.

The iterations start with an initial codebook, which is in the consecutive steps denoted by: C_m. Then, according to the nearest neighbor rule, there is realized partitioning of the entire training sequence into partitions R_i:

$$R_i = \{\mathbf{y} \in R : d(\mathbf{y}, \mathbf{c}_i) \leq d(\mathbf{y}, \mathbf{c}_j), \text{for} j \neq i\} \quad (3)$$

where:

$$d(\mathbf{y}, \mathbf{c}) = \sum_{m=1}^{M} \sum_{n=1}^{N} (y_{m,n} - c_{m,n})^2 \quad (4)$$

After that there are computed the centroids for each partition, the components of a new codebook C_{m+1} :

$$c_i = cent(R_i) = \frac{1}{|R_i|} \sum_{k=1}^{|R_i|} y_k \tag{5}$$

The value of an average global error (for coding all the training sequence) is controlled in each iteration step:

$$D_m = \frac{1}{|C|} \sum_{i=1}^{|C|} \sum_{k=1}^{|R_i|} d(y_k, c_i) \tag{6}$$

where $|C|$ - denotes codebook size, and $|R_i|$ - the size of i-th training sequence partition.

The Lloyd procedure is a kind of descent algorithm, meaning that each iteration always reduces average distortion but the new codebook is generally not very much different from the old one. So that, once an initial codebook is chosen, the algorithm will lead to the nearest local minimum. Since a cost function will generally have many local minima, better or worse, it is clear that the GLA does not have the ability to locate an optimal codebook. There are some methods reducing or eliminating the dependence of the solution on the initial codebook. In order to avoid local minima in a codebook design procedure, in this project, there have been used a simulated annealing procedure with the reaction to decoder (SA- D). There were introduced some perturbations (noise ξ_i) to each codebook component in each iteration step:

$$c_i^m = c_i^m + \xi_i(T_m) \tag{7}$$

The parameter called "temperature" (T_m) is used to refer to the noise variance and a "cooling schedule" to refer to the sequence of the "temperature" values as a function of the iteration number m.

$$T_m = \sigma_m^2 = \frac{\sigma_M^2}{(m+1)^p} \tag{8}$$

where:
σ_M^2 - beginning value of the noise variance averaged for all vector components,
σ_m^2 - current value of the noise variance averaged for all vector components,
m - iteration number,
p - integer value of power.

The LBG algorithm for the training sequence including a simulated annealing process (SA-D type) is presented below:

1. Start with an initial codebook $C = C_0, m = 0$.
2. Traditional Lloyd iteration step: $C_m \rightarrow C_{m+1}$.
3. Introducing perturbations to code vectors: $c_i^m + \xi_i(T_m)$
4. Global average distortion computing: D_{m+1}
5. A comparison of the error reduction level $\Delta D/D$ to the threshold value ϵ : $\Delta D/D < \epsilon$?, Decision: stop or go to step 2.

In this project the "temperature" was defined as:

$$T_m = \sigma_x^2 (1 - \frac{m}{I})^3 \qquad (9)$$

Simulated annealing establishes rather complex and very time consuming process. A novel approach to the non convex optimization of VQ, deterministic annealing (DA), appears to capture the benefits of SA without adding any randomness in the design process. It is conceptually similar to the technique of fuzzy clustering. In DA the statistical description of randomness is incorporated into the cost function.

3 Simulation Results

The lena and baboon were implemented in a testing procedure. The first results, describing the influence of codebook sizes (CB) on the PSNR and the bit rate (BR) values versus vector dimensions (block sizes) are presented in table 1.

TABLE 1. PSNR and BR versus block sizes for different codebook sizes: DCT-VQ,lena

Codebook	2x2	4x4	8x8
size	PSNR[dB]/BR[bpp]		
64	32.56 / 1.400	26.53 / 0.375	19.76 / 0.094
128	34.73 / 1.748	28.18 / 0.437	21.05 / 0.109
256	36.81 / 2.000	29.67 / 0.500	22.19 / 0.125

The influence of vector sizes on the PSNR value (and other parameters) for a standard (CB=256) codebook size is presented in table 2. The presentation does not include the size 16x16 because it was not coded in a direct manner.

TABLE 2. Coding parameters versus codebook sizes: DCT-VQ, CB=256, lena

Block size	BR [bpp]	CR	MSE	RMSE	NMSE [%]	SNR [dB]	PSNR [dB]
2x2	2.0	4	13.55	3.68	0.36	24.41	36.81
4x4	0.5	16	70.02	8.36	1.87	17.23	29.67
8x8	0.125	64	392.54	19.81	10.50	9.78	22.19

The similar results were achieved for baboon (table 3).

TABLE 3. DCT-VQ, lena, baboon

Block size	BR [bpp]	lena PSNR[dB]	baboon PSNR[dB]
2x2	2.0	36.81	34.87
4x4	0.5	29.67	27.83
8x8	0.125	22.19	20.34

The block of size 16x16 is definitely to large for a direct vector quantization. So that there is recommended a procedure of "zig-zag" scanning and partitioning of the 1D block. In this project the entire block has been partitioned into 16 vectors with different sizes (16 points in average) resulting from the implementation of the allocation procedure in a vector area. That case (16 vectors) gave in average bit rate of 0.5bpp, it means the same like in a case of 4x4 with direct VQ, described above. The simulations have proved that the coding quality of the system 1D-VQ (16x16) is better then in a 2D case of 4x4. As an explanation it is enough to remind that in a case of larger data blocks there appears the dipper data mixing effect (data decorrelation) in the transformation process. The vector quantization "has less to do". The similar experiment has been done for the case of the size 8x8. Each block has been partitioned into 16 vectors with sizes 4 points in average. The achieved bit rate was 2bpp. The simulation results seemed to be very promising too. Table 4 includes the final results.

TABLE 4. A comp. of DCT-VQ-2D to DCT-VQ-1D, lena

Block size	BR [bpp]	CR	MSE	RMSE	NMSE [%]	SNR [dB]	PSNR [dB]
2x2	2.0	4	13.55	3.68	0.36	24.41	36.81
4x4	0.5	16	70.02	8.36	1.87	17.23	29.67
8x8	0.125	64	392.54	19.81	10.50	9.78	22.19
8x8 (1D)	2.0	4	11.33	3.36	0.30	25.18	37.59
16x16 (1D)	0.5	16	34.28	5.85	0.91	20.37	32.78

Very important and difficult problems to overcome or eliminate are the blocky effects. Even the high resolution digital images are blocky when viewed at the large scale. The VQ image is however much worse. It could be partly seen in the set of pictures demonstrated in fig.3. The blockness and the "sawtooth" effects along diagonal edges are apparent. They are especially meaningful in a case of large block sizes: 8x8 and 16x16. The implementation of a very simple filtering (smoothing by averaging pixel values) process around the block edges can improve the quality good enough. This was also introduced into the project but it did not influence the value of the PSNR and the results were examined in a subjective manner.

The original image and the ones achieved after vector quantization for different vector (block) sizes and a codebook of 64 code vectors (CB=64) are presented in fig.3.

4   Conclusion

An algorithm of vector quantization of images in Discrete Cosine Transform domain was presented. The two-dimensional array of DCT coefficients was "zig-zag" scanning before the Vector Quantization. There were used large data blocks in order to achieve a better VQ performance. The encoding complexity of DCTVQ was reduced drastically by dividing the 1D block into small vectors (product code with a bit allocation). In order to avoid local minima there was implemented simulated annealing in a codebook design procedure. The achieved performance gain was about 1dB in SNR.

The majority of papers in the area of DCT VQ of images were published at the conference ICASSP and symposium PCS organized in Japan. The main representatives of them are papers [2] and [3]. Work [2] is devoted to the independent vector quantization of RGB components of color images. There is implemented a decomposition of constant size block (8x8) into 15 diagonally taken 1D - vectors and an adaptive bit allocation. Such a system is much more complex but is adequate to a direct comparison with the one proposed above. Generally there exists equality in the area of PSNR (+/-1dB). But the original system 1D (16x16), proposed above, gives the results about 2dB better.

References

1. Abdelwahab, A.A., Kwatra, S.C.: Image Data Compression with Vector Quantization in the Transform Domain. IEEE Int. Conf. on Commun. ICC'86, Toronto, 1986
2. Aizawa, K.,Harashima, H., Miakawa, H.: Adaptive Discrete Cosine Transform Coding with Vector Quantization for Color Images. Proc. ICASSP '86, Tokyo, Japan, 1986
3. Aizawa, K.,Harashima, H., Miakawa, H.: Adaptive Discrete Cosine Transform Coding with Vector Quantization. PCS'86 Picture Coding Symp., Tokyo, Japan, 1986
4. Bellifemine, F., Picco, R.: 2D-DCT coding with Pyramidal Vector Quantization. Picture Coding Symp., Torino, Italy, 1988
5. Cho, N.I., Lee, S.U.: A fast 4x4 DCT for the recursive 2-D DCT. IEEE Trans. Sign. Processing vol.40, Sept.1992
6. Clarke, R.J.: Digital Compression of Still Images and Video. Academic Press, 1996
7. Flanagan, J.K., Morrell, D.R., Frost, R.L., Read, C.J., Nelson, B.E.: Vector Quantization Codebook Generation Using Simulated Annealing. ICASSP, Glasgow, Scotland, May 1989
8. Gersho, A., Gray, R.M.: Vector Quantization and Signal Compression. Kluwer Academic Publishers,1992
9. Gotze, M.: Adaptive Vector Quantization of Images in the Discrete Cosine Transform Domain. Picture Coding Symp. PCS'86, Tokyo, Japan, 1986
10. Marescq, J.P., Labit, C.: Vector Quantization in Transformed Image Coding. Int. Conf. on Acoust. Speech and Sgn. Proc., ICASSP'86, Tokyo, Japan, 1986.
11. Rabbani, M., Jones P.W.: Digital Image Compression Techniques. SPIE Optical Engineering Press, 1991
12. Rak, R.J.: Signal Compression based on Fourier Transform Vector Quantization. Mediterranean Electrotechnical Conf. MELECON'94, Antalya, Turkee, 1994

13. Rak, R.J. : A System For Transform Vector Coding of Images. 3rd International Conference on Signal Processing ICSP'96, Bejjing, China, 1996
14. Rak, R.J.: Wavelet Transform Vector Quantization of Images. 13th International Conference on Signal Processing DSP97, Santorini, Greece, 1997
15. Rao, K.R., Yip, P.: Discrete Cosine Transform. Academic Press 1990.
16. Saito, T., Takeo, H., Aizawa, K., Harashima, H., Miyakawa, H.: Discrete Cosinte Transform Coding System Using Gain/Shape Vector Quantizers and its application to Image Coding. Picture Coding Symposium, PCS'86, Tokyo, Japan, 1986

Fig. 3. The original image and the ones achieved after vector quantization for different vector (block) sizes top-left: original image; top-right: recovered for the size 2x2 (BR=1.5bpp, PSNR=32.56 dB); bottom-left: recovered for the size 4x4 (BR=0.375bpp, PSNR=24.96 dB); bottom-right: recovered for the size 8x8: (BR=0.09375bpp, PSNR=19.76 dB)

Fig. 1 The model image and the ... achieved after an optimization for three ...

Daubechies Filters for 2D Wavelet Transforms

Waldemar Rakowski[1] and Zbigniew Bartosiewicz[2]

[1] Instytut Informatyki, Politechnika Białostocka
Wiejska 45, 15-351 Białystok, Poland
waldi@ii.pb.bialystok.pl
[2] Instytut Matematyki i Fizyki, Politechnika Białostocka
Wiejska 45, 15-351 Białystok, Poland
bartos@cksr.ac.bialystok.pl

[Abstract.] **The wavelet compression method is one of the most effective techniques of digital image compression. Efficiency of this method strongly depends on the filters used to two-dimensional wavelet transform. The fundamental way of construction of finite impulse response filters was given by I. Daubechies. The paper presents a new proof of the fact that the Daubechies filters satisfy the power complementary condition, which is one of the conditions for perfect image reconstruction.**

In [Dau1], [Dau2] I. Daubechies proposed a method of a design of filters for two-dimensional wavelet transform, which is used in image compression [Mal], [Str], [Rak]. Such filters must satisfy the power complementary or Smith–Barnwell condition [Vai], [Str]:

$$|H_0(\omega)|^2 + |H_0(\omega + \pi)|^2 = 2, \tag{1}$$

where $H_0(\omega)$ is the Fourier transform of the impulse response of the filter, i.e.

$$H_0(\omega) = \sum_n h_0(n)e^{-i\omega n}. \tag{2}$$

In this paper we give a new proof of the fact that the polynomial

$$|H_0(\omega)|^2 = 2 \left(\frac{1 + \cos \omega}{2} \right)^p \sum_{n=0}^{p-1} \binom{p+n-1}{n} \left(\frac{1 - \cos \omega}{2} \right)^n. \tag{3}$$

is a solution of the equation (1).

To show that (3) satisfy in fact (1) let us introduce a new variable x:

$$x = \frac{1 - \cos \omega}{2}. \tag{4}$$

It is easy to see that $x \in [0, 1]$. Thus it is enough to prove the following theorem

---

* This work was supported in part by ESPRIT and INCO EC programmes under grant CRIT2
** This work was supported in part by ESPRIT and INCO EC programmes under grant CRIT2

L. Polkowski and A. Skowron (Eds.): RSCTC'98, LNAI 1424, pp. 369–372, 1998.
© Springer-Verlag Berlin Heidelberg 1998

[**Theorem 1.**] *For $x \in [0,1]$ and $p \geq 1$ we have*

$$(1-x)^p \sum_{n=0}^{p-1} \binom{p+n-1}{n} x^n + x^p \sum_{n=0}^{p-1} \binom{p+n-1}{n} (1-x)^n = 1. \qquad (5)$$

[*Proof.*]Let us first observe that (5) holds trivially for $x = 0$ and $x = 1$. So let us assume that $x \in (0,1)$. For such x the formula (5) is equivalent to

$$\frac{1}{x^p(1-x)^p} = \sum_{n=0}^{p-1} \binom{p+n-1}{n} \left(\frac{1}{x^{p-n}} + \frac{1}{(1-x)^{p-n}} \right). \qquad (6)$$

Before we show (6) let us first prove a simpler statement.

[**Lemma 1.**] *For $x \in (0,1)$ and $p \geq 1$ it holds*

$$\frac{1}{x^p(1-x)} = \frac{1}{x^p} + \cdots + \frac{1}{x} + \frac{1}{1-x}.$$

[*Proof.*]The formula clearly holds for $p = 1$. Assume that it holds for some $p \geq 1$. Then

$$\frac{1}{x^{p+1}(1-x)} = \frac{1}{x^{p+1}} + \cdots + \frac{1}{x^2} + \frac{1}{x(1-x)}$$

$$= \frac{1}{x^{p+1}} + \cdots + \frac{1}{x^2} + \frac{1}{x} + \frac{1}{1-x}.$$

Thus, by induction, it holds for all $p \geq 1$.

Switching x and $1-x$ in Lemma 1 we get

$$\frac{1}{(1-x)^p x} = \frac{1}{(1-x)^p} + \cdots + \frac{1}{1-x} + \frac{1}{x}. \qquad (7)$$

The proof of the following identity can be found in [GKP]. It holds for $k \geq 0$:

$$\sum_{n=0}^{k} \binom{m+n}{n} = \binom{m+k+1}{k}. \qquad (8)$$

Now we are going to show (6). It holds for $p = 1$, so let us assume that (6) holds for some $p \geq 1$. We shall show that it holds for $p + 1$. From the assumption, using Lemma 1 and (7), we get

$$
\frac{1}{x^{p+1}(1-x)^{p+1}} = \sum_{n=0}^{p-1} \binom{p+n-1}{n} \left(\frac{1}{x^{p+1-n}(1-x)} + \frac{1}{(1-x)^{p+1-n}x} \right)
$$

$$
= \sum_{n=0}^{p-1} \binom{p+n-1}{n} \left(\frac{1}{x^{p+1-n}} + \cdots + \frac{1}{x} + \frac{1}{1-x} \right.
$$

$$
\left. + \frac{1}{(1-x)^{p+1-n}} + \cdots + \frac{1}{1-x} + \frac{1}{x} \right)
$$

$$
= \sum_{n=0}^{p-1} \binom{p+n-1}{n} 2 \left(\frac{1}{x} + \frac{1}{1-x} \right)
$$

$$
+ \sum_{n=0}^{p-1} \binom{p+n-1}{n} \sum_{k=n}^{p-1} \left(\frac{1}{x^{p+1-k}} + \frac{1}{(1-x)^{p+1-k}} \right)
$$

$$
= 2 \sum_{n=0}^{p-1} \binom{p+n-1}{n} \left(\frac{1}{x} + \frac{1}{1-x} \right)
$$

$$
+ \sum_{k=0}^{p-1} \left(\sum_{n=0}^{k} \binom{p+n-1}{n} \right) \left(\frac{1}{x^{p+1-k}} + \frac{1}{(1-x)^{p+1-k}} \right).
$$

Now we can use (8) and the simple identity

$$
2 \binom{2p-1}{p-1} = \binom{2p}{p}
$$

to get

$$
\frac{1}{x^{p+1}(1-x)^{p+1}} = \binom{2p}{p} \left(\frac{1}{x} + \frac{1}{1-x} \right)
$$

$$
+ \sum_{k=0}^{p-1} \binom{p+k}{k} \left(\frac{1}{x^{p+1-k}} + \frac{1}{(1-x)^{p+1-k}} \right)
$$

$$
= \sum_{k=0}^{p} \binom{p+k}{k} \left(\frac{1}{x^{p+1-k}} + \frac{1}{(1-x)^{p+1-k}} \right).
$$

Thus (6) holds for $p+1$ and, by induction, for every $p \geq 1$.

References

Dau1. Daubechies, I.: Orthonormal bases of compactly supported wavelets, *Comm. Pure Appl. Math.* 41 (1988), 909–996.

Dau2. Daubechies, I.: Ten Lectures on Wavelets. SIAM, 1992.

GKP.   Graham, R. L., Knuth, D. E., and Patashnik, O.: Concrete Mathematics. A Foundation for Computer Science. Addison-Wesley, Reading, MA, 1994.

Mal.   Mallat, S. G.: A Theory for Multiresolution Signal Decomposition: The Wavelet Representation. *IEEE Transactions on Pattern Analysis and Machine Intelligence* 11, No. 7 (1989), 674–693.

Rak.   Rakowski, W.: Wavelet representation of digital images — theory and practice (in Polish). ICS PAS Reports, No. 817, Warsaw, 1996.

Str.   Strang, G., and Nguyen, T.: Wavelets and Filter Banks. Wellesley-Cambridge Press, 1996.

Vai.   Vaidyanathan, P.P.: Multirate Systems and Filter Banks. Prentice-Hall, Englewood Cliffs, N.J., 1992.

Some Heuristics for Default Knowledge Discovery

Tor-Kristian Jenssen, Jan Komorowski and Aleksander Øhrn

Knowledge Systems Group
Dept. of Computer and Information Science
Norwegian University of Science and Technology
N-7034 Trondheim, Norway
{tkj, janko, aleks}@idi.ntnu.no

[Abstract.] In this paper discovery of *default knowledge* as proposed by Mollestad [7], [8], [9], [10] is further investigated. Mollestad's algorithm, as described in [9], is refined and extended in several ways. In particular, new heuristics guiding the search for *default decision rules* are proposed and evaluated. The results so far have been encouraging when the (modified) framework is compared to other rough set methods.

1 Introduction

Knowledge Discovery in Databases (KDD) [2], [15] is motivated by the need for *automated*, *efficient*, and *intelligent* methods for data analysis (summarization, clustering, classification etc.). The ultimate goal of KDD is to discover *knowledge* in the form of useful patterns from raw data.

Rough set theory has been used in KDD and machine learning [21] to find decision rules from data tables [14], [19], [20]. In an abstracted form, these rules are written like $A \Rightarrow B$, and interpreted as "A implies B", or equivalently "if A then B". Such rules are appealing as they closely resemble the way humans represent knowledge in everyday parlor.

In [7], [8], [9], [10], Mollestad et al. proposed to synthesize *default decision rules*, i.e. rules that have exceptions, by systematically removing attributes from the given data. The motivation for inducing rules in this way is twofold. First, data is in general uncertain (due to noise and errors) and inconsistent. This implies that even correct rules in general are approximate. Second, by removing information in the induction, the resulting rules describe more general trends in the data, and equally important, these rules are particularly suitable for classification of *new* objects with missing values (situations with limited knowledge).

Mollestad's algorithm is, to the best of our knowledge, the first approach to automatic synthesis of large knowledge bases for default reasoning. We propose heuristics for an improved version of his algorithm and test the extended framework on real data. The quality of our classifiers seems to be better than those obtained by other rough set methods, and though the costs remain high, the heuristics have greatly contributed in making the approach feasible. The reported work is in progress, and further validation is required, both for the framework in general and in particular for the heuristics.

L. Polkowski and A. Skowron (Eds.): RSCTC'98, LNAI 1424, pp. 373–380, 1998.
© Springer-Verlag Berlin Heidelberg 1998

In the following section, ideas and terminology from default reasoning is reviewed. Section 3 recapitulates rule generation in the rough set approach, and Mollestad's algorithm is briefly recalled in Sect. 4. Heuristics are proposed and discussed in Sect. 5. Results are presented and commented in Sect. 6, and we give preliminary conclusions based on our results in the last section.

2   Default Reasoning

Very informally, default reasoning is about "jumping to conclusions" on the basis of limited information, and is omnipresent in common sense reasoning with inconclusive evidence. Default reasoning is a case of *nonmonotonic reasoning*: The set of consequences does not necessarily increase as new axioms are added, and the addition may invalidate previous consequences.

Several formal approaches to default reasoning have been proposed. The Default Logic (DL) of Reiter [17] (also [1], [6]) formalizes default reasoning by the use of *default rules*. A DL-theory consists of two disjoint sets of formulae, the defaults and the *facts* (axioms). The facts are assumed to be logically valid pieces of knowledge, whereas the defaults describe relations between facts which in general have exceptions. Reiter's approach has two disadvantages: 1) It is necessary to a priori know all exceptions to a default rule and 2) In the case of conflicting defaults, DL offers no means for deciding which is correct (cf. the *multiple extensions problem*).

An alternative formalization is offered by the Preferred Subtheories of Brewka [1]. His ideas generalize the Theorist system of Poole [16]. Also in Theorist, the formulae are divided into a set of *facts*, F, and *defaults*, Δ. Facts are assumed to be irrefutable while the defaults may be counter-proven. Brewka proposed two generalizations to this two-level prioritization. The first is to partition all formulae T (corresponding to $F \cup \Delta$) into k mutually disjoint sets $T_0, \ldots, T_k$, where a formula in T_i is considered more reliable than a formula in any T_j when $j > i$. This approach has two drawbacks: 1) Any formula must be judged against all others, and 2) There may still be conflicts between two formulae (rules) with the same "reliability". The other generalization amends this by defining a partial order between all and only those rules that may conflict.

3   Rule Generation in the Rough Set Approach

Rough set (RS) theory was introduced by Pawlak [12], [13] as a formal tool for approximate reasoning. The basic construct is an *information system* defined as a pair $\mathcal{A} = (U, A)$, where U is a finite, nonempty set of objects, and A is a finite, nonempty set of attributes. Each $a \in A$ is a total function, $a : U \rightarrow V_a$, where V_a is the value-set of a. A *decision table* (DT) is an information system where the set of attributes A is divided into two disjoint, nonempty subsets C (conditions) and D (decisions). Hence forward, D will be assumed to be a singleton, i.e. $D = \{d\}$. For a DT $\mathcal{A} = (U, C \cup \{d\})$, the *indiscernibility relation* with respect to $B \subseteq C$ in $\mathcal{A}$, $IND_{\mathcal{A}}(B)$, is defined as $IND_{\mathcal{A}}(B) = \{(x, y) \in U^2 \mid \forall a \in B \; a(x) = a(y)\}$. The equivalence class of $x \in U$ in $IND_{\mathcal{A}}(B)$ is denoted $[x]_B$. A *reduct* in $\mathcal{A}$ is a minimal (with respect to inclusion) set of attributes $R \subseteq A$ such that $IND_{\mathcal{A}}(R)$

$= IND_\mathcal{A}(C)$. The set of all reducts of $\mathcal{A}$ is denoted $RED_\mathcal{A}$. The *discernibility matrix* of $\mathcal{A} = (U, C \cup \{d\})$, $\mathcal{M}_\mathcal{A}$, is defined as the $n \times n$, matrix $[m_{ij}]$, where $n = |U|$ and $m_{ij} = \{a \in C \mid a(x_i) \neq a(x_j)\}$. For the discernibility matrix *modulo decision* of $\mathcal{A}$, $\mathcal{M}_\mathcal{A}^D$, m_{ij} is defined as $m_{ij} = \{a \in C \mid a(x_i) \neq a(x_j)\}$ if $d(x_i) \neq d(x_j)$ otherwise $m_{ij} = \emptyset$.

Given a DT $\mathcal{A}$, an expression of the form $a = v$, where $a \in (C \cup \{d\})$ and $v \in V_a$ is called a *descriptor* in $\mathcal{A}$. Let $\mathcal{C}_\mathcal{A}$ denote the set $\{\tau \mid \tau = (c_1 = v_1 \wedge \cdots \wedge c_k = v_k)\}$, where $k \geq 0, c_i \in C, v_i \in V_{c_i}, i = 1, \ldots, k$ and $c_i \neq c_j, i \neq j$, and let $\mathcal{D}_\mathcal{A}$ denote the set $\{\tau' \mid \tau' = (d = v_1 \vee \cdots \vee d = v_l)\}$, where $l \geq 1, v_1, \ldots, v_l \in V_d$. The decision rules of interest is then given by the set $RULES_\mathcal{A} = \{\tau \Rightarrow \tau' \mid \tau \in \mathcal{C}_\mathcal{A}, \tau' \in \mathcal{D}_\mathcal{A}\}$. For $\tau \in \mathcal{C}_\mathcal{A} \cup \mathcal{D}_\mathcal{A}$, $[\tau]$ denotes those $x \in U$ satisfying the conditions stated in τ, i.e. if τ is a descriptor $a = v$, $[\tau] = \{x \in U \mid a(x) = v\}$, if $\tau = \tau_1 \wedge \tau_2$, $[\tau] = [\tau_1] \cap [\tau_2]$, if $\tau = \tau_1 \vee \tau_2$, $[\tau] = [\tau_1] \cup [\tau_2]$, and, if $\tau = \neg \tau_1$, $[\tau] = U - [\tau_1]$. For a rule $r = (\tau \Rightarrow \tau') \in RULES_\mathcal{A}$, where $\tau' = d = v_1, \ldots, d = v_l$, the *probability* of r in $\mathcal{A}$, $p_\mathcal{A}(r)$, is defined as $p_\mathcal{A}(r) = max_{i=1,\ldots,l}\{|[d = v_i] \cap [\tau]|/|[\tau]|\}$. The *support* of r in $\mathcal{A}$, $\sigma_\mathcal{A}(r)$, is defined as $\sigma_\mathcal{A}(r) = |[\tau]|$.

4 Default Rules

This section reviews Mollestad's framework for generation of default rules (as presented in [9]). For a DT $\mathcal{A}$, a rule $r \in RULES_\mathcal{A}$ is said to be *definite* if $p_\mathcal{A}(r) = 1$ (aka *true* in $\mathcal{A}$). In [9], a default rule is defined as a rule with $p_\mathcal{A}(r) \geq \mu_t$, where $0 \leq \mu_t \leq 1$. In this paper, the term default rule will mostly be used for rules with probability strictly less than 1.

Mollestad's algorithm may in loose terms be described as a search for default rules over the power set of the condition attributes. A DT $\mathcal{A}' = (U, C' \cup \{d\})$ obtained from a DT $\mathcal{A} = (U, C \cup \{d\})$ by deleting a (nonempty) set $C - C'$, is called a *variant* of $\mathcal{A}$. Mollestad proposed to create new variants by removing attributes so that *indeterminacies* are introduced, and subsequently generate rules from all resulting variants. A (new) indeterminacy is introduced by removing attributes discerning objects with different decisions. To formalize, let $\Phi_\mathcal{A}^D = \{\phi \mid \phi \in \mathcal{M}_\mathcal{A}^D, \phi \neq \emptyset\}$, i.e. the set of nonempty discernibility factors modulo decision. Furthermore, let $\hat{\Phi}_\mathcal{A}^D = \{\phi \in \Phi_\mathcal{A}^D \mid \forall \phi' \in \Phi_\mathcal{A}^D \ \phi' \not\subset \phi\}$. Then a *constructed* variant is defined in the following inductive way: i) (Base case) $\mathcal{A}$ is constructed, ii) (Induction step) if $\mathcal{A}' = (U, C' \cup \{d\})$ is a constructed variant, then the DT $\mathcal{A}'' = (U, C'' \cup \{d\})$ is constructed if and only if $C' - C'' \in \hat{\Phi}_{\mathcal{A}'}^D$.

5 Heuristics

Even the set of constructed variants may have size exponential in the number of condition attributes. This forces heuristics to be employed in order to further restrict the set of variants from which to generate rules. The following heuristics can be divided into two categories: 1) Restrictions on the set of discernibility factors used when generating new variants, and 2) Performance-based heuristics combined with thresholds for cutting search.

5.1 Factor Filtering Heuristics

The notion of a constructed variant is based on the set of discernibility factors that are not supersets of any other factor. This implies that at each step a "minimal" number of attributes are removed. In order to faster find the most general rules, and also to find a larger number of new rules in each new variant it is better to remove larger sets of attributes. In this respect, using the "longest" factors in the set $\Phi_{\mathcal{A}}^{D}$ seems a better option. To this end, we propose the following heuristics for filtering the set $\Phi_{\mathcal{A}}^{D}$: 1) Use only factors that are longer than average[1], 2) Use the $c \times |\Phi_{\mathcal{A}}^{D}|$ longest factors ($c \in (0,1]$ is a parameter), and 3) Use the N longest factors (parameter N).

Basing filtering criteria on factor length is a subcase of a more general strategy employing attribute costs (length corresponds to uniform costs). As a refinement to the above heuristics we propose to assign costs to each attribute and calculate costs for each factor as the cost of the attribute sets of the resulting variant. The cost of an attribute may reflect resource-requirements with respect to acquiring information on an attribute or represent an inverted measure of the attribute's information content. Strategies 1) to 3) may thus be rephrased in terms of least cost factors.

5.2 Performance-Based Heuristics

These heuristics are based on the assumption that removing attributes from a variant with high "performance" along various dimensions results in new variants which also perform well. The purpose of the discovery process may be to find strong individual patterns, to find compact, yet accurate models of the data, or to find classifiers for prediction of new objects.

Assume a DT $\mathcal{A} = (U, C \cup \{d\})$ is given, and that $\mathcal{A}' = (U, C' \cup \{d\})$ is a variant of $\mathcal{A}$. In [9], the following two measures of variant interestingness are proposed: $glued(\mathcal{A}') = |\{x \in U \mid [x]_C \subset [x]_{C'}\}|$, and $kept(\mathcal{A}') = |\{x \in U \mid d(x) = d([x]_{C'})\}|$, where $d([x]_{C'})$ is the decision supported by the most objects in $[x]_{C'}$. Larger equivalence classes results in more general rules (high value of $glued$), and variants with a high $kept$-value result in rules with high probabilities.

If the purpose is to find classifiers or accurate models, the rule accuracy of a variant (the variant's rule set) is an obvious heuristic. Another measure of classificational power is found in the area under the *Receiver Operating Characteristic* (ROC) curve [3]. If classifier performance is the only issue, it seems as a good idea to maximize performance on one (or both) of these two measures. As a description of the given data, model transparency (comprehensibility) is important. When smaller models are desired, the complexities of resulting sets of rules and reducts may be suitable heuristics. Furthermore, measures of interestingness of individual rules should be employed when the aim is to find the strongest patterns.

The proposed measures of variant "quality" can be combined with thresholds to decide if further search should be stopped, or together with a maximum, N, for the total number of variants in order to guide search to the N most

---

[1] This was initially proposed in [9].

promising variants based on performance on any of the proposed measuring functions. These heuristics are readily employed in conjunction with the notion of constructibility, but may also be used alone.

6   Experiments and Results

To give an indication of the value of synthesizing default rules, we compare the rule sets found with our algorithm with models obtained by "standard" RS methods. All experiments have been carried out with the ROSETTA [18], [22] toolkit wherein the framework for default rules generation has been implemented. The available reduct calculation algorithms are: Dynamic, Exhaustive, Genetic, and Johnson. In all experiments, object-related reducts were calculated, and within the default rules framework the Exhaustive (exact reduct calculations) algorithm was invoked for each variant[2]

The following data sets, all from the *UCI Repository of Machine Learning and Domain Theories* [11], were chosen for analysis: Australian Credit Card Approval (AUS), Cleveland Heart Disease (CLE), Hepatitis (HEP), and Breast Cancer (BCO) and Lymphography (LYM) from the Oncology Institute.[3] The test strategy was single train-test validation by randomly splitting the original data set into two disjoint sets; training on one and validating against the other. For domains with continuous attributes, scaling was carried out after splitting by discretizing the training table and subsequently the test table was scaled with the same cuts. The best accuracies obtained with the default rules algorithm are contrasted with other rough set methods in Tab. 1. Note, for the Dynamic and Genetic algorithms all parameters were kept at their default values (cf. [18]).

Table 1. *Comparison of default rules synthesis to other rough set methods. Number in parenthesis refers to percentage of objects (of total data set) used for training*

Method	AUS (20)	CLE (30)	HEP (30)	BCO (30)	LYM (40)
Default	.8714	.8374	.9074	.7600	.8315
Dynamic	.8315	.8079	.8241	.7300	.7416
Exhaustive	.8424	.7783	.8148	.7150	.7416
Genetic	.8406	.7734	.7963	.7150	.7640
Johnson	.7736	.7734	.7870	.6600	.6854

In order to assess the heuristics, train-test pairs from three of the domains were taken for further analysis using the factor filtering heuristics (uniform costs) with the default rules algorithm. Table 2 summarizes results from this part of the experiments.

---

[2] Employing approximate reduct calculations may reduce time requirements, but the current implementations of the Johnson and the Genetic algorithms had deficiencies rendering this impossible.

[3] For detailed information on these domains, the reader is referred to [11].

Table 2. *Factor filtering heuristics. AVG refers to the strategy of using factors longer than average, R (c) refers to using the $c \times |\Phi_{\mathcal{A}}^D|$ longest factors, and N (n) refers to using the n longest factors*

			Heuristic			
Domain	None	AVG	R (.40)	R (.60)	N (100)	N (300)
AUS	.8714	.8659	.8659	.8659	.8587	.8659
CLE	.8374	.8177	.8177	.8227	.7783	.7980
HEP	.9074	.8704	.8704	.8704	.8704	.9074

6.1 Comments and Observations

A caveat is required before commenting the results: The accuracy of the best variant's rule set cannot immediately be recognized as the performance of the algorithm as a classifier inducer (in that case, it would be necessary to predict this variant from the training table). However, what the results in Tab. 1 do suggest is that valuable information is missed when rule generation is based on the discernibility of all attributes, and hence, that default rules synthesis actually is worthwhile. From all data sets, a large number of variants giving rule accuracy better than the input tables were found. Considering model size, the advantage of these variants is further strengthened, and even if the rule sets of variants did not obtain better accuracy than the standard RS methods, they are valuable in offering more compact and thus more comprehensible models.

Filtering the factor set significantly reduced the total number of variants and in particular variants with many attributes. This resulted in a considerable boost of efficiency, though, as indicated by Tab. 2, with some loss of information. Removing all factors with less than k attributes, for some $1 \leq k \leq |C|$ from a DT $\mathcal{A} = (U, C \cup \{d\})$ effectively excludes all variants with more than $|C| - k$ attributes. If the optimal, in some sense, variant has more than this number of attributes, a filtering strategy removing all factors with less than k attributes is bound to miss this variant. A priori finding the correct value of this k is clearly not trivial, and in using any filtering strategy this aspect should be kept in mind.

Comparing the obtained results with previously reported studies is difficult. Actual observed rule-accuracy is a product of a number of steps each having several options, and without detailed knowledge of the constituent steps in other studies, we refrain from making any concrete comparisons. In the experiments we used very small training tables compared to, for instance, what would be the case in a 10-fold cross-validation setting, and though our results are based on a single partitioning split this fact suggests that the reported accuracies are fairly conservative estimates. Even so, the obtained accuracies, also for the heuristics, compare quite well with numbers reported elsewhere (cf. [4], [9], [11]), which indicate that defaults generated from a small number of training examples generalizes well to new data. It is interesting to observe that the default rules perform particularly well on the HEP and LYM data sets which both has "many" attributes. A plausible, albeit ad hoc, explanation for this is that there may be several irrelevant attributes which do not contribute in prediction but confuse the standard RS approaches.

7   Conclusions

The heuristics significantly reduced time requirements and gave encouraging results with respect to performance of the best variant. Among the factor-filtering heuristics, the strategy of using the $c \times |\varPhi_{\mathcal{A}}^D|$ longest factors seems more promising than the other two. This strategy is more flexible than considering only factors longer than average, and though the N-longest strategy also can be adjusted it is absolute in that it does not consider the actual number of factors. Filtering on attribute costs have not been tested yet, but is an aspect that should be given high priority in future experiments. The performance measuring heuristics have not been assessed in these experiments. In [5], promising results were obtained combining performance heuristics with a maximum number of variants in a bottom-up search.

For all practical purposes the collection of all synthesized rule sets is unwieldy, and there is a high degree of overlap in that several rules are repeated. We propose the following three ways to utilize the result of default rules generation: 1) Extract a specified number of the most interesting rules, defined by performance on some rule-interestingness measure, 2) Collect the union of all rules and construct a prioritized knowledge base in which a partial order among rules is defined by ordering on rule-specificity (cf. partially ordered Preferred Subtheories, Brewka), and 3) Refine the algorithm by defining a predictor-function in order to predict a variant and return its rule set as the result of the algorithm as a classifier inducer.

Acknowledgments

Torulf Mollestad was always ready to discuss this work and provided several useful comments. We also thank the repository administrators at [11] and the domain donors.
This research has been supported in part by the NFR grants #74467/410 and #107409/320.

References

1. Brewka, G. *Nonmonotonic Reasoning: Logical Foundations of Commonsense*, Cambridge University Press, 1991
2. Fayyad, U. M., Piatetsky-Shapiro, G., Smyth, P., and Uthurasamy, R. (eds.) *Advances in Knowledge Discovery and Data Mining*, AAAI Press / MIT Press, 1996
3. Hanley, J. A. and McNeil, B. J. *The Meaning and Use of the Area under a Receiver Operating Characteristic (ROC) Curve*, Radiology, 143, pp. 29-36, 1982
4. Holte, R. C. *Very simple classification rules perform well on most commonly used datasets*, Machine Learning, 11, pp. 63-90
5. Jenssen, T.-K. *Refinements to Mollestad's Algorithm for Synthesis of Default Rules*, MSc Thesis, Norwegian University of Science and Technology, 1998
6. Lukaszewics, W. *Non-Monotonic Reasoning - formalization of commonsence reasoning*, Ellis Horwood, 1990
7. Mollestad, T. *Learning Propositional Default Rules using the Rough Set Approach*, In: Aamodt, A. and Komorowski, J. (eds.), Scandinavian Conference on Artificial Intelligence, pp. 208-219, IOS Press, 1995

8. Mollestad, T. and Skowron, A. *A Rough Set Framework for Data Mining of Propositional Default Rules*, In: Michalewicz, Z. and Ras, Z. R. (eds.), Proc. of the 9th Intl. Symposium on Intelligent Systems, ISMIS '96, pp. 448-457, Springer Verlag, 1996

9. Mollestad, T. *A Rough Set Approach to Data Mining: Extracting a logic of default rules from data*, PhD Thesis, Norwegian University of Science and Technology, 1997

10. Mollestad, T. and Komorowski, J. *A Rough Set Framework for Propositional Default Rules Data Mining*, To appear in: Pal, S. K. and Skowron, A. (eds.), Fuzzy Sets, Rough Sets and Decision Making Processes, Springer-Verlag Singapore Pte Ltd, 1998

11. Murphy, P. M. *UCI Repository of machine Learning and Domain Theories*, At: http://www.ics.uci.edu/~mlearn/MLRepository.html

12. Pawlak, Z. *Rough Sets*, In: Intl. J. of Information and Computer Science, 11(5) pp. 341-356, 1982

13. Pawlak, Z. *Rough Sets: Theoretical Aspects of Reasoning about Data*, Kluwer Academic Publisher, 1991

14. Pawlak, Z. and Skowron A. *A Rough Set Approach to Decision Rules Generation*, Technical Report, Warzaw University of Technology, 1993

15. Piatetsky-Shapiro, G. and Frawley, W. J. (eds.) *Knowledge Discovery in Databases*, AAAI Press/MIT Press, 1991

16. Poole, D. L. *A Logical Framework for Default Reasoning*, J. of Artificial Intelligence, 36, pp. 27-47, 1988

17. Reiter, R. *A Logic for Default Reasoning*, Computational Intelligence, 13, pp. 81-132, 1980

18. The ROSETTAWWW homepage At: http://www.idi.ntnu.no/~aleks/rosetta/

19. Skowron, A. *Boolean Reasoning for Decision Rules Generation* In: Komoroski, J. and Ras, Z. W. (eds.), 7th Intl. Symposium for Methodologies for Intelligent Systems (ISMIS '93), pp. 295-305, Springer Verlag, 1993

20. Skowron, A. *Synthesis of Adaptive Decision Systems from Experimental Data*, In: Aamodt, A. and Komorowski, J. (eds.), Proc. 5th Scandinavian Conference on Artificial Intelligence, Trondheim, Norway, May 29–31, Frontiers in Artificial Intelligence and Applications, Vol. 28, pp. 220–238, IOS Press, 1995

21. Ziarko, W. P. (ed.) *Rough Sets, Fuzzy Sets, and Knowledge Discovery - Proc. of the Intl. Workshop on Rough Sets and Knowledge Discovery (RSKD '93)* , Springer Verlag, 1993

22. Øhrn, A., Komorowski, J., Skowron, A. and Synak, P. *The Design and Implementation of a Knowledge Discovery Toolkit Based on Rough Sets - The ROSETTA System*, To appear in: Skowron, A. and Polkowski, L. (eds.), Rough Sets in Knowledge Discovery, Physica Verlag

Fuzzy Partitions II: Belief Functions
A Probabilistic View

T. Y. Lin

Department of Mathematics and Computer Science, San Jose State University,
San Jose, California 95192 , USA
tylin@cs.sjsu.edu

1. Discrete Random Variables

In this section, we will examine the classical random variables from the point of view of belief functions. Let (C, Θ, Pr) be a probability space, where C is the universe, Θ is a σ-algebra, and Pr is a probability measure [2]; we will be interested in the case C is a finite set.

1.1. Suppose C is a finite set. *A real-valued function X: $C{\rightarrow}R$, is a measurable function*, if all its inverse images are measurable (this definition is equivalent to that of [2] only when C is finite). Let the set of distinct values be $S = \{a_1, a_2,..., a_n\}$. We will be interested in the inverse image $X^{-1}(a_i)$ of a_i under X; for convenience we will write

$$X_i = X^{-1}(a_i) = \{c \mid c \in C \text{ and } X(c) = a_i \}$$

By abuse of notation, we also use X to denote the collection of inverse images

$$X = \{X_i : i = 1, 2, ..., n\}$$

So X *is a random variable as well as the collection* of inverse images, which forms a partition on U. The collection X generates a σ-algebra $\Theta(X) \subseteq \Theta$.

1.2. A new probability P_X can thus be defined by setting $P_X(X_i) = Pr(X_i)$, and extending it to $\Theta(X)$ additively, namely, for $Y = \cup_j X_j$,

$$P_X(Y) = \Sigma_j P_X(X_j), \text{ where j varies through a finite subset of } \{1, 2, .., n\}.$$

In particular, the total sum $P_X(C) = \Sigma^n_{i=1} P_X(X_i) = 1$. So $(C, \Theta(X), P_X)$ is a new probability space.

1.3. The inner probability can be expressed by

$$P_{X*}(A) = SUP\{\Sigma_j P_X(Y_j) \mid A \supseteq Y = \cup_j Y_j \text{ and } Y_j \in X\},$$

where SUP is the least upper bound.

L. Polkowski, A. Skowron (Eds.): RSCTC'98, LNAI 1424, pp. 381–386, 1998.
© Springer-Verlag Berlin Heidelberg 1998

1.4. So given a random variable, we have

(S1) a partition X on C, and
(S2) a numerical value, $m(X_i) = P_X(X_i)$, $i = 1, 2, \ldots, n$, such that

(S3) $\sum_{i=1}^{n} m(X_i) = 1$.

and *the inner X-probability is the belief function.*

$$Bel(A) = P_{X*}(A) = SUP\{\sum_j m(Y_j) \mid A \supseteq Y = \cup_j Y_j \text{ and } Y_j \in X\}$$

It is clear Shafer theory is a generalization of random variables, in which (S1) is weakened to a partial covering, yet the theory computes belief functions as if X were disjoint. This observation spells out the precise point of our motivation in re-interpreting Shafer theory [6], [7], [8]. The main goal of this paper is to re-interpret the notion of focal elements: even though they appear to have non-empty intersections, *their basic probability assignments are, in fact, "measures" of disjoint fuzzy sets. Focal elements are merely the support of these fuzzy sets.*

1.5. Let A be any subset of C. The lower and upper approximation are defined by

$$L[A] = \{Y_j \in X : \exists \ Y_j \subseteq A\}; \quad H[A] = \{Y_j \in X : \forall \ Y_j, \ Y_j \cap A \neq \varnothing\}$$

It is clear that $L[A] \in \Theta(X)$, and we have

Proposition $Bel_X(A) = P_{X*}(A) = P_X(L[A])$

Similar results, mainly on concrete counting measures, are obtained by many rough setters; e.g. see [9] for references and exposition. There are generalizations ([6], [8], [10]).

2. Belief Functions -- Shafer Theory

2.1. Let Ω be a finite set and 2^Ω be its power set. A unit interval valued function $m : 2^\Omega \rightarrow [0, 1]$, is a *basic probability assignment* (**bpa**) if

$m(\varnothing) = 0$ and

$\sum_n m(E_n) = 1$, where E_n varies through 2^Ω.

A set E_n is a focal element, if $m(E_n) \neq 0$.

2.2. Belief function Bel can be defined by basic probabilities:

Bel $(A) = \Sigma \{m(B): B \subseteq A\}$, where the summation runs through all subsets of A.

2.3. Let $C = \{1, 2, 3, 4, 5, 6, 7, 8\}$. Two focal elements and their bpa's are

$X = \{1,2,3,4,5\}$ and $Y = \{2,3,4,5,6\}$;
$m(X) = 2/3$, $m(Y) = 1/3$, and other bpa's are zero.

Then, by definition,

$Bel(X) = 2/3$, $Bel(Y) = 1/3$, and $Bel(X \cap Y) = 0$.

The belief measure of atomic intersection is zero; an intersection is called atomic if it does not contain any focal element. Intuitively, it implies that the evidences of any two atomic events are always "independent;" atomic event is an event that has no sub-events with positive evidences.

2.4. Let $C = \{1, 2, 3, 4, 5\}$ be the universe. Let B be a binary relation $B \subseteq C \times C$, which is defined by

$B_1 = \{x \mid (x, 1) \in B\} = \{1, 2, 3\}$
$B_2 = \{x \mid (x, 2) \in B\} = B_3 = \{x \mid (x, 3) \in B\} = \{1, 2, 3, 4\}$
$B_4 = \{x \mid (x, 4) \in B\} = \{2, 3, 4\}$, $B_5 = \{x \mid (x, 5) \in B\} = \varnothing$

Each B_i consists of those points that are B-related to the element i; they are referred to as elementary B-neighborhoods. We will show that "direct" generalizations are invalid: Let $\Theta(B)$ be the σ-ring generated by $\{B_1, B_2, B_3, B_4, B_5\}$. Suppose a measure is defined on $\Theta(B)$ as follows:

$$Pr_B(B_1) = Pr_B(B_4) = 3/10, \ Pr_B(B_2) = Pr_B(B_3) = 2/5, \ Pr_B(B_5) = 0.$$

Its inner measure is:

$$Pr_B*[\{1, 2, 3, 4, 5\}] = SUP\{Pr_B(B_1 \cup B_4 \cup B_5), Pr_B(B_2 \cup B_5), Pr_B(B_3 \cup B_5)$$
$$= SUP\{Pr_B(B_1 \cup B_4), Pr_B(B_2), Pr_B(B_3)\} = 2/5, \text{ since } B_5 = \varnothing.$$

Next, we take the measure of B_i as bpa 's,

$$m_B(B_i) = Pr_B(B_i), \ i = 1, 2, 3, 4, \text{ and other bpa's are zero,}$$

and we have

$$Bel_B(\{1, 2, 3, 4, 5\}) = m_B(\{1, 2, 3\}) + m_B(\{2, 3, 4\}) + m_B(\{1, 2, 3, 4\})$$
$$= Pr_B(B_1) + Pr_B(B_4) + Pr_B(B_1 \cup \{4\}) \ = 1$$

The difference results from the fact that belief value sums up the multiple measures of the *overlapping areas*; B_1 is measured twice in belief value, only once in the inner probability. We believe that bpa's should not be interpreted as measurements of focal elements, they are measurements of fuzzy sets whose supports are focal elements. With such interpretations, belief functions are the inner probabilities of granular fuzzy sets [3], [4], [5]; this is the subject of next section.

3. Belief Functions – A New View

3.1. Suppose we are given a bpa, $m : 2^C \rightarrow [0, 1]$. Let $Y_i : i = 1, 2, ..., s$ be the focal elements. Let χ_i be the characteristic function of Y_i. Then we define our *target* membership functions (fuzzy sets and membership functions are synonyms) by

(1) $TY_i(c) = m(Y_i)\chi_i(c), i = 1, 2, .., s.$

We write $n = s+1$ and define: $TY_n(c) = 1 - \sum_{i=1}^{s} TY_i(c)$, or equivalently,

(2) $\sum_{i=1}^{n} TY_i(c) = 1.$

It is clear that $TY_n(c) \geq 0$; it is a legitimate membership function. Next we set bpa of TY_i, $i = 1, 2, ..., n$:

(3) $m(TY_i) = m(Y_i), i = 1, 2, .., s,$ and $m(TY_n) = 0.$

(4) $\sum_{i=1}^{n} m(TY_i) = \sum_{i=1}^{s} m(Y_i)\chi_i(c) + m(TY_n) = 1$

3.2. Let us recall the notion of fuzzy partitions introduced in [1]. A family of fuzzy sets FX_i, $i = 1, 2, ..., t$ is said to be a BH- partition, if $\sum_{i=1}^{t} FX_i,(c) = 1$

Proposition $\{TY_i \mid i = 1, 2, .., n\}$ is a BH-partition.

3.3. Let $U = C \times [0, 1]$ be called the total space [5]. Consider the natural projection NP: $U \rightarrow C$. We will show that there is a crisp partition $\{X_i\}$ on U such that

$TY_i(c) = \alpha ([c] \cap X_i) / \alpha([c]) = \alpha ([c] \cap X_i)$

where $[c] = c \times [0, 1]$, and $\alpha (c \times S) = \alpha(S)$ is the Lesbegue measure of $S \subseteq [0,1]$. Given a membership function TY_i we will define X_i as follows: At each point $c \in C$, we set $c_0 = 0, c_n = 1$. For $i = 1, 2,.. , n$, we define half-open intervals,

$I_c = [c_i, c_{i+1})$ such that $TY_i(c) = c_i - c_{i-1}$, and set

$$X_i = \cup_c \{ c \times [c_{i-1}, c_i) : c \in C \}$$

Such an X_i is called a *realization* of TY_i ; it is a crisp set representation of a fuzzy set TY_i . The 6-tuple

$$(U, NP, C, \alpha, \{TY_i: i = 1, 2,.., n\}, \{X_i: i = 1, 2,.., n\})$$

is called the context of $\{TY_i : i = 1, 2,.., n\}$. Since $\sum^n_{i=1} TY_i(c) = 1$,

$I_c = \{[c_{i-1}, c_i) : i = 1, 2, ..., n\}$ is a crisp partition on $[0, 1]$, and

$X = \{X_i : i = 1, 2, ..., n\}$ is a crisp partition on $U = C \times [0, 1]$.

So, we have

(S1) a partition $X = \{X_i : i = 1, 2, ..., n\}$ on U, and

(S2) a numerical value, $m(X_i) \equiv m(TY_i) = m(Y_i)$, $i = 1, 2, ..., n$, such that

(S3) $\sum^n_{i=1} m(X_i) = 1$

These three conditions clearly say that X can be treated as a random variable on the total space U. Next, we will define a belief function on C by X.

$$Bel_X(A \times [0, 1]) = P_{X*}(A \times [0, 1]) = SUP\{\Sigma_j m(X_j) \mid A \times [0, 1] \supseteq \cup_j X_j \}$$

Observe that $A \times [0, 1] \supseteq \cup_j X_j$ if and only if $A \supseteq \cup_j Y_j$, so, we have

$$P_{X*}(A \times [0, 1]) = SUP\{\Sigma_j m(X_j) \mid A \times [0, 1] \supseteq \cup_j X_j \}$$
$$= SUP\{\Sigma_j m(Y_j) \mid A \supseteq \cup_j Y_j \} = Bel_Y(A)$$

So we have defined $Bel_Y(A)$ in terms of the probability P_X of random variable X. Therefore Bel_Y is a well-defined *set function*.

5. Conclusions

In this paper, we show that we can view the bpa of focal elements as probabilities of fuzzy sets that have those focal element as their supports. Using the notion of granular fuzzy sets [3], [4], these fuzzy sets on base space C can be realized by crisp sets on the total space U. Thus we have the setting of classical random variables on the total space. We show that the belief functions of fuzzy sets on C are the inner probabilities of the random variables on U. So belief functions is a sound and well-defined set function. These findings also imply that we can introduce an additive fuzzy measure theory, in fact it is a generalized function on membership functions; this will be reported in near future.

References

1. Bezdek, J. and Harris, J. (1978), Fuzzy Partitions and Relations: An Axiomatic Basis for Clustering, *Fuzzy Sets and Systems* 1, 112-127, 1978
2. Halmos, P. (1950), *Measure Theory*, Van Nostrand, 1950.
3. Lin, T. Y. (1997), Rough Sets, Fuzzy Sets and Their Interactions. In: Proceedings of Fifth National Conferences on Fuzzy Theory and Applications (Fuzzy 97), Dec 19-20, Tainan, Taiwan, 1997
4. Lin, T. Y. (1998a), Sets with Partial Memberships. In: Proceedings of 1998 IEEE World Congress on Computational Intelligence, Anchorage, Alaska, May 4-9, 1998
5. Lin, T. Y. (1998b) Fuzzy Partitions: Rough Set Theory. In : Proceedings of the Seventh Conference on Information Processing and Management of Uncertainty in Knowledge-Based Systems, July 6 - 10, Paris, France, 1998
6. Lin, T. Y.(1998c) Granular Computing of Binary relations II: Rough Set Representation and Belief Functions. In: Rough Sets and Knowledge Discovery, Polkowski and Skowron (eds), Springer-Verlag, 1998 (to appear).
7. Lin, T. Y. and Liau, C. J. (1997) Belief Functions Based on Probabilistic Multivalued Random Variables. In: Proceedings of Joint Conference of Information Science, Research Triangle Park, North Carolina, March 1-5, 1997, 269-272.
8. Lin, T. Y. and Yao, Y. Y., Neighborhoods Systems and Belief Functions Proceedings of The Fourth Workshop on Rough Sets, Fuzzy Sets and Machine Discovery, Tokyo, Japan, November 8-10,1996, 202-207.
9. Skowron A., and Grzymala-Busse, J.W., (1994), From rough set theory to evidence theory. In: Advances in the Dempster Shafer Theory of Evidence, R.R Yaeger, M. Fedrizzi and J. Kacprzyk (eds.), John Wiley & Sons, Inc., New York, Chichester, Brisbane, Toronto, Singapore, 193--236.
10. Yao and Lingras (1998, Interpretations of belief functions in the theory of rough sets. Information Sciences, Vol. 104, No. 1-2, pp. 81-106, 1998)
11. Zadeh, L. A.(1965), Fuzzy sets, *Information and Control*, 8, 338-353, 1965.

Frameworks for Mining Binary Relations in Data

T.Y. Lin [*,1,2], N. Zhong & J. J. Dong[3] and S. Ohsuga[4]

[1] Department of Mathematics and Computer Science
San Jose State University, San Jose, California 95192
[2] Department of Electric Engineering and Computer Science
University of California, Berkeley, California 94720
tylin@cs.sjsu.edu, tylin@cs.berkely.edu
[3] Department of Computer Science and Systems Engineering
Yamaguchi University, Tokiwa-Dai, 2557. Ube 755, Japan
{zhong,dong}@ai.csse.yamaguchi-u.ac.jp
[4] Department of Information and Computer Science
Waseda University, 3-4-1 Okubo Shinjuku, Tokyo 169. Japan
ohsuga@ohsuga.info.waseda.ac.jp

Abstract. This paper extends the notion of information tables and concept hierarchies of equivalence relations to binary relations. So extended rough set theory and attribute oriented generalization techniques can be used to mining binary relations in data.

1 Introduction

By giving each elementary set (equivalence class) a name, we represent a universe by an information table. If we apply the same naming procedure to a nested sequence of equivalence relations, we get a concept hierarchy [2,4]. So traditional data mining is essentially a mining of equivalence relations.

In [7], Lin showed that by naming each granule of binary relations (elementary neighborhoods), we represent the universe by, roughly speaking, a "topological" relational data model, or precisely speaking an information tables with neighborhood systems [15,14,3]. Similarly concept hierarchies can be extended to nested sequences of binary granulations [16]. Thus we have an environment of computing binary relations, and hence we can extend the traditional data mining techniques to mining binary relations [2,5,4,10,18,11,6].

2 Binary Relations and Binary Neighborhood Systems

We will recall few relevant terms from the theory of neighborhood systems [15,9,6]. All results are valid for both crisp and fuzzy worlds. For simplicity, we will state the results in crisp terms. Let U and V be two crisp sets; V is called an object space, U a data space.

* partially supported by EPRI, SJSU, NASA NCC2-275, ONR N00014-96-1-0556, LLNL 442427-26449, ARO DAAH04-961-0341, and BISC at UC-Berkeley

1. *Binary neighborhood system*: With each object $p \in V$, we associate a (possibly empty)subset B_p, called the *elementary, binary or basic* neighborhood at p. The map

$$B : V \longrightarrow 2^U; p \longrightarrow B_p \subseteq U, \text{ or the collection } \{B_p\}$$

   is referred to as the *binary neighborhood system*.

2. *Binary relations*: B is said to be a binary relation on U and V if $B \subseteq V \times U$. A binary relation defines a binary neighborhood system by setting

$$B_p \equiv \{u \mid (p, u) \in B\}.$$

   If the binary relation is an equivalence relation ($V = U$), elementary neighborhoods are pairwise disjoint; so they form a partition. We will regard an elementary set as the neighborhood of its points. In this paper, we will use elementary sets in this sense. See example below.

3. A subset $X \subseteq U$ is a definable neighborhood/set, if X is a union of elementary neighborhoods/sets. If a definable neighborhood/set X contains a elementary neighborhood/elementary set B_p, it is a definable neighborhood of p. In fuzzy world, the union is expressed by means of a t-norm, for example, max operation.

4. For simplicity, a space with a neighborhood system is called a NS-space; it is a mild generalization of Frechet(V) space.

5. *Clump System*: A neighborhood system is called a clump system , if there are additional information structures imposed on the neighborhood system [6,7,8]. In this paper, a concept hierarchy is the additional information structure; see Section 3. In general, information structure is an intuitive, not a formalized, notion;

3   Concept Hierarchies of Binary Relations

A concept hierarchy is best represented best as a tree. A node represents a concept c. Each parent node represents the concept immediately more general than(strongly depend on; see below) c and each child node represents a concept immediately more specific than c. The leaf nodes have no children and represent the base concepts. Traditional concept hierarchies require that all sets of siblings form elementary sets. In this section, we relax the equivalence relation to a general binary relation (among sibling concepts). Siblings form elementary neighborhoods, not necessarily elementary sets [16].

1. Let B^2 and B^1 be two binary neighborhood systems. We say B^2 is *strongly depended* on B^1, denoted by

$$B^1 \Longrightarrow B^2,$$

   iff every B^2-neighborhood is a definable B^1-neighborhood.

2. If $B^1 \Longrightarrow B^2$, we will say B^2 is *definably finer* than B^1 or B^1 is *definably coarser* than B^2.

A *concept hierarchy* is a nested sequence of neighborhood systems, in which each elementary neighborhood is given a name, called an elementary concept or simply concept:

1. $B_{j,i}^0$ is an elementary neighborhood $\subseteq U$;
   $C_{j,i}^0 = NAME(B_{j,i}^0)$ is a 0th level elementary concept.
2. $B_{j,i}^1 = \bigcup\{B_{j,i}^0 \mid B_{j,i}^0$ is in a neighborhood subsystem $\}$ is a 1th level definable neighborhood;
   $C_{j,i}^1 = NAME(B_{j,i}^1)$ is a 1th level elementary concept.
3. $B_{j,i}^{(k+1)} = \bigcup\{B_{j,i}^k \mid B_{j,i}^k$ is in a neighborhood subsystem $\}$ is a $(k+1)^{\text{th}}$ level definable neighborhood;
   $C_{j,i}^{(k+1)} = NAME(B_{j,i}^{(k+1)})$ is a $(k+1)^{\text{th}}$ level elementary concept.

4 Granular Structures and Representations

An equivalence relation decomposes the universe into pairwise disjoints elementary sets. A binary relation decomposes the universe into elementary neighborhoods that are not necessarily disjoint. The decomposition is called a binary granulation or a binary neighborhood system [6,7,8].

1. A *binary granular structure* consists of 4-tuple

$$(V, U, B, C),$$

   where (1) V is the object space, (2) U is the data space (V and U could be the same set), (3) B is the binary neighborhood system or binary clump system (see Section 2,Item 5), and (4)C the concept space which consists of names of elementary neighborhoods.
2. A binary granular structure is called a rough structure, denoted by (U, E, C), if $V = U$, E is a partition.
3. A *binary knowledge base* is a finite collection of binary granular structures; it will be denoted by $(V, U, \{B_i, \ C_i, i = 1, 2, \ldots, n.\})$ or simply $\{B_i, i = 1, 2, \ldots, n.\}$. We will assume all granular structures have additional structures, namely, concept hierarchies. If $V = U$ and B is a partition (with no-additional structure) Pawlak call it knowledge base [20].
4. A *knowledge* is referred to a granular structure with one single binary neighborhood system or clump system, for example (V, U, B_{i_o}, C_{i_o}) or simply B_{i_o}.

4.1 Extended Tables of Binary Granulations

Zdzislaw Pawlak showed that a knowledge base can be represented by an information table and vice versa [20]. T. Y. Lin extended his notion to binary relations [7]; in this paper, we consider a mild generalization, namely, to binary granular structures.

   Let $(V, U, \{B_i, \ C_i, i = 1, 2, \ldots, n.\})$ be a given binary granular structures. For simplicity, we may suppress the index i, namely, $B_i = B, C_i = C, LEARN_i = LEARN$. At level i, we consider the mapping $LEARN$ [17], that is,

$$LEARN : V \longrightarrow C;$$

that maps an object $p \in V$ to its unique elementary neighborhood $B_p \subseteq U$, then to its name $C_p = NAME(B_p) \in C$ (called elementary concept). In notations,

$$p \longrightarrow B_p \longrightarrow NAME(B_p).$$

Intuitively, the mapping, $LEARN$, generalizes a data to a i-th level concept. A collection of such LEARN's forms an extended information table [7]. We should like to point out that, if the binary granular structure is a rough structure, then the extended table reduces to Pawlak's information table. However, we should caution readers that unlike classical rough set theory, the entries in the extended table are not semantically independent; there are inter-relationships among data; see Table 2. Rough set theory is a theory of "name(word) processing" on discrete words (a crisp set), while ours is on clustering words (a NS-space).

It is interesting to observe that we reached the same conclusion from pure database point of view about a decade ago; see earlier works in [15,13] and later references and exposition [3,19,6].

Examples

To avoid unnecessary complex notations, we will skip the concept hierarchies in the example. Let the object and data space be the same, $U = V = \{1, 2, 3, 4\}$.

1. B is a binary neighborhood system defined by

$$B_1 = \{2, 3, 4\}; \quad B_2 = \{1, 2\}; \quad B_3 = B_4 = \{3, 4\}$$

Equivalently, it can be expressed as a binary relation:

$$B = \{(1, 2), (1, 3), (1, 4), (2, 1), (2, 2), (3, 3), (3, 4), (4, 3), (4, 4)\}$$

Elementary concepts of B are:

$$NAME(B_1) = all; \quad NAME(B_2) = middle;$$
$$NAME(B_3) = NAME(B_4) = large$$

2. E is an equivalence relation defined by

$$E_1 = E_2 = \{1, 2\}; \quad E_3 = E_4 = \{3, 4\}$$

These neighborhoods (elementary sets)forms a partition:

$$\{1, 2\}; \quad \{3, 4\}.$$

Elementary concepts of E are:

$$NAME(E_1) = NAME(E_2) = low; \quad NAME(E_3) = NAME(E_4) = high.$$

Objects	(V, U, B, C)-attribute	(U, U, E, C)-attribute
ID_1	all	low
ID_2	middle	low
ID_3	large	high
ID_4	large	high

Table 1. Binary granulations from Example; Entries are semantically interrelated; see Table 2

(V, U, B, C_B)-attribute	(V, U, B, C_B)-attribute
all	large
all	middle
middle	all
middle	middle
large	large

Table 2. Exemplary semantic relations between entries in Table 1

4.2 Knowledge Bases

Pawlak introduced the notions of cores and reducts for knowledge base. These notion can be extended to binary granular structures; the detail will appear elsewhere. We summarize the notions in Table 3 [7].

5 Mining Binary Relations in Data and Conclusions

I. Mining Environment:

1. Domain: Extended information tables. Unlike rough set theory, these entries in the extended tables are semantically related; see Table 2.
2. Background Knowledge: For each attribute, there is a nested sequence of binary relations; nested by strong dependencies.
3. Level of Target Knowledge: By forming extended information table on different levels, we will have different level of rules. For example, let us form an extended table with k^{th} level concepts and $(k+j)^{th}$ level concepts. If the knowledge level is at 0th level, we call them hard rules. If $k > 1$ and $j = 0$, we get high level rules of the same level. If $j > 1$, we get different level of soft rules [18,6]

II. Mining Methodology:

1. Rough Set Approach: By applying table processing techniques of rough set theory, we get the rules from different levels [18,16].
2. Other approaches: For example, association rules or its combination with rough set theory [1,10,11,12].

knowledge oriented terms	rough set theory	single level granulation	multilevel granulation
knowledge (geometric)	partition (classification)	binary granular structure	granular structure
knowledge (algebraic)	equivalence relations	binary relations	
granule	elementary set (equivalence class)	elementary neighborhood	fundamental neighborhood
concept space	elementary concept space	elementary concept space	fundamental concept space
knowledge base	Pawlak knowledge base	binary knowledge base	formal word knowledge base
knowledge Representation	information table	extended information table	formal word table [8]

Table 3. Knowledge Bases of Granulations

6   Conclusion

Table processing methodology of rough set theory and attribute oriented generalization (AOG) of data mining are powerful techniques and notions, they are extended to binary relations. The techniques of traditonal data mining of discrete data (a crisp set) can be extended to that of clustered data (A NS-space). We are exploring more, in fact, extending them to general neighborhood systems; we have some success in fuzzy world [8]. Applications are on the way; we will report them in the near future.

References

1. R. Agrawal, T. Imielinski A. Swami, Mining Association Rules Between Sets of Items in Large Databases. In Proceeding of ACM-SIGMOD international Conference on Management of Data, pp. 207-216, Washington, DC, June, 1993
2. Y.D. Cai, N. Cercone, and J. Han. Attribute-oriented induction in relational databases. In *Knowledge Discovery in Databases*, pages 213–228. AAAI/MIT Press, Cambridge, MA, 1991.
3. W. Chu, Neighborhood and associative query answering, *Journal of Intelligent Information Systems*, 1, 355-382, 1992.
4. H. J. Hamilton and D. R. Fudger. Estimating DBLEARN's potential for knowledge discovery in databases. *Computational Intelligence*, 11(2), 1995.
5. M. Hadjimichael and A. Wasilewska. Interactive inductive learning. *International Journal of Man-Machine Studies*, 38:147–167, 1993.
6. T. Y. Lin, Granular Computing of Binary relations I: Data Mining and Neighborhood Systems. In:Rough Sets and Knowledge Discovery, L. Polkowski and A. Skowron (Editors), Springer-Verlag (to appear).

7. T. Y. Lin, Granular Computing of Binary relations II: Rough Set Representations and Belief Functions. In:Rough Sets and Knowledge Discovery, L. Polkowski and A. Skowron (Editors), Springer-Verlag (to appear).

8. T. Y. Lin, Granular Computing: Fuzzy Logic and Rough Sets. In *Computing with words information/intelligent systems*, L.A. Zadeh and J. Kacprzyk (Editors), Springer-Verlag (to appear)

9. T. Y. Lin, Neighborhood Systems - A Qualitative Theory for Fuzzy and Rough Sets. In: *Advances in Machine Intelligence and Soft Computing*, Volume IV. Ed. Paul Wang, 132-155, 1997.

10. T. Y. Lin, Rough Set Theory in Very Large Databases, Symposium on Modeling, Analysis and Simulation, IMACS Multi Conference (Computational Engineering in Systems Applications), Lille, France, July 9-12, 1996, Vol. 2 of 2, 936-941.

11. T. Y. Lin and R. Chen, Finding Reducts in Very Large Databases, Proceedings of Joint Conference of Information Science,Research Triangle Park, North Carolina, March 1-5, 1997, 350-352.

12. Supporting Rough Set Theory in very large Databases using ORACLE RDBMS. In : *Soft Computing in Intelligent Systems and Information Processing*, Proceedings of 1996 Asian Fuzzy Systems Symposium, Kenting, Taiwan, December 11-14, 1996, 332-337

13. T. Y. Lin , Neighborhood Systems and Approximation in Database and Knowledge Base Systems. In:*Proceedings of the Fourth International Symposium on Methodologies of Intelligent Systems*, Poster Session, October 12-15, 1989, 75-86.

14. T. Y. Lin, "Topological Data Models and Approximate Retrieval and Reasoning. In: Proceedings of 1989 ACM Seventeenth Annual Computer Science Conference, February 21-23, Louisville, Kentucky, 1989, 453.

15. T. Y. Lin, Neighborhood Systems and Relational Database. In: *Proceedings of 1988 ACM Sixteen Annual Computer Science Conference*, February 23-25, 1988, 725

16. T. Y. Lin and M. Hadjimichael, Non-Classificatory Generalization in Data Mining. In: *Proceedings of The Fourth International Workshop on Rough Sets, Fuzzy Sets, and Machine Discovery*, November 6-8, 1996, Tokyo, Japan, 404–411

17. T. Y. Lin and Y. Y. Yao, Neighborhoods Systems and Belief Functions. In: *Proceedings of The Fourth Workshop on Rough Sets, Fuzzy Sets and Machine Discovery*, Tokyo, Japan, November 8-10,1996, 202–207.

18. T. Y. Lin, and Y. Y. Yao, Mining Soft Rules Using Rough Sets and Neighborhoods. In: Symposium on Modeling, Analysis and Simulation, CESA'96 IMACS Multiconference (Computational Engineering in Systems Applications), Lille, France, 1996, Vol. 2 of 2, 1095-1100, 1996.

19. B. Michael and T.Y. Lin, Neighborhoods, Rough sets, and Query Relaxation, Rough Sets and Data Mining: Analysis of Imprecise Data, Kluwer Academic Publisher, 1997, 229 -238. (Final version of paper presented in Workshop on Rough Sets and Database Mining , March 2, 1995

20. Z. Pawlak, *Rough Sets (Theoretical Aspects of Reasoning about Data)*. Kluwer Academic, Dordrecht, 1991.

21. W. Sierpenski and C. Krieger, General Topology, University of Toronto Press 1952.

Handling Continuous Attributes in Discovery of Strong Decision Rules

Jerzy Stefanowski

Institute of Computing Science, Poznań University of Technology,
Piotrowo 3A, 60-965 Poznań, Poland
Jerzy.Stefanowski@cs.put.poznan.pl

[Abstract.] We consider discovery of strong decision rules from
information systems defined by continuous attributes. The rule
discovery algorithm Explore is extended by introducing a new
method that handles continuous attributes while constructing
the elementary conditions in the way corresponding to requirements for getting strong decision rules. The usefulness of this
method is evaluated in an experiment.

1 Introduction

The one of purposes of rule discovery systems is to extract from data sets information patterns and regularities which are interesting and useful to different
kinds of users. Discovery of such decision rules is a difficult problem depending
on the context of application and requires close interaction with the user as he
has to define requirements to the derived knowledge, should be able to direct
the knowledge discovery process and validate its results (cf. [5]). Such postulates
led the author to introduce the algorithm *Explore* [7] that induces *all decision
rules* which *satisfy user defined requirements*. This algorithm takes into account
requirements related to the various criteria of rule evaluation, e.g. the *strength* of
the rules (representing the relative number of learning examples supporting the
rule), the length of condition part of rules, the level of discrimination. In current
experiments with the algorithm we focused mainly on the strength of decision
rules (for more details see [7]). The motivations for discovery all strong rules
according to the user's requirements are also typical for data mining algorithms
(cf. Apriori algorithm [1] for mining association rules in large databases - which
are in fact another kind of rules than considered in our approach).

The first version of the algorithm Explore was limited to analyze only data
sets defined using *qualitative* attributes. The *continuous* attributes could be handled by *discretization* techniques which convert them into *discrete* attributes.
There are already proposed several discretization methods which are applied as
a *preprocessing step* before the phase of rule induction (see, e.g. reviews in [2,3]).
The main difficulty with these methods is that they determine discretization
independently from the underlying concept of applied further rule induction algorithm. It is possible that the set of discretized subintervals being candidates
for elementary conditions may not satisfy requirements for getting strong rules.
For instance, in some specific data sets the discretization may produce a very

L. Polkowski and A. Skowron (Eds.): RSCTC'98, LNAI 1424, pp. 394–401, 1998.
© Springer-Verlag Berlin Heidelberg 1998

limited numbers of subintervals while in other cases they could give too much week conditions. As a result the algorithm may be unable to discover sufficient number of strong decision rules. Therefore, it is necessary to look for yet another discretization technique that should be more deeply integrated with the algorithm Explore and its requirements.

The aim of the following paper is to extend the algorithm Explore by introducing a phase of direct discretizing continuous attributes while creating elementary conditions. The method should try to generate such conditions that correspond to requirements for extracting the strongest decision rules. The proposed method will be compared experimentally with the method of preliminary discretization to evaluate which of them allows to induce better rules.

The paper is organized as follows. The next section describes shortly the algorithm Explore. Then, the method of handling continuous attributes is introduced. Summary of results (restricted due to the paper size) is given in section 4. Discussion of the results and conclusions are drawn in the final section.

2   Algorithm Explore as a tool for inducing strong decision rules

We give only basic information about the algorithm Explore (for more details see [7]). The decision rules are *iteratively* induced for each decision class. If the input data contain *inconsistencies* they can be handled by means of the *rough set theory*, i.e. *exact* (certain) and *approximate* (possible) decision rules are induced from lower and upper approximations of decision classes [6,9]. Another way of handling inconsistencies is to induce *partly discriminating* decision rules from each decision class, i.e. such rules besides learning examples from a given class could cover a limited number of examples from other classes (similar motivations are used in *variable precision rough sets model* [10]). The second way will be used in our experiment. Below we introduce some formal definitions.

Let $(U, A \cup \{d\})$ denote the *information system* where U is a finite set of objects, A is a finite set of *condition attributes*, $d \notin A$ is a distinguished *decision attribute* that defines partition of objects into a set of *decision classes* $Y_1, Y_2, \ldots, Y_k$. Let K will represent the decision concept to be described by rules (i.e. decision class Y_j or its approximation depending on the way of handling inconsistencies). An *elementary condition* c is defined as an expression $(a \ rel \ v)$ where $a \in A$ and v is its value (or a set of values) and rel stands for relational operator e.g. $=, \leq, \geq, \in$. Let C be a *conjunction* of q elementary conditions, i.e. $C = c_1 \wedge c_2 \wedge \cdots \wedge c_q$. Then $[C]$ is the subset of examples which satisfy the conditions represented by C. Considering the concept K to be described, this subset is divided into the *positive cover* $[C]_K^+ = [C] \cap K$ and the *negative cover* $[C]_K^- = [C] \cap (U \setminus K)$.

A *decision rule* r is an assertion of the form *if R then K*, where R is a minimal conjunction $c_1 \wedge c_2 \wedge \cdots \wedge c_q$, satisfying $[R]_K^+ \neq \emptyset$. A rule r is *discriminant* (exact) if $[R]_K^- = \emptyset$ otherwise is *partly discriminant*. The rules are characterized by two measures. The *strength* of the rule r is defined as $Strength(r) = |\ [R]_K^+\ | \ / \ |\ [K]\ |$. The *level of discrimination* $D(R)$ is defined as $|\ [R]_K^+\ | \ / \ |\ [R]\ |$.

In the algorithm Explore the search for rules is controlled by parameters called *stopping conditions* **SC** that reflect the user's requirements. As the main attention is put to the strength of the rules, the definition of SC is connected with determining the threshold value for the minimal strength of the conjunction being candidate for the condition part of the rule. If its strength is lower than SC it is discarded otherwise it can be further evaluated. Additionally, one can define a threshold d expressing the minimum value of the level of discrimination $D(R)$ of the rules to be generated. The algorithm *Explore* is based on a *breadth-first* strategy which generates rules of increasing size starting from the shortest ones. The main part of the algorithm is presented in pseudo-code below:

Procedure Explore(SC: stopping_conditions; d: discrimination threshold;
        L: list_of_conditions; var $\mathcal{R}$: set_of_rules)
begin
    $\mathcal{R} \leftarrow \emptyset$;
    for each condition c from list L **do**
    begin
        if $[c]_K^+ = \emptyset$ or c satisfies SC **then** discard c;
        if $D(c) \geq d$ **then** $\mathcal{R} \leftarrow \mathcal{R} \cup \{c\}$ and discard c
    end;
    form a queue Q with all the remaining elementary conditions $c_1, \ldots, c_n$
    (ordered according to the decreasing strength);
    while $Q \neq \emptyset$ **do**
    begin
        remove the first conjunction C from the queue Q;
        generate the set $\mathcal{C}$ of all the conjunctions $C \wedge c_{h+1}, C \wedge c_{h+2}, \ldots, C \wedge c_n$
        where h is the highest index of the condition involved in C;
        for each $C' \in \mathcal{C}$ **do**
        begin
            if $[C']_K^+ = \emptyset$ or C' satisfies SC **then** $\mathcal{C} \leftarrow \mathcal{C} \setminus \{C'\}$;
            if $D(c) \geq d$ **then**
            begin
                if C' is *minimal* **then** $\mathcal{R} \leftarrow \mathcal{R} \cup \{C'\}$;
                $\mathcal{C} \leftarrow \mathcal{C} \setminus \{C'\}$
            end;
        end;
        place all the conjunctions from $\mathcal{C}$ at the end of the queue
    end
end

The crucial issue refers to the creation of the list L of allowed elementary conditions. In the previous experiments [7] only nominal or preliminary discretized attributes were considered. For given decision concept K, the elementary conditions were created as attribute-value pairs $(a = v)$ such that $[(a = v)]_K^+ \neq \emptyset$. As for considered discrete attributes the list of their values is rather small, these conditions are quite fast to detect by scanning respective attribute values for objects in the input information system.

Let us stress that the Explore algorithm guaranties getting the set of *all* decision rules that satisfy the user's requirements. As this perspective is quite different from the rule learning methods that induce the minimum set of classification rules (as e.g. C4.5, or LEM2), the algorithm Explore is more demanding from the time and memory complexity point of view. Its computational complexity is exponential in the worst case. However, in practice the users are interested in discovery strong rules of small size. According to the way the search is performed (see the concept of the ordered queue) such rules are induced first. Therefore, if the user sets proper stopping conditions it efficiently reduces the computational costs.

3   Discretizing continuous attributes while creating elementary conditions

We propose to discretize continuous attributes directly at the moment when elementary conditions are created. As decision rules are mainly evaluated by their strength and secondly by their level of discrimination, the searched conditions should cover the largest number of examples from the given decision class and at the same moment cover the smallest number of examples from other classes. If one starts from strong enough and partly discriminating elementary conditions, the algorithm Explore performing its breadth-first search strategy could combine to sufficiently strong rules. Due to the strength requirements, the discussed below method will be called *Maximal Strength Partitioning* and denoted shortly MSP.

The created elementary conditions c will be evaluated by two measures: their strength $Strength(c)$ and level of discrimination $D(c)$. Similarly to the Explore requirements these measures are controlled by two threshold parameters: *Min_Strength* and *Min_Discr*.

As the Explore generates rules iteratively for each decision class, the MSP method also generates elementary conditions independently for each class. The generation of selectors is done *locally* for each attribute. The conditions are represented in the form either $(a \leq x)$ or $(a \geq x)$ where a is attribute and x is a cut point. The candidates for the cut point are locally scanned for the range of each continuous attribute. It means that values of the attribute for objects in input data are sorted in the increasing sequence. With each value it is also stored information about the decision class assignments of objects having this value. Candidate cut points are computed as mid-points between successive value points in the sorted sequence. We consider only mid-points between points that are characterized by the change of class assignment. If for candidate cut-point x any of conditions $(a \leq x)$ or $(a \geq x)$ satisfies both criteria $Strength \geq Min\_Strength$ and $D(c) \geq Min\_ Discr$, it is temporary added to L – the list of allowed conditions for the algorithm Explore. The time complexity of this technique is linear in $O(nm)$ where n is a number of examples, and m candidate cut points in the worst case restricted by the number of values for a given attribute.

If the list L at the end of discretizing the attribute contains too many potential conditions it is possible to restrict their number. It needs to define the input parameter called *Maximum_conditions*. For its given value the best conditions from the list L are selected. Half of them is chosen according to the criterion of the highest value of discrimination level while the rest according to the highest value of the strength. This differentiation results from the possible trade off between both criteria (i.e. the conditions having the highest discrimination level may not be the strongest ones).

If the list L for given requirements contains less elementary conditions than the required *Maximum_conditions* number it is possible to extend their number. We propose to additionaly use minimal entropy partitioning method (ME) [4] that preliminary divides the range of values of the attribute into two 'purer' subintervals. Then, cut points inside these subintervals are scanned using our MSP approach. Due to preliminary sub-division the tested conditions are now represented in the form $(a \in [x, b])$ or $(a \in [b, x])$ where b is boundary generated by ME and x is a tested cut-point. If it is possible to identify the conditions that satisfy the tests for *Min_Strength* and *Min_ Discr* they are added to L the list of allowed conditions.

4   Experiment

In our experiment, we compared the proposed discretization method MSP to an approach where *Minimal Entropy Partitioning* method [4], denoted as ME, is used in preprocessing phase. The ME method is one of the most often applied approaches to discretize attributes in preprocessing phase before the rules induction. Moreover it was chosen because of some similarity to scanning the histogram of attribute value proposed in the MSP method.

The aim of the comparison was to evaluate which of the methods allows the algorithm Explore to produce the better set of decision rules. The average strength of the rule was the main criterion (the higher the better). We also analyzed the number of rules and as a supporting criterion we used the classification accuracy estimated in the 10-fold cross validation test. In the experiment we used three data sets from U.C. Irvine repository [8], i.e. *bank, bupa* and *iris*. The fourth data *buses* was coming from our previous experiments with technical diagnostics. Each of the data sets was defined using continuous attributes only. Let us shortly characterize them: *bank* concerns two decision classes (cardinality of objects in classes 33/33) described by 5 continuous attributes; *bupa* - two classes (201/85 objects) described by 6 attributes; *buses* - two classes (46/30 objects) by 8 attributes and *iris* - three classes (50/50/50) by 4 attributes.

Table 1. The characteristics of sets of elementary conditions obtained by MSP and ME discretization methods for BUPA data set; */* denotes results in each decision class, '–' means that it was not possible to obtain any elementary conditions.

Min_Discr [%]	MSP discretization			ME discretization		
	no_cond [%]	av_strength [%]	av_discr [%]	no_cond [%]	av_strength [%]	av_discr [%]
30	35/22	81/73	45/63	7/8	78/66	48/59
40	35/22	81/73	45/63	6/8	80/66	47/59
50	11/20	50/70	52/63	2/8	54/66	56/59
60	5/18	21/59	66/65	–/3	–/59	–/69

For the proposed MSP discretization method we decided to test several values of control parameters. We assumed that the maximal number of elementary

Table 2. The Sets of elementary conditions created for all data sets (Min_Str = 10% and Min_Discr = 30%).

Data	MSP discretization			ME discretization		
	no_cond	av_strength	av_discr	no_cond	av_strength	av_discr
Bank	22/19	88/90	80/76	5/5	89/78	76/80
Bupa	35/22	81/73	45/63	7/8	78/66	48/59
Buses	14/20	91/89	88/80	9/8	89/86	88/90
Iris	11/20/16	81/88/86	77/49/65	4/4/4	88/84/97	91/46/48

conditions for one attribute will be restricted to 10. We tested systematically the following values of *Min_Strength* for elementary conditions: 10%, 20%, 30%, 40%, 50%. For each of them we checked the following values *Min_Discr*: 30%, 40%, 50%, 60% and 70%.

For all data sets we noticed that various values of *Min_Strength* did not influence too much the quality of created elementary conditions, assuming only that the strength is defined as at least 10%. Moreover, the values of minimal level of discrimination *Min_Discr* did not influence the created elementary condition for Bank and Buses data while for two other data Bupa and Iris increasing their value over 50% led to difficulties in getting a sufficient number of strong enough elementary conditions. This tendency is illustrated by the results for Bupa data presented in Table 1. In Tables 1 – 2 we will use the following abbreviations: no_cond – number of elementary conditions, av_strength – average strength of elementary conditions [%] , av_discr. – average level of discrimination of elementary condition [%]. The summarized results evaluating quality of elementary conditions for all data sets are given in Table 2 - they refer to the parameters *Min_Strength* = 10% and *Min_Discr* = 30%.

Table 3. The characteristics of decision rules; */* denotes results in each decision class, '–' means that it was not possible to obtain any decision rule

Data set	MSP discretization				ME discretization			
	no_rules	av_strength [%]	av_discr [%]	accuracy [%]	no_rules	av_strength [%]	av_discr [%]	accuracy [%]
Bank	45/34	87/84	94/95	91.67	2/3	84/74	95/96	92.33
Bupa	3/11	16/17	85/91	62.85	2/–	28/–	85/–	27
Buses	11/16	94/88	96/96	98.57	8/6	92/85	97/95	97.56
Iris	2/11/6	72/71/78	99/94/92	94.14	3/–/–	85/–/–	99/–/–	32.34

These sets of elementary conditions were the basis for using the algorithm Explore. We tested four different stopping conditions SC: 15%, 20%, 25%, 30%. Only the first threshold led to satisfactory results, in particular for higher values we met difficulties with identifying strong decision rules from conditions produced by ME discretization. We also noticed that it is interesting to look for partly discriminant rules (with the requirement to the level of discrimination 85% - 95 %) as strictly discriminant rules were weaker and more numerous. Ta-

Table 4. Comparison of decision rules induced by Explore and LEM2 algorithms.

Data set	Explore			LEM2			
	no_rules	av_strength	accuracy		no_rules	av_strength	accuracy
Buses	27	91	98.57		3	93	98.7
Bank	79	85.5	91.67		4	83.1	93.9
Bupa	14	16.5	62.85	2ME	9	3.05	52.4
				kME	30	2.3	62.85
Iris	37	77	94.14	2ME	2	–	33.33
				kME	14	22	94.33

ble 3 summarizes information characterizing discovered set of rules - the rules were induced for minimal strength SC = 15% and d=85%. The abbreviations used in Table 3 refer to average strength and discrimination levels of rules; accuracy means classification accuracy estimated in 10-fold cross validation test. Let us notice that for ME discretization it was not possible to induce decision rules for some classes in case of Bupa and Iris data sets.

Then, we compared the satisfactory set of rules induced by the algorithm Explore with rules obtained by means of the classification-oriented rule induction system LEM2 [6] which was used on elementary conditions prepared by ME preliminary discretization phase. As this discretization was quite weak for Bupa and Iris data sets we additionally used its version with larger numbers of discrete subintervals (denoted by *kME*). Comparative results are given in Table 4.

5 Discussion of results and final remarks

Let us summarize the results of the experiment. The introduced MSP discretization method produces greater number of elementary conditions characterized by higher average strength than the typical preprocessing discretization method based on minimal entropy partitioning (ME). It could be considered as the better basis for the further induction of decision rules. This was particularly observed for difficult data sets like Bupa.

The MSP method has to be parametrized by two parameters: minimal strength and minimal discrimination level of the elementary condition. We noticed for analyzed data sets that good MSP results are quite stable while changing these parameters. On the other hand increasing these values for filtering ME discretization deteriorates the quality of elementary conditions for two data sets Iris and Bupa. It seems also that minimal level of discrimination has greater influence on the final results assuming that minimal strength is not lower than 10%. Although both parameters could be tuned by the user depending on the context of application, the experiments indicates the default values for them: strength - 10% and discrimination level - 30%.

The MSP discretization helped the Explore in discovering more decision rules characterized by higher average strength. In particular, its advantage could be noticed for more difficult data sets (Bupa and Iris) where ME pre-discretization failed in discovery strong decision rules.

Finally, the comparison of rules derived by Explore and LEM2 algorithms showed that by using of the former algorithm it was always possible to discover set of better rules according to the criteria: number of rules, average rule strength without decreasing too much the classification accuracy.

Acknowledgments

Research on this paper was supported by the grant KBN no. 8T11C 013 13. The author is grateful to Daniel Vanderpooten and Robert Mienko for discussions and joint work on a first version of the Explore algorithm.

References

1. Agrawal R., Mannila H., Sirkant R., Toivonen H., Verkamo A.: Fast discovery of association rules. In: *Advances in Knowledge Discovery and Data Mining*, MIT Press, (1996) 307–328.
2. Chmielewski M. R., Grzymala–Busse J. W.: Global discretization of continuous attributes as preprocessing for machine learning. In: T. Y. Lin and Wilderberger A. M. (eds.), *Soft Computing* (1995) 294–297.
3. Dougherty J., Kohavi R., Sahami M.: Supervised and unsupervised discretizations of continuous features. In: *Proc. 12th Int. Conf. on Machine Learning*, 1995, 194–202.
4. Fayyad U.M., Irani K.B.: Multi–interval discretization of continuous-valued attributes for classification learning. In: *Proc. of 13th Int. Conf. on Machine Learning*, Morgan Kaufmann, (1993) 1022–1027.
5. Fayyad U.M., Piatetsky-Shapiro G., Smyth P.: From data mining to knowledge discovery: an overview. In: *Advances in Knowledge Discovery and Data Mining*, MIT Press, (1996), 1–36.
6. Grzymala–Busse J.W.: LERS - a system for learning from examples based on rough sets. In: R. Słowiński, (ed.) *Intelligent Decision Support*, Kluwer, (1992) 3–18.
7. Mienko R., Stefanowski J., Toumi K., Vanderpooten D.: Discovery-Oriented Induction of Decision Rules. *Cahier du Lamsade* no. 141, Universite de Paris Dauphine, Septembre 1996.
8. Murphy P.M., Aha D.W.: UCI Repository of machine learning databases. University of California at Irvine, Dept. of Computer Science.
9. Stefanowski J., Vanderpooten D.: A general two stage approach to rule induction from examples. In: W. Ziarko (Ed.), *Rough Sets, Fuzzy Sets and Knowledge Discovery*, Springer-Verlag, 1994, 317–325.
10. Ziarko W.: Variable precision rough sets model. *Journal of Computer and Systems Sciences*, vol. 46. no. 1, (1993) 39–59.

Covering with Reducts - A Fast Algorithm for Rule Generation

Jakub Wróblewski

Institute of Mathematics
Warsaw University
Banacha 2, 02-097 Warsaw, Poland
e-mail: jakubw@alfa.mimuw.edu.pl
http://alfa.mimuw.edu.pl/~jakubw

Abstract. In a rough set approach to knowledge discovery problems, a set of rules is generated basing on training data using a notion of reduct. Because a problem of finding short reducts is NP-hard, we have to use several approximation techniques. A covering approach to the problem of generating rules based on information system is presented in this article. A new, efficient algorithm for finding local reducts for each object in data table is described, as well as its parallelization and some optimization notes. A problem of working with tolerances in our algorithm is discussed. Some experimental results generated on large data tables (concerned with real applications) are presented.

1 Introduction

Rough set expert systems base on the notion of a *reduct* ([7], [8]), a minimal subset of attributes which is sufficient to discern between objects with different decision values. A set of short reducts can be used to generate rules ([1]). A problem of short reducts generation is NP-hard, but an approximate algorithm (like the genetic one described in [9], [4] and implemented successfully - see [6]) can be used to obtain reducts in reasonable time. On the other hand, rules generated basing on reducts are often too specific and cannot classify new objects. Another types of reducts have been considered to improve efficiency on new objects (see [2]). One of the methods is to calculate reducts basing on a single object.

Let $A = (U, A \cup \{d\})$ be an *information system* (see [8]), where U - set of objects, A - set of attributes, d - decision.

Definition: A *local reduct* $R(o_i) \subseteq A$ (or a *reduct relative to decision and object* $o_i \in U$; o_i is called a *base object*) is a subset such that:

a) $\forall\, o_j \in U,\ d(o_i) \neq d(o_j) \implies \exists\, a_k \in R$: $a_k(o_i) \neq a_k(o_j)$

b) R is minimal with respect to inclusion.

A classical reduct will be referred to as *global reduct*. A rule generated by a local reduct is concerned with the base object and may not recognize any other object from U. To assure that a set of rules will recognize (at least) all objects from the training set, we have to generate a local reduct for every object. A simple algorithm checking whether a subset is a local superreduct works at a

L. Polkowski and A. Skowron (Eds.): RSCTC'98, LNAI 1424, pp. 402–407, 1998.
© Springer-Verlag Berlin Heidelberg 1998

time complexity of $O(mn)$: we have to compare our base object with all other objects and check whether condition a) (see local reduct definition) holds. It takes $O(mn^2)$ time to do this for all objects (when we are looking for a local reduct for every object), where n = number of objects, m = number of attributes. This time complexity is not acceptable for large data tables. A fast approximation algorithm for local reducts generation is presented in the next section. In sections 3 and 4 some related topics, concerned with parallelization of the algorithm and dealing with tolerance are discussed. In section 5 some experimental results are presented.

2   Covering algorithm

An algorithm presented below realizes the following objective: assuming the information system is consistent, find a family of subsets R_1, R_2,... R_k such that for any object o_i from U at least one R_j is a local reduct (we will say, that R_j covers o_i). We will look for possibly small family R_1,... R_k, i.e. we will prefer these subsets which cover possibly many objects. We assume, that these subsets reflect regularities in data and generate more general rules - it means better classification of new samples and less memory required to store rules.

1. Let σ be a random permutation of attributes.
2. Let $R = A$ and N_1,...N_n - a table of numbers of local reducts found for each object. Set $N_j = 0$.
3. Test whether R is a local reduct for any object. If so, increment N_i for these objects and store rules.
4. Let $R = R - a_i$, where a_i - the first attribute from R. Calculate a number M_i of these objects, for which R is a (super)reduct, and which are not covered by reducts found previously. Let $R = R + a_i$.
5. Continue step 4. with the next attribute from R. Finish after collecting numbers M_i for all attributes.
6. Find the maximal number among M_i; if there are more than one such a number - get the first one with respect to the permutation σ. Let a_j - an attribute associated with this maximum. Let $R = R - a_j$.
7. Continue from 3. until R is empty.
8. If there is at least one uncovered object - let $R = A$, continue from 4.

   Lemma. The algorithm described above generates a covering for all objects in at most $n = |U|$ cycles (by "cycle" we mean one sequence of steps from 2 to 8).
   Proof. We will prove, that in one cycle at least one uncovered object is covered by newly produced reduct. When we find out, that a set R is a local superreduct for a number of objects not covered so far in step 6. of the algorithm, there are two possibilities: a) all M_i are equal to 0, but it means that R is a local reduct for all these objects (because it is a superreduct and none of its subsets is a superreduct) so we have covered some new objects; b) there exists an $M_i > 0$, i.e. at least one subset of R is a superreduct for M_i objects not covered so far -

we continue from step 4. If our subset R has two attributes, possibility b) means, that there exists a local reduct with one attribute (a superreduct with only one attribute must be a reduct). So, in one cycle (starting from $R = A$, which is a local superreduct for all objects) we either realize possibility a) or, in the worst case, achieve a reduct containing only one attribute.

We need to have a method of determining whether a subset is a superreduct to complete our algorithm.

1. Sort a table of objects using attribute values (for attributes belonging to R).
2. Scan the table of sorted objects one by one. Our objects are divided into groups with equal values on attributes (abstract classes of indiscernibility relation generated by R, see [8]).
3. If a group has an uniform value of decision - it means that R is a local superreduct for objects belonging to this group. If not - R is not a local superreduct for these ones.

Since we may use a fast method of sorting, our algorithm has the complexity of $mn\ log\ (n)$, where n = number of objects, m = number of attributes.

3   Parallel algorithm and practical notes

The algorithm described in the previous section covers all objects by at least one reduct. On the other hand, the more reducts for each object we find, the more rules we can generate. Since the algorithm is deterministic for a given permutation σ, we have the following possibilities:

1. We may choose a set of p permutations $\sigma_1, \ldots \sigma_p$ and generate p coverings using this algorithm in parallel on p machines. When permutations are different, the obtained coverings usually are different too.
2. We may do the same on one machine in sequential way. In this case we can perform an additional optimization: at each stage of algorithm we look for covering for only these objects which are covered in minimal degree during the previous stages.
3. We may use an evolutionary algorithm to find the best permutation - i.e. the permutation generating a covering using a minimal set of possibly short reducts. An order-based genetic algorithm (see [3], [10]) can be used in case of sequential as well as parallel computations.

When we check whether a subset R is a local superreduct for any object, we can easily check whether it is a global superreduct (R is a global superreduct $\Longleftrightarrow R$ is a local superreduct for all objects). On the other hand, the algorithm of finding global reducts described in [4] and [11] uses a structure called "reduct store" containing all known global superreducts of information system. Thus, we can check whether R is known as a global superreduct before we start to sort our object set, as well as we can add R to this structure when we find out that R is a global superreduct. Moreover, the same structure can be used by many agents in parallel implementation (each agent calculates covering for different permutation) and by one specialized agent calculating global reducts.

4   Tolerance

We use local reducts to generate rules which are more general than these generated basing on general reducts. On the other hand, these rules may still be too specific - especially when we work on numerical data. One of the ways to manage this problem is to use a discretization technique (see e.g. [5]). Alternatively, we can use a *tolerance measure*, which allows us to treat two different (but close) values as equal.

An algorithm presented in section 2. can be easily adopted to this new situation, in case a tolerance relation is transitive. In this case we can sort a set of objects and divide it into classes of this relation - then continue with the standard algorithm. Alternatively, we can initially replace attributes' values with their representatives (found by e.g. methods of scalar quantization or discretization).

Unfortunately, many tolerance relations are not transitive, and we cannot simply sort data and check adjacent pairs of objects. More research is needed to use our algorithm in this case.

5   Experimental results

The algorithm described in section 2. was implemented and tested on several information systems used in real applications - results are shown in the table presented below.

Size: obj× attr	#red	Time [sec]
4,492× 36	1	13
	10	49
24,000× 10	1	25
22,000× 27	1	90
	5	1600
47,000× 28	1	360

Calculations were performed on Pentium-200 machine. The first data set is the "Satellite image" database, the second is the "Shuttle" database. The column "#red" indicates how many reducts (at least) we found for each object.

The results show, that our new method is relatively fast, even for large data tables (finding local reducts for each object using the previous methods takes many hours for tables with number of objects greater than 20,000), especially when we are interested in just a covering of objects. Actually, when we cover objects by at least one reduct, an average number of reducts covering an object is usually equal to about 3.5.

6   Conclusions and future work

We have presented a covering approach to rule generation problem and an efficient algorithm for finding local reducts for a set of objects. A computation time for this algorithm is close to the time of global reduct finding ([4]). Our new method is fast, and it should generate more general rules - a comparison of efficiency of these rules generated in classical and a new way will be performed in the future. Another direction of future research is to implement and test an evolutionary algorithm (see section 3.) and a tolerance-based techniques. Moreover, a parallel system has been not implemented so far.

Acknowledgment

This work was supported by Polish State Committee for Scientific Research grant #8T11C01011 and Research Program of European Union - ESPRIT-CRIT2 No. 20288

References

1. Bazan J., Skowron A., Synak P., 1994. *Dynamic reducts as a tool for extracting laws from decision tables*, Proc. of the Symp. on Methodologies for Intelligent Systems, Charlotte, NC, October 16-19, 1994, Lecture Notes in Artificial Intelligence 869, Springer-Verlag, Berlin 1994, 346–355, also in: ICS Research Report 43/94, Warsaw University of Technology.
2. Bazan J., 1998. *A Comparison of Dynamic and non-Dynamic Rough Set Methods for Extracting Laws from Decision Tables.* In: L. Polkowski, A. Skowron (eds.). Rough Sets in Knowledge Discovery. Physica Verlag, 1998.
3. Goldberg D.E., 1989. *GA in Search, Optimisation, and Machine Learning.* Addison-Wesley.
4. Nguyen S. H., Skowron A., Synak P., Wróblewski J., 1997. *Knowledge Discovery in Databases: Rough Set Approach.* Proc. of The Seventh International Fuzzy Systems Association World Congress, vol. II, pp. 204–209, IFSA97, Prague, Czech Republic.
5. Nguyen H. S., Nguyen S. H., 1998. *Discretization Methods in Data Mining.* In: L. Polkowski, A. Skowron (eds.). Rough Sets in Knowledge Discovery. Physica Verlag, 1998.
6. Øhrn A., Komorowski J., 1997. *Rosetta - A rough set toolkit for analysis of data.* Proc. of Third International Join Conference on Information Sciences (JCIS97), Durham, NC, USA, March 1 - 5, 3 (1997), pp. 403–407.
7. Pawlak Z., 1991. *Rough sets: Theoretical aspects of reasoning about data.* Kluwer: Dordrecht 1991.
8. Skowron A., Rauszer C., 1992. *The Discernibility Matrices and Functions in Information Systems.* In: R. Slowiński (ed.): Intelligent Decision Support. Handbook of Applications and Advances of the Rough Sets Theory. Kluwer: Dordrecht 1992, pp. 331–362.
9. Wróblewski J., 1995. *Finding minimal reducts using genetic algorithms.* Proc. of the Second Annual Join Conference on Information Sciences, pp.186-189, September 28-October 1, 1995, Wrightsville Beach, NC. Also in: ICS Research report 16/95, Warsaw University of Technology.

10. Wróblewski J., 1996. *Theoretical Foundations of Order-Based Genetic Algorithms.* Fundamenta Informaticae, vol. 28 (3, 4), pp. 423–430. IOS Press, 1996.

11. Wróblewski J., 1998. *Genetic algorithms in decomposition and classification problem.* In: L. Polkowski, A. Skowron (eds.). Rough Sets in Knowledge Discovery. Physica Verlag, 1998.

Syntactical Content of Finite Approximations of Partial Algebras

Wiktor Bartol[1], Xavier Caicedo[2], Francesc Rosselló[3]

[1] Institute of Mathematics, Warsaw University, 02-097 Warszawa (Poland)
bartol@mimuw.edu.pl
[2] Dep. Matemáticas,Univ. de los Andes, Apartado Aéreo 4976, Bogotá (Colombia)
xcaicedo@zeus.uniandes.edu.co
[3] Dep. Matemàtiques i Informàtica, Univ. Illes Balears, 07071 Palma (Spain)
dmifrl0@ps.uib.es

[Abstract.] In this paper we give a syntactical answer to the
following question: What do we actually know about a partial
algebra when we know its set of weak or relative subalgebras
with cardinal smaller than a fixed bound, if we do not have any
information on how they are linked to each other within the
algebra?

1   Introduction

The "roughness" of a theory consists essentially in the fact that, under some data
system, objects can only be described approximately. Thus, different objects may
be indistinguishable from each other by the means available in the system. This
situation is made precise by the notion of indiscernibility.

This paper is intended to pursue the idea of structural approximation for
algebras. Given a weak or relative subalgebra of an algebra (total or partial),
it can be seen as an approximation of the structure of the latter. This leads to
the following notion of indiscernibility. Let $\mathcal{K}$ be a class of algebras of a given
signature. We define $S_w(\mathcal{K})$ as the class of all weak subalgebras of algebras in
$\mathcal{K}$. Then for any class $\mathcal{M}$ of algebras we say that two algebras $\mathbf{A}$ and $\mathbf{B}$ in $\mathcal{K}$
are $\mathcal{M}$-*indiscernible* iff for every $\mathbf{D} \in \mathcal{M}$, $\mathbf{D} \in S_w(\mathbf{A})$ if and only if $\mathbf{D} \in S_w(\mathbf{B})$.
Clearly, a similar notion can be defined for relative subalgebras, too. Thus two
algebras are indiscernible when they cannot be distinguished from each other by
the available approximations.

In this paper we are concerned with approximations of an algebra $\mathbf{A}$ given
by finite weak or relative subalgebras (or even by subalgebras of at most some
fixed finite cardinality). Notice that such finite approximations are ubiquitous in
$\mathbf{A}$ (each finite subset of the carrier of $\mathbf{A}$ supports at least one weak subalgebra
and exactly one relative subalgebra of $\mathbf{A}$), and they determine completely $\mathbf{A}$,
provided we know the way they glue (a partial algebra is the direct limit of

---

* This work has been partially supported by the KBN grant 8-T11C-01011, and the
DGCIyT grant PB96-0191-C02-02

L. Polkowski and A. Skowron (Eds.): RSCTC'98, LNAI 1424, pp. 408–415, 1998.

its directed system of finite weak, or finite relative, subalgebras [2, Cor. 4.4.7]). But what do we actually know about a partial algebra when we know its finite approximations (up to isomorphisms), if no information on how they are linked to each other within the algebra is available?

We give here a syntactical answer to this question. We define the *syntactical content* of a type of finite approximations as, roughly, a set of formulas of the form "conjunction implies disjunction" that are 'captured' by these approximations (see Def. 2 below), and we determine it for weak and relative subalgebras (with a fixed bound on the cardinality of their carrier) of partial and total algebras.

Notice that partial algebras are often used as models of objects appearing in soft computing, such as graphs, relational systems and data bases. For instance, a binary relation on a set can be understood as a partial binary operation given by the first projection defined only on the pairs in the relation. Our results allow us to characterize syntactically, for instance, the knowledge of all its weak subsystems (weak subalgebras of that binary algebra) with less than a fixed number of elements. Another way of looking at the problem considered in this paper is to ask what knowledge on an algebraic structure can be derived from finite experiments, which thus brings us close to machine learning.

This note is born in part from the desire to better understand some of the results on similar problems obtained in [1], to be published elsewhere.

To simplify things, we only deal here with partial algebras over *finite* (i.e., with finitely many operation symbols) and *homogeneous* (i.e., one-sorted) signatures, but the results we obtain in this case are easily generalized to more general cases; cf. again [1].

2   Preliminaries and notations

For the convenience of the reader, in this section we recall the basic definitions on partial algebras, used in this paper (except for weak and relative subalgebras defined at the beginning of the next section); for any notion not defined here, as well as for more details about those defined, see [2]. In this section we also fix some notation and conventions to be used throughout the paper.

We fix for the rest of this paper a *signature* $\Sigma = (\Omega, \eta)$, where Ω is a finite set of *operation symbols* and $\eta : \Omega \to \mathbb{N}$ is the *arity mapping*. We set $\Omega^{(n)} = \{\varphi \in \Omega \mid \eta(\varphi) = n\}$ for every $n \in \mathbb{N}$.

A *partial Σ-algebra* (an *algebra*, for short) is a structure $\mathbf{A} = (A, (\varphi^{\mathbf{A}})_{\varphi \in \Omega})$, where A is a set, called the *carrier* of the algebra, and for every $\varphi \in \Omega$, $\varphi^{\mathbf{A}} : A^{\eta(\varphi)} \to A$ is a partial mapping with domain $\operatorname{dom} \varphi^{\mathbf{A}} \subseteq A^{\eta(\varphi)}$. We denote the class of all such algebras by Alg_Σ.

Given an algebra denoted by a capital letter in boldface type ($\mathbf{A}$, $\mathbf{B}$, etc.), we always denote, unless otherwise stated, its carrier set by the same capital letter in slanted type (A, B, etc.).

An algebra is *finite* when its carrier is finite. The *cardinal* $|\mathbf{A}|$ of a finite algebra $\mathbf{A}$ is the cardinal of its carrier.

An algebra $\mathbf{A}$ is *total* when $\varphi^{\mathbf{A}}$ is a total mapping, for every $\varphi \in \Omega$; we denote by TAlg_Σ the class of all total algebras.

Two algebras $\mathbf{A} = (A, (\varphi^{\mathbf{A}})_{\varphi \in \Omega})$ and $\mathbf{B} = (B, (\varphi^{\mathbf{B}})_{\varphi \in \Omega})$ are *isomorphic* when there exists a bijection $h : A \to B$ (an *isomorphism*) such that for every $\varphi \in \Omega$ and for every $a, a_1, \ldots, a_{\eta(\varphi)} \in A$, $\varphi^{\mathbf{A}}(a_1, \ldots, a_{\eta(\varphi)}) = a$ iff $\varphi^{\mathbf{B}}(h(a_1), \ldots, h(a_{\eta(\varphi)})) = h(a)$.

We fix henceforth a countably infinite *set of variables* $\mathcal{X} = \{x_i \mid i \in \mathbb{N}\}$, disjoint from Ω. The set $\mathrm{T}_\Sigma(\mathcal{X})$ of *(Σ-)terms with variables in* $\mathcal{X}$ is defined as the least set T such that $\mathcal{X} \cup \Omega^{(0)} \subseteq \mathrm{T}$ and, if $\varphi \in \Omega$ and $\mathbf{t}_1, \ldots, \mathbf{t}_{\eta(\varphi)} \in \mathrm{T}$, then $\varphi(\mathbf{t}_1, \ldots \mathbf{t}_{\eta(\varphi)}) \in \mathrm{T}$.

Given a term $\mathbf{t} \in \mathrm{T}_\Sigma(\mathcal{X})$ and an algebra $\mathbf{A}$, we define the (partial) *term function* $\mathbf{t}^{\mathbf{A}} : A^{\mathcal{X}} \to A$ (where $A^{\mathcal{X}}$ denotes the set of all *valuations* $v : \mathcal{X} \to A$) as follows:

- If $\mathbf{t} = x_i \in \mathcal{X}$, then $\mathbf{t}^{\mathbf{A}}(v) = v(x_i)$ for every $v : \mathcal{X} \to A$.
- If $\mathbf{t} = \varphi \in \Omega^{(0)}$, then $\mathbf{t}^{\mathbf{A}}(v) = \varphi^{\mathbf{A}}$ for every $v : \mathcal{X} \to A$ if $\varphi^{\mathbf{A}}$ is defined[1], and $\mathrm{dom}\, \mathbf{t}^{\mathbf{A}} = \emptyset$ otherwise.
- If $\mathbf{t} = \varphi(\mathbf{t}_1, \ldots \mathbf{t}_n)$ for some $\varphi \in \Omega^{(n)}$ and terms $\mathbf{t}_1, \ldots, \mathbf{t}_n$, then $v \in \mathrm{dom}\, \mathbf{t}^{\mathbf{A}}$ iff $v \in \bigcap_{i=1}^n \mathrm{dom}\, \mathbf{t}_i^{\mathbf{A}}$ and $(\mathbf{t}_1^{\mathbf{A}}(v), \ldots, \mathbf{t}_n^{\mathbf{A}}(v)) \in \mathrm{dom}\, \varphi^{\mathbf{A}}$, and if $v \in \mathrm{dom}\, \mathbf{t}^{\mathbf{A}}$ then $\mathbf{t}^{\mathbf{A}}(v) = \varphi^{\mathbf{A}}(\mathbf{t}_1^{\mathbf{A}}(v), \ldots, \mathbf{t}_n^{\mathbf{A}}(v))$.

Notice that the definedness and value of $\mathbf{t}^{\mathbf{A}}(v)$ only depend on the images under v of the variables appearing in $\mathbf{t}$. Moreover, if $\varphi \in \Omega^{(n)}$, then the term function associated to $\varphi(x_1, \ldots, x_n)$ is (essentially) the operation $\varphi^{\mathbf{A}} : A^n \to A$.

To simplify the notation, and unless otherwise stated, when we write in the sequel $\mathbf{t}^{\mathbf{A}}(v)$, we always assume that it is defined, i.e., that $v \in \mathrm{dom}\, \mathbf{t}^{\mathbf{A}}$.

An *existence equation*, an *equation* for short, is a pair $(\mathbf{p}, \mathbf{q}) \in \mathrm{T}_\Sigma(\mathcal{X})^2$ of terms, and will be written $\mathbf{p} \approx \mathbf{q}$ in the sequel.

Given an algebra $\mathbf{A}$ and a valuation $v : \mathcal{X} \to A$, the equation $\mathbf{p} \approx \mathbf{q}$ is *satisfied* in $\mathbf{A}$ w.r.t. v, in symbols $(\mathbf{A}, v) \models \mathbf{p} \approx \mathbf{q}$, when $v \in \mathrm{dom}\, \mathbf{p}^{\mathbf{A}} \cap \mathrm{dom}\, \mathbf{q}^{\mathbf{A}}$ and $\mathbf{p}^{\mathbf{A}}(v) = \mathbf{q}^{\mathbf{A}}(v)$. So, for instance, $(\mathbf{A}, v) \models \mathbf{p} \approx \mathbf{p}$ means simply that $\mathbf{p}^{\mathbf{A}}(v)$ is defined; therefore, we denote the equation $\mathbf{p} \approx \mathbf{p}$ by $\exists \mathbf{p}$.

Using equations as atoms, and the connectives $\neg, \vee, \wedge, \Rightarrow, \ldots$ (with their usual logical meaning), we can build up formulas and define their satisfaction in a partial algebra w.r.t. a given valuation; see [2, §7.1] for details. In this paper we are only interested in a special type of such formulas.

A *quasi-existence equation of type Σ* is a formula of the form $\left(\bigwedge_{i \in I} \mathbf{p}_i \approx \mathbf{q}_i\right) \Rightarrow \mathbf{p} \approx \mathbf{q}$ with I a finite set. A *disjunctive quasi-existence equation*, a $\vee$-*equation* for short, *of type Σ* is a formula of the form

$$\left(\bigwedge_{i \in I} \mathbf{p}_i \approx \mathbf{q}_i\right) \Rightarrow \left(\bigvee_{j \in J} \mathbf{p}_j' \approx \mathbf{q}_j'\right)$$

with I and J finite sets; so, $\vee$-equations include, as special cases, quasi-existence equations and disjunctions of equations (taking $|J| = 1$ and $I = \emptyset$, respectively).

---

[1] If $\varphi \in \Omega^{(0)}$, we say that $\varphi^{\mathbf{A}}$ is *defined* when $\varphi^{\mathbf{A}} : A^0 \to A$ is total, and then we use the same symbol $\varphi^{\mathbf{A}}$ to denote the image of this mapping.

To simplify the notation, we usually omit the brackets embracing the premise and the conclusion in $\vee$-equations.

We denote by $\mathcal{L}$ the set of all $\vee$-equations of some previously fixed type.

Let now $\Phi = \bigwedge_{i \in I} \mathbf{p}_i \approx \mathbf{q}_i \Rightarrow \bigvee_{j \in J} \mathbf{p}'_j \approx \mathbf{q}'_j$ be a $\vee$-equation. Then Φ is *satisfied* in an algebra $\mathbf{A}$ w.r.t. a valuation $v : \mathcal{X} \to A$, in symbols $(\mathbf{A}, v) \models \Phi$, iff the following condition holds:

If $(\mathbf{A}, v) \models \mathbf{p}_i \approx \mathbf{q}_i$ for every $i \in I$, then $(\mathbf{A}, v) \models \mathbf{p}'_j \approx \mathbf{q}'_j$ for some $j \in J$.

So, $(\mathbf{A}, v) \not\models \Phi$ iff $\mathbf{p}_i^{\mathbf{A}}(v) = \mathbf{q}_i^{\mathbf{A}}(v)$ for every $i \in I$ but, for every $j \in J$, either $v \notin \operatorname{dom} \mathbf{p}'_j{}^{\mathbf{A}}$, or $v \notin \operatorname{dom} \mathbf{q}'_j{}^{\mathbf{A}}$, or $\mathbf{p}'_j{}^{\mathbf{A}}(v) \neq \mathbf{q}'_j{}^{\mathbf{A}}(v)$.

Now, an algebra $\mathbf{A}$ *(globally) satisfies* a $\vee$-equation Φ, in symbols $\mathbf{A} \models \Phi$, when $(\mathbf{A}, v) \models \Phi$ for every $v : \mathcal{X} \to A$.

It is clear that two isomorphic algebras satisfy exactly the same $\vee$-equations (as we will see later, the converse implication is false, even for total algebras).

We say that an equation $\mathbf{p} \approx \mathbf{q}$ is a *consequence* of a finite set of equations $\{\mathbf{p}_i \approx \mathbf{q}_i\}_{i \in I}$ when the quasi-equation $\bigwedge_{i \in I} \mathbf{p}_i \approx \mathbf{q}_i \Rightarrow \mathbf{p} \approx \mathbf{q}$ is a *tautology* (i.e., it is satisfied by *all* algebras); it is equivalent to say that $\mathbf{p} \approx \mathbf{q}$ is deduced from $\{\mathbf{p}_i \approx \mathbf{q}_i\}_{i \in I}$ through Burmeister's deduction rules for existence equations [2, §6.4.8].

3 Main results

We begin by recalling the definitions of weak and relative subalgebras.

[**Definition 1.**] *Let* $\mathbf{A} = (A, (\varphi^{\mathbf{A}})_{\varphi \in \Omega})$ *and* $\mathbf{B} = (B, (\varphi^{\mathbf{B}})_{\varphi \in \Omega})$ *be two algebras, with* $B \subseteq A$.

i) $\mathbf{B}$ *is a weak subalgebra of* $\mathbf{A}$ *when, for every* $\varphi \in \Omega$, *if* $\underline{b} \in \operatorname{dom} \varphi^{\mathbf{B}}$ *then* $\underline{b} \in \operatorname{dom} \varphi^{\mathbf{A}}$ *and* $\varphi^{\mathbf{B}}(\underline{b}) = \varphi^{\mathbf{A}}(\underline{b})$.

ii) $\mathbf{B}$ *is a relative subalgebra of* $\mathbf{A}$ *when it is a weak subalgebra and, for every* $\varphi \in \Omega$, *if* $\underline{b} \in \operatorname{dom} \varphi^{\mathbf{A}} \cap B^{n(\varphi)}$ *and* $\varphi^{\mathbf{A}}(\underline{b}) \in B$ *then* $\underline{b} \in \operatorname{dom} \varphi^{\mathbf{B}}$.

Notice that every subset B of the carrier of an algebra $\mathbf{A}$ supports (in principle) many weak subalgebras of $\mathbf{A}$, but only one such relative subalgebra, namely the greatest possible weak subalgebra of $\mathbf{A}$ supported on B.

Given an algebra $\mathbf{A}$, let $S_w^{(n)}(\mathbf{A})$ and $S_r^{(n)}(\mathbf{A})$ be the classes of all algebras of cardinal at most n that are isomorphic to weak and relative subalgebras of $\mathbf{A}$, respectively. Let also $S_w^f(\mathbf{A})$ and $S_r^f(\mathbf{A})$ be the classes of all finite algebras that are isomorphic to weak and relative subalgebras of $\mathbf{A}$, respectively. These are the *finite approximations* of $\mathbf{A}$ we consider in this paper.

[**Definition 2.**] *Let* $\mathcal{C}$ *be a class of algebras and let* $\tilde{S}$ *be an algebraic operator corresponding to some type of finite subalgebras (for instance, one of those defined above).*

A set $\tilde{\mathcal{L}}$ *of* $\vee$-*equations is the* syntactical content *of* $\tilde{S}$ *for* $\mathcal{C}$ *when it is the greatest subset of* $\mathcal{L}$ *satisfying the following three conditions:*

[i)] *If* $\bigwedge_{i \in I} \mathbf{p}_i \approx \mathbf{q}_i \Rightarrow \bigvee_{j \in J} \mathbf{p}'_j \approx \mathbf{q}'_j$ *belongs to* $\tilde{\mathcal{L}}$ *then* $\tilde{\mathcal{L}}$ *also contains every* $\vee$*-equation of the form* $\bigwedge_{i \in I} \mathbf{p}_i \approx \mathbf{q}_i \Rightarrow \bigvee_{j \in J'} \mathbf{p}'_j \approx \mathbf{q}'_j$ *with* $J' \subseteq J$ *(we say then that* $\tilde{\mathcal{L}}$ *is* well-formed*)*.

[ii)] *For every formula* $\Phi \in \tilde{\mathcal{L}}$ *there exists a non-empty finite set* $C_{\tilde{S},\mathcal{C}}(\Phi)$ *of finite algebras such that, for every algebra* $\mathbf{A} \in \mathcal{C}$,

$$\mathbf{A} \not\models \Phi \text{ iff there exists some } \mathbf{A}_0 \in C_{\tilde{S},\mathcal{C}}(\Phi) \cap \tilde{S}(\mathbf{A}).$$

[iii)] *For every finite algebra* $\mathbf{A}_0$, *there exists a formula* $\Phi_{\tilde{S},\mathcal{C}}(\mathbf{A}_0) \in \tilde{\mathcal{L}}$ *such that, for every algebra* $\mathbf{A} \in \mathcal{C}$,

$$\mathbf{A}_0 \in \tilde{S}(\mathbf{A}) \text{ iff } \mathbf{A} \not\models \Phi_{\tilde{S},\mathcal{C}}(\mathbf{A}_0).$$

Thus, a well-formed set $\tilde{\mathcal{L}}$ of $\vee$-equations is the syntactical content of $\tilde{S}$ for a class $\mathcal{C}$ when it is the greatest such set such that, for every $\mathbf{A} \in \mathcal{C}$, the knowledge of $\tilde{S}(\mathbf{A})$ is equivalent to the knowledge of

$$\tilde{\mathcal{L}}(\mathbf{A}) = \{\Phi \in \tilde{\mathcal{L}} \mid \mathbf{A} \models \Phi\}.$$

Indeed, notice that, for every $\mathbf{A} \in \mathcal{C}$:

- To know whether $\mathbf{A}$ satisfies a formula $\Phi \in \tilde{\mathcal{L}}$, one has only to check whether some algebra in the finite set $C_{\tilde{S},\mathcal{C}}(\Phi)$ belongs to $\tilde{S}(\mathbf{A})$;
- To know whether a given finite algebra $\mathbf{A}_0$ belongs to $\tilde{S}(\mathbf{A})$, one only has to check the non-satisfaction of $\Phi_{\tilde{S},\mathcal{C}}(\mathbf{A}_0)$ by $\mathbf{A}$.

In particular, the following result holds.

[**Proposition 1.**] *Let* $\tilde{\mathcal{L}}$ *be the syntactical content of the operator* $\tilde{S}$ *for a class* $\mathcal{C}$ *of algebras. Then, given any two algebras* $\mathbf{A}, \mathbf{B} \in \mathcal{C}$, $\tilde{S}(\mathbf{A}) = \tilde{S}(\mathbf{B})$ *iff* $\tilde{\mathcal{L}}(\mathbf{B}) = \tilde{\mathcal{L}}(\mathbf{A})$.

Consider now the following definition.

[**Definition 3.**] *Given a non-tautological* $\vee$*-equation* Φ, *we shall call its* complexity *the greatest cardinal* $\kappa(\Phi)$ *of an algebra* $\mathbf{A}$ *such that* $\mathbf{A} \not\models \Phi$ *but* $\mathbf{A}' \models \Phi$ *for every strict weak subalgebra of* $\mathbf{A}$. *We adopt the convention that tautological* $\vee$*-equations have complexity* 0.

Let $\mathcal{L}^{(n)}$ be the set of all $\vee$-equations of complexity at most n, and, for every $\tilde{\mathcal{L}} \subseteq \mathcal{L}$, set $\tilde{\mathcal{L}}^{(n)} = \tilde{\mathcal{L}} \cap \mathcal{L}^{(n)}$.

Notice that the complexity of a $\vee$-equation Φ is always smaller or equal than the cardinal of the least initial segment[2] of $T_\Sigma(\mathcal{X})$ containing all terms appearing in it.

---

[2] A subset $Y \subseteq T_\Sigma(\mathcal{X})$ is an *initial segment* when, for every $\varphi \in \Omega$ and $\mathbf{t}_1, \dots \mathbf{t}_{\eta(\varphi)} \in T_\Sigma(\mathcal{X})$, $\varphi(\mathbf{t}_1, \dots \mathbf{t}_{\eta(\varphi)}) \in Y$ implies $\mathbf{t}_1, \dots \mathbf{t}_{\eta(\varphi)} \in Y$.

[**Proposition 2.**] *Let $\mathcal{L}_w$ denote the set of all $\vee$-equations of the form*

$$\bigwedge_{i \in I} \mathbf{p}_i \approx \mathbf{q}_i \Rightarrow (\bigvee_{(j_1,j_2) \in J} x_{j_1} \approx x_{j_2}) \vee (\bigvee_{k \in K} \mathbf{p}'_k \approx \mathbf{q}'_k)$$

such that, for every $k \in K$, $\exists \mathbf{p}'_k, \exists \mathbf{q}'_k$ are consequences of $\bigwedge_{i \in I} \mathbf{p}_i \approx \mathbf{q}_i$.

  i) For every $n \in \mathbb{N}$, $\mathcal{L}_w^{(n)}$ is the syntactical content of $S_w^{(n)}$ for Alg_Σ.

  ii) $\mathcal{L}_w$ is the syntactical content of S_w^f for Alg_Σ.

[*Proof.*]We will prove only (i), since (ii) follows immediately. To do that, $\mathcal{L}_w^{(n)}$ being clearly well-formed, we check points (ii) and (iii) in the definition of syntactical content, and then we show that $\mathcal{L}_w^{(n)}$ is the greatest well-formed set of $\vee$-equations satisfying point (ii) therein.

  a) For every $\Phi = \bigwedge_{i \in I} \mathbf{p}_i \approx \mathbf{q}_i \Rightarrow (\bigvee_{(j_1,j_2) \in J} x_{j_1} \approx x_{j_2}) \vee (\bigvee_{k \in K} \mathbf{p}'_k \approx \mathbf{q}'_k)$ in $\mathcal{L}_w^{(n)}$, let $C_{S_w^{(n)}, \mathrm{Alg}_\Sigma}(\Phi)$ be a (minimal) set containing one, and only one, representative of every isomorphism class of algebras $\mathbf{A}'$ such that $\mathbf{A}' \not\models \Phi$ and $|\mathbf{A}'| \leq \kappa(\Phi) \leq n$. This set is clearly finite (and it is empty iff Φ is a tautology). We will show that it satisfies the property required in Definition 2.

  Let $\mathbf{A}$ be an algebra such that $\mathbf{A} \not\models \Phi$, and let $v : \mathcal{X} \to A$ be a valuation such that $(\mathbf{A}, v) \not\models \Phi$, i.e., such that $\mathbf{p}_i^{\mathbf{A}}(v) = \mathbf{q}_i^{\mathbf{A}}(v)$ for every $i \in I$ but $\mathbf{p}'_k{}^{\mathbf{A}}(v) \neq \mathbf{q}'_k{}^{\mathbf{A}}(v)$ for every $k \in K$ (notice that all $\mathbf{p}'_k{}^{\mathbf{A}}(v)$ and $\mathbf{q}'_k{}^{\mathbf{A}}(v)$ are defined because $\exists \mathbf{p}'_k$ and $\exists \mathbf{q}'_k$ are consequences of $\bigwedge_{i \in I} \mathbf{p}_i \approx \mathbf{q}_i$) and $v(x_{j_1}) \neq v(x_{j_2})$ for every $(j_1, j_2) \in J$ with $j_1 \neq j_2$.

  Let V be the set of all variables appearing (explicitly) in the terms of Φ, and let $\mathbf{A}'$ be the least finite weak subalgebra of $\mathbf{A}$ containing $v(V)$ and such that $\mathbf{p}_i^{\mathbf{A}'}(v)$ and $\mathbf{q}_i^{\mathbf{A}'}(v)$ are defined for every $i \in I$. Then, for any valuation $v' : \mathcal{X} \to A'$ that coincides with v on V we have $(\mathbf{A}', v') \not\models \Phi$, hence $\mathbf{A}' \not\models \Phi$.

  Since any strict weak subalgebra of $\mathbf{A}'$ satisfies Φ, we have that $|\mathbf{A}'| \leq \kappa(\Phi)$ and thus it has an isomorphic copy $\mathbf{A}'_0$ in $C_{S_w^{(n)}, \mathrm{Alg}_\Sigma}(\Phi)$. This shows that if $\mathbf{A} \not\models \Phi$ then there exists some $\mathbf{A}'_0$ in $C_{S_w^{(n)}, \mathrm{Alg}_\Sigma}(\Phi) \cap S_w^{(n)}(\mathbf{A})$.

  Conversely, let $\mathbf{A}'$ be a finite algebra of cardinality less then $\kappa(\Phi)$ such that $\mathbf{A}' \not\models \Phi$ and let $\mathbf{A}$ be an algebra containing $\mathbf{A}'$ as a weak subalgebra. Let $v : \mathcal{X} \to A'$ be a valuation such that $(\mathbf{A}', v) \not\models \Phi$: then $\mathbf{p}_i^{\mathbf{A}'}(v) = \mathbf{q}_i^{\mathbf{A}'}(v)$ for every $i \in I$ but $\mathbf{p}'_k{}^{\mathbf{A}'}(v) \neq \mathbf{q}'_k{}^{\mathbf{A}'}(v)$ for every $k \in K$ (remember that $\exists \mathbf{p}'_k$ and $\exists \mathbf{q}'_k$ are consequences of $\bigwedge_{i \in I} \mathbf{p}_i \approx \mathbf{q}_i$) and $v(x_{j_1}) \neq v(x_{j_2})$ for every $(j_1, j_2) \in J$ with $j_1 \neq j_2$.

  Taking as $v : \mathcal{X} \to A$ the same valuation with target set A, we also have $\mathbf{p}_i^{\mathbf{A}}(v) = \mathbf{q}_i^{\mathbf{A}}(v)$ for every $i \in I$ (because they are already defined, and equal, in $\mathbf{A}'$), $\mathbf{p}'_k{}^{\mathbf{A}'}(v) \neq \mathbf{q}'_k{}^{\mathbf{A}'}(v)$ for every $k \in K$ (because they are already defined, and different, in $\mathbf{A}'$), and $v(x_{j_1}) \neq v(x_{j_2})$ for every $(j_1, j_2) \in J$ with $j_1 \neq j_2$, so $(\mathbf{A}, v) \not\models \Phi$ and consequently $\mathbf{A} \not\models \Phi$.

  b) Every partial algebra has an empty weak subalgebra; thus, we can take as $\Phi_{S_w^{(n)}, \mathrm{Alg}_\Sigma}(\emptyset)$ the equation $x_1 \approx x_1$.

Assume now that $\mathbf{A}_0$ is a non-empty partial algebra of cardinality $m \geq 1$, with carrier $A_0 = \{a_1, \ldots, a_m\}$. Let $I(\mathbf{A}_0)$ be the set of equations

$$I(\mathbf{A}_0) = \{\varphi(x_{i_1}, \ldots, x_{i_n}) \approx x_{i_0} \mid \varphi \in \Omega^{(n)}, n \geq 0, i_0, \ldots, i_n \in \{1, \ldots, m\},$$
$$\varphi^{\mathbf{A}_0}(a_{i_1}, \ldots, a_{i_n}) = a_{i_0}\}$$

and take as $\Phi_{S_w^{(n)}, \mathrm{Alg}_\Sigma}(\mathbf{A}_0)$ the $\vee$-equation

$$\bigwedge I(\mathbf{A}_0) \Rightarrow \bigvee_{1 \leq j_1 < j_2 \leq m} x_{j_1} \approx x_{j_2}$$

Notice that $(\mathbf{A}_0, v) \not\models \Phi_{S_w^{(n)}, \mathrm{Alg}_\Sigma}(\mathbf{A}_0)$ for any valuation $v : \mathcal{X} \to A_0$ such that $v(x_i) = a_i$, $i = 1, \ldots, m$. Therefore, if $\mathbf{A}_0 \in S_w^f(\mathbf{A})$, then $\mathbf{A} \not\models \Phi_{S_w^{(n)}, \mathrm{Alg}_\Sigma}(\mathbf{A}_0)$.

Conversely, assume that $\mathbf{A} \not\models \Phi_{S_w^{(n)}, \mathrm{Alg}_\Sigma}(\mathbf{A}_0)$, and let $v : \mathcal{X} \to A$ be a valuation such that $(\mathbf{A}, v) \not\models \Phi_{S_w^{(n)}, \mathrm{Alg}_\Sigma}(\mathbf{A}_0)$. Taking $a_i' = v(x_i)$ for every $i = 1, \ldots, m$, we have that

- if $\varphi(x_{i_1}, \ldots, x_{i_n}) \approx x_{i_0} \in I(\mathbf{A}_0)$, i.e., if $\varphi^{\mathbf{A}_0}(a_{i_1}, \ldots, a_{i_n}) = a_{i_0}$, then $\varphi^{\mathbf{A}}(a_{i_1}', \ldots, a_{i_n}') = a_{i_0}'$;
- if $1 \leq j_1 < j_2 \leq m$ then $a_{j_1}' \neq a_{j_2}'$.

Let $A_0' = \{a_1', \ldots, a_m'\}$, and take the weak subalgebra $\mathbf{A}_0' = (A_0', (\varphi^{\mathbf{A}_0'})_{\varphi \in \Omega})$ of $\mathbf{A}$ with

$$\mathrm{dom}\,\varphi^{\mathbf{A}_0'} = \{(a_{i_1}', \ldots, a_{i_n}') \mid (a_{i_1}, \ldots, a_{i_n}) \in \mathrm{dom}\,\varphi^{\mathbf{A}_0}\}, \quad \varphi \in \Omega^{(n)}, n \geq 0.$$

Then the mapping $h : A_0 \to A_0'$ defined by $h(a_i) = a_i'$, $i = 1, \ldots, m$, is an isomorphism of $\mathbf{A}_0$ onto $\mathbf{A}_0'$, and therefore $\mathbf{A}_0 \in S_w^{(n)}(\mathbf{A})$.

Notice that the complexity of $\Phi_{S_w^{(n)}, \mathrm{Alg}_\Sigma}(\mathbf{A}_0)$ is exactly $|\mathbf{A}_0| = m$. Therefore, for every algebra $\mathbf{A}_0$ of cardinality at most n we have constructed a $\vee$-equation $\tilde{\Phi}_{S_w^{(n)}, \mathrm{Alg}_\Sigma}(\mathbf{A}_0)$ of complexity at most n such that, for every $\mathbf{A} \in \mathcal{C}$, $\mathbf{A}_0 \in S_w^{(n)}(\mathbf{A})$ iff $\mathbf{A} \not\models \tilde{\Phi}_{S_w^{(n)}, \mathrm{Alg}_\Sigma}(\mathbf{A}_0)$, as we wanted.

c) Assume that there is some well-formed set of equations $\tilde{\mathcal{L}}$, not contained in $\mathcal{L}_w^{(n)}$, and satisfying (ii) in Definition 2 w.r.t. the operator $S_w^{(n)}$ and the class Alg_Σ. Then, $\tilde{\mathcal{L}}$ will contain a formula Φ of the form $\bigwedge_{i \in I} \mathbf{p}_i \approx \mathbf{q}_i \Rightarrow \mathbf{p} \approx \mathbf{q}$, where, say, $\exists \mathbf{p}$ is not a consequence of the premise.

Let $\mathbf{A}$ be a minimal algebra not satisfying $\bigwedge_{i \in I} \mathbf{p}_i \approx \mathbf{q}_i \Rightarrow \mathbf{p} \approx \mathbf{q}$, and therefore Φ either. Then by (ii) $\mathbf{A}$ has a finite weak subalgebra $\mathbf{A}_0$ that belongs to $C_{S_w^{(n)}, \mathrm{Alg}_\Sigma}(\Phi)$; however, $\mathbf{A}$ and thus also $\mathbf{A}_0$ has an extension satisfying Φ, in contradiction with (ii). $\square$

In the proofs of the next propositions, we shall only give the corresponding sets $C_{\tilde{S}, \mathcal{C}}(\Phi)$ and $\vee$-equations $\Phi_{\tilde{S}, \mathcal{C}}(\mathbf{A}_0)$; the proofs of the desired properties are similar to those in the previous proposition, already presented in detail.

[**Proposition 3.**] a) $\mathcal{L}^{(n)}$ is the syntactical content of $S_w^{(n)}$ for TAlg_Σ.

  b) $\mathcal{L}$ is the syntactical content of S_w^f for TAlg_Σ.

[*Proof.*]Given an arbitrary $\vee$-equation $\Phi = \bigwedge_{i \in I} \mathbf{p}_i \approx \mathbf{q}_i \Rightarrow \bigvee_{j \in J} \mathbf{p}_j' \approx \mathbf{q}_j$ in $\mathcal{L}^{(n)}$, let $C_{S_w^{(n)}, \mathrm{TAlg}_\Sigma}(\Phi)$ be a (minimal) set containing one, and only one, representative of every isomorphism class of algebras $\mathbf{A}'$ of cardinality at most $\kappa(\Phi)$ that do not satisfy Φ for some valuation $v : \mathcal{X} \to A'$.

  Moreover, given a finite algebra $\mathbf{A}_0$, let $\Phi_{S_w^{(n)}, \mathrm{TAlg}_\Sigma}(\mathbf{A}_0) = \Phi_{S_w^{(n)}, \mathrm{Alg}_\Sigma}(\mathbf{A}_0)$. □

[**Proposition 4.**] Let $\mathcal{L}_r$ denote the set of all $\vee$-equations of the form

$$\bigwedge_{i \in I} \mathbf{p}_i \approx \mathbf{q}_i \Rightarrow \left(\bigvee_{j \in J} \mathbf{t}_j \approx x_{i_j} \right) \vee \left(\bigvee_{k \in K} \mathbf{p}_k' \approx \mathbf{q}_k' \right)$$

where every $\mathbf{t}_j$ is either a variable or a term of the form $\varphi(x_{i_1}, \ldots, x_{i_{n(\varphi)}})$ for some $\varphi \in \Omega$, and, for every $k \in K$, $\exists \mathbf{p}_k', \exists \mathbf{q}_k'$ are consequences of $\bigwedge_{i \in I} \mathbf{p}_i \approx \mathbf{q}_i$.

  a) For every $n \in \mathbb{N}$, $\mathcal{L}_r^{(n)}$ (resp. $\mathcal{L}^{(n)}$) is the syntactical content of $S_r^{(n)}$ for Alg_Σ (resp. TAlg_Σ).

  b) $\mathcal{L}_r$ (resp. $\mathcal{L}$) is the syntactical content of S_w^f for Alg_Σ (resp. TAlg_Σ).

[*Proof.*]For every $\Phi \in \mathcal{L}_r^{(n)}$ take $C_{S_r^{(n)}, \mathrm{Alg}_\Sigma}(\Phi) = C_{S_w^{(n)}, \mathrm{Alg}_\Sigma}(\Phi)$ and $C_{S_r^{(n)}, \mathrm{TAlg}_\Sigma}(\Phi) = C_{S_w^{(n)}, \mathrm{TAlg}_\Sigma}(\Phi)$. Moreover, for every finite algebra $\mathbf{A}_0$:

- $\Phi_{S_r^{(n)}, \mathrm{Alg}_\Sigma}(\emptyset) = \Phi_{S_r^{(n)}, \mathrm{TAlg}_\Sigma}(\emptyset) = x_1 \approx x_1$;
- If $\mathbf{A}_0$ has carrier $A_0 = \{a_1, \ldots, a_m\}$, $m \geq 1$, then let $I(\mathbf{A}_0)$ be the set of equations associated to $\mathbf{A}_0$ as in the proof of Proposition 2, and let

$$I^c(\mathbf{A}_0) = \{ \varphi(x_{i_1}, \ldots, x_{i_n}) \approx x_{i_0} \mid \varphi \in \Omega^{(n)}, n \geq 0, i_0, \ldots, i_n \in \{1, \ldots, m\},$$
$$(a_{i_1}, \ldots, a_{i_n}) \notin \mathrm{dom}\, \varphi^\mathbf{A} \text{ or }$$
$$\varphi^\mathbf{A}(a_{i_1}, \ldots, a_{i_n}) \neq a_{i_0} \}$$

  Then take $\Phi_{S_r^{(n)}, \mathrm{Alg}_\Sigma}(\mathbf{A}_0) = \Phi_{S_r^{(n)}, \mathrm{TAlg}_\Sigma}(\mathbf{A}_0)$ to be the $\vee$-equation

$$\bigwedge I(\mathbf{A}_0) \Rightarrow \left(\bigvee I^c(\mathbf{A}_0) \vee \bigvee_{1 \leq j_1 < j_2 \leq m} x_{j_1} \approx x_{j_2} \right)$$

□

References

1. W. Bartol, X. Caicedo, F. Rosselló, "Local equivalence of partial algebras." Preprint (1997).
2. P. Burmeister, *A Model Theoretic Oriented Approach to Partial Algebras*. Mathematical Research vol. 32, Akademie–Verlag (1986).

A New Approach to Linguistic Negation based upon Compatibility Level and Tolerance Threshold

Daniel Pacholczyk

LERIA, University of Angers, 2 Bd Lavoisier, F-49045 Angers Cedex 01, FRANCE
Email: pacho@univ-angers.fr

Abstract. *In this paper, we focus our attention on the representation of linguistic negation of nuanced information in knowledge-based systems. The linguistic negation presented here uses a compatibility level and tolerance threshold based upon neighborhood and similarity relations. Their combination allows us to choose the reference frame from which the possible values of a linguistic negation of A appearing in the statement "x is not A" will be extracted. This new approach to negation takes into account linguistic analysis of negation and leads to intended properties in common sense reasoning.*

1 Introduction

In this paper, we focus our attention on the representation of nuanced information expressed in *affirmative form* like "the weather is *really very* wet" or in *negative form* like "the weather is *not* dry". Our main goal has been to create a new symbolic model dealing with this information, but more particularly with negative information referring to linguistic negation, and this, within the context of fuzzy set theory [27]. It is obvious that it is not easy to solve the problem of representation of *nuances* ([18,20,19]), *modifiers* ([28], [2,4], [3], [5], [7], [16], [13]), or *linguistic labels* ([25]) in terms of membership functions. In Section 2, we present the initial representation of nuanced information based on an *automatic process* defining the L-R functions ([10]) associated to *nuances of basic properties* ([8,9]). Section 3 is devoted to the presentation of a new approach to linguistic negation which, *(1) alleviates difficulties* of logical negation based on fuzzy complementation ([27]) and of qualitative negation [25,16] used in symbolic models, and *(2) improves* the approach to linguistic negation proposed by Pacholczyk in ([18,17,20]). We first define a more general concept of *linguistic negation* depending on *compatibility level* ρ and *tolerance threshold* ε. These parameters define linguistic negation through *neighbourhood* and *similarity* relations. Then, a combination of ρ and ε values allows us to choose the *reference frame*, a set of nuanced properties, denoted as $\mathrm{Neg}_{\rho,\varepsilon}(A)$, from which the possible values of linguistic negation of A will be extracted. Each element of $\mathrm{Neg}_{\rho,\varepsilon}(A)$ is said to be a *linguistic negation ρ-compatible with A with a tolerance threshold ε*. Then, we can define a reference frame subset, denoted as $\mathrm{neg}_{\rho,\varepsilon}(A, x)$, which consists of the *intended meanings of the linguistic negation ρ-compatible with A at x with a tolerance threshold ε*. This new linguistic negation leads to the intended meaning resulting from the analysis proposed by linguists. Finally, we present properties of this new linguistic negation.

L. Polkowski and A. Skowron (Eds.): RSCTC'98, LNAI 1424, pp. 416–423, 1998.
© Springer-Verlag Berlin Heidelberg 1998

2   The Initial Frame of Information Representation

In many domains a part of knowledge is represented by facts, denoted as "x is A", and rules, denoted as if "x is A" then "y is B". So, their representation can be handled with Fuzzy Sets associated with respective properties. Moreover, this knowledge based upon *fuzzy properties*, *(1)* can be expressed in *natural language with the aid of nuanced expressions*, and *(2)* can refer to *linguistic negations of properties*. The knowledge base can be characterized by a finite number of distinct concepts C_j. The set $\mathcal{I}$ denotes the objects (or individuals) belonging to the discourse universe. A finite set of distinct *basic properties* denoted as P_{ik}, defined in the same *domain* D_j, is associated with each concept C_j. The user applies *linguistic modifiers* to these basic properties to express his *knowledge*.
- Example. A knowledge base can contain rules like R1: if Jack is not small then he is visible in the crowd, R2: if the wage is not high then the summer holidays are not long, R3: if the weather is not wet then the tourist season is not bad. The user can introduce facts like F1: Jack is really very tall, F2: the wage is really low, F3: the weather is dry.

The model proposed by Desmontils & Pacholczyk in [8,9] allows us to refer to affirmative information like "x is $f_\alpha m_\beta P_{ik}$" or negative information like "x is not $f_\alpha m_\beta P_{ik}$". A property like "$f_\alpha m_\beta P_{ik}$" which requires for its expression a list of linguistic terms is called a *nuanced property*. Let us denote as $\mathcal{N}$ the set containing the nuances of all basic properties P_{ik}. We have selected two sets of fuzzy modifiers:
- The first one consists of *translation modifiers*: $M_7=$\{extremely little, very little, rather little, moderately (Ø), rather, very, extremely\} (Fig. 1). M_7 is supposed to be totally ordered by the relation: $m_\alpha < m_\beta \Leftrightarrow \alpha < \beta$.
- The second one consists of *precision modifiers*: $F_6=$\{vaguely, neighboring, more or less, moderately (Ø), really, exactly\} (Fig. 2). In the same way, F_6 is supposed to be totally ordered: $f_\alpha < f_\beta \Leftrightarrow \alpha < \beta$.

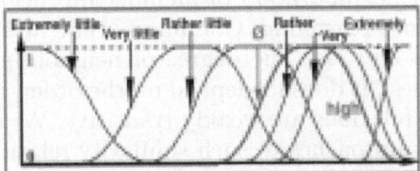 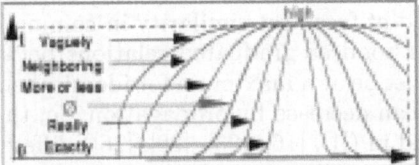

Fig. 1. Translation modifiers   **Fig. 2.** Precision modifiers

3   Negation ρ–compatible with A with Tolerance Threshold ε

As already pointed out by Pacholczyk in [18,19,20], *it is necessary* to leave a representation of linguistic negation in terms of a one-to-one correspondence between the elements of L *and to turn towards a one-to-one correspondence between an element of L and a subset of L*. More generally, any function associating with each nuanced property A a subset of nuanced properties defined on the same domain as A, will be called a *multiset function*.

Basing one's argument on linguistic analysis (Muller [15], Culioli [6], Horn [12], Ducrot and Schaeffer [11], Ladusaw [14]), Pacholczyk has pointed out in [18,20] that when one asserts that "x is not A" then, (1) one *rejects* a reference to "x is A", and (2) if necessary, *one refers either* to the *logical negation of* A, or to *another property* P different from A but defined in the same domain, or sometimes to a *nuance* $f_\alpha m_\beta A$ of A, or finally to *a new basic property* denoted as not-A.

The following examples illustrate these different cases: Asserting that "Smith is not guilty", Sherlock Holmes can only reject the hypothesis "Smith is guilty" ; Saying that "my hat is not blue", the clown can signify that his hat has another colour ; The statement of an employer "the wage is not very high" can receive as an intended meaning "the wage is really low" ; Saying "this wine is not bad" the restaurant owner can signify to his customer that "it is not-bad ". So, "not-bad" is a new basic property associated with wine quality.

The statement "x is A" may be interpreted as one of the following statements ([24]): "x is ØA", "x is really A" or "x is more or less A". In the set F_6 of precision modifiers, we can put: F_{61}={more or less, Ø, really} and F_{62} ={vaguely, neighboring, exactly}. So, linguistically speaking we put: "x is A" ⇔ { "x is $f_\alpha A$" with $f_\alpha \in F_{61}$}. In other words: Rejection of "x is A" ⇒ Rejection of { "x is $f_\alpha A$" with $f_\alpha \in F_{61}$}. Moreover, the assertion "x is not A" can in addition imply a reference to a nuanced property P in order to define "x is P" as the intended meaning of "x is not A". Intuitively the speaker understands this real difference in terms of a weak neighborhood between the membership degrees to A and P for their significant values, that is to say: $\mu_A(x)$ or $\mu_P(x)$ *is rather close to* $1 \Rightarrow \mu_A(x)$ and $\mu_P(x)$ are *weakly neighboring*. It is obvious that the expressions "rather close to" and "weakly neighboring" can receive different translations, each of one defining a linguistic negation by the strength with which the assertion is denied. *That's why we have introduced a first parameter ρ, with $1 \geq \rho \geq 0$, to take into account this negation strength.*

Let us now talk about the notions of neighborhood of membership degrees and of fuzzy set similarity in common sense reasoning. Commonly they are understood as graduated relations between entities. The degree of neighborhood must be at a maximum for identical entities ; it doesn't depend on the order, and it can decrease by propagation. So, these relations are weakly transitive. We can find in ([1], [16], [23], [29], [26]) different approaches to such similarity relations. Let us point out a final detail: the strength of linguistic negation depends also on the precision of each degree of neighborhood or similarity. In other words, a speaker attaches a tolerance threshold to the precision of each degree. *That's why we use a second parameter, denoted as ε, with $\rho \geq \varepsilon \geq 0$, to define the negation strength according to a tolerance threshold.*

[**Definition 1.**]*A neighborhood relation V is defined in [0, 1] as follows:*
VN1: *V(x, x)=1 (Reflexivity) ;* **VN2**: *V(x, y)=V(y, x) (Symmetry) ;*
VN3: *V(x, z)$\geq$T(V(x, y), V(y, z)), where T is a T-norm (Pseudo-transitivity).*

[**Definition 2.**]*Given neighborhood relation V defined in [0, 1], x and y are said to be (at least) α-neighboring if V(x, y)$\geq \alpha$.*

- Example. $V_L(x, y)=\text{Min}(x\rightarrow_L y, y\rightarrow_L x)$ with $x\rightarrow_L y=1$ if $x\leq y$ else $1-x+y$ (Lukasiewicz' implication) is a neighborhood relation.

Let us now denote by $\mathcal{F}$ the set of all fuzzy sets defined in the domain $\mathcal{D}$.

[**Definition 3.**]*Let* $V: \mathcal{F} \times \mathcal{F} \rightarrow [0, 1]$. *Then* V *defines a similarity relation in* $\mathcal{F}$ *if and only if:* **S1**: $V(A, A)=1$ *(Reflexivity)* ; **S2**: $V(A, B)=V(B, A)$ *(Symmetry)* ; **S3**: $V(A, C)\geq T(V(A, B),V(B, C))$, *T being the previous T-norm (Pseudotransitivity)*

[**Definition 4.**]*Given that the fuzzy sets A and B of* $\mathcal{F}$ *are said to be (at least)* α *similar, we denote this as* $V(A, B) \geq \alpha$, *iff* $\forall$ *x*, $V(\mu_A(x), \mu_B(x)) \geq \alpha$.

In order to define the *linguistic nuanced similarity* of fuzzy sets, we can introduce a set of linguistic expressions $\mathcal{L}=\{\theta_1,\ldots,\theta_7\}=\{$not at all, very little, rather little, moderately (Ø), rather, very, totally$\}$ totally ordered as follows: $i>j \Leftrightarrow \theta_i > \theta_j$. Moreover, we suppose that they are defined as *fuzzy subintervals* ([10]) of [0, 1] (Fig. 3).

[**Definition 5.**]*Given* V, *two fuzzy sets A and B of* $\mathcal{F}$ *are said to be* θ_i *similar if and only if,* $\mu_{\theta_i}(\alpha) = Max_j\{\mu_{\theta_j}(\alpha)\}$ *knowing that* $\alpha = Max\{\delta|V(A, B) \geq \delta\}$.

Then, we can prove the following properties.

[*Property 1.*]The linguistic similarity relation possesses the following properties:
LS1: A is totally similar to A (Reflexivity)
LS2: if A and B are θ_i similar then B and A are θ_i similar (Symmetry)
LS3: if A and B are θ_i similar and B and C are θ_j similar then A and C are θ_k similar with $\theta_k \geq \theta_3$ knowing that θ_3 is such that $\mu_{\theta_3}(T(\alpha_1,\alpha_2)) = Max_l\{\mu_{\theta_l}(T(\alpha_1,\alpha_2))\}$, $\alpha_1 = Max\{\delta|V(A, B) \geq \delta\}$ and $\alpha_2 = Max\{\delta|V(B,C) \geq \delta\}$ (Weak transitivity).

[*Property 2.*]There exists a multiset function n: $\mathcal{N} \rightarrow \mathcal{P}(\mathcal{N})$ such that $n(A)$ is the set of all nuanced properties associated with the concept defining the nuanced property A.

This being done, we can go into the details of the new concept of linguistic negation.

[**Definition 6.**]*Let* ρ, ε *such that:* $0 \leq \varepsilon \leq \rho \leq 1$. *The multiset function* $Neg_{\rho,\varepsilon}$: $\mathcal{N} \rightarrow \mathcal{P}(\mathcal{N})$ *defined as follows:* $\forall A \in \mathcal{N}, \forall P \in n(A), P \in Neg_{\rho,\varepsilon}(A)$ *if and only if :* **N1**: *P and A are* θ_i-*similar with* $\theta_i <moderately$, **N2**: $\mu_A(x) \geq \rho \Rightarrow V(\mu_A(x), \mu_P(x)) \leq 1 - \rho + \varepsilon$, **N3**: $\mu_P(x) \geq \rho \Rightarrow V(\mu_P(x), \mu_A(x)) \leq 1 - \rho + \varepsilon$, *is said to be the linguistic negation* ρ-*compatible with a tolerance threshold* ε. *Moreover, any* $P \in Neg_{\rho,\varepsilon}(A)$ *is said to be a linguistic negation* ρ-*compatible with A with the tolerance threshold* ε.

We can specify the functions of parameters ρ and ε (conditions **N2** and **N3**). For given ε, we can illustrate the function of ρ within the application domain of linguistic negation of A. When ρ creases, its application domain, the strong

ρ-cut of A (resp. P) ([10]), decreases by inclusion in the domain D to the strong 1-cut. So, when $\rho > 0$, the conditions **N2** and **N3** define the threshold ρ of minimal compatibility between elements of D with A regarding linguistic negation of A. *So, ρ defines among possible reference frames of the negation of A in the domain D, the strong ρ-cut of A as the reference frame for linguistic negation ρ-compatible with A.* For given ρ, the parameter ε induces for any $\mu_P(x) \geq \rho$, the local neighborhood condition $V(\mu_A(x), \mu_P(x)) \leq 1 - \rho + \varepsilon$. Its translation implies a maximal threshold for the values of $\mu_P(x)$ (resp. $\mu_A(x)$. Moreover, when ε decreases, the local neighborhood decreasing to $(1 - \rho)$, this threshold also decreases. In other words, *ε defines a maximal tolerance threshold to which we refer to modulate the precision of linguistic negation ρ-compatible with A.* As a conclusion, *the combination of ρ and ε allows the choice of reference frame from which the possible values of linguistic negation will be extracted.*
- Example. Using $V = V_L, \rho = 0.97$ and $\varepsilon = 0.3$, we have collected in Figure 4, $\text{Neg}_{0.97,0.3}$("low") the set of linguistic negations 0.97-compatible with "low" with tolerance threshold 0.3.

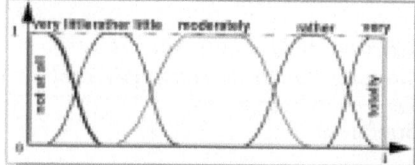 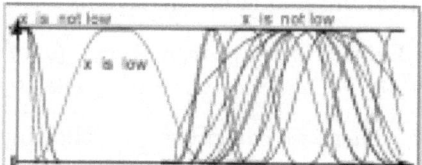

Fig. 3. Linguistic similarity degrees   **Fig. 4.** Plausible linguistic negations
- Remark: In [18,17,20,19], the linguistic negation was based upon the previous particular neighborhood relation V_L. Moreover, the condition **N2** was basically different: $\mu_A(x) = \xi \geq 0.67 + \varepsilon \Rightarrow \mu_P(x) \leq \xi - 0.67$. It is easy to establish that this linguistic negation satisfies the new property **N2**. *In other words, our new approach to linguistic negation improves the previous models.*

The set $\text{Neg}_{\rho,\varepsilon}(A)$ defines the reference frame from which we have to extract the values of the linguistic negation. This comes down to define explicitly a subset of $\text{Neg}_{\rho,\varepsilon}(A)$, consisting of the intended meaning of this occurrence of "x is not A". More formally, we can denote as (A, x) the pair representing "x is A". The previous process leads to the definition of *choice function* $\text{neg}_{\rho,\varepsilon} : \mathcal{N} \times \mathcal{I} \to \mathcal{P}(\mathcal{N})$, which associates with (A, x) the set $\text{neg}_{\rho,\varepsilon}(A, x)$ of nuances accepted as intended meanings of "x is not A".

[**Definition 7.**]*Any $P \in \text{neg}_{\rho,\varepsilon}(A, x)$ is called an intended meaning in $\mathcal{I}$ at x of the linguistic negation ρ-compatible with A with tolerance threshold ε. We say also that "x is P" is an intended meaning in $\mathcal{I}$ with the tolerance threshold ε of the linguistic negation ρ-compatible of "x is A". If no confusion is possible, we simply say that P is an intended meaning of the linguistic negation of A at x.*

We can now present without proofs some properties of this linguistic negation.

[*Property 3.*]For many properties A, "x is A" does not automatically define the knowledge about "x is not A". More precisely, for any A we only know that "x is P" can be an intended meaning of "x is not A" if P belongs to $\text{neg}_{\rho,\varepsilon}(A, x)$.

[*Property 4.*]If $P \in \text{Neg}_{\rho,\varepsilon}(A)$ then $A \in \text{Neg}_{\rho,\varepsilon}(P)$. But, if $P \in \text{neg}_{\rho,\varepsilon}(A, x)$ we cannot assert that $A \in \text{neg}_{\rho,\varepsilon}(P, x)$. In other words, the double negation of A belongs to a frame reference of the linguistic negation, but the intended meaning of "x is not P" does not generally lead to "x is A".

[*Property 5.*]$\text{Neg}_{1,\varepsilon}(A) = \{P \in n(A), P$ is less than moderately similar to $A, \mu_A(x)$ (resp. $\mu_P(x)) = 1 \Rightarrow V(1, \mu_P(x)(\text{resp. } \mu_A(x)) \leq \varepsilon\}$.

[*Property 6.*]$0 \leq \varepsilon \leq \rho \leq 1, \text{Neg}_{1,\varepsilon}(A) \subseteq \text{Neg}_{\rho,\varepsilon}(A) \subseteq \text{Neg}_{0,0}(A)$.

[*Property 7.*]$0 \leq \varepsilon \leq \rho \leq 1$, if $\text{Neg}_{1,\varepsilon}(A) \neq \emptyset$ then $\text{Neg}_{\rho,\varepsilon}(A) \neq \emptyset$.

[*Property 8.*]$\{\rho \leq \rho', \varepsilon' \leq \varepsilon \leq \rho\} \Rightarrow \text{Neg}_{1,0}(A) \subseteq \text{Neg}_{\rho',\varepsilon'}(A) \subseteq \text{Neg}_{\rho,\varepsilon}(A) \subseteq \text{Neg}_{0,0}(A)$.

[*Property 9.*]If we choose the neighboring relation: $V_L(x,y) = \text{Min}\{x \rightarrow_L y, y \rightarrow_L x\}$, then $\text{Neg}_{1,0}(A)$ and $\text{Neg}_{\rho,\varepsilon}(A)$ are generally not empty.

Let us recall conditions satisfied by modifiers ([8,9]).
H1: A and $f_\alpha A$ are *less than moderately* similar, if and only if, $f_\alpha \in F_{62}$,
H2: The *translation modifiers* are defined in such a way that the *resulting nuances cover the corresponding domain*.
We suppose that the properties fulfil also the following conditions :
H3: For any basic property P_{ik}, there exists one translation modifier defining a nuance which differs from P_{ik} for the significant values : $\forall P_{ik}, \forall \rho_k, \exists \beta_k, 0 \leq \beta_k \leq \rho_k, \exists m_\alpha \in M_7, \{\forall x, \mu_{P_{ik}}(x) \geq \rho_k \Rightarrow \mu_A(x) \leq \beta_k$ with $A = m_\alpha P_{ik}\}$.
H4: For any concept C_i, the associated basic properties differ basically for significant values: $\forall k, \exists \sigma_k, 0 \leq \sigma_k \leq 1, \exists \gamma_k, 0 \leq \gamma_k \leq \sigma_k, \{\forall x, \mu_{P_{ik}}(x) \geq \sigma_k \Rightarrow \forall j \neq k, \mu_{P_{ij}}(x) \leq \gamma_k\}$.
Then, it is easy to prove the following result.

[*Property 10.*]There exists neighborhood and similarity relations leading to a concept of linguistic negation taking into account all linguistic interpretations of "x is not A".

4   Deductive Process dealing with Linguistic Negation

The presence of linguistic negations in the knowledge base does not generally modify the use of the existing deductive process. For example, we can suppose that for given ρ and ε, the intended meaning of negations in the initial knowledge base lead to R'1: if Jack is very tall then Jack is visible in the crowd, R'2: if the wage is low then the summer holidays are Q$\in$ {short, average}, R'3: if the weather is dry then the tourist season is not-bad. The precision modifiers give us: Jack is really very tall $\Rightarrow$ Jack is very tall and the wage is really low $\Rightarrow$ the wage is low. By using facts F1-F3, we can deduce that: Jack is visible in the crowd, the summer holidays are short or average, and the tourist season is not-bad.

By using classical systems, deductive process based upon logical negation cannot be applied to facts and rules which include linguistic negations. But, it is not the case with our approach to linguistic negation, since the same deductive process can be apply to equivalent rules referring only to affirmative information. Moreover, a lack of information about the exact intended meaning does not necessary prohibit the use of rules containing linguistic negation in its premise. So, our approach to linguistic negation improves the abilities in the management of knowledge base in that facts or rules can include linguistic negations.

5 Conclusion

We have defined a general concept of *linguistic negation of nuanced property* based upon *neighborhood* and *similarity* relations and depending on *compatibility level* and *tolerance threshold*. A combination of these values allows the choice of *reference frame* from which the possible values of linguistic negation of "x is A" are extracted. This new approach to linguistic negation can be presented as a *generalization of previous models* taking into account linguistic analysis of negation. It appears clearly that this approach to linguistic negation improves the abilities in the management of a knowledge base in that *facts or rules can include linguistic negations*.

References

1. J. F. Baldwin and B. W. Pilsworth. Axiomatic approach to implication for approximate reasoning with fuzzy logic. *Fuzzy Sets and Systems* **3**, pp. 193–219, 1980.
2. B. Bouchon. *Stability of Linguistic Modifiers compatible with a Fuzzy Logic*, **313**. Lecture Notes in Computer Science, 1988.
3. B. Bouchon and S. Desprès. *Acquisition numérique/symbolique de connaissances graduelles*. Actes des 3èmes journées du PRC-GDR Intelligence Artificielle, Paris, Hermès, 1990.
4. B. Bouchon–Meunier. *Fuzzy Logic and Knowledge Representation using Linguistic Modifiers*. Fuzzy Logic for the Management of Uncertainty, Zadeh L. A., Kacprzyk J. Eds, 1992.
5. B. Bouchon–Meunier and J. Yao. Linguistic modifiers and gradual membership to a category. *Int. Jour. of Intelligent Systems*, **7**, pp. 25–36, 1992.
6. A. Culioli. *Pour une linguistique de l'énonciation : Opérations et Représentations*. **1** Ophrys Eds. Paris, 1991.
7. J. Delechamp and B. Bouchon–Meunier. Graduality by means of analogical reasoning. *Proc. Int. Conf. ECSQARU-FAPR'97, Lecture Notes in Artificial Intelligence*, **1244**, pp. 210–222, 1997.
8. E. Desmontils and D. Pacholczyk. Modélisation déclarative en synthèse d'images: traitement semi-qualitatif des propriétés imprécises ou vagues. In *Proceedings of AFIG'96*, pp. 173–181, Dijon, 1996.
9. E. Desmontils and D. Pacholczyk. Towards a linguistic processing of properties in declarative modeling. *Int. Journal of CADCAM and Comp. Graphics*, **12:4**, pp. 351–371, 1997.
10. D. Dubois and H. Prade. *Théorie des Possibilités*. Masson, 1985.

11. O. Ducrot and J.-M. Schaeffer. *Nouveau dictionnaire encyclopédique des sciences du langage*. Seuil, Paris, 1995.

12. L.R. Horn. *A Natural History of Negation*. The University of Chicago Press, Chicago, 1995.

13. J. Kacprzyk *Fuzzy logic with linguistic quantifiers in inductive learning*. Fuzzy Logic for the Management of Uncertainty. L.A. Zadeh and J. Kacprzyk, J. Wiley, pp. 465–478, 1992.

14. W.A. Ladusaw. *Negative Concord and " Made of Judgement "*. a notion in Focus. H. Wansing eds. Berlin, 1996.

15. C. Muller. *La négation en français*. Publications romanes et françaises, Genève, 1991.

16. D. Pacholczyk. *Contribution au traitement logico-symbolique de la connaisance*. Thèse d'état, Paris 6, Avril 1992.

17. D. Pacholczyk. About the representation of negative information. *Proc. of Int. Conf. EUFIT'97*, **2**, pp. 877–881, 1997.

18. D. Pacholczyk. A fuzzy analysis of linguistic negation of nuanced property in knowledge-based systems. *Proc. of Int. Conf. ECSQARU-FAPR'97, Lecture Notes in Art. Int*, **1244**, pp. 451–465, 1997.

19. D. Pacholczyk. An intelligent system dealing with negative information. *Proc. of Int. Conf. ISMIS'97, Charlotte, U. S. A, Lecture Notes in A. I.*, **1325**, pp. 467–476, 1997.

20. D. Pacholczyk. Towards a qualitative representation of linguistic negation of nuanced properties. *Proc. of AIIA'97, Roma, Italy, Lecture Notes in Art. Int.*, **1321**, pp. 393–404, 1997.

21. D. Pacholczyk and E. Desmontils. A qualitative approach to fuzzy properties in scene description. *Proc. of Int. Conf CISST'97, Las Vegas, U. S. A.*, pp. 139–148, 1997.

22. D. Pacholczyk and E. Desmontils. Vers une description nuancée des scènes en synthèse d'images. *to appear in Proc. LFA'97, Lyon, France*, 1997.

23. E.H. Ruspini. The semantics of vague knowledge. *Rev. int. De Systémique*, **3:4**, pp. 387–420, 1989.

24. P. Scheffe. On foundations of reasoning with uncertain facts and vague concepts. *fuzzy reasoning and its applications*, pp. 189–216, 1981.

25. V. Torra. Negation functions based semantics for ordered linguistic labels. *Int. Jour. of Intelligent Systems*, **11**, pp. 975–988, 1996.

26. A. Tversky. Features of similarity. *Psychological Review*, **4**, 1977.

27. L.A. Zadeh. Fuzzy sets. *Information and Control*, **8**, pp. 338–353, 1965.

28. L.A. Zadeh. *PRUF-A meaning representation language for natural languages*, **10:4**. Int. J. Man-Machine Studies, 1978.

29. L.A. Zadeh. *Similarity relations and Fuzzy orderings*, **3**. Selected papers of L. A. Zadeh, 1987.

Some Issues on Nondeterministic Knowledge Bases with Incomplete and Selective Information

Hiroshi Sakai

Department of Computer Engineering
Kyushu Institute of Technology
Department of Computer Engineering
Tobata Kitakyushu 804, Japan
Tel(Fax) +81-93-884-3258, e-mail:sakai@comp.kyutech.ac.jp

Abstract. Rough set theory depending upon deterministic information systems or knowledge bases is now becoming a mathematical foundation of soft computing. In this paper, we pick up nondeterministic knowledge bases with incomplete and selective information. The both information are given as a set of attribute values, whose difference comes from the temporal concept. If the information is referring the past information then we see it incomplete information. On the other hand, selective information means that the real attribute value is not decided in a set, i.e., we can select the most proper value from this set. By introducing these two information into knowledge bases, we develop another framework for nondeterministic knowledge bases. Namely, we discuss question-answering, approximation, rough set concept and dependencies of attributes on this nondeterministic knowledge bases.

1 Introduction

Rough set theory is seen as a mathematical foundation of soft computing, which covers some areas of research in AI, i.e., knowledge, imprecision, vagueness, learning, induction[1,2]. We recently see some applications of this theory to knowledge discovery and data mining[3,4,5]. The rough set theory historically seems to depend upon the information systems including information retrieval systems[6,7,8]. As for information systems, the main purpose seems to establish the theory for question-answering. We know some famous works. However rough set theory handles not only question-answering but also other several issues for soft computing, like approximation, dependencies of attributes and decision rules.

In this paper we basically depend upon the definitions in [1,2]. Furthermore, we deal with a case that *every attribute value for every object is unique and not multivalued*. As for dealing with multivalued attributes, we need more research. According to [1], we use term *'knowledge bases'* instead of *'information systems'*. For these knowledge bases, we introduce the following incomplete information and selective information into them.

(1) *Incomplete information: For object x, there is a real attribute value in a set, but we can not decide which is the real one for the lack of information.*

L. Polkowski and A. Skowron (Eds.): RSCTC'98, LNAI 1424, pp. 424–431, 1998.
© Springer-Verlag Berlin Heidelberg 1998

(2) *Selective information: For object x, we can select its attribute value from a set. In this case, the selection is done by every user of these knowledge bases. The issue for handling selective information is to find the most proper attribute value for every user's purpose.*

The incomplete information causes modal concept '*possibility*' and '*necessity*', and there exists unknown real attribute value. However in selective information, there is no modal concept and no real attribute value. This is the big difference of two information. If we see $affiliation(tom, \{management, development\})$ as incomplete information then we see that Tom's affiliation is either *management* or *development*. Therefore, we see that $affiliation(tom, management)$ may be true. On the other hand, if we see it selective information then we see that we can select his affiliation from *management* or *development*. The user of this information will decide Tom's affiliation by checking the expectation value in every case. Therefore, the introduction of selective information supports us to discuss the issues like planning and decision making, too. Here, the purpose is to find the most proper attribute value for every user.

We call knowledge bases with two above information '*Nondeterministic knowledge bases*' from now on. We mainly discuss the question-answering and variational rough set theory in nondeterministic knowledge bases.

2   Nondeterministic Knowledge Bases : NKBs

Now in this section, we give some basic definitions according to [1]. Let $KB = (X, A, V, f)$ be a knowledge base, where X is a finite set of objects, A is a finite set of attributes, V is the set-theoretical union of domains of attributes from A and f is a classification function such that $f : X * A \rightarrow V$. In every KB, the value of classification function is definite.

Then, we give the definition of *nondeterministic knowledge base(NKB) by incomplete and selective information*. We need to specify three kinds of information for every NKB, namely definite, incomplete or selective information. We name a set $S = \{d(definite), i(incomplete), s(selective)\}$ state set. In this case, we define an $NKB = (X, A, V, S, g)$ where g is a classification function which satisfies $g : X * A \rightarrow 2^V * S$. We usually use the attribute value instead of a singleton set in case of definite information.

Example 1. Let's consider a party planning issue. We have to decide place and date for party. Here $X = \{plan1, plan2, plan3\}$, $A = \{place, date, estimated\_cost\}$, $V_{place} = \{place1, place2\}$, $V_{date} = \{fri, sat, sun\}$ and $V_{estimated\_cost} = Set\_of\_natural\_number$. As for the details, we know the following table by classification function g. This $NKB_1 = (X, A, V, S, g)$ shows that we have three plans with incomplete and selective information. For example, *plan1* shows we can reserve *place1* on Saturday or Sunday and the estimated cost is between 12 to 15, i.e., estimated cost is either 12, 13, 14 or 15. When we see this NKB_1 as a knowledge base, we may think the following queries according to NKB_1.
(1) *How much is necessary at least for selecting place1?*
(2) *How much is necessary at least for selecting Sunday?*

X	place	date	estimated_cost
plan1	place1	$(\{sat, sun\}, s)$	$([12, 15], i)$
plan2	place1	fri	$([7, 9], i)$
plan3	place2	$(\{fri, sat, sun\}, s)$	$([5, 7], i)$

Table 1. Knowledge base NKB_1

These are the aspect of question-answering in NKB_1. However, we have another aspect in NKB_1, namely the aspect of planning. Every user has the most proper selection according to his criteria from this NKB_1. For example, if the attendance is usually maximum on Sunday and a user wants to gather a large attendance, then he will select Sunday from Table 1.

3 Question-Answering in NKBs

Now in this section, we refer to question-answering in every NKB. In incomplete information systems, two valuational concepts '*may hold*' and '*surely hold*' come from the incomplete information[7]. However, NKBs have not only two valuational concepts but also concept of optimal selection.

3.1 Extensions in Every NKB

We first give definitions in every NKB. Let $NKB = (X, A, V, S, g)$ be an nondeterministic knowledge base, where $g : X * A \rightarrow 2^V * S$. In this case, we call a classification function g' which satisfies the following (1) and (2) an *extensional classification function with selective information*.

(1) *For every object $x(\in X)$ and attribute $a(\in A)$ which satisfy $g(x, a) = (subset\_SV\_of\_V, i)$, $g'(x, a) = (element\_of\_SV, d)$.*

(2) *For every object $x(\in X)$ and attribute $a(\in A)$ which satisfy $g(x, a) = (element\_of\_V, d)$ or $g(x, a) = (subset\_SV\_of\_V, s)$, $g'(x, a) = g(x, a)$.*

In the above (1), every incomplete information is removed. In this case, we call an $NKB' = (X, A, V, \{d, s\}, g')$ an *extension of NKB with selective information*. Of course, the extension of every NKB may not be unique. So, we define $EXT_s(NKB) = \{NKB'|\ NKB'$ *is an extension of NKB with selective information* $\}$. For $NKB' = (X, A, V, \{d, s\}, g')(\in EXT_s(NKB))$, we call a classification function g^* which satisfies the following (1) and (2) an *extensional classification function*.

(1) *For every object $x(\in X)$ and attribute $a(\in A)$ which satisfy $g'(x, a) = (subset\_SV\_of\_V, s)$, $g^*(x, a) = (element\_of\_SV, d)$.*

(2) *For every object $x(\in X)$ and attribute $a(\in A)$ which satisfy $g'(x, a) = (element\_of\_V, d)$, $g^*(x, a) = g'(x, a)$.*

Here, we call an $NKB^* = (X, A, V, \{d\}, g^*)$ an *extension of NKB via NKB'*. We also define $EXT(NKB, NKB') = \{NKB^*|\ NKB^*$ *is an extension of NKB via NKB'* $\}$ and $EXT(NKB) = \bigcup_{NKB' \in EXT_s(NKB)} EXT(NKB, NKB')$.

3.2   Question-Answering with Some Conditions in NKBs

According to $EXT_s(NKB)$ and $EXT(NKB, NKB')$, we give the following definition. Let q be a query, where q is a conjunction of atoms. Because, we depend upon prolog in implementation. We did not discuss the equivalence translation for queries like [8] yet.

(1) *For a query q, the query q holds in $NKB'(\in EXT_s(NKB))$ under selective values, if there is an $NKB^*(\in EXT(NKB, NKB'))$ such that NKB^* satisfies q.*

(2) *For a query q, the query q may hold in NKB, if there is an $NKB'(\in EXT_s(NKB))$ such that q holds in NKB' under selective values.*

(3) *For a query q, the query q surely holds in NKB, if q holds in every $NKB'(\in EXT_s(NKB))$.*

In this definition, there is no uncertainties for $EXT_s(NKB)$, so there is no concept of *may* and *sure*. In (1), our purpose is to select the $NKB^*(\in EXT(NKB, NKB'))$ which satisfies query q. We have better to use term 'constraints' instead of term 'query'. In (2) and (3), we are handling incomplete information, so there is concept of *may* and *sure*.

3.3   Implemented Question-Answering System in NKBs

Now we briefly show the implemented question-answering system for NKBs. We have also been discussing logic programs handling incomplete information[9]. We revised this prover for handling not only incomplete information but also selective information, which can solve queries in the above (1), (2) and (3). The following is the preliminary prolog program for query interpreter.

```
dmprob(X,CHYPO,AHYPO):-functor(X,Y,Num),Y/==(,),!,dmatom(X,CHYPO,AHYPO).

dmprob(X,CHYPO,AHYPO):-functor(X,(,),2),arg(1,X,X1),dmatom(X1,CHYPO,AHYPO1),

    arg(2,X,X2),dmprob(X2,AHYPO1,AHYPO).

dmatom(X,CHYPO,AHYPO):-clause(X,true),AHYPO=CHYPO.

dmatom(X,CHYPO,AHYPO):-clause(X,Y),Y/==true,dmprob(Y,CHYPO,AHYPO).

dmatom(X,CHYPO,AHYPO):-hypo(M1,M2,X),iccheck(hp(M1,M2),CHYPO),add(hp(M1,M2),CHYPO,AHYPO).

dmatom(X,CHYPO,AHYPO):-hypo(M1,M2,(X:-GOAL)),GOAL/==true,iccheck(hp(M1,M2),CHYPO),

    add(hp(M1,M2),CHYPO,CHYPO1),dmprob(GOAL,CHYPO1,AHYPO).
```

We omit the details of implementation for reducing the length of this paper. The details of the previous version of this prover is in [9].

4   Approximations and Rough Sets in NKBs

Now in this section, we give definitions for the approximation and roughness in every $NKB = (X, A, V, S, g)$. Let's suppose $NKB^*(\in EXT(NKB))$ and $R(\subset A)$. The NKB^* is a typical knowledge base, so we know that $X'(\subset X)$ is R-definable or R-rough in NKB^* according to [1]. We use R-equivalence relation on X for deciding it. First we give a definition in $NKB'(\in EXT_s(NKB))$, where there is no concept of incompleteness.

(1) *For subsets $X'(\subset X)$ and $R(\subset A)$, X' is R-definable in NKB', if X' is R-definable in an $NKB^*(\in EXT(NKB, NKB'))$.*

(2) *For subsets $X'(\subset X)$ and $R(\subset A)$, X' is R-rough in NKB', if X' is not R-definable in every $NKB^*(\in EXT(NKB, NKB'))$.*

The above definition seems to be natural by the definition of selective information. Then we go to the definition in NKB, which has the incompleteness. We need the concept 'may' and 'sure'.

(1) *For subsets $X'(\subset X)$ and $R(\subset A)$, X' may be R-definable(or R-rough) in NKB, if X' is R-definable(or R-rough) in an $NKB'(\in EXT_s(NKB))$.*
(2) *For subsets $X'(\subset X)$ and $R(\subset A)$, X' is surely R-definable(or R-rough) in NKB, if X' is R-definable(or R-rough) in every $NKB'(\in EXT_s(NKB))$.*

Example 2. Let's consider the following $NKB_2 = (X, A, V, S, g)$. In this NKB_2,

X	a_1	a_2
1	$(\{1,2\}, i)$	2
2	1	$(\{1,2\}, s)$
3	2	2

Table 2. Knowledge base NKB_2

a set $\{1, 3\}(\subset X)$ is surely A-definable. Because, the following holds.

(Case1) *If we deal with $g'(1, a_1) = (1, d)$ then we select $g^*(2, a_2) = (1, d)$. In this case, $(a_1 = 1 \wedge a_2 = 2) \vee (a_1 = 2 \wedge a_2 = 2)$ expresses the set $\{1, 3\}$.*
(Case2) *If we deal with $g'(1, a_1) = (2, d)$ then we select $g^*(2, a_2) = (1, d)$. In this case, $a_1 = 2 \wedge a_2 = 2$ expresses the set $\{1, 3\}$.*

As for the set $\{1\}$, it may be A-definable and also may be A-rough, because in (Case1) $a_1 = 1 \wedge a_2 = 2$ expresses $\{1\}$, but in (Case2) we can not discriminate 1 and 3.

If $X'(\subset X)$ is surely R-definable, then the X' is not influenced by the incomplete information. However if X' is not surely R-definable, then the X' is influenced by them. Suppose X' is R-rough in $NKB'(\in EXT_s(NKB))$. In this case, the inclusion relation $inf_{NKB^*}(X') \subset X' \subset sup_{NKB^*}(X')$ depends upon $NKB^*(\in EXT(NKB, NKB'))$. Especially if the best inclusion relation uniquely exists in every $EXT(NKB, NKB')$, then we get the following proposition.

$$\bigcap_{EXT_s(NKB)}(inf_{NKB^*}(X')) \subset X' \subset \bigcup_{EXT_s(NKB)}(sup_{NKB^*}(X')).$$

This inclusion relation is not influenced by the incomplete information.

5   An Effective Procedure for Checking Rough Sets in NKBs

In the previous section, we defined R-rough set concept in an NKB, whose definition depends upon the extensions in every NKB. Now in this section, we show an effective procedure to check whether X' is R-definable or R-rough in $NKB'(\in EXT_s(NKB))$. We can easily revise this procedure to check whether X' may be R-definable(or R-rough) or is surely R-definable(or R-rough) in every NKB.

Suppose $R(\subset A)$, and we consider $NKB' = (X, R, V_R, \{d, s\}, g')$. For every object $x_i(\in X)$, if $g'(x_i, a) = (\_, d)$ for every $a(\in R)$ then we call x_i an *object with fixed value*. Otherwise, we call it an *object with selection*. We also define $X_{fixed} = \{x_i \in X | x_i$ *is an object with fixed value* $\}$. We continue the definition. For every object with selection x_i, we can make an object with fixed value $x_{i,\theta}$ by a selection θ, which we call an *object with selection* θ *from* x_i. Finally we give the following definitions:

(1) $inf(x_{i,\theta}) = \{x_i\} \bigcup \{x_j \in X_{fixed}|$ *every attribute value in* x_j *and* $x_{i,\theta}$ *is the same* $\}$.

(2) $sup(x_{i,\theta}) = \{x_j \in X|$ *there exists an object* $x_{j,\theta'}$ *with selection* θ' *whose attributes values are the same as* $x_{i,\theta}$ $\}$.

Then, we get the following proposition. Here $[x_i]$ implies an R-equivalence relation including object x_i.

Proposition 1.

(1) *The* $inf(x_{i,\theta})$ *is the minimal R-equivalence relation including object* $x_{i,\theta}$, *which is not influenced by the selective information.*

(2) *The* $sup(x_{i,\theta}) - inf(x_{i,\theta})$ *is a set of objects, where every element* x_j *satisfies* $x_{j,\theta'} \in [x_{i,\theta}]$ *for some* θ' *and* $x_{j,\theta''} \notin [x_{i,\theta}]$ *for some* θ''.

(3) X' *which satisfies* $inf(x_{i,\theta}) \subset X' \subset sup(x_{i,\theta})$ *for some* x_i *and selection* θ *is R-definable in* NKB'.

We have to remark that $inf(x_{i,\theta})$ and $sup(x_{i,\theta})$ are not independent in every x_i. The $inf(x_{i,\theta})$ and $sup(x_{i,\theta})$ are mutually related to other $inf(x_{j,\theta'})$ and $sup(x_{j,\theta'})$. Now we show the overview of the procedure for checking R-rough set by the Example 3.

Example 3. Let's consider the following $NKB_3 = (X, A, V, \{d, s\}, g')$. According

X	a_1	a_2
1	$(\{1,2\}, s)$	2
2	1	2
3	1	$(\{1,2\}, s)$
4	2	2

Table 3. Knowledge base NKB_3

to the Table 3, we first prepare the following inf and sup relations for every object.

(a) $\{1,2\} \subset [1_{a_1=1}] \subset \{1,2,3\}$,   (b) $\{1,4\} \subset [1_{a_1=2}] \subset \{1,4\}$,
(c) $\{2\} \subset [2] \subset \{1,2,3\}$,   (d) $\{3\} \subset [3_{a_2=1}] \subset \{3\}$,
(e) $\{2,3\} \subset [3_{a_2=2}] \subset \{1,2,3\}$,   (f) $\{4\} \subset [4] \subset \{1,4\}$.

We show how we check the R-definability by using a set $\{1,2,4\}$. We use (a) and conclude $\{1,2\}$ is R-definable. In this case, we implicitly rejected (b) and selected $3 \notin [1]$ and $3 \notin [2]$. Namely we can not use (e), because (e) implies $[3] = [2]$. So we need to solve $\{4\}$ is R-definable or not by (c), (d) and (f). If we use $[4] = \{4\}$ in (f) then $1 \notin [4]$. It contradicts (b). Therefore we try another search path. We use (b) or (f) and we conclude $\{1,4\}$ is R-definable, here we rejected (a). So we need to solve $\{2\}$ is R-definable or not by (c), (d) and (e). If we use $[2] = \{2\}$ in (c) then $3 \notin [2]$. Finally we check $\{3\}$ by (d) and (e). We can select $[3] = \{3\}$ by (d), which makes no contradiction. Therefore $\{1,2,4\}$ is R-definable in NKB_3. This is the nondeterministic procedure with backtracking, which seems to be a kind of the resolution procedure. After checking the R-definability, we also get the selected values from selective information. The following is the overview of this procedure.

[Overview for checking R-definability in NKB']
    Suppose we are given $X'(\subset X)$, $inf(x_{i,\theta})$ and $sup(x_{i,\theta})$ for every $x_{i,\theta}(\in X)$.
(1) *Set $X^* = X'$.*
(2) *Pick up the first element $x_j(\in X^*)$ and find $X''(\subset X^*)$ such that $inf(x_{j,\theta'}) \subset X'' \subset sup(x_{j,\theta'})$ for some θ'. If we can not find such X'' then X' is R-rough.*
(3) *The usable $inf(x_{i,\theta})$ and $sup(x_{i,\theta})$ are restricted by selecting X'' in (2). So, check the usable inf and sup, and go to (4).*
(4) *If there is no contradiction in (3), then set $X^* = X^* - X''$ and go to (2). Especially if $X^* = \emptyset$ then X' is R-definable. If there is contradiction in (3), then backtrack to (2) and try another X''.*

We are now formally refining the above procedure. As for the roughness of subset X' in an NKB is as follows. If X' is R-definable(R-rough) in $\forall NKB'(\in EXT_s(NKB))$, then X' is surely R-definable(R-rough) in NKB. If X' is R-definable(R-rough) in $\exists NKB'(\in EXT_s(NKB))$, then X' may be R-definable(R-rough) in NKB.

We can also apply the above procedure to find all R-equivalence relations in every $NKB'(\in EXT_s(NKB))$. The following is an overview.

[Overview for finding all R-equivalence relations in NKB']
    Suppose we are given $inf(x_{i,\theta})$ and $sup(x_{i,\theta})$ for every $x_{i,\theta}(\in X)$.
(1) *Set $X^* = X$.*
(2) *Pick up the first element $x_j(\in X^*)$ and find $X''(\subset X^*)$ such that $inf(x_{j,\theta'}) \subset X'' \subset sup(x_{j,\theta'})$ for some θ'.*
(3) *The usable $inf(x_{i,\theta})$ and $sup(x_{i,\theta})$ are restricted by selecting X'' in (2). So, check the usable inf and sup, and go to (4).*
(4) *If there is no contradiction in (3), then set $[x_j] = X''$, $X^* = X^* - X''$ and go to (2). Especially if $X^* = \emptyset$ then we got an R-equivalence relation. To find other relations, backtrack to (2). If there is contradiction in (3), then backtrack to (2) and try another X''.*

We can also revise this procedure for every NKB with incomplete information.

6   Concluding Remarks

In this paper, we discussed nondeterministic knowledge bases $NKBs$ with incomplete and selective information. The introduction of two information made knowledge bases more powerful and caused new several issues on knowledge bases. Our framework will be an advancement from rough set theory on deterministic knowledge bases. However, we have just prepared a tool, i.e., analytic procedure by $inf(x_i)$ and $sup(x_i)$ for every object $x_i(\in X)$. ¿From now on we will discuss several issues on $NKBs$, for example,

(1) *Handling of multivalued attributes in $NKBs$,*
(2) *Significance of attributes in $NKBs$,*
(3) *Simplification of decision tables in $NKBs$,*
(4) *Reduction of decision rules in $NKBs$,*
(5) *Real applications in $NKBs$,*
(6) *Learning in $NKBs$.*

References

[1] Z. Pawlak: Rough Sets, Kluwer Academic Publisher, 1991.
[2] Z. Pawlak: Deta versus Logic A Rough Set View, Proc. 4th Int'l. Workshop on Rough Set, Fuzzy Sets and Machine Discovery, pp. 1–8, 1996.
[3] A. Nakamura, S. Tsumoto, H. Tanaka and S. Kobayashi: Rough Set Theory and Its Applications, Journal of Japanese Society for AI, Vol.11, No.2, pp. 209–215, 1996.
[4] J. Grzymala–Busse: A New Version of the Rule Induction System LERS, Fundamenta Informaticae, Vol.31, pp. 27–39, 1997.
[5] J. Komorowski and J. Zytkow (Eds.): Principles of Data Mining and Knowledge Discovery, Lecture Notes in AI, Vol. 1263, 1997. pp. 162–167, 1996.
[6] W. Marek and Z. Pawlak: Information Storage and Retrieval Systems: Mathematical Foundations, Theoretical Computer Sciences, Vol.1, No.4, pp. 262–269, 1979.
[7] W. Lipski: On Semantic Issues Connected with Incomplete Information Databases, ACM Trans. on Database Systems, Vol.4, pp. 269–296, 1979.
[8] Z. Ras and S. Joshi: Query Approximate Answering System for an Incomplete DKBS, Fandamenta Informaticae, Vol.30, pp. 313–324, 1997.
[9] H. Sakai: Another Fuzzy Prolog, Proc. 4th Int'l. Workshop on Rough Set, Fuzzy Sets and Machine Discovery, pp. 261–268, 1996.

The OI-Resolution of Operator Rough Logic

Qing Liu

Department Of Computer Science
NanChang University, NanChang, JiangXi 330029,China

[**Abstract.**] Based on rough set theory, this paper establishes operator space $[\xi_*, \xi^*]$. It is also a subset on truth value interval $[0,1]$. The operators is put in the front of the formulas to produce the many-valued logic called operator rough logic(ORL). It defines OI-valid and OI-inconsistent, OI-resolution of the logic, where OI is an abbreviation of Operator Interval. And it also proves the soundness theorem of the logic resolution.

Keywords: Rough Set Theory, Operator Rough Logic, OI-Resolution.

1   The Notions in Operator Rough Logic

Let U be a non-empty set, and R be an equivalence relation on U. If $X \subseteq U$ is a subset on U, then the lower approximate set: $R_*(X) = \{ x \epsilon U \colon R(x) \subseteq X \}$; the upper approximate set: $R^*(X) = \{ x \epsilon U \colon R(x) \cap X \neq \emptyset \}$, where $\emptyset$ is an empty set, $R(x)$ is an equivalent class to include x. The qualities of lower and upper approximation are denoted by
$\xi_* = K(R_*(X))/K(U)$ and $\xi^* = K(R^*(X))/K(U)$ respectively, where K(S) denotes the cardinal number of S which is limited as finite for short, Obviously, $0 \leq \xi_* \leq \xi^* \leq 1$. The qualities of lower and upper approximations, ξ_* and ξ^* are constructed as the operator interval $[\xi_*, \xi^*]$. Obviously, it is also a subset on $[0,1]$.

Definition 1. Let $\xi, \xi_1 \epsilon [\xi_*, \xi^*]$ be the operators, $t \epsilon [0,1]$ be the truth value of the formulas in ORL. Then
(1).Composite operation of operators $\bigcirc \colon \xi \bigcirc \xi_1 = (\xi + \xi_1)/2$, where "+" and "/" are plus and division symbols in algorithm respectively;
(2).Taking negation operation of truth values $\sim \colon \sim t = 1\text{-}t$, where "-" is a subtraction operation symbol in algorithm.

Definition 2. Let P be a n-place predicate symbol, $\xi \epsilon [\xi_*, \xi^*]$ be an operator, $P(x_1, \ldots, x_n)$ be an atom of first order logic (FOL), then ξP is a rough atom of ORL.

Definition 3. Let $\xi \epsilon [\xi_*, \xi^*]$ be a rough operator, then the formulas of ORL are defined recursively as follows:
(1). ORL atom is a formula in ORL;

---

* This study is supported by national natural science fund and natural science fund of JiangXi Province in China.

L. Polkowski and A. Skowron (Eds.): RSCTC'98, LNAI 1424, pp. 432–435, 1998.
© Springer-Verlag Berlin Heidelberg 1998

(2). if W and W_1 are the formulas in ORL, then ξW, $\sim$W, W$\vee W_1$, W$\wedge W_1$, W $\rightarrow W_1$, W $\leftrightarrow W_1$ are the formulas in ORL;

(3). If W is a formula of ORL, x is a free variable in W, then $(\forall x)$W(x),$(\exists x)$W(x), $(\xi\forall x)$W(x) and $(\xi\exists x)$W(x) are the formulas in ORL;

(4). The obtainable formulas what (1)-(3) are quoted finite times are the formulas in ORL.

Definition 4. Let A=(U,R) be an approximate space[2], I_R and u_R be rough interpretation of the formulas and rough assignment to individual variables in the formulas in ORL respectively. The term assignment $T_{I_R u_R}$ corresponding to I_R and u_R is a map from term to objects, then

(1). If the term τ is an object constant symbol to occur in W, then $T_{I_R u_R}(\tau)=I_R(\tau)=$e, where e is an entity on U;

(2). If the term τ is a variable symbol to occur in W, then $T_{I_R u_R}(\tau)=u_R(\tau)=$c, where c is a constant on U;

(3). If the term τ is a n-place function of the form $\pi(\tau_1,\ldots,\tau_n)$, $I_R(\pi)=$g and $T_{I_R u_R}(\tau_i)=x_i$, then $T_{I_R u_R}(\tau)=g(x_1,\ldots,x_n)$, where g is a map from U^n to U;

(4). If the term τ is a n-place predicate of the form $\rho(\tau_1,\ldots,\tau_n)$, $I_R(\rho)=$P and $T_{I_R u_R}(\tau_i)=x_i$, then $T_{I_R u_R}(\tau)=$P$(x_1,\ldots,x_n)$, where P is a relation on U.

The predicates to occur in the formulas of the logic are viewed as a relation. For any relation R, we might find out the lower approximation $R_*(X)$ and the upper approximation $R^*(X)$ with respect to R. Hence, the truth values of the formulas in ORL can be computed by mathematical *formulas*[1,6].

The truth value of a formula, $T_{I_R u_R}$(W) is uniquely determined through the following definition.

Definition 5. Let W and W_1 be formulas in ORL, I_R and u_R are interpretation and assignment of the formulas respectively, then

(1). $T_{I_R u_R}(\xi W)=\xi T_{I_R u_R}(W)$;

(2). $T_{I_R u_R}(\sim W)=1-T_{I_R u_R}(W)$;

(3). $T_{I_R u_R}(W \vee W_1)= \max\{T_{I_R u_R}(W),T_{I_R u_R}(W_1)\}$;

(4). $T_{I_R u_R}(W \wedge W_1)= \min\{T_{I_R u_R}(W),T_{I_R u_R}(W_1)\}$;

(5). $T_{I_R u_R}((\forall x)W(x))= \min\{T_{I_R u_R}(W(x_1)),\ldots,T_{I_R u_R}(W(x_n))\}$,

where $x \epsilon U$ and U are limited as finite.

2 The OI-Resolution of ORL

The truth values of the formulas are taken on [0,1], and the resolutions of ORL with respect to operator set $[\xi_*, \xi^*]$ are called OI-resolution of ORL.

Definition 6. Let $[\xi_*, \xi^*]$ be an operator set, [0,1] be the truth value set of formulas in ORL, if $(\xi_* + \xi^*)/2 \geq 0.5$, and $T_{I_R u_R}(W) \geq \xi^*$, then W is called $[\xi_*, \xi^*]$-valid, in brief written by OI-valid; $T_{I_R u_R}(\sim W)=1-T_{I_R u_R}(W) \leq \xi_*$, then $\sim$W is called $[\xi_*, \xi^*]$-inconsistent, in brief written by OI-inconsistent.

For any formula in FOL, there is a set of clauses that is equivalent to the original formula in that the formula is satisfiable iff the corresponding set of clauses is satisfiable. By the *properties*[1], we have following proposition in ORL.

Proposition. Formula in ORL, W can be transformed into a *Conjunctive Normal Form(CNF)*:

$$W'=C_1\wedge\ldots\wedge C_m$$

where $m\geq 1$ and each C_i, $i=1,\ldots,m$, is a *disjunction*[3].

Definition 7. Let $\xi\epsilon[\xi_*,\xi^*]$ be an operator, then ξL and $\sim\xi L$ are literals in ORL. The former is a position literal; The latter is a negation litaral.

Definition 8. Let ξ, $\xi_1\epsilon[\xi_*,\xi^*]$ and $(\xi_*+\xi^*)/2\geq 0.5$, then ξL and $\sim\xi L$ is a complementary pair of literals with respect to $[\xi_*,\xi^*]$ in ORL, in brief written by OI-complementary literal; And if $\xi=_R\xi_1$, where $=_R$ is rough equal sign with respect to the error θ then ξL and $\xi_1 L$ are called similar literals, for short denoted by OI-similar literal.

The oprator ξ can only be limited to move into before predicates and to have no concern with individual variables and items within the predicates, hence the unification algorithm in ORL is similar as in FOL[5].

Definition 9. Let C_1 and C_2 be two clauses with no individual variable in common in ORL, $\xi_1 L_1$ and $\xi_2 L_2$ be two literals in C_1 and C_2 respectively. If C_1 and C_2 have a most general unifier(mgu) σ in the algorithm of FOL, and $\xi_1 L_1 \circ \sigma$ and $\xi_2 L_2 \circ \sigma$ is a complementary pair of literals, where $\circ$ is an operation sign of substituting composite, then the clause $C_1 - \xi_1^s L_1^{\sigma s} \cup (C_2 - \xi_2^s L_2^{\sigma s})$ is called binary resolvent of C_1 and C_2 in ORL, denoted by $OI(C_1,C_2)$, where $L_i^{\sigma s}$ is a set of OI-similar literals in the C_i.

The resolution is sound in that any clause that can be derived from a set of clauses using resolution is logically implied by that set of clauses in FOL[3,5]. Thus we have following OI-resolution sound theorem in ORL.

Theorem. Let S be a set of clauses in ORL, if there is a resolution reasoning of the clause C from the set S of clauses, then S logically implies C.

Proof: It is finished by simple induction on the length of resolution reasoning. For the induction, we need to show only that any given resolution step is sound. Suppose, then, that C_1 and C_2 are arbitrary clauses in ORL, that resolve to produce a new clause:

$$(C_1-\{\xi_1 L_1,\ldots,\xi_n L_n\})\circ\sigma\cup(C_2-\{\xi_1'L_1'\ldots,\xi_m'L_m'\})\circ\sigma \quad (1)$$

where σ is mgu of $L_1,\ldots,L_n$ and $L_1',\ldots,L_m'$,$\xi_i L_i$($i=1,\ldots,n$) are the OI-simiral literal; $\xi_j'L_j'$($j=1,\ldots,m$) are the simiral literal. We prove the (1) to be sound, only need to show that

$$(C_1-\{\xi_1 L_1,\ldots,\xi_n L_n\})^\sigma\cup(C_2-\{\xi_1'L_1',\ldots,\xi_m'L_m'\})^\sigma \quad (2)$$

is OI-valid for any I_R and u_R.

Because σ is a mgu of $L_1,\ldots,L_n$ and $L_1',\ldots,L_m'$, so it can set $L=L_i^\sigma=L_j'^\sigma$, and

$\xi =_R \xi_i =_R \xi_j'$,i=1,...,n,j=1,...,m, thus (2) can be written by

$$(C_1^\sigma \text{-}\{\xi L\}) \cup (C_2^\sigma \text{-}\{\sim \xi L\}) \ (3)$$

By inductive assumption, we have C_1 and C_2 which are OI-valid, therefore, if ξL is OI-valid then $\sim \xi L$ is OI-inconsistent, thus $C_2^\sigma \text{-}\{\sim \xi L\}$ is OI-valid; Again, if ξL is OI-inconsistent then $\sim \xi L$ is OI-valid, hence $C_1^\sigma \text{-}\{\xi L\}$ is OI-valid. Therefore, the truth value of (3), namely $(C_1^\sigma \text{-}\{\xi L\}) \cup (C_2^\sigma \text{-}\{\sim \xi L\})$ is OI-valid. Hence, the original resolvent of C_1 and C_2

$$(C_1 \text{-}\{\xi_1 L_1,...,\xi_n L_n\}) \circ \sigma \cup (C_2 \text{-}\{\xi_1' L_1',...,\xi_m' L_m'\}) \circ \sigma$$

is OI-valid. The proof is finished.

3 Conclusion

It proposes the system of ORL in the paper. The operator interval $[\xi_*, \xi^*]$ is constructed by the quality of lower and upper approximations based on rough set theory. It can be computed by mathematical $formulas$[1,6]. But the operator λ in [4] is an approximate degree of artificial forgery, its operator interval Ω is also imginary. These show that ORL is different from other many-valued logic. On the other hand, the relations are used as propositions and predicates in ORL, while the relations are not easily defined, maybe, this is the complexity of ORL.

References

[1] Q. Liu, Operator Rough Logic and Its Resolution Principles, *Chinese Journal of Computer*,5 (1998), (in Chinese).

[2] Z.Pawlak, Rough Set Theory, Theoretical Aspects of Reasoning about Data, *Kluwer Academic Publishers*, (1992).

[3] Q.Liu, The Course of Atificial Itelligence,*High Colloge Jion Publisher in JiangXi*, 9 (1992), 75-76,(in Chinese).

[4] X.H. Liu,Fuzzy Logic and Fuzzy Reasoning,*Publisher of JiLin University in China*, 5 (1989), 145-151, (in Chinese).

[5] C.L. Chang, and R.C.T. Lee, *Symbolic Logic and Its Machine Theorem Proving*,(1973)

[6] Q. Liu, Accuracy Operator Rough Logic and Its Resolution Reasoning, *The Proceedings of RSFD'96 International Conference by The University of Tokoy in Japan*, 11 (1996), 45-63.

Pedagogical Method for Extraction of Symbolic Knowledge

Krzysztof Krawiec, Roman Słowiński, Irmina Szcześniak

Institute of Computing Science
Poznań University of Technology
Piotrowo 3A, 60-965 Poznań, Poland

Abstract. This paper addresses the extraction of symbolic knowledge from trained artificial neural networks. Specifically, for that purpose the so-called *pedagogical* approach is incorporated, where the trained network is used as an oracle when inducing the symbolic description. We present an essential extension of the TREPAN algorithm proposed originally by Craven and Shavlik [4][5]. The crucial modification concerns the way of generating artificial training instances. The paper ends with an empirical verification of the proposed method on popular machine learning benchmarks and comparison with the original TREPAN.

1 Introduction

It is commonly recognized, that artificial neural networks (ANNs) became in the last decade one of the most promising computational paradigms in Computer Science, especially in Artificial Intelligence (AI) and related disciplines. Besides being a subject of extensive theoretical studies, they often outperform other approaches in various practical applications, related to machine learning, pattern recognition, signal processing, to list only a few.

Many controversies concerning ANNs were risen and discussed in the literature already in late 80's [2][7]. One of the most important and still unsolved problems related to ANNs is their *opacity*, also referred to as *black-box* problem. These qualifications refer to the inherent inability of ANNs to explain, in terms of languages legible to human beings, decisions they make and knowledge they posses. This is due to the *distributed* model of processing implemented by networks, where the final decision is a resultant of many processing elements (PEs) working simultaneously. The explanation is hidden in many synaptic connections between particular PEs. That shortcoming prevents ANNs from being used in many real-world domains, where explanation is an essential point, due to, for instance, importance of the decisions being made.

Thus, from the beginning of 90's, there has been an increasing interest in the literature in the topics related to the above-mentioned problem. The early proposals [9][6] were mostly focused on the explanation ability of an ANN concerning justification of the decision taken by the network for a particular example. There have been also some research concerning hybrid systems, like [19], where an ANN is used to refine the pre-defined symbolic knowledge. However, most of the work done in the area addresses the more general problem of extracting the knowledge

L. Polkowski and A. Skowron (Eds.): RSCTC'98, LNAI 1424, pp. 436–443, 1998.
© Springer-Verlag Berlin Heidelberg 1998

in symbolic form from an ANN, shortly referred to as *knowledge extraction*. As a language for the symbolic representation the decision rules [18][8][3] or decision trees [5] are usually chosen.

This paper addresses the extraction of decision trees from ANNs by means of the so-called *pedagogical* approach. Specifically, we present an extension of the TREPAN algorithm proposed originally by Craven and Shavlik [4]. The crucial modification concerns the way of generating artificial training instances.

2   Decompositional and Pedagogical Methods

In the literature, many different formulations of the problem have been proposed, however, the most suitable for the purpose of the considered research is the one given by Craven and Shavlik [3]:

> "Given a trained neural network and the examples used to train it, produce a *concise* and *accurate* symbolic description of the network"

It is assumed in the above definition, that the set of training examples is accessible beside the trained network. Following that fact, we assume (which is not explicitly stated in the definition) that the symbolic description will be extracted from the network *in context of the data set*. Thus, we should not be interested in the knowledge which is irrelevant from the viewpoint of the training examples (and, thus, for the considered real-world classification problem). Secondly, two criteria are taken into account when evaluating the resulting description: *conciseness* (length), which should be minimized to ensure legibility, and *fidelity* with respect to the network, obviously maximized (expressed, for instance, as a percentage of decisions consistent with the network). Unfortunately, these criteria are usually conflicting - high fidelity often requires extensive description. Their importances depend on the final use of the extracted knowledge; for instance, for explanation purposes, mostly the conciseness matters.

In late 80's and early 90's, most of the proposals concerning the knowledge extraction from ANNs were based on the so-called *decompositional paradigm*, where the description is built by decomposing the network into particular PEs and analyzing their weights (e.g. [10][8]). The exploration of the network is here formulated in terms of a search problem. Their main deficiency is high computational complexity, resulting from taking into account all (or many) combinations of signals coming into a particular PE. Moreover, to limit the search space, they usually discretize weights or signals of the network, which results in questionable fidelity of the obtained description.

To cope with the above-mentioned problems, Craven and Shavlik introduced in 1994 [3] a completely different approach to the knowledge extraction from ANN, which incorporates the so-called *pedagogical* principle (known also as an *oracle-based* approach). The main idea relies on employing some symbolic machine learning algorithm (inducer), producing, for instance, decision rules or decision trees. In general, it is possible to incorporate some existing system, like C4.5 [14]. Then, knowledge extraction consists in the induction of a classifier, however, using data set which is fundamentally modified in comparison to the original set of examples used for ANN training.

The above modifications are twofold and rely on using the trained network as an *oracle*, i.e. on testing its answers on examples prepared in a way described below. Firstly, for each example, its (original) class label is replaced by classification suggested for it by the network. This ensures the fidelity of the resulting description with respect to the considered network. Secondly, the original training set is being enriched by additional, 'artificial', examples. The values of condition attributes for an artificial example are generated at random according to the probability distributions of particular attributes in the original data set. The class label is induced using the network as an oracle, as it is the case for the original examples. The introduction of artificial examples allows for more precise examination of the behavior of the network. Specifically, this technique is extremely useful for induction of decision trees, being a remedy for the well-known problem of rapidly decreasing number of cases supporting the splitting of tree nodes on its lower levels. With that extension, the inducer is able to supplement the original examples supporting the node with the artificial ones.

Thus, the main difference between decompositional and oracle-based methods is that the former consider the network as a whole. As a consequence, one avoids to some degree an oversimplification (e.g. due to discretization) of the network, which should lead to better fidelity of the induced description. The distributed nature of the network processing is also preserved by the oracle-based method. Moreover, the computational complexity of the approach is basically determined only by the complexity of the inducer, which is usually polynomial. However, it should be stressed, that the pedagogical approach extracts only the *processing* (*behavior*) of the network and omits its *structure*, while the decompositional methods consider both those aspects.

In this paper we present a significant extension of the TREPAN algorithm [4][5], which is, in our opinion, the most interesting representative of the pedagogical paradigm. The next sections outline the algorithm and the proposed extensions.

3   The TREPAN Algorithm

The TREPAN algorithm (TREes PArroting Networks), proposed originally in [3], is probably the most comprehensive representant of the oracle-based approaches. It chooses decision trees as a language for a symbolic representation of the extracted knowledge. A thorough description of the method can be found in [5]. The aim of this section is merely to outline it as an introduction to our work.

As mentioned before, the main idea of TREPAN is to view the knowledge extraction from a trained network as an inductive learning task. Since the chosen knowledge representation is a decision tree, the authors refer to popular tree induction algorithms, such as CART [1], ID3 [13], and C4.5 [14]. The remainder of this section presents the similarities and differences between conventional decision-tree inducers and TREPAN.

The decision-tree inducing methods proceed in general as follows. Basing on a set of training examples, the input space, spanned over the attributes, is recursively partitioned in order to separate examples with different class labels.

Each inner node of the induced tree represents some partition of the input space and each leaf is labelled by the class predicted for those examples which reach this leaf. Every partition of the input space involves a test on the attribute values. Such a test may be a condition placed on a single attribute (as in [14]). In that case, a test on a symbolic (nominal) attribute with n possible values gives n outcomes (child nodes), one for each value. A test on a continuous attribute is specified by a threshold v on its values and results in 2 outcomes (one for examples with the attribute's value $\leq v$, and the other for the remaining ones). The tests originally used in TREPAN were more complex M-of-N conditions, introduced to decision trees in [12].

Since the objective is to build compact yet accurate decision trees, choosing the best of all possible tests at each tree node is a nontrivial problem. TREPAN, as many other tree inducers, uses *information gain* [13] as the criterion for evaluating candidate tests. According to this entropy-based criterion a test is chosen, which maximizes the information gained about the class labels of the examples.

Considering that, one major inconvenience of conventional decision-tree building algorithms becomes apparent: the number of training examples available at a tree-node decreases with the depth of the tree. Thus, tests near the bottom of the tree may often be poorly chosen due to insufficient number of examples. TREPAN overcomes this by randomly generating additional, artificial training examples when needed. In fact, additional training examples are generated for every tree node in which the number of original examples is less then S_{min}, where S_{min} is a parameter of the algorithm. It is worth emphasizing, that such a solution would not be possible when learning a concept underlying some given set of examples, since the class labels for artificial examples would be unknown. In TREPAN however, the target function is the concept given by an ANN - the *oracle* which is used to assign class labels to artificial examples as well as to the original ones.

In generating additional training examples the authors of TREPAN use a fairly simple approach based on modeling the *marginal distributions* of individual attribute values. Such an approach suffers from the fact, that it does not take into account possible dependencies among the attributes. The authors of TREPAN are aware of that problem and try to overcome it by estimating the marginal distributions locally for the particular nodes as the decision tree is being built. However, this solution is not fully satisfactory. In our opinion the method used for generation of training examples has a major impact on the quality of the extracted trees, so it is worth further investigation. Thus, in the next section we propose an entirely new method, which should result in generation of artificial examples preserving the interdependencies among attributes.

4   Attribute-Interdependency Preserving Method

As already mentioned in Section 3, we are strongly convinced that the method of generating artificial training data is crucial. Ignoring the existence of dependencies among features often results in obtaining 'nonsense' examples, which do not have a sensible interpretation in the problem domain. This is particularly true for continuous features, since the examples are generated randomly from within a hypercube extending from the minimal to the maximal values of every attribute. The problem aggravates with the increase in space dimensionality, i.e. the number of the attributes.

As a further consequence, the induced decision tree approximates the behavior of the considered ANN also in those parts of the problem space where real-world examples are unlikely to be found. Taking into account that our aim is to built a concise, and thus legible description of an ANN, it becomes obvious, that considering nonsense examples may lead to an unsatisfactory exploration of relevant parts of the problem space. Taking the above into account, we propose herein a simple heuristic, which regards the actual distribution of the instances in the problem space to produce artificial examples.

In order to generate a single artificial training example a' at a specified tree-node N our algorithm, called hereafter TREPAN+, employs the following steps:

1. Randomly choose an example $a = \langle v_1^a, v_2^a, ..., v_m^a \rangle$ of those from the original training set, which have reached N.
2. For every attribute $i = 1, 2, ..., m$ generate a random value $v_i^{a'}$ according to a Gaussian distribution around the value v_i^a (the variances σ_i^N of the distributions are estimated individually for each attribute i and examples in node N).
3. Use the *oracle* (i.e. query the ANN) to determine the class label for the newly generated example $a' = \left\langle v_1^{a'}, v_2^{a'}, ..., v_m^{a'} \right\rangle$.

Thus, an artificial example a' is being created by introducing perturbations into values of attributes of a randomly chosen original example a. We claim that such a technique preserves the dependencies among the attributes better than the technique using marginal distributions. As an illustrative example, let us consider a data set described by a pair of attributes (x, y), where the examples constitute 3 groups of equal quantity. The attribute values are $(0.25, 0.25)$, $(0.25, 0.75)$, and $(0.75, 0.25)$ for the first, second and third group of examples, respectively. As it can be seen in Fig. 1, the original TREPAN generates artificial examples based on marginal distributions, which results in an extra group around the values $(0.75, 0.75)$, which does not exist in the original data set. Moreover, the probability around $(0.25, 0.25)$ is overestimated at the expense of the two remaining groups. On the contrary, TREPAN+ is free of these deficiencies.

Our method uses the well-known Box-Müller algorithm for generating random numbers according to a Gaussian distribution, and thus has another major advantage of being efficient in comparison to, e.g., Monte Carlo technique used in [17]. This is particularly important considering the usually large number of examples generated at every tree node.

5   Empirical Evaluation

The main aim of the experiment was to compare the fidelity and size of trees generated using the original TREPAN and TREPAN+. To obtain comparable results, the computational experiments have been carried out on well-known reference data sets. As the ANNs prove their usefulness especially on continuous (quantitative, real-valued) attributes, and extraction of knowledge for such attributes is

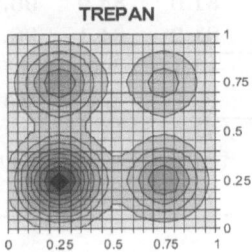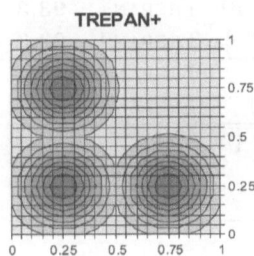

Fig. 1. Probability distributions of artificial examples for both compared methods.

in general harder than for nominal ones, we focused our attention on the former type of attributes. Thus, from the Repository of Machine Learning Databases [11] a few domains composed solely of quantitative attributes have been chosen: *Iris* (Fisher's iris plant database), *Glass* (glass identification database), *Pima* (Pima Indians diabetes database), and *Bupa* (liver disorders). The computations have been performed also for the *Busses* data set, collected in our environment [16]. Table 1 outlines the profiles of the data sets.

Standard feed-forwarded multi-layer perceptrons with sigmoidal transfer function were subject of the knowledge extraction. However, to speed up the training, instead of backpropagation, we used much more efficient RPROP algorithm [15] with a threshold on MSE error as the stopping condition. The number of hidden layers and their sizes have been estimated experimentally to maximize the accuracy of classification. The topologies of the networks are shown in Table 1.

The experiments have been carried out in the cross-validation framework, to ensure the robustness of results. Each of the 10 cross-validation iterations is composed of the following steps: (i) network training on the training set, (ii) network verification on the testing set, and (iii) extraction of symbolic representation in context of the training set, using original TREPAN and our modified version TREPAN+ with improved generation of artificial examples. Each resulting description (decision tree) is then verified on the testing set with respect to the fidelity to the ANN it has been extracted from, and with respect to the classification accuracy. It should be stressed, that both TREPAN and TREPAN+ work on precisely the same networks.

The parameters of both the algorithms have been adjusted to obtain optimal and similar (as far as it was possible) fidelity with respect to the network. The generation of M-of-N conditions [12] has been disabled in the original TREPAN, as it did not improve the results and would have made difficult the comparison between both algorithms, because TREPAN+ does not implement that feature.

Table 1 presents the above-mentioned results of the experiments, with fidelity, tree size, and test set accuracy averaging over 10 iterations of cross validation.

Table 1. Data sets, ANN architectures, and properties of extracted decision trees.

Data set		Iris	Glass	Pima	Bupa	Busses
Examples/Attributes/Classes		150/4/3	214/9/6	768/8/2	345/6/2	76/8/2
Network architecture		4-5-3	9-10-6	8-10-2	6-5-2	8-5-2
Test set fidelity with	TREPAN	93.3	56.3	81.0	84.9	96.1
respect to ANN [%]	TREPAN+	93.3	71.5	81.2	84.1	96.1
Tree size (no.	TREPAN	3.8	7.2	13.1	5.1	1.8
of internal nodes)	TREPAN+	2.5	6.1	4.4	6.5	1.0
Test set accuracy [%]	ANN	95.3	65.0	73.6	67.5	98.7
	TREPAN	92.7	49.4	71.7	64.1	94.7
	TREPAN+	96.0	64.5	73.8	64.1	94.7

6 Conclusions

The results of computational evaluation confirm basically the main thesis of the study. As it is shown in Table 1, both the algorithms yield decision trees of comparable fidelity with respect to the ANN, however, the knowledge extracted by TREPAN+ is in most cases much more compact, and thus more comprehensible. That advantage is especially impressive in case of the *Pima* set, where the trees built by TREPAN+ are on average about three times smaller than those built by TREPAN (4.4 vs. 13.1 internal nodes), preserving at the same time the fidelity of description. The trees built by both algorithms for the *Glass* data set are similar in size, however, we were unable to force the original TREPAN to achieve fidelity comparable to TREPAN+. The only exception from that tendency is the *Bupa* data set, where TREPAN yields smaller trees with slightly better fidelity. This may be due to more sophisticated tree induction strategy used by TREPAN. It should be also mentioned that the trees built by TREPAN+, beside being smaller, obtain better accuracy of classification.

Thus, the above experiment shows, that taking into account the interdependencies between attributes when generating the artificial examples improves significantly the quality (especially size) of the resulting description. Ignoring those dependencies may lead to decision trees which are in some part redundant, as they are partially founded on examples which have no interpretation in terms of the domain knowledge.

Acknowledgments

First of all, we would like to thank Mark Craven and Jude Shavlik for making the source code of original TREPAN available to us. The authors wish also to acknowledge the financial support from State Committee for Scientific Research, KBN research grant no. 8T11C 013 13 and from CRIT 2 - Esprit Project no. 20288. We also express our gratitude to Poznań Supercomputing and Networking Center for making the computational part of this research possible.

References

1. Breiman, L., Friedman, J.H., Olshen, R.A., Stone, C.J.: Classification and Regression Trees. Wadsworth and Brooks (1984)
2. Partridge, D., Wilks, Y. (eds.): The Foundations of Artificial Intelligence. A Sourcebook. Cambridge Univ. Press (1990)
3. Craven, M.W., Shavlik, J.W.: Using Sampling and Queries to Extract Rules from Trained Neural Networks. In: Proc. Eleventh Int. Conf. on Machine Learning, Morgan Kaufmann (1994)
4. Craven, M.W., Shavlik, J.W.: Extracting Tree-Structured Representations of Trained Networks. In: Touretzky, D., Mozer, M., Hasselmo, M. (eds.): Advances in Neural Information Processing Systems (vol. 8), MIT Press, Cambridge (1996)
5. Craven, M.W.: Extracting Comprehensible Models from Trained Neural Networks. Ph.D. Thesis, Comp. Sci. Dept., Univ. of Wisconsin-Madison (1996)
6. Diederich, J.: Explanation and artificial neural networks. Int. J. Man-Machine Studies 37 (1992) 335-355
7. Dinsmore, J. (ed.): The Symbolic and Connectionist Paradigms: Closing the Gap. Lawrence Erlbaum Associates (1992)
8. Fu, L.: Rule Generation from Neural Networks. IEEE Trans. SMC-24 (1994)
9. Gallant, S.I.: Connectionist expert systems. Comm. of the ACM 31 (1988) 152-169
10. Hayashi, Y.: A neural expert system with automated extraction of fuzzy if-then rules and its application to medical diagnosis. In: Advances in Neural Information Processing Systems (vol. 3). Morgan Kaufmann (1990)
11. Merz, C.J., Murphy, P.M.: Repository of machine learning databases. Univ. of California, Dept. of Information and Comp. Sci. (1996)
12. Murphy, P.M., Pazzani, M.J.: ID2-of-3: Constructive induction of M-of-N concepts for discriminators in decision trees. Proc. Eighth Int. Machine Learning Workshop, Morgan Kaufmann (1991)
13. Quinlan, J.R.: Induction of Decision Trees. Machine Learning 1 (1986) 81-106
14. Quinlan, J.R.: C4.5: Programs for Machine Learning. Morgan Kaufmann (1993)
15. Riedmiller, M., Braun, H.: RPROP - A Fast Adaptive Learning Algorithm. Technical Report, Univ. Karlsruhe (1992)
16. Słowiński, R., Stefanowski, J., Susmaga, R.: Rough set analysis of attribute dependencies in technical diagnosis. Proc. 4th Intl. Workshop on Rough Sets, Fuzzy Sets, and Machine Discovery, Univ. of Tokyo (1996)
17. Szcześniak, I.: Acquisition of symbolic knowledge from neural networks. M.Sc. Thesis, Inst. of Comp. Sci., Poznań Univ. of Tech. (1997)
18. Thrun, S.B.: Extracting provably correct rules from artificial neural networks. Technical Report IAI-TR-93-5, Institut für Informatik, Univ. Bonn (1993)
19. Towell, G.G., Shavlik, J.W.: Knowledge-Based Artificial Neural Networks. Artificial Intelligence 70 (1994) 119-165

Wavelets, Rough Sets and Artificial Neural Networks in EEG Analysis

Piotr Wojdyłło

Institute of Mathematics
Warsaw University
Banacha 2, 02-097 Warsaw
e-mail: pwoj@mimuw.edu.pl

[Abstract.] We present a method for processing of EEG signals by means of wavelets, rough set based algorithms and neural networks. The hybrid approach makes problem of discerning between posttraumatic epilepsy and other causes of epilepsy solvable. Experimental results are showing that proposed approach is promising.

1 Introduction

The problem of epilepsy diagnosis is one of important problems in EEG analysis. The research is carried out in two directions. The first one is related to a detection of characteristic patterns in epileptic EEG. This automated analysis of signals helps physicians to look quickly throughout great amount of data. Some results of research in this direction are reported in [8]. They are devoted to differentiating singlefocal and multifocal epilepsy. Another approach is based on search for relevant features which could support diagnosis. Our paper is relevant to the second approach. In this paper we propose to use the wavelet analysis and to apply to the data that is received from this analysis some rough set methods or neural networks' based ones. Wavelets have already proved their efficiency in the field (see [9]).

The objective of this paper is to solve the following problem for EEG signals. There are two groups of children suffering from epilepsy. The first, as clinical treatment says, suffers posttraumatic epilepsy (denoted B) and numbers 11. Epilepsy of children in the second group (denoted A) is due to other facts and there are 25 children in this group. There are also two additional groups *probably* belonging to the basic ones. For the purpose of this research, they were added to appropriate basic groups. For every patient we have at hand an EEG score of 21 electrodes of 2,5 second sampled at frequency 102,4 Hz. The problem is: Can one, using only these EEG scores and no other clinical information, distinct between posttraumatic epilepsy and the other group?

The experiments are showing that hybrid methods using wavelets in combination with rough set methods and neural networks can give satisfactory results. The initial data has 5736 real-valued attributes for 44 objects. This large number of attributes with many values causes that direct application of methods

L. Polkowski and A. Skowron (Eds.): RSCTC'98, LNAI 1424, pp. 444–449, 1998.
© Springer-Verlag Berlin Heidelberg 1998

for decision rule generation can not give a satisfactory solution. Hence, in pre-processing compression of data by wavelet methods is used. Data received from wavelet analysis are used as input for rough set or neural network methods. In many cases hybrid methods can give better results (see for example [10]).

The paper is organized as follows: In Section 2 the basis of applied wavelet analysis is presented.

In Section 3 we present methods used for further processing of the data. After their application one treats data with two classification methods. These are :

(i) The rough set methods implemented in RSES library (see [7],[1]) by the team supervised by A. Skowron in Institute of Mathematics in Warsaw University and

(ii) methods based on artificial neural networks by means of system Neural Works for Professional 2 Plus.

We have used Cross-Validation scheme. For rough set methods the classification was done in 5-Cross-Validation scheme, while for ANN - 2-Cross-Validation scheme was applied. Dividing data into two subgroups for ANN is better setup than 5-Cross-Validation what was confirmed experimentally. Both methods analyze one subgroup and then classify the other.

Section 4 summarizes our results.

Let us note that a small cardinality of group B makes mean efficiency useless, so we give classification efficiency for both groups. As group B is more difficult to classify, we interpret the result as best, if it is best for group B.

2   Wavelet analysis of the signal

We construct a wavelet in a way presented in [5]. For a certain sequence of coefficients (c_n) we build a function satisfying the, so called, scaling equation

$$\varphi(x) = \sum_{n \in Z} c_n \varphi(2x - n).$$

Then we define a function

$$\psi(x) = \sum_{n \in Z} (-1)^n c_{-n+1} \varphi(2x - n).$$

For an appropriate choice of (c_n) we get $\psi(x)$ being a wavelet i.e. functions $\psi_{jk}(x) = 2^{j/2} \psi(2^j x - k)$ with $j, k \in Z$ form an orthonormal basis for $L^2(R)$. It means that the expression $\sum \langle f, \psi_{jk} \rangle \psi_{jk}$ gives a perfect reconstruction of f. Moreover, rejecting coefficients of expansion satisfying $|\langle f, \psi_{jk} \rangle| < \varepsilon$, we obtain a function different from f by $C\varepsilon$ in L^∞-norm. Therefore, to compress the signal f we replace it with

$$\tilde{f} = \sum_{(j,k) \in A_\vartheta} \langle f, \psi_{jk} \rangle \psi_{jk}$$

where $A_\vartheta = \{(j, k) \in Z^2 : |\langle f, \psi_{jk} \rangle| > \vartheta\}$. These facts are connected with the fact that wavelets form unconditional bases for spaces L^p [6]. These bases are optimal for compression as is proved in [3].

For the purpose of this paper we choose Daubechies wavelet of the 5th order [2]. It is a function of C^1 class with a compact support contained in the interval $[0, 9]$. Fig. 1 and 2 present respectively a scaling function and a wavelet.

Fig. 3 presents the dependence between compression ratio and $\left\| f - \tilde{f} \right\|_{L^\infty}$. We have chosen for our purposes $\vartheta_0 = 1, 0$ as coefficient ensuring enough compression ratio and signal quality. Fig. 4 shows the accuracy of wavelet compression.

3   Data processing

3.1   Frequential analysis

Wavelet analysis of the signal resulted in a quite small number of coefficients $\langle f, \psi_{jk} \rangle$ having absolute value bigger than $\vartheta_0 = 1, 0$, namely about 15. However, for application of discretization procedures it is important to get the same coefficient for every patient and electrode that could serve as an attribute. The straightforward procedure of taking all coefficients with absolute value bigger than $\vartheta_0 = 1, 0$ at least for one patient and one electrode would result in $86 \times 21 = 1806$ attributes for every patient which is too big number. Therefore, we have decided to apply frequential analysis. The frequential analysis consists in choosing wavelet coefficients $\langle f, \psi_{jk} \rangle$ with numbers (j, k) such that the number of their representatives greater than ϑ_0 in all electrodes and for all patients is greater than a certain threshold M.

For threshold M=200 we have obtained 29 coefficients creating our basic data D. For any signal these 29 coefficients are considered and these with absolute value smaller than ϑ_0 are replaced by 0. This data gives a result of big dispersion, which can be seen from the efficiency of classification (see Table 1). ANN even failed to solve this problem because of its time complexity.

In next experiments no. 3-5 we took the 4 most frequent coefficients with M=660.

The last case (experiments no. 6-7) belongs to the most complicated ones. We performed it to find features of the signal translated in time. We did it considering 13 coefficients from 5 levels (5 different values of j) and consecutively located in time (subsequent values of k). These coefficients have M$\approx$600. As an input data we took averages of these 13 coefficients in levels.

Results of these experiments are presented in Table 1. The best efficiency was obtained in 4th experiment with M=660 analyzed by RSES with dynamic scaling: **75%** (A) and **56%** (B).

3.2   Global cuts

The basic data D was analyzed by means of RSES procedures for discretization [4]. The interesting fact is that only 4 cuts is enough to split all data D into subsets of decision classes. Therefore, efficiency of this classification is 100% (A) and 100%(B). We claim that they may be relevant features of EEG. The

No.	Experiment	Efficiency (A)	Efficiency (B)
1	M=200 (RSES)	60-71 %	27-58 %
2	M=200 (ANN)	*	*
3	M=660 (RSES)	80-82 %	33-38 %
4	M=660 (RSES dynamic scaling)	**75 %**	**56 %**
5	M=660 (ANN)	66 %	54 %
6	means M≈ 600 (RSES)	68 %	40 %
7	means M≈ 600 (ANN)	63 %	54 %

Table 1. Application of frequential analysis - the efficiency classification

following experiment confirms it. We translated the cuts from their attributes to the appropriate attributes from the next electrode. Explicitly, we changed the cuts (no. of coefficient, no. of electrode, threshold) to (no. of coefficient, no. of electrode+1, threshold). It resulted in good efficiency of classification. Namely, 93% (A) and 73% (B).

3.3 Best cuts

Our next approach was to scale data before analysis to ensure its greater stability. Using discretization procedures from RSES library 20, 50 and 100 cuts discerning the largest possible number of pairs for population were generated. Data has been scaled using these cuts and then analyzed by RSES system and ANN. Results are presented in Table 2.

No.	Experiment	Efficiency (A)	Efficiency (B)
1	20 best cuts (RSES)	75 %	50 %
2	20 best cuts (ANN)	62 %	52,5 %
3	50 best cuts (RSES)	**84 %**	**64 %**
4	50 best cuts (ANN)	73 %	46,5 %
5	100 best cuts (RSES)	100 %	3 %
6	100 best cuts (RSES with additional scaling)	85 %	69 %
7	100 best cuts (ANN)	89,5 %	47 %

Table 2. Application of best cuts - the efficiency of classification

It can be seen that they are good for smaller numbers of cuts. The optimal result was obtained for 50 best cuts with RSES (3rd experiment) : **84%** (A) and **64%** (B). The result of 6th experiment is robust - **85%** (A) and **69%** (B). However, one should remember that in this experiment data was scaled twice.

3.4   Pattern identification

This experiment comes from medical intuition about the problem. It consists in seeking for a pattern in EEG score or for its elements. We have chosen four triples (j, k, ϑ_{jk}) and data was scaled putting 1 if $\langle f, \psi_{jk} \rangle > \vartheta_{jk}$ and 0 otherwise. For each electrode we count ones and their number is one of 21 attributes. Thus, we compare wavelet components of the signal with the pattern and the number of ones is the quality of their matching. These four triples correspond to the pattern $\sum_{j,k} \vartheta_{jk} \psi_{jk}$ given in Fig.5.The classification quality in this case was **68,5%** (A) and **67%** (B) by means of ANN.

4   Conclusions

1. Best results are: **75%** (A) and **56%** (B) for frequential analysis, **84%** (A) and **64%** (B) for best cuts' approach, **85%** (A) and **69%** (B) for best cuts' approach with additional scaling and **68,5%** (A) and **67%** (B) for pattern identification. The obtained results allow us to suggest that described hybrid method can be treated as a promising tool for classification of objects. However, further work in this direction is needed.

2. The comparison of results shows that improvement of classification in group B yields worse results on group A. One of possible causes may be due to the fact that part of data from group B does not include information necessary for efficient classification.

3. The whole population can be divided by using only 4 cuts to subsets of decision classes. Hence the decision classes can be defined by these 4 cuts. Our experiments with translated cuts are showing that they carry a real piece of information from EEG.

4. Further works on rough set methods should allow to discover features of signals discriminating groups. Then wavelet analysis may be concentrated on preprocessing efficient in denoising to make these features detectable by rough set methods.

5. The analysis of attribute dissimilarities for patients from different decision classes is a promising field of investigation.

6. Next direction of further research is optimization of wavelets with respect to essential signal features finding, compression ratio or reconstruction accuracy. It requires interpolating wavelets. This enables us to use genetic algorithms for wavelet optimization.

Acknowledgments

I gladly acknowledge help and cooperation of Sinh Hoa Nguyen, Jan Bazan and Marcin Szczuka in experiments with application of RSES library and neural networks and for sharing their new methods.

The work was partially supported by KBN Research Grant 8T11C01011 and ESPRIT project 20288 CRTI-2.

References

1. J. Bazan "A Comparison of Dynamic and Non-Dynamic Rough Set Methods for Extracting Laws from Decision Table" in : L. Polkowski, A. Skowron (eds.) "Rough Sets in Knowledge Discovery" Physica Verlag, to appear
2. I. Daubechies "Orthonormal bases of compactly supported wavelets" Comm. Pure Appl. Math., 41 (1988) p. 909-996
3. D.L.Donoho "Unconditional bases are optimal bases for data compression and for statistical estimation" Applied and Computational Harmonic Analysis, 1 (1993) pp. 100–115
4. S.H. Nguyen, H. S. Nguyen "From optimal hyperplanes to optimal decision trees: Rough Set and Boolean reasoning approach" in: RSFD'96 pp. 82–88
5. W. Lawton "Tight frames of compactly supported wavelets" J.Math.Phys., 31 (1990) pp. 1898–1901
6. Y. Meyer "Wavelets and Operators" Cambridge University Press, Cambridge 1992
7. Z. Pawlak "Rough Sets - Theoretical Aspects of reasoning about data", Kluwer Academic Publishers, Dordrecht 1991
8. E. Rodin, M. Litzinger, J. Thompson "Complexity of focal spikes suggests relative epileptogenicity" Epilepsia. 36(11):1078-83, 1995 Nov.
9. L. Senhadji, J. L. Dillenseger, F. Wendling, C. Rocha, A. Kinie "Wavelet analysis of EEG for three-dimensional mapping of epileptic events" Annals of Biomedical Engineering. 23(5):543–52, 1995 Sep-Oct.
10. R.W. Swiniarski "Rough Sets and Principal Component Analysis and Their Applications in Data Model Building and Classification" in S.K. Pal, A. Skowron (eds.) "Fuzzy Sets, Rough Sets and Decision Making Processes" Springer, to appear

Parallel Computation of Reducts

Robert Susmaga

Institute of Computing Science, Poznań University of Technology,
Piotrowo 3A, 60-965 Poznań, Poland.
Robert.Susmaga@CS.PUT.Poznan.PL

Abstract. The paper addresses the problem of reduct generation, one of the key issues in the rough set theory. A considerable speed up of computations may be achieved by decomposing the original task into subtasks and executing these as parallel processes. This paper presents an effective method of such a decomposition. The presented algorithm is an adaptation of the reduct generation algorithm based on the notion of discernibility matrix. The practical behaviour of the parallel algorithm is illustrated with a computational experiment conducted for a real-life data set.

1 Introduction

This paper addresses the problem of reduct computation in information systems. The reduct is a notion that has been given much attention in numerous papers within the Rough Set community [4,5,6,7,9]. The idea of reduct and attribute reduction in information tables is, in general, related to a broader problem of feature selection, which has been the focus of many papers in the area of Machine Learning (see [1] for a comprehensive review of feature selection methods). A successful application of rough set reducts has been reported e.g. in [3].

One of the most challenging issues related to reducts is the problem of reduct generation. This is because generating reducts is a computationally complex task. The problem of generating a minimal reduct has been proved to be NP-hard in [7]. This result directs the line of research towards different methods of handling the problem of computational complexity. As a result, the reduct generating procedures employ both exact and heuristic methods.

This paper deals with exact methods. The forthcoming sections provide motivation for parallelization of reduct generation and introduce a parallel algorithm for the computation of all exact reducts. The algorithm is based on the decomposition of an exact algorithm from [7]. Because introducing the framework for formal definition of reducts is impossible due to the paper size restrictions, the reader is referred to [4,5,7,9].

The presented methodology is certainly not the only existing approach to the problem of decomposition in reduct generation. It focuses on a single data set and the parallelization is introduced mainly to speed up the computations for this data set. An earlier approach from [4] considers information systems in which the data are distributed among several locations. The process of reduct

L. Polkowski and A. Skowron (Eds.): RSCTC'98, LNAI 1424, pp. 450–458, 1998.
© Springer-Verlag Berlin Heidelberg 1998

computation may then proceed independently in all of those locations and the results may be combined together to produce the final set of reducts.

The rest of the paper is organized as follows. Section 2 presents a sequential reduct generation algorithm. Section 3 introduces the parallelization of this algorithm. In Section 5 results of the experiment aimed at providing practical evaluation of the parallel algorithm are presented and discussed. The last section draws attention to some unsolved problems and outlines directions of the future research on the subject.

2   The Sequential Implementation of the Algorithm

The original algorithm from [7] is based on the notion of the discernibility matrix. The result of the algorithm is the set of all exact reducts. The main notion of the algorithm, the discernibility matrix, is defined as follows. Given the information table $IT =< U, Q, V, d >$, $|U| = N$, the discernibility matrix is defined as the matrix $[C_{i,j}]$, $i = 1..N$, $j = 1..N$, where:
$C_{i,j} := \{q \in Q : d(u_i, q) \neq d(u_j, q), \; for \; each \; pair \; i, j\}$.

Input: A set of objects U ($|U|=N$) described by values of attributes from the set Q.

Output: The set K of all (absolute) reducts for the set U.

PHASE I
Step 1
Create the absorbed discernibility list ADL for $i=1..N$, $j=1..N$, eliminating empty and non-minimal elements:
$ADL:=\{C_{ij}: C_{ij}\neq\emptyset$ and for no $C_{lm}\in ADL: C_{lm}\subset C_{ij}\}$, where:
$C_{ij}:=\{q\in Q: \delta(u_i,q)\neq\delta(u_j,q)$, for each pair $i, j\}$.
The resulting absorbed discernibility list contains elements $(C_1, C_2, ..., C_d)$, where $d\in [1, N(N-1)/2]$.

Step 2
Sort the ADL in the ascending order of the cardinality of its elements.

PHASE II
Step 1
$R_0:=\{\emptyset\}$.

Step 2
For $i=1..d$ compute:
$S_i:=\{R\in R_{i-1} : R\cap C_i\neq\emptyset\}$.
$T_i:= \bigcup\limits_{q\in C_i} \bigcup\limits_{R\in R_{i-1}: R\cap C_i\neq\emptyset} \{R\cup\{q\}\}$.
$MIN_i:=\{R\in T_i: Min(R,ADL,i)=true\}$.
$R_i:=S_i\cup MIN_i$.

The final result is $K:=R_d$.

Fig. 1. The Modified Reduct Generation Algorithm (MRGA)

The main idea of the method is to find all minimal attribute subsets that have non-empty intersections with all (non-empty) elements of the discernibility matrix. The algorithm consists of two phases: the first phase creates the elements of the discernibility matrix, and the second one generates reducts using these elements. To improve efficiency, the elements of the discernibility matrix are stored in form of an absorbed lists. In the last step of the first phase, what is a modification of the original algorithm, the absorbed list is sorted in the ascending order of the cardinality of its elements. The modified reduct generating algorithm (MRGA) is presented in Figure 1.

In the algorithm, Min performs an operation that corresponds to checking for prime implicants of a Boolean function. The returned value is $true$ if the argument R does not contain redundant attributes. $ADL_{1..i}$ is a list of i initial elements of ADL: $ADL_{1..i} = (C1, C2, ..., Ci)$. Thanks to the links established in [7], the algorithm may be also directly applied to the problem of searching for all prime implicants of Boolean functions.

In the form given above the algorithm searches for so called absolute reducts. This is determined by the definition of the discernibility matrix, which is computed with a uniform treatment of all attributes from Q. If the split of Q into condition (C) and decision (D) attributes is to be taken into account and so called relative reducts are to be produced, the discernibility matrix should be computed as follows:

$C_{i,j} := \{q \in Q : d(u_i, q) \neq d(u_j, q) \text{ and } (u_i, u_j) \notin IND(D), \text{ for each pair } i, j\}$,

where $(u_i, u_j) \notin IND(D)$ means that the objects u_i and u_j belong to different classes defined by the decision attributes.

3   The Parallel Implementation of the Algorithm

The most general idea of parallelization is decomposing the computing task into several subtasks which may be then processed at the same time on different processors. This requires, first of all, a specialized algorithm. The main objective of parallelization is reducing the computing time. Another important benefit may be the reduction of memory requirements (see [2] for a detailed description of different aspects of parallel algorithms).

Parallelization of MRGA is based on its following characteristics.

<u>Observation</u> In each iteration of the *Step 2 (PHASE II)* of the algorithm the following holds: $S_i \cap T_i = \emptyset$; every $R \in S_i$ is a copy of an $R' \in R_{i-1}$; every $R \in T_i$ is created from a single $R' \in R_{i-1}$ by augmenting it by an element $q \in Q$; MIN_i is a subset of T_i created by discarding some of its elements — the process is conducted on the basis of the index i and the value of $ADL_{1..i}$.

<u>Conclusion 1</u> Any two $R, R' \in R_i$ are independent in the sense that no information on R or on how it has been created is required for producing R'.

<u>Conclusion 2</u> The process of subset creation has a tree-like structure: in each iteration an element $R \in R_i$ may be either copied to the next iteration without changes or be deleted, giving rise to a collection of its proper supersets, which are subsequently processed independently of one another and of the other elements of R_i.

Input: A set of objects U ($|U|=N$) described by values of attributes from the set Q;
the subtask count SC, the branching factor BF (it is assumed that $BF \geq SC$).

Output: The set K of all (absolute) reducts for the set U.

PHASE I – <u>As in Modified Reduct Generating Algorithm</u>

PHASE II
Step 1
$R_0 := \{\emptyset\}$.
$i := 1$.

Step 2
While ($|R_0| < BF$) *and* ($i \leq d$) do loop :
$\quad S_i := \{R \in R_{i-1} : R \cap C_i \neq \emptyset\}$.
$\quad T_i := \bigcup_{q \in C_i} \bigcup_{R \in R_{i-1}, R \cap C_i \neq \emptyset} \{R \cup \{q\}\}$.
$\quad MIN_i := \{R \in T_i : Min(R, ADL, i) = true\}$.
$\quad R_i := S_i \cup MIN_i$.
$\quad i := i+1$.

Step 3
If ($i > d$) then set $K := R_d$ and stop the algorithm.
Otherwise proceed with *Step 4*.

Step 4
Partition the set R_i into SC subsets R_i^j, ($j = 1..SC$), such that $R_i^k \cap R_i^l = \emptyset$ for each $k \neq l$,
$\bigcup_{j=1}^{SC} R_i^j = R_i$ and the subsets R_i^j are of approximately equal size.

Step 5
Execute SC parallel subtasks, each performing the procedure $K^j := SUB(i, d, R_i^j, ADL)$, for $j = 1..SC$.
Set $K := \bigcup_{j=1}^{SC} K^j$ and stop the algorithm.

Procedure SUB (*InitialIndex* , *FinalIndex* , *PartialSolution* , *AbsorbedDiscernibilityList*)

Step 1
$R_{InitialIndex-1} := PartialSolution$.

Step 2
For $i = InitialIndex..FinalIndex$ compute:
$\quad S_i := \{R \in R_{i-1} : R \cap C_i \neq \emptyset\}$.
$\quad T_i := \bigcup_{q \in C_i} \bigcup_{R \in R_{i-1}, R \cap C_i \neq \emptyset} \{R \cup \{q\}\}$.
$\quad MIN_i := \{R \in T_i : Min(R, ADL, i) = true\}$.
$\quad R_i := S_i \cup MIN_i$.

Return $R_{FinalIndex}$ as the result of the procedure.

Fig. 2. The Parallel Implementation of MRGA

The conclusions lead to a formulation of the parallel algorithm for reduct computation. After completing the *PHASE I* of the sequential algorithm and performing several iterations of the *Step 2* (*PHASE II*), the resulting R_i may be split into a family of subsets which create a partition of the set R_i, i.e. all of them are pairwise disjoint and they all sum up to the set R_i. From now on the computations may continue in form of several subtasks, each of which proceeds with exactly one partition substituted for the whole set R_i. It is important that the resulting subtasks are absolutely independent of one another. The final set of reducts is the sum of the sets of reducts generated by each subtask.

The implementation of the parallel algorithm requires two additional parameters: the number of subtasks which are to be initiated (subtask count, SC) and the cardinality of the set R_i which is to be reached before the algorithm branches into multiple subtasks (branching factor, BF). The branching factor must be equal to or greater than the number of subtasks so that each subtask may start with at least one element of R_i. The role of this parameter is closely discussed in Section 5.

There are no formal restrictions as to the method of partitioning the set R_i into disjoint subsets. In the described experiments this was implemented simply as assigning the attribute subset number n to the subtask number $m = (n \bmod SC) + 1$.

4   Experimental Evaluation of the Parallel Implementation

The data set used in the experiment was a medical file containing descriptions of 343 patients after the ESWL treatment [8]. Each patient is described by a vector of 33 condition attributes and one decision attribute (defining two decision classes). The file was selected for presenting the results of the experiments only because of its computational characteristics: a large number of relative reducts (38207), a moderate computing time (327 seconds) and small memory requirements (about 380 kB). It is stressed that this paper contains no claims as to any potential suitability of the selected data file for reduct generation or the usefulness of the generated reducts in further analyses.

The practical behaviour of the parallel implementation may be best characterized by plotting the computing times (understood as the maximal computing time of all of the subtasks + the time required to reach the branching point) against the increasing number of subtasks employed (see Figure 3). The same applies to the maximal memory requirements (Figure 4).

The main computing platform in the experiments was a SUN Workstation equipped with a Sparc4 processor running at 110 MHz.

Computing Times

Inspection of Figure 3 allows to conclude that the parallel implementation of the algorithm reduces the computing time considerably. The fact that the decrease of the computing time is worse than linear may be explained as follows.

The problem of decomposing the original problem into subtasks is unbalanced in the sense that the number of reducts generated by each of the subtasks is highly variable. It results from the fact that it is impossible to determine in advance how many reducts will be finally generated by a given subtask, what, in turn, results from the highly variable 'productivity' of different attribute subsets initially assigned to the subtask. Experiments indicate that running the algorithm with the value of the branching factor equal to the number of subtasks (which is the lowest possible value of this parameter) is highly unbalanced, including situations where the maximal subtask computing times are close to the computing time of the sequential implementation. A solution to this seems to be increasing the value of the branching factor. This results, first of all, in a direct increase of the initial computing time (it takes longer to reach the branching point of the algorithm) but it has also a positive effect on balancing the subtasks and, in longer perspective, proves to be beneficial.

The general conclusion is that the computing time decreases satisfactorily only after the branching factor is set to a value which enables each of the subtasks to start with at least several attribute subsets. This is, however, only an improvised, temporary principle — discovering optimal (provided there exists one) or close to optimal value of the branching factor needs further research.

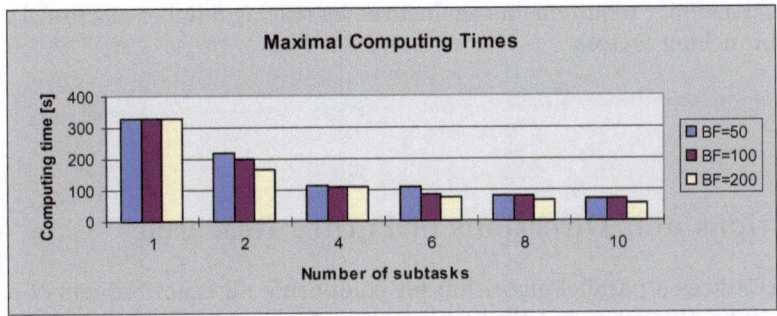

Fig. 3. Maximal computing times against the increasing number of subtasks and different branching factors

Memory Requirements

Another very important feature of the parallel implementation is the characteristics of its memory requirements. Inspection of Figure 4 reveals that the maximal memory requirements of the sequential algorithm are considerably reduced by its decomposition into subtasks, the reduction being most evident for a small number of subtasks and becoming less evident with more subtasks. The fact that the decrease of memory requirements is worse than linear may be explained as follows. There are two main memory-consuming structures of the algorithm: the set R_i and the initial (unabsorbed) form of $ADL_{1..i}$. Usually the size of R_i exceeds considerably that of the unabsorbed list, but after partitioning of R_i the size of the unabsorbed list starts to dominate, obliterating the effects of partitioning.

Applying parallelization with small numbers of subtasks, however, still seems to be a very good alternative, especially in computing environments where the memory proves to be the bottle-neck of the process.

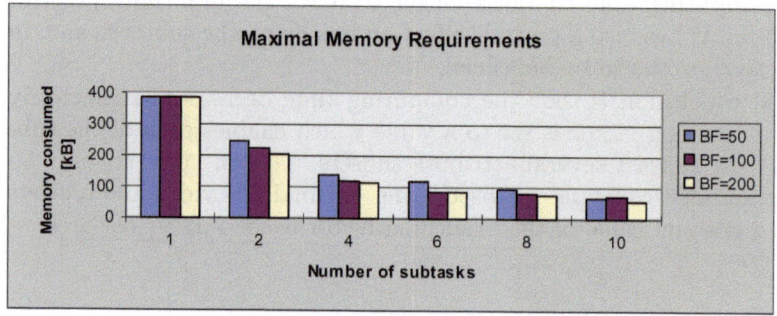

Fig. 4. Maximal memory requirements against the increasing number of subtasks and different branching factors

5   Conclusions and Directions of Future Research

This paper introduces a parallel algorithm for computing all exact reducts of a decision table. The algorithm is based on an algorithm described originally in [7]. Results of experiments aimed at verifying the algorithm's practical behaviour are also presented and discussed. The results indicate usefulness of the parallel implementation, which allows for the reduction of both the computing time and memory requirements of the reduct generating process.

There still remain some issues which need further research. The main of them are: establishing the number of subtasks to be started and balancing the subtasks so that they generate approximately equal numbers of reducts and exhibit similar computing times and memory requirements.

In its current version, the parallel algorithms branches at one point into an arbitrary number of subtasks. Such a solution does not take into account the computational needs of the particular data set for which the algorithm has been invoked, because it is not possible, in general, to determine the final number of reducts in advance. In an alternative approach, the number of subtasks would not have to be specified — the only control parameter being the branching factor. Upon reaching it, the algorithm would branch into two parallel subtasks, with each of them proceeding with a half of the current load. The same branching procedure would be subsequently applied recursively in both resulting subtasks.

A related problem is that of balancing the resulting subtasks. Presently, the parallel algorithm assigns the initially created attribute subsets one by one to consecutive subtasks. Because the original ordering of the attribute subsets is random, this partitioning may also be viewed as random. It must be stressed, however, that the ordering of the absorbed discernibility list, from which the attribute subsets are created, is not random — the list is ordered in ascending order of its elements. As a result, the initial attribute subsets contain those attributes that occurred in low-cardinality elements of the discernibility list. A better balancing scheme could incorporate some attribute counts, so that the numbers of different attributes in the initial subsets would be approximately the same.

Acknowledgements

This paper has been supported by the grant KBN no 8T11C–013–13. Parts of the reported computations were conducted at the Supercomputing and Networking Centre of Poznań, Poland. Special thanks are due to Darek Wawrzyniak from the Institute of Computing Science for his eager help on parallelization issues.

References

1. Dash M. & Liu H. 'Feature Selection for Classification', *Intelligent Data Analysis* (on-line journal), Vol. 1, No 3, (1997), http://www-east.elsevier.com/ida.
2. Jaja J. *An Introduction to Parallel Algorithms*, Addison Wesley, (1992).
3. Jelonek J., Krawiec K. & Slowinski R. 'Rough set reduction of attributes and their domains for neural networks', *Computational Intelligence*, Vol. 11, No 2, (1995), pp. 339–347.
4. Kryszkiewicz M. & Rybinski H. 'Finding Reducts in Composed Information Systems', *Fundamenta Informaticae*, Vol. 2, No 2/3, (1996), pp. 183–196.
5. Orlowska M. & Orlowski M. 'Maintenance of Knowledge in Dynamic Information Systems', In: Slowinski R., (ed.), *Intelligent Decision Support. Handbook of Applications and Advances of the Rough Set Theory*, Kluwer Academic Publishers, Dordrecht, (1992), pp. 315–330.
6. Romanski S. 'Operations on Families of Sets for Exhaustive Search Given a Monotonic Function', In: Beeri, C., Smith, J.W., Dayal, U., (eds), *Proceedings of the 3rd International Conference on Data and Knowledge Bases*, Jerusalem, Israel, (1988), pp. 28–30.
7. Skowron A. & Rauszer C. 'The Discernibility Matrices and Functions in Information Systems', In: Slowinski R., (ed.), *Intelligent Decision Support. Handbook of Applications and Advances of the Rough Set Theory*, Kluwer Academic Publishers, Dordrecht, (1992), pp. 331–362.

8. Slowinski K., Stefanowski J., Antczak A. and Kwias Z. 'Rough Set Approach to the Verification of Indications for Treatment of Urinary Stones by Extracorporeal Shock Wave Lithotripsy (ESWL)', In: Lin T.Y., Wildberger A.M., (eds.), *Soft Computing*, Society for Computer Simulation, San Diego, California, (1995), pp. 142–145.
9. Ziarko W. & Shan N. 'Data-Based Acquisition and Incremental Modification of Classification Rules', *Computational Intelligence*, Vol. 11, No 2, (1995), pp. 357–370.

Rough Rules in Prolog

Allan Tony Ath and Krzysztof Cielak

Institute of Theoretical and Applied Computer Science
Polish Academy of Sciences
Ul. Bałtycka 5, 44-100 Gliwice, Poland

Abstract. In this paper we present an approach to develop decision support systems. We focus on knowledge acquisition and processing with the use of rough set theory. The rule based approach implementation is also considered. We discuss the use of Prolog as a tool for knowledge representation and processing. Finally, the way of embedding Prolog code in procedural language processing is presented. Our work is illustrated with an exemplary system supporting credit decisions.

1 Introduction

Systems for supporting decisions are one of the artificial intelligence domains widely applied in practical industries. By systems of this type we mean systems that perform features of spectrum of specific problems. Typically, they are transparent on explicit knowledge base, and control mechanisms mutually bound. Another advantage is the possibility of giving explanations for the solutions found by the inference engine.

The technology of expert systems in the field of economy has many applications, e.g. in banking [BI9, SE9], management [SH9], hotels, e.b. [VE9], insurance, investments and marketing [Bas19, VE9].

The main components of an expert system are as follows: a knowledge base, an inference engine, and a user interface (see [LI9]). The synthesis of an expert system should be preceded by an analysis of three problems: knowledge acquisition, knowledge representation and knowledge utilisation.

One of the knowledge acquisition methods — obtaining knowledge from procedural systems — analysis of decisions made by experts. Such means of learning is also defined as an inductive method [LI9]. From the source techniques of knowledge processing compiled in the article [LI9], we have chosen and applied the result set theory introduced by Pawlak [Pas9, Pa9] with extensions described in the book [Li9]. We used Pathobasis system [LI9] to transform the obtained knowledge from the database into the form of the set. Knowledge base. The reduced knowledge base consists of its kind rules which, in turn, can be directly presented in the form of Horn clause rules [Me9]. Prolog is a language of logic based on predicate calculus. In fact, it is a natural tool for knowledge representation and for developing an inference engine. In the end, it was instead of Prolog apposed to do, shifts of expert systems. The interconnection and the knowledge base were implemented in the language of logic, whereas the user interface and

Rough Rules in Prolog

Adam Mrózek and Krzysztof Skabek

Institute of Theoretical and Applied Computer Science
Polish Academy of Sciences
ul. Bałtycka 5, 44–100 Gliwice, Poland

[Abstract.] **In this paper we present an approach to develop decision support systems. We focus on knowledge acquisition and processing with the use of rough set theory. The rule knowledge representation is also considered. We discuss the use of Prolog as a tool for knowledge representation and processing. Finally, the way of embedding Prolog code in procedural language programs is presented. Our work is illustrated with an exemplary system supporting credit decisions.**

1 Introduction

Systems for supporting decisions are one of the artificial intelligence domain widely applied in practical solutions. By systems for supporting decisions we mean programs that perform inquiries of specific problems. Typically these systems gather an explicite knowledge base and control mechanisms mutually separated. Another advantage is the possibility of giving explanations for the solutions found by the inference engine.

The technology of expert systems in the field of economy has many applications, i.e. in banking [Sl95,Sk96], management [Sr94], finances [Kr95], insurance, investments and marketing [Ba94a,Po97].

The main components of an expert system are as follows: a knowledge base, an inference engine and a user interface (see Fig. 1). The synthesis of an expert system should be proceded by an analysis of three problems: knowledge acquisition, knowledge representation and knowledge utilization.

One knowledge acquisition methods is obtaining knowledge from previously existing examples of decisions made by experts. Such tactics of learning is often defined as an inductive method [Mi83]. From the various techniques of knowledge processing, compared in the article [Po97], we have chosen and applied the rough set theory introduced by Pawlak [Pa82,Pa91] with extensions described in [Pl,Sl92,Ba94b]. We used DataLogic system [Dl92] to transform the obtained knowledge from the database into the form of the rule knowledge base. The resulting knowledge base consists of decision rules which, in turn, can be directly presented in the form of Prolog rules [Me89]. Prolog as a language of logic based on predicate calculus [Ch73,Cl84] is a perfect tool for knowledge representation and for developing an inference engine. In the article we presented a hybrid approach to developing of expert systems. The inference engine and the knowledge base were implemented in the language of logic, whereas the user interface and

L. Polkowski and A. Skowron (Eds.): RSCTC'98, LNAI 1424, pp. 458–466, 1998.
© Springer-Verlag Berlin Heidelberg 1998

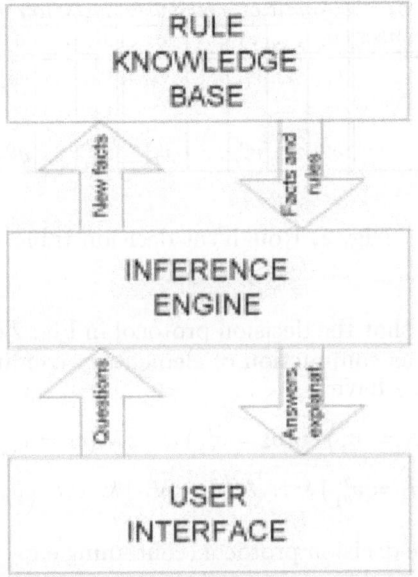

Fig. 1. Main components of the expert system

database modules were developed using classical procedure programming. Such connection was possible due to *Amzi!* Logic Server [Am95].

This article presents a practical implementation of an expert system for supporting credit decisions and application of mentioned above mechanisms.

2 Knowledge Acquisition - Rough Approach

It is accepted that a decision problem can be described by: finite parameter set $C = \{c_1, c_2, \ldots, c_k\}$ called further *condition attributes*, finite parameter set $D = \{d_1, d_2, \ldots, d_p\}$ called further *decision attributes*, and determined *domains* (ranges of values) respectively:

$$V_C = \bigcup_{c_i \in C} V_{c_i} \text{ for condition attributes} \tag{1}$$

and

$$V_D = \bigcup_{d_j \in D} V_{d_j} \text{ for decision attributes.} \tag{2}$$

Such description of a decision problem causes that its solution consists in determination of condition attribute values and, on this ground, setting values on particular decision attributes. Such a process of solving decision problems can be represented as a decision protocol. Fig. 2 shows the structure of this decision protocol.

Rule number	Condition attributes					Decision attributes				
	c_1	...	c_j	...	c_k	d_1	...	d_j	...	d_n
1	$v_{c_1}^1$	...	v_{cj}^1	...	$v_{c_k}^1$	$v_{d_1}^1$	...	$v_{d_j}^1$	...	$v_{d_n}^1$
$\vdots$	$\vdots$	$\vdots$	$\vdots$	$\vdots$	$\vdots$	$\vdots$	$\vdots$	$\vdots$	$\vdots$	$\vdots$
N	$v_{c_1}^N$	...	v_{cj}^N	...	$v_{c_k}^N$	$v_{d_1}^N$	...	$v_{d_j}^N$	...	$v_{d_n}^N$

Fig. 2. Rough set decision table

It is easy to notice that the decision protocol in Fig. 2 contains decision rules. They can be denoted as conjunction of elementary conditions. For j-th decision we have:

$$\text{if } \left\{ \left(c_1 = v_{c_1}^j\right) \& \left(c_2 = v_{c_2}^j\right) \& \ldots \& \left(c_k = v_{c_k}^j\right)\right\}$$
$$\text{then } \left\{ \left(d_1 = v_{d_1}^j\right) \& \ldots \& \left(d_l = v_{d_l}^j\right) \& \ldots \& \left(d_p = v_{d_p}^j\right)\right\} \tag{3}$$

Having the suitable decision protocol (containing e.g. expert decisions solving definite problem) the problem of knowledge acquisition can be reduced to the problem of generation all different decision rules in the form (3) based on such protocol.

We can use for this purpose efficient software tools based on rough set theory as e.g. LERS [Gr92], DataLogic [Dl92]. The main function of these tools is the transformation of data bases (expressed in a form of decision protocols or semantically equivalent decision tables) into the form of a reduced decision rule base (a knowledge base).

3   Rule Knowledge Base Representation

Knowledge representation is a formal way of knowledge projection in order to efficiently store and process it in a computer memory.

In Prolog the knowledge base consist of a set of facts and rules. Prolog makes possible recording in the knowledge base not only the facts but also the information about relations between the facts in the form of rules that are called Horn clauses [Cl84].

Prolog Horn clauses accept also the usage of condition alternatives and disjunction at the same time [Ch73]. During the decision rule analysis in the form (3) it is not difficult to notice the possibility of its natural representation in the form of an adequate Prolog Horn clause.

4   Prolog Inference Engine

Prolog as a language of logic is favorable for implementations of inference systems. Prolog inference engines are based on search strategies. A built-in unification mechanism as well as inference based on the resolution principle [Ch73] are also helpful during the inference process.

The most important features of expert systems are easy to implement in Prolog [Me89]. These are: backward chaining, forward chaining, rule representation of data and explanations.

Although the expert systems written by means of conventional languages, such as C, have often a better performance, the Prolog code makes expert systems close to a logical specification of a program and thus easy to modify.

5   User Interface

For a user, the most important part of a computer system supporting decisions is a user interface. Interactive program environments become nowadays an indispensable tool for solving complicated problems related to the systems development.

For this purpose, either intelligent communication algorithms or modern interactive methods can be used. Intelligent communications algorithms make it possible to transform computer requirements into the human way of reasoning in the process of making decisions whereas modern interactive methods allow a user an efficient communication with a computer. In this way a human is able to watch particular stages of the decision process performed by computers, influence its run and obtain, except of the final problem solution, intermediate information of various type.

6   Logic Server

Logic server enables Prolog predicates to be integrated with conventional programming languages, databases and other programming tools. In our research we use Logic Server produced by *Amzi!*. It supports the memory management separated from a base application. It includes full graphic interface that makes possible to access Prolog module from outside. Interface libraries are supported to many popular languages such as: C++, Delphi, Visual Basic, etc.

The structure of the complete decision supporting system is presented in Fig. 3.

7   An Example of the Decision Support System

An example of an economic decision problem presented below intends to point at the elements of rough set theory as a formal tool that can be efficiently used in synthesis of appropriate computer systems supporting their solution.

Granting credits to individuals or businesses belongs to the fundamental duties and functions of modern banks. Such activity includes a certain level of risk. That risk results from the difficulties of explicit determination of so called *credit capacity* of a debtor, i.e. the possibility of credit repayment including payable interest [Si93]. At the stage of credit terms negotiations, the contrary interests of banks and debtors occur.

This contradiction of bank and debtor interests as well as incompleteness, inexplicitness, uncertainty of available information and difficulties with selection

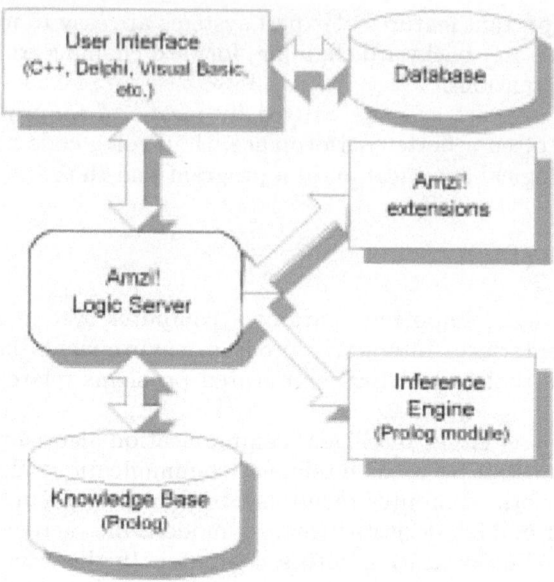

Fig. 3. The complete decision supporting system

of the parameters and criteria allowing for objective credit capacity evaluation, make the credit decision problem difficult to formalize.

This fact influenced the attempt of synthesis of computer system supporting economic decision based on rough set theory elements at the stage of knowledge acquisition and representation. Particular description of such synthesis was presented in [Sk96].

7.1 Knowledge Sources and Acquisition

From the formal point of view, bank crediting process consists of two partial problems:

- preparing the premises for decision making i.e. honest complete and credible valuation of debtor's credit capacity;
- opening the credit up to the certain limit and on the condition that minimize a risk.

The resolution of this problem must concern the solutions of both partial problems. Taking advantage of accessible publications, possibilities of discussion with experts and observations of credit decisions made by banks in real conditions makes possible the knowledge acquisition. It was in details described in [Sk96].

The following economic indexes useful in valuation of credit capacity have been chosen [Si93]:

– **Net Profitability Ratio**

$$\text{Ind1} = \frac{\text{Net Profit}}{\text{Sale Value}} 100\%$$

– **Current Ratio**

$$\text{Ind2} = \frac{\text{Current Assets}}{\text{Current Liabilities} + \text{Short-term Credit}}$$

– **Quick Ratio**

$$\text{Ind3} = \frac{\text{Current Assets} - \text{Stock}}{\text{Current Liabilities}}$$

– **Accounts Receivable Turnover Ratio**

$$\text{Ind4} = \frac{\text{Average Accounts Receivable}}{\text{Net Sale}} 365$$

– **Inventory Turnover Ratio**

$$\text{Ind5} = \frac{\text{Average Stock}}{\text{Net Sale}} 365$$

– **Exceeded Payables Ratio**

$$\text{Ind6} = \frac{\text{Exceeded Accounts Payables}}{\text{Total Payables}}$$

– **Equity Ratio**

$$\text{Ind7} = \frac{\text{Outside Capital}}{\text{Ownership Capital}}$$

– **Ownership Capital Ratio**

$$\text{Ind8} = \frac{\text{Ownership Capital}}{\text{Total Assets}}$$

– **Interest Coverage Ratio**

$$\text{Ind9} = \frac{\text{Interest}}{\text{Sale}} 100\%$$

Condition and decision attributes as well as their domains were determined with a help of experts' suggestions and economic indexes described above. The data were recorded in a relevant economic decision protocol which makes the basis of the knowledge acquisition process. General scheme of decision protocol was presented in Fig. 2.

Condition Attributes. Here are the selected condition attributes:

$[c_1]$ – Net Profitability Ratio – noted as *Ind1*, domain: $\{acceptable, unacceptable\}$
$[c_2]$ – Net Profitability Ratio tendency – domain: $\{increase, decrease\}$
$[c_3]$ – Net Profitability Ratio in comparison with the other companies of a branch – domain: $\{high, low\}$
$[c_4]$ – Current Ratio – noted as *Ind2*, domain: $\{acceptable, unacceptable\}$
$[c_5]$ – Quick Ratio – noted as *Ind3*, domain: $\{acceptable, unacceptable\}$
$[c_6]$ – Accounts Receivable Turnover Ratio – *Ind4*, domain: $\{acceptable, unacceptable\}$
$[c_7]$ – Inventory Turnover Ratio – noted as *Ind5*, domain: $\{acceptable, unacceptable\}$

$[c_8]$ – Exceeded Payables Ratio – noted as *Ind6*, domain: {*acceptable, unacceptable*}
$[c_9]$ – Equity Ratio – noted as *Ind7*, domain: {*acceptable, unacceptable*}
$[c_{10}]$ – Ownership Capital Ratio – noted as *Ind8*, domain: {*acceptable, unacceptable*}
$[c_{11}]$ – Ownership Capital Ratio tendency – domain: {*increase, decrease*}
$[c_{12}]$ – Interest Coverage Ratio – noted as *Ind9*, domain: {*acceptable, unacceptable*}

In this way we have obtained the set C of condition attributes

$$C = \{c_1, c_2, \ldots, c_{12}\}.$$

Decision Attributes. Risk rating applied in practice by crediting banks allows us to classify credits into the following groups [Sk96]:

[Group 1] – ordinary credit.
[Group 2] – observed credit.
[Group 3] – doubtful credit.

Accordingly to the above classification, it was assumed that the credit risk group is the only decision attribute.

Finally we obtained the single element set D of decision attributes ($D = \{d_1\}$). The above mentioned groups of risk make the domain of the set D.

Establishing of condition and decision attribute sets has explicitly determined the structure of the relevant decision protocol. This protocol was helpful in data acquisition process.

Credit decisions recorded in protocol cases have arisen from practical bank consultations (they took place during student internships) and from availiable specialistic publications [Si93]. The complete data set contains 512 rules and is described in [Sk96].

7.2 Knowledge Reduction

The data set has been reduced by means of DataLogic [Dl92].

The main function of this program is the reduction of data sets into the form of decision rules. The process of generating decision rules consists of the following stages: reduct searching, redundant attributes exclusion, redundant record reduction.

The complete decision table consisted of 512 items. After the reduction process the knowledge base included 150 decision rules.

Three condition attributes (c_2, c_3 and c_{11}) were found unneccessary and could be removed. The reason of it is that these attributes consider the tendency and the comparison with the other companies of the branch, values of which already exist in the set of attributes. However, because of a possibility of the knowledge base extension, they remained as parameters in the program. As the application makes possible to record new cases in the knowledge base, these parameters may become useful during the system exploitation.

The particular decisions included the following numbers of rules:

- *DOUBTFUL* — 55 rules,
- *OBSERVED* — 81 rules,
- *ORDINARY* — 14 rules.

The complete set of decision rules is published in [Sk96].

In the rough analysis the attribute strength report is also very important. It appears that the attributes remaining in the knowledge base have approximately equal volume (the highest difference reaches 17%). It means that particular attributes have a similar influence on the final decision.

7.3   Implementation of the Inference Engine

The credit decisions support system includes the inference engine which performs the following tasks:

- reads contents of the credit database;
- accepts and removes facts about economic indicators of an analyzed company;
- performs the inference;
- gives the explanation "*HOW?*";
- displays all facts currently available in the system;
- displays all rules from the knowledge base.

7.4   Modular Structure of the System

The complete system for supporting credit decisions consists of the following modules:

- module for data processing – used for edition of credit applications, filling information about companies and their financial reports;
- module for monitoring of economic indicators – used for current analysis of economic indicators; it provides an instant overview of indicator values;
- interface engine with a rule knowledge base – performs credit risk estimation on the basis of economic ratios of the company being examined; the module includes a built-in explanation mechanism and also allows a user the knowledge base modifications; it was developed in Prolog.

References

Am95. *Amzi!Prolog. User's Guide & Reference ver. 3.3.* Amzi! Inc., Stow, Massachusetts, U.S.A., 1995.

Ba94a. Bazan, J., Skowron, A., Synak, P.: *Market Data Analysis: A Rough Set Approach.* ICS Research Report 6/94, Warsaw University of Technology, 1994.

Ba94b. Bazan, J., Skowron, A., Synak, P.: *Discovery of Decision Rules from Experimental Data.* The Third International Workshop on Rough Sets and Soft Computing, San Jose, 1994.

Kurtys, E., Waśniewski, T., Wersty, B.: *Economic analysis in a company* [in Polish]. Wydawnictwo Akademii Ekonomicznej we Wrocławiu, Wrocław, 1993.

Ch73. Chang, C.R., Lee, R.C.T.: *Symbolic Logic and Mechanical Theorem Proving.* Academic Press, 1973.

Cl84. Clocksin, W., Mellish, C.: *Programming in Prolog,* 2nd ed. Berlin B.R.D.: Springer-Verlag, 1984.

De94. Debski, W.: *Bank risk* [in Polish]. "Bank i Kredyt", no. 10, 1994.

Gr92.   Grzymała-Busse, J.: *LERS—A System for Learning from Examples Based on Rough Sets*. In "Intelligent Decision Support. Handbook of Applications and Advances of Rough Sets Theory." Słowiński, R. (ed), Kluwer Academic Publishers, Dordrecht, 1992, pp. 3–18.

Kr95.   Krawczyk, R.: *Computer system supporting companies valuation* [in Polish]. Master Thesis, Silesian Technical University, Gliwice, 1995.

Le94.   Lenarcik, A., Piasta, Z.: *Rough Classifiers* in W. Ziarko (ed.): "Rough Sets, Fuzzy Sets and Data Mining", pp. 298–316, Springer–Verlag, London, 1994.

Me89.   Merritt, D.: *Building Experts Systems in Prolog*. Springer–Verlag New York Inc., 1989.

Mk93.   Michalik, K.: *Intelligent Systems for Supporting Economic Decisions* [in Polish]. Wydawnictwo Akademii Ekonomicznej w Katowicach, Katowice 1993.

Mi83.   Michalski, R.: *A theory and methodology of inductive learning* in W: Michalski R., Charbonell J., Mitchell T.M. (eds.): "Machine Learning". Tioga, Palo Alto 1983.

Pa82.   Pawlak, Z.: *Rough Sets*. International Journal of Information and Computer Science, Vol. 11, No. 341, 1982.

Pa91.   Pawlak, Z.: *Rough Sets: Theoretical Aspects of Reasoning about Data*. Kluwer Academic Publishers, Dordrecht, The Neatherlands, 1991.

Pi.   Piasta, Z.: *Data Mining and Knowledge Discovery in Marketing and Financial Databases with Rough Classifiers* [in Polish]. Wydawnictwo Akademii Ekonomicznej we Wrocławiu, Wrocław, (to appear).

Pl.   Piasta, Z., Lenarcik, A.: *Learning Rough Classifiers from Large Databases with Missing Values* in A. Skowron and T. Polkowski (eds.): "Rough Sets and Knowledge Discovery", Physica–Verlag, (to appear).

Dl92.   REDUCT System, Inc. *DataLogic/R Reference Manual*. Regina, 1992.

Ro65.   Robinson, J.A.: *A machine oritnted logic based on the resolution principle*. J. Assoc. Comput. Mach., vol. 12, pp. 23–41, 1965.

Si93.   Sierpińska, M., Jachna, T.: *The company evaluation according to international standards* [in Polish]. Wydawnictwo Naukowe PWN, Warszawa, 1993.

Sk96.   Skabek, K.: *Computer system supporting credit decisions* [in Polish]. Master Thesis, Silesian Technical University, Gliwice, 1996.

Sl92.   Słowiński, R. (ed.): *Intelligent Decision Support: Handbook of Applications and Advances of the Rough Sets Theory*. Kluwer Academic Publishers, Dordrecht, 1992.

Sl95.   Słowiński, R., Zopounidis, C.: *Application of the Rough Set Approach to Evaluation of Bankrupcy Risk*. Intelligent Systems in Accounting and Management, Vol. 4, John Wiley & Sons, 1995.

Sr94.   Sroka, H.: *Expert Systems: Computer Decision Support in Management and Finance* [in Polish]. Wydawnictwo Akademii Ekonomicznej w Katowicach, Katowice 1994.

Po97.   Van den Poel, D. *Rough Sets for Database Marketing*. Department of Applied Economic Sciences, Catholic University Leuven, Naamsestraat 69, B-3000 Leuven, Belgium, 1997.

Tuning the Perceptual Noise Reduction Algorithm Using Rough Sets

A. Czyzewski, B. Kostek

Technical University of Gdansk, Sound Engineering Dept., 80-952 Gdansk, Poland
{andrzej, bozenka}@sound.eti.pg.gda.pl

Abstract. As is shown by the results of some recently proposed methods, the perceptual coding of audio can be used for suppressing the noise affecting transmitted audio signals. The process of tuning the perceptual audio coding algorithm demands finding the relations between masking algorithm parameters and their influence on the subjective quality of processed audio. The rough set method was employed to discover ill-defined relations underlying the implemented perceptual model of hearing.

1 Introduction

Presently, the majority of audio signals transmitted or recorded in digital systems is perceptually encoded. However, as is shown by the results of the lately proposed methods, the perceptual coding of audio can be also used for suppressing the noise affecting audio signals [1]. Such a noise can be caused by the recording procedures or by the transmission of audio signals through telecommunication channels. Thus, the application of the perceptual coding may not only prevent the original quality of audio, but it can also subjectively improve it. The masking curves providing the basic mechanism of this algorithm divide the spectral magnitudes of audio signal to two categories: audible components (stretched beyond masking curves) and not-audible ones (remaining below these curves). When determining the settings of the masking model, we have to base on the experimental procedures. The parameter values to be optimized could be discerned only by listening tests, yet their results do not reveal clearly the sought dependencies between parameters and audio quality. Hence, the rough set method was employed by the authors to facilitate the process of optimizing the perceptual algorithm for noise reduction. A special test procedure was elaborated for this purpose, which uses both: principles of psychometric scaling and the rule base building method based on rough sets. The perceptual coding algorithm and the proposed soft computing method for its optimization are presented in this paper.

2 Outline of the Perceptual Method for Noise Reduction

The method consists mainly in the constant analysis of the signal-to-noise relations in the consecutive sample packets, calculating the optimum shape of masking curves and processing the noisy audio using the perceptual coding algorithm. The noise sample is always taken from a silence passage nearest to the

L. Polkowski and A. Skowron (Eds.): RSCTC'98, LNAI 1424, pp. 467–474, 1998.
© Springer-Verlag Berlin Heidelberg 1998

currently processed fragment of the signal. On the basis of the spectral power density of noise in neighbor "silence passages" and the spectral power density of noise in the current signal packet, the masking threshold is raised separately in each critical band in order to keep the noise below the masking curves. The determination of the absolute threshold of hearing is calculated as in the literature [2] :

$$T_q = 3.64 \cdot f^{-8} - 6.5 \cdot \exp[-0.6 \cdot (f - 3.3)^2] + 10^{-3} \cdot f^4 \tag{1}$$

where: T_q - level of hearing threshold $[dB]$, f - frequency $[kHz]$.

The linear frequency scale is transformed into the critical band-related Bark scale on the basis of the dependencies found by Zwicker [2]:

$$b = 13 \cdot arctg(0.76 \cdot f) + 3.5 \cdot arctg[(f/7.5)]^2 \tag{2}$$

where: b - frequency $[Bark]$, f - frequency $[kHz]$.

The approximation of the masking phenomenon caused by the excitation component on the Basiliar membrane in the inner ear may be approximated by two segments inclined at angles S_1, S_2 (see Fig.1):

$$\begin{aligned} S_1 &= 27 \\ S_2(E, i) &= 24 + 0.23 \cdot f_c^{-1}(i) - 0.2 \cdot E(i) \end{aligned} \tag{3}$$

where: S_1 , S_2 - inclination measures expressed in $[dB/Bark]$,
  i - critical band number, $i = 0, 1, ..., 24$,

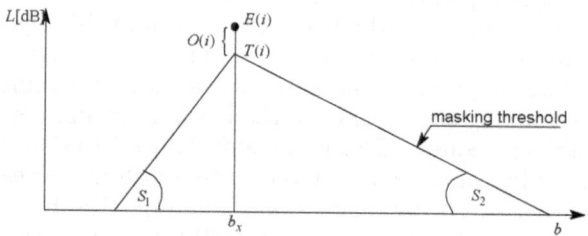

Fig. 1. Approximation of the masking threshold

According to Johnston [3] the distance between the excitation level and masking level $O(i)$ can be determined on the basis of the following relationship:

$$O(i) = \alpha(14.5 + i) + 5.5 \cdot (1 - \alpha) \tag{4}$$

where: $\alpha(0 \leq \alpha \leq 1)$ - is the so called *tonality index* which may be computed on the basis of the N-point Fast Fourier Transform Algorithm [3].

For the need to hide noise affecting the useful signal, the masking thresholds T(i) should be raised properly, what may be achieved numerically through setting appropriate values to them. This task can be done, when additional variable $\beta(i)$ is added to the formula (4) as follows:

$$O(i) = \alpha(14.5 + i) + 5.5 \cdot (1 - \alpha) + \alpha_n \beta(i) \tag{5}$$

where: α_n − is the tonality index for a sample of noise taken from a silence passage nearest to the currently processed fragment of the signal

Since masking is not a local phenomenon (it influences a certain bandwidth), thus the level of masking $E_{x,k}(E_x, b_x, b_k)$ for the frequency $b_k[Bark]$ caused by the presence of the component of the frequency b_x is derived, allowing to define masking threshold as follows:

$$T(i) = 10^{\log E'(i) - O(i)/10} \tag{6}$$

where:

$$E'(i) = B_{ij} \cdot S_p(i) \tag{7}$$

and: $E'(i)$- integrated excitation in i-th critical band $B_{ij}(i, j)$ - spreading function for the distance between critical bands $\Delta b = i - j$, (this function can be given in the form of a look-up table [3]) $S_P(i)$ - spectral power density in the i-th critical band.

Another phenomenon which should be taken into the consideration is the post-masking (the influence of previously occurring excitements to the current masking effect). The post-masking can be modeled by the dependence:

$$S_t(i, k) = S_a(i, k) + T_\tau(i) \cdot S_t(i, k - 1) \tag{8}$$

where: $S_t(i, k), S_t(i, k-1)$- spectral power densities for k-th and $(k-1)$-th sample packet $S_a(i, k)$ - transformed spectral power density in i-th critical band:

$$S_a(i, k) = a_0(b_i) \cdot \sum_{\Omega = \Omega_{di}}^{\Omega_{gi}} [S_p(e^{i\Omega}, k)] \tag{9}$$

where: $a_0(b_i)$ - coefficients of transformation, $S_P(e^{j\Omega}, k)$ - spectral power density for the k-th sample packet

The energy transmission coefficient $T(i)$ used in the equation (8) is given by the relationship:

$$T_\tau(i) = e^{\frac{-d}{\tau(i)}} \tag{10}$$

where: d - time lag between consecutive sample packets $[ms]$, $\tau(i)$- time constant $[ms]$ depended on the i-th critical band. The post-masking features can be supported by an algorithm, provided $S_P(i)$ in the equation (7) is replaced by $S_t(i, k)$calculated basing on eq. (8).

3   Tuning the Model

As results from the previous paragraph, in order to determine the masking effect, some values have to be computed for each signal packet, among others $S_1(i), S_2(i), O(i), E(i), T(i)$. In order to calculate these values within a given critical band, three parameters should be defined for each critical band $i = 1, 2, ..., 24$, namely $\beta(i), \tau(i), and\ a_0(i)$

This results in the set containing $(3 \bullet 24 = 72)$ parameters to be tuned. A proper selection of these parameter values allow one to control the process of noise removal. Practically 24 critical bands were grouped in three regions: low frequency (1 to 8 band), mid-frequency (9-16 band) and high frequency (17-24 band). The three parameters to be optimised, namely $\beta(i), \tau(i), and\ a_0(i)$ were set identical in each group of critical bands. Consequently, $(3 \bullet 3 = 9)$ parameters were subjected to the optimisation procedure supported by the soft computing data processing, as is described in the next paragraphs.

4   Subjective Testing Procedure and Soft Processing of Results

The most popular subjective quality testing method is the paired comparison test. The goal of this method is to compare objects ordered to pairs and to assess them on the basis of a two-level scale (better/worse attribute scale). Technically, signal samples are presented in A-B order. The expert task is to choose the better one from a pair of sound samples that differ in acoustic features. As a result a certain number is assigned to each compared sound sample. That number reflects the experts' overall preference. The statistical analysis of test results cannot reveal hidden relations between tested parameters nor give the rules instructing one on how to tune a system based on such parameters. That is why the soft computing rule-based system (the rough set method) was employed to this task. There are many other effective machine learning algorithms, however, in the studied case there is a necessity to fulfil some special demands to make the acquired knowledge base applicable to the task of tuning the audio processing algorithm. The demands are as follows:

  - the knowledge should be presented in the form of a set of readable rules
  - the rules should be associated with a belief measure, allowing to rank them, because it is not probable to induce only certain rules on the basis of subjective opinions of many experts (there is expected a wide margin of uncertainty)
  - the system should be able to deal with values expressed by ranges

Above considerations led authors' choose the rough set decision system as a tool matching the demands related to the processing data acquired on the basis of subjective opinions. Let's assume that the number of assessed audio patterns (related to a single tuned parameter of the algorithm) is set to X, the number of subjects involved in the subjective listening session equals Y, and finally, the number of test series equals Z. Hence, the maximum number of answers in a paired comparison test equals:

$$N = N_1 \cdot Y \cdot Z \tag{11}$$

where N_1- is the number of pairs assessed by an individual expert in one series of a test calculated according to the following formula:

$$N_1 = \binom{X}{2} = \frac{X!}{(X-2)! \cdot 2!} \tag{12}$$

In this way testing 10 audio patterns by 5 experts in 2 series will result in 450 votes which are distributed between objects. The test was performed 9 times, because it was 9 parameters: $\beta(1), \beta(2), \beta(3), \tau(1), \tau(2), \tau(3), a_0(1), a_0(2), a_0(3)$, which were related to 3 frequency regions. The preference diagrams were obtained on the basis of experts' answers (see the example in Fig. 2). The individual objects (A to J) were assessed when paired with all others. Because each object was ranked higher certain times by some experts when compared to another objects, the previously mentioned 450 votes are distributed among objects, as is seen in Fig. 2. The preference curve reveals some maxima for the objects processed with certain parameter settings. Thus, the plot indirectly provides information on perceptible differences between parameter values. In this way the so-called perceptual quantization can be done. Usually, for practical needs, the number of preference ranges could be diminished, so the scale of number of votes reflecting this preference could be for example decimated. Hence, in the showed example (Fig. 2) the vertical axis can be re-scaled, reflecting preferences in the range of 0 to 5.

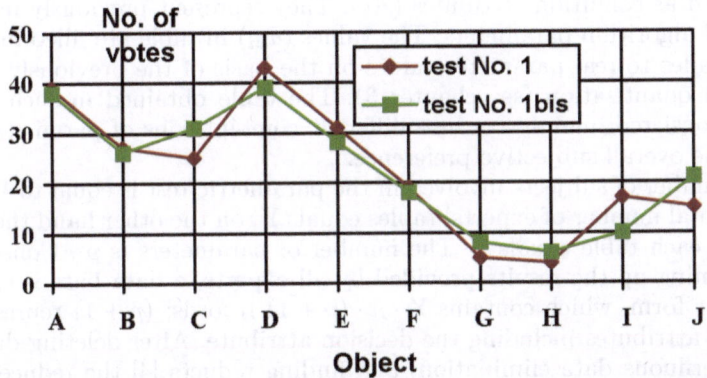

Fig. 2. Number of votes given to each tested object when assessed in pairs, reflecting the degree of subjective preference of objects by experts (the test procedure was repeated twice). $A - J$ are audio patterns related to 10 values of one parameter (in the case presented in figure it was parameter $a_0(i)$)

Having the parameter values discretized to 5 ranges, it is possible to generate $5^9 = 1953125$ audio patterns related to all combination of values of 9 parameters. Obviously, such number of tests would be completely impractical. Thus, the number of combinations was randomly decreased to 4096 (data compression ratio about 500). This operation makes the subjective testing realizable, however imposes a wide margin of uncertainty to the data processing algorithm. A special computer program was prepared to generate and play patterns automatically, without the need to store the resulting audio files. The grades were acquired from experts using special keyboards interfaced to the computer. Each audio pattern contained $5[s]$ of mixed music and speech. This resulted in $4096 \bullet 5[s] = 6[h]$ of music assessed by the experts (plus $3[h]$ for pauses between fragments). The sessions were organized during 6 days in order to prevent experts' from getting tired. After listening to each pattern, each of experts proposed a grade of overall preference according to the 5-point preference scale. The experts were asked to assess the noise reduction effect and the audible distortion level, simultaneously. Since it were 5 experts employed, thus a database containing $5 \bullet 4096 = 20480$ records was created. Because the number of test results can be so large, thus the relations between parameters remain hidden until they are discovered by an automatic rule induction algorithm. The large portion of missing combinations and inconsistency of the database left a wide margin of uncertainty to be managed by the soft computing algorithm.

4.1   Rough set-based analysis of test results

In order to discover tendencies underlying the choice of the overall quality assigned by experts to particular combinations of parameter values, the rough set-based analysis was performed. The perceptual quantization of parameter values was used which was introduced in the previous paragraph. The Quality was defined as the decision attribute (D), all other attributes included in table are used as condition attributes (A_k). They represent previously introduced perceptual algorithm parameters. The values (a_{ki}) in table are filled in after assigning grades to real parameter values on the basis of the previously executed perceptual quantization (see chapter 3). The table obtained in such a way is highly inconsisted, mainly because different combinations of parameters result in the same overall subjective preference.

The number of subjects involved in the parametric test is equal to Y. Therefore, the total number of experts' tables equals Y, on the other hand the number of rows in each table equals n. The number of parameters is p . Consequently, after summing up the results provided by all experts, a data base is created in the tabular form, which contains $Y \cdot n \cdot (p + 1)$ records; $(p + 1)$ represents the number of attributes including the decision attribute. After deleting duplicated rows (superfluous data elimination) and finding reducts [4] the reduced sets of rules is obtained which contains the knowledge of the masking algorithm parameters values which are preferred by the experts. The final step is generating rules from the reducts. However, in the specific case related to the discussed field of application, the whole set of rules should be used initially as a guidance for tuning the perceptual coding system. That is caused by the fact, that the reduced rules may not show how to set the values of some parameters. Since some rules contain only some parameters to be set, thus the sum of rules generated from reducts should be taken into consideration. The practical way of tuning the system using the set of rules is described in the next chapter.

5   Experimental Procedures

The 10 values of each parameter within each range was selected for testing. In this way total 90 parameter values were defined in each of 3 frequency ranges. Then, 10 speech and music patterns (each of them of time duration approximately 5[s]) were processed using all 10 previously defined values of perceptual coding algorithm parameters. Consequently, after the processing, 9 sets of 10 objects (audio patterns) were obtained. Subsequently, each 10-element set of objects related to a defined parameter was transformed to the set of $\binom{10}{2} = 45$ pairs to be assessed subjectively. Then, the experts provided their opinions and the number of votes given by them to each object was computed. The scale of subjective grades was decimated, so the perceptual quantization of parameter values was obtained. After listening tests the decision table was built up according to the previously discussed scheme (see Tab. 1). The rough set algorithm elaborated at the Sound Engineering Department of the TU Gdansk was used [1]. This algorithm generated rules from this table, of various rough measure values (within the range of $< 0.5, 1.0 >$). After the calculation of reducts the new rule set was generated (basing on reducts), containing 36 strongest ($\mu_{RS} > 0.8$) rules of the length from 2 to 9. The rules were ordered in such a way that the shorter and the stronger were listed first. The strongest rules were used during the experiments as a guidance for tuning the perceptual noise reduction system. The exemplary rules obtained in this way were as follows:

$(\beta(2) = 3)$**and**$(\beta(3) = 1) => (Overall\_Quality = 4)$, $\mu_{RS} = 1$

$(\beta(3) = 0)$**and**$(\tau(2) = 4) => (Overall\_Quality = 1)$, $\mu_{RS} = 0.816$

$(\beta(1) = 4)(\beta(2) = 3(\beta(3) = 1)(\tau(2) = 3)(\tau(3) = 2) => (5)$, $\mu_{RS} = 1$

$(\beta(1) = 1)(\beta(2) = 1)(\tau(3) = 1)a_0(3) = 0) => (2)$, $\mu_{RS} = 0.803$

$(\beta(1) = 4)(\beta(2) = 3)(\beta(3) = 1)(\tau(3) = 1)(a_0(2) = 1)(a_0(3) = 3) => (5)$, $\mu_{RS} = 0.911$

$(\beta(1) = 4)(\beta(2) = 3)(\beta(3) = 1)(\tau(1) = 2)(a_0(1) = 2)(a_0(2) = 1)a_0(3) = 3) => (5)$, $\mu_{RS} = 0.81$

No one rule, which employs all 9 conditional attributes was found associated with the decision showing the highest grade of subjective preference ($Overall\_Quality = 5$). Consequently, using the induced set of rules the parameters of the perceptual coding system were set in such a way that first were considered the shortest and the strongest rules. After some additional listening the first rule showing the best grade ($= 5$) was applied to set $\beta(1)$ to 4 and $\beta(2)$ to 3 and $\beta(3)$ to 1 and $\tau(2)$ to 3 and $\tau(3)$ to 2. The remaining parameters values ($\tau(1), a_0(1), a_0(2)$ and $a_0(3)$) were set on the basis of the two last rules listed above, namely $\tau(1)$ to 2 and $a_0(1)$ to 2 and $a_0(2)$ to 1 and $a_0(3)$ to 3. Final listening test showed that these settings were acceptable by the experts.

6   Conclusions

The concept of perceptual quantization has been introduced in this paper, providing a usable method of replacing values by ranges in psychometric scaling. Despite the data reduction obtained with the use of the perceptual quantization, the number of combinations of parameter settings was too large for practical listening tests. The reduction of the number of such combinations made the testing procedure uncertain and the table of subjective preference highly incomplete. In such conditions the statistical processing of data cannot help to find the optimum values of tested parameters. The rough set algorithm dealing with inconsistency and missing data allowed to generate the rules which were applied for tuning the perceptual noise reduction system.

Acknowledgements

Research sponsored by the State Committee for Scientific Research, Warsaw, Poland. Grant No. 8T11D 021 12.

References

1. 1. Czyzewski, A., Królikowski R.: Application of Intelligent Decision Systems to the Perceptual Noise Reduction of Audio Signals. Proc. EUFIT'97, Aachen, vol. 1, 188–192 (1997).
2. 2. Shlien, S.: Guide to MPEG-1 Audio Standard, IEEE J. Trans. on Br., vol. 40, No. 4, (1994).
3. 3. Beerends J., Stemerdink, J., A Perceptual Audio Quality Measure Based on a Psychoacoustic Sound Representation, J. Audio Eng. Society, vol. 40, No. 12, pp. 963–978, (1992).
4. 4. Pawlak, Z.: Data versus Logic A Rough Set View, Proc. 4th Int. Workshop RSFD'96, Tokyo, 1–8, (1996).

Modelling Medical Diagnostic Rules Based on Rough Sets

Shusaku Tsumoto

Department of Information Medicine, Medical Research Institute,
Tokyo Medical and Dental University
1-5-45 Yushima, Bunkyo-ku Tokyo 113 Japan
E-mail: tsumoto@computer.org

Abstract. *This paper discusses the characteristics of medical reasoning and shows the representation of these diagnostic models by the use of rough set theory. The key ideas are both a variable precision rough set model, which corresponds to an ordinal positive reasoning, and an upper approximation of a target concept, which corresponds to a focusing procedure. Acquired representation suggests that rough set model should be closely related with medical diagnosis.*

1   Introduction

Medical reasoning always includes uncertainty[1], which is caused by the limitations of medical knowledge, available data and our recognition, compared with the complexities of human body. Thus, medical databases also have a certain degree of uncertainty: rules extracted from databases are also incomplete, which suggests that rule induction method should deal with uncertain rules.

According to this motivation, rule induction based on rough set theory have been applied to medical databases empirically[6–8,10], the results of which show that rough-set-based methods are very useful to extract medical diagnostic rules.

This paper presents how medical diagnostic rules are modeled by the concepts of rough sets[5] in a more theoretical way. The key ideas are both a variable precision rough set model, which corresponds to an ordinal positive reasoning, and an upper approximation of a target concept, which corresponds to a focusing mechanism in medical reasoning. Acquired models show that the characteristics of medical reasoning reflect the concepts on approximation of rough sets, which explains why rough sets work well in medical domains. The paper is organized as follows: in Section 2, two important measures, accuracy and coverage are defined and a probabilistic rule is defined. Section 3 to 5 presents description of three types of medical reasoning: simple differential diagnosis, focusing mechanism and $m-$of$-n$ criteria, respectively. Section 6 concludes our paper.

2   Probabilistic Rules

In this section, a probabilistic rule, which is a basis for describing diagnostic rules, is defined by the use of the following three notations of rough set theory[5]. In

L. Polkowski and A. Skowron (Eds.): RSCTC'98, LNAI 1424, pp. 475–482, 1998.

the following, these notations are illustrated by a small database shown in Table 1.

First, a combination of attribute-value pairs, corresponding to a complex in AQ terminology[4], is denoted by a formula R. For example, $[age = 50 - 59]\&[loc = occular]$ will be one formula, denoted by $R = [age = 50 - 59]\&[loc = occular]$. Secondly, a set of samples which satisfies R is denoted by $[x]_R$, corresponding to a star in AQ terminology. For example, when $\{2, 3, 4, 5\}$ is a set of samples which satisfies $[age = 40 - 49]$, $[x]_{[age=40-49]}$ is equal to $\{2, 3, 4, 5\}$. [1] Finally, U, which stands for "Universe", denotes all training samples.

Table 1. An Example of Database

	age	loc	nat	prod	nau	M1	class
1	50-59	occ	per	0	0	1	m.c.h.
2	40-49	who	per	0	0	1	m.c.h.
3	40-49	lat	thr	1	1	0	migra
4	40-49	who	thr	1	1	0	migra
5	40-49	who	rad	0	0	1	m.c.h.
6	50-59	who	per	0	1	1	psycho

DEFINITIONS: loc: location, nat: nature, prod: prodrome, nau: nausea, M1: tenderness of M1, who: whole, occ: occular, lat: lateral, per: persistent, thr: throbbing, rad: radiating, m.c.h.: muscle contraction headache, migra: migraine, psycho: psychological pain, 1: Yes, 0: No.

2.1   Classification Accuracy and Coverage

Definition of Accuracy and Coverage According to the notations above, classification accuracy and coverage (true positive rate) is defined as:

$$\alpha_R(D) = \frac{|[x]_R \cap D|}{|[x]_R|}, \text{ and } \kappa_R(D) = \frac{|[x]_R \cap D|}{|D|},$$

where $|A|$ denotes the cardinality of a set A, $\alpha_R(D)$ denotes a classification accuracy of R as to classification of D, and $\kappa_R(D)$ denotes a coverage, or a true positive rate of R to D, respectively. In the above example, when R and D are set to $[nau = 1]$ and $[class = migraine]$, $\alpha_R(D) = 2/3 = 0.67$ and $\kappa_R(D) = 2/2 = 1.0$.

It is notable that $\alpha_R(D)$ measures the degree of the sufficiency of a proposition, $R \to D$, and that $\kappa_R(D)$ measures the degree of its necessity. For example, if $\alpha_R(D)$ is equal to 1.0, then $R \to D$ is true. On the other hand, if $\kappa_R(D)$ is equal to 1.0, then $D \to R$ is true. Thus, if both measures are 1.0, then $R \leftrightarrow D$.

---

[1] In this notation, "n" denotes the nth sample in a dataset (Table 1).

2.2 Probabilistic Rules

By the use of accuracy and coverage, a probabilistic rule is defined as follows:

Definition 1 ((Probabilistic Rules)). *Let R be a formula (conjunction of attribute-value pairs), D denote a set whose elements belong to a class d, or positive examples in all training samples (the universe), U. Finally, let $|D|$ denote the cardinality of D. A probabilistic rule of D is defined as a tripule, $< R \overset{\alpha,\kappa}{\to} d, \alpha_R(D), \kappa_R(D) >$, where $R \overset{\alpha,\kappa}{\to} d$ satisfies $\alpha_R(D) > 0.$*[2]

In the following sections, all the diagnostic rules are represented as special types of the probabilistic rule above.

3 Simplest Diagnostic Rules

3.1 Representation of Diagnostic Rules

The simplest probabilistic model is that which only uses classification rules which have high accuracy and high coverage. Such rules can be defined as:

$$R \overset{\alpha,\kappa}{\to} d \text{ s.t. } \quad R = \vee_i R_i = \vee_i \wedge_j [a_j = v_k],$$
$$\alpha_{R_i}(D) \geq \delta_\alpha \text{ and } \kappa_{R_i}(D) \geq \delta_\kappa,$$

where δ_α and δ_κ denote given thresholds for accuracy and coverage, respectively. For the above example shown in Table 1, probabilistic rules for m.c.h. are given as follows (both δ_α and δ_κ are set to 0.75):

$$[prod = 0] \to m.c.h. \ \alpha = 3/4 = 0.75, \ \kappa = 1.0,$$
$$[nau = 0] \to m.c.h. \ \alpha = 3/3 = 1.0, \ \kappa = 1.0,$$
$$[M1 = 1] \to m.c.h. \ \alpha = 3/4 = 0.75, \ \kappa = 1.0,$$

3.2 An Rule Induction Algorithm

An rule induction algorithm is defined as Figure 1, which is discussed precisely in [9]. It is notable that rule induction of other type rules is derived by simple modification of this algorithm.

4 Focusing Mechanism

One of the characteristics in medical reasoning is a focusing mechanism, which is used to select the final diagnosis from many candidates[8]. For example, in differential diagnosis of headache, more than 60 diseases will be checked by

---

[2] It is notable that this rule is a kind of probabilistic proposition with two statistical measures, which is one kind of an extension of Ziarko's variable precision model(VPRS) [11].

procedure *Induction of Classification Rules*;
  var
    $i : integer$; $M, L_i : List$;
  begin
    $L_1 := L_{er}$; /* L_{er}: List of Elementary Relations */
    $i := 1$; $M := \{\}$;
    for $i := 1$ **to** n **do**   /* n: Total number of attributes */
      begin
        while ($L_i \neq \{\}$) **do**
          begin
            Sort L_i with respect to the value of coverage;
            Select one pair $R = \wedge[a_i = v_j]$ from L_i,
            which have the largest value on coverage;
            $L_i := L_i - \{R\}$;
            if $(\kappa_R(D) \geq \delta_\kappa)$
              then do
                if $(\alpha_R(D) \geq \delta_\alpha)$
                  then do $S_{ir} := S_{ir} + \{R\}$; /* Include R as Classification Rule */
                      $M := M + \{R\}$;
        end
        $L_{i+1} :=$ (A list of the whole combination of the conjunction formulae in M);
      end
  end {*Induction of Classification Rules* };

Fig. 1. An Algorithm for Classification Rules

present history, physical examinations and laboratory examinations. In diagnostic procedures, a candidate is excluded if a symptom necessary to diagnose is not observed.

This style of reasoning consists of the following two kinds of reasoning processes: exclusive reasoning and inclusive reasoning. First, exclusive reasoning excludes a disease from candidates when a patient does not have a symptom which is necessary to diagnose that disease. Secondly, inclusive reasoning suspects a disease in the output of the exclusive process when a patient has symptoms specific to a disease. These two steps are modeled as usage of two kinds of rules, negative rules (exclusive rules) and positive rules, the former of which corresponds to exclusive reasoning and the latter of which corresponds to inclusive reasoning. In the next two subsections, these two rules are represented as special kinds of probabilistic rules.

4.1   Positive Rules

A positive rule can be defined as a rule supported by only positive examples, which means that the classification accuracy of a rule is equal to 1.0. Thus, a positive rule is represented as:

$$R \to d \quad s.t. \quad R = \wedge_j[a_j = v_k], \quad \alpha_R(D) = 1.0$$

In the above example, one positive rule of "m.c.h." is:

$$[nau = 0] \rightarrow m.c.h. \quad \alpha = 3/3 = 1.0.$$

This positive rule is often called deterministic rules. However, in this paper, we use a term, positive (deterministic) rules, because deterministic rules which are supported only by negative examples, called negative rules, is introduced as in the next subsection.

4.2 Negative Rules

Before defining a negative rule, let us first introduce an exclusive rule, the contrapositive of a negative rule[8]. An exclusive rule can be defined as a rule including all the positive examples, which means that the coverage of a rule is equal to 1.0.[3] Thus, an exclusive rule is represented as:

$$R \rightarrow d \quad s.t. \quad a \quad R = \wedge_j [a_j = v_k], \quad \kappa_R(D) = 1.0.$$

In the above example, exclusive rule of "m.c.h." is:

$$[prod = 0] \wedge [nau = 0] \wedge [M1 = 1] \rightarrow m.c.h. \quad \kappa = 1.0,$$

It is notable that exclusive rule corresponds to an upper approximation of a target concept. For example, the set which supports the exclusive rule above is an upper approximation of m.c.h.

¿From the viewpoint of propositional logic, an exclusive rule uniquely corresponds to

$$d \rightarrow \wedge_j [a_j = v_k],$$

because the condition of an exclusive rule correspond to the necessity condition of conclusion d. Thus, it is easy to see that a negative rule is defined as the contrapositive of an exclusive rule:

$$\vee_j \neg [a_j = v_k] \rightarrow \neg d,$$

which means that if a case does not satisfy any attribute value pairs in the condition of a negative rules, then we can exclude a decision d from candidates. For example, the negative rule of m.c.h. is:

$$\neg [prod = 0] \vee \neg [nau = 0] \vee \neg [M1 = 1] \rightarrow \neg m.c.h.$$

In summary, a negative rule is defined as:

$$\vee_j [a_j = v_k] \rightarrow \neg d \quad s.t. \quad \forall [a_j = v_k] \, \kappa_{[a_j = v_k]}(D) = 1.0,$$

where D denotes a set of samples which belongs to a class d.

---

[3] Exclusive rules represent the necessity condition of a decision.

4.3   Rule Induction Algorithm

An algorithm for induction of positive and negative rules is derived by simple modification of the algorithm in Figure 1: if the thresholds of accuracy and coverage is set to 0.0 and 1.0, respectively, the algorithm for negative rules will be obtained. On the other hand, if the thresholds of accuracy and coverage are set to 1.0 and 0.0, respectively, the algorithm for negative rules will be obtained.

It is notable that positive and negative rules can be extended to probabilistic versions, which is discussed precisely in [8].

5   Criteria Tables

5.1   Representation of Rules

Another characteristic reasoning in medicine is $m-$of$-n$ concepts, or a criteria table, which is discussed in [2]. The criteria table for a disease d is described by n attributes, which are enough to make its diagnosis. If at least m attributes are observed in a patient, d should be suspected.

Langley discusses that this $m-$of$-n$ description can be rewritten as a simple linear combination of attribute-value pairs. Thus, he implements an induction of this description as an induction of threshold concepts.

However, a $m-$of$-n$ rule in medicine is not equivalent to a linear combination rule, which is a special kind of statistical discriminant functions[3]. Rather, this type of rule is based on relations between sets as follows.

1. If total n attributes are observed, a disease d is suspected with the highest accuracy. (The coverage is equal to 1.0).

2. If m attributes are satisfied, a disease d should be suspected with high accuracy. (The coverage is equal to 1.0).

3. If less than m attributes are satisfied, the probability of d is low. However, the coverage is equal to 1.0. Thus, $m-$of$-n$ concept is described as combination of exclusive rules (below, we call them *unit rules*) with the constraint that their accuracies are high:

$$R \rightarrow d \ s.t. \ R = \wedge_{j=1}^{i}[a_j = v_k](m \le i \le n)$$
$$\alpha_R(D) \ge \delta_\alpha$$
$$\forall[a_j = v_k], \kappa_{[a_j=v_k]}(D) = 1.0,$$
$$(\alpha_{[a_j=v_k]}(D) \ge \delta_2),$$

which also satisfies that: if R is represented as $\wedge_{j=1}^{i}(i < m)$, then $\alpha_R(D) < \delta_\alpha$ holds.

For the above example in Table 1, exclusive rule of m.c.h. is:

$$[prod = 0] \wedge [nau = 0] \wedge [M1 = 1] \rightarrow m.c.h. \quad \kappa = 1.0, \alpha = 1.0$$

This attains the highest accuracy. If the threshold for accuracy is set to 0.75, then

$$[prod = 0] \rightarrow m.c.h. \ \kappa = 1.0, \alpha = 0.75,$$
$$[nau = 0] \rightarrow m.c.h. \ \kappa = 1.0, \alpha = 0.75, \ and$$
$$[M1 = 1] \rightarrow m.c.h. \ \ \kappa = 1.0, \alpha = 1.0.$$

So, diagnostic rules for m.c.h. can be viewed as $1-of-3$ concept. In this way, combination of accuracy and coverage is also important to represent $m-of-n$ type rules.

5.2 Rule Induction Algorithm

An algorithm for induction of unit rules is derived by simple modification of the algorithm in Figure 1: if the thresholds of accuracy and coverage are set to δ and 1.0, respectively, then the algorithm for induction of each unit rule will be obtained. In this model, we should only add integration of unit rules after rule induction to obtain the total algorithm, which is not shown for the limitation of the space.

6   Conclusion

In this paper, rough set framework is introduced to model medical diagnostic rules. Acquired models show that the characteristics of medical reasoning reflect the concepts on approximation of rough sets, which explains why rough sets work well in medical domains. These results have not been validated by experiment results yet, which will be reported in the near future.

References

1. Buchnan, B. G. and Shortliffe, E. H.(eds.) (1984). *Rule-Based Expert Systems*, Addison-Wesley.
2. Langley, P. (1996). *Elements of Machine Learning*, Morgan Kaufmann, CA.
3. Mclachlan, G.J. (1992). *Discriminant Analysis and Statistical Pattern Recognition*. John Wiley and Sons, NY.
4. Michalski, R. S. (1983). A Theory and Methodology of Machine Learning. Michalski, R.S., Carbonell, J.G. and Mitchell, T.M., *Machine Learning - An Artificial Intelligence Approach*. Morgan Kaufmann, CA.
5. Pawlak, Z. (1991). *Rough Sets*. Kluwer Academic Publishers, Dordrecht.
6. Slowinski, K. et al. (1988). Rough sets approach to analysis of data from peritoneal lavage in acute pancreatitis. *Medical Informatics*, **13**, 143–159.
7. Tsumoto, S. and Tanaka, H. (1995). PRIMEROSE: Probabilistic Rule Induction Method based on Rough Sets and Resampling Methods. *Computational Intelligence*, **11**, 389–405.

8. Tsumoto, S. and Tanaka, H. (1996). Automated Discovery of Medical Expert System Rules from Clinical Databases based on Rough Sets. Proceedings of the Second International Conference on Knowledge Discovery and Data Mining 96, pp. 63–69, AAAI Press.
9. Tsumoto, S. and Tanaka, H. (1998). Automated Knowledge Acquisition from Medical Databases and Its Evaluation, *MEDINFO-98*, IMIA(in press).
10. Wakulicz-Deja, A., Paszek, P. (1997). Optimization on decision problems on medical knowledge bases. in: *Proceedings of EUFIT-97.*
11. Ziarko, W (1993). Variable Precision Rough Set Model. *Journal of Computer and System Sciences*, **46**, 39–59.

Discretization of Continuous Attributes on Decision System in Mitochondrial Encephalomyopathies *

Alicja Wakulicz–Deja, Mariusz Boryczka, Piotr Paszek

Institute of Computer Science,
University of Silesia,
41–200 Sosnowiec, Poland

Abstract. In this work we check how the automatic discretization algorithms generate decision rules for the concrete medical problem – diagnosing mitochondrial encephalomyopathies (MEM).
We describe several algorithms for discretization – local and global – of continuous attributes obtained in the second stage of diagnosing MEM. All of these algorithms act together with the data analysis method based on the rough sets theory.
This work compares results — quality of classification rules — which were obtained using different discretization methods of the continuous attributes.

1 Discretization of the continuous attributes in the decision systems

Data describing characteristics of real objects, which are a base for collecting the data in the decision systems, are usually represented by real numbers. In connection with it, the continuous attributes, what means, such of which values are the real numbers of a certain range $[a, b]$, appear in such systems. The number of values of a continuous attribute is the higher the higher is the accuracy of the measurement from which those values come. The information (decision) system having continuous attributes is characterized by a great number of equivalence classes (atomic sets) - a number of the equivalence classes is frequently of order of a number of objects, particularly when there are many continuous attributes in the system.

The precise values of the continuous attributes cause the formation of a great number of the equivalence classes, which results in generating a lot of decision rules. A great number of the decision rules gives the chance that the rules will be deterministic, but on the other hand too high number of decision rules hinders their verification by an expert. Additionally, those rules can classify poorly new cases, because matching the attributes values of new objects with the attribute

---

* This work was supported by the Committee for Scientific Research, Warsaw, Poland, Grant No. 8T11C 005 12.

L. Polkowski and A. Skowron (Eds.): RSCTC'98, LNAI 1424, pp. 483–490, 1998.
© Springer-Verlag Berlin Heidelberg 1998

values in the rules may be hampered. For that reason it is necessary to conduct the discretization process of the continuous attributes.

Discretization is based on dividing the continuous attribute range into a certain number of intervals and assigning determined different values to those intervals. The transformation of an attribute with real values into an attribute with discrete values takes place. Let A be a continuous attribute and the $[a, b]$ interval be a range of the A attribute value. Then, the following k set of intervals is called the π_A partition of the $[a, b]$ interval:

$$\pi_A = \{[a_0, a_1), [a_1, a_2), \ldots, [a_{k-1}, a_k]\}$$

where: $a_0 = a$, $a_1 < a_i$ for $i = 1, 2, \ldots, k$ and $a_k = b$.

Thus, the discretization is a process which creates a set of the π_A intervals in the value range of the A the $[a, b]$ interval attribute .

After performing the discretization process of the continuous attribute, values of this attribute are transformed into the discrete values of the A^D attribute of which a set of values is defined in the following way:

$$V_{A_D} = \{1, 2, \ldots, k\}$$

Those values determine interval numbers of the π_A partition.

In the discretization process of the continuous attributes, a value determined as level of consistency is used to evaluate consistency of the decision system based on a theory of rough sets:

$$\frac{\sum\limits_{X \in \{d\}^*} |\underline{\mathbf{A}}X|}{|U|}$$

where:

- A - a set of the system attributes
- U - a set of the system objects
- X - a set of objects connected with a given decision

The simplest discretization method is based on division of the continuous attribute values into two intervals. Two values of a discrete attribute ($\pi_A = 2$) are obtained. From considerations a case when a number of intervals of the discretized attribute is one, is excluded because an attribute having only one value does not contribute any essential information into the system.

If we consider discretization leading to the division of the continuous attribute value into two intervals, considering m different values of this attribute describing objects of a certain decision system, there will be $m - 1$ ways of the division of those values. This number increases with geometric progression together with an increase in a number of intervals of the discretized attribute values.

Since there is a great number of possibilities of finding the division points in the assigned set of the continuous attribute values, the process based on searching the best way of discretization by checking results for all possible cases of the division is impossible to be carried out. For that reason different discretization algorithms are used, suggesting one way of division, which gives the results close to the optimum.

The simplest algorithm divides a set of the continuous attribute values into intervals equal in width and it is called Equal Interval Width Method. This algorithm does not require to consider the attribute values for given objects an because of it, the result of its activity depends mainly on the assumed number of the A^D discrete attribute value and boundaries of the continuous attribute value interval.

Another algorithm, using so called the adaptive discretization [2] acts in such a way that first the $[a, b]$ interval is divided into two intervals of equal width, and then, using discrete values, it induces decision rules. The induced rules are verified. If the non–deterministic rules have been induced or the consistency level of the system is lower than assumed, one of those two intervals is divided into two and the whole process of the creation of rules and their verification is repeated. This process is repeated until the assumed consistency level of the decision system is reached.

The successive discretization algorithms use the term of entropy, calculated in the following way:

$$\text{Ent}(U) = -\sum_{j=1}^{k} p_j \log p_j$$

Where the p_j values are connected with a number of the continuous attribute values being in respective intervals.

Equal Frequency per Interval Method - EF is also called Maximum Entropy Discretization [13]. It is based on the fact that in the continuous attribute value set, an assigned number of intervals is created; the intervals are of that type that in each of them, in approximation, there is an equal number of the attribute values for the chosen set of objects.

Another method, called Minimal Class Entropy Method - MCE [4], is based on the minimal entropy. Entropy is a criterion for searching a list of the best division points, which together with the boundaries of the continuous attribute value interval create the searched model.

The presented discretization methods are local methods. They act each time on one continuous attribute. The global methods find partitions for all continuous attributes at the same time, using tools connected with the analysis method of the decision system in question. So called globalisation of local discretization methods can be made at that place.

Cluster Analysis Method - CA [3] may be an example of a global discretization method. This method begins from a cluster analysis, and then when clustering cannot be performed any longer (because of reaching the assumed level of consistency), linking of adjacent value interval of particular continuous attributes is made. The entropy value is also taken as a criterion for the choice of the attribute and intervals for linking.

In this work for the discretization of the continuous attributes the following methods: EF, MCE and CA were used. The experiment results will presented in the next points.

2 Medical problem

Our problem was to develop and describe a knowledge base for progressive encephalopathy (PE). PE is a progressive loss of psychomotor and neuromuscular functions occurring in the infancy or in older children. The disease is grave and life threatening [1,8,12].

The real reasons of PE are metabolic diseases. Metabolic processes in a cell are catalyzed with enzymes and enzyme systems which are in nucleus and in various cytoplasmic organelles such as: lysosome, mitochondria and peroxysome.

In the work we have paid attention to encephalopathy in which respiratory enzymes of the cell located in mitochondria's are impaired. Mitochondrial encephalomyopathies (MEM) occur with elevated levels of lactic and pyruvic acid in the blood serum and the cerebrospinal fluid (CSF). MEM are the heterogenic group of disorders in which function disturbances may concern many organs, particularly the brain.

The disease detection requires a series of tests, of which some are typically invasive ones and they are not indifferent to a child's health. Invasive tests are divided into two groups: testing levels of pyruvic and lactic acids in the blood serum, and cerebrospinal fluid (CSF) in the first stage and the examinations of a nerve or muscle segment to determine the enzyme levels, which are the final tests confirming the disease. As they are most threatening to a child's health, they are made as the last resort for a small group of children.

In such a way we create a three stage classification where a set of objects (patients) on each classification level is smaller. The most important problem is to create an appropriate classification system of patients on each level. This consists of an appropriate choice of attributes for the classification process and the generation of a set of rules, a base to make decisions in new cases.

3 Description of data and neurological tests used in the system

In the first stage we made first selection. Based on clinical symptoms we created rules that classify the ill children into 2 groups: the suspicion of PE or other diseases [7].

The second stage consist of further elimination of children from the first group, who do not suffer from PE. The stage consists of taking blood samples and CSF and then on making biochemical examinations of pyruvate and lactate levels in the samples.

Finally, on the basis on these results rules qualifying to the suitable group were created [9].

Clinical material consists of 114 patients in the age between 3 months -15 years (the patients were: 60 boys and 54 girls) suspected of MEM because of the 7 following symptoms: – lactate level in blood;

– pyruvate level in blood;
– ratio of lactate to pyruvate level in blood;
– lactate level in CSF;
– pyruvate level in CSF;
– ratio of lactate to pyruvate level in blood;
– changes of acids level in blood and CSF.

The values of six first attributes were real. The last one has tree value set.

In this work we describe the discretization process of the continuous attributes obtained in the second stage of diagnosing MEM.

This work compares results - quality of classification rules - which were obtained using different discretization methods of the continuous attributes.

Table 1. Decision table before discretization – fragment

Object	\multicolumn{8}{c}{Attribute}

Object	1	2	3	4	5	6	7	Decision
1150	2.17	1.00	2.17	5.37	0.23	23.35	1	3
1160	0.55	0.07	7.86	1.24	0.09	13.78	1	2
1170	0.60	0.04	15.00	1.20	0.09	13.33	1	2
1190	0.87	0.48	1.81	0.91	0.24	3.79	1	1

In the second stage, for rule generation we used machine learning program LERS (Learning From Examples based on Rough Sets) [5,6]. The system handles inconsistencies using rough set theory [10,11].

4 Experiment

4.1 Discretization method

During the second stage of diagnosis MEM we get six continuous attributes.

We used four discretization methods of the continuous attributes

Discretization on the basis of a control group. Making measurements of acid levels in the control group in blood and CSF, norms for particular attribute were calculated.

To calculate the norms the T-student's test was used.

Table 2. Table with norm in control group method

\multicolumn{8}{c}{Attribute}							
\multicolumn{2}{c}{1}	2	3	\multicolumn{2}{c}{4}	5	6
norm	pathology	norm	norm	norm	pathology	norm	norm
1.96	2.5	0.43	33.1	2.47	3.0	0.3	24.1

Using these norms values of real attributes were digitized.

Equal Frequency per Interval Method. In this method we determined a number of intervals for 3, in order to enable a comparison with the method based on the control group.

The table with the limit values for six continuous attributes is following.

Table 3. The limit values in EF method

Attribute	Interval		
1	(0, 1.07)	[1.07, 1.9)	[1.9, $\max_1$]
2	(0, 1.2)	[1.2, 2.8)	[2.8, $\max_2$]
3	(0, 5.625)	[5.625, 11.935)	[11.935, $\max_3$]
4	(0, 0.71)	[0.71, 1.425)	[1.425, $\max_4$]
5	(0, 0.065)	[0.065, 0.17)	[0.17, $\max_5$]
6	(0, 3.045)	[3.045, 12.06)	[12.06, $\max_6$]

where $max_i = \max\{a_i(x) : x \in U\}$

Minimal Class Entropy Method. As in the previous case a number of intervals was determined for 3.

The table with the limit values for six continuous attributes is following .

Table 4. The limit values in MCE method

Attribute	Interval		
1	(0, 1.335)	[1.335, 1.35)	[1.35, $\max_1$]
2	(0, 1.95)	[1.95, 2.05)	[2.05, $\max_2$]
3	(0, 8.315)	[8.315, 8.43)	[8.43, $\max_3$]
4	(0, 1.16)	[1.16, 1.19)	[1.19, $\max_4$]
5	(0, 0.085)	[0.085, 0.095)	[0.095, $\max_5$]
6	(0, 7.265)	[7.265, 7.32)	[7.32, $\max_6$]

where $max_i = \max\{a_i(x) : x \in U\}$

Cluster Analysis Method The table with the limit values for six continuous attributes is following:

Table 5. The limit values in CA method

Attribute	Begin of interval						
1	0.36	2.02	3.78				
2	0.02	0.15	0.17	0.21	0.34	0.43	2.30
3	0.88	12.18	69.33				
4	0.0	2.54	8.6				
5	0.0	0.29	0.82				
6	0.0	11.43	20.75	45.26			

4.2   Checking quality of rules

Rule sets, induced from training data by a LERS system, were used to classify new examples.

Training set consists 114 examples.

Unseen set consists 50 examples - new patients.

For these sets we are going to use naive approach to classification new examples, where an attempt is made to classify an example using all possible rules. In case a bad result we are going to use classification scheme from LERS (complete and partial matching).

The table compares results obtained by the methods described above.

Table 6. Results of four discretization method

	Naive classification scheme		
Discretization method	Correctly classified examples	Unclassified examples	Error Rate
Control group	32	14	4/50 = 0.08
EF	20	16	14/50 = 0.28
MCE	13	22	15/50 = 0.30
CA	29	14	7/50 = 0.14

5   Conclusion

This work compares results — quality of classification rules — which were obtained using different discretization methods of the continuous attributes obtained in the second stage of diagnosing MEM.

Results obtained in this work suggest explicitly that the method based on evaluation of norms on basis of a control group leads to the best results. Method based on control group has the smallest error rate.

Error rates in the Equal Frequency per Interval Method and Minimal Class Entropy Method are much bigger then in control group method.

Cluster Analysis method gives compare value of error rate and the same quantity of unclassified examples.

However the division of continuous attributes into three or more intervals caused objection and incomprehension of physicians - experts.

The relatively big number of unclassified new examples in each method is the result of incomplete data – missing attribute values.

Those results are preliminary ones. Value of discretization methods will be eventually confirmed when a number of new cases will be comparable with a training set.

References

1. Barkovich A. J.: "Toxic and metabolic brain disorders." *In: Pediatric Neuroimaging*. Raven Press Ltd, New York, 1995.
2. Chan, C–C., Batur, C., Srinivassan, A.: "Determination of quantization intervals in rule based model for dynamic systems", *Proc. of the IEEE Conference on Systems, Man and Cybernetics*, VA Oct.13–16, 1991, 1719–1723.
3. Chmielewski, M. R., Grzymala–Busse, J. W.: "Global Discretization of Continuous Attributes as Preprocessing for Inductive Learning", *Technical Report TR-92-7*, Department of Computing Science, University of Cansas, Lawrence, USA, 1992.
4. Fayyad, U. M., Irani, K. B.: "On the handling of continuous–valued attributes in decision tree generation", *Machine Learning*, 8(1992), 87–102.
5. Grzymala–Busse J.: "Managing Uncertainty in Expert Systems." Kluwer Academic Publishers, 1992.
6. Grzymala–Busse J.: "LERS - a system for learning from examples based on Rough Sets. *In R. Slowinski R. (ed.), In intelligent decision support. Handbook of Applications and Advances of the Rough Sets Theory*, 3, Kluwer Academic Publishers, 1992.
7. Marszal–Paszek B., Paszek P., Wakulicz–Deja A.: "Applying Rough Sets to diagnose in Children's Neurology." *Sixt International Conference Information Processing and Management of Uncertainty in Knowledge-Base System*, Granada, Spain, 1996, Vol. 3, 1463–1468.
8. Matthews P. M., Anderman F., Silver K., Karpati G., Arnold D. L.: "Proton MR spectroscopic characterization of differences in regional brain metabolic abnormalities in mitochondrial encefalomyopathies." *Neurology* 43 (1993), 2484-2490.
9. Paszek P., Wakulicz–Deja A.:"Optimalization Diagnose In Progressive Encephalopathy Applying The Rough Set Theory." *Four European Congress on Inteligent Techniques and Soft Computing*, Aachen, Germany, 1996, Vol. 1, 192–196.
10. Pawlak Z.: "Rough Sets: Theoretical aspects of reasoning about data", Boston: Kluwer Academic Publishers, 1991.
11. Skowron A., Rauszer G.:"The discernibility matrices and functions in information systems," *In R. Slowinski R. (ed.), In intelligent decision support. Handbook of Applications and Advances of the Rough Sets Theory*, 331-336, Kluwer Academic Publishers, 1992.
12. Tulinius M. H., Holme E., Kristianson B., Larsson N., Oldfors A.: "Mitochondrial encephalomyopathies in childhood", *I. Biochemical and morphologic investigations. J. Pediatrics*, 119, 242-50, 1991.
13. Wong, A. K. C., Chiu, D. K. Y.: "Synthesizing statistical knowledge from incomplete mixed–mode data", *IEEE Transaction on Pattern Analysis and Machine Intelligence*, 9(1987), 796–805.

Approximate Time Rough Control: Concepts and Application to Satellite Attitude Control*

J.F. Peters[1], K. Ziaei[2], S. Ramanna[3]

[1] Department of Electrical and Computer Engineering,
jfpeters@ee.umanitoba.ca
[2] Department of Mechanical Engineering,
University of Manitoba,
Winnipeg, Manitoba R3T 2N2 Canada,
kamyar@ee.umanitoba.ca
[3] Department of Business Computing,
University of Winnipeg,
Winnipeg, R3B 2E9, Canada
ramanna@ee.umanitoba.ca

Abstract. This paper presents an approach to the design of approximate time rough controllers. In this paper, the clocks used to measure durations required to achieve controller objectives are modeled as approximate time windows. An approximate time window (atw) partitions time relative to granules (clumps of similar timing measurements) such as early, ontime, late. An atw determines the degree of membership of each observed duration in each of its temporal partitions. Based on observations of the degree of overshoot, rise time, and settling time during the operation of a control system, the architecture of an approximate time rough control system is established. The rough controller is guided by rules derived from a real-time decision-making system. Rough sets theory is used to derive controller rules. Fuzzy sets theory is used to model decision system sensors as fuzzy implications. A roughly fuzzy Petri net model of an approximate controller is given. The approach taken in approximate time rough control is illustrated relative to controlling the pitch angle in an attitude control system for a small satellite.

1 Introduction

Considerable work in control technology has been reported on a rapidly growing range of problems being solved with a decisions systems approach [1]. This occurs especially in those niches of control engineering where the classic approaches have been faced with some difficulties or have appeared to perform inherently weakly. The objectives of this study are twofold. First, we introduce an approach to the design of approximate real-time controllers which utilize a combination of rough sets and fuzzy sets. Second, we illustrate the approximate real-time rough control design methodology relative to attitude control system for small satellites. Experiments with this form of rough control have been carried out in the

* Research supported by NSERC Industrial-Research Partnership Grant , Bristol Aerospace Ltd. Grant and Canadian Space Agency Grant

L. Polkowski and A. Skowron (Eds.): RSCTC'98, LNAI 1424, pp. 491–498, 1998.
© Springer-Verlag Berlin Heidelberg 1998

design of an attitude control system for a small, scientific satellite. This research has been carried out over the past two years by a research group consisting of two space systems engineers from a local aerospace industry as well as two professors and several graduate students in the faculty of engineering at the University of Manitoba working in cooperation with the Canadian Space Agency.

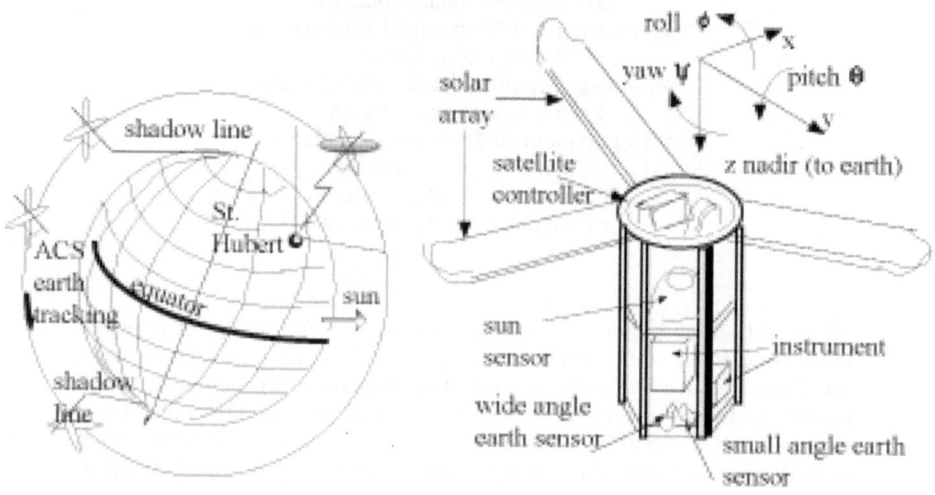

Fig. 1 Orbit of Satellite   Fig. 2 Structure of Satellite

For simplicity, the discussion is limited to a linearized pitch control strategy for nadir-pointing, momentum bias satellites (see Figs. 1 and 2) using sensors, momentum wheel, and torquers to achieve attitude control [2]. The focus of this paper is a description of how rough control rules derived from a real-time decision system table have been used in fine-pointing for attitude control of a small satellite. The contribution of this paper is the application of rough sets, fuzzy sets and approximate time windows in the design of approximate time rough control systems.

2 Knowledge Discovery Process

The problem in this research is instrumenting sensors to obtain knowledge of the interdependencies of controller overshoot ov, rise time rt and settling time st to achieve fine-pointing in selected classes of controllers. More deeply, the selection of the "mechanism" which defines the operation of a sensor is a nontrivial task, and depends on engineering skills, intuition, and an understanding of the behavior of a controller. It is our knowledge of the "usual" distribution of controller rise time and settling time which underlies design choices concerning sensors.

2.1   Approach

In the context of controllers, the design of an approximate-time rough control system is accomplished by (a) selecting a universe consisting of objects which are observations of overshoot as well as durations s_t (settling time) and r_t (rise time) and (b) instrumenting what is known as an approximate time window with attributes identified with sensors used to evaluate durations as well as sensors to evaluate overshoot. Overshoot (ov) is the biggest deviation of step response from a particular steady state after the step response has reached a tolerance band for the first time. Rise time rt is the time when a step response reaches 90% of its steady-state value for the first time, and settling time st is measured relative to rise time (i.e., the clock for s_t is reset at $t = r_t$). Time itself is measured relative to a clock which measures durations in the context of fuzzy sets named early, ontime, and late.

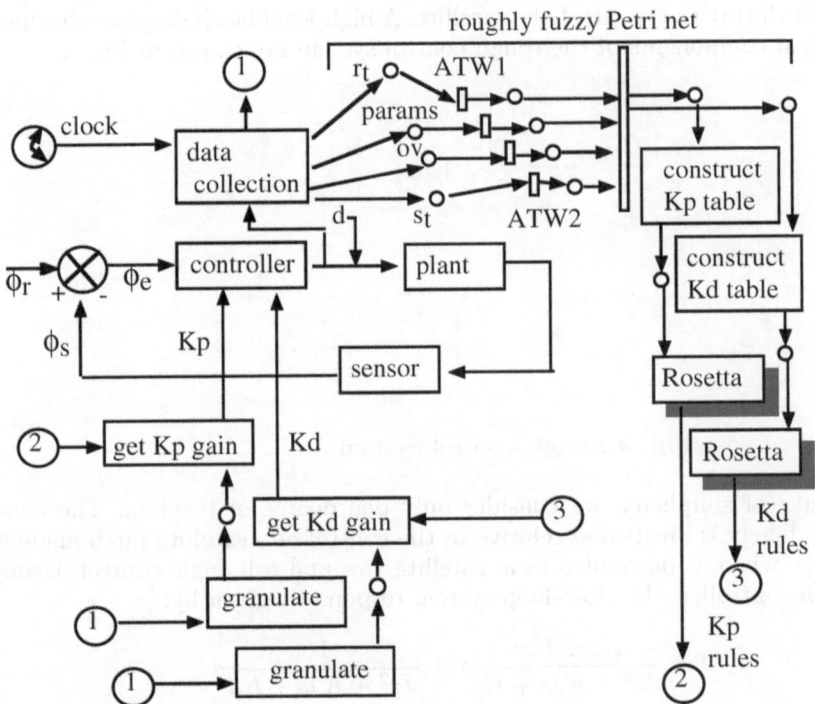

Fig. 3 Architecture of Approximate Time Rough Control System

In other words, based on observed rise time and settling time, approximate time windows (designated ATW1 and ATW2 in Fig. 3) are designed relative to fuzzy partitions of temporal intervals. Approximate time window measurements are defined relative to durations between firings of transitions in roughly fuzzy Petri nets, which were introduced in [3,6]. Approximate time windows are an extension of the fuzzy clock model introduced in [4]. Overshoot is conceptualized in terms of fuzzy sets relative to linguistic labels acceptable, big, and very big.

2.2 Background Knowledge

The rough sets approach to decision systems, especially in the context of real-time decision-making and the representation of decisions with Petri nets, has been investigated in [5,6]. In deriving decision system rules, the discernibility matrix and discernibility function are essential. Precise conditions for decision rules can be extracted from a discernibility matrix. The application of rough sets theory in control systems has been investigated by [1].

3   Controller Example

As a real-time application of the approximate time rough control methodology, we consider the attitude control of a satellite. A high-level block diagram showing the principal components of the rough control system are shown in Fig. 4.

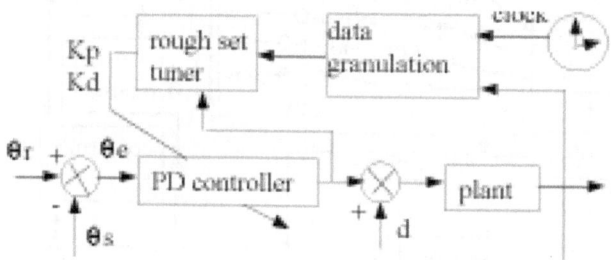

Fig. 4 Rough Control System

For the sake of simplicity, we consider only one degree of freedom. The control methodology is illustrated relative to the control of a satellite pitch angle θ (see Fig. 2), which is decoupled from satellite yaw and roll angle control. Using a PD pitch controller, the close-loop system response is given by:

$$\theta = \frac{K_p + K_d s}{Js^2 + K_d s + K_p} r + \frac{1}{Js^2 + K_d s + K_p} d$$

where θ is the pitch angular position; r, setpoint; d, disturbance; J, moment of inertia of the plant; and K_d and K_p are the controller differential and proportional gain parameters. To build an information system, we define the following features.

- Overshoot ov: The biggest deviation of step response from steady state q_{ss} after the step response reached the tolerance band $q_{ss} \pm \in$ for the first time. Overshoot is divided by the height of the step demand to obtain a relative quantity.

- Rise time r_t: The time when the step response reaches 90% of its steady-state value for the first time.
- Settling time s_t: For $t > s_t$ representing deviations from a steady state relative to a tolerance band $\pm \in$.

Here we measure the settling time relative to the rise time, i.e. the clock for s_t is reset at $t = r_t$. Our set of objects are the step responses of the system for different controller gains. For each object (observed step response of the system), we decide on correction factors for both proportional and derivative gain, which change the controller parameters in order to improve the performance. It should be also be noted that changes in controller parameter values (K_d and K_p) in a decision table are inserted by an experienced control engineer using a form of pattern recognition. Each decision value inducing changes in K_d and K_p is a judgment about controller performance from a measured step response: the observed response is compared to the picture of an ideal response. A decision table is constructed with nine condition attributes: a_1, a_2, a_3 for granulations of overshoot measurement, a_4, a_5, a_6 for rise time granulations, and a_6, a_7, a_8 for settling time granulations. Sample rows from a rough controller tuning information table are given in Table 1.

$\{o_v, \mathsf{r}, \mathsf{s}\} \in U \setminus A$		a_1	a_2	a_3	a_4	a_5	a_6	a_7	a_8	a_9	d_1	d_2
x_1	{3.5,1.80,2.60}	0.9	0.1	0.0	0.60	1.00	0.93	0.62	0.86	0.90	2.00	1.80
x_2	{0.50,3.10,1.40}	1.0	0.0	0.0	0.60	1.00	1.00	0.81	1.00	0.90	2.00	1.50
x_3	{17.50,1.50,4.80}	0.0	0.0	0.8	0.60	0.96	0.92	0.60	0.80	0.91	1.80	2.00
x_4	{0,2.10,1.10}	1.0	0.0	0.0	0.60	1.00	0.96	0.91	1.00	0.90	1.50	1.20
x_5	{4.50,1.10,1.60}	0.7	0.3	0.0	0.72	0.90	0.90	0.76	1.00	0.90	0.80	1.20

Table 1 Decision Table(s)

The distribution of degree of membership values in a granule associated with a sensor a_i, $1 \le i \le 9$, is assumed to be approximately normal in a Gaussian distribution with mean (modal point) m and standard deviation s (spread). Let g, x be the name of a granule associated with sensor a_i and measurement (e.g. overshoot at given instant in time), respectively. Hence, the membership function used in modeling sensor is given by

$$g(x) = \exp\left(\frac{-(x-m)^2}{s^2}\right)$$

In modeling sensors a_4, a_5, a_6 (rise time sensors) and a_6, a_7, a_8 (settling time sensors), we also introduce a modulator r and strength-of-connection w. Taken collectively, each trio of sensors constitutes an atw (approximate time window). A modulator imposes a threshold on stimuli, and a strength-of-connection raises or lowers the impact of an input in an atw. Then sensor a_i in Table 1 is modeled as an aggregation of a fuzzy implication value and strength-of-connection w:

$a_i(x) = (r \rightarrow g(x))sw$, where $(r \rightarrow g(x) = \min\left(1, \frac{g(x)}{r}\right)$

In this research, the operator s (s-norm operator) computes a probabilistic sum. It has been shown that the modulator and strength-of-connection parameters in approximate time windows can be calibrated [3]. Table 1 is processed as two separate decision tables in Rosetta, one to derive rules for changing proportional gain ($d_1 = v_{K_p}$) and a second table to derive rules for changing differential gain ($d_2 = v_{K_d}$). It should also be noted that each application of a rule relative to an observed step response of the control system results in changes in both Kp and Kd. The reducts $\{a_3, a_5, a_6, a_7\}$, $\{a_5, a_7, a_9\}$ were derived with Rosetta. A sampling of the controller rules is given as follows:

K_p : [a_3(0.0) AND a_5(1.00) AND a_6(1.00) AND a_7(0.81)] OR
[a_3(0.0) AND a_5(1.00) AND a_6(0.93) AND a_7(0.62)]$\Longrightarrow d_1$(2.00)
K_d : [a_2(0.1) AND a_5(1.00) AND a_7(0.81)] $\Longrightarrow d_2$(1.80)

Such rules are derived from a real-time decision system table based on a sufficient number of prototypical experimental measurements of controller performance and the granulation of these measurements.

Rule Firing Algorithm

step 1. Let x, a_i, a_j, a_k, v_{a_i}, v_{q_j}, v_{a_k} be an experimental value observed during actual operation of a control system, sample decision system condition sensors for a sample control rule r $\in$ D(S), and sensor values from decision system table (U, A $\cup$ {d}, V), respectively. Let s be defined as a sum s = (a_i(x) - v_{a_i}) + (a_j(x) - v_{q_j}) + (a_k(x) - v_{a_k}) where x is an input value (e.g observed overshoot, rise time, or settling time) evaluated with sensor a_i in A (for example) to produce a particular value v_{ai}.

step 2. Let n, m be the number of K_p, K_d rules, respectively. Let s_i, $1 \le i \le n$, s_j, $1 \le j \le m$ be sums of the form introduced in step 1 relative to n rules for K_p and m rules for K_d, respectively. Then let m_{K_p}, m_{K_d} be functions defined as follows as follows:

m_{K_p}: s_1, ..., s_i, ..., s_n $\rightarrow$ i such that s[i] = $\min(s_1, ..., s_i, ..., s_n)$
m_{K_d}: s_1, ..., s_j, ..., s_m $\rightarrow$ j such that s[j] = $\min(s_1, ..., s_j, ..., s_m)$

In other words, m_{K_p}, m_{K_d} each finds the index of the smallest sum, which identifies the premise of a rule which is closest to the measured condition during the operation of a controller.

step 3. Let K_p, K_d be the current values of proportional and differential coefficients. Then compute

$K_p := K_p * d[i]$ and $K_d := K_d * d[j]$

At this point, it should observed that variations of step 3 of the rule-firing algorithm are possible. First, it has been found that it is helpful to change K_p only if the percent of overshoot exceeds some k (e.g. k = 0.1). Second, performance of the controller can be improved by forming {derived rules} ∪ {default rules), constituting a population which evolves. Such refinements of this algorithm are outside the scope of this paper.

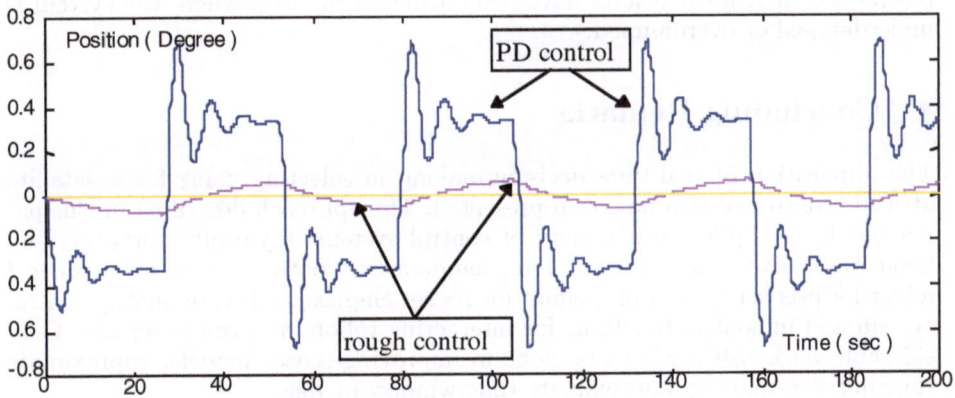

Fig. 5 Closed-loop Response to Rectangular Disturbance

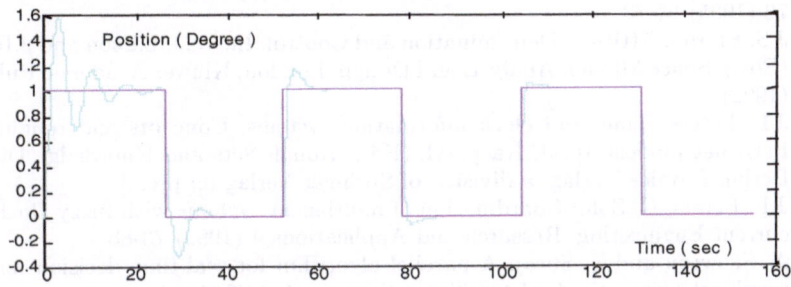

Fig. 6 Closed-loop response during self-tuning

K_p and K_d correction rules for a PD controller have been derived using a combination of a roughly fuzzy Petri net and Rosetta. Tuning information from a number of controller simulations was used to build an information system and to generate some tuning decisions. These rules can be used to tune a satellite pitch controller on-line. The new information collected after each tuning were added to the rough control system and dynamic reducts were employed to modify decision rules periodically. A comparison of the performance of rough control with a PD control of the pitch angle in the presence of a rectangular (square wave form of) disturbance is shown in Figs. 5 and 6. Fine-pointing is achieved rapidly. Each step response of the rough controller is due to the firing of a pair of rules (see Fig. 4) used to select appropriate changes in proportional and differential gains of the PD controller. This approach differs from classical PD control, since the gains are changing dynamically depending on the degree of disturbance. Other forms of disturbance have also be investigated with similar fine-pointing results. Further, good results have been achieved in cases where the system is underdamped or overdamped.

4   Concluding Remarks

The application of real-time decision-making in selecting gains for a satellite attitude control system has been presented. The approach described in this paper has been applied in a variety of control systems (hydraulic servo system, flood water diversion control system, temperature control system for a vertical mixer for preparing solid propellant for rocket engines), software quality control system and in control functions for interacting robots in a computer zoo. Considerable work still needs to be done in improving sensor models, approximate time decision-making, approximate time window models.

References

1.   E. Czogala, A. Mrozek, Z. Pawlak: The idea of a rough fuzzy controller and its application to the stabilization of a pendulum-car system. Fuzzy Sets and Systems 72 (1995) 61-73
2.   J.S. Eterno: Attitude Determination and Control. In: W.J. Larson and J.R. Wertz (Eds.), Space Mission Analysis and Design, London, Kluwer Academic Publishers, (1992)
3.   J.F. Peters :Time and clock information systems: Concepts and roughly fuzzy Petri net models. In: J. Kacprzyk (Ed.), Rough Sets and Knowledge Discovery. Berlin, Physica Verlag, a division of Springer Verlag [in press]
4.   J.F. Peters, N. Sohi: Coordination of multiagent systems with fuzzy clocks. Concurrent Engineering: Research and Applications 4 (1996) 73-88
5.   A. Skowron and Z. Suraj: A parallel algorithm for real-time decision making: a rough set approach. J. of Intelligent Systems 7, 5-28, 1996
6.   J.F. Peters, A. Skowron, Z. Suraj and S.Ramanna: Approximate Real-Time Decision Making: Concepts and Roughly Fuzzy Petri Net Models. Internation Journal of Intelligent Systems [To appear].

Some Relationships between Decision Trees and Decision Rule Systems

Mikhail Moshkov

Research Institute for Applied Mathematics and
Cybernetics of Nizhni Novgorod State University
10, Uljanova St., Nizhni Novgorod, 603005, Russia
moshkov@nnucnit.unn.ac.ru

[Abstract.] **Relationships between parameters of a decision rule system and the minimal depth of a decision tree which solves the problem of the search of all realizable rules from the system are considered. Unimprovable upper and close to unimprovable lower bounds on the minimal depth of a decision tree are obtained.**

1 Introduction

Decision trees and decision rule systems are widely used in different applications as algorithms and as a form of knowledge representation. Problems of comparative analysis of decision trees and decision rule systems are interesting both for theory and for practice [2, 6].

In this paper we consider relationships between parameters of a decision rule system and the minimal depth of a decision tree which solves the problem of the search of all realizable rules from the system. The necessity to find all realizable rules arises, for example, if we consider problems which can have simultaneously many decisions, and the number of realizable rules with the same decision characterizes the importance of this decision. The main question considered in the paper is to clarify can a decision tree restricts oneself to recognition values of only a part of attributes from a decision rule system.

As the parameters of a decision rule system we consider the number of different attributes in the system, the maximal number of values of an attribute and the maximal length of a rule. Unimprovable upper and close to unimprovable lower bounds on the minimal depth of a decision tree are obtained. The main consequences of these results are the following: in the worst case for the solving of the considered problem for a decision rule system by a decision tree we must recognize values of all attributes from the system. However, there exist systems such that for the considered problem solving by a decision tree it is sufficient to recognize values of only a small part of attributes from the system.

In proofs we use methods of test theory [1, 3, 4, 5] and rough set theory [7, 8].

---

* This work was supported by Russian Foundation of Fundamental Research (grant # 96-01-00428) and by Polish State Research Committee (grant # 8T11C01011).

L. Polkowski and A. Skowron (Eds.): RSCTC'98, LNAI 1424, pp. 499–505, 1998.

2   Main Definitions and Results

2.1   Decision Rule Systems

Let $\omega = \{0, 1, 2, \ldots\}$ and $A = \{a_i : i \in \omega\}$. Elements of the set A will be called *attributes*. *Decision rule* is an expression of the kind

$$a_{i_1} = \delta_1 \wedge \ldots \wedge a_{i_m} = \delta_m \Rightarrow \sigma,$$

where $m \in \omega$, $a_{i_1}, \ldots, a_{i_m}$ are pairwise different attributes from A and $\delta_1, \ldots, \delta_m$, $\sigma \in \omega$. Denote this decision rule by r. The expression $a_{i_1} = \delta_1 \wedge \ldots \wedge a_{i_m} = \delta_m$ will be called *the left part* of the rule r. The number m will be called *the length* of the rule r. Denote $A(r) = \{a_{i_1}, \ldots, a_{i_m}\}$ and $K(r) = \{a_{i_1} = \delta_1, \ldots, a_{i_m} = \delta_m\}$ (if $m = 0$ then $A(r) = K(r) = \emptyset$).

A *decision rule system* S is a finite nonempty set of decision rules. Let $A(S) = \bigcup_{r \in S} A(r)$, $n(S) = |A(S)|$ and $d(S)$ is the maximal length of a decision rule from S. For $a_i \in A(S)$ let $V_S(a_i) = \{\delta : \delta \in \omega, (a_i = \delta) \in \bigcup_{r \in S} K(r)\}$. Denote $k(S) = \max\{|V_S(a_i)| : a_i \in A(S)\}$. Denote Σ the set of all decision rule systems.

Let $S \in \Sigma$ and $A(S) = \{a_{j_1}, \ldots, a_{j_n}\}$ where $j_1 < \ldots < j_n$. Denote $V(S) = V_S(a_{j_1}) \times \ldots \times V_S(a_{j_n})$. For $\bar{\delta} = (\delta_1, \ldots, \delta_n) \in V(S)$ denote $K(S, \bar{\delta}) = \{a_{j_1} = \delta_1, \ldots, a_{j_n} = \delta_n\}$. We will say that a decision rule $r \in S$ is *realized* for the tuple $\bar{\delta}$ if $K(r) \subseteq K(S, \bar{\delta})$.

We define the problem *All Realizable Rules* for a system $S \in \Sigma$ as follows: for a given tuple $\bar{\delta} \in V(S)$ it is required to find all rules from S which are realized for the tuple $\bar{\delta}$. Denote this problem by $\mathrm{ARR}(S)$.

2.2   Decision Trees

A *finite oriented tree with the root* is a finite oriented tree containing exactly one node with no entering edges. This singular node is called the *root*. The nodes of the tree having no issuing edges are called *terminal* nodes. The non-terminal nodes of the tree are called *working* nodes. A *complete path* of a finite oriented tree with the root is a sequence $\xi = v_1, d_1, \ldots, v_m, d_m, v_{m+1}$ of nodes and edges of the tree such that v_1 is the root, v_{m+1} is a terminal node and for $i = 1, \ldots, m$ the edge d_i issues from the node v_i and enters the node v_{i+1}.

Let S be a decision rule system. A *decision tree over* S is a labeled finite oriented tree with the root which satisfies the following conditions:

a) every working node is labeled with an attribute from $A(S)$;

b) a working node which is labeled with an attribute a_i has $|V_S(a_i)|$ issuing edges, and these edges are labeled with pairwise different elements from $V_S(a_i)$;

c) every terminal node is labeled with a subset of the set S.

Let Γ be a decision tree over S. Denote by $CP(\Gamma)$ the set of all complete paths of Γ. Let $\xi = v_1, d_1, \ldots, v_m, d_m, v_{m+1}$ be a complete path of Γ. We will associate with ξ a set of attributes $A(\xi)$ and a system of equations $K(\xi)$. If $m = 0$ then $A(\xi) = \emptyset$ and $K(\xi) = \emptyset$. Let $m > 0$ and for $j = 1, \ldots, m$ let the node v_j be labeled with the attribute a_{i_j}, and the edge d_j be labeled with the number δ_j. Then $A(\xi) = \{a_{i_1}, \ldots, a_{i_m}\}$ and $K(\xi) = \{a_{i_1} = \delta_1, \ldots, a_{i_m} = \delta_m\}$. Denote by $\tau(\xi)$ the set of decision rules which is the label of the node v_{m+1}.

A systems of equations $\{a_{i_1} = \delta_1, \ldots, a_{i_m} = \delta_m\}$ where $a_{i_1}, \ldots, a_{i_m} \in A$ and $\delta_1, \ldots, \delta_m \in \omega$ will be called *inconsistent* if there exist $l, k \in \{1, \ldots, m\}$ such that $l \neq k$, $i_l = i_k$ and $\delta_l \neq \delta_k$. If a system of equations is not inconsistent then it will be called *consistent*.

Let S be a decision rule system and Γ be a decision tree over S. We will say that Γ solves the problem $\mathrm{ARR}(S)$ if for each path $\xi \in CP(\Gamma)$ with consistent system of equations $K(\xi)$ the following conditions hold:

a) $K(r) \subseteq K(\xi)$ for each rule $r \in \tau(\xi)$;
b) for any rule $r \in S \setminus \tau(\xi)$ the system of equations $K(r) \cup K(\xi)$ is inconsistent.

For an arbitrary complete path $\xi \in CP(\Gamma)$ denote by $h(\xi)$ the number of working nodes in the path ξ. The value

$$h(\Gamma) = \max\{h(\xi) : \xi \in CP(\Gamma)\}$$

will be called the *depth* of the decision tree Γ.

Denote by $h(S)$ the minimal depth of a decision tree over S which solves the problem $\mathrm{ARR}(S)$.

2.3   Main Results

One can show that $\{(n(S), d(S), k(S)) : S \in \Sigma\} = \{(0, 0, 0)\} \cup \{(n, d, k) : n, d, k \in \omega \setminus \{0\}, d \leq n\}$.

Let $n, d, k \in \omega \setminus \{0\}$ and $d \leq n$. Denote

$$h(n, d, k) = \min\{h(S) : S \in \Sigma, n(S) = n, d(S) = d, k(S) = k\},$$
$$H(n, d, k) = \max\{h(S) : S \in \Sigma, n(S) = n, d(S) = d, k(S) = k\}.$$

Considered values are the unimprovable lower (the value $h(n, d, k)$) and the unimprovable upper (the value $H(n, d, k)$) bounds on minimal decision tree depth for systems $S \in \Sigma$ such that $n(S) = n$, $d(S) = d$ and $k(S) = k$.

[**Theorem 1.**] *Let $n, d, k \in \omega \setminus \{0\}$ and $d \leq n$. Then $H(n, d, k) = n$.*

This theorem shows that in the worst case for a problem $\mathrm{ARR}(S)$ solving by a decision tree we must recognize values of all attributes from the set $A(S)$.

[**Theorem 2.**] *Let $n, d, k \in \omega \setminus \{0\}$ and $d \leq n$. If $k = 1$ or $d = 1$ then $h(n, d, k) = n$. If $k \geq 2$ and $d \geq 2$ then*

$$\max\left\{d, \frac{n(k-1)}{k^d}\right\} \leq h(n, d, k) \leq d + \frac{n}{k^{d-1}}.$$

This theorem shows that there exist systems S such that for the problem $\mathrm{ARR}(S)$ solving by a decision tree we must recognize values of attributes from a small part of the set $A(S)$ only.

3   Auxiliary Statements

Let $S \in \Sigma$ and $\alpha = \{a_{i_1} = \delta_1, \ldots, a_{i_m} = \delta_m\}$ be a consistent system of equations such that $a_{i_1}, \ldots, a_{i_m} \in A(S)$ and $\delta_1 \in V_S(a_{i_1}), \ldots, \delta_m \in V_S(a_{i_m})$. Define a decision rule system S_α. Let r be a decision rule from S such that the system of equations $K(r) \cup \alpha$ is consistent. Denote by r_α the decision rule obtained by removing from the left part of r all equations contained in α. Then S_α is the set of all rules r_α such that $r \in S$ and the system $K(r) \cup \alpha$ is consistent.

It is not difficult to prove the following two statements.

[**Lemma 1.**] *Let $S \in \Sigma$ and $\alpha = \{a_{i_1} = \delta_1, \ldots, a_{i_m} = \delta_m\}$ be a consistent system of equations such that $a_{i_1}, \ldots, a_{i_m} \in A(S)$ and $\delta_1 \in V_S(a_{i_1}), \ldots, \delta_m \in V_S(a_{i_m})$. Then $h(S) \geq h(S_\alpha)$.*

[**Lemma 2.**] *Let $S \in \Sigma$ and S' be a subsystem of S. Then $h(S) \geq h(S')$.*

Now we consider some lower bounds on the value $h(S)$.

[**Lemma 3.**] *Let S be a decision rule system. Then $h(S) \geq d(S)$.*

[*Proof.*]Let r be a rule from S such that the length of r is equal to $d(S)$. Let Γ be a decision tree over S which solves the problem $\mathrm{ARR}(S)$ and for which $h(\Gamma) = h(S)$. It is clear that there exists a complete path ξ of Γ such that the system $K(r) \cup K(\xi)$ is consistent. Since Γ solves the problem $\mathrm{ARR}(S)$ we obtain that $K(r) \subseteq K(\xi)$. Therefore $h(\xi) \geq d(S)$, $h(\Gamma) \geq d(S)$ and $h(S) \geq d(S)$.

[**Lemma 4.**] *Let S be a decision rule system such that $d(S) = 1$. Then $h(S) \geq n(S)$.*

[*Proof.*]Let Γ be a decision tree over S which solves the problem $\mathrm{ARR}(S)$ and for which $h(\Gamma) = h(S)$. Evidently there exists a complete path ξ of Γ such that the system $K(\xi)$ is consistent. It is clear that for each decision rule $r \in S$ either $K(r) \subseteq K(\xi)$ or the system $K(r) \cup K(\xi)$ is inconsistent. Therefore $A(\xi) \cap A(r) \neq \emptyset$ for any rule $r \in S$, $A(r) \neq \emptyset$. Taking into account that $|A(r)| \leq 1$ for any rule $r \in S$ and $|\bigcup_{r \in S} A(r)| = n(S)$ we obtain that $|A(\xi)| = n(S)$. Therefore $h(\xi) \geq n(S)$, $h(\Gamma) \geq n(S)$ and $h(S) \geq n(S)$.

[**Lemma 5.**] *Let S be a decision rule system such that $k(S) = 1$. Then $h(S) \geq n(S)$.*

[*Proof.*]Let Γ be a decision tree over S which solves the problem $\mathrm{ARR}(S)$ and for which $h(\Gamma) = h(S)$. Let $\bar{\delta} = (\delta_1, \ldots, \delta_{n(S)}) \in V(S)$ and ξ be a complete path of Γ such that $K(\xi) \subseteq K(S, \bar{\delta})$. Since Γ solves the problem $\mathrm{ARR}(S)$ one can show that the terminal node of this path is labeled with the set S and $K(r) \subseteq K(\xi)$ for any $r \in S$. Therefore $K(\xi) = K(S, \bar{\delta})$ and $h(\xi) \geq n(S)$. Consequently $h(\Gamma) \geq n(S)$ and $h(S) \geq n(S)$.

4   Proofs of Theorems

Proof of Theorem 1. Consider the decision rule system $S = S_1 \cup S_2$ where $S_1 = \{a_1 = 0 \wedge \ldots \wedge a_d = 0 \Rightarrow 0, a_{d+1} = 0 \Rightarrow d + 1, \ldots, a_n = 0 \Rightarrow n\}$ and $\{a_1 = 1 \Rightarrow n + 1, \ldots, a_1 = k - 1 \Rightarrow n + k - 1\}$. If $k = 1$ then $S_2 = \emptyset$. It is clear that $n(S) = n$, $d(S) = d$ and $k(S) = k$.

Let Γ be a decision tree over S which solves the problem $\mathrm{ARR}(S)$ and for which $h(\Gamma) = h(S)$. Let $\bar{\delta} = (0, \ldots, 0) \in V(S)$ and ξ be a complete path of Γ such that $K(\xi) \subseteq K(S, \bar{\delta})$. Since Γ solves the problem $\mathrm{ARR}(S)$ one can show that the terminal node of this path is labeled with the set S_1, and $K(r) \subseteq K(\xi)$ for any $r \in S_1$. Therefore $K(\xi) = K(S, \bar{\delta})$ and $h(\xi) \geq n$. Consequently $h(\Gamma) \geq n$ and $h(S) \geq n$. Hence $H(n, d, k) \geq n$. It is clear that $H(n, d, k) \leq n$.

Proof of Theorem 2. Let $k = 1$ or $d = 1$. Using Lemmas 4 and 5 we obtain $h(n, d, k) \geq n$. It is clear that $h(n, d, k) \leq n$. Therefore $h(n, d, k) = n$.

Let $k \geq 2$ and $d \geq 2$.

At first we consider the lower bound on $h(n, d, k)$. Using Lemma 3 we obtain that $h(n, d, k) \geq d$. We prove by induction on d that for any $n, d, k \in \omega \setminus \{0\}$ such that $d \leq n$ the following inequality holds:

$$h(n, d, k) \geq \frac{n(k - 1)}{k^d}. \tag{1}$$

Since $h(n, 1, k) = n$ we have that (1) holds if $d = 1$. Suppose that for some $d \geq 1$ the inequality $h(n, t, k) \geq \frac{n(k-1)}{k^t}$ holds for any integer t, $1 \leq t \leq d$. We prove that (1) holds for $d + 1$ as well. Let S be a decision rule system such that $n(S) = n$, $d(S) = d + 1$ and $k(S) = k$. Let S' be a subsystem of the system S such that $A(S') = A(S)$ and $A(S'') \neq A(S)$ for any system $S'' \subset S'$. It is clear that $n(S') = n$, $d(S') \leq d + 1$ and $k(S') \leq k$. Using Lemma 2 we obtain

$$h(S) \geq h(S'). \tag{2}$$

Let $S' = \{r_1, \ldots, r_p\}$. It is clear that for $j = 1, \ldots, p$ the system of equations $K(r_j)$ contains some equation $a_{i_j} = \sigma_j$ such that $a_{i_j} \notin A(S') \setminus A(r_j)$. Therefore

$$|S'| \leq n. \tag{3}$$

Suppose $|S'| \leq \frac{(k-1)n}{k}$. Denote $\alpha = \{a_{i_1} = \sigma_1, \ldots, a_{i_p} = \sigma_p\}$. It is clear that α is a consistent system. Consider the system S'_α (see definition before Lemma 1). Denote $n_0 = n(S'_\alpha)$, $d_0 = d(S'_\alpha)$ and $k_0 = k(S'_\alpha)$. One can show that $n_0 \geq \frac{n}{k}$, $1 \leq d_0 \leq d$ and $1 \leq k_0 \leq k$. Let $d_0 = 1$ or $k_0 = 1$. Using Lemmas 4 and 5 we obtain that $h(S'_\alpha) \geq n_0 \geq \frac{n}{k} \geq \frac{n}{k} \frac{(k-1)}{k^d} = \frac{n(k-1)}{k^{d+1}}$. Let $d_0 \geq 2$ and $k_0 \geq 2$. Using the inductive hypothesis we obtain that $h(S'_\alpha) \geq \frac{n_0(k_0-1)}{k_0^{d_0}} \geq \frac{n}{k} \frac{(k_0-1)}{k_0^d}$.

One can show that $\frac{k_0-1}{k_0^d} \geq \frac{k-1}{k^d}$. Therefore $h(S'_\alpha) \geq \frac{n(k-1)}{k^{d+1}}$. Using Lemma 1 we obtain that $h(S') \geq h(S'_\alpha)$. From these relations and from (2) follows that $h(S) \geq \frac{n(k-1)}{k^{d+1}}$.

Suppose now $|S'| > \frac{(k-1)n}{k}$. Denote $m = n - |S'|$. From (3) follows that $m \geq 0$.

Let $m = 0$. One can show that in this case $d(S') = 1$. Using Lemma 4 we obtain $h(S') \geq n(S') = |S'| > \frac{(k-1)n}{k} \geq \frac{(k-1)n}{k^{d+1}}$. Using (2) we obtain $h(S) \geq \frac{(k-1)n}{k^{d+1}}$.

Now let $m > 0$. Denote $B = A(S) \setminus \{a_{i_1}, \ldots, a_{i_p}\}$. It is clear that $|B| = m$. Let $B = \{a_{l_1}, \ldots, a_{l_m}\}$ and $l_1 < \ldots < l_m$. Let $j \in \{1, \ldots, m\}$. Define a set V_j. If $|V_{S'}(a_{l_j})| = k$ then $V_j = V_{S'}(a_{l_j})$. If $|V_{S'}(a_{l_j})| < k$ then V_j is a subset of the set ω possessing the following properties: $|V_j| = k$ and $V_{S'}(a_{l_j}) \subset V_j$. Denote $V = V_1 \times \ldots \times V_m$. Let $q = d(S') - 1$. It is clear that $q \leq m$ and $q \leq d-1$. One can show that for any decision rule $r \in S'$ there exist at least k^{m-q} tuples $\bar{\delta} = (\delta_1, \ldots, \delta_m) \in V$ such that the system of equations $K(r) \cup \{a_{l_1} = \delta_1, \ldots, a_{l_m} = \delta_m\}$ is consistent. For each $\bar{\delta} = (\delta_1, \ldots, \delta_m) \in V$ let $N(\bar{\delta})$ be the number of decision rules $r \in S'$ such that the system $K(r) \cup \{a_{l_1} = \delta_1, \ldots, a_{l_m} = \delta_m\}$ is consistent. Denote $N = \sum_{\bar{\delta} \in V} N(\bar{\delta})$. It is clear that $N \geq |S'| \cdot k^{m-q} \geq \frac{k^m (k-1)n}{k^{q+1}}$. It is clear also that there exists a tuple $\bar{\delta}' \in V$ such that $N(\bar{\delta}') \geq \frac{N}{|V|} = \frac{N}{k^m} \geq \frac{(k-1)n}{k^{q+1}} \geq \frac{(k-1)n}{k^{d+1}}$. Let $\bar{\delta}' = (\delta_1', \ldots, \delta_m')$. Define a tuple $\bar{\delta} = (\delta_1, \ldots, \delta_m)$. Let $j \in \{1, \ldots, m\}$. If $\delta_j' \in V_{S'}(a_{l_j})$ then $\delta_j = \delta_j'$. If $\delta_j' \notin V_{S'}(a_{l_j})$ then δ_j is the minimal number from the set $V_{S'}(a_{l_j})$. One can show that $N(\bar{\delta}) \geq N(\bar{\delta}') \geq \frac{(k-1)n}{k^{d+1}}$. Denote $\alpha = \{a_{l_1} = \delta_1, \ldots, a_{l_m} = \delta_m\}$ and consider the system S_α'. One can show that $k(S_\alpha') = 1$, $d(S_\alpha') = 1$ and $n(S_\alpha') \geq \frac{(k-1)n}{k^{d+1}}$. Using Lemma 4 we obtain that $h(S_\alpha') \geq \frac{(k-1)n}{k^{d+1}}$. Using Lemma 1 we obtain that $h(S') \geq \frac{(k-1)n}{k^{d+1}}$. From this inequality and from (2) follows that $h(S) \geq \frac{(k-1)n}{k^{d+1}}$. Thus (1) holds for $d + 1$ and (1) is proved.

Now we consider the upper bound on $h(n, d, k)$. Define a decision rule system S such that $A(S) = \{a_1, \ldots, a_n\}$. Denote $E_k = \{0, 1, \ldots, k - 1\}$. Consider a partition $\{a_d, \ldots, a_n\} = \bigcup_{\bar{\delta} \in E_k^{d-1}} B(\bar{\delta})$ such that $|B(\bar{\delta})| \leq \lceil \frac{n-d+1}{k^{d-1}} \rceil$ for any $\bar{\delta} \in E_k^{d-1}$ (we suppose that $B(\bar{\delta}_1) \cap B(\bar{\delta}_2) = \emptyset$ for any $\bar{\delta}_1, \bar{\delta}_2 \in E_k^{d-1}$, $\bar{\delta}_1 \neq \bar{\delta}_2$). Some sets in this partition are possibly empty but at least one of these sets is nonempty. Let $\bar{\delta} = (\delta_1, \ldots, \delta_{d-1}) \in E_k^{d-1}$. Define a decision rule system $S(\bar{\delta})$. If $B(\bar{\delta}) = \emptyset$ then $S(\bar{\delta}) = \{a_1 = \delta_1 \wedge \ldots \wedge a_{d-1} = \delta_{d-1} \Rightarrow 0\}$. If $B(\bar{\delta}) \neq \emptyset$ then $S(\bar{\delta}) = \{a_1 = \delta_1 \wedge \ldots \wedge a_{d-1} = \delta_{d-1} \wedge a_i = 0 \Rightarrow 0 : a_i \in B(\bar{\delta})\}$. Denote $S = \bigcup_{\bar{\delta} \in E_k^{d-1}} S(\bar{\delta})$. It is clear that $n(S) = n$, $k(S) = k$ and $d(S) = d$.

Describe a decision tree Γ which solves the problem $\mathrm{ARR}(S)$. At first we compute values of the attributes $a_1, \ldots, a_{d-1}$. Let $a_1 = \delta_1, \ldots, a_{d-1} = \delta_{d-1}$. Denote $\bar{\delta} = (\delta_1, \ldots, \delta_{d-1})$. If $B(\bar{\delta}) = \emptyset$ then $S(\bar{\delta})$ is the solution of the problem $\mathrm{ARR}(S)$. Let $B(\bar{\delta}) \neq \emptyset$ and $B(\bar{\delta}) = \{a_{i_1}, \ldots, a_{i_m}\}$. Now we compute values of attributes $a_{i_1}, \ldots, a_{i_m}$. Let $a_{i_1} = \sigma_1, \ldots, a_{i_m} = \sigma_m$. Then the set $\{a_1 = \delta_1 \wedge \ldots \wedge a_{d-1} = \delta_{d-1} \wedge a_{i_j} = 0 \Rightarrow 0 : j \in \{1, \ldots, m\}, \sigma_j = 0\}$ is the solution of the problem $\mathrm{ARR}(S)$. It is not difficult to show that the decision tree Γ solves the problem $\mathrm{ARR}(S)$ and $h(\Gamma) \leq d - 1 + \lceil \frac{n-d+1}{k^{d-1}} \rceil \leq d + \frac{n}{k^{d-1}}$. Therefore $h(n, d, k) \leq d + \frac{n}{k^{d-1}}$.

5 Conclusion

In the paper we obtained unimprovable upper and close to unimprovable lower bounds on the minimal depth of a decision tree which solves the problem of the search of all realizable rules from a decision rule system. These bounds allow to compare the efficiency of decision trees and decision rule systems. The lower bound is nontrivial and has some independent theoretical interest.

Acknowledgment

We would like to thank the anonymous referees for helpful comments.

References

1. Chegis, I.A., Yablonskii, S.V.: Logical methods of electric circuit control. Trudy MIAN SSSR **51** (1958) 270–360 (in Russian)
2. Michalski, R.S., Imam, I.F.: Learning problem-oriented decision structures from decision rules: the AQDT-2 system. Proceedings of 8th International Symposium Methodologies for Intelligent Systems. Lecture Notes in Artificial Intelligence **869**. Springer Verlag, Heidelberg (1994) 416–426
3. Moshkov, M.Ju.: On conditional tests. DAN SSSR **265**(3) (1982) 550–552 (in Russian)
4. Moshkov, M.Ju.: Conditional tests. Problemy Kybernetiki **40**. Nauka Publishers, Moscow (1983) 131–170 (in Russian)
5. Moshkov, M.Ju.: Decision Trees. Theory and Applications. Nizhni Novgorod University Publishers, Nizhni Novgorod (1994) (in Russian)
6. Moshkov, M.Ju.: Decision trees and decision rule systems. Proceedings of International Workshop on Rough Sets and Soft Computing, San Jose (1994) 578–585
7. Pawlak, Z.: Rough Sets - Theoretical Aspects of Reasoning about Data. Kluwer Academic Publishers, Dordrecht (1991)
8. Skowron, A., Rauszer, C.: The discernibility matrices and functions in information systems. Intelligent Decision Support. Handbook of Applications and Advances of the Rough Set Theory. Kluwer Academic Publishers, Dordrecht (1992) 331–362

On Decision Trees with Minimal Average Depth

I. Chikalov

Faculty of Calculating Mathematics and
Cybernetics of Nizhni Novgorod State University
23, Gagarina Av., Nizhni Novgorod, 603600, Russia

[**Abstract.**] Decision trees are studied in rough set theory [6],[7] and test theory [1], [2], [3] and are used in different areas of applications. The complexity of optimal decision tree (a decision tree with minimal average depth) construction is very high. In the paper some conditions reducing the search are formulated. If these conditions are satisfied, an optimal decision tree for the problem is a result of simple transformation of optimal decision trees for some problems, obtained by decomposition of the initial problem. The decompostion properties are used to show that bounds given in [4] are unimprovable bounds on minimal average depth of decision tree.

1 Basic Notions

Let A be a nonempty set, F be some set of functions from A to $\{0,1\}$ and for an arbitrary function $f \in F$ let the relation $f \not\equiv const$ hold. Functions from F will be called *attributes* and the pair $U = (A, F)$ will be called *an information system*.

Problem over the information system $U = (A, F)$ is any $(n + 1)$-tuple $z = (\nu, f_1, \ldots, f_n)$ where $f_1, \ldots, f_n \in F$, $\nu : \{0,1\}^n \to \omega$ and $\omega = \{0, 1, \ldots\}$. The problem z may be interpreted as a problem of searching for the value $z(a) = \nu(f_1(a), \ldots, f_n(a))$ for an arbitrary element $a \in A$. Different problems of pattern recognition [3], fault diagnosis [1] and discrete optimization [3] can be represented in such form.

Two elements a and b from A will be called *equivalent for the problem z* if $f_i(a) = f_i(b)$ for $i = 1, \ldots, n$. This equivalence relation divides the set A onto nonempty equivalence classes $A_1, \ldots, A_s$. Denote by Δ_z the set $\{\bar{d}_1, \ldots, \bar{d}_s\} \subseteq \{0,1\}^n$, where $\bar{d}_i = (f_1(a_i), \ldots, f_n(a_i))$ and $a_i \in A_i$, $i = 1, \ldots, s$. A problem z will be called *diagnostic* if for any $\bar{d}_i, \bar{d}_j \in \Delta_z$, $i \neq j$, the relation $\nu(\bar{d}_i) \neq \nu(\bar{d}_j)$ holds.

Probability distribution for the problem z is a mapping $P : \Delta_z \to (0,1]$ such that $\sum_{\bar{d} \in \Delta_z} P(\bar{d}) = 1$. For $\bar{d} \in \Delta_z$ we will interpret the number $P(\bar{d})$ as the probability of the event $(f_1(a), \ldots, f_n(a)) = \bar{d}$ for an element a from A.

A decision tree for the problem z is a finite oriented tree with the root satisfying the following conditions:

L. Polkowski and A. Skowron (Eds.): RSCTC'98, LNAI 1424, pp. 506–512, 1998.
© Springer-Verlag Berlin Heidelberg 1998

a) each nonterminal vertex has assigned an attribute from $\{f_1, \ldots, f_n\}$ (i.e. only those attributes are used in the decision tree which are listed in the problem z description);

b) from each nonterminal vertex exactly two edges leave which have assigned numbers 0 and 1, respectively;

c) each terminal vertex has assigned a number from ω.

A path from the root of the tree to a terminal vertex will be called *complete*.

Let Γ be a decision tree for z and ξ be a complete path in Γ. Assume ξ contains $t \geq 1$ nonterminal vertices $v_1, \ldots, v_t$, for $j = 1, \ldots, t$ the vertex v_j has assigned the attribute f_{i_j} and the edge, which leaves the vertex v_j and enters to the vertex v_{j+1}, has assigned the number $\delta_j \in \{0, 1\}$. Then we will say that *the system of equations* $\{f_{i_1}(x) = \delta_1, \ldots, f_{i_t}(x) = \delta_t\}$ *corresponds to the path* ξ. The set of solutions in A of the system of equations, which corresponds to the path ξ, will be denoted by $A(\xi)$. We assume $A(\xi) = A$ for a path ξ which consists of a terminal vertex only. The set of complete paths in Γ will be denoted by $\Xi(\Gamma)$. It can be shown that $\bigcup_{\xi \in \Xi(\Gamma)} A(\xi) = A$, and for any two different complete paths ξ_1, ξ_2 the relation $A(\xi_1) \cap A(\xi_2) = \emptyset$ holds. Moreover, for an arbitrary complete path ξ in Γ either $A_i \subseteq A(\xi)$ or $A_i \cap A(\xi) = \emptyset$ for $i = 1, \ldots, s$.

We will say that *the decision tree* Γ *solves the problem* z if for $i = 1, \ldots, s$ to the terminal vertex of the path ξ such that $A_i \subseteq A(\xi)$ is assigned the number $z(a_i)$ where a_i is an element from A_i.

For $i = 1, \ldots, s$ we denote by h_i the length of the path ξ_i such that $A_i \subseteq A(\xi_i)$. The value $\sum_{i=1}^{s} h_i P(\bar{d}_i)$ will be called *the average depth of decision tree* Γ *relatively to the probability distribution* P (or, in short, P-*average depth of* Γ). A decision tree Γ for the problem z solving z and having minimal P-average depth, will be called *optimal for* z *and* P.

2   On Decomposition of Problem

2.1   Auxiliary Notions

Let U be an information system, $z = (\nu, f_1, \ldots, f_n)$ be a problem over U and P be a probability distribution for z. For arbitrary attribute $f_i \in \{f_1, \ldots, f_n\}$ and arbitrary number $\delta \in \{0, 1\}$ we put $\Delta_z(f_i, \delta) = \{\bar{d} \in \Delta_z, \bar{d} = (d_1, \ldots d_m), d_i = \delta\}$.

For each subset $T \subseteq \Delta_z$ we denote

$$N(T, P) = \sum_{\bar{\delta} \in T} P(\bar{\delta}).$$

2.2   Proper Decomposition of the Problem

Let $U = (A, F)$ be an information system, $z_0 = (\nu_0, f_1^0, \ldots, f_{n_0}^0)$ be a diagnostic problem over U with m equivalence classes $A_1, \ldots, A_m$ and $\Delta_{z_0} = (\bar{\delta}_1^0, \ldots, \bar{\delta}_m^0)$ where $\bar{\delta}_i^0 = (f_1^0(a_i), \ldots, f_{n_0}^0(a_i))$ and $a_i \in A_i$, $i = 1, \ldots, m$. For $i = 1, \ldots, m$ let

$z_i = (\nu_i, f_1^i, \ldots, f_{n_i}^i)$ be a problem over the information system (A_i, F) with s_i equivalence classes and $\Delta_{z_i} = \{\bar{\delta}_1^i, \ldots, \bar{\delta}_{s_i}^i\}$. Let P_i be a probability distribution for the problem z_i, $i = 0, \ldots, m$. Let $\Sigma = \{\bar{\sigma}_1^1, \ldots, \bar{\sigma}_{s_1}^1, \ldots, \bar{\sigma}_1^m, \ldots, \bar{\sigma}_{s_m}^m\} \subseteq \{0, 1\}^{n_0 + \cdots + n_m}$ where $\bar{\sigma}_j^i = (\bar{\alpha}_0^{ij} \bar{\alpha}_1^{ij} \ldots \bar{\alpha}_m^{ij})$, $\bar{\alpha}_k^{ij} \in \{0, 1\}^{n_k}$, $k = 0, \ldots, m$, and

$$\bar{\alpha}_k^{ij} = \begin{cases} \bar{\delta}_i^0, & \text{if } k = 0, \\ \bar{\delta}_j^i, & \text{if } k = i, \\ (0, \ldots, 0), & \text{if } k \in \{1, \ldots m\} \setminus \{i\}, \end{cases}$$

$j = 1, \ldots, s_i$, $i = 1, \ldots, m$. Define a function $\nu : \{0, 1\}^{n_0 + \cdots + n_m} \to \omega$ as follows :

$$\nu(\bar{\delta}) = \begin{cases} \nu_i(\bar{\delta}_j^i), & \text{if } \bar{\delta} \in \Sigma \text{ and } \bar{\delta} = \bar{\sigma}_j^i, \\ 0, & \text{if } \bar{\delta} \notin \Sigma. \end{cases}$$

Consider the problem $z = (\nu, f_1^0, \ldots, f_{n_0}^0, \tilde{f}_1^1, \ldots, \tilde{f}_{n_1}^1, \ldots, \tilde{f}_1^m, \ldots, \tilde{f}_{n_m}^m)$ over U where

$$\tilde{f}_j^i(a) = \begin{cases} f_j^i(a), & \text{if } a \in A_i, \\ 0, & \text{if } a \notin A_i, \end{cases}$$

for $j = 1, \ldots, n_i$, $i = 1, \ldots, m$ and $a \in A$. One can show that $\Delta_z = \Sigma$. Define a probability distribution P for the problem z as follows: $P(\bar{\sigma}_j^i) = P_0(\bar{\delta}_i^0) P_i(\bar{\delta}_j^i)$ for $j = 1, \ldots, s_i$ and $i = 1, \ldots, m$. The $(m+1)$-tuple $((z_0, P_0), (z_1, P_1), \ldots, (z_m, P_m))$ will be called *a proper decomposition of the pair* (z, P) if the following conditions hold:

1) $P_0(\bar{\delta}_i^0) N(\Delta_{z_i}(f_j^i, 1), P_i) \leq \frac{1}{2} \min_{\bar{\delta} \in \Delta_{z_0}} P_0(\bar{\delta})$ for $j = 1, \ldots, n_i$, $i = 1, \ldots, m$;

2) for any $i, j \in \{1, \ldots, m\}$, $i \neq j$, and $c \in \omega$ such that $q_i = \sum\limits_{\bar{\delta} \in \Delta_{z_i}, \nu_i(\bar{\delta}) = c} P_i(\bar{\delta})$ > 0 and $q_j = \sum\limits_{\bar{\delta} \in \Delta_{z_j}, \nu_j(\bar{\delta}) = c} P_j(\bar{\delta}) > 0$ the following inequalities $\min(q_i, q_j) < 1/2$ and $\max(q_i, q_j) < 1$ hold.

Let $((z_0, P_0), (z_1, P_1), \ldots, (z_m, P_m))$ be a proper decomposition of the pair (z, P) and for $i = 0, \ldots, m$ let Γ_i be a decision tree for the problem z_i, which solves z_i. For $i = 1, \ldots, m$ assign each nonterminal vertex of the decision tree Γ_i instead of the attribute f_j^i the attribute $\tilde{f}_j^i$. Denote by $\tilde{\Gamma}_i$ the obtained decision tree. For $i = 1, \ldots, m$ find a complete path ξ_i in Γ_0 such that $A_i \subseteq A(\xi_i)$ and change the terminal vertex of the path ξ_i to the root of the decision tree $\tilde{\Gamma}_i$. Denote the obtained tree by $\Omega(\Gamma_0, \Gamma_1, \ldots, \Gamma_m)$.

2.3   Main Result

[**Theorem 1.**]*Let z be a problem over an information system U, P be a probability distribution for z and $((z_0, P_0), (z_1, P_1), \ldots, (z_m, P_m))$ be a proper decomposition of the pair (z, P). Let Γ_i be a decision tree for z_i solving the problem z_i and optimal for z_i and P_i, $i = 0, \ldots, m$. Then the tree $\Omega(\Gamma_0, \Gamma_1, \ldots, \Gamma_m)$ is a tree for the problem z, which solves z and which is optimal for z and P.*

We omit the proof of Theorem 1 because it is too long. Further we will consider some application of this result.

3   Some Application of Decomposition

Theorem 1 allows us to construct optimal decision trees for some problems over information systems. In this section we show that the announced in [5] result about closeness to unimprovable upper bound on minimal average depth of decision trees from [4] can be obtained as a consequence of Theorem 1.

3.1   Parameters of Problems and Probability Distributions

At first we define the parameter $M(z)$ for a problem $z = (\nu, f_1, \ldots, f_n)$ over an information system U. If $z(x) \equiv const$ on A then $M(z) = 0$. Let $z(x) \not\equiv const$ on A. For an arbitrary n-tuple $\bar{\delta} = (\delta_1, \ldots, \delta_n) \in \{0,1\}^n$ we denote by $M(z, \bar{\delta})$ the minimal natural number m such that there exist numbers $i_1, \ldots, i_m \in \{1, \ldots, n\}$ which satisfy the following condition: either the set of solutions on A of the system of equations $\{f_{i_1}(x) = \delta_{i_1}, \ldots, f_{i_m}(x) = \delta_{i_m}\}$ is empty or $z(x) \equiv const$ on this set. Then

$$M(z) = \max_{\bar{\delta} \in \{0,1\}^n} M(z, \bar{\delta}).$$

As a parameter of a probability distribution P for the problem z we will consider the value

$$H(P) = -\sum_{\bar{\delta} \in \Delta_z} P(\bar{\delta}) \log_2 P(\bar{\delta})$$

which is called *the entropy of the probability distribution P*.

For a problem z and a probability distribution P for z we denote by $h(z, P)$ the minimal P-average depth of a decision tree for z solving the problem z.

3.2   Close to Unimprovable Upper Bound on Minimal Average Depth

Following statement gives us the upper bound on the minimal average depth of a decision tree for an arbitrary problem over an information system.

[**Theorem 2.**][4] *Let z be a problem over an information system U and P be a probability distribution for z. Then*

$$h(z, P) \leq \begin{cases} M(z), & \text{if } M(z) \leq 1, \\ M(z) + 2H(P), & \text{if } 2 \leq M(z) \leq 3, \\ M(z) + \frac{M(z)}{\log_2 M(z)} H(P), & \text{if } M(z) \geq 4. \end{cases}$$

Following statement characterizes the quality of upper bound from Theorem 2.

[**Theorem 3.**][5] *For arbitrary natural numbers $m \geq 2$, n there exist an information system U_m^n, a problem z_m^n over U_m^n with m^n equivalence classes and the probability distribution $P_m^n \equiv \frac{1}{m^n}$ such that $H(P) = n \log_2 m$,*

$$M(z) = \begin{cases} m - 1, & \text{if } n = 1, \\ m, & \text{if } n = 2, \\ m + 1, & \text{if } n \geq 3, \end{cases} \quad \text{and } h(z, P) = \frac{(m+2)(m-1)}{2m} n.$$

3.3 Proof of Theorem 3

Let $m \geq 2, n$ be arbitrary natural numbers. At first we describe the information system U_m^n. Define the system of circles B_m^n on the plane. By definition, B_m^1 consists of m circles on the plane pairwise disjoint. Let the system B_m^{i-1} has been already defined. Then the system B_m^i consists of m circles pairwise disjoint, and each of them contains the system of circles B_m^{i-1}. We will say, that a circle from B_m^n is *a circle of zero kind*, if it does not contain circles from B_m^n. Let the circles of kinds from 0 to $i - 1$ has been defined, where $i < n$. We will say that a circle from B_m^n is a circle of i-th kind if the kind of each circle, which embedded in it is not greater then $i - 1$, and at least one circle, which embedded in it, is a circle of $(i - 1)$-th kind. One can show that B_m^n contains $s = m^n$ circles of zero kind. Denote them $C_1, \ldots, C_s$. Denote a_i the set of points on a plane, which is situated inside the circle C_i and denote $A = \{a_1, \ldots, a_s\}$. Set into correspondence to the each circle $C \in B_m^n$ the function $f : A \to \{0, 1\}$. The function f takes the value 1 on the element a_i if the set of points a_i is situated inside the circle C and otherwise it takes the value 0. Denote $F = \{f_1, \ldots, f_t\}$ the set of functions, which corresponds to all circles from B_m^n. Then $U_m^n = (A, F)$.

Let $z_m^n = (\nu, f_1, \ldots, f_t)$ be a diagnostic problem over U_m^n. The following statement gives us the value of parameter $M(z_m^n)$ for the problem z_m^n.

[**Lemma 1.**]*Let $m \geq 2, n$ be arbitrary natural numbers. Then*

$$M(z_m^n) = \begin{cases} m - 1, \text{ if } n = 1, \\ m, \quad \text{ if } n = 2, \\ m + 1, \text{ if } n \geq 3. \end{cases}$$

[*Proof.*]Consider the case $n \geq 3$. Let us estimate the parameter $M(z_m^n, \bar{\delta})$ for an arbitrary $\bar{\delta} \in \{0, 1\}^t$. Let $\bar{\delta} = (0, \ldots, 0)$. The system of equations $\{f_{i_1}(x) = 0, \ldots, f_{i_m}(x) = 0\}$ does not have solutions on the set A, if $f_{i_1}, \ldots, f_{i_m}$ are pairwise different attributes from F which correspond to circles of the $(n - 1)$-th kind. Therefore $M(z_m^n, \bar{\delta}) \leq m$. Let $\bar{\delta} \neq (0, \ldots, 0)$. Denote by C_0 the circle of the least kind such that in $\bar{\delta}$ the value of the attribute corresponding to C_0 is equal to 1. Let C_0 be a circle of n_0-th kind and f_{i_0} be an attribute corresponding to C_0. If $n_0 = 0$ then the system of equations $\{f_{i_0}(x) = 1\}$ has only solution in the set A and $M(z_m^n, \bar{\delta}) = 1$. Let $n_0 \geq 1$. Denote by $f_{i_1}, \ldots, f_{i_m}$ the attributes corresponding to the circles of $(n_0 - 1)$-th kind, which embedded in C_0. By the choice of the circle C_0, in $\bar{\delta}$ values of attributes $f_{i_1}, \ldots, f_{i_m}$ are equal to 0. The system of equations $\{f_{i_0}(x) = 1, f_{i_1}(x) = 0, \ldots, f_{i_m}(x) = 0\}$ does not have solutions in the set A and $M(z_m^n, \bar{\delta}) \leq m + 1$. Therefore

$$M(z_m^n) \leq m + 1. \tag{1}$$

We will show that the value $m + 1$ is obtained on the t-tuple $\bar{\delta} = (\delta_1, \ldots, \delta_t)$, such that $n_0 = 2$, the values of the attributes corresponding to circles, which includes the circle C_0, is equal to 1, and the values of other attributes is equal to 0. Let $S = \{f_{j_1}(x) = \delta_{j_1}, \ldots, f_{j_k}(x) = \delta_{j_k}\}$ be an arbitrary system of equations such that the set of solutions of this system in A is empty or $z_m^n(x) \equiv const$ on this set. Let $l \in \{1, \ldots, k\}$. Change the equation $f_{j_l}(x) = \delta_{j_l}$ to the equation $f_{i_r}(x) = 0$ if the circle corresponding to the attribute f_{j_l} is embedded into the circle corresponding to the attribute f_{i_r} or equals to this circle for some $r \in \{1, \ldots, m\}$. Otherwise, change the equation $f_{j_l}(x) = \delta_{j_l}$ to the equation $f_{i_0}(x) = 1$. Make this changes for $l = 1, \ldots, k$ and denote the obtained system by S_1. It is clear that S_1 has at most k equations, the set of solutions of S_1 in A is a subset of the set of solutions of S in A, and S_1 is a subsystem of the system $S_2 = \{f_{i_0}(x) = 1, f_{i_1}(x) = 0, \ldots, f_{i_m}(x) = 0\}$. Suppose that $S_1 \neq S_2$. One can show that in this case $z_m^n(x) \not\equiv const$ on the set of solutions of S_1 in A. But it is impossible. Therefore $k \geq m + 1$ and $M(z_m^n, \bar{\delta}) \geq m + 1$. From this inequality and from (1) follows that $M(z_m^n) = m + 1$. The cases $n = 1$ and $n = 2$ are considered similarly. □

Proof of Theorem 3. We will proceed by induction on n. Let $n = 1$. Define the decision tree Γ for the problem z_m^1. The decision tree Γ contains $m - 1$ nonterminal vertices $v_1, \ldots, v_{m-1}$, which assigned pairwise different attributes $f_1, \ldots, f_{m-1}$ respectively and m terminal vertices $v_m, w_1, \ldots, w_{m-1}$. For $i = 1, \ldots, m - 1$ the vertex v_i leaves two edges, which assigned the numbers 0 and 1 respectively. The edge, which is assigned the number 0, enters to the vertex v_{i+1}, and the edge, which is assigned the number 1, enters to the vertex w_i. For $i = 1, \ldots, m - 1$ the vertex w_i is assigned the number $z_m^1(a_i)$ where a_i is the element from the set A such that $f_i(a_i) = 1$. The vertex v_m is assigned the number $z_m^1(a_0)$ where a_0 is the element from the set A such that $f_i(a_0) = 0$ for $i = 1, \ldots, m - 1$. The decision tree Γ does not contains other vertices and edges. One can show that the decision tree Γ solves the problem z_m^1 and Γ is optimal for z_m^1 and P_m^1. Let us evaluate average depth of Γ: $h(\Gamma, P_m^1) = \sum_{i=1}^{m-1} i \frac{1}{m} + (m-1)\frac{1}{m} = \frac{(m+2)(m-1)}{2m}$. Then $h(z_m^1, P_m^1) = \frac{(m+2)(m-1)}{2m}$.

Consider now a natural number $n \geq 2$ such that the considered statement is true for any natural number less then n. Consider a decomposition $((z_0, P_0), (z_1, P_1), \ldots, (z_m, P_m))$ of the pair (z_m^n, P_m^n). The diagnostic problem z_0 contains attributes corresponding to all circles of $(m-1)$-th kind from the system B_m^n. Each of the diagnostic problems $z_1, \ldots, z_m$ contains attributes corresponding to all circles from one of the system B_m^{n-1}, which is contained into B_m^n, and these systems are pairwise different. The problem z_i is defined such that $z_i(x)$ is a restriction of the mapping $z : A \to \omega$ on the set A_i, $i = 1, \ldots, m$. For $i = 0, \ldots, m$ let P_i be an uniform probability distribution for the problem z_i. One can show that $((z_0, P_0), (z_1, P_1), \ldots, (z_m, P_m))$ is proper decomposition for the pair (z_m^n, P_m^n). One can show that $h(z_0, P_0) = h(z_m^1, P_m^1)$ and $h(z_i, P_i) = h(z_m^{n-1}, P_m^{n-1})$ for $i = 1, \ldots, m$. Using the inductive hypothesis we obtain that $h(z_0, P_0) = \frac{(m+2)(m-1)}{2m}$ and $h(z_i, P_i) = \frac{(m+2)(m-1)}{2m}(n-1)$ for $i = 1, \ldots, m$. Let Γ_i be a decision tree for the problem z_i, which solves z_i and which is optimal for z_i and P_i, $i = 0, \ldots, m$. Let $\Omega = \Omega(\Gamma_0, \Gamma_1, \ldots, \Gamma_m)$. From the definition of the tree Ω it follows that $h(\Omega, P_m^n) = h(\Gamma_0, P_0) + \sum_{i=1}^{m} \frac{h(\Gamma_i, P_i)}{m} = \frac{(m+2)(m-1)}{2m}n$. Using Theorem 1 we obtain $h(z_m^n, P_m^n) = \frac{(m+2)(m-1)}{2m}n$. □

References

1. Chegis, I., Yablonskii, S.: Logical methods for electric circuit control. Trudy MIAN SSSR **51** (1958) 270–360 (in Russian).
2. Moshkov, M.: Conditional tests. Problemy Cybernetici **40** (1983) 131–170 (in Russian).
3. Moshkov, M.: Decision Trees. Theory and Applications. Nizhni Novgorod University Publishers, Nizhni Novgorod (1994) (in Russian).
4. Moshkov, M., Chikalov, I.: Bounds on average weighted depth of decision trees. Fundamenta Informaticae (1997). **31** 145–157
5. Moshkov, M., Chikalov, I.: Bounds on average depth of decision trees. Proceedings of the Fifth European Congress on Intelligent Techniques and Soft Computing, Aachen (1997) 226–230
6. Pawlak, Z.: Rough Sets - Theoretical Aspects of Reasoning about Data. Kluwer Academic Publishers, Dordrecht (1991)
7. Skowron, A., Rauszer, C.: The discernibility matrices and functions in information systems. Intelligent Decision Support. Handbook of Applications and Advances of the Rough Set Theory. Kluwer Academic Publishers, Dordrecht (1992) 331–362

On Diagnosis of Retaining Faults in Circuits

Albina Moshkova*

Faculty of Calculating Mathematics and
Cybernetics of Nizhni Novgorod State University
23, Gagarina Av., Nizhni Novgorod, 603600, Russia

Abstract. Diagnosis of faults in circuits is important field of applications of rough set theory and test theory. The problem of search an optimal circuit basis for an arbitrary closed class of Boolean functions is considered. The basis should be optimal in the sense of simplicity of diagnosis of so-called retaining faults in iteration-free circuits over the basis. This problem is solved for all closed classes. In the paper the complexity of diagnosis of retaining faults in iteration-free circuits over optimal bases is studied.

1 Introduction

The structure of the system of all classes of Boolean functions closed over substitution operation has been described by Post in [7] and [8]. Yablonskii, Gavrilov and Kudriavtzev in [10] studied the structure (slightly different from that of Post) of all classes of Boolean functions closed over substitution operation with the assumption that if a Boolean function is given then all functions which differ from this function by unessential variables are given. The latter structure will be used in the paper.

We will consider the realization of functions from closed classes by combinatorial circuits and we will study the depth of decision trees for diagnosis of so-called retaining faults in these circuits. The faults under consideration consist of the change of the function realized by a gate such that there exist two tuples on which values of the function are invariable and are equal to 0 and 1 respectively (if a gate realizes a constant then this gate has no faults).

The problem of diagnosis of arbitrary circuits is a complicated problem. Therefore a nonstandard approach to design and diagnosis of circuits is considered [4, 5]. Only formula-like circuits over special chosen bases are used. In addition to the usual work mode of the circuit there exists the diagnostic mode in which the circuit transforms to so-called iteration-free circuit for which the diagnosis problem solves efficiently.

The main problem considered in the paper is to find for an arbitrary closed class of Boolean functions a circuit basis for this class (an optimal basis) for which the minimal depth of decision trees for diagnosis of iteration-free circuits over this

---

* This work was supported by Russian Foundation of Fundamental Research (grant # 96-01-00428).

L. Polkowski and A. Skowron (Eds.): RSCTC'98, LNAI 1424, pp. 513–516, 1998.

basis grows most slowly with growth of the number of gates in circuits. Optimal bases are found for all closed classes of Boolean functions. In the paper the complexity of diagnosis of retaining faults in iteration-free circuits over optimal bases is studied. Note that analogous problem for constant faults on gate inputs was solved in [5] for 40 closed classes.

In proofs we use methods and results of test theory [1, 2, 3] and rough set theory [6, 9].

2   On Diagnosis of Iteration-Free Circuits

Let $f(x_1, \ldots, x_n)$ be a Boolean function. The variable x_i of the function f will be called *essential*, if there exist two n-tuples $\bar{\delta}$ and $\bar{\sigma}$ from $\{0,1\}^n$ which differ only in the i-th digit and such that $f(\bar{\delta}) \neq f(\bar{\sigma})$. The variables of function f which are not essential will be called *unessential*.

A *circuit basis* is a finite nonempty set B of Boolean functions for each of which all variables are essential. Divide the basis B into two parts: $B = B_C \cup B_F$ where $B_C = B \cap \{0,1\}$ and $B_F = B \setminus \{0,1\}$. Let S be a combinatorial circuit over the basis B. We will assume that circuit gates can have the following faults which will be called *retaining faults*. Let a gate realizes a constant from B_C. Then this gate has no faults. Let a gate realizes a function $f(x_1, \ldots, x_n) \in B_F$. Then there exist two n-tuples α_f^0 and α_f^1 from $\{0,1\}^n$ such that $f(\alpha_f^0) = 0$, $f(\alpha_f^1) = 1$ and the gate with a fault can realize an arbitrary Boolean function $f'(x_1, \ldots, x_n)$ such that $f'(\alpha_f^0) = 0$ and $f'(\alpha_f^1) = 1$.

For example, let a gate realizes the function $f(x, y) = x \vee y$, $\alpha_f^0 = (0,0)$ and $\alpha_f^1 = (1,1)$. Then the considered gate with a fault can realize an arbitrary Boolean function from the set $\{x \vee y, x \cdot y, x, y\}$.

The diagnosis problem for a circuit consists in the recognition of the function realized by the circuit which, possibly, has retaining faults of gates. For this problem solving we use decision trees. Each check of a decision tree consists in observing the output of the circuit at the inputs of which a binary tuple is supplied. As a complexity measure we consider the depth of a decision tree (the maximal number of checks which this tree realizes).

The number of gates in the circuit S will be denoted by $L(S)$ and the minimal depth of a decision tree which solves the diagnosis problem for the circuit S will be denoted by $h(S)$.

A combinatorial circuit will be called *iteration-free* if each node (input or gate) of it has at most one issuing edge. It is not difficult to prove the following statement.

Proposition 1. *Let B be a circuit basis for which $B_F \neq \emptyset$. Then there exists a constant c such that $h(S) \leq c \cdot L(S)$ for any iteration-free circuit S over B_F.*

We denote by $\mathcal{C}(B)$ the minimal natural number c for which the inequality $h(S) \leq c \cdot L(S)$ holds for any iteration-free circuit S over B_F. If $B_F = \emptyset$ then let $\mathcal{C}(B) = 0$. For a Boolean function f we denote by $\rho(f)$ the number of essential variables of the function f. For a nonempty finite set of Boolean functions D let $\rho(D) = \max\{\rho(f) : f \in D\}$. The following statement describes the behavior of the parameter $\mathcal{C}(B)$.

Theorem 2. *Let B be a circuit basis. Then*

$$\mathcal{C}(B) = \begin{cases} 0 & if \ \rho(B) = 0 \\ 2^{\rho(B)} - 2 & if \ \rho(B) > 0. \end{cases}$$

3   On Optimal Bases

We will say that two Boolean functions are *equal* if they differ on unessential variables. As in [10] we will assume that if a Boolean function f is given, then all functions which are equal to f are given. Let U be a nonempty set of Boolean functions. The closure of the set U over operation of substitution will be denoted by $[U]$. The set U will be called a *closed class* if $U = [U]$.

Let U be a closed class of Boolean functions and B be a circuit basis. We will say that B *correctly generates* the class U if $U = [B_F] \cup B_C$.

Let B correctly generates the class U. This basis may be used for constructing circuits which realize functions from U and have effective algorithms for diagnosis of retaining faults of gates. Let $f \in U$.

Assume that $f \in [B_F]$. Let φ be a formula over B_F realizing the function f. Then the circuit S over B_F is constructed according to the formula φ and satisfying the following conditions:

a) the circuit S realizes the function f;

b) $L(S) = L(\varphi)$ where $L(\varphi)$ is the number of functional symbols in the formula φ;

c) from any gate of the circuit S issues at most one edge.

In addition to the usual work mode of the circuit S there exists the *diagnostic* mode in which the inputs of the circuit S are "split" so that it becomes the iteration-free circuit $\tilde{S}$. The relations $h(\tilde{S}) \leq \mathcal{C}(B) \cdot L(\tilde{S}) = \mathcal{C}(B) \cdot L(\varphi)$ hold for the circuit $\tilde{S}$.

Assume now that f is a constant from B_C. Then we can realize the function f by a circuit S which has one gate and for which $h(S) = 0$.

One can prove the following statement.

Proposition 3. *For any closed class U of Boolean functions there exists a circuit basis B which correctly generates the class U.*

Denote $\mathcal{C}(U) = \min \mathcal{C}(B)$ where B is a circuit basis which correctly generates the class U. A circuit basis B will be called an *optimal circuit basis for the class U* if B correctly generates the class U and $\mathcal{C}(B) = \mathcal{C}(U)$.

A nonempty finite set of Boolean functions D will be called a *basis* of the class U if $[D] = U$ and for any set $D' \subset D$ the relation $[D'] \neq U$ holds. Denote $\rho(U) = \min \rho(D)$ where D is a basis of the class U. For each closed class U the value $\rho(U)$ can be found in [10].

Optimal circuit bases were constructed for each closed class U. The following theorem describes the behavior of the parameter $\mathcal{C}(U)$.

Theorem 4. *Let U be a closed class of Boolean functions. Then*

$$C(U) = \begin{cases} 0 & if\ \rho(U) = 0 \\ 2^{\rho(U)} - 2 & if\ \rho(U) > 0. \end{cases}$$

Further we will consider examples of some closed classes of Boolean functions.

1. The class C_1 comprises all Boolean functions. For this class $C(C_1) = 2$ and $\{\bar{x} \vee \bar{y}\}$ is an optimal circuit basis.
2. The class L_1 comprises all linear functions. For this class $C(L_1) = 2$ and $\{x + y, x + y + 1\}$ is an optimal circuit basis.
3. The class D_3 comprises all self-dual functions. For this class $C(D_3) = 6$ and $\{x \cdot \bar{y} \vee x \cdot \bar{z} \vee \bar{y} \cdot \bar{z}\}$ is an optimal circuit basis.
4. The class A_1 comprises all monotone functions. For this class $C(A_1) = 2$ and $\{x \cdot y, x \vee y, 0, 1\}$ is an optimal circuit basis.

References

1. Chegis, I.A., Yablonskii, S.V.: Logical methods for electric circuit control. Trudy MI AN SSSR **51** (1958) 270–360 (in Russian)
2. Moshkov, M.: Conditional tests. Problems of Cybernetics **40**. Nauka Publishers, Moscow (1983) 131–170 (in Russian)
3. Moshkov, M.: Decision Trees. Theory and Applications. Nizhni Novgorod University Publishers, Nizhni Novgorod (1994) (in Russian)
4. Moshkov, M.: Diagnosis of constant faults of circuits. Proceedings of the Fourth International Workshop on Rough Sets, Fuzzy Sets and Machine Discovery, Tokyo (1996) 325–327
5. Moshkov, M., Moshkova, A.: Optimal bases for some closed classes of Boolean functions. Proceedings of the Fifth European Congress on Intelligent Techniques and Soft Computing, Aachen (1997) 1643–1647
6. Pawlak, Z.: Rough Sets - Theoretical Aspects of Reasoning about Data. Kluwer Academic Publishers, Dordrecht (1991)
7. Post, E.: Introduction to a general theory of elementary propositions. Amer. J. Math. **43** (1921) 163–185
8. Post, E.: Two-Valued Iterative Systems of Mathematical Logic. Annals of Math. Studies 5. Princeton Univ. Press, Princeton-London (1941)
9. Skowron, A., Rauszer, C: The discernibility matrices and functions in information systems. Intelligent Decision Support. Handbook of Applications and Advances of the Rough Set Theory. Kluwer Academic Publishers, Dordrecht (1992) 331–362
10. Yablonskii, S.V., Gavrilov, G.P., Kudriavtzev, V.B.: Boolean Functions and Classes of Post. Nauka Publishers, Moscow (1966) (in Russian)

On the Depth of Decision Trees for Diagnosing of Nonelementary Faults in Circuits

V. Shevtchenko

Research Institute for Applied Mathematics and
Cybernetics of Nizhni Novgorod State University
10, Uljanova St., Nizhni Novgorod, 603005, Russia

Introduction

Diagnosing of faults in circuits is a significant area of applications for test theory [1, 3, 4, 5, 6] and rough set theory [2]. The circuits considered in the paper realize Boolean functions and use gates each of which realizes a function from some finite set of Boolean functions (basis for circuits). We will consider the class of so-called nonelementary faults of these circuits. This class contains different combinations of $\wedge$- and $\vee$- types of faults, constant faults and "negation"- type of faults. We will consider the problem of construction of decision trees with minimal depth which recognize the function realizable by a given circuit probably with faults. We will consider not the set of all possible faults of a circuit, but some subsets containing at most $k \geq 2$ functionally distinguishable faults. In the circuits faults are represented by gates which realize functions from some finite set of Boolean functions (basis of faults). We will consider nonelementary bases of faults. A nonelementary basis of faults contains function which is neither conjunction nor constant, function which is neither disjunction nor constant, function which is not linear function.

It is shown that for an arbitrary nonelementary basis of faults for any circuit in an arbitrary finite basis with at least 2 gates and $n \geq 4$ inputs, and for any k, $2 \leq k \leq \binom{n-2}{\lfloor (n-2)/2 \rfloor}$, there is a set of k functionally distinguishable faults of circuit for diagnosing of which it is necessary to use decision trees which depth is at least $k - 1$.

1 Definitions and Notation

A circuit will be called a finite oriented contour-free graph where

1) each vertex without incoming arcs is assigned the constant 0 or 1, or a variable from the alphabet $X = \{x_1, x_2, \ldots\}$; here the variables assigned to different vertices are different;

2) to each vertex, which has $\tau \geq 1$ incoming arcs, there is assigned some Boolean function being essentially dependent on τ variables;

3) all the arcs are assigned numbers such that if some vertex has τ incoming arcs then these arcs are assigned numbers from 1 through τ;

L. Polkowski and A. Skowron (Eds.): RSCTC'98, LNAI 1424, pp. 517–520, 1998.

4) exactly one vertex is labeled by *.

The vertices, which are labeled by variables, are called circuit inputs; all other vertices are called gates; and the vertex labeled by * is called an output of the circuit.

We assume that a) the circuit contains at least one input and one gate; and b) a gate is taken as the circuit output.

We will speak that in the circuit a vertex v_1 precedes a vertex v_2 if there exists an oriented path from the vertex v_1 to the vertex v_2. Sometimes we will speak that the vertex v_1 immediately precedes the vertex v_2 if v_1 and v_2 are connected by the arc (v_1, v_2).

To determine a function realizable by the circuit, each vertex of the circuit can be associated with a function in the following way:

1) the vertex which is labeled by a variable (constant) will be associated with the function being equal to this variable (constant);

2) let v_0 be a vertex which is labeled by a Boolean function $\phi(y_1, \ldots, y_\tau)$, and $v_1, \ldots, v_\tau$ be the vertices immediately preceding v_0. Let the arc (v_i, v_0) be labeled by the number $i, i = 1, \ldots, \tau$, and let the vertices $v_1, \ldots, v_\tau$ be already associated with the functions $g_1, \ldots, g_\tau$ respectively. Then, let the vertex v_0 be associated with the function $\phi(g_1, \ldots, g_\tau)$.

The circuit will realize the function associated with the vertex labeled by *.

An arbitrary finite nonempty set of Boolean functions P containing function which is neither conjunction nor constant, function which is neither disjunction nor constant, function which is not linear function, will be called a nonelementary basis of faults or simply basis of faults.

Let $\psi(\tilde{x}_t)$ be some Boolean function from P, $\tilde{x}_t = (x_1, \ldots, x_t)$. Now let us determine an operation of introducing a ψ-fault into the circuit S.

1) Let $t > 0$. Add to S the vertex e and assign to it the function ψ, and then draw the arcs $(v_1, e), \ldots, (v_t, e)$ where $v_1, \ldots, v_t = \bar{v}$ is an arbitrary sequence of vertices in S. Upon it, let us assign the number $i, i = 1, \ldots, t$, to the arc (v_i, e). The obtained circuit will be denoted by U'; and here, if $\bar{v}$ contains the output of S, then the vertex e will be labeled as the output of U'. Now, in U' let us take the arc (v_i, ω) such that v_i is a vertex from $\bar{v}$, and the vertex ω differs from e, w is not contained in $\bar{v}$, and w does not precede any vertex from $\bar{v}$. In U', let us draw the arc (e, ω), eliminate the arc (v_i, ω), and assign its number to the arc (e, ω). The obtained circuit will be denoted by U''. Concerning the circuits U' and U'' we will speak that they have been obtained from the circuit S by introducing the ψ-fault.

2) Let $t = 0$. In this case, ψ is the constant 0 or 1. Now, let us add the vertex e to the circuit S, assign to it the function ψ, and mark it as the circuit output. The obtained circuit will be denoted by Γ'. Further, let us arbitrarily choose an arc (v_1, v_2) in S, add the vertex e to S, assign to it the function ψ, draw the arc (e, v_2), assign to it the number of the arc (v_1, v_2), and eliminate the arc (v_1, v_2). The obtained circuit will be denoted by Γ''. Concerning the circuits Γ' and Γ'' we will speak that they have been obtained from the circuit S by introducing the ψ-fault.

Now determine a set of circuits $H_P(S)$ in the following way:

1) $S \in H_P(S)$;

2) Let $U \in H_P(S)$ and $\psi(\tilde{x}_t) \in P$. Then all the circuits, possibly obtained from S by introducing the ψ-fault, belong to $H_P(S)$ as well.

3) The set $H_P(S)$ contains no other circuits.

Let F be some set of Boolean functions. Then by $\bar{F}$ we will denote the set of all superpositions of functions from F. Note that the set $\bar{P}$ is some closed class of Boolean functions, containing function which is neither conjunction nor constant, function which is neither disjunction nor constant, function which is not linear function.

1.1 Problem of Diagnosing

Divide the set of circuits $H_P(S)$ into subsets $H_P^1(S), H_P^2(S), \ldots, H_P^m(S)$ such that all circuits from the same subset realize the same Boolean function and circuits from different subsets realize different functions. Let $i_1, \ldots, i_k \in \{1, \ldots, m\}$.

Now we define the problem of diagnosing S with respect to subsets $H_P^{i_1}(S)$, $\ldots, H_P^{i_k}(S)$. For an arbitrary circuit $U \in \bigcup_{j=1}^{k} H_P^{i_j}(S)$ it is required to find the subset which contains the circuit U.

For solving this problem we will use decision trees.

The set of all Boolean functions realizable by circuits from $\bigcup_{j=1}^{k} H_P^{i_j}(S)$ will be denoted by $F_P^{i_1,\ldots,i_k}(S)$ $(F_P(S) = F_P^{1,\ldots,m}(S)$ is the set of Boolean functions realizable by circuits from $H_P(S))$. A circuit S with $n \geq 1$ inputs will be denoted by S_n if convenient.

A decision tree Y for the circuit S_n with respect to subsets $H_P^{i_1}(S_n)$, $\ldots,$ $H_P^{i_k}(S_n)$ is a finite oriented rooted tree where each nonterminal vertex is labeled by a tuple from $\{0,1\}^n$, and each terminal vertex is labeled by some number from the set $\{i_1, \ldots, i_k\}$. ¿From each nonterminal vertex there run out exactly two arcs which are labeled by numbers 0 and 1 respectively. For any function $f_{i_j} \in F_P^{i_1,\ldots,i_k}(S_n)$, which is realized by a circuit from $H_P^{i_j}(S_n)$, there exists some complete path $\gamma = v_1, u_1, \ldots, v_r, u_r, v_{r+1}$ (from the root to a terminal vertex) such that the vertex v_{r+1} is labeled by the number i_j; and if for $q = 1, \ldots, r$ the vertex v_q is labeled by the tuple $\alpha_q \in \{0,1\}^n$, and the arc u_q is labeled by the number $\delta_q \in \{0,1\}$, then the function f_{i_j} is the single function from $F_P^{i_1,\ldots,i_k}(S_n)$ which on the tuples $\alpha_1, \ldots, \alpha_r$ has the values $\delta_1, \ldots, \delta_r$ respectively.

The maximal length of a complete path is called the depth of a decision tree Y and is denoted by $h(Y)$.

Let $h_P^{i_1,\ldots,i_k}(S_n) = \min h(Y)$, where minimum is found among all decision trees for S_n with respect to subsets $H_P^{i_1}(S_n), \ldots, H_P^{i_k}(S_n)$.

In this paper we study the value $h_P^k(S_n) = \max h_P^{i_1,\ldots,i_k}(S_n)$ where maximum is found among all tuples from the set $\{(i_1, \ldots, i_k) : 1 \leq i_1 < \ldots < i_k \leq m\}$.

2   Results

Let

$$\alpha_n^k = \min\{\lambda_n, k-1\},$$
$$\beta_n^k = \min\{2^n, k-1\},$$

where $n \geq 1$, $k \geq 2$, and

$$\lambda(n) = \binom{n}{\lfloor n/2 \rfloor} + \binom{n}{\lfloor n/2 \rfloor + 1}.$$

Denote by $G(S)$ the set of functions assigned to vertices of the circuit S.

[**Theorem 1.**]

Let P be an arbitrary nonelementary basis of faults and S_n be some circuit in an arbitrary basis which has n inputs and at least two gates.

a) If the set $P \cup G(S_n)$ contains only monotone Boolean functions and $n \geq 4$, then for any $k \geq 2$

$$\alpha_{n-1}^k \leq h_P^k(S_n) \leq \alpha_n^k.$$

b) If the basis P contains only monotone self-dual Boolean functions, the set $G(S_n)$ contains at least one nonmonotone function, and $n \geq 4$, then for any $k \geq 2$,

$$\beta_{n-2}^k \leq h_P^k(S_n) \leq \beta_n^k.$$

c) If the set $P \cup G(S_n)$ contains a nonmonotone Boolean function and P contains a function which is not monotone self-dual function then for $n \geq 1$ and $k \geq 2$,

$$\beta_{n-1}^k \leq h_P^k(S_n) \leq \beta_n^k.$$

References

1. Chegis, I., Yablonskii, S.: Logical methods for electric circuit control. Trudy MIAN SSSR **51** (1958) 270–360 (in Russian).
2. Pawlak, Z.: Rough Sets - Theoretical Aspects of Reasoning about Data. Kluwer Academic Publishers, Dordrecht (1991).
3. Shevtchenko, V.: On complexity of diagnosing "⊕"-type faults in circuits. Kombinatorno-Algebraitcheskie i Verojatnostnye Metody v Prikladnoi Matematike, Gorky (1989) 129–140 (in Russian).
4. Shevtchenko, V.: On complexity of diagnosing "0"-, "1"-, "∧"- and "∨"- types of faults in circuits. Kombinatorno-Algebraitcheskie i Verojatnostnye Metody v Prikladnoi Matematike, Gorky (1990) 125–150 (in Russian).
5. Shevtchenko, V.: On the depth of decision trees for diagnosing faults in circuits. Proceedings of The Third International Workshop on Rough Sets and Soft Computing, San Jose (1994) 594–601.
6. Shevtchenko, V.: On the depth of decision trees for control faults of circuits. Proceedings of the Fourth International Workshop on Rough Sets, Fuzzy Sets and Machine Discovery, Tokyo (1996) 328–330.

Discovery of Decision Rules by
Matching New Objects Against Data Tables

Jan G. Bazan

Institute of Mathematics, Pedagogical University of Rzeszów
Rejtana 16A, 35-310 Rzeszów, Poland
E-mail: bazan@univ.rzeszow.pl

Abstract. In this paper we present an exemplary algorithm classifying new objects by matching them directly against data table to generate relevant decision instead of matching it against all rules generated from data table (see [1]). We report results of experiments on three medical data sets, concerning lymphography, breast cancer and primary tumor (see [8]). We compare standard methods for extracting laws from decision tables (see e.g. [17], [1]), based on rough set (see [13]) and boolean reasoning (see [2]), with the method based on algorithms calculating relevant decision rules for new objects. We also compare the results of computer experiments on those data sets obtained by applying our system based on rough set methods with the results on the same data sets obtained with help of several data analysis systems known from literature.

1 Introduction

A *classification algorithm* is an algorithm which permits us to repeatedly make a forecast on the basis of accumulated knowledge in new situations (see e.g. [9]). We consider here a classification related to construction of a classifying algorithm which on the basis of current knowledge will be applied to a number of cases to classify objects previously unseen; each new object will be assigned to a class belonging to a predefined set of classes on the basis of observed values of suitably chosen attributes (features).

Many approaches have been proposed for constructing classification algorithms, among them we would like to point out classical and modern statistical techniques (see e.g. [9]), neural networks (see e.g. [9]), decision trees (see e.g. [3], [15], [9]), decision rules (see e.g. [4], [8], [6] [18], [12]) inductive logic programming (see e.g. [5]).

The most popular method for classification algorithms construction is based on learning rules from examples. One can use rough set methods to discover rules from data sets (see e.g. [17], [1]). The methods based on calculation of all reducts allow to compute, for given data, the descriptions of concepts by means of decision rules (see [13]). Unfortunately, the searching problem for reduct of minimal length (minimal number of attributes) is *NP*-hard (see [16]). Therefore we often apply approximation algorithms to obtain some knowledge about the reduct set (see [11], [19]). Another approach can be based on construction of

L. Polkowski and A. Skowron (Eds.): RSCTC'98, LNAI 1424, pp. 521–528, 1998.
© Springer-Verlag Berlin Heidelberg 1998

algorithms not requiring calculation of the decision rule set before new objects are classified. These algorithms classify any new object by generation of relevant rules for this object only. The aim of the paper is to present an example of such algorithm. We compare the results of classification algorithm based on standard rough set methods with methods based on presented algorithm and we also compare rough set methods with classification algorithms known from literature.

From experiments it follows, that algorithms generating only decision rules relevant to new cases are faster than algorithm based on decision rule generation especially when they are applied to large decision tables. The results suggest that methods using rough set techniques can be treated as a promising tool for extracting laws from experimental data sets and their performance is fully comparable with the performance of other classification systems (see also [1] for more complete discussion).

2 Classification Algorithms based on Decision Rules

We assume that the reader is familiar with basic notions of the rough set theory (see [13]) and methods of decision rules generation based on boolean reasoning (see [2]). In particular by $\mathbf{A} = (U, A \cup \{d\})$ we denote the decision table and by $RUL(\mathbf{A})$ we denote the set of all optimal basic decision rules of $\mathbf{A}$ (i.e. decision rule with minimal number of descriptors in predecessor and only one decision descriptor in successor - see [13], [1]). The cardinality of the image $d(U) = \{k : d(s) = k$ for some $s \in U\}$ is called the *rank of d* and is denoted by $r(d)$. We assume that the set $V_{\mathbf{A}}^d$ of values of the decision d is equal to $\{v_d^1, ..., v_d^{r(d)}\}$. The decision d determines partition $CLASS_{\mathbf{A}}(d) = \{X_{\mathbf{A}}^1, ..., X_{\mathbf{A}}^{r(d)}\}$ of the universe U, where $X_{\mathbf{A}}^k = \{x \in U : d(x) = v_d^k\}$ for $1 \le k \le r(d)$. The set $X_{\mathbf{A}}^i$ is called the *i-th decision class of $\mathbf{A}$*. By $X_{\mathbf{A}}(u)$ we denote the decision class $\{x \in U : d(x) = d(u)\}$, for any $u \in U$. If $r \in RUL(\mathbf{A})$, then by $Pred(r), Succ(r)$ we denote the predecessor of r and the successor of r, respectively, and by $d(r)$ we denote the decision value specified by $Succ(r)$. An object $u \in U$ is *matched* by a decision rule $r \in RUL(\mathbf{A})$, iff u belongs to the set describing the *meaning* (see e.g. [13]) of $Pred(r)$. Then we will say that the rule is *classifying u* to the decision class $d(r)$. The set of all objects matched by decision rule r is denoted by $Match_{\mathbf{A}}(r)$. An object $u \in U$ *supports* a decision rule $r \in RUL(\mathbf{A})$ iff u belongs to the meaning of $Pred(r)$ and u belongs to the meaning of $Succ(r)$. The set of all objects supporting a decision rule r is denoted by $Supp_{\mathbf{A}}(r)$. If r is a decision rule in $\mathbf{A}$, then the number $\mu_{\mathbf{A}}(r) = \frac{card(Supp_{\mathbf{A}}(r))}{card(Match_{\mathbf{A}}(r))}$ is called *the coefficient of consistency* of the rule r. If $\mu_{\mathbf{A}}(r) = 1$ then we will say that the decision rule r is *consistent* in $\mathbf{A}$, otherwise the decision rule r is *inconsistent* or *approximate* in $\mathbf{A}$.

Let $\mathbf{W} = (W, A \cup \{d\})$ be a hypothetical *universal decision table* (including known and unknown objects describing an actual considered aspect of reality - see [1]) and by $\mathbf{A} = (U, A \cup \{d\})$ we denote a given subtable of the universal decision table. Let $u \in W$ be a so called *tested object*. Our task consists in assigning the value $d(u)$ of the decision d to the tested objects u, basing only on values of condition attributes of u, and relying on a given decision table $\mathbf{A}$. A solution of this problem is a classification algorithm sufficiently approximating the decision function d. There are many methods for decision rules generation from data (see e.g. [6], [7], [8], [17], [1]).

When a set of decision rules has been computed then it is necessary to use some methods to resolve conflicts between rule sets classifying tested objects to different decision classes. Therefore we use some measures of the strength of rule set matched by a given tested object and classifying this object to decision classes pointed out by the rules from this set. In this paper we consider *the global strength of rule set* presented in [1] and defined by

$$
GlobalStrength(X_i, u_t) = \frac{card\left(\displaystyle\bigcup_{r \in MRul(X_i, u_t)} Match_{\mathbf{A}}(r) \cap X_i\right)}{card(X_i)},
$$

where $u_t \in W$ is a tested object, $MRul(X_i, u_t) \subseteq RUL(\mathbf{A})$ is a set of all calculated basic decision rules for $\mathbf{A}$, classifying objects to the decision class X_i and matching tested objects u_t. The global strength of decision rule set is similar to strength of rule presented in [8] and [6].

Sometimes the decision rules generated by applying rough set methods cannot be accepted as laws valid for data encoded in a given decision table. This happens, e.g. when the number of examples supporting the decision rule is relatively small. Therefore one can use approximate decision rules instead of consistent decision rules to construct the classification algorithm for a decision table $\mathbf{A}$. Different methods (see [1], [10], [14], [20]) are now widely used to generate approximate decision rules. In our experiments (see Section 4) we used the method of approximate rules generation described in [1].

3 Decision rule synthesis by matching new objects against data tables

The method of classification algorithms construction mentioned in the previous section is based on methods for decision rules generation from decision tables. We have applied (see e.g. [1]) an algorithm for reduct set computation to obtain decision rule set. However, the time cost of the reduct set computation can be too high when the decision table has too many attributes or/and different values of attributes or objects. Therefore we often apply some approximate algorithms to extract some knowledge about the reduct set (see [11], [19]). Another approach can be based on construction of algorithms not requiring calculation of the decision rule set before new objects are classified. These algorithms classify any new object by generation of relevant rules for this object only. In this paper we present an example of such algorithm.

Let $\mathbf{W} = (W, A \cup \{d\})$ be a universal decision table and let $\mathbf{A} = (U, A \cup \{d\})$ be a given decision table $(U \subseteq W)$. By $EQL_{\mathbf{A}}(u_1, u_2)$ we denote the set $\{a \in A : a(u_1) = a(u_2)\}$ for any $u_1, u_2 \in W$.

An object $u_r \in U$ *classifies* a tested object $u_t \in W$ to the decision class $X_{\mathbf{A}}(u_r)$ iff $card(EQL_{\mathbf{A}}(u_r, u_t)) > 0$.

An object $u_r \in U$ *exactly classifies* a tested object $u_t \in W$ to the decision class $X_{\mathbf{A}}(u_r)$ iff u_r classifies u_t to the decision class $X_{\mathbf{A}}(u_r)$ and for any $u \in U$:

$$EQL_{\mathbf{A}}(u_r, u_t) \subseteq EQL_{\mathbf{A}}(u, u_r) \Rightarrow d(u) = d(u_r).$$

An object $u_r \in U$ does *not classify* a tested object $u_t \in W$ *exactly* to the decision class $X_{\mathbf{A}}(u_r)$ iff u_r classifies u_t to the decision class $X_{\mathbf{A}}(u_r)$ and two conditions are satisfied: $EQL_{\mathbf{A}}(u_r, u_t) \subseteq EQL_{\mathbf{A}}(u'_r, u_r)$ and $d(u'_r) \neq d(u_r)$ for some object u'_r.

Let $u_r \in U$ and $u_t \in W$. By $\alpha_{\mathbf{A}}(u_r, u_t)$ we denote the number

$$\frac{card(REC_{\mathbf{A}}(u_r, u_t) \cap X_{\mathbf{A}}(u_r))}{card(REC_{\mathbf{A}}(u_r, u_t))}$$

called *the coefficient of classification of* u_t by object u_r, where $REC_{\mathbf{A}}(u_r, u_t)$ is the set

$$\{u \in U : card(EQL_{\mathbf{A}}(u, u_t)) > 0 \wedge EQL_{\mathbf{A}}(u_r, u_t) \subseteq EQL_{\mathbf{A}}(u, u_r)\}.$$

It is easy to see that the object u_r exactly classifies an object u_t iff $\alpha_{\mathbf{A}}(u_r, u_t) = 1$.

If $u_t \in W, v \in V_{\mathbf{A}}^d$ and $\alpha \in [0,1]$ then by $REC_{\mathbf{A}}^v(\alpha, u_t)$ we denote the set $\{u \in U : d(u) = v \wedge card(EQL_{\mathbf{A}}(u, u_t)) > 0 \wedge \alpha_{\mathbf{A}}(u, u_t) \geq \alpha\}$.

Now we present an example of classification algorithm. It is based on the value $REC_{\mathbf{A}}^v(\alpha, u_t)$ for any tested objects u_t, the fixed threshold α and decision values $v \in \{v_d^1, ..., v_d^{r(d)}\}$.

Algorithm A2. *Classification by generation of relevant decision rules*
Input:

1. $\mathbf{A} = (U, A \cup \{d\})$ *is a given decision table, where* $U = \{u_1, ..., u_n\}$ *and* $A = \{a_1, ..., a_m\}$; $\mathbf{A}$ *is a subtable of a universal decision table* $\mathbf{W} = (W, A \cup \{d\})$ *(i.e.* $U \subseteq W$),
2. *a tested object* $u_t \in W$.
3. *a classification threshold* α_0 *(e.g.* $\alpha_0 = 0.9$).

Output: decision value for object u_t.

Data structure:
OA : array 1...n of boolean values (0 or 1),
OL : integer list of object numbers (maximum size: n),
AL : integer list of attribute numbers (maximum size: m),
DVQ : integer array 1...card($V_{\mathbf{A}}^d$) of decision value weights.

Method:

For $i = 1$ to n do $OA[i] = 0$.

For $i = 1$ to n do

   If $OA[i] = 0$ then

   begin

      $AL = EQL_{\mathbf{A}}(u_t, u_i)$

      If $card(AL) > 0$ than

      begin

         $OL = \emptyset$ *and* $count = 0$

         For $j = 1$ to n do

         begin

            If $AL \subseteq EQL_{\mathbf{A}}(u_t, u_j)$ than

            begin

               $count = count + 1$

               If $d(u_i) = d(u_j)$ than $OL = OL \cup u_j$

            end

         end

         If $\frac{card(OL)}{count} \geq \alpha$ than

            for any $u_l \in OL \subseteq \{u_1, ..., u_n\} = U$ do $OA[l] = 1$

      end

   end

For $i = 1$ to $card(V_{\mathbf{A}}^d)$ calculate $DVQ[i] = \frac{card(\{u_j \in U : d(u_j) = v_i \wedge OA[j] = 1\})}{card(\{u_j \in U : d(u_j) = v_i\})}$.

Extract $MDVQ = \left\{ v_i \in V_{\mathbf{A}}^d : DVQ[i] = max_{j \in \{1, ..., card(V_{\mathbf{A}}^d)\}} \{DVQ[j]\} \right\}$

Randomly choose decision value v_g from $MDVQ$.

Return v_g □

It is easy to see that the time and space complexity of Algorithm A2 are of order $O(n^2 \cdot m)$ and $O(n + m + card(V_{\mathbf{A}}^d))$, respectively.

One can treat our method as a method of generation of those decision rules only (with coefficient of classification not lower than the fixed threshold) which can be involved in the classification of a given tested object u_t. It is not necessary to compute all decision rules and then to match of u_t against all of them.

One can show that results of the above algorithm are equivalent, in a sense, to the results of the algorithm based on calculating all decision rules with consistency coefficient not lower than the fixed threshold and the global strength of rule set (see Section 2) as a strategy for conflict resolving.

Let us observe that the presented method generates only rules relevant to tested objects. This can save time of computation because it is not necessary to match a new object against all decision rules generated for decision table.

4   Experiments with Data

We present the results of experiments performed on the following three medical data sets: lymphography, breast cancer and primary tumor (see [8]). The medical

Data	Algorithms	Coefficients μ_0 or α_0	Number of rules	Time Train	Time Test	Error rate Train	Error rate Test
Lymphography	A1	0.9	175	57.32	0.05	0.085	0.193
	A2	0.9	—	54.18	0.38	0.095	0.200
Breast cancer	A1	0.75	647	50.33	0.58	0.261	0.271
	A2	0.75	—	10.80	1.16	0.261	0.271
Primary tumor	A1	0.75	6599	625.78	8.42	0.366	0.679
	A2	0.75	—	35.62	3.77	0.392	0.697

Table 1. Comparison of selected methods with rules and without rules

data used in our experiments were obtained from the University Medical Centre, Institute of Oncology, Ljubljana, Yugoslavia. Thanks go to M. Zwitter and M. Soklic for providing the data (see [8]).

The lymphography data consists of 18 condition attributes (nominal attributes with small numbers of values), one decision attribute (4 decision classes) and 148 objects.

The breast cancer data consists of 9 condition attributes (nominal attributes with small numbers of values), one decision attribute (2 decision classes) and 286 objects.

The primary tumor data consists of 17 condition attributes (nominal attributes with small numbers of values), one decision attribute (22 decision classes) and 339 objects.

We have applied the cross-validation method of estimating error rate (see e.g. [9]). Ten-fold cross validation was performed by using tested classification algorithms.

The algorithm presented in this paper is implemented in object-oriented programming library: "RSES-lib", creating the computational kernel of the system "ROSETTA" (see [12] for more details). The computers used for experiments were HP Workstations (series 9000, model 712/60MHz, 12.9 MFlops, 79 MIPS).

We report results of experiments performed with help of classification algorithm allowing to predict decision for new objects by computing only decision rules relevant for new objects classification (algorithm A2 - see Section 3) We compare there results with those obtained by algorithm A1 based on calculating all decision rules with consistency coefficient for rules not lower than the fixed threshold μ_0 and the global strength strategy of rule set (see Section 2). Table 1 shows the results of the considered classification algorithms for the medical data sets. In case of the cross-validation method we present the average (from all folds) values of error rate, time of computation and number of rules.

Table 2 shows error rates for other classification systems obtained using medical data sets. The tested in this paper learning systems fall into two categories:

1. methods based on Decision Trees: Assistant Professional (see [3]), C4.5 (see [15]).
2. methods based on Decision Rules: AQ15 (see [8]), CN2 (see [4]), Naive LERS and New LERS (see [6]).

	Error rate		
System	Lymphography	Breast cancer	Primary tumor
Assistant Professional	0.240	0.220	0.560
C4.5	0.230	—	0.600
AQ15	0.180-0.200	0.320-0.340	0.590-0.710
CN2	0.180	0.320-0.340	0.550
Naive LERS	0.380	0.490	0.790
New LERS	0.190	0.300	0.670
RSES-lib	0.193	0.271	0.679

Table 2. A comparison results for medical data

It is easy to see that the our results are fully comparable with results obtained when using other systems.

5 Summary

We have presented an example of classification algorithm for computing decision rules relevant to particular new objects. The experiments show that the results of these algorithm are similar to results of the algorithm using the whole set of decision rules generated on training cases. From experiments it follows, that algorithms generating only decision rules relevant to new cases are faster than algorithm based on decision rule generation especially when they are applied to large decision tables. Performed experiments have also proved that results of classification algorithms based on rough set theory are fully comparable with results of known from literature alternative classification systems.

Acknowledgment: This work was supported by the Polish State Committee for Scientific Research grant No. 8T11C01011 and the Research Program of the European Union: ESPRIT-CRIT2 No. 20288.

References

1. Bazan, J.: A Comparison of Dynamic and non-Dynamic Rough Set Methods for Extracting Laws from Decision Table. To appear in Rough Sets in Knowledge Discovery, L. Polkowski and A. Skowron (eds.), Physica Verlag, 46 pages.
2. Brown, E., M.: Boolean reasoning. Kluwer Academic Publishers, Dordrecht (1990)
3. Cestnik, B., Kononenko, I., Bratko, I.: ASSISTANT 86: A knowledge elicitation tool for sophisticated users. In: Proceedings of EWSL-87, Bled, Yugoslavia (1987) 31–47
4. Clark, P., Niblett, T.: The CN2 induction algorithm. Machine Learning **3** (1989) 261–284
5. Dzeroski, S.: Handling noise in inductive logic programming. MSc Thesis, Dept. of EE and CS, University of Ljubljana, Slovenia (1991)
6. Grzymała-Busse, J. W.: A new version of the rule induction system LERS. Fundamenta Informaticae **31** (1997) 27–39

7. Michalski, R., Carbonell, J., G. and Mitchel, T., M. (eds.): Machine learning 1, Tioga/Morgan Kaufmann, Los Altos, CA (1983)
8. Michalski, R.,S., Mozetic, I., Hong, J. and Lavrac, N.: The multi-purpose incremental learning system AQ15 and its testing to three medical domains. In: Proceedings of AAAI-86, Morgan Kaufmann, San Mateo, CA (1986) 1041–1045
9. Michie, D., Spiegelhalter, D., J., Taylor, C., C.: Machine learning, neural and statistical classification. Ellis Horwood, New York (1994)
10. Mollestad, T.: A rough set approach to default rules data mining. PhD Thesis, supervisor J. Komorowski, Norvegian Institute of Technology, Trondheim, Norway (1996)
11. Nguyen, S. Hoa, Nguyen, H. Son: Some efficient algorithms for rough set methods. In: Proceedings of the Sixth International Conference, Information Procesing and Management of Uncertainty in Knowledge–Based Systems (IPMU'96), July 1–5, Granada, Spain (1996) 2 1451–1456
12. Øhrn, A., Komorowski, J.: Rosetta – A rough set toolkit for analysis of data. In: Proceedings of Third International Joint Conference on Information Sciences (JCIS'97), Durham, NC, USA, March 1–5, 3 (1997) 403–407
13. Pawlak, Z.: Rough sets: Theoretical aspects of reasoning about data. Kluwer Academic Publishers, Dordrecht (1991)
14. Piasta, Z., Lenarcik, A., Tsumoto S.: Machine discovery in databases with probabilistic rough classifiers. In: S. Tsumoto, S. Kobayashi, T. Yokomori, H. Tanaka, and A. Nakamura (eds.): Proceedings of the Fourth International Workshop on Rough Sets, Fuzzy Sets, and Machine Discovery (RSFD'96), The University of Tokyo, November 6–8 (1996) 353–359
15. Quinlan, J., R.: C4.5: Programs for machine learning. Morgan Kaufmann, San Mateo, California (1993).
16. Skowron , A., Rauszer, C.: The discernibility matrices and functions in information systems. In: R. Słowiński (ed.): Intelligent Decision Support – Handbook of Applications and Advances of the Rough Sets Theory, Kluwer Academic Publishers, Dordrecht (1992) 331–362
17. Skowron, A.: Boolean reasoning for decision rules generation. In: J. Komorowski, Z.W. Ras (eds.), Proceedings of of the Seventh International Symposium on Methodologies for Intelligent Systems (ISMIS'93), Trondheim, Norway, June 15–18, 1993, Lecture Notes in Computer Science 689 (1993) 295–305
18. Słowiński, R., Stefanowski, J.: RoughDAS and roughClass' software implementations of the rough sets approach. In: R. Słowiński (ed.): Intelligent Decision Support – Handbook of Applications and Advances of the Rough Sets Theory, Kluwer Academic Publishers, Dordrecht (1992) 445–456
19. Wróblewski, J.: Finding minimal reducts using genetic algorithm (extended version). In: P.P. Wang (ed.), Second Annual Joint Conference on Information Sciences (JCIS'95), September 28 – October 1, Wrightsville Beach, North Carolina, USA (1995) 186–189
20. Ziarko, W.: Variable precision rough set model. Journal of Computer and System Sciences 40 (1993) 39–59

Rule+Exception Modeling Based on Rough Set Theory

Yujian Zhou and Jue Wang

AI Lab, Institute of Automation, Chinese Academy of Sciences, Bejing 100080
CHINA
wangj@sunserver.ia.ac.cn

[Abstract.] In this paper Rough Set Theory (RS) is employed to discuss "rule+exception" modeling, which will have fewer rules compared with rule-based modeling and fewer exceptions compared with example-based modeling. An attribute reduction strategy based on discernibility matrix is described. We attempt to consider what kind of data sets are suitable for the model, and how to distinguish exceptions within the data sets. To illustrate the principle the psychological model of Nosofsky's category learning is simulated, and three more complex examples are provided.

1 Introduction

In Shepard, Hovland, and Jenkins' psychological study [4], six types of classification problems on category learning were presented (see sec. 5 in this paper). Nosofsky, Gluck, and Glauthier [1] conducted more complicated replications and extensions of this classic study. Later Nosofsky, Palmeri, and McKinley [2] explained the set of learning data by the RULEX model based on the "Rules plus Exceptions" model. They claimed that for problems of type 1 and 2, concise rules are available; for type 3 , 4 and 5, concise rules and a few additional exceptions are needed; for type 6 however, the only way is to store all of the examples in the memory. This seems to imply three different types of data sets which correspond to three different models under the consistent classification constraints: (1) can be modeled as concise rules; (2) can be modeled as concise rules plus some exceptions; (3) can not be modeled as concise rules and need to store examples. We believe it is important that different data sets require different models. Especially, if some data sets adopt a "rule+exception" model, they will need fewer rules compared with the rule-based model, and will have fewer exceptions compared with the example-based model.

In order to discuss the above three models, the following two questions have to be addressed first: (1) For a given data set, which model is suitable, rule-based, example-based, or "Rule+Exception"-based? (2) Given a data set which is suitable for "Rule+Exception," how to distinguish exceptions within data sets?

L. Polkowski and A. Skowron (Eds.): RSCTC'98, LNAI 1424, pp. 529–536, 1998.

2 An Example

Table 1 is the "Car Classification" data set taken from [7], where mileage is the dependent variable.

Table 1. The car example

No.	Size	Cyl	Turbo	Fuelsys	Displace	Comp	Power	Trans	Weight	Mileage
1	comp	6	y	EFI	med	high	high	auto	med	med
2	comp	6	n	EFI	med	med	high	manual	med	med
3	comp	6	n	EFI	med	high	high	manual	med	med
4	comp	4	y	EFI	med	high	high	manual	light	high
5	comp	6	n	EFI	med	med	med	manual	med	med
6	comp	6	n	2-BBL	med	med	med	auto	heavy	low
7	comp	6	n	EFI	med	med	high	manual	heavy	low
8	subcomp	4	n	2-BBL	small	high	low	manual	light	high
9	comp	4	n	2-BBL	small	high	low	manual	med	med
10	comp	4	n	2-BBL	small	high	med	auto	med	med
11	subcomp	4	n	EFI	small	high	low	manual	light	high
12	subcomp	4	n	EFI	med	med	med	manual	med	high
13	comp	4	n	2-BBL	med	med	med	manual	med	med
14	subcomp	4	y	EFI	small	high	high	manual	med	high
15	subcomp	4	n	2-BBL	small	med	low	manual	med	high
16	comp	4	y	EFI	med	med	high	manual	med	med
17	comp	6	n	EFI	med	med	high	auto	med	med
18	comp	4	n	EFI	med	med	high	auto	med	med
19	subcomp	4	n	EFI	small	high	med	manual	med	high
20	comp	4	n	EFI	small	high	med	manual	med	high
21	comp	4	n	2-BBL	small	high	med	manual	med	med

RS reduction [3] is employed to get the rule set, which is shown in the left side of table 2. If the 20th example in table 1 is considered as an exception, another set of rules, which has fewer and more concise rules, is found by RS reduction, as shown in the right side of table 2 (In table 2 an asterisk means that feature can be ignored in the corresponding rule.) "Rule+Exception" model needs two attributes and four rules compared to four attributes and six rules needed by the rule only model, with one exception (the 20th example conflicts with rule 1 in the right side of table 2). For roughness, rule model is 1, while rule+exception model is about 0.95.

3 Principle

The key idea of the "rule+exception" model is how to select the set of exceptions from a given data set. First we describe our algorithm, we will justify it later in this section.

Table 2. The complete rule set and the reduced rule set.

Rule					Rule + Exception		
size	fuelsys	displace	weight	mileage	size	weight	mileage
comp	*	med	med	med	comp	med	med
comp	2-BBL	small	*	med			
*	*	*	light	high	*	light	high
subcomp	*	*	*	high	subcomp	*	high
*	EFI	small	*	high			
*	*	*	heavy	low	*	heavy	low

(1) Reduce a given database W, obtaining a set of rules R. Create a frequency histogram of the rules.

(2) Select exceptions E according to the criterion described below.

(3) Delete E from W and form a new data set $W' = W - E$.

(4) Repeat step (1) for database W' and find a new rule set R'.

(5) Use R' to test the given database W. Examples conflicting with R' constitute an exception set E'. Generally, $E' \subseteq E$.

$R' + E'$ is the "rule+exception" model of the given data set. If E is empty after step (2), it means the given data set is not suitable for a "rule+exception" model.

In step (1), the set of rules is a reduction based on RS for the given data set. In view of the importance of attribute reduction for obtaining more concise rules and a smaller rule set, we present an attribute reduction strategy with polynomial time complexity $O(n^2)$ based on a discernibility matrix as follows.

Let M be the Discernibility Matrix of a decision table D. An element of M, T_{ij}, is called a *term*. It is the set of all attributes which discern the ith and j^{th} examples in D. $A = \{a_1, a_2, ..., a_n\}$ is the set of all attributes in D. $T_{ij} \subseteq A$ [5].

Let $p(a_k)$ be the number of all terms in M containing attribute a_k, called the *attribute frequency function* of a_k. Then $p(a_k)$ is the significance of the attribute a_k. By this definition, the following reduction strategy can be designed:

Let C_0 be the set of cores on M, and let $R = C_0$.

(1) Let $Q = \{T_{ij} : T_{ij} \cap R \neq \emptyset, i \neq j, i, j = 1, 2, ..., n\}$, $M = M - Q$, and $B = A - R$;

(2) For all $a_k \in B$, compute $p(a_k)$ in M, and let $p(a_q) = \max_k(p(a_k))$;

(3) Let $R = R \cup \{a_q\}$;

(4) Repeat the above steps till $M = \emptyset$;

The set R is an attribute reduction of decision table D.

Although we have proved in theory that this strategy is incomplete for minimal attribute reduction [6], it is very efficient in our experiment.

4   The suitability of the "Rule+Exception" model

Given a consistent data set W, a set of rules R can be obtained by reduction under the consistent decision constraint. If this data set can be modeled as "Rule+Exception," then by applying the above principle, this data set can be divided into a set of exceptions E and the remaining data set W'. Reducing W' again, another set of rules R' can be acquired. Generally, R' can not ensure the correct classification of the exceptions E. Therefore to make the above procedure nontrivial, R' should be more concise than R and the set E should be small. This is the key idea that decides whether or not a data set can be modeled as "Rule+Exception."

Let A and A' be the condition attribute sets of the rules R and R' produced by reducing W and W', respectively. E is a set of exceptions. Let $CARD(.)$ be the function that computes the number of the elements of a given set. Function $f(.)$ is a function for computing the frequency of a rule. For a given data set to be modeled as "Rule+Exception," the following conditions should be satisfied:

$$\max\{f(R_r) : R_r \in R\} \div \min\{f(R_r) : R_r \in R\} > \theta \tag{1}$$

$$CARD(A) - CARD(A') > \alpha \tag{2}$$

$$CARD(R) - CARD(R') > \gamma \tag{3}$$

$$CARD(E) < CARD(R) - CARD(R') + \beta \tag{4}$$

In the above conditions, (1) is a necessary condition for a given data set to be modeled as "Rule+Exception," where θ is an integer larger than 1 (generally θ is a very large positive integer). It means that, if data set W can be divided into rules and exceptions, then there must exist at least one rule whose frequency is fewer than other rules. This condition can also be written as:

$$E = \{R_r : f(R_r) < \theta, R_r \in R\} \tag{5}$$

That is, if the frequency of a rule is below θ, then the examples corresponding to it can be regarded as exceptions. In this paper we use

$$E = \{R_r : f(R_r) = 1, R_r \in R\} \tag{6}$$

as the exception criterion.

If W satisfies conditions (2) and (3), then R' is relatively concise . If condition (4) is satisfied, then there will be few exceptions. Here α, γ , and β are constants related with the given database.

If all the above conditions are satisfied by W, we say W is suitable to be modeled as "Rule +Exception." That is, the rule+exception model is more effective for W than the rule-based or the example-based models.

It is interesting that the above principle also implies the conditions for a data set to be modeled by rule-based or example-based model. If (1) is not satisfied, and

$$\max\{f(R_r) : R_r \in R\} \cong \min\{f(R_r) : R_r \in R\} \tag{7}$$

and (2) and (3) are satisfied, then a rule-based model is recommended for this data set. If none of conditions (1), (2) or (3) are satisfied, and $CARD(A') \cong CARD(A)$ or $CARD(R') \cong CARD(R)$, an example-based model is recommended for the data set.

Table 3. Shepard's six problems

	P_1	P_2	P_3	P_4	P_5	P_6
	a b c d	a b c d	a b c d	a b c d	a b c d	a b c d
1	0 0 0 A	0 0 0 A	0 0 0 A	0 0 0 A	0 0 0 A	0 0 0 A
2	0 1 0 A	0 0 1 A	0 0 1 A	0 0 1 A	0 0 1 A	1 0 1 A
3	0 1 1 A	1 1 1 A	1 0 1 A	0 1 0 A	0 1 0 A	0 1 1 A
4	0 0 1 A	1 1 0 A	0 1 0 A	1 0 0 A	1 1 1 A	1 1 0 A
5	1 0 0 B	1 0 0 B	1 0 0 B	1 1 0 B	1 0 0 B	0 0 1 B
6	1 1 0 B	1 0 1 B	1 1 0 B	1 1 1 B	1 1 0 B	1 0 0 B
7	1 0 1 B	0 1 0 B	1 1 1 B	1 0 1 B	1 0 1 B	0 1 0 B
8	1 1 1 B	0 1 1 B	0 1 1 B	0 1 1 B	0 1 1 B	1 1 1 B

5 Shepard's six problems

In this section, the above principle is used to repeat Nosofsky, Palmeri, and McKinley's study [2] on Shepard's six problems, shown in table 3.

(1) Reduce the above data sets P_i and obtain the rule sets $R_i(i = 1, 2, \ldots, 6)$, as shown in the left side of table 4.

(2) Create the frequency table of $R_i(i = 1, 2, \ldots, 6)$, as shown in the right side of table 4.

Table 4. The rule sets of Shepard's six problems and its frequency table

	The rule set						The frequency table					
	R_1	R_2	R_3	R_4	R_5	R_6		R_1	R_2	R_3	R_4	R_5	R_6
	a d	a b d	a b c d	a b c d	a b c d	a b c d							
1	0 A	0 0 A	0 0 * A	0 0 * A	0 0 * A	0 0 0 A	f(1)	4	2	2	2	1	1
2						1 0 1 A	f(2)						1
3		1 1 A	* 0 1 A	0 * 0 A	0 * 0 A	0 1 1 A	f(3)		2	1	1	1	1
4			0 * 0 A	* 0 0 A	1 1 1 A	1 1 0 A	f(4)			1	1	1	1
5	1 B	1 0 B	1 * 0 B	1 1 * B	1 0 * B	0 0 1 B	f(5)	4	2	1	2	1	1
6			1 1 * B		1 * 0 B	1 0 0 B	f(6)			2		2	1
7		0 1 B		1 * 1 B		0 1 0 B	f(7)		2		1		1
8			* 1 1 B	* 1 1 B	0 1 1 B	1 1 1 B	f(8)			1	1	1	1

(3) According to conditions (2), (3) and (6), exceptions E_i are selected, and $P_i' = P_i - E_i$.

Problems 1 and 2 do not satisfy condition (6), therefore they don't have any exceptions; for problem 3, $E_3 = \{3, 4, 5, 8\}$, $P_3' = \{1, 2, 6, 7\}$; for problem 4, $E_4 = \{3, 4, 7, 8\}$, $P_4' = \{1, 2, 5, 6\}$; for problem 5, $E_5 = \{3, 4, 5, 8\}$, $P_5' = \{1, 2, 6, 7\}$. Because problem 6 does not satisfy conditions (1), (2) and (3), it has no exceptions (or it has no rules).

(4) Reduce the databases P_j' again and obtain rule sets $R_j'(j = 3, 4, 5)$, see the left of table 5.

(5) Compute the frequency table of the reduced rule set $R'_j (j = 3, 4, 5)$, see the right of table 5.

Table 5. The reduced rule set and its frequency table.

The reduced rule set						The frequency of each rules				
$R_3 I$		$R_4 I$		$R_5 I$		$R_3 I$		$R_4 I$		$R_5 I$
Rule	a d	Rule	a d	Rule	a d					
1	0 A	1	0 A	1	0 A	f(1)	3	f(1)	3	f(1) 3
6	1 B	5	1 B	6	1 B	f(6)	3	f(5)	3	f(6) 3

(6) Produce the final exceptions E'_j, which are examples conflicting with R'_j. For problem 3, $E'_3 = \{3, 8\}$; for problem 4, $E'_4 = \{3, 8\}$; for problem 5, $E'_5 = \{4, 8\}$.

Since Problems 1 and 2 have $\max\{f(R_r) : R_r \in R\} = \min\{f(R_r) : R_r \in R\}$, they do not satisfy condition (1), therefore these problems can not be modeled as "Rule+Exception." However, they satisfy conditions (2) and (3), so a rule-based model is suggested for them. Problems 3, 4 and 5 satisfy all the above conditions, so they can use the "Rule+Exception" model. Problem 6 does not satisfy conditions (1), (2) and (3), so an example-based model is suggested.

6   Three more complex databases

In this section, three databases from the UCI repository are selected to illustrate the principle described in this paper. The problems are Voting, which is suitable for a "Rule+Exceptions" model, Moral-Reasoner, suitable for a rule-based model, and Soybean, which is suitable for an example-based model.

The Voting data set is the 1984 questionnaire on the 435 congressmen in the U. S. House of Representatives about 16 key problems. Using above method we can analyze this database as below:

(1) Through reduction, 44 rules with 9 attributes are produced. The frequency histogram of the rules is shown in figure 1.

(2) Delete all the examples (21 in all) that are used only once. Reduce the remaining database again, and 6 rules with 4 attributes are obtained.

(3) Use the 6 rules to test all of the 435 examples. 13 examples are found conflicting with the rules. Thus six rules and thirteen exceptions are a solution to the Voting problem. Obviously, this database satisfies all the conditions for using the "Rule+Exception" model, and is a typical database for it. Its roughness is about 0.97. The 6 rules are shown in table 6. It can be interpreted in natural language as:

The main divergences between Republicans and Democrats are: Republicans agree on the physician-fee-freeze (PFF) and anti-satellite-test-ban (ASTB) problems, while Democrats disagree or keep silence. On the adoption-of-the-budget-resolution (ABR) and synfuels-corporation-cutback (SCC) problems, Democrats agree while Republicans disagree.

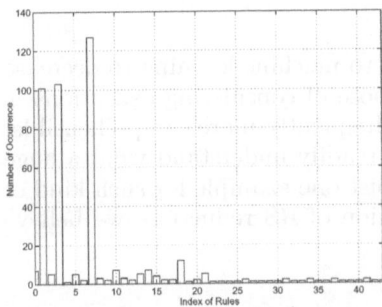

Fig. 1. The frequency histogram of the Voting experiment

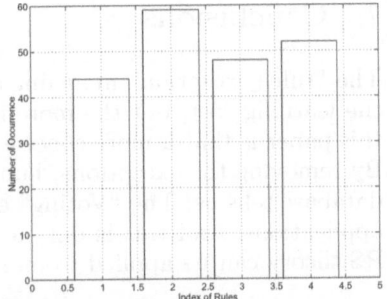

Fig. 2. The frequency histogram of the Moral-reasoner experiment

Table 6. The rule set of the Voting data set

Rule	ABR	PFF	ASTB	SCC	PARTY
0	n	y	*	*	republican
1	*	n	*	*	democrat
2	*	y	y	*	republican
3	y	*	n	y	democrat
4	*	?	*	*	democrat
5	*	y	*	n	republican

If the effect of exceptions is ignored, the above description can be regarded as the conclusion of the questionnaire. If the effect of the exceptions has to be considered, it is not difficult to analyze their difference from the final rules and translate them into natural language.

If the given data set is more suited for a rule-based model, the method in this paper will recommend a rule-based model for the given data set rather than a "Rule+Exception" model. Let's see the next example.

The Moral-Reasoner database has 202 examples and 23 attributes. After reduction, only 4 rules with 1 attribute remain. Its corresponding histogram is shown in figure 2.

As Shepard's problem 2, this database does not satisfy condition (1), none of the rules are infrequent, so it is recommended to be modeled as rule-based.

The Soybean database contains 307 examples and 35 condition attributes. Through RS reduction, 116 rules with 9 attributes are produced. Since rules of different classes have quite different histograms, only those rules which belong to the alternarialeaf-spotand and frog-eye-leaf-spot classes satisfy condition (1). And when the "Rule+Exception" principle is applied to the examples of the two classes, the numbers of attributes and rules become 8 and 111 respectively, with 14 exceptions. As Shepard's problem 6 it can not perfectly satisfy condition (2), (3) and (4), and an example-based model is recommended.

7   Conclusions

The "rule+exception" modeling can be applied to machine learning to decrease the learning cost, but the more important purpose of considering exceptions in this paper is that a more concise set of rules is frequently more comprehensible. By removing the exceptions, humans can more readily understand what a huge database tells us. The "Voting" experiment is just one example for such kind of applications. And this is only a direct application of RS reduction, we believe RS theory can be applied to more broad areas.

Acknowledgments Research supported by "863" program and NSF of China. The authors are grateful to all the people who contributed to UCI Repository used in the empirical study.

References

1. Nosofsky, R., Gluck, M., Glauthier, P.: Comparing models of rule-based classification learning: A replication and extension of Shepard, Hovland, and Jenkins(1961). Memory & Cognition. 22,(3), (1994) 352–369
2. Nosofsky, R., Palmeri, J., McKinley, C.: Rule-Plus-Exception Model of Classification Learning. Psychological Review. Vol. 101, No. 1, (1994) 53–79
3. Pawlak, Z.: Rough Sets. Int. J. Comput. Inf. Sci. **11** (1982) 341–356
4. Shepard, N., Hovland, C., Jenkins, H.: Learning and memorization of classification. Psychological Monographs. 75 (13, Whole No.517), (1961)
5. Skowron, A., Rauszer, C.: The Discernibility Matrices and Functions in Information Systems, Intelligent Decision Support. Handbook of Applications and Advances of the Rough Sets Theory. (1992) 331–362.
6. Wang, J., Miao, D.: Analysis on attribute reduction strategies of rough set. Journal of Computer Science and Technology. Vol 13 No.2, (1997)
7. Ziarko, W.: Discovery in Databases. Edited by Piatetsky-Shapiro, G. and Frawley, W. AAAI/MIT press. (1990) 213–228

On Finding Optimal Discretizations
for Two Attributes*

Bogdan S. Chlebus   and   Sinh Hoa Nguyen

Instytut Informatyki
Uniwersytet Warszawski
ul. Banacha 2, 02-097 Warszawa, Poland
e-mail: {chlebus,hoa}@mimuw.edu.pl

Abstract. We show that finding optimal discretization of instances of
decision tables with two attributes with real values and binary decisions
is computationally hard. This is done by abstracting the problem in such
a way that it regards partitioning points in the plane into regions, subject
to certain minimality restrictions, and proving them to be NP-hard. We
also propose a new method to find optimal discretizations.

1   Introduction

Discretization of attributes with real values is an important problem in knowledge discovery and data mining. It is an indispensable tool in data analysis and
extraction of decision rules from decision tables. A lot of effort has been spent
to find effective methods for real value attribute discretization (see [1,2,3,10]).
We show that certain optimization problems motivated by the discretization
problem are NP-hard.

To facilitate extracting decision rules from a decision table with real value
attributes, the decision tables are usually discretized. Among the well known
discretization methods are those based on the equal width and equal frequency
intervals, statistical tests [7], entropy [1,2,3], adaptive quantizers and dynamic
programming. All these methods are heuristics for discretization of data.

We restrict our attention to the discretization that selects a set of cut points of
attributes, which determine a partition of the real value attributes into intervals.
The set of cuts determines a grid in k-dimensional space with $\prod_{i=1}^{k} n_i$ regions,
where k is the number of attributes and n_i is the number of intervals of the ith
attribute. It was shown in [9,10] that the problem Optimal Discretization, to
find a consistent discretization of a given decision table with minimal number
of cuts, is NP-hard. In this paper we discuss the computational complexity of
the problem of finding a consistent discretization with the minimum number of
regions. The main result of this paper is showing that the decision problem of
checking if there is a consistent set of cuts such that the grid defined by them
contains at most K regions, for a given K, is NP-complete. We also improve the

---

* This research was partly supported by Polish State Committee for Scientific Research
grant No. 8T11C03614, No. 8T11C01011 and Research Program of European Union
- ESPRIT-CRIT2 No. 20288.

result of [10] by showing that the problem Optimal Discretization remains NP-hard even if a given decision table is restricted to two attributes. These results justify the search for approximation algorithms yielding as good discretizations as possible.

2   Basic notions

An *information system* [11] is a pair $\mathbf{A} = (U, A)$, where U is a non-empty finite set called the *universe* and A is a non-empty finite set of *attributes*, i.e. $a : U \rightarrow V_a$ for $a \in A$, where V_a is called *the value set of a*. Elements of U are called *objects*. An information system of the form $\mathbf{A} = (U, A \cup \{d\})$ is called a *decision table*, where $d \notin A$ is called the *decision attribute* and the attributes from A are called *condition attributes*. Let $V_d = \{1, \dots, r(d)\}$. The decision d determines a partition of U into decision classes: $\{C_1, ..., C_{r(d)}\}$, where $C_k = \{x \in U : d(x) = k\}$ for $1 \leq k \leq r(d)$. Any non-empty set $B \subseteq A$ defines a *B-information* function by $Inf_B(x) = \{(a, a(x)) : a \in B$ for $x \in U\}$. A decision table $\mathbf{A}$ is called *consistent* if $Inf_A(x) \neq Inf_A(y)$ for any $x, y \in U$ such that $d(x) \neq d(y)$.

Let $\mathbf{A} = (U, A \cup \{d\})$ be a decision table with real value attributes $A = \{a : a : U \rightarrow V_a, V_a \subset \mathbb{R}\}$ and $d : U \rightarrow \{1, ..., r(d)\}$. Any pair (a, c) where $a \in A$ and $c \in \mathbb{R}$ will be called a *cut on V_a*. For $a \in A$, any set of cuts: $\{(a, c_1^a), (a, c_2^a), \dots, (a, c_{k_a}^a)\}$ on V_a defines a partition $\mathbf{P}_a$ of V_a into sub-intervals $\mathbf{P}_a = \{[c_0^a, c_1^a), [c_1^a, c_2^a), \dots, [c_{k_a}^a, c_{k_a+1}^a)\}$ where $l_a = c_0^a < c_1^a < c_2^a < \dots < c_{k_a}^a < c_{k_a+1}^a = r_a$ and $V_a = [c_0^a, c_1^a) \cup [c_1^a, c_2^a) \cup \dots \cup [c_{k_a}^a, c_{k_a+1}^a)$. Therefore, any set of cuts $\mathbf{P} = \bigcup_{a \in A} \mathbf{P}_a$ defines new a decision table $\mathbf{A}^{\mathbf{P}} = (U, A^{\mathbf{P}} \cup \{d\})$, where $A^{\mathbf{P}} = \{a^{\mathbf{P}} : a^{\mathbf{P}}(x) = i$ iff $a(x) \in [c_i^a, c_{i+1}^a)$, for $x \in U$ and $i \in \{0, .., k_a\}\}$. A set of cuts $\mathbf{P}$ is $\mathbf{A}$-*consistent* if it discerns all pairs of objects with different decisions.

3   Optimal splitting in $\mathbb{R}^2$

In this section we consider a consistent decision table $\mathbf{A} = (U, \{a, b\} \cup \{d\})$ with two real value condition attributes a, b and binary decision $d : U \rightarrow \{0, 1\}$. Such a decision table is a representation of the set of points $S = \{(a(u_i), b(u_i)) : u_i \in U\}$ in the plane $\mathbb{R}^2$ partitioned into two disjoint categories $S = S_1 \cup S_2$. We can determine such a partition by assigning black and white colors to the points. Any cut (a, c) on a (or (b, c) on b), where $c \in \mathbb{R}$, can be represented by a vertical (or horizontal) line. The set of cuts is $\mathbf{A}$-consistent if the set of lines representing them defines a partition of the plane into regions in such a way that if any two points are in the same region then they have the same color. Such a set of lines is said to be *consistent*.

PROBLEM 2-OS: Optimal Splitting in $\mathbb{R}^2$
Input: A set S of points $P_1, ..., P_n$ in the plane $\mathbb{R}^2$, partitioned into two disjoint
   categories S_1, S_2 and a natural number T.
Question: Is there a consistent set of lines such that the partition of the plane
   into regions defined by them consists of at most T regions?

Theorem 1. *2-OS is NP-complete.*

Proof. It is clear that 2-OS is in NP. The NP-hardness part of the proof is done
by reducing 3SAT to 2-OS (cf. [5]).
   Let $\Phi = C_1 \wedge ... \wedge C_k$ be an instance of 3SAT. We construct an instance I_Φ
of 2-OS such that Φ is satisfiable iff there is a sufficiently small consistent set
of lines for I_Φ. The description of I_Φ will specify a set of points S, which will
be partitioned into two subsets of white and black points. A pair of points with
equal horizontal coordinates is said to be *vertical*, similarly, a pair of points
with equal vertical coordinates is *horizontal*. If a configuration of points includes
a pair of horizontal points p_1 and p_2 of different colors, then any consistent
set of lines will include a vertical line L separating p_1 and p_2, which will be
in the vertical strip with p_1 and p_2 on its boundaries. Such a strip is referred
to as a *forcing strip*, and the line L as *forced* by points p_1 and p_2. Horizontal
forcing strips and forced lines are defined similarly. The instance I_Φ has an
underlying grid-like structure consisting of vertical and horizontal forcing strips.
The rectangular regions inside the structure and consisting of points outside the
strips are referred to as *f-rectangles* of the grid. In the figures that follow and
illustrate the construction of I_Φ, the forcing strips are depicted simply as lines.
These lines make a grid of rectangles which represent the f-rectangles.
   For each propositional variable p occurring in C use one special row and
one special column of f-rectangles. In the f-rectangle that is at the intersection
of the row and column place an instance R_p of the *variable configuration*, as
depicted in Figure 1 (a). Notice that the variable configuration requires at least
one horizontal and one vertical line to separate the white points from the black
ones. If only one such vertical line occurs in a consistent set of lines, then it
separates either the left or the right white point from the central black one,
what we interpret as an assignment of the value *true* or *false* to the propositional
variable.

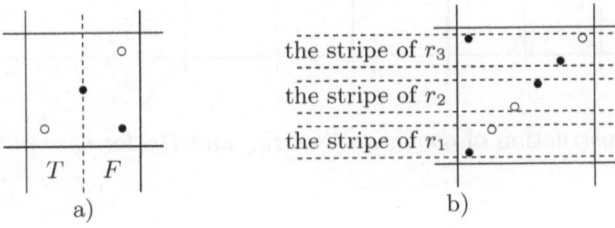

Fig. 1. a) The variable configuration. b) The clause configuration.

For each clause C in Φ use one special row and one special column of f-rectangles. In the region of the intersection of this row and column place an instance R_C of the *clause configuration*, which is depicted in Figure 1 (b). Notice that R_C requires at least three lines to separate the white from the black points, and among them at least one vertical. Let C be of the form $C = r_1 \vee r_2 \vee r_3$, where the variables in the literals r_i are all different. Subdivide the row of f-rectangles of C into three strips corresponding to the literals. For each r_i create an instance $R_{C,i}$ of the *literal configuration*, which consists of one black and one white points, of distinct vertical and horizontal coordinates. Place $R_{C,i}$ at the intersection of the horizontal strip of r_i and the column of the variable of r_i; if the variable is p, then either in the 'true' part of R_p, if $r_i = p$, or in the 'false' part of R_p, if $r_i = \neg p$. The pair of points in $R_{C,i}$ has their vertical coordinates equal to the vertical coordinates of the pair of points in R_C in the strip of r_i. An example of this construction is depicted in Figure 2. Column x_{p_i} and row y_{p_i} correspond to variable p_i, row y_C corresponds to clause C.

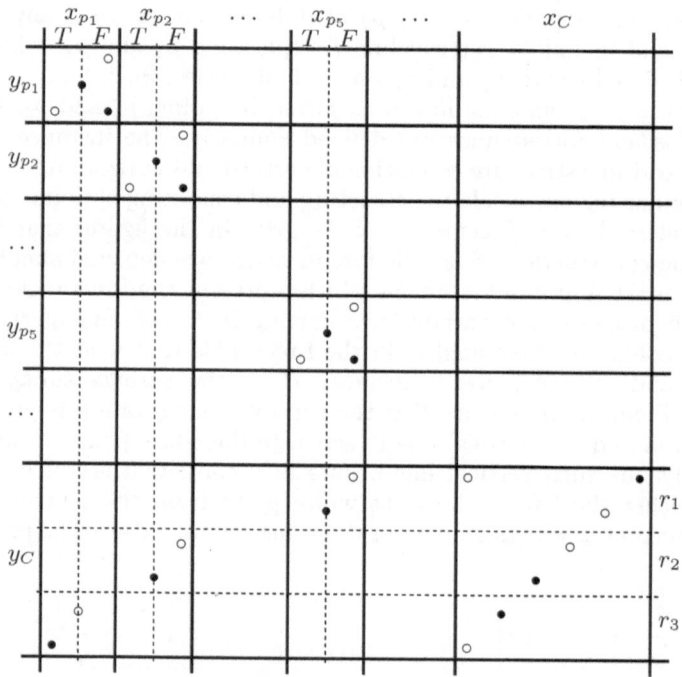

Fig. 2. Construction of configurations R_{p_i} and R_C for $C = p_1 \vee \neg p_2 \vee \neg p_5$

Let the underlying grid of of f-rectangles be minimal to accommodate this construction. We need to add a number of horizontal rows of f-rectangles, so that each vertical line will contribute sufficiently more regions than a horizontal line, here we mean lines different from those defining the grid of f-rectangles. To be specific, let the number of these rows of f-rectangles be equal to the size of Φ plus 1 (size is the number of occurences of symbols).

Suppose conceptually that a consistent set of lines W includes exactly one vertical and one horizontal line per each R_p, and exactly one vertical and two horizontal lines per each R_C, let L_1 be the set of all these lines. There is also the set L_2 of lines inside the forcing strips, precisely one line per each strip. We have $W = L_1 \cup L_2$. Let the number of horizontal lines in W be equal to l_h and vertical to l_v. That many lines create $T = (l_h - 1) \cdot (l_v - 1)$ regions, and this number is the last component of I_Φ.

Next we show the correctness of the reduction. Suppose first that Φ is satisfiable, let us fix a satisfying assignment of logical values to the variables of Φ. The consistent set of lines is determined as follows. Place one line into each forcing strip. For each variable p place one vertical and one horizontal line to separate points in R_p, the vertical line determined by the logical value assigned to p. Each configuration R_C is handled as follows. Let C be of the form $C = r_1 \vee r_2 \vee r_3$. Since C is satisfied, at least one $R_{C,i}$, say $R_{C,1}$, is separated by the vertical line that separates also R_p, where p is the variable of r_1. Place two horizontal lines to separate the remaining $R_{C,2}$ and $R_{C,3}$. They also separates two pairs of points in R_C. Add one vertical line to complete separation of the points in R_C. All this means that there is a consistent set of lines which creates T regions.

On the other hand, suppose that there is a consistent set of lines for I_Φ, which determines at most T regions. The number T was defined in such a way that two lines must separate each R_p and three lines each R_C, in the latter case at least one of them vertical. Notice that a horizontal line contributes fewer regions than a vertical one because the grid of splitting strips contains much more rows than columns. Hence one vertical line and two horizontal lines separate each R_C, because changing horizontal to vertical would increase the number of regions beyond T. It follows that, for each clause $C = r_1 \vee r_2 \vee r_3$, at least one $R_{C,i}$ is separated by such a vertical line which also separates R_p, where p is the variable of r_i, and this yields a satisfying truth assignment.

4   Optimal discretization in $\mathbb{R}^2$

In this section we consider the problem of finding a consistent partition of the plane using minimal number of cuts. We analyze a following decision problem:

PROBLEM k-D2: Discretization in $\mathbb{R}^2$ by at most k cuts.
Input: Set S of points $P_1, ..., P_n$ in the plane, partitioned into two disjoint categories S_1, S_2 and a natural number k.
Question: Is there a consistent set of at most k lines?

We show that the problem *Set Cover* (see [5]) is reducible to k-D2.

Theorem 2. *k-D2 is NP-complete.*

Proof. An instance of I of Set Cover consists of $S = \{u_1, u_2, ..., u_n\}$, $\mathcal{F} = \{S_1, S_2, ..., S_m\}$, where $S_j \subseteq S$ and $\bigcup_{i=1}^{m} S_i = S$, and an integer M, and the question is if there are M sets from $\mathcal{F}$ whose union contains all elements of

S. We need to construct an instance I' of k-D2 such that I has a positive answer iff I' has a positive answer. The construction of I' is quite similar to the construction described in the previous section. We start by building a grid-like structure consisting of vertical and horizontal forcing stripes. The regions are in rows labeled by $y_{u_1}, ..., y_{u_n}$ and columns labeled by $x_{S_1}, ..., x_{S_m}, x_{u_1}, ..., x_{u_n}$ (see Figure 3). The region (x_{S_j}, y_{u_i}) represents $u_i \in S_i$ and the region (x_{u_i}, y_{u_i}) represents the sets which include u_i. First, for any element $u_i \in S$, define a family $\mathcal{F}_i = \{S_{i_1}, S_{i_2}, ..., S_{i_{m_i}}\}$ of all subsets containing u_i. Then subdivide the row y_{u_i} into m_i strips, corresponding to the subsets from $\mathcal{F}_i$. In the strip labeled by $u_i \in S_j$ place one pair of black and white points inside the region (x_{u_i}, y_{u_j}) and another pair in the column labeled by x_{S_j} (see Figure 3). The points in adjacent strips in the region (x_{u_i}, y_{u_j}) have the same color. In each region (x_{u_i}, y_{u_j}) add a special point in the top left corner with a color different from the color of the point in the top right corner. This point is introduced to force at least one vertical line across a region. Place the configuration R_{u_i} for u_i in the region labeled by (x_{u_i}, y_{u_i}). Examples of R_{u_1} and R_{u_2} where $\mathcal{F}_{u_1} = \{S_1, S_2, S_4, S_5\}$ and $\mathcal{F}_{u_2} = \{S_1, S_3, S_4\}$, are depicted in the Figure 3.

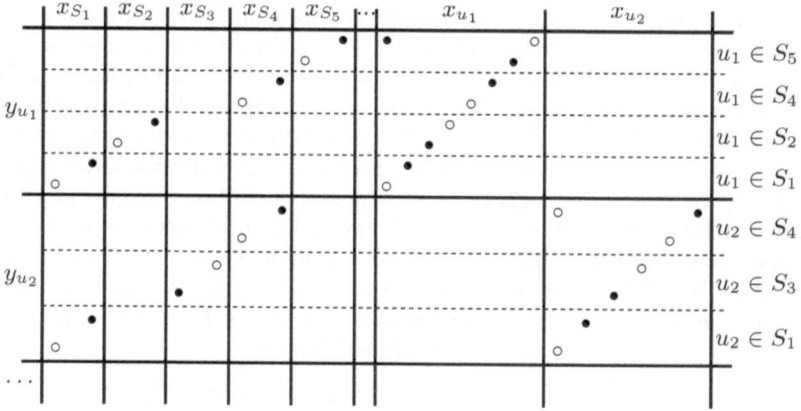

Fig. 3. Construction of configurations R_{u_1} and R_{u_2} where $\mathcal{F}_{u_1} = \{S_1, S_2, S_4, S_5\}$ and $\mathcal{F}_{u_2} = \{S_1, S_3, S_4\}$

The configuration R_{u_i} requires at least m_i lines to be separated, among them at least one vertical. Thus, the whole construction for u_i requires at least $m_i + 1$ lines. Let I' be an instance of k-D2 defined by the set of all points forcing the grid and all configurations R_{u_i} with $k = M + \sum_{i=1}^{n} m_i + (2n + m + 2)$ as the parameter, where the last component $(2n + m + 2)$ is the number of lines defining the grid. If there is a covering of S by M subsets $S_{j_1}, S_{j_2}, ..., S_{j_M}$, then we can construct k lines that separate well the set of points, namely $(2n + m + 2)$ grid lines, M vertical lines in the columns corresponding to $S_{j_1}, S_{j_2}, ..., S_{j_M}$ and m_i lines per element u_i, for $1 \leq i \leq n$.

On the other hand, let us assume that there are k lines separating the points from instance I'. We show that there exists a covering of S by M subsets. There is a set of lines such that, for any $i \in \{1, ..., n\}$, there are exactly m_i lines passing across the configuration R_{u_i}, which is the region labeled by (x_{u_i}, y_{u_i}), among them exactly one vertical line. Hence, there are at most M vertical lines on rows labeled by $x_{S_1}, ..., x_{S_M}$. These lines determine M subsets that cover the whole set S.

5   Methods to learn optimal partitions

In the previous section we have shown NP-hardness of discretization problems for the case of a decision table with two attributes. This implies the hardness of discretization problems with more attributes. In this section we present a heuristic for semi-optimal partition generation for an arbitrary dimensional space. An efficient algorithm for generating a minimal consistent set of cuts was presented in [8], which takes $O(kn \log n)$ time to extract one cut, where n is the number of objects and k is the number of attributes. Below we present an algorithm finding a consistent set of cuts which minimizes the *number of regions*. To simplify the exposition, we present the algorithm for the case of two attributes. The algorithm can be generalized to be able to handle more attributes.

Let $\mathbf{A} = (U, \{a, b\} \cup \{d\})$ be a consistent decision table with two real attributes a and b, where $Card(U) = n$. We assume that the decision d classifies the universe U into m decision classes. Such a decision table can be represented as a set of points $S = \{(a(u_i), b(u_i)) : u_i \in U\}$ in the plane $\mathbb{R}^2$, painted by m colors. The task is to find a set of vertical and horizontal lines that divides the plane into a minimum number of rectangular regions, such that every one contains points of the same color.

Let L be a set of all possible horizontal and vertical lines for the set of points S. The main idea of the algorithm is to reduce L by removing from it many useless lines without loss of consistency. The use of a partition line l is characterized by *a number of regions defined by l* and *a number of point pairs discerned by l*.

The line l is *useless* if both numbers are "small". A given region is *adjacent* to line l if l is one of its boundaries. For every line l we introduce a function $Density(l)$ being a density of regions adjacent to l. Let R^l_{left} and R^l_{right} are the sets of points belonging to regions adjacent to the line l on the left and right, respectively. Let $N(l)$ be the number of regions adjacent to line l. The *density function* is defined as follows: $Density(l) = \frac{Card(R^l_{left}) + Card(R^l_{right})}{N(l)}$

For every partition line l, define two functions measuring its global and local discernibility degree as follows:

$$GlobalDisc(l) = Card\{(u, v) : d(u) \neq d(v), u \in R^l_{left}, v \in R^l_{right}\}$$

$$LocalDisc(l) = Card\{(u, v) : d(u) \neq d(v) \wedge u, v \text{ are discerned by } l \text{ only}\}.$$

A line l can be rejected if $LocalDisc(l) = 0$. A line l is a preferable candidate to be rejected if both $Density(l)$ and $GlobalDisc(l)$ are "small". The algorithm is in four steps:

Step 1. Start with the set L of all possible lines.
Step 2. Find a partition line l with $LocalDisc(l) \neq 0$ of minimum value

of *Density(l)*. If there are several such lines then select the one with the minimum value *GlobalDisc(l)*.

Step 3. Set L to $L - \{l\}$. Update the structure of the regions.

Step 4. If the set of lines L is reducible then **go to** Step 1, otherwise **stop**.

The algorithm can be implemented to run in time $O(n)$ per one line reduction. Therefore a reduction of R lines can be done in time $O(|R|.n)$. The time performance of the algorithm for a k-dimensional space will be $O(|R|.kn)$.

6   Conclusions

We have shown that the problems Optimal Splitting and Optimal Discretization are NP-complete. We have also proposed an approximation algorithm for the optimalization version of Optimal Splitting. The algorithm can be treated as a new discretization method, and it can be used in the preprocessing step for other classification methods.

References

1. Catlett J. (1991). On changing continuous attributes into ordered discrete attributes. In Y. Kodratoff, (ed.), *Machine Learning-EWSL-91*, Porto, Portugal, March 1991, LNAI, pp. 164–178.
2. Chmielewski M.R., Grzymala–Busse J.W. (1994). Global discretization of attributes as preprocessing for machine learning. *Proc. of the III International Workshop on RSSC94*, November 1994, pp. 294–301.
3. Fayyad U. M., Irani K.B. (1992). The attribute selection problem in decision tree generation. *Proc. of AAAI-92*, July 1992, San Jose, CA. MIT Press, pp. 104–110.
4. Fayyad, U. M., Irani, K.B. (1993). Multi-interval discretization of continuous-valued attributes for classification learning. In *Proc. of the 13th International Joint Conference on Artificial Intelligence*, Morgan Kaufmann, pp. 1022-1027.
5. Garey, M.R., Johnson, D.S. (1979). *Computers and Intractability, A guide to the theory of NP-completeness*, W.H. Freeman, San Francisco.
6. Holt R.C. (1993). Very simple classification rules perform well on most commonly used datasets, *Machine Learning* **11**, pp. 63-90.
7. Kerber R. (1992), Chimerge, Discretization of numeric attributes. *Proc. of the Tenth National Conference on Artificial Intelligence*, MIT Press, pp. 123–128.
8. Nguyen S. H., Nguyen H. S.(1996), Some efficient algorithms for rough set methods. *Proc. of the Conference of Information Processing and Management of Uncertainty in Knowledge-Based Systems*, Granada, Spain, pp. 1451–1456.
9. Nguyen, H.S., Nguyen, S.H.(1997). Discretization methods for data mining. In A.Skowron and L.Polkowski (Eds.), *Rough Set in Data Mining and Knowledge Discovery (in preparation)*. Berlin, Springer Verlag.
10. Nguyen H. S, Skowron A. (1995). Quantization of real values attributes, rough set and boolean reasoning approaches. *Proc. of the Second Joint Annual Conference on Information Sciences*, Wrightsville Beach, NC, 1995, USA, pp. 34–37.
11. Pawlak Z.(1991). *Rough sets: Theoretical aspects of reasoning about data*, Kluwer Dordrecht.

Discretization Problem for Rough Sets Methods

Hung Son Nguyen

Institute of Mathematics,
Warsaw University
02-097, Banacha Str. 2, Warsaw Poland
e-mail: son@alfa.mimuw.edu.pl

Abstract. We study the relationship between reduct problem in Rough Sets theory and the problem of real value attribute discretization. We consider the problem of searching for a minimal set of cuts on attribute domains that preserves discernibility of objects with respect to any chosen attributes subset of cardinality s (where s is a parameter given by a user). Such a discretization procedure assures that one can keep all reducts consisting of at least s attributes. We show that this optimization problem is NP-hard and it is interesting to find efficient heuristics for solving this problem.

1 Introduction

Discretization of real value attributes is an important task in data mining, particularly for the classification problem. Empirical results are showing that the quality of classification methods depends on the discretization algorithm used in preprocessing step. In general, discretization is a process of searching for partition of attribute domains into intervals and unifying the values over each interval. Hence discretization problem can be defined as a problem of searching for a *suitable* set of cuts (i.e. boundary points of intervals) on attribute domains.

Usually we are looking for *consistent set of cuts* i.e. preserving the discernibility relation between objects from different decision classes [10, 8]. In previous papers we considered the problem of searching for optimal consistent set cuts, where optimization criteria were defined by number of cuts (OD-problem). It has been shown that such optimization problem was NP-hard [8]. Any discretization algorithm reduces the information on decision tables. Our heuristic for OD-problem called *MD-algorithm* produces a new decision table with *"one reduct"* only. In some applications based on Rough Sets (e.g. dynamic reduct and dynamic rule methods [1]), where reducts are an important tool, it is not enough to obtain the strong rules.

Hence, in some sense we would like to obtain more excessive set of cuts producing new discretized decision table containing more reducts, and, at the same time, reducing the superfluous information. In this paper we consider the problem of searching for minimal set of cuts such that it saves the discernibility between objects with respect to any subset of s attributes. One can show that this problem, called *s-optimal discretization problem* (*s*-OD problem), is also

L. Polkowski and A. Skowron (Eds.): RSCTC'98, LNAI 1424, pp. 545–552, 1998.
© Springer-Verlag Berlin Heidelberg 1998

NP-hard, where the heuristic algorithm is more complicated than in case of OD-problem. Similarly to the case of OD-problem, we propose the method based on Boolean reasoning to solve the s-OD problem.

2   Preliminaries

An *information system* [11] is a pair $\mathbf{A} = (U, A)$, where U is a non-empty, finite set called the *universe* and A is a non-empty, finite set of *attributes*, i.e. $a : U \rightarrow V_a$ for $a \in A$, where V_a is called *the value set of a*. Elements of U are called *objects*. Any non-empty set $B \subseteq A$ defines a B-*information* function by $Inf_B(x) = \{(a, a(x)) : a \in B$ for $x \in U\}$.

For any subset of attributes $B \subseteq A$, an equivalence relation called the B-indiscernibility relation [11], denoted by $IND(B)$, is defined by

$$IND(B) = \{(x, y) \in U \times U : \forall_{a \in B} (a(x) = a(y))\}$$

Objects x, y satisfying relation $IND(B)$ are indiscernible by attributes from B. By $[x]_{IND(B)}$ we denote the equivalence class of $IND(B)$ defined by x. A minimal (in sense of inclusion) subset B of A such that $IND(A) = IND(B)$ is called a reduct of $\mathbf{A}$.

Any information system of the form $\mathbf{A} = (U, A \cup \{d\})$ is called a decision table where $d \notin A$ is called the decision and the elements of A are called conditions. Let $V_d = \{1, \ldots, r(d)\}$. The decision d determines the partition $\{C_1, \ldots, C_{r(d)}\}$ of the universe U, where $C_k = \{x \in U : d(x) = k\}$ for $1 \leq k \leq r(d)$. The set C_k is called the kth decision class of $\mathbf{A}$.

For any subset of attributes $B \subseteq A$ we define an equivalence relation, called the *relative B-indiscernibility relation* and denoted by $IND(B, d)$, as follows

$$IND(B, d) = \{(x, y) \in U \times U : (d(x) = d(y)) \vee (Inf_B(x) = Inf_B(y))\}$$

A minimal (in sense of inclusion) subset B of attributes from A such that $IND(B, d) = IND(A, d)$ is called *a relative reduct of* $\mathbf{A}$.

Let $\mathbf{A} = (U, A \cup \{d\})$ be a decision table and $B \subseteq A$. We define a function $\partial_B : U \rightarrow 2^{\{1, \ldots, r(d)\}}$, called the generalized decision in $\mathbf{A}$, by $\partial_B(x) = d\left([x]_{IND(B)}\right)$. A decision table $\mathbf{A}$ is called *consistent (deterministic)* if $card(\partial_A(x)) = 1$ for any $x \in U$. Otherwise $\mathbf{A}$ is *inconsistent (non-deterministic)*.

3   Optimal Reduct-Discretization Problems

Let $\mathbf{A} = (U, A \cup \{d\})$ be a decision table where $U = \{x_1, \ldots, x_n\}$; $A = \{a_1, \ldots, a_k : a_m : U \rightarrow V_{a_m}\}$ and $d : U \rightarrow \{1, \ldots, r(d)\}$. Any pair (a, c) where $a \in A$ and $c \in \Re$ will be called a *cut on* V_a. For $a \in A$, any set of cuts: $\{(a, c_1^a), (a, c_2^a), \ldots, (a, c_{k_a}^a)\}$ on $V_a = [l_a, r_a) \subset \Re$ defines a partition $\mathbf{P}_a$ of V_a onto subintervals $\mathbf{P}_a = \{[c_0^a, c_1^a), [c_1^a, c_2^a), \ldots, [c_{k_a}^a, c_{k_a+1}^a)\}$ where $l_a = c_0^a < c_1^a < c_2^a < \ldots < c_{k_a}^a < c_{k_a+1}^a = r_a$ and $V_a = [c_0^a, c_1^a) \cup [c_1^a, c_2^a) \cup \ldots \cup [c_{k_a}^a, c_{k_a+1}^a)$. Therefore, any set of cuts $\mathbf{P} = \bigcup_{a \in A} \mathbf{P}_a$ defines new decision table $\mathbf{A^P} = (U, A^\mathbf{P} \cup \{d\})$, where $A^\mathbf{P} = \{a^\mathbf{P} : a^\mathbf{P}(x) = i \Leftrightarrow a(x) \in [c_i^a, c_{i+1}^a)$ for $x \in U$ and $i \in \{0, .., k_a\}\}$.

Two sets of cuts $\mathbf{P'}, \mathbf{P}$ are equivalent, i.e. $\mathbf{P'} \equiv_A \mathbf{P}$, iff $\mathbf{A^P} = \mathbf{A^{P'}}$. The equivalence relation $\equiv_A$ has a finite number of equivalence classes. In the sequel we will not discern equivalent sets of cuts. The set of cuts $\mathbf{P}$ is $\mathbf{A}$-*consistent* if $\partial_A = \partial_{A\mathbf{P}}$, where ∂_A and $\partial_{A\mathbf{P}}$ are generalized decisions of $\mathbf{A}$ and $\mathbf{A^P}$, respectively. The $\mathbf{A}$-consistent set of cuts $\mathbf{P}^{irr}$ is $\mathbf{A}$-*irreducible* if $\mathbf{P}$ is not $\mathbf{A}$-consistent for any $\mathbf{P} \subset \mathbf{P}^{irr}$. The $\mathbf{A}$-consistent set of cuts $\mathbf{P}^{opt}$ is $\mathbf{A}$-*optimal* if $card(\mathbf{P}^{opt}) \leq card(\mathbf{P})$ for any $\mathbf{A}$-consistent set of cuts $\mathbf{P}$.

Definition 1. *The set $\mathbf{P}$ of cuts is called s-consistent with $\mathbf{A}$ (or s-consistent in short) where $1 \leq s \leq card(A)$ iff for any decision subtable $\mathbf{B} = (U, B \cup \{d\})$ with s conditional attributes, the set of cuts $\mathbf{P} \cap (B \times \Re)$ is $\mathbf{B}$-consistent.*

The 1-consistent set of cuts will be called locally consistent and the k-consistent set of cuts (where $k = card(A)$) will be simply called consistent.

Definition 2. *The s-consistent set of cuts $\mathbf{P}$ is called s-irreducible iff $\mathbf{Q}$ is not s-consistent for any proper subset $\mathbf{Q} \subset \mathbf{P}$.*

Definition 3. *The s-consistent set of cuts $\mathbf{P}$ is called s-optimal iff $card(P) \leq card(Q)$ for any s-consistent set of cuts $\mathbf{Q}$.*

When $s = k = card(A)$, instead of saying that the set of cuts is k-irreducible or k-optimal, we say that it is irreducible or optimal. We show that any s-consistent set of cuts saves all relative reducts with cardinality not smaller than s in the following sense:

Proposition 1. *If $\mathbf{P}$ is a s-consistent set of cuts in the decision table $\mathbf{A} = (U, A \cup \{d\})$ then for any relative (super-) reduct B of $\mathbf{A}$ such that $|B| \geq s$ the set of discretized attributes $B^\mathbf{P}$ is also a relative (super-) reduct in discretized decision table $\mathbf{A^P}$.*

We illustrate our concepts by Table 1. One can see that the set of all relative reducts of Table 1 is equal to $R = \{\{a_1, a_2\}, \{a_2, a_3\}\}$. The set of all possible cuts is equal to $\mathbf{C}_A = C_{a_1} \cup C_{a_2} \cup C_{a_3}$ where

$$C_{a_1} = \{(a_1, 1.5), (a_1, 2.5), (a_1, 3.5), (a_1, 4.5), (a_1, 5.5), (a_1, 6.5), (a_1, 7.5)\}$$
$$C_{a_2} = \{(a_2, 1.5), (a_2, 3.5), (a_2, 5.5), (a_2, 6.5)(a_2, 7.5)\}$$
$$C_{a_3} = \{(a_3, 2.0), (a_3, 4.0), (a_3, 5.5)(a_3, 7.0)\}$$

The s-optimal sets of cuts for $s = 1, 2, 3$ are the following

$$\mathbf{P}_1 = \{(a_1, 1.5), (a_1, 2.5), (a_1, 3.5), (a_1, 4.5), (a_1, 5.5), (a_1, 6.5), (a_1, 7.5)\}$$
$$\cup \{(a_2, 1.5), (a_2, 3.5), (a_2, 5.5), (a_2, 6.5)\} \cup \{(a_3, 2.0), (a_3, 4.0), (a_3, 7.0)\}$$
$$\mathbf{P}_2 = \{(a_1, 3.5), (a_1, 4.5), (a_1, 5.5), (a_1, 6.5)\} \cup \{(a_2, 3.5), (a_2, 6.5)\}$$
$$\cup \{(a_3, 2.0), (a_3, 4.0)\}$$
$$\mathbf{P}_3 = \{(a_1, 3.5)\} \cup \{(a_2, 3.5), (a_2, 6.5)\} \cup \{(a_3, 4.0)\}$$

A	a_1	a_2	a_3	d
u_1	1.0	2.0	3.0	0
u_2	2.0	5.0	5.0	1
u_3	3.0	7.0	1.0	2
u_4	3.0	6.0	1.0	1
u_5	4.0	6.0	3.0	0
u_6	5.0	6.0	5.0	1
u_7	6.0	1.0	8.0	2
u_8	7.0	8.0	8.0	2
u_9	7.0	1.0	1.0	0
u_{10}	8.0	1.0	1.0	0

$\implies$

$a_1^{\mathbf{P_2}}$	$a_2^{\mathbf{P_2}}$	$a_3^{\mathbf{P_2}}$	d
0	0	1	0
0	1	2	1
0	2	0	2
0	1	0	1
1	1	1	0
2	1	2	1
3	0	2	2
4	2	2	2
4	0	0	0
4	0	0	0

Table 1. The exemplary decision table with ten objects, three attributes and three decision classes. The discretized table $\mathbf{A^{P_2}}$ is presented in the right hand size.

One can see that in the table $\mathbf{A^{P_2}}$ we still have both reducts: $\{a_1^{\mathbf{P_2}}, a_2^{\mathbf{P_2}}\}$ and $\{a_2^{\mathbf{P_2}}, a_3^{\mathbf{P_2}}\}$, but in the table $\mathbf{A^{P_3}}$ we have only one reduct: $\{a_1^{\mathbf{P_3}}, a_2^{\mathbf{P_3}}, a_3^{\mathbf{P_3}}\}$.

4   Boolean reasoning for discretization problems

Consider a decision table $\mathbf{A} = (U, A \cup \{d\})$ where $U = \{u_1, u_2, \ldots, u_n\}$ and $A = \{a_1, \ldots, a_k\}$. By $C_{a_m} = \{(a_m, v_m^1), \ldots, (a_m, v_m^{n_m})\}$ we denote the set of all possible cuts on the attribute a_m for $m = 1, ..., k$ and by $P_{a_m} = \{p_m^1, \ldots, p_m^{n_m}\}$ we denote the set of Boolean variables corresponding to them.

For any objects $u_i, u_j \in U$ such that $d(u_i) \neq d(u_j)$ and for any attribute $a_m \in A$ we define the set $C_{a_m}^{i,j}$ of cuts which discern u_i and u_j as follows:

$$C_{a_m}^{i,j} = \left\{ (a_m, v_m^l) \in C_{a_m} : \left(a_m(u_i) - v_m^l \right) \left(a_m(u_j) - v_m^l \right) < 0 \right\}.$$

By $C_B^{i,j}$, where $B \subset A$, we denote the set of cuts on attributes from B which discern u_i and u_j as $C_B^{i,j} = \bigcup_{a_m \in B} C_{a_m}^{i,j}$.

The Boolean function $\psi_B^{i,j}$ called discernibility function of objects u_i and u_j over B is defined by $\psi_B^{i,j} = \bigvee P_B^{i,j}$ where $P_B^{i,j}$ is the set of Boolean variables corresponding to the cuts $C_B^{i,j}$. We can say that the set of cuts $\mathbf{P}$ satisfies the function $\psi_B^{i,j}$ iff $\mathbf{P} \cap C_B^{i,j} \neq \emptyset$.

The *discernibility Boolean function* for the set of attributes B is defined by:

$$\Phi_B = \bigwedge_{d(u_i) \neq d(u_j)} \psi_B^{i,j}.$$

We say that the set of cuts $\mathbf{P}$ satisfies the function Φ_B iff $\mathbf{P}$ satisfies functions $\psi_B^{i,j}$ for all i, j such that $d(u_i) \neq d(u_j)$. In previous papers we have considered the discernibility Boolean function for the table $\mathbf{A}$ defined by: $\Phi_{\mathbf{A}} = \Phi_A$ and it has been shown that any set of cuts is consistent if it satisfies the function $\Phi_{\mathbf{A}}$. We have also shown the following:

Theorem 1. *The set of cuts* **P** *is optimal iff the set of Boolean variables corresponding to* **P** *is a prime implicant of the function* $\Phi_{\mathbf{A}}$.

This fact allows us to construct an efficient algorithm searching for semi-optimal set of cuts by applying the approximation algorithm (greedy algorithm) for computing the minimal prime implicant [14] of the function $\Phi_{\mathbf{A}}$.

5   Properties and generalizations

In this section we explore some interesting properties of s-consistent sets of cuts. Firstly, we discuss the computational complexity of the optimization problem related to s-optimal sets of cuts.

Theorem 2. *For a given decision table* $\mathbf{A} = (U, A \cup \{d\})$ *and an integer* s, *the problem of searching for* s-*optimal set of cuts is*

1. $DTIME(kn \log n)$ *for* $s = 1$;
2. *NP-hard for any* $s \geq 2$.

Proof. The Fact 1. is obvious. To prove the Fact 2. we recall the following Theorem presented in [5]: *The problem of searching for optimal set of cuts in decision tables with two condition attributes is NP-hard.*   □

Let us generalize Φ_B introduced in previous section and define Boolean function Φ_s such that any s-optimal set of cuts corresponds to the minimal implicant of Φ_s i.e.

$$\Phi_s = \bigwedge_{|B|=s} \Phi_B.$$

One can see that the number of clauses $\psi_B^{i,j}$ in the Boolean function Φ_s is equal to $\binom{k}{s}$ times $conflict\,(\mathbf{A}) = card\,\{(u_i, u_j) : d\,(u_i) \neq d\,(u_j)\} = O\,(n^2)$ being the number of pairs of objects with different decision values. Thus, any greedy algorithm searching for minimal reduct of the function Φ_s needs $O\left(n^2 \cdot \binom{k}{s}\right)$ steps to compute the number of clauses satisfied by any given cut. The function Φ_s can be rewritten as follows:

$$\Phi_s = \bigwedge_{d(u_i) \neq d(u_j)} \bigwedge_{|B|=s} \psi_B^{i,j}$$

For any pair of objects u_i, u_j such that $d\,(u_i) \neq d\,(u_j)$ we denote the set of attributes which discern u_i and u_j by $A_{i,j} = \{a \in A : a\,(u_i) \neq a\,(u_j)\}$. We have

Proposition 2. *If* $(|A_{i,j}| \leq k - s + 1)$ *then* $\bigwedge_{|B|=s} \psi_B^{i,j} = \bigwedge_{a_m \in A_{i,j}} \psi_{a_m}^{i,j}$.

In the consequence we have:

Theorem 3. *Let $k_{i,j} = \min\{|A_{i,j}|, k - s + 1\}$. The set of cuts $\mathbf{P}$ satisfies the function Φ_s if for any pair of objects u_i, u_j such that $d(u_i) \neq d(u_j)$, there are $k_{i,j}$ attributes $a_{m_1}, \ldots a_{m_{k_{i,j}}}$ such that $\mathbf{P}$ satisfies all functions $\psi_{i,j}^{m_1}, \ldots, \psi_{i,j}^{m_{k_{i,j}}}$, i.e. $\mathbf{P} \cap C_{i,j}^{m_l} \neq \emptyset$ for $l = 1, \ldots, k_{i,j}$.*

The last theorem implies that one can construct more efficient discretization algorithm then the greedy heuristics. It will be presented in the next section. Now we conclude with the fact that s-consistency is monotone with respect to s. In particular, it implies that one can reduce the s-optimal set of cuts to obtain $(s + 1)$-optimal set of cuts.

Proposition 3. *For any decision table $\mathbf{A} = (U, A \cup \{d\})$, $card(A) = k$, and for any integer $s \in \{1, \ldots, k - 1\}$, if the set of cuts $\mathbf{P}$ is s-consistent with $\mathbf{A}$, then $\mathbf{P}$ is also $(s + 1)$-consistent with $\mathbf{A}$.*

6   The Algorithm Framework

In previous papers [8, 9] we presented a discretization algorithm called *MD-heuristic* [1] for the total optimal (k-optimal) discretization problem. This is a version of the greedy algorithm applying to the Boolean function Φ_k. The main idea of this method is based on a construction and an analysis of a new table $\mathbf{A}^* = (U^*, A^*)$ where

- $U^* = \{(u_i, u_j) \in U^2 : d(u_i) \neq d(u_j)\}$
- $A^* = \{c : c \text{ is a cut on } \mathbf{A}\}$, where $c((u_i, u_j)) = \begin{cases} 1 & \text{if } c \text{ discerns } u_i, u_j \\ 0 & \text{otherwise} \end{cases}$

This table consists of $O(nk)$ attributes (cuts) and $O(n^2)$ objects (see Table 2). We denote by $Disc(a, c)$ the *discernibility degree* of the cut (a, c) which is defined as a number of pairs of objects from different decision classes (or number of objects in table $\mathbf{A}^*$) discerned by c. The *MD-heuristic* is searching for a cut $(a, c) \in A^*$ with the largest discernibility degree $Disc(a, c)$. Then we move the cut c from A^* to the result set of cuts $\mathbf{P}$ and remove from U^* all pairs of objects discerned by c. Our algorithm is continued until $U^* = \emptyset$. It has been shown in [9] that MD-heuristic is very efficient, because it determines the best cut in $O(kn)$ steps using $O(kn)$ space only.

One can modify this algorithm in case of s-optimal discretization problem by applying Theorem 3. At the beginning, we confer *required cut number $k_{i,j}$* and *set of discerning attributes $L_{i,j} := \emptyset$* upon every pair of objects $(u_i, u_j) \in U^*$ (see Theorem 3). Next we look for a cut $(a, c) \in A^*$ with the largest discernibility degree $Disc(a, c)$ and remove (a, c) from A^* to the result set of cuts $\mathbf{P}$. Then we insert the attribute a into lists of attributes of all pairs of objects discerned by (a, c). We also delete from U^* such pairs (u_i, u_j) that $|L_{i,j}| = k_{i,j}$. This algorithm is continued until $U^* = \emptyset$.

---

[1] Abbreviation of "Maximal Discernibility Heuristic"

A^*	a_1							a_2				a_3		
	1.5	2.5	3.5	4.5	5.5	6.5	7.5	1.5	3.5	5.5	6.5	2.0	4.0	7.0
(u_1, u_2)	1							1				1		
(u_1, u_3)	1	1						1	1	1	1			
(u_1, u_4)	1	1						1	1			1		
(u_1, u_6)	1	1	1	1				1	1			1		
(u_1, u_7)	1	1	1	1	1			1					1	1
(u_1, u_8)	1	1	1	1	1	1		1	1	1			1	1
(u_2, u_3)		1							1	1	1	1		
(u_2, u_5)		1	1						1			1		
...	...	...	...	...	...	...	...	...	...	...	...	...	...	...
(u_3, u_4)													1	
...	...	...	...	...	...	...	...	...	...	...	...	...	...	...
(u_8, u_{10})							1	1	1	1	1	1	1	1

Table 2. A fragment of the temporary table A^* constructed from A (see Table 1)

In Figure 1 we show all possible cuts for decision table A (presented in Table 1). Table A^* (see Table 2) consists of 33 pairs of objects from different decision classes. For $s = 2$, the required numbers of cuts $k_{i,j}$ for all $(u_i, u_j) \in U^*$ (see Theorem 3) are equal to 2 except $k_{3,4} = 1$. Our algorithm begins by choosing the best cut $(a_3, 4.0)$ discerning 20 pairs of objects from A. In the next step the cut $(a_1, 3.5)$ will be chosen because of 17 pairs of objects discerned by this cut. After this step one can remove 9 pairs of objects from U^* e.g. $(u_1, u_6), (u_1, u_7), (u_1, u_8), (u_2, u_5), \ldots$ because they are discerned by two cuts on two different attributes. When the Algorithm stops one can eliminate some superfluous cuts to obtain the set of cuts P_2 presented in Section 3.

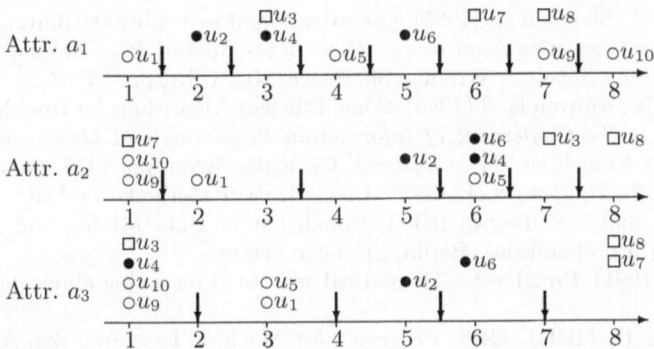

Fig. 1. Illustration of cuts on the table A. Objects are marked by three labels with respect to their decision values.

7   Conclusions

We presented the discretization method with respect to the discernibility between objects and relative reducts of cardinality $\geq s$ for a parameter s given by a user. We proposed the method solving this problem based on Boolean reasoning approach. Also the initial approximation algorithm framework was presented. For continuation we plan to adopt very efficient MD-heuristic presented in [9].

Acknowledgment: The work was supported by Polish State Committee for Scientific Research grant #8T11C01011 and Research Program of European Union - ESPRIT-CRIT2 No. 20288.

References

1. Bazan, J., Skowron, A., Synak, P. (1994). Dynamic reducts as tool for extracting laws from decision table. Proc. of the Symp. on Metodologies for Intelligent System, Charllotte, NC, Lecture Notes in Artificial Intelligence **869**, Springer- Verlag, Berlin, 346–355.
2. Brown F.M., 1990. Boolean reasoning, Kluwer, Dordrecht.
3. Catlett J. (1991). On changing continuous attributes into ordered discrete attributes. In Y. Kodratoff, (ed.), Machine Learning-EWSL-91, Porto, Portugal, LNAI, pp. 164–178.
4. Chmielewski M.R., Grzymala–Busse J.W. (1994). Global Discretization of Attributes as Preprocessing for Machine Learning. Proc. of the III International Workshop on RSSC94, pp. 294–301.
5. Chlebus B., Nguyen S. Hoa. On finding good discretizations for two attributes (also composed to RSCTC'98).
6. Fayyad U. M., Irani K.B. (1992). The attribute selection problem in decision tree generation. Proc. of AAAI-92, July 1992, San Jose, CA.MIT Press, pp. 104-110.
7. Kondratoff Y., Michalski R. (1990): *Machine learning: An Artificial Intelligence approach*, vol.3, Morgan Kaufmann.
8. Nguyen H. S., Skowron A. (1995). Quantization of real value attributes, Rough set and Boolean Reasoning Approaches. *Proc. of the Second Joint Annual Conference on Information Sciences,* Wrightsville Beach, NC, USA, pp. 34–37.
9. Nguyen S. H., Nguyen H. S.(1996), Some Efficient Algorithms for Rough Set Methods. *Proc. of the Conference of Information Processing and Management of Uncertainty in Knowledge-Based Systems*, Granada, Spain, pp. 1451–1456.
10. Nguyen, H.S., Nguyen, S.H.(1997). Discretization Methods for Data Mining. In A.Skowron and L.Polkowski (eds.), Rough Set in Data Mining and Knowledge Discovery (in preparation). Berlin, Springer Verlag.
11. Pawlak Z.(1991): Rough sets: Theoretical aspects of reasoning about data, Kluwer Dordrecht.
12. Quinlan, J. R. (1993). C4.5: Programs for Machine Learning. San Mateo. CA: Morgan Kaufmann Publishers.
13. Skowron A., Rauszer C.(1992). The Discernibility Matrices and Functions in Information Systems. In: Intelligent Decision Support-Handbook of Applications and Advances of the Rough Sets Theory, Słowiński R.(ed.), Kluwer Dordrecht, 331–362.
14. Wegener I. (1987). The Complexity of Boolean Functions. Stuttgart: John Wiley & Sons.

Rough Mereology for Industrial Design

Julia Johnson

Department of Computer Science, University of Regina,
Regina, SK. S4S 0A2 Canada

Abstract. A rough sets approach to the optimisation of product schedules is proposed. This approach provides a general method for finding approximate solutions to a certain class of NP-hard mathematical programming problems that arise in product scheduling in manufacturing.

1 Introduction

By industrial design we mean the development of intelligent solutions for industrial and business problems. Rough mereology offers an new paradigm for applications in industrial design because our solutions are being expressed in approximate terms, that is "in acceptable degree", unlike previous approaches in which an optimal solution was expected. An industrial problem which shows most immediate benefit from the rough mereology approach is that of production scheduling in manufacturing. This problem needs to be re-examined for the possibility of better solutions that the new paradigm makes possible.

In this paper I will discuss the relevance of the rough mereology approach to production scheduling in manufacturing. In Section 2 the production scheduling problem is described. In Section 3 the rough mereology is reviewed. In Section 4 we demonstrate that the production scheduling problem can be formulated within a rough mereology approach. Conclusions are presented in Section 5.

2 The Production Scheduling Problem

Products are scheduled for production on several different production lines (processors). Product demands over a certain number of time periods (called the planning horizon) are known. Each processor is capable of producing more than one product, but only one at a time. A production schedule is a week by week plan showing which products are produced on which processors in which time periods.

The event of switching a processor from producing one product to producing another is known as a change-over. Its cost is called *changeover cost*. The cost of holding a product in the warehouse after it is produced is called holding cost. Total production cost is the sum of manufacturing, holding and changeover cost.

A feasible schedule satisfies demands. An optimal schedule minimises total production cost while satisfying demands. Our objective is to produce a close to optimal schedule.

L. Polkowski and A. Skowron (Eds.): RSCTC'98, LNAI 1424, pp. 553-556, 1998.
© Springer-Verlag Berlin Heidelberg 1998

3   What is Rough Mereology?

Rough mereology [3, 4] is an approach to problems of approximate reasoning based on rough set theory extended so that the problems can be allocated to a system of agents. It comprises a process of learning about how objects are constructed and a process of classifying objects based on the knowledge acquired in the learning process. The basic idea is embodied in the meaning of the statement $\mu(x, y) = \varepsilon$ which is read *x is a part of y in degree at least ε*. But what metric do we use to make precise the notion of x being a part of y to a degree? One answer would be to look at the similarity between the two objects: Information about objects is expressed by attributes applicable to objects of a given class and values for those attributes for each object of that class. An approximate measure for $\mu(x, y)$ is the proportion of attribute values that the two objects x and y share in common.

We assume a hierarchically structured agent system in which an agent synthesises objects of a particular class from sub-objects sent to it by its children. Consider the following example from [3]: We are constructing a man from subparts of a body (which includes head, arms and trunk) and a pair of legs. There are two types of bodies, a skinny body $B1$ and a fat body $B3$, and two types of legs, skinny legs $L2$ and fat legs $L3$. A consistent set of values for the μ's is as follows: $\mu_{body}(B3, B1) = \varepsilon_{body} \geq 0.25$, $\mu_{legs}(L2, L2) = \varepsilon_{legs} \geq 1$, $\mu_{man}(B3L2, B1L2) = \varepsilon_{man} \geq 0.28$, $\mu_{legs}(L3, L2) = \varepsilon_{legs} \geq 0.4$, $\mu_{man}(B3L3, B1L2) = \varepsilon_{man} \geq 0.14$. Men $B3L2$ and $B1L2$ who have the same legs share at least 28% of their attribute values in common while men $B3L3$ and $B1L2$ who have neither the same legs nor the same bodies share only at least 14% of their attribute values in common.

The ε's for a particular object class are treated as values of a decision attribute and the respective μ's as values of respective condition attributes. Decisions propagate from the lowest level agents to the top level agent during synthesis of an object. Each non-leaf agent chooses from among the possible objects that can be constructed from parts sent to it by its children the object that most closely matches the desirable specification that has been sent to it from above.

4   Rough Mereology Production Scheduling

A mapping between the production scheduling domain and the blocks world setting to facilitate user acceptance of the decision support system has been provided in [1, 2]. Thereafter, we can operate in a two dimensional world inhabited by a robot that moves blocks around to develop preferable schedules.

The top level agent constructs a preferable schedule by expanding itself to a system of agents as illustrated in Fig. 1. Multiple goals are achieved including those of satisfying customer demands and running the products on a given processor in a predefined order. Regarding the latter, it might not be a good idea to produce widgets before gadgets on a given processor because widgets are black and gadgets are white. The processor would have to be shut down

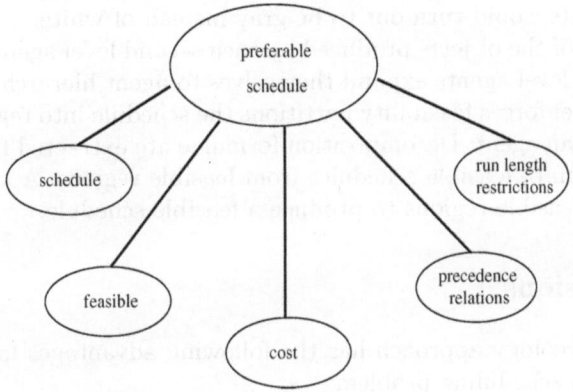

attributes of a preferable schedule

      * simplicity   * user - satisfaction

      * learnability   * ease of use

      * comprehensibibly   * user - friendliness

Fig. 1. Agent Based Scheduling

attributes of a schedule
    * number of processors
    * number of products
    * number of time periods
    * which processors can produce which products

attributes of a feasible schedule

    * weekly production
    * weekly consumption

attributes of a cost effective schedule

    * holding cost
    * manufacturing cost
    * change over cost

attributes of a schedule with the inclusion of precedence relations

    * partial ordering on product numbers

attributes of a schedule with the inclusion of run length restrictions

    * maximum run length
    * number of idle time periods

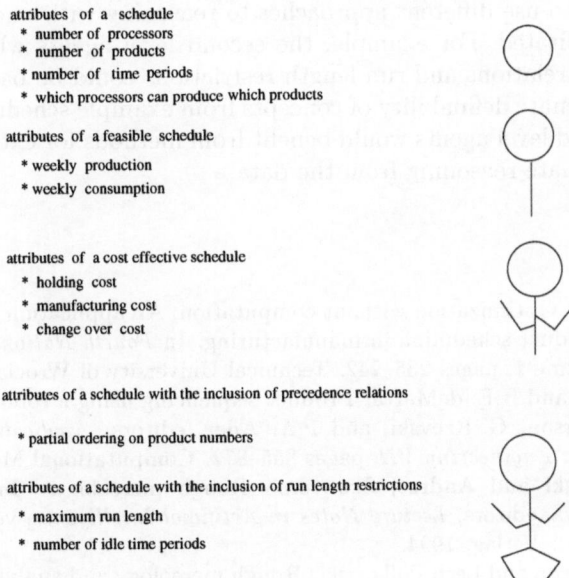

Fig. 2. Attributes of Second Level Agents

for cleaning after producing widgets before producing gadgets. Otherwise that batch of gadgets would turn out to be gray instead of white.

Attributes of the objects produced by each second level agent are provided in Fig. 2. Second level agents expand themselves to agent hierarchies. For example the agent that enforces feasibility partitions the schedule into regions and assigns each region to an agent. Decomposition formulae are extracted from examples of the construction of feasible schedules from feasible regions. A synthesis process combines the feasible regions to produce a feasible schedule.

5   Conclusions

The rough mereology approach has the following advantages for application to the production scheduling problem:

1. The problem of constructing a mathematical programming (MP) model can be formulated as a problem of learning about how objects are constructed from sub-objects.
2. The problem of solving an MP model can be formulated as the problem of synthesising objects from sub-objects within an agent based framework.
3. An infeasible solution to an MP model corresponds to a schedule that fails to satisfy customer demands while keeping the cost of the schedule at an acceptable level.
4. General constraints of the rough mereology force a proper design for partitioning a problem into subproblems for allocation to a hierarchy of agents.
5. Flexibility to use different approaches to reasoning with uncertain information is facilitated. For example, the second level agents which provide for precedence relations and run length restrictions could be based on methods for approximate definability of concepts from example schedules whereas the other second level agents would benefit from methods for extracting schemes of approximate reasoning from the data.

References

1. J.A. Johnson. Optimization without computation: An application of artificial intelligence to product scheduling in manufacturing. In *Fourth National Conference on Robotics*, volume 1, pages 235–242. Technical University of Wroclaw, 1993.
2. J.A. Johnson and R.E. deMatta. Product sequencing using a robot problem solver. In D.E. Grierson, G. Rzevski, and P.A. Adey, editors, *Applications of Artificial Intelligence in Engineering VII*, pages 855–872. Computational Mechanics, 1992.
3. Lech Polkowski and Andrzej Skowron. Rough mereology. In Z.W. Raś and M. Zemankowa, editors, *Lecture Notes in Artificial Intelligence*, volume 869, pages 85–94. Springer Verlag, 1994.
4. Andrzej Skowron and Lech Polkowski. Rough mereology and analytical morphology. In E. Orlowska, editor, *Incomplete Information: Rough Set Analysis*, chapter 13, pages 399–437. Physica Verlag, 1998.

Optimal Stochastic Scaling of CAE Parallel Computations[*]

Mariusz Flasiński[1], Robert Schaefer[1], Wojciech Toporkiewicz[2]

[1]Institute of Computer Science, Jagiellonian University, Cracow, Poland.
[2]PhD student, Cracow University of Technology, Cracow, Poland.
(mflas,schaefer,toporkiewicz)@ii.uj.edu.pl

Abstract. The meshless strategy of CAE problem partitioning is presented. It offers significant decrement of time complexity in comparison with usual mesh decomposition algorithms, and therefore may be applied in on-line processing. The effective IE graph solid representation is applied as well as the stochastic performance forecast system for parallel MIMD computer nodes. The main scaling problem is formulated as the optimal fuzzy graph matching and it is proposed to be solved by efficient ETPL(k) parsers.

1 Introduction

A successful utilization of multiprocessor computer installations imposes parallelization of most complex problems that appear in the discrete mechanical analysis of structures (CAE). A suggested course of parallelization is to decompose the whole domain and then process each subdomain concurrently. A particular decomposition is as good as the resulting tasks are scaled to the available performance distribution and as it decreases communication between processors. Although an accurate scaling of CAE problem can be obtained only by a decomposition of a computational mesh, a rough decomposition may provide a similar speedup, because of an inherent uncertainty of an evaluation of virtual computer performance parameters. The new, meshless strategy proposed in this paper takes into account such an uncertainty. We undertake the following assumptions:

- Partitioning is performed on a feature-oriented, unambiguous IE graph (see [4,6]) representation of the solution domain before the computational mesh is generated. It allows us to minimize an interprocess communication involving maximum information about solid topology.

- We make use of synthetic information about computational complexity introduced by each part of the problem's domain, e.g. the space density of *degrees of freedom* (d.o.f), and we take into account uncertainty of their distribution. In particular, we know the error distribution for evaluated computational complexity of each task and communicational complexity resulting from assembling partial results obtained on

---

[*] The research was supported from the ESPRIT Project 20288-13.

interfaces. This information is contained in the *density of d.o.f function* ρ. It may be defined by designer's introductory heuristic analysis of a CAE problem, or basing on an earlier solution, using a'posteriori error estimation.

- We dedicate our strategy to a distributed environment (a computer network) with an asynchronous communication medium (e.g. such as in Ethernet Lan's). A stochastic forecast is used to predict a distribution of a computer performance over a network (cf. [8]).

2 IE-graph Solid Representation Preliminaries

We aim to represent all kinds of structures like machine parts, buildings etc. and call them *Engineering Solids - ES*. These are compact, bounded and closed subsets of $\Re^3$ having a Lipshitz boundary.

Definition 1. (cf. [4,5]) An *Indexed Edge-unambiguous graph, (IE-graph)*, over Σ and Γ is a quintuple $G = (V, E, \Sigma, \Gamma, \Phi)$, where V is a finite, nonempty set of nodes, to which indices have been ascribed in an unambiguous way on the basis of properties of a represented object, Σ is a finite nonempty set of node labels, Γ is a finite nonempty set of edge labels, E is a set of edges, of the form (v,λ,w) where $v,w \in V$, $\lambda \in \Gamma$, such that the index of v is less than the index of w, and $\Phi : V \to \Sigma$ is a *node labeling function*. The *family of all the IE graphs* will be denoted with *IEG*.

Definition 2. (*Graph inclusion*) Let there are two given IE graphs $A = (V_A, E_A, \Sigma, \Gamma, \Phi_A)$, and $B = (V_B, E_B, \Sigma, \Gamma, \Phi_B)$. When there is an injection $i : V_A \to V_B$, such that $\Phi_B \circ i = \Phi_A$ and $E_B = \{(i(v),\lambda,i(w)) : (v,\lambda,w) \in E_A\}$, we say that A is a *subgraph* of B or that A is *included* in B and we denote it by $A \subset B$.

We distinguish two separate classes of node labels: *Basic Constructive Solids, BCS*, (for any *ES*, *BCS* is its convex hull; faces of such a convex hull are indexed unambiguously (see [5])) and *Features* $\Sigma = BCS \cup FEATURE$.

IE graphs are used for representing *Engineering Solids - ES* according to the following scheme (see [5]). Any engineering solid is modelled with subtracting from *BCS* volumes chosen from a predefined set of (primitive) *Features*. A subtraction consists in placing the so-called sweep base (a contour) on some *BCS* faces and cutting a volume along some direction (see Fig. 1a). The IE graph representation is defined in such a way that node

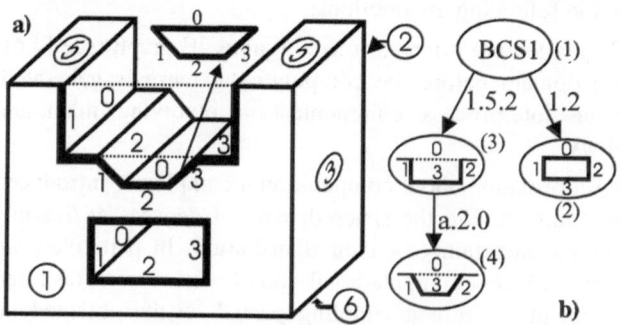

Fig. A Initial solid representation

labels describe types of *BCS*-es and types of sweep bases, whereas edge labels define positioning of a modelled feature (i.e. placing a sweep base on the face and a way of its sweeping in relation to the *BCS*). For example, let us look at the IE graph shown in Fig. 1b. The edge connecting a node indexed with 1 (*BCS1*) and a node indexed with 2 (corresponding to the „square" through pocket in a solid shown in Fig. 1a) is labelled with 1.2, because the feature interacts with faces 1 and 2 of *BCS* (see Fig. 1a). The edge connecting a node 1 with a node 3 (a square slot) is labelled with 1.5.2, because the feature „starts" from a face 1 „ends" at a face 2 and additionally interacts with a face 5 of *BCS* (the „upper" one). With the scheme we can also represent adjacency of features. For example, a square slot represented with a node 3 is adjacent with a side (of its sweep base) indexed with **2** to a V-slot (a graph node 4), in particular to its (sweep base's) side indexed with **0** - cf. Figs: 1a and 1b. Therefore, the corresponding edge is labelled with **a.2.0** („a" means „adjacent", **2** and **0** are sides of adjacent sweep bases). This scheme has been defined and discussed in a detailed way in [5]. In [5] we have also proved that engineering solids treated in such a way can be represented with the family of IE graphs in a unique and unambiguous way, i.e. there is a one-to-one mapping between *ES* and *IEG*.

3 Decomposition Graps, Computational Complexity Graphs and Computer Network Graphs.

A *decomposition graph* is an IE graph derived from a graph representing an engineering solid S and reflects geometry and topology of the particular partitioning of such a solid. The image of a decomposition graph throughout the representation mapping is the same as the initial image, but the solid's body is treated as splited into a given number of parts. Thus, within a decomposition graph we can distinguish several subgraphs.

Definition 3. A d*ecomposition graph* of an engineering solid $S \in ES$ is an IE graph $G=(V,E,\Sigma,\Gamma,\Phi)$ that consists of at least two nonoverlapping subgraphs $G_i=(V_i,E_i,\Sigma,\Gamma,\Phi_i)$, | $V_i = \varnothing$, $i = 1,...,M$, $M > 1$ such that in each of them we may distinguish a 'representative' *BCS*-labeled node v: $\forall i=1,...,M$, $\exists v \in V_i$: $\Phi_i(v) \in BCS$, and the only graph edges linking separate subgraphs, connect their *BCS*-labeled nodes, thus: $\forall v \in V_i, w \in V_j, i, j = 1,...,M, i \neq j$

$$\exists (v,\lambda,w) \in E \;\Rightarrow\; \left(\{\Phi_i(v) \in BCS\} \, and \, \{\Phi_j(w) \in BCS\}\right).$$

Let us denote the set of such edges by :

$$\acute{E} = \left\{ \begin{array}{l} (v,\lambda,w) \in E: \; v \in V_i, w \in V_j, \; \lambda \in \Gamma, \; i,j = 1,...,M, \; i \neq j, \\ \Phi_i(v) \in BCS \, and \, \Phi_j(w) \in BCS \end{array} \right\}$$

A decomposition graph G of an engineering solid S is denoted with $d(S)$. The *family of all the decomposition graphs* over an engineering solid S is denoted with $DG_S = \{G : G = d(S)\}$.

For example, if we decompose a solid shown in Fig. 1a in a way shown in Fig. 2a, then we obtain a decomposition graph shown in Fig. 2b.

To be able to forecast a computational complexity of a CAE task we have to estimate the number of d.o.f falling into particular subparts of a partitioned volume, as well as the number of d.o.f contained in interface surfaces between subdomains. Their

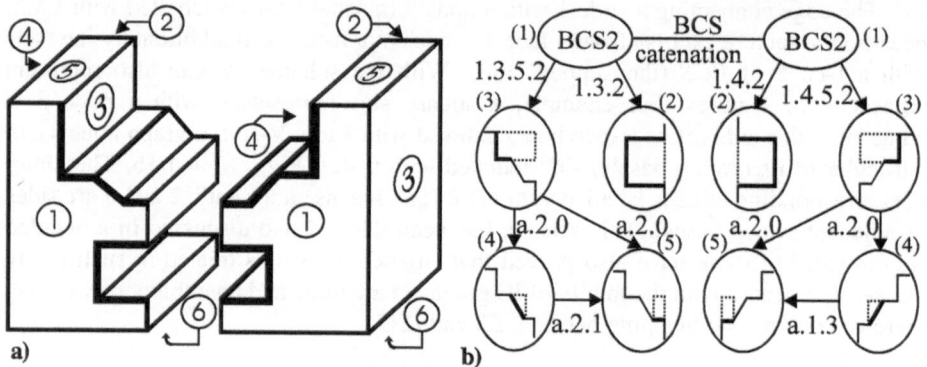

a)   b)

Fig. B Solid after decomposition

evaluation will be expressed with the *computational complexity graph*. In order to provide its formal description let us define, firstly, a set of random variables: $\Xi_i : \Omega_\Xi^i \to N$, $I_j : \Omega_I^j \to N$, $i=1,...,M, j=1,...,m \leq M^2$ which values are number of degrees of freedom in the i-th subdomain, and number of degrees of freedom on the j-th interface, where j is an index that enumerates interfaces. Means of Ξ_i, I_j are usually determined by integrals of the degrees of freedom density function ρ, mentioned in the introduction. The integrals are taken over the i-th subdomain and the j-th interface, respectively. Means may be randomized using a simple Gauss error distribution, or some specific distribution derived from adoption characteristics.

Definition 4. A *computational complexity graph* $\overline{G} = \left(\overline{V}, \overline{E}, \overline{\Sigma}, \overline{\Gamma}, \overline{\Phi} \right)$ corresponding to a decomposition graph $G=(V,E,\Sigma,\Gamma,\Phi)$ is such an IE-graph that: $\overline{V} = \left\{ \overline{v}_i : \exists v \in V \wedge \Phi(v) \in BCS \right\}$ is the node set, $\overline{\Sigma} = \left\{ \Xi_i : \Omega_\Xi \to N, i = 1,...,M \right\}$ is a nonempty set of node labels, $\overline{\Gamma} = \left\{ I_j : \Omega_I \to N, \ j = 1,...,m \leq M^2 \right\}$ is a nonempty set of edge labels, $\overline{E} = \left\{ \left(\overline{v}, I_j, \overline{w} \right) : \exists (v, \lambda, w) \in \overline{E} \subset E, j = 1,...,m \leq M^2 \right\}$ is a nonempty set of graph edges, and $\overline{\Phi} : \overline{V} \ni \overline{v}_i \to \Xi_i \in \overline{\Sigma}, \ i = 1,...,M$ is the node labeling function and j is an index that enumerates interfaces. The *family of all computational complexity graphs* resulting from a decomposition of an engineering solid S is denoted with $L_S = \left\{ \overline{G} : G \in DG_S \right\}$.

Nodes of a computational complexity graph correspond to subdomains of a particular decomposition and they reflect a complexity of computations combined with solving a problem within them. Edges of such a graph relate to predicted interprocess communication resulting from assembling partial results obtained on interfaces of subdomains.

The last of three graphs introduced in the paragraph is a *computer network graph*. In our approach a physical heterogeneous parallel computer consists of computer nodes v_i $i=1,...,M$ and an *asynchronous communication medium* which connects all nodes (e.g. Ethernet LAN technology). We assume that this computer is *nondecomposable* what means, that v_i can perform only one operation of our CAE application at a time. We assume that each task of our application is *atomic i.e.* should be evaluated sequentially, on a single machine, without any communication with other tasks. Moreover, we assume that there exists the *pattern task* for the considered class of CAE applications. Let us introduce (see [8]) the stochastic vector $\left\{ \overline{T_n} \right\}_{n=\mu,\mu+1,...}$, $\overline{T_n} = \left\{ \overline{T_n^1},...,\overline{T_n^M} \right\}$, that refers to the virtual computer. $\overline{T_n^i}$ is the execution time of the pattern task on the i-th machine in the n-th time step, while μ denotes the starting point index. $\overline{T_n}$ may be evaluated as a discrete, Markow, periodic process. Its evolution determines the formula $\Pi^j(n+1) = P_n^j \Pi^j(n)$, $j = 1,...,M, n = \mu,\mu+1,...$ where $\Pi^j(n)$ is the probability distribution for $\overline{T_n^i}$ and P_n^j is a $k \times k$ transition probability matrix identified for k distinct states of i-th machine loading, and for finite set of time steps (due to process periodicity). Let us define new random vectors:

$$\mathfrak{I}_{\mu,Z} = \frac{1}{Z}\sum_{n=\mu}^{\mu+Z} \overline{T_n}, \ \mathfrak{I}_{\mu,Z} = \left\{\mathfrak{I}_{\mu,Z}^j\right\}, \ \chi_{\mu,Z} = \left\{\chi_{\mu,Z}^j\right\}, \ \chi_{\mu,Z}^j = \frac{1}{\mathfrak{I}_{\mu,Z}^j}, \ j=1,...,M$$

$\mathfrak{I}_{\mu,Z}$ expresses the mean execution time of the pattern task during time horizon Z, while $\chi_{\mu,Z}$ is a random vector of *power coefficients*, components of which will serve for labels of the computational complexity graph nodes. In [3] we have shown that computer network structure can be represented with the family of IE graphs, which means that such a representation model preserve good computational properties of its processing schemes. Therefore, we define a computer network graph as an IE graph.

Definition 5. A *computer network graph* is a complete IE graph $H = \left(V_H, E_H, \Sigma_H, \Gamma_H, \Phi_H\right)$, where $V_H = \left\{v_i : i=1,...,M\right\}$ is the node set, $\Sigma_H = \left\{\chi_{\mu,Z}^i, i=1,...,M\right\}$ is the set of node labels, Γ_H is a nonempty set of edge labels, $E_H = \left\{\left(v_i, \lambda, v_j\right) : v_i, v_j \in V_H, \lambda \in \Gamma, i,j=1,...,M, i<j\right\}$ is a nonempty set of graph edges, $\Phi_H : V_i \ni v_i \rightarrow \chi_{n,Z}^i \in \Sigma_H$, $j=1,...,M$ is the node labeling function.

4 Scaling of CAE Tasks as a Stochastic Control Problem

Having introduced the necesseary formalism we may formulate scaling of a CAE work as a stochastic control problem. To factorize adjustment of a particular decomposition to the presumed distribution of computer performance over a network we will utilize the functional:

$$F\left(\overline{G}, \chi_{\mu,z}, \sigma\right) = E\left(\sqrt{\sum_{i \neq j,\ i,j=1,\dots,M} \left(\frac{\chi^i_{\mu,z}}{\chi^j_{\mu,z}} - \frac{\Xi_{\sigma(i)}}{\Xi_{\sigma(j)}}\right)^2}\right)$$

where σ is a permutation of an M-element node set of a computational complexity graph. E denotes the expected value operator.

For a given computational complexity graph $\overline{G} \in L_S$ we denote:

$$\overline{F}\left(\overline{G}, \chi_{\mu,z}\right) = \min_{\sigma \in L_M} F\left(\overline{G}, \chi_{\mu,z}, \sigma\right)$$

where L_M is a group of permutations of an M-element set, and L_S is the family of all computational complexity graphs resulting from decomposition of an engineering solid S. $\overline{F}$ can be understood as the distance function (metrix) between $\overline{G} \in L_S$ and H.

The *optimal stochastic scaling problem* consists in finding such a $G_{opt} \in DG_S$ for which $\overline{G}_{opt} \in L_S$ satisfies the minimum property and the constraint:

$$\overline{F}\left(\overline{G}_{opt}, \chi_{\mu,z}\right) = \min_{G \in L_S} \overline{F}\left(\overline{G}, \chi_{\mu,z}\right), \qquad E\left(\frac{I}{\Xi}\right) \leq ComRange,$$

where: $I = \sum_{j=1}^{m} I_j$, $\Xi = \sum_{i=1}^{M} \Xi_i$ are random variables expresing communicational complexity and approximating computational load of an entire work for $\overline{G}$. *ComRange* stands for the maximum admissible communication complexity contribution.

5 Syntactic Pattern Recognition for Efficient Decomposition and Allocation

As one can easily notice, we have been able to define the optimal stochastic scaling problem as a problem of finding a graph that satisfies a certain property in a set of allowable graphs. Since we have been able to define all the representation models (decomposition graphs, computational complexity graphs, and computer networks graphs) as the special cases of the family of IE graphs, we can make use of very good computational properties of this family. It is well-known that, in general, graph processing schemes are based on graph matching, which makes them inefficient (a non-polynomial time complexity), due to NP-completeness of the graph isomorphism problem. Fortunately, IE graphs used in our model belong to a class of ETPL(k) graph languages with a polynomial membership problem, which means that processing schemes based on graph matching are computationally efficient.

Theorem (Theorem 5.1 in [3]) The algorithm of parsing ETPL(k) graph grammars has a time complexity $O(n^2)$, where n is a number of IE graph nodes.

It means that searching of any IE graph in a set of IE graphs (treated here as the ETPL(k) language) made with the ETPL(k) parser in $O(n^2)$ time (see [4]). However, in order to apply such an efficient scheme, we have to discuss two key issues that condition a successful use of such a syntactic pattern recognition-based processing scheme. The first issue consists in defining an efficient processing scheme for generating a decomposition graph from the initial one (cf. Figs: 1b and 2b). From the syntactic pattern recognition point of view, this problem resolves itself into the problem of translating one formal language into another according to a set of predefined rules.

Fortunately, for ETPL(k) graph languages used for a solid representation, this „translation" problem is solved by defining an efficient *Syntax-Directed Translation Scheme, SDTS* (cf. [5], pp 416-422). In [5], we have shown also that the IE-graph based representation scheme is constructed in a way allowing us to derive „translating" rules in an easy and intuitive way. For example, let us come back to our examples shown in Figures 1 and 2. Let us notice that we have decomposed the solid symetrically with a plane parallel to its faces 3 and 4 (cf. Figs: 1a and 2a). Then, instead of a graph shown in Fig. 1b we have a (decomposition) graph shown in Fig. 2b. The graph edges labelling exhibits this symmetry. Instead of the edge (1, 1.2, 2) pointing a square through pocket we have received two edges labelled with 1.3.2 and 1.4.2 pointing to two square slots being a result of the pocket decomposition. These labels (1.3.2 and 1.4.2) can be „computed" knowing that we „decompose" the label 1.2 with the plane parallel to faces 3 and 4 of the original solid. A reader easily can see other possibilities of such computations made for our example as well as a possibility of defining „general" rules translating a graph before a succeeding decomposition into a graph after it with such a representation scheme. These „general" rules can be then formalized in the form of an efficient syntax-directed translation scheme.

The second problem concerns an extension of a „pure" IE graph, for which an efficient syntactic pattern recognition schemes has been defined (see [4]) into a representation that allows one to find an optimal graph basing on certain measurements added to its structural elements, i.e. its nodes and edges. (In our case these measurements are of the stochastic nature.) The key questions are as follows:

• *Is it possible to add such measurements to IE graph structures that can be used for comparing „distances" between graphs and this way decide which one is „better" according to some optimalization criteria ?*

• *If we add such information to IE graphs and we use parsing processing scheme, does the computational complexity increases?*

Answers for both questions are satisfactory. In [2] „fuzzy" (in particular random) IE graphs have been defined and a way of their comparison based on metric measuremets has been suggested. For processing such fuzzy IE graphs, the (efficient, $O(n^2)$) error-correcting parser finding the „best matched" (according to predefined criteria, in particular in case to the distance defined by $\overline{F}$) graph out of the set (language) of IE graphs has been constructed.

6 Concluding remarks

• Some phases of the presented optimal partitioning strategy have been succesfully implemented. The rough solid partitioning algorithm using IE graph representation was described in [7]. Error-correcting parsers for fuzzy IE graphs were implemented in [2]. The Markov performance forecast system was tested for a network of SUN workstations (see [8]). Its advantage was confirmed for simple (homogenous) decomposition and for task migration strategy (see [9]).

• The computational efficiency of such a meshless strategy, being its biggest advantage, results from the use of a class of IE graphs as a formalism of a representation of both: decomposed solids and computer network structures. The IE graphs belong to the family of ETPL(k) languages with a polynomial membership problem (see [6]), which results in $O(n^2)$ time complexity of underlying processing schemes (see [4]). The rough partitioner mentioned in [4] is $O(n)$.

• The proposed strategy is dramatically less complex then mesh partitioners (see e.g. [10]), moreover it allows to generate the computational network in parallel which additionally increases the overal speedup of a parallel CAE computation. It also restricts an interprocess communication to the acceptable fraction of the entire effort.

• The presented approach is adjusted for the problem partitioning in case of discrete meshless solving methods (see e.g. [1]).

References

1. Belytschko, Y. Krongauz, D. Organ, M. Fleming, P. Krysl, Meshless methods: An overview and recent developments, *Comp. Meth. in Applied Mech. and Eng.* **139** (1996), 3-47.
2. M. Flasiński, Distorted pattern analysis with the help of node label controlled graph languages, *Pattern Recognition* **23** (1990), 765-774.
3. M. Flasiński, L. Kotulski, On the use of graph grammars for the control of a distributed software allocation, *Comp. J.* **35** (1992), A165-A175.
4. M. Flasiński, On the parsing of deterministic graph languages for syntactic pattern recognition, *Pattern Recognition* **26** (1993), 1-16.
5. M. Flasiński, Use of graph grammars for the description of mechanical parts, *Computer Aided Design* **27** (1995), 403-433.
6. M. Flasiński, Power properties of NLC graph grammars with a polynomial membership problem, *Theoretical Computer Science* **203** (1998), in print.
7. M. Flasiński, R.F. Schaefer, W. Toporkiewicz, Supporting CAE Parallel Computations with IE Graph Solid Representation, *J. for Geom. and Graph.*, VOL 1 (1997), No. 1, 23-29.
8. R.F. Schaefer, Z. Onderka, Markov Chain Based Management of Large Scale Distributed Computations of Earthen Dam Leakages, *L.N.C.S* **1215**, Springer 1997, 49-64.
9. R.F. Schaefer, J. Sipowicz, G. Myoliwiec, Control Activities in Message Passing Environment, *L.N.C.S* **1332**, Springer 1997, 143-151.
10. B.H.V. Toping, A.I. Khan, J.K. Wilson, Parallel Dynamic Relaxation and Domain Decomposition, *Advances in Parallel and Vector Processing for Structural Mechanics*, CIVIL-COMP, 215-232.

CBR for Complex Objects Represented in Hierarchical Information Systems

Jan Wierzbicki

Polish-Japanese Institute of Computer Techniques
Koszykowa 86, 02-008 Warsaw, Poland
and Center for Computer Science Education
Raszyńska 8/10, 02-026 Warsaw, Poland
e-mail: jwierzbi@oeiizk.waw.pl

Abstract. We discuss how to use Case-Based Reasoning (CBR) philosophy for solving various problems specified by complex objects represented
by means of hierarchical information systems [3]. We show how to use
this kind of knowledge base for the recognition of novel cases. Next we
show how to identify new problems and how to use and adapt methods
which were successful in past situations to the new ones. All issues are
illustrated by examples, which are here some elementary mathematical
tasks.

1 Introduction

The idea of Case-Based Reasoning (CBR) systems is to solve new problems by
adapting previously successful solutions for similar problems. The process is represented by the four steps: *Retrieve, Reuse, Revise, Retain* creating a schematic
cycle [1]. The most difficult problem is related to discovery of cases similar to
a given one, so that on the basis of algorithms corresponding to the extracted
similar cases an algorithm for the given case can be constructed. Hence a problem arises how to construct a knowledge base relevant for this task. Here we
have two pragmatic measures: the functionality and the ease of acquisition of
the information represented in the case [2].

In this paper cases are represented by means of hierarchical complex objects
together with algorithms (strategies) for transforming them. Any complex object
is defined by hierarchically structured attributes, on the basis of which (and using
expert knowledge), some additional characteristics of the object are created.
Any case (object) has assigned a strategy (algorithm) or a family of algorithms
working on it. The algorithm can start computation if values of its input variables
(parameters) are specified properly by the object representation. Algorithms
create a hierarchical structure, starting from the very precise ones referring to
the objects defined exactly to general ones for the objects defined more generally
or in a vague way. To extract successfully an algorithm for the new objects
(cases) they must be properly decomposed using attributes. This causes that
even if the algorithm for the similar object is not to be reused exactly, it can be
adapted. Adaptation of algorithm is a transformation, which let to replace some

L. Polkowski and A. Skowron (Eds.): RSCTC'98, LNAI 1424, pp. 565-572, 1998.
© Springer-Verlag Berlin Heidelberg 1998

its segments by the proper ones for the case, taking characteristics of such object into account. The idea of this paper is to show a method for construction of such hierarchical complex object knowledge base by proper case decomposition, and adapting methods of algorithms known for some cases similar to a new identified case, to receive the algorithm corresponding to that case.

2   Information systems for complex objects

An information system is a pair $\mathbf{A} = (U, A)$ where U is a set called the universe of objects (cases) and A is a set of attributes, any attribute $a \in A$ is a mapping on the universe U. With every attribute $a \in A$ we associate a set of its values V_a - domain of a. We let $V = \cup\{V_a : a \in A\}$ [4].

The task (problem) can be characterized as an object. Information vector of any object $O \in U$ is defined by: $Inf_A(O) = \{(a, a(O)) : a \in A\}$. For example when we consider the problems of elementary mathematics, the attributes can be found as types of geometrical figures, edge lengths, relations between figures, equation or inequality variables etc. The attribute values are the specific values of the attributes taken from the task contents. For example for the attribute 'type of a geometrical figure' its value can be - 'circle', for the attribute 'radius' its value can be - '2 cm', while for the attribute 'area' its value can be - 'unknown'. One can consider a new information system $\mathbf{A}' = (U, A')$ where some attributes corresponding to attributes from A can have values Yes, No or Alg. If for example $a(O) =$Yes, $b(O) =$Yes and $c(O) =$Alg then it means that the dependency $a, b \rightarrow c$ is true in $\mathbf{A}$ and Alg is a pointer to the algorithm which returns the value c if the values of a,b are given.

Example 1. Let us to consider simple geometrical problems concerning the circle. They can be represented by a table (Table 1), columns of which can be seen as attributes, rows as objects and table entries are attribute values. Elementary objects are the elementary problems (like compute circumference having the radius length), attributes are *radius, diameter, circumference, area*; attribute values are Yes - (the attribute has a value) or Alg (one can obtain the value of attribute by the algorithm pointed to).

Table 1. (*Circle*)

object	a_1 (radius)	a_2 (diameter)	a_3 (circumference)	a_4 (area)
O_1	Yes	Alg	Alg	Alg
O_2	Alg	Yes	Alg	Alg
O_3	Alg	Alg	Yes	Alg
O_4	Alg	Alg	Alg	Yes

For any object from Table 1 one can obtain values of all its attributes using known dependencies. An attribute value can be equal to Alg only when other attributes have values allowing to compute the unknown value of the attribute using the algorithm pointed by Alg. For objects $O_1, ..., O_4$ from Table 1 the same algorithm for solving equations: $Area = \Pi(Radius)^2$; $Circumference = 2\Pi(Radius)$; $Diameter = 2(Radius)$ can be used. We have three equations, to solve them we may have three unknowns. If the value of the attribute is Yes we may substitute to the equation a specific attribute value taken form the case description.

This is an example of objects which we call simple or elementary objects. Any simple object is represented by the elementary attributes. Elementary attributes are the primary concepts of each field, for example primary concepts of elementary mathematics (geometry). Obviously, among elementary mathematical problems we can find other such simple objects, for example concerned with other geometric figures (triangles, tetragons etc.). The representation of such objects is made in this same way as in the above example. Simple objects described by the same set of attributes are represented by means of information system. Example 1 shows information system $A_c = (U_c, A_c)$ for the problems dealing with circle. Index $'c'$ represents the type of object from the system. Tables which represent information systems are labeled also by the object type of the system. For example objects in Table 1 have type $Circle$.

Taking two (or more) simple objects into account one can construct new object called a complex object. Such a complex object is described by the attributes representing information about simple objects used in its construction, attributes describing relations between these simple objects, and some extra attributes specifying the so called $characteristics$ of such object, defined by the expert.

Let us consider two information systems for objects of some types $A_1 = (U_1, A_1)$, $A_2 = (U_2, A_2)$, where $A_1 \cap A_2 = \emptyset$. The indices 1 and 2 represent the types of the objects. We may construct a new information system $A_{(1,2)}$ by composition of objects from information systems A_1 and A_2 in the following way: $A_{(1,2)} = (U_{(1,2)}, A_{(1,2)}, f)$ where f represents a decomposition function $f : U_{(1,2)} \to U_1 \times U_2$ (with projection functions f_1, f_2 for f defined by $f_1(O) = O'$, $f_2(O) = O''$; where $f(O) = (O', O'')$; the complex object O is constructed out of objects O', O'').

Information system $A_{(1,2)}$ points to the information systems A_1, A_2 corresponding to types of components (of type 1,2) of objects in $A_{(1,2)}$. Links from subset of the universe $U_{(1,2)}$ of the information system $A_{(1,2)}$ to the subsets of the universes U_1, U_2 of the information systems A_1, A_2 are specified by attribute value vectors. Sets of attributes $A' = \{a_1, ..., a_n\} \subseteq A_1, B' = \{b_1, ..., b_m\} \subseteq A_2$ specify subsets of the universes U_1, U_2 proper for the decomposition of complex object, while the set of attributes C specifies relations between objects from which the complex object has been constructed. An information system for complex objects can be also represented by a table (see Table 2). Any value vector of attributes $a_1, \ldots, a_n$ specifies a subset of the universe U_1, defined by all objects from A_1 consistent with this value vector; any value vector of attributes $b_1, \ldots, b_m$ specifies a subset of the universe U_2, defined by all objects from A_2 consistent with this value vector; and any value vector of attributes $c_1, \ldots, c_k$ specifies the relations between objects from the universe U_1 and U_2. The set C of attributes is a set of relational attributes c of the form $c : U_{(1,2)} \to V_c = \{0, 1\}$. For example such attribute can be defined by relations $r \subseteq U_1 \times U_2$ in the following way $c(\mathbf{O}) = 1 \leftrightarrow (f_1(\mathbf{O}), f_2(\mathbf{O})) \in r$. Any value vector of attributes $d_1, \ldots, d_l$ is defined by the expert and specifies the characteristics of the complex object, taken from the values of relational attributes, types and special attributes of objects from which complex object has been constructed. The value vector of such attributes can be found as a description of such characteristic, concrete attribute value, or pointer to the algorithm which can compute concrete value of such attribute.

Table 2. (1&2)

object	a_1	$\ldots$	a_n	b_1	$\ldots$	b_m	c_1	$\ldots$	c_k	d_1	$\ldots$	d_l
$\mathbf{O_1}$	v_1	$\ldots$	v_n	w_1	$\ldots$	w_m	u_1	$\ldots$	u_k	s_1	$\ldots$	s_l
$\mathbf{O_2}$	v_1'	$\ldots$	v_n'	w_1'	$\ldots$	w_m'	u_1'	$\ldots$	u_k'	s_1'	$\ldots$	s_l'

Information system $\mathbf{A_{(1,2)}}$ can be generalized to a new information system $\mathbf{A_{(G1,\ G2)}}$ by the generalization of types and attributes of objects represented by the system, and generalization of relations between these objects.

For example information system describing objects constructed by an object of the type *Triangle* and an object of the type *Rectangle*, can be generalized to information system describing objects constructed by two objects of the type *Figure*. We may note that after such generalization, information system $\mathbf{A_{(1,2)}}$ can be linked to new information systems for objects of generalized type.

Expert can make compositions of various defined objects (simple and complex). In each case the method is the same, as it was described for composition of two simple objects. A complex object constructed by a composition of some other complex objects can be decomposed in various ways if we do not have any information about possible methods of such decomposition. If the decomposition of the complex object is not appropriate, the algorithm assigned for it will fail. That is why formulation of relational attributes and its values has influence on a proper decomposition. For example for the complex object $\mathbf{O}$ constructed from two other complex objects $\mathbf{O_1, O_2}$, the only information on relations of objects $\mathbf{O_1, O_2}$ is not sufficient. To make a proper decomposition of object $\mathbf{O}$, we should have additional information about objects $\mathbf{O_1, O_2}$ and their interrelations as well.

Figure 1.

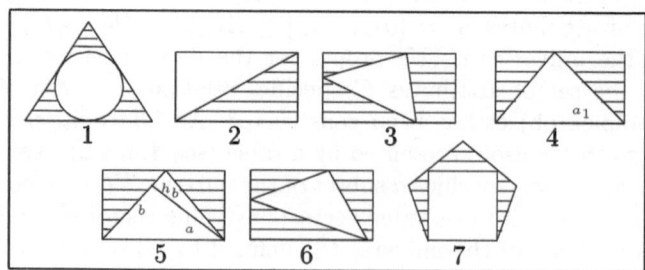

Example 2. What is the area of the figure created by inscribing a figure A into a figure B? Figures A and B are any of the geometrical figures (like circles,

triangles, tetragons, pentagons) represented by different values of attributes. The inscribing types of a figure A into a figure B can be different as well. Some possible examples are given in Figure 1.

To simplify the task we will focus only on cases 2-6 (Fig. 1). To solve such kinds of problems we need to have in the knowledge base a description of simple objects, more complex objects and finally the description of the main complex object (general contents of the problem). An example of simplified (with reduction of the amount of possible attributes) collection of such simple objects and relations among them is presented in Tables 3,4,5; labeled by types *Triangle*, *Rectangle*, *Triangle&Rectangle* of objects.

Table 3. (*Triangle*)

object	a'_1 (subtype of object)	a'_2 (edge a)	a'_3 (edge b)	a'_4 (edge c)	a'_5 (height h_a)	a'_6 (height h_b)	a'_7 (height h_c)	a'_8 (area)
O'_1	equilateral	Yes	Alg'	Alg'	Alg'	Alg'	Alg'	Alg'
O'_2	ordinary	Yes	No	No	Yes	No	No	Alg'
O'_3	ordinary	No	Yes	No	No	Alg'	No	Yes
O'_4	ordinary	Yes	Yes	No	No	Yes	No	Alg'

Table 4. (*Rectangle*)

object	a''_1 (subtype of object)	a''_2 (edge a)	a''_3 (edge b)	a''_4 (diagonal d)	a''_5 (area)
O''_1	square	Yes	Alg''	Alg''	Alg''
O''_2	ordinary	Yes	Yes	Alg''	Alg''
O''_3	ordinary	Yes	Alg''	Yes	Alg''
O''_4	ordinary	Alg''	Yes	Yes	Alg''

Table 5. (*Triangle&Rectangle*)

object	a_1 edge	a_2 height	a_3 common edge	b_1 edge	b_2 height	c_1	c_2 (vertex A)	c_3 (vertex B)	c_4 (vertex C)	d_1	d_2
O_1	No	Yes	Yes	No	No	inscr	Yes	Yes	Yes	char$_1$	Alg$_1$
O_2	Yes	Yes	Yes	No	No	inscr	Yes	No	Yes	char$_2$	Alg$_2$
O_3	No	Yes	No	No	No	inscr	Yes	No	No	char$_3$	Alg$_3$
O_4	Yes	No	Yes	No	No	inscr	No	No	No	char$_4$	Alg$_4$
O_5	No	Yes	Yes	No	Yes	circum	Yes	No	No	char$_5$	Alg$_5$

Attribute c_1 describes the type of relation, attributes $c_2 - c_4$ specify whether the vertices, of figures extracted from a given object, coincide. Any value vector of attributes $a_1 - a_3$ points to the object subset of Table 3 (for example the subset pointed by object O_2 is $\{O'_1, O'_4\}$); any value vector of attributes b_1, b_2 points to the subset of Table 4 ; attributes $c_1 - c_4$ are relational ones (specify for example possible methods of inscribing a triangle into a rectangle); and finally attributes d_1, d_2 precise a characteristic of a complex object. For example attribute d_1 for object O_2, has a value (char$_2$) which, among others, describes that there is a common edge (between coinciding vertices of components), one rectangle edge is a height of the triangle to the common edge and some consequences of this fact. Attribute d_2 specifies an area of the figure created by inscribing a triangle into a rectangle. Value of attribute $d_2 - Alg_i$ is a pointer to the algorithm, which returns a concrete value of the area for the created figure. Algorithms Alg_i refer to complex objects. These algorithms are described by sequences of operations $Alg_i = (op_1, \ldots, op_n)$. Some of these operations refer to subproblems dealing with concrete component objects of a complex object. To solve these subproblems complex object must be properly decomposed to its component objects, characterized by special attributes and their value vectors. That is why such algorithm is characterized by the type of component objects, value vector of attributes specifying sets of component objects, relations between them, and known characteristics of the complex object. We call such type of algorithm a detailed algorithm. The value vector of attributes specifying sets of proper component objects is pointing to the subset of the universe for the information systems corresponding for component objects. Table 5 can be generalized by generalization of the object type *Triangle&Rectangle* and attributes $a_1 - a_3$; b_1, b_2 to the a general object type - *Figure&Figure*, while attributes $c_1 - c_4$; d_1, d_2 are generalized to attributes c, d. Table 6 presents the description of main complex object of a general type defined by attributes c, d.

Table 6. (*Figure&Figure*)

object	c (relation)	d (area of the figure created by inscribing two figures)
O_1^i	inscribe	$\mathbf{ALG_1}$
O_2^i	circumscribe	$\mathbf{ALG_2}$

This general table can be linked not only to Tables 3, 4 as Table 5 do, but to any other tables for the objects whose type can be generalized to the type *Figure*.

Value $\mathbf{ALG}_i$ of attribute d is a pointer to the algorithm, which returns a concrete value of an area of the figure created by inscribing two other figures. Algorithms corresponding to $\mathbf{ALG}_i$, refer to complex objects of a general type. These algorithms are in fact collections of equivalent algorithms corresponding to different decompositions of the complex object. Any algorithm choice from these collections can require different input information about components of complex object. We call algorithms of such type general algorithms. If decomposition of the complex object is not appropriate, the detailed algorithm assigned for such decomposition fails. In such situation the general algorithm let to assign another different corresponding algorithm appropriate to other possible decomposition of the complex object. In Example 2 object O_1^i of the type *Figure&Figure* (Table 6) can be decomposed into two objects or to few objects of type *Figure*. For example case 3 (Fig. 1) can be decomposed to one triangle and one rectangle; or to three triangles and one tetragon etc. If first decomposition is not allowing to compute area, one may try an algorithm for the second decomposition.

3 Identification of the new object with the similar ones coded in the knowledge base

The problem of object identification in the CBR cycle is called *Retrieve*. The *Retrieve* process must be proceeded by proper representation of the new object. During implementation process the new object is matched against the constructed knowledge base. To do this it should be decomposed up to some level. Next one may find descriptions of similar known objects and adapt their algorithms to our case. Decomposition to more general modules, for example to the general object type and relations of its components, gives a little information about such object, and it can be difficult to assign correct detailed algorithm due to the lack of some attributes values. Decomposition to the detailed attributes, may cause difficulties with adaptation of general strategies. The retrieving starts from the identification of proper information system by comparing of the new object type with types of objects from the information systems. Next we specify subset of the universe U_i for chosen information system $\mathbf{A_i} = (U_i, A_i)$ by comparing the attribute value vector of the new object with objects from the universe U_i. We select a similar object or a subset of similar objects with maximal attribute value vector consistent with value vector of the new object. One can find here a problem of selecting of proper algorithm for the case, when the subset of similar objects with different algorithms have been chosen. For example for Table 5 we may obtain a subset of objects $\{O_1, O_3, O_5\}$. For such situation the most promising object for the success, must be selected. This allows to reduce nondeterministic choices in object identification. This reduction can be done by using weights of attributes defined by experts.

4   The reuse strategies of similar objects to the new object

This problem in the *CBR* cycle is called *Reuse*. One may find here a few situations of reusing and adaptation of algorithms: (i) the exact matching of the object (case) from the information system and the new one; (ii) the new object (case) is similar to a case from knowledge base, i.e. they have the same values on sufficiently many attributes; (iii) the new object (case) is similar to a few cases from the same or different information systems. In the first situation one can reuse algorithm without any adaptation. In the second situation algorithm has to be adapted, taking the characteristics of such object into account. The method of algorithm adaptation by analysis of object characteristics is a transformation, which let to substitute some operations of the algorithm with different ones or to rearrange these operations to the more proper ones for the case by taking these characteristics into account. The rearranging process replaces or modifies the chosen operation to the new one by taking some additional relations between object components specified by the characteristics. For example we may count first the area of the tetragon instead the area of the triangle, if this two figures are well related. In Example 2 we may easy adapt, by the use of this method, algorithm for case 5 (Fig. 1) to case 4 (Fig. 1). In the third situation algorithm has to be adapted as well, but the methods of adaptation are much more complicated here, due to fact that characteristics of many objects must be taken into account.

The algorithm adaptation by analysis of object characteristics returns a solution, when the given new object was properly decomposed. For the case of unproperly decomposition such adaptation method fails, and we have to refer to the general algorithm. From the collection of equivalent detailed algorithms we may choose new algorithm corresponding to different decomposition of the complex object. If the decomposition is proper algorithm can be adapted to the new object. Specification of the object is done by given attributes and its value vector. Components given by the formulated problem, not in all cases are the ones to be matched for in knowledge base. Sometimes the proper decomposition elements can be hidden in the contents of the task. In Example 2 we may note that for case 3 (Fig. 1) characterized by some attributes, case 7 (Fig. 1) can be more similar than case 6 (Fig. 1). During *Reuse* process, the operations of a new algorithm are temporarily stored, and the operations which were undertaken as wrong are not stored. The final reduction of some left wrong operations is done during process of renewal execution of a new algorithm operations, which is called in the CBR cycle *Revise*. In *Revise* process we check whether the result of each operation is needed to execute some of the next operations, otherwise such operation will not be stored. New algorithm is stored by *Retain* process.

Conclusions. In the paper a construction of a hierarchical knowledge base which seems to be well suited for CBR systems was shown. In implementation we use the object oriented methodology. The main problem is a proper decomposition and definition of the object which reflects the case. Another problem is the ease of acquisition of the algorithm, which could be achieved by the right structure of algorithms construction, and right object decomposition. Mentioned methods of algorithm adaptation were described just to show some abilities of such approach to the problems dealing with complex objects, and are not the only ones. According to these ideas the system is constructed, which contains much more methods of algorithm adaptation and object decomposition as well as recognition. Simple mathematical examples were for the simplicity of understanding the methods taken, however the system constructed in the way presented can be adapt to other fields.

Acknowledgments. The author is due to thank Professor Andrzej Skowron for formulating the subject and for his numerous discussions and helpful critical remarks throughout the investigation.

References

1. A.Aamodt & E.Plaza (1994). Case Based Reasoning: Foundational Issues, Methodological Variations, and System Approaches. AI Communications.
2. J.Kolodner (1993). Case Based Reasoning. Morgan Kaufmann.
3. Z.Pawlak (1991). Rough Sets. Theoretical Aspects of Reasoning about Data. Kluwer Academic Publishers, Boston, London, Dordrecht.
4. L.Polkowski, A.Skowron, J.Komorowski (1996). Approximate case-based reasoning: A rough mereological approach. In: H.D.Burkhard, M.Lenz (eds.), Fourth German Workshop on Case-Based Reasoning. System Development and Evaluation, Informatik Berichte 55, Humboldt University, Berlin, 144-151.
5. M.M.Veloso & J.G.Carbonell (1993). Derivational analogy in PRODIGY: Automating case acquisition, storage and utilization. Machine Learning, Vol.10(iii).
6. M.M.Veloso, H.Munioz, R.Bergmann (1996). Case-based planing: Selected methods and systems. AI Communications.

Application of the Information Measures to Input Support Selection in Functional Decomposition

M.Rawski[1], L.Jóźwiak[2], and A.Chojnacki[2]

[1] Warsaw University of Technology, Faculty of Electronics and Inf. Tech.
ul.Nowowiejska 15/19 00-665 Warsaw, Poland
[2] Eindhoven University of Technology, Faculty of Electrical Engineering
P.O.Box 513, EH 10.25 5600 MB Eindhoven, The Netherlands

[Abstract.]
General functional decomposition has important application in many fields of modern engineering and science. Its practical usefulness for very complex systems is however limited by lack of an effective and efficient method for selection of the appropriate input supports for sub-systems. In this paper, an effective and efficient heuristic method for input support selection is proposed and discussed. The experimental results demonstrate that the method is able to construct optimal or near optimal supports efficiently even for large systems.

1   Introduction

Decomposition is a central activity in analysis and design of complex systems. It is fundamental to many fields of modern engineering and science. Strong stimuli for developing decomposition techniques come from such areas as pattern analysis, knowledge discovery, machine learning, data mining and decision making, but also from logic synthesis in computer-aided design of very large integrated circuits (VLSI-CAD) [2].

Functional decomposition consists of breaking down a complex system into a network of smaller and relatively independent co-operating sub-systems, in such a way that the original systems behavior is preserved. It can be used in all fields mentioned above. The motivation for using it in system analysis and design is to reduce the problem complexity and to find well structured network of coherent sub-systems. A complex system is decomposed into a network of smaller subsystems, such that each of them is easier to analyze, understand or synthesize. Although the multi-level functional decomposition gives very good results [7][8], its practical usefulness for very complex systems is limited by lack of an efficient method for the construction of sub-systems. The decompositions quality heavily depends on the effectiveness and efficiency of the input support selection for subsystems. However, the commonly used systematic method of input support selection, based on checking of all possible supports, is inefficient for larger problem instances.

In this paper, an efficient heuristic method for input support selection is proposed. It is based on the analysis of information relationships in a considered

L. Polkowski and A. Skowron (Eds.): RSCTC'98, LNAI 1424, pp. 573–580, 1998.

system. Application of information relationship measures for construction of the input supports allows us to reduce the search space to a manageable size while keeping the high-quality solutions in the reduced space.

After introducing some basic theory, the proposed input support selection method is presented. Subsequently, some experimental results are discussed, which are reached with a prototype tool that implements the method.

The experimental results demonstrate that the method is able to construct optimal or near optimal supports efficiently, even for large systems. It is much faster than the systematic method while delivering results of comparable quality.

2   Functional decomposition

"Partitions" with non-disjoint blocks are referred to as rough partitions [5][6] (r-partitions) or set systems [1]. We recall some information on partition-based modeling that is necessary for understanding of the paper. For example, the function F of Table 1 can be represented by the following set of r-partitions:

$P(x_1) = \{\overline{1,2,4,5,8,9}; \overline{3,6,7}; \overline{10}\}$, $P(x_2) = \{\overline{1,2,3,4,7,8}; \overline{4,5,6,7,9}; \overline{4,7,10}\}$,
$P(x_3) = \{\overline{1,5,9}; \overline{2,4,8}; \overline{3,6,7,10}\}$, $P(x_4) = \{\overline{1,3,4}; \overline{2,8}; \overline{5,7}; \overline{6,9,10}\}$,
$PF = \{\overline{9,10}; \overline{3,4,6}; \overline{5}; \overline{5,8}; \overline{1,7}; \overline{2,8}\}$.

	x_1	x_2	x_3	x_4	y_1	y_2	y_3
1	0	0	0	0	1	1	0
2	0	0	1	1	1	1	1
3	1	0	2	0	0	1	1
4	0	-	1	0	0	1	1
5	0	1	0	2	1	0	-
6	1	1	2	3	0	1	1
7	1	-	2	2	1	1	0
8	0	0	1	1	1	-	1
9	0	1	0	3	0	1	0
10	2	2	2	3	0	1	0

Table 1. Function table of the multiple-valued, multiple-output, incompletely specified discrete function F.

The product of r-partitions $P = P(x_2) \bullet P(x_3)$ (computed by finding products of the r-partitions' blocks) is as follows: $P = \{\overline{1}; \overline{2,4,8}; \overline{3,7}; \overline{5,9}; \overline{4}; \overline{6,7}; \overline{7,10}\}$.

In this way, various information streams in discrete information systems can be modeled using r-partitions.

Let A and B be two subsets of X such that $A \cup B = X$. Assume that the variables $x_1, \ldots, x_n$ have been relabeled in such a way that $A = \{x_1, \ldots, x_r\}$ and $B = \{x_{n-s+1}, \ldots, x_n\}$. For an n-tuple x, the first r components are denoted by x^A, and the last s components, by x^B.

Let F be a Boolean function, with $n > 0$ inputs and $m > 0$ outputs, and let (A, B) be as above. Assume that F is specified by a set F of the function's cubes. Let G be a function with s inputs and p outputs, and let H be a function with $r + p$ inputs and m outputs. The pair (G, H) represents a serial decomposition of F with respect to (A, B), if for every minterm b relevant to F, $G(b^B)$ is defined, $G(b^B) \in \{0, 1\}^p$, and $F(b) = H(b^A, G(b^B))$. G and H are called blocks of the decomposition.

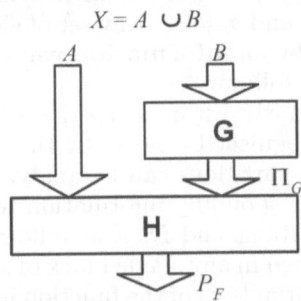

Fig. 1. Schematic representation of serial decomposition.

[**Theorem 1.**] *If there exists an r-partition Π_G on F such that $P(B) \leq \Pi_G$, and $P(A) \bullet \Pi_G \leq P_F$, then F has a serial decomposition with respect to (A, B) [5].*

[*Example 1.*] Let's consider a functional decomposition with $A = \{x_3, x_4\}$ and $B = \{x_1, x_2\}$ for the function of Table 1 specified by the set F of its cubes numbered 1 through 10. For these A and B: $P(A) = \{\overline{1}; \overline{5}; \overline{9}; \overline{4}; \overline{2, 8}; \overline{3}; \overline{7}; \overline{6, 10}\}$, and $P(B) = \{\overline{1, 2, 4, 8}; \overline{4, 5, 9}; \overline{4}; \overline{3, 7}; \overline{6, 7}; \overline{7}; \overline{10}\}$, and $\Pi_G = \{\overline{1, 2, 4, 5, 6, 7, 8, 9}; \overline{3, 7, 10}\}$ satisfies the conditions of Theorem 1.

If k denotes the number of blocks in Π_G then the number of outputs from block G is: $p = \lceil \log_2 k \rceil$. Outputs of G constitute a part of the input support for H. Thus, the size of G and the size of H grow both with the number of blocks in partition Π_G. The number of blocks in Π_G strongly depends on the input support chosen for G. Therefore, the decomposition's quality strongly depends on the input support chosen. Most of the known functional decomposition algorithms use systematic input support selection, which finds partition Π_G for all possible input supports [5]. These algorithms are inefficient for large problem instances. Therefore, we present below an efficient heuristic method of input selection based on information relationships and information relationship measures [3][4].

3   Information relationships and measures

Analysis of information relationships between various information streams is of primary importance for analysis and synthesis of information systems. The theory of information relationships and measures is presented in a separate paper of this conference [3] and extensively discussed in [4]. Below, only some information is recalled that is necessary for understanding of this paper.

Information on symbols from a certain set S means the ability to distinguish certain symbols from some other symbols. An **elementary information** describes the ability to distinguish a certain single symbol s_i from another single symbol s_j, where: $s_i, s_j \in S$ and $s_i \neq s_j$. Any set of elementary portions of information can be represented by an **information relation** I **or an information set** IS defined on $S \times S$ as follows:

$I = \{(s_i, s_j) | s_i$ is distinguished from s_j by the modeled information$\}$,
$IS = \{\{s_i, s_j\} | s_i$ is distinguished from s_j by the modeled information$\}$.

Relationships between r-partitions can be analyzed by considering the relationships between their corresponding information relations and sets. The correspondence between r-partitions and IS is as follows: IS contains the pairs of symbols that are not contained in any single block of a corresponding r-partition.

For instance, for input variable x_2 of the function in Table 1 the corresponding r-partition and information set are as follows:
$P(x_2) = \{\overline{1, 2, 3, 4, 7, 8}; \overline{4, 5, 6, 7, 9}; \overline{4, 7, 10}\}$,
$IS(x_2) = \{1|5, 1|6, 1|9, 1|10, 2|5, 2|6, 2|9, 2|10, 3|5, 3|6, 3|9, 3|10, 5|8, 5|10, 6|8,$
$\qquad\qquad 6|10, 8|9, 8|10, 9|10\}$,
Symbol "|" in pairs $s_i | s_j$ from $IS(x_2)$ is used to stress that the elements s_i and s_j of a certain pair $\{s_i, s_j\}$ are distinguished from each other.

In [3] and [4], we defined the following **relationships between information** of two r-partitions P_1 and P_2:

- **common information** $CI(P_1, P_2) = IS(P_1) \cap IS(P_2)$,
- **extra information** $EI(P_1, P_2) = IS(P_2) \setminus IS(P_1)$.

For an r-partition P the **information quantity** IQ: $IQ(P) = |IS(P)|$ is defined in [3] and [4]. For two r-partitions P_1 and P_2, the following **relationship measures** are defined in [3] and [4]:

- **information similarity** $ISIM$: $ISIM(P_1, P_2) = |CI(P_1, P_2)|$,
- **information increase** $IINC$: $IINC(P_1, P_2) = |EI(P_1, P_2)|$.

4   Input support selection

Each realization of a discrete function, thus also each decomposition, must be able to compute information required by the output variables of the function from information provided by the input variables of the function. Theorem 1 describing the conditions for serial decomposition, can be re-expressed in terms of the information sets in the following way:

[**Theorem 2.**] *If there exists an r-partition Π_G on F such that $IS(P(B)) \supseteq IS(\Pi_G)$, and $IS(P(A) \bullet \Pi_G) \supseteq IS(P_F)$, then F has a serial decomposition with respect to (A, B).*

In general, information that is necessary for computing values of a certain output is distributed on a number of inputs, the inputs also deliver some information that is not needed for the output, and information on the inputs is represented otherwise than on the considered output. Moreover, decomposition is nontrivial if the number of block's G outputs is smaller than the number of its inputs.

Block G can be described as a block where an intermediate information transformation is performed. This transformation consists of construction of an appropriate partition Π_G from a selected partition $P(B)$. Partition Π_G should carry a part of the information delivered by partition $P(B)$, which in combination with information delivered by partition $P(A)$, is essential to compute the required output information. The number of block's G outputs is in practical cases much smaller than the number of its inputs. Partition Π_G is created by merging the blocks of partition $P(B)$. To avoid big loss of information in this process or creation of partitions Π_G with big number of blocks, which would require many physical logic blocks in implementation, partitions generated by each of the input variables from set B should be similar each to another. Set B should also contain variables that carry relatively much unimportant information. A part of the important information delivered by variables from set B is in most cases also delivered by some variables from set A. Therefore it is not necessary to transfer this information to the output of G.

Based on these sorts of observations we developed the following rules of input variable selection for set B. Set B should contain variables which:

- carry relatively much unnecessary information for computing output information,
- carry relatively much information delivered also by variables from set A,
- carry quite much common information.

The above rules are expressed below using the information relationship measures introduced earlier in this paper. Variable x_i should be included into set B if:

- $EI(P_F, P(x_i))$ is relatively big, i.e. $IINC(P_F, P(x_i))$ is relatively high,
- $CI(P(A), P(x_i))$ is relatively big, i.e. $ISIM(P(A), P(x_i))$ is relatively high,
- $CI(P(B'), P(x_i))$ is relatively big, i.e. $ISIM(P(B'), P(x_i))$ is relatively high,

where: x_i - candidate to set B, B' - partially crated set of indirect variables.

Using these rules, we developed and implemented an algorithm for near optimal input support selection in serial decomposition. In the algorithm, set B is constructed step by step. First, a pair of variables (x_i, x_j) is chosen that maximizes **ISIM**(x_i, x_j). Next such variables are added to B for which **IINC**$(P_F, P(x_i))$ and **ISIM**$(P(A), P(x_i))$ are possibly highest. This increases a chance of constructing of a partition Π_G with a small number of blocks without loss of substantial information. The algorithm performs a beam-search what allows to control both the quality of results and the computation time.

5   Experimental results

This section compares the systematic input selection method with the proposed heuristic selection method that is based on information relationship measures by, applying both methods to several small, medium and large benchmarks from the international logic synthesis benchmark set [9]. Tables 2 and 3 report results of the input support selection for set B in a single serial decomposition step, as illustrated in Fig. 1 and described in Section 2. Table 2 shows comparison of the minimum number of blocks of partition Π_G. The results were obtained for decompositions with 3, 4, 5, and 6 input variables in set B. The method based on information measures, despite of its heuristic character, produces the optimal or near optimal results.

Benchmark	inputs	outputs	cubes	Systematic method ($|B|$) 3	4	5	6	Heuristic method ($|B|$) 3	4	5	6
z4	7	4	128	4	6	8	12	4	6	8	12
5xp1	7	10	126	8	16	32	64	8	16	32	64
misex1	8	7	18	4	6	7	9	4	6	7	9
root	8	5	71	5	9	15	17	5	9	15	17
opus	9	10	23	4	6	8	10	5	6	8	10
9sym	9	1	191	4	5	6	7	4	5	6	7
alu2	10	3	391	6	12	24	43	6	12	24	43
sao2	10	4	60	4	6	9	11	4	6	9	11
sse	11	11	39	4	6	8	11	4	6	9	11
keyb	12	7	147	6	9	13	19	6	10	14	19
s1	13	11	110	5	8	13	19	6	10	15	19
plan	13	25	115	5	7	11	17	5	9	14	19
styr	14	15	140	4	6	9	13	5	7	10	13
ex1	14	24	127	4	6	8	11	4	6	8	11
kirk	16	10	304	4	4	5	6	4	5	5	6

Table 2. Comparison of the number of blocks in partition Π_G obtained by the systematic and heuristic method.

Table 3 shows comparison of the computation time. For large benchmarks, the systematic method is very slow in comparison to the new method based on information relationships. For functions with more than 10 input variables, the new method is many times faster. The difference in processing time between these two methods grows very fast with the size of function. For the largest of the tested functions the heuristic method is more than 50 times faster.

Table 4 shows the results of decomposition of some benchmark functions into a network of 4-input, 1-output logic cells, obtained by repeating the serial decomposition step from Fig. 1 a number of times. The decomposition aims in minimal number of cells. In the table the number of logic cells in decomposition is presented for systematic and heuristic method. The results show that the heuristic character of proposed method has almost no influence on the number of logic blocks obtained in decomposition. In two cases (misex1 and alu2) the results from the heuristic method are even better than from the systematic search. This results from the fact that in the systematic method the first found

Benchmark	Size			Systematic method ($|B|$)				Heuristic method ($|B|$)			
	inputs	outputs	cubes	3	4	5	6	3	4	5	6
z4	7	4	128	1	1	2	2	1	2	5	8
5xp1	7	10	126	0	2	1	2	2	2	5	7
misex1	8	7	18	0	1	1	3	1	1	2	5
root	8	5	71	1	2	2	3	1	2	5	12
opus	9	10	23	1	2	3	8	0	1	3	7
9sym	9	1	191	10	20	41	79	5	8	16	34
alu2	10	3	391	33	51	64	63	32	40	60	98
sao2	10	4	60	3	6	14	36	1	2	6	21
sse	11	11	39	2	7	20	63	1	2	8	25
keyb	12	7	147	17	52	156	503	7	11	24	51
s1	13	11	110	13	46	137	552	7	10	17	33
plan	13	25	115	16	51	184	595	6	9	19	42
styr	14	15	140	31	109	404	1453	12	17	32	58
ex1	14	24	127	24	91	333	1377	8	12	26	58
kirk	16	10	304	108	528	2125	11234	55	69	119	230

Table 3. Comparison of the computation time (in seconds) for the systematic and heuristic method.

solution with the minimal number of blocks in partition Π_G is selected and in the heuristic method the solution with the minimal number of blocks which is the best from the information measures viewpoint.

6   Conclusions

The proposed heuristic method of input support selection is very efficient. The method delivers decompositions of similar quality as decompositions obtained from the systematic method. In some cases, the results from the heuristic method are even better than from the systematic method. For largest of the tested benchmarks, the method based on information relationships is more than 50 times faster than the systematic method.

Benchmark	Size			Systematic method	Heuristic method
	inputs	outputs	cubes		
z4	7	4	128	7	7
5xp1	7	10	126	20	21
misex1	8	7	18	19	18
root	8	5	71	46	47
alu2	10	3	391	116	114

Table 4. Comparison of number of logic cells in decomposition.

These features make the proposed heuristic method very useful for decomposition-based analysis and synthesis and demonstrate high usefulness of the information relationships and measures to the analysis and synthesis of information systems.

References

1. Hartmanis, J., Stearns R.E.: *Algebraic Structure Theory of Sequential Machines* Prentice-Hall, 1966.
2. Jóźwiak, L.: *General Decomposition And Its Use in Digital Circuit Synthesis, VLSI Design* An International Journal of Custom-Chip Design Simulation , and Testing, Special Issue on Decomposition in VLSI Design, vol.3, Nos.3-4, 1995, pp. 225–228.
3. Jóźwiak, L.: *Analysis and Synthesis of Information Systems with Information Relationship Measures* First Int. Conference on Rough Sets and Current Trends in Computing (RSCT'98), Warsaw, Poland, June 22-26, 1998.
4. Jóźwiak, L.: *Information Relationships and Measures: An Analysis Apparatus for Efficient Information System Synthesis,* Proc. of the EUROMICRO'97 Conference, Budapest, Hungary, Sept. 1-4, 1997, pp. 13–23, IEEE Computer Society Press.
5. Łuba, T., Selvaraj, H.: *A General Approach to Boolean Function Decomposition and its Applications in FPGA-based Synthesis* VLSI Design, Special Issue on Decompositions in VLSI Design, vol. 3, Nos. 3-4, pp. 289–300, 1995.
6. Rawski, M., Jóźwiak, L., Nowicka, M., Łuba, T.: *Non- Disjoint Decomposition of Boolean Functions and Its Application in FPGA- oriented Technology Mapping* Proc. of the EUROMICRO'97 Conference, Budapest, Hungary, Sept. 1-4, 1997, pp. 24–30, IEEE Computer Society Press.
7. Volf, F.A.M., Jóźwiak, L., Stevens, M.: *General Decomposition And Its Use in Digital Circuit Synthesis* VLSI Design: An International Journal of Custom-Chip Design Simulation , and Testing, Special Issue on Decomposition in VLSI Design, vol.3, Nos.3-4, 1995, pp. 225–228.
8. Volf, F.A.M., Jóźwiak, L.: *Decompositional Logic Synthesis Approach for Look-up Table Based FPGAs* 8-th IEE International ASIC Conference - ASIC'95, Austin, Texas, September, 1995.
9. Collaborative Benchmarking Laboratory, Department of Computer Science at North Carolina State University, http://www.cbl.ncsu.edu/.

Modelling Social Game Systems by Rule Complexes

Tom R. Burns[1] and Anna Gomolińska[2]*

[1] University of Uppsala, Dept. of Sociology, Box 821, 75108 Uppsala, Sweden
[2] University of Białystok, Dept. of Mathematics,
Akademicka 2, 15267 Białystok, Poland

Abstract. In the paper we present the notion of rule complex as a promising tool to formalize social game systems. We also hope to arouse the interest of the computer science community in application of the rough-set and other current computing methods to the social game theory.

1 Introduction

Systems of rules guiding social actors in their activities and interactions as well as social game orders may be formalized by means of rule complexes in a uniform and clear though fairly general way[1]. Rule and rule complex are key concepts for us. Our rule is a kin to default rule [3] while our rule complex is not merely a set of rules. Using the latter notion to formalize social game systems is absolutely novel[2]. We develop the idea presented in [1] that social organization may be seen as a certain system of rules. Our framework essentially extends the classical von Neumann-Morgenstern game theory [2], where a game order is a finite set of pre-determined rules. According to our approach, rules may be imprecise and open to innovation and modification. Game orders viewed as rule complexes are subject to transformation. Social organization and, in particular, social games form a system the complexity of which far exceeds the actual capabilities of humans to create a uniform formal model with full particulars. This and the difficulties with communication between researchers in computer and social sciences discourage many to apply the current computing methods in social sciences. Our framework is to explain in a formal way what the social game theory is about and to facilitate the application of rule-oriented technics, e.g., the rough-set methods to certain problems in social game theory, e.g., to generate judgment complexes.

---

* The author expresses her gratitude to Andrzej Skowron for deep and useful comments.
[1] For simplicity, the term "rule complex" will denote both a social system of rules and its formal counterpart in our theory.
[2] Informally speaking, the notion of rule complex is related to that of a set of rules as the notion of a program with procedures to that of a program with instructions only.

L. Polkowski and A. Skowron (Eds.): RSCTC'98, LNAI 1424, pp. 581–584, 1998.

Briefly speaking, we are mainly concerned with social rule complexes (i.e., rule complexes shared by a group or population of actors), actors' complexes (i.e., social roles of actors), and social game orders. Actors use social rule complexes, a type of collective knowledge, to constitute and regulate their interactions or game processes. Each type of social relationship has its corresponding rule based roles derived from cultural frameworks. The roles (i.e., actors' complexes) vary because actors play different roles in social relationships and because actors often differ somewhat in the ways they have learned and developed roles through their personal histories and continuing practice. Social life is often characterized by ambiguity and the underspecification of options, i.e., game information is not only incomplete; it remains to be generated. In general, interaction situations are typically not fully specified. The social constructions of games and game processes provide the specification. Among other things, the participating actors use their social structural knowledge and previous experience with one another, possibly in similar game situations to fill in undefined or uncertain "spaces" of action opportunities and outcomes. Given that social actors tend to acquire and develop different rule complexes, we find conflicts and struggles over rule complexes. That is, rule complexes – and their articulation in institutions and games – are historical products of interactions among social actors.

2   Rules and Rule Complexes

meta levels[3]. The main assumption is that actors organize rules in rule complexes. In consequence, we can uniformly and relatively easy investigate various sorts of rules (e.g., evaluative rules, norms, judgment rules, and action rules), complex objects consisting of rules (e.g., roles, routines, procedures, algorithms, game orders, and models of the reality and the actors), and interdependencies among them. Given a language where the object and meta levels are not separated, let us denote the set of all formulas by FOR, while rules (resp., rule complexes) by r (resp., C) with sub/superscripts if necessary. Formally, by a *rule* we mean a triary relation $r \subseteq \mathcal{P}(FOR)^2 \times FOR$, where there exist $m, n \in \mathbb{N}$ such that for any $(X, Y, \gamma) \in r$, $card(X) = m$ and $card(Y) = n$. Where possible, rules will be written by schema, viz., $r : \frac{\alpha_1,\ldots,\alpha_m : \beta_1,\ldots,\beta_n}{\gamma}$, where $X = \{\alpha_1, \ldots, \alpha_m\}$ and $Y = \{\beta_1, \ldots, \beta_n\}$. *Axiomatic* rules, where $X = Y = \emptyset$, represent facts.

A *rule complex* is a set obtained according to the following formation rules: (1) Any finite set of rules is a rule complex; (2) If C_1, C_2 are rule complexes, then $C_1 \cup C_2$, $C_1 \cap C_2$, $C_1 - C_2$, and $\mathcal{P}(C_1)$ are rule complexes; (3) If $C_1 \subseteq C_2$ and C_2 is a rule complex, then C_1 is a rule complex. Thus, a rule complex C is a set $C = \{r_1, \ldots, r_m, C_1, \ldots, C_n\}$, where $m, n \in \mathbb{N}$, $r_1, \ldots, r_m$ are some rules and $C_1, \ldots, C_n$ are some rule complexes. All rules constituting a rule complex C form its *rule base*, $rb(C)$. Similarly, rule complexes constituting C form its *complex base*, $cb(C)$. The meaning of the two notions is explained by the example.

---

[3] In practice, people carry on both the operative and the meta levels; in part, they switch back and forth or operate simultaneously on both levels. Nevertheless, we are aware of the circularity which may occur.

Example 1. Where r_1, r_2, r_3, r_4 are rules, sets $C_1 = \{r_2, r_3\}$, $C_2 = \{r_4\}$, $C_3 = \{r_1, C_1\}$, and $C_4 = \{r_1, C_2, C_3\}$ are rule complexes. $rb(C_4) = \{r_1, r_2, r_3, r_4\}$ and $cb(C_4) = \{C_1, C_2, C_3\}$.

Example 2. Algorithms as collections of instructions may be seen as rule complexes.

By a *subcomplex* of C we mean an element of $cb(C)$ or a rule complex obtained from C by dropping some elements of $rb(C) \cup cb(C)$. In particular, any subset of C is a subcomplex of C as well. However, C_2 and $C_6 = \{r_1, C_2, C_5\}$, where $C_5 = \{r_1\}$, are subcomplexes but not subsets of C_4.

The notion of rule complex may be used to formalize game (or interaction) orders as well as actors' systems of rules: normative orders, value complexes, judgment systems, action modules, and models of the reality and the actors.

Example 3. Given an interaction situation S_t at time t, consider an actor acting as an employee and a mother in S_t. The roles are formalized by rule complexes $ROL_E(t)$ and $ROL_M(t)$, respectively. The actor's role in S_t, $ROL(t)$, may be defined as

$$ROL(t) = \{ROL_E(t), ROL_M(t), R_R(t)\},$$

where $R_R(t)$ is a complex of extra rules describing, among others, interdependencies between $ROL_E(t)$ and $ROL_M(t)$. Now consider subcomplex $ROL_E(t)$ which describes the actor's activities as an employee in S_t and, in particular, norms, values (and hence goals), judgment system(s), action modules (routines, procedures, etc.) in S_t, and her beliefs and knowledge about S_t (i.e., about the reality, herself, and other actors involved). The norms form a rule complex, $NO_E(t)$, called a *normative order* associated with the role of employee in S_t. Similarly, the evaluative rules form a *value complex*, $VAL_E(t)$, the judgment rules constitute a *judgment complex*, $JUDG_E(t)$, the action modules form an *action complex*, $ACT_E(t)$, and the actor's beliefs and knowledge about S_t form her model of S_t, $MOD_E(t)$. Hence, $ROL_E(t)$ may be written as

$$ROL_E(t) = \{NO_E(t), VAL_E(t), JUDG_E(t), ACT_E(t), MOD_E(t), R_E(t)\},$$

where $R_E(t)$ is a complex of other relevant rules. Going on along these lines, one can obtain specific rules, routines, and procedures associated with subcomplexes of $ROL_E(t)$; and analogously for $ROL_M(t)$. On the other hand, if consider particular questions like judgment making, one may investigate the actor's judgment complex in S_t, $JUDG(t)$, which may be defined as

$$JUDG(t) = \{JUDG_E(t), JUDG_M(t), R_J(t)\},$$

where $R_J(t)$ is a complex of some extra rules taken into account. Thus, the actor's role $ROL(t)$ may be also written as

$$ROL(t) = \{NO(t), VAL(t), JUDG(t), ACT(t), MOD(t), R'_R(t)\},$$

where $R'_R(t)$ is a complex of other rules relevant for $ROL(t)$.

3   Final Remarks

For lack of space, our presentation is limited to the definitions of rule and rule complex and to a simple example on modeling social roles by means of rule complexes. Nevertheless, we have investigated such problems as application of rules and rule complexes, consistency of rule complexes, and transformations of rule complexes (in particular, compositions and decompositions).

In spite of its conceptual simplicity, the notion of rule complex is powerful and flexible enough to model systems of social actors, games, and interactions. Apart from modeling social game systems in a novel way, we aim at building a bridge between the computer and social sciences to facilitate application of new computing ideas and technics to the contemporary problems investigated in the social sciences.

References

1. Burns, T. R., Flam, H.: The Shaping of Social Organization: Social Rule System Theory with Applications. Sage Publications (1987, reprinted 1990)
2. von Neumann, J., Morgenstern, O.: Theory of Games and Economic Behaviour. Princeton University Press (1953)
3. Reiter, R.: A logic for default reasoning. Artificial Intelligence **13** (1980) 81–132

Analysis and Synthesis of Information Systems with Information Relationships and Measures

Lech Jóźwiak

Eindhoven University of Technology, Faculty of Electrical Engineering
P.O.Box 513, EH 10.25 5600 MB Eindhoven, The Netherlands

[**Abstract.**]
Analysis of relationships between information streams is of primary importance for analysis and synthesis of discrete information systems. This paper defines and discusses various information relationships and measures for the strength and importance of the information relationships.

1 Introduction

Analysis of information and information relationships is of primary importance for analysis and synthesis of discrete information systems. This paper aims to introduce and discuss the fundamental apparatus for analysis and evaluation of information and information relationships.

2 Representation of information in information systems

Information is represented in discrete systems by values of some discrete signals or variables. Lets consider a certain finite set of elements S called symbols. Knowing a certain value of a certain signal or variable x, it is possible to distinguish a certain subset B of elements from S from all other elements of S, but it is impossible to distinguish between

| term | inputs | | | | | | outputs | | | |
symbols	x_1	x_2	x_3	x_4	x_5	x_6	y_1	y_2	y_3	y_4
1	0	-	0	0	0	1	0	1	1	0
2	0	0	0	1	0	-	1	1	-	0
3	1	-	0	0	1	-	1	1	1	0
4	1	1	1	1	1	-	0	-	0	1
5	-	0	1	-	0	0	-	0	-	0

Fig. 1. A multi-output Boolean function $y = f(x_1, x_2, x_3, x_4, x_5, x_6)$.

the elements from B. For example, for Boolean function in Fig.1, where symbols $1 - 5$ represent terms on input variables $x_1 - x_6$, different values of the input variable x_3 enable us to distinguish between the subset $\{1, 2, 3\}$ and $\{4, 5\}$. For 1, 2 and 3: $x_3 = 0$ and for 4 and 5: $x_3 = 1$. In such a way information is modeled with set systems [1].

A **set system** [1] (r-partition [4]) SS on a set S is defined as a collection of subsets B_1, B_2,..., B_k of S such that $\bigcup_i B_i = S$ and $B_i \not\subseteq B_j$ for $i \neq j$.

An elementary information describes the ability to distinguish a certain single symbol s_i from another single symbol s_j, where: $s_i, s_j \in S$ and $s_i \neq s_j$. Any set of

L. Polkowski and A. Skowron (Eds.): RSCTC'98, LNAI 1424, pp. 585–588, 1998.
© Springer-Verlag Berlin Heidelberg 1998

elementary portions of information can be represented by an information relation I, set IS and graph IG defined on $S \times S$ as follows:
$I = \{(s_i, s_j)|s_i$ is distinguished from s_j by the modeled information$\}$,
$IS = \{\{s_i, s_j\}|s_i$ is distinguished from s_j by the modeled information$\}$, and
$IG = \{S, \{\{s_i, s_j\}|s_i$ is distinguished from s_j by the modeled information$\}\}$.

Relationships between set systems can be analyzed by considering relationships between their corresponding information relations, sets and graphs. The correspondence between SS and IS is as follows: IS contains the pairs of symbols that are not contained in any single block of a corresponding SS.

[*Example 1.*] (information modeling with set systems and information sets)
For the function from Fig.1, the appropriate set systems and information sets are listed below:

$$SS(x_1) = \{\{1,2,5\}; \{3,4,5\}\} \quad IS(x_1) = \{1|3, 1|4, 2|3, 2|4\}$$
$$SS(x_2) = \{\{1,2,3,5\}; \{1,3,4\}\} \quad IS(x_2) = \{1|4, 2|4, 4|5\}$$
$$SS(x_3) = \{\{1,2,3\}; \{4,5\}\} \quad IS(x_3) = \{1|4, 1|5, 2|4, 2|5, 3|4, 3|5\}$$
$$SS(x_4) = \{\{1,3,5\}; \{2,4,5\}\} \quad IS(x_4) = \{1|2, 1|4, 2|3, 3|4\}$$
$$SS(x_5) = \{\{1,2,5\}; \{3,4\}\} \quad IS(x_5) = \{1|3, 1|4, 2|3, 2|4, 3|5, 4|5\}$$
$$SS(x_6) = \{\{1,2,3,4\}; \{2,3,4,5\}\} \, IS(x_6) = \{1|5\}$$
$$SS(y_1) = \{\{1,4,5\}; \{2,3,5\}\} \quad IS(y_1) = \{1|2, 1|3, 2|4, 3|4\}$$
$$SS(y_2) = \{\{1,2,3,4\}; \{4,5\}\} \quad IS(y_2) = \{1|5, 2|5, 3|5\}$$
$$SS(y_3) = \{\{1,2,3,5\}; \{2,4,5\}\} \quad IS(y_3) = \{1|4, 3|4\}$$
$$SS(y_4) = \{\{1,2,3,5\}; \{4\}\} \quad IS(y_4) = \{1|4, 2|4, 3|4, 4|5\}$$
$$SS(y) = SS(y_1, y_2, y_3, y_4) = \{\{1\}; \{2,3\}; \{4\}; \{5\}\},$$
$$IS(y) = \{1|2, 1|3, 1|4, 1|5, 2|4, 2|5, 3|4, 3|5, 4|5\}$$

3   Information relationships

Information relationships between two set systems SS_1 and SS_2 are defined as follows:

- **common information** CI (information that is present in both SS_1 and SS_2): $CI(SS_1, SS_2) = IS(SS_1) \cap IS(SS_2)$
- **total (combined) information** TI (information that is present either in SS_1 or in SS_2): $TI(SS_1, SS_2) = IS(SS_1) \cup IS(SS_2)$
- **missing information** MI (information that is present in SS_1, but missing in SS_2): $MI(SS_1, SS_2) = IS(SS_1) \setminus IS(SS_2)$
- **extra information** EI (information that is missing in SS_1, but present in SS_2): $EI(SS_1, SS_2) = IS(SS_2) \setminus IS(SS_1)$
- **different information** DI (information present in one and missing in the other set system): $DI(SS_1, SS_2) = MI(SS_1, SS_2) \cup EI(SS_1, SS_2)$.

[*Example 2.*] (some information relationships for function from Fig.1)

$$CI(SS(y_1), SS(x_4)) = \{1|2, 3|4\} \, MI(SS(y_1), SS(x_4)) = \{1|3, 2|4\}$$
$$EI(SS(y_1), SS(x_4)) = \{1|4, 2|3\} \, DI(SS(y_1), SS(x_4)) = \{1|3, 1|4, 2|3, 2|4\}$$
$$TI(SS(y_1), SS(x_4)) = \{1|2, 1|3, 1|4, 2|3, 2|4, 3|4\}$$

With the relationship apparatus defined above, we can have such questions answered as: what information required to compute values of a certain variable is present in another variable, what information is missing etc. However, to take appropriate decisions, some quantitative relationship measures are often necessary.

4   Information relationship measures

For set system SS **information quantity** IQ: $IQ(SS) = |IS(SS)|$ is defined. For two set systems SS_1 and SS_2, the following **information relationship measures** are defined:

- **similarity measure ISIM**: $ISIM(SS_1, SS_2) = |CI(SS_1, SS_2)|$
- **dissimilarity measure IDIS**: $IDIS(SS_1, SS_2) = |DI(SS_1, SS_2)|$
- **decrease measure IDEC**: $IDEC(SS_1, SS_2) = |MI(SS_1, SS_2)|$
- **increase measure IINC**: $IINC(SS_1, SS_2) = |EI(SS_1, SS_2)|$
- **total information quantity TIQ**: $TIQ(SS_1, SS_2) = |TI(SS_1, SS_2)|$.

It is also possible to define some relative measures, by normalizing the above absolute measures and weighted measures by associating a certain importance weight $w(s_i|s_j)$ with each elementary information.

5   Application of the relationships and measures

The information relationships and measures introduced above are of primary importance for an effective and efficient analysis and synthesis of information systems. They provide designers and tools with data necessary for decision making. Results of the relationship analysis make it possible to discover the nature of a considered system or to decide its structure. Below, we will illustrate with an example the way of use of the information relationships and measures.

[*Example 3.*] (application to input support minimization) Input support minimization consists of finding a minimal sub-set of inputs that still enables unambiguous computation of values for a specified function [2]. Let's find the minimum support for the example function from Fig.1, by analyzing the relationships between the information required for computing the output values and information provided by particular input variables.

$$CI(y, x_1) = \{1|3, 1|4, 2|4\} \qquad\qquad |CI(y, x_1)| = 3$$
$$CI(y, x_2) = \{1|4, 2|4, 4|5\} \qquad\qquad |CI(y, x_2)| = 3$$
$$CI(y, x_3) = \{1|4, 1|5, 2|4, 2|5, 3|4, 3|5\} \quad |CI(y, x_3)| = 6$$
$$CI(y, x_4) = \{1|2, 1|4, 3|4\} \qquad\qquad |CI(y, x_4)| = 3$$
$$CI(y, x_5) = \{1|3, 1|4, 2|4, 3|5, 4|5\} \quad |CI(y, x_5)| = 5$$
$$CI(y, x_6) = \{1|5\} \qquad\qquad\qquad |CI(y, x_6)| = 1$$

$|CI(y, x_3)| > |CI(y, x_5)| > |CI(y, x_1)| = |CI(y, x_2)| = |CI(y, x_4)| > |CI(y, x_6)|$.
Thus, x_3 delivers more information for y than x_5, x_5 more than x_1 , etc. Furthermore, x_3 delivers a unique information, namely: $2|5$. Therefore, x_3 must be in any minimal support. The symbols from the pairs given by $IS(y) - CI(y, x_3) =$

$\{1|2, 1|3, 4|5\}$ are not distinguished from each other by x_3, and therefore at least one extra input variable is necessary. Because $CI((IS(y)-CI(y, x_3)), x_1) = \phi$, $CI((IS(y)-CI(y, x_3)), x_2) = \{4|5\}$, $CI((IS(y)-CI(y, x_3)), x_4) = \{1|2\}$, $CI((IS(y)-CI(y, x_3)), x_5) = \{1|3, 4|5\}$, $CI(y, x_6) = \phi$, no single variable delivers the lacking information, but x_4 and x_5 together provide it. This means that x_3, x_4 and x_5 constitute the minimum input support.

6   Conclusions

Some of the relationships and measures have already been applied in CAD tools for finding optimal decompositions of combinational functions [6], and for finding minimal input support [2]. Application of the relationships and measures to input support selection in functional decomposition is presented in a separate paper of this conference [5]. Their more extensive discussion can be found in [3]. The theory of information relationships and measures introduced briefly in this paper makes operational the famous theory of set systems of Hartmanis [1]. Set systems enable us to model information. The relationships and measures enable us to analyze and measure the modeled information and the relationships between the modeled information streams. They form a fundamental apparatus for analysis and design of information systems, and in particular for logic design, database design, pattern analysis, machine learning etc. [1]-[6].

References

1. J. Hartmanis, R.E. Stearns: *Algebraic Structure Theory of Sequential Machines*, Englewood Cliffs, N.J.: Prentice-Hall, 1966.
2. L. Jóźwiak, N.Ederveen: *Genetic Algorithm for Input Support Minimization*, International Conference on Computational Intelligence and Multimedia Applications (ICCIMA'98), Melbourne, Australia, Febr. 9-11, 1998.
3. L. Jóźwiak: *Information Relationships and Measures: An Analysis Apparatus for Efficient Information System Synthesis*, Proc. of the EUROMICRO'97 Conference, Budapest, Hungary, Sept. 1-4, 1997, pp. 13–23, IEEE Computer Society Press.
4. T. Łuba and J. Rybnik: *Rough Sets and Some Aspects of Logic Synthesis, in R. Slowinski (Ed.): Intelligent Decision Support - Handbook of Applications and Advances of the Rough Sets Theory*, pp. 181–199, Kluwer, 1993.
5. M. Rawski, L. Jóźwiak, A. Chojnacki: *Application of the Information Relationship Measures to Input Support Selection in Functional Decomposition*, The First International Conference on Rough Sets and Current Trends in Computing (RSCT'98), Warsaw, Poland, June 22-26, 1998.
6. F.A.M. Volf, L. Jóźwiak: *Decompositional Logic Synthesis Approach for Look Up Table Based FPGAs*, 8th IEEE International ASIC Conference, Austin, Texas, 18-22 September, 1995.

Approximations in Data Mining

Wieslaw Traczyk

Institute of Control and Computation Engineering
Warsaw University of Technology

Abstract. There are many different ways of discovering knowledge in large databases but in all of them the same problem arises: how rigorously the results of discovery answer the purpose? In the paper we discuss some sources of vagueness in data mining, and present non-statistical method of inaccuracy evaluation.

1 Introduction

Question how rigorously the results of a work answer the purpose, is common for all kinds of our activity. It is particularly important in knowledge discovery from large databases, because requirements in this domain can be diverse, and algorithm efficiency limitations are frequently accepted.

It will be assumed that the process of discovery starts from user query, which contains some information on the *domain* of search and more or less precise specification of the *concept* in request. Simple query to database may look as follows:

WHAT ARE THE FACTORS THAT LEAD TO *concept(s)* IN *domain?*

(for instance *concept* = STUDY PROLONGATION, *domain* = MY DEPARTMENT)

Inductive learning is essential for generating hypotheses from data automatically, and therefore our interest will be focused on logical descriptions of hypotheses, obtained by induction.

2 Basic notions

Relational database collects attribute values of the set of objects. Let $X = \{x_1, x_2, \ldots, x_M\}$ be the set of *objects* in the problem domain, described by the set $A = \{a_1, a_2, \ldots, a_N\}$ of *attributes*. By V_i we denote the *domain (value set)* of the attribute a_i i.e. $V_i = \{v_1, v_2, \ldots, v_K\}_i$. Equation $a_i = v_{ji}$ is called *selector*. *Possible values space* $\mathbf{V}' = V_1 \times V_2 \times \ldots \times V_N$ has elements $\mathbf{v}$ named *characteristic* or *tuple*. Each object x_m has characteristic $\mathbf{v}_m$. *Real values space* is defined by $\mathbf{V} = \bigcup_{m=1}^{M} \mathbf{v}_m$. *Data matrix* $\mathbf{M} = \langle \mathbf{v}_1, \mathbf{v}_2, \ldots \mathbf{v}_M \rangle^T$ has rows determined by objects, and columns corresponding to attributes.

Attribute values v_i (*original values*) may have different domains: *continuous* ($V_i \subseteq \mathcal{R}$), *ordinal* or *nominal*. These domains are frequently modified, in order to reduce its cardinality or with the intention to adapt attribute values to external requirements. Some more frequently used *attribute modifications* are listed below.

L. Polkowski and A. Skowron (Eds.): RSCTC'98, LNAI 1424, pp. 589–592, 1998.
© Springer-Verlag Berlin Heidelberg 1998

Discretization (coding) of continuous values produces discrete values;
Partitioning merges some values of the ordinal attribute into intervals, intro-
ducing a new form of selectors; e.g. $a \in [v_\square, v^\square]$ or $a \geq v$.
Assembling selects (by *or* and *not* several discrete values,
Generalization joints similar discrete values into one group.

Since selectors with modified attributes can include different relations and simple
values or intervals, we will use generally the symbol s_i for all kinds of selectors
with an attribute a_i.

Two or more original or modified attributes (from the same or distinct tables)
can be used for construction of *virtual attribute a^**, with a domain determined by
operations on domains of component attributes. In the case of logical dependence
each value v of a^* is defined by a conjunction of selectors (with original or
modified attributes $a_i, a_j, \ldots$ and relations $=, \leq, \ldots$):
$$(a^* = v) \text{ if } (a_i = v') \text{ AND } (a_j = v'') \text{ AND } \ldots$$
We will also use the equivalent notation
$$Def(a^* = v) = Def(s^*) = \{s_i, s_j, \ldots\}$$
It will be useful to extend the definition and assume that virtual attribute
is a generalization, covering one or more original or modified attributes of the
given data matrix.

Concept name, introduced in a user query, can be related to the database if
there exists an virtual attribute with a value that strictly match the meaning
of a concept. We will use symbols σ for a concept name, v^σ – for corresponding
attribute value and a^σ – for the virtual attribute with value v^σ ($a^\sigma = a^*$).

Intension Int(σ) of a concept σ is a definition of appropriate virtual attribute
and may be used as a definition of a concept (for logical dependencies):
$$Int(\sigma) = Def(a^\sigma = v^\sigma)$$
Extension Ext(σ) (or *range $X(\sigma)$*) of a concept σ contains all objects from
X for which an attribute a^σ has value v^σ: $Ext(\sigma) = X(\sigma) = \{x \mid a^\sigma(x) = v^\sigma\}$.

Attributes from an intension of a concept define it, but frequently some
other attributes describe the causes or conditions of a concept appearance.
Then, for a given concept σ the set of all original or modified attributes can
be divided into three parts: *defining* attributes A_D^σ (represented by a^σ), *explaining (interpretative)* attributes A_E^σ and *irrelevant* attributes A_I^σ, such that
$A = A_D^\sigma \cup A_E^\sigma \cup A_I^\sigma$. For example the concept STUDY PROLONGATION is *defined*
by the attribute SEMESTER-OF-THE-STUDY (modified by coding "long" study),
explained by attributes EXTERNAL-JOB, AVERAGE-GRADE, ..., but attributes
like NAME, ID-NUMBER are *irrelevant*.

Defining, explaining and irrelevant attributes determine appropriate partition
of object characteristic: $\mathbf{v}_m = \langle \mathbf{v}_{Dm}, \mathbf{v}_{Em}, \mathbf{v}_{Im} \rangle$. We will call them D-, E- and
I-characteristics of an object x_m. Let
$$\mathbf{V}_D = \bigcup_{m=1}^M \mathbf{v}_{Dm}, \quad \mathbf{V}_E = \bigcup_{m=1}^M \mathbf{v}_{Em} \text{ and } \rho: X \to \mathbf{V}_E.$$
All objects from X represented by E-characteristic $\mathbf{v}_E$ are then defined by
$$X(\mathbf{v}_E) = \{x \mid \rho(x) = \mathbf{v}_E\}$$
In many cases the main goal of data mining is to find the dependency be-
tween defining attributes (as dependent variables) and explaining or remaining

attributes (as independent variables). Explanation of a concept σ, represented by virtual attribute $a^* = a^\sigma$ with discrete value, can be stated as a predicate formula Φ, with selectors used as arguments $(a_k, a_l, \ldots \in A_E^\sigma)$:

$$Expl(a^* = v) = Expl(s^*) = \Phi(s_k, s_l, \ldots)$$

This equation is usually presented as a production rule. Explanations generated in this way designate the answer to the user query, but frequently a query concerns not only one concept but a set of similar concepts, exhaustive and disjoint, which form *classes* of the domain:

$$X = X(\sigma_1) \cup X(\sigma_2) \cup \ldots \cup X(\sigma_p)$$

In this case explanation formulae are used as a tool for *classification*.

3 Approximations

In a process of reply generation for the user query we should be able to evaluate the quality of result, mainly determined by univocal and precise final explanations or interpretations. Precise concept(s) explanation (or interpretation) by means of logical formula assumes univocal relationship between values of explaining attributes A_E^σ and values of defining attributes A_D^σ. This is equivalent to the assumption that the set of objects relevant to σ (i.g. $X(\sigma)$) covers all objects represented by E-characteristics explaining a concept σ (i.g. $X(\mathbf{v}_E)$). Appropriate formula (valid for all σ) looks as follows:

$$(X(\mathbf{v}_E) \cap X(\sigma) \neq \emptyset) \Rightarrow (X(\mathbf{v}_E) \subseteq X(\sigma)).$$

If reality is not so ideal – some characteristics $\mathbf{v}_E$ refer not only to considered concept but also to another notions (then explanation of the concept is not correct), or objects with equal characteristics $\mathbf{v}_E$ are associated with different concepts (giving wrong classification, if based on $\mathbf{v}_E$). If we are interested in the degree of ambiguity – *rough sets* [3] can help to evaluate it.

E-characteristics that always properly explain the concept – define the set of objects being *lower approximation* of the concept (exactly – of its extension), calculated as:

$$X_\sigma = \{x_m \mid X(\mathbf{v}_{Em}) \subseteq X(\sigma)\}$$

E-characteristics that explain the concept, but can effect also some other notions, define the set of objects being *upper approximation* of the concept:

$$X^\sigma = \{x_m \mid X(\mathbf{v}_{Em}) \cap X(\sigma) \neq \emptyset\}$$

A *quality of interpretation* may now be defined as the ratio of univocally interpreted objects to all objects referred to the concept:

$$\eta = \frac{|X_\sigma|}{|X(\sigma)|}$$

In the ideal case $X_\sigma = X^\sigma = X(\sigma)$ and $\eta = 1$.

For complex explanations it is important to know that some set of E-characteristics marks *certain but incomplete* explanation:

$$\mathbf{V}_\sigma = \{\mathbf{v}_E \mid X(\mathbf{v}_E) \subseteq X(\sigma)\}$$

and another set defines *redundant* explanation:

$$\mathbf{V}^\sigma = \{\mathbf{v}_E \mid X(\mathbf{v}_E) \cap X(\sigma) \neq \emptyset\}$$

When more than one concept should be interpreted by data mining, evaluation of interpretation can be done by means of the *evidence theory*. In this theory, the *basic probability assignment* $m : \theta \to [0, 1]$ is interpreted as the degree of evidence that a specific element of X belongs to the set of concepts θ (subset of all concepts considered), but not to any special subset of θ.

Notions of lower and upper approximations can be extended to more than one concept in the set θ:

$$X_\theta = \bigcap_{\sigma \in \theta} X^\sigma \quad X^\theta = \bigcup_{\sigma \in \theta} X^\sigma.$$

Having sets X_θ one can easily calculate $m(\theta)$ from the equation

$$m(\theta) = \frac{|X_\theta|}{|X|}$$

If θ contains one, two or more concepts and Δ is the set of concepts, then the *belief function* is defined as

$$Bel(\Delta) = \textstyle\sum_{\theta \in \Delta} m(\theta)$$

and the *plausibility function* – as

$$Pl(\Delta) = \textstyle\sum_{\theta \cap \Delta \neq \emptyset} m(\theta)$$

$Bel(\Delta)$ represents the total evidence or believe that the object from X belongs to Δ as well as to the various subsets of Δ. $Pl(\Delta)$ represents all this and additional evidence associated with sets that overlap with Δ.

The pair $\langle Bel(\Delta), Pl(\Delta) \rangle$ for different Δ constitutes a measure of approximation for the set Δ: if $Bel(\Delta) = Pl(\Delta)$ – concepts in Δ are interpreted precisely (as a set), the difference between $Bel(\Delta)$ and $Pl(\Delta)$ shows the level of approximation, and values of $Bel(\Delta)$ and $Pl(\Delta)$ qualify contribution of Δ members in the total set of objects.

References

1. Fayyad U.M., Piatetsky–Shapiro G., Smyth P., Uthurosamy R., *Advances in Knowledge Discovery and Data Mining*, AAAI Press / The MIT Press, 1996.
2. Kloesgen W., Zytkow J.M., *Knowledge Discovery in Databases Terminology*, in [1], pp 573–593.
3. Pawlak Z., *Rough Sets – Theoretical Aspects of Reasoning about Data*, Kluwer Academic Publ., 1991.

Purchase Prediction in Database Marketing with the ProbRough System

Dirk Van den Poel[1] and Zdzisław Piasta[2]

[1] Catholic University Leuven
Department of Applied Economic Sciences
Naamsestraat 69, B-3000 Leuven, Belgium
e-mail: Dirk.VandenPoel@econ.kuleuven.ac.be

[2] Kielce University of Technology
Mathematics Department
Al. 1000-lecia P.P. 5, 25-314 Kielce, Poland
e-mail: zpiasta@sabat.tu.kielce.pl

[Abstract.] We describe how probabilistic rough classifiers, generated by the rule induction system ProbRough, were used for purchase prediction and discovering knowledge on customer behavior patterns. The decision rules were induced from the mail-order company database. Construction of ProbRough is based on the idea of the attribute space partition and was inspired by the rough set theory. The system's beam search strategy in a space of models is guided by the global cost criterion. The system accepts noisy and inconsistent data with missing attribute values. Background knowledge is used in the form of prior probabilities of decisions and different costs of misclassification. ProbRough provided a lot of useful information about the problem of customer response modeling, and demonstrated its usefulness and efficiency as a data mining tool.

Introduction

Predicting future purchase behavior at the level of the individual consumer has become a key issue in database marketing, which is defined as a method of analyzing customer data to look for patterns among existing customer preferences and to use these patterns for more targeted selection of customers (Fayyad et al.; 1996). The prediction and targeting of the individual consumer was made possible by the capability to individually address every single customer by direct marketing media such as direct mail, catalogs, and more recently the internet (Petrison et al.; 1993). This is in contrast to the traditional advertising media such as print and television, which do not offer this opportunity. In the paper we describe how probabilistic rough classifiers, generated by the ProbRough system

L. Polkowski and A. Skowron (Eds.): RSCTC'98, LNAI 1424, pp. 593–600, 1998.
© Springer-Verlag Berlin Heidelberg 1998

(Piasta and Lenarcik; 1996, 1998) were used for purchase prediction, discovering knowledge on customer behavior patterns, and presenting this knowledge in the form transparent to the user. ProbRough is a system based on the idea of the attribute space partition. Its construction was inspired by the rough set theory (Pawlak, 1991). ProbRough beam search strategy in a space of models is guided by the global cost criterion. The system accepts noisy and inconsistent data with missing attribute values. Background knowledge is used in the form of prior probabilities of decisions and different costs of misclassification. ProbRough provided a lot of useful information about the problem of customer response modeling, and demonstrated its usefulnes and efficiency as a data mining tool (see also, Kowalczyk and Piasta, 1998; Lenarcik and Piasta, 1997). This paper is structured as follows: in the next section we describe the process of knowledge discovery, including a detailed description of the business problem, the data mining method, characterization of the data and a discussion of the results. We conclude with a summary of our findings.

The KDD process

Business problem

Every mailing period, mail-order companies are confronted with the decision problem whether or not they have to mail a catalog to a particular customer. The importance of this task is enhanced by the tendency of rising mailing costs and increasing competition (Hauser, 1992). In this case study, we formulated this decision problem as a classification task, i.e., we tried to predict, based on all available data, whether a customer would (re)purchase during the next mailing period. Hence, the response variable in the specification which we investigated was binary (0/1). The marketing manager would then typically use the prediction of the classification technique to rank the total customer base and mail to the best part of its customer base. How many customers would receive a particular catalog is determined by the mailing company based on the cost/profit trade-off or budget constraints. Two important issues in this decision problem arose, which have not been dealt with extensively by previous discussions of this topic of response modeling: (1) the prior probabilities and (2) the misclassification costs. The former issue is concerned with the fact that frequently the proportion of buyers to non-buyers in the data do not reflect the true proportion in the whole population of customers. The latter problem deals with the fact that the cost of incorrectly assigning a buyer as a non-buyer (i.e., the opportunity cost of missing a sale) is much higher than the situation in which a non-buyer is predicted to be a buyer (i.e., the cost of sending a catalog to somebody who will not purchase).

Data Mining Method: ProbRough

We selected the ProbRough system for generating probabilistic rough classifiers as a data mining tool. ProbRough has the ability to handle the key features of the problem at hand: (1) prior probabilities and (2) unequal misclassification costs. In the sequel by standard priors and costs we mean the prior probabilities set to the frequencies of classes in the training data and equal costs of misclassification. Now, we present a general idea of ProbRough. The detailed description of the algorithm is given in Piasta and Lenarcik (1996).

Domains of the decision rules generated by ProbRough are of the form

$$\Delta = \Delta_1 \times \ldots \times \Delta_m \, ,$$

where Δ_q is an interval, when the values of the q-th attribute are ordered, or an arbitrary subset, otherwise. Such Δ_q-s are referred to as the segments. The sets Δ are called the feasible subsets in the attribute space. Because of the specific form of the domains, the decision rules can be expressed in a transparent format that can be easily understood by users.

The algorithm of rough classifier generation consists of the two fundamental phases:

- [(i)] the global segmentation of the attribute space,
- [(ii)] the reduction of the number of decision rules.

In the first phase ProbRough tries to minimize the average global cost of the decision-making. In this phase a number of divisions of the whole attribute space is peformed. This number is one of the parameters and is referred to as the number of iterations of the algorithm. The number of iterations can be optimized in order to improve the predictive properties of the algorithm (see, Piasta and Lenarcik, 1996). Each single iteration consists in dividing one of the attribute value sets (or its subset) into two segments. When the attribute is continuous then a finite set of intermediate values has to be used. This set should be given in advance or can be obtained from the data. The attributes which are not involved in the partition process are eliminated. The resulting partitions determine the unique partition of the attribute space into feasible subsets. Each partition element is associated with a set of equally important decisions. In the second phase we consider only those partitions which are associated with the minimum value of the cost criterion. In this phase the partition elements are joined into the bigger feasible subsets, provided that the sets of equally important decisions have a non-empty intersection. During the above procedure the number of rules of the resultant classifier is minimized.

We refer to Van den Poel (1998) for a comparison study between several different data mining methods for response modeling.

Dataset

A random sample from the mail-order company database from an anonymous European mail-order company containing 6.800 observations was taken in such a way that 50 % of the customers in the sample responded to the offers during a 6 month period and 50 % did not respond to the offer. This sample was randomly split to obtain a learning sample and a test sample. All models were built on the basis of the learning sample. The predictive performance was judged on the basis of the test sample, i.e. observations which the system had not used during the learning phase. Among the many attributes that could be constructed based on past transaction data, three variables have been identified by Cullinan (Petrison et al.; 1993) to be of particular importance in database marketing modeling: recency, frequency and monetary value. Several instances of these RFM-variables were included in this study as shown in Table 1. All variables were already categorized and were provided by the mail-order company at the level of the individual customer.

Table 1. Description of the attributes in the database.

Name	Type	Description
Buy_{t-1}	0/1	Did the customer buy during the previous 6 months?
Buy_{t-2}	0/1	Did the customer buy in the period: 1 year ago - 6 months ago?
Buy_{t-3}	0/1	Did the customer buy during the period: 1.5 years - 1 year ago?
Buy_{t-4}	0/1	Did the customer buy during the period: 2 years - 1.5 years ago?
Customer	6 cat.	For how long is this person a customer?
LastFreq	9 cat.	What was the purchasing frequency during the last 6 months?
LastSales	5 cat.	Sales generated by the customer during the last 6 months?
LastProfit	10 cat.	Profit generated by the customer during the last 6 months?
DaysSince	7 cat.	No. days since the last purchase
Unimulti	3 cat.	Does the customer live in a stand-alone home or an appartment?
Socclass	6 cat.	Social class of the customer
VAT	0/1	Is the person self-employed?
Household	4 cat.	Type of household the customer belongs to
Family	4 cat.	Number of families living at the address
Natclass	6 cat.	Nationality distribution in the street of the customer
State	9 cat.	Province of Belgium the customer lives in

Results

This section discusses the results which were obtained from applying the Prob-Rough system to the problem at hand. We present the resulting rough classifiers obtained with the standard priors and costs of misclassification, together with two alternative results with non-standard priors and costs of misclassification, which were typical for the mail-order company.

Standard priors and costs of misclassification. Table 2 contains three equivalent sets of decision rules, created by the ProbRough system for the case of equal prior probabilities of purchase and no purchase, and equal costs of misclassification. A decision for every (new) customer could be taken based on each of the three rulesets. The classification of cases to either the *Do mail* or *Do* *not mail* category was then complemented by the strengths of each of the rules. The latter criterion was used to assign an order among the segments.

Table 2. Results with standard priors and costs of misclassification

1. if $LastFreq > 0$ then $d = Do\,mail$,
2. if $LastFreq = 0$ and $Buy_{t-2} = Yes$ then $d = Do\,mail$,
3. if $LastFreq = 0$ and $Buy_{t-2} = No$ and $DaysSince < 180$ then $d = Do\,mail$,
4. if $LastFreq = 0$ and $Buy_{t-2} = No$ and $DaysSince \geq 180$ then $d = Do\,not\,mail$.

1. if $Buy_{t-1} = Yes$ then $d = Do\,mail$,
2. if $Buy_{t-1} = No$ and $Buy_{t-2} = Yes$ then $d = Do\,mail$,
3. if $Buy_{t-1} = No$ and $Buy_{t-2} = No$ and $DaysSince < 180$ then $d = Do\,mail$,
4. if $Buy_{t-1} = No$ and $Buy_{t-2} = No$ and $DaysSince \geq 180$ then $d = Do\,not\,mail$.

1. if $LastProfit > 0$ then $d = Do\,mail$,
2. if $LastProfit = 0$ and $Buy_{t-2} = Yes$ then $d = Do\,mail$,
3. if $LastProfit = 0$ and $Buy_{t-2} = No$ and $DaysSince < 180$ then $d = Do\,mail$,
4. if $LastProfit = 0$ and $Buy_{t-2} = No$ and $DaysSince \geq 180$ then $d = Do\,not\,mail$.

We observed that the structure of the three sets of decision rules was very similar even though different attributes were used. These predictors were close substitutes because the correlations between LastFreq and Buy_{t-1} (0.80) and LastFreq and LastProfit (0.91) were high. This caused the results in Table 2 to be very closely together. The decision rules which we obtained closely reflected database marketing theory that recency (attribute DaysSince), frequency (Last-Freq, Buy_{t-1}, Buy_{t-2}) and monetary value (LastProfit) are the best predictors for future purchasing behavior, even though these measures are known to be highly intercorrelated. The findings also show that only very recent transactional data (up to one year before the period of consideration) were useful for predictive purposes.

Table 3. Classification results (standard priors and costs of misclassification).

	Ruleset 1	Ruleset 2	Ruleset 3
Prob(Decision=$Do\,mail$ if Reality=no purchase)	0.25	0.22	0.25
Prob(Decision=$Do\,not\,mail$ if Reality=no purchase)	0.75	0.78	0.75
Prob(Decision=$Do\,mail$ if Reality=purchase)	0.74	0.72	0.74
Prob(Decision=$Do\,not\,mail$ if Reality=purchase)	0.26	0.28	0.26

Table 3 contains the results of the use of the rough classifiers shown in Table 2 on the test sample (i.e., cases which have not been submitted to the data mining method before). The classification results exhibited a favorable picture (about 3/4th of the cases were correctly classified and neither of the two decisions were favored) and revealed that all three rulesets performed very similarly, which showed that they were also in terms of a test sample truly substitutable. These results were similar to the findings by Van den Poel (1998).

Unequal priors and costs of misclassification 2:1. In database marketing applications, misclassification costs are typically not equal. Usually, costs of classifying a buyer as a non-buyer are much higher than classifying a non-buyer as a buyer, because the cost in the latter case is often limited to the cost related to the mailing piece, whereas in the former case foregone profits are much more important. Therefore, we reran the ProbRough algorithm with a ratio of misclassification costs 2:1. Moreover, we adjusted the prior probabilities to 0.4 for the probability of a purchase and hence, 1 - 0.4 or 0.6 for the probability of no purchase. The search process of ProbRough is directed now by the average cost of misclassification that includes unequal priors and costs of misclassification. The problems of unequal priors and unequal costs of misclassification are related (see, Piasta and Lenarcik, 1998). The ratio of misclassification costs 2:1 is partly compensated by the assumed prior probabilities.

When we compared the generated rough classifiers to those in Table 2, the only difference was in the value for the attribute of the number of days since the last purchase. We could therefore conclude that the change in the costs of misclassification led to an increase in the period of consideration (value of 180 days in Table 2 versus a value of 365 days now).

Unequal priors and costs of misclassification 3:1. Now we present results obtained when we increased the ratio of misclassification costs from 2:1 to 3:1. The purpose was to investigate the impact on both the types of attributes used and the classification results. The use of the induced rough classifier shown in Table 4) on the test sample exhibited a decrease in the probability of wrongfully assigning a buyer to the non-buyer category (as compared to Table 3) and an increase in the probability of incorrectly assigning a non-buyer to the buyer category. This is a consequence of the fact that the search process in the space of models is guided by the criterion which depends on the assumed ratio of misclassification costs.

From a marketing perspective, it was remarkable that a geographic element became an important segmentation variable in the prediction of specific consumer behavior. However, this finding was confirmed by the mail-order company managers, who pointed to substantial differences in consumer behavior between different regions of Belgium. Moreover, the number of families living at a certain address became important in the decision whether or not to mail. Nevertheless, we have to mention that the strength of rules containing these additional non-RFM variables is not all that large. With respect to the period of consideration

Table 4. Results with unequal prior probabilities and costs of misclassification 3:1

1. **if** $LastFreq > 0$ & $State \notin \{4,5\}$
then $d = Do\,mail$,
2. **if** $LastFreq > 0$ & $State \in \{4,5\}$ & $\qquad\qquad$ $Family > 2$
then $d = Do\,not\,mail$,
3. **if** $LastFreq > 0$ & $State \in \{4,5\}$ & $\qquad\qquad$ $Family \leq 2$
then $d = Do\,mail$.
4. **if** $LastFreq = 0$ & $State \notin \{4,5\}$ & $Buy_{t-2} = Yes$
then $d = Do\,mail$.
5. **if** $LastFreq = 0$ & $State \notin \{4,5\}$ & $Buy_{t-2} = No$ & $\qquad\qquad$ $DaysSince \leq 365$
then $d = Do\,mail$.
6. **if** $LastFreq = 0$ & $\qquad\qquad$ $Buy_{t-2} = No$ & $\qquad\qquad$ $DaysSince > 365$
then $d = Do\,not\,mail$.
7. **if** $LastFreq = 0$ & $State \in \{4,5\}$ & $Buy_{t-2} = Yes$ & $Family \leq 2$
then $d = Do\,mail$.
8. **if** $LastFreq = 0$ & $State \in \{4,5\}$ & $Buy_{t-2} = Yes$ & $Family > 2$
then $d = Do\,not\,mail$.
9. **if** $LastFreq = 0$ & $State \in \{4,5\}$ & $Buy_{t-2} = No$ & $Family > 2$ & $DaysSince \leq 365$
then $d = Do\,not\,mail$.
10. **if** $LastFreq = 0$ & $State \in \{4,5\}$ & $Buy_{t-2} = No$ & $Family \leq 2$ & $DaysSince \leq 365$
then $d = Do\,mail$.

of the RFM variables, the same duration of one year was discovered to be most important as for the ratio of misclassifications 2:1.

Conclusions and future research

The results obtained from the application of the ProbRough algorithm rediscovered the RFM variables (known from theory) as most significant predictors of future mail-order buying behavior. However, the data mining method highlighted the fact that only rather recent transactional data (up to one year before the period of consideration) was useful in the prediction of future purchase behavior. A change in the costs of misclassification was reflected in an increase of the period of consideration (to a one-year period). A significant finding was the importance of non-RFM variables in predicting purchasing behavior as the ratio of misclassification costs became larger. This corresponded to prior beliefs of marketing managers but has not yet been revealed by other data mining efforts. Moreover, the application of a data mining method which was capable of handling misclassification costs clearly showed a decrease in the overall cost due to a lower percentage in foregone profits. We used predictors which had already been categorized. This is not a necessity for the ProbRough algorithm, since it features a built-in discretization process. However, we could not investigate this feature because only categorical data was available. Therefore, we leave this issue as a topic for future research.

Acknowledgments

This work was supported by grant no. 8 T11C 010 12 from the State Committee for Scientific Research (KBN).

References

1. Fayyad U.M., Piatetsky–Shapiro G., Smyth P. (1996), From data mining to knowledge discovery: an overview, in *Advances in Knowledge Discovery and Data Mining*, AAAI Press, CA.
2. Hauser B. (1992), List segmentation, in Nash, E. L. (ed.), *The Direct Marketing Handbook*, 233–247.
3. Kowalczyk, W., Piasta, Z. (1998), Rough-set inspired approach to knowledge discovery in business databases, *The Second Pacific-Asia Conference on Knowledge Discovery and Data Mining*, (PAKDD-98), Melburne, Australia, April 15-17, 1998, (accepted).
4. Lenarcik, A., Piasta, Z. (1997), Probabilistic rough classifiers with mixtures of discrete and continuous condition attributes, in T. Y. Lin, N. Cercone (eds), *Rough Sets and Data Mining*, Kluwer Academic Publishers, Boston, London, Dordrecht, 373–383.
5. Pawlak, Z. (1991), *Rough Sets: Theoretical Aspects of Reasoning About Data*, Kluwer Academic Publishers, Dordrecht.
6. Petrison, L.A., Blattberg, R.C., Wang, P. (1993), Database marketing: past, present, and future, *Journal of Direct Marketing*, 7, 27–43.
7. Piasta, Z., Lenarcik, A. (1996), Rule induction with probabilistic rough classifiers, ICS Research Report 24/96, Warsaw University of Technology, to appear in *Machine Learning*.
8. Piasta, Z., Lenarcik, A. (1998), Learning rough classifiers from large databases with missing values, in L. Polkowski, A. Skowron (eds): *Rough Sets in Knowledge Discovery*, Physica-Verlag (Springer), forthcoming.
9. Van den Poel, D. (1998), Rough sets for database marketing, in L. Polkowski, A. Skowron (eds), *Rough Sets in Knowledge Discovery*, Physica Verlag (Springer), forthcoming.

Rough Sets in Optical Character Recognition

Witold Czajewski

Warsaw University of Technology, Electrical Engineering Faculty, Institute of Control
and Industrial Electronics, 00-661 Warszawa, Poland

Abstract. The paper presents an attempt to apply the Rough Sets Theory to Optical Character Recognition. In this approach specific characters' features are referred to as an information system, from which the most important information is being extracted by the Rough Sets Theory. This process is fully automatic and does not require any human decision in the area of usefulness of certain characters' features. A discernibility matrix which is built in this way constitutes a reduced database for classification algorithms. A brief description of Classical Optical Character Recognition Theory and Rough Sets Theory as well as some selected research and experimental results are also presented.

1 Introduction

Automatic identification of characters is getting more and more important in our modern civilization (e.g. recognition of addresses, zip codes, signatures). A recognition algorithm must be based on a certain, earlier acquired knowledge on the objects it is about to identify. The more objects there are and the more complex they are, the bigger the knowledge base of the algorithm is. Therefore it seems important to find an efficient way of data reduction, providing the same (or lower but acceptable) quality of identification. The solution to this problem may be the Rough Sets described in [1].

2 Classical Optical Character Recognition Theory

Optical Character Recognition (OCR) is a field of science dealing with pattern analysis, especially with identification of characters. It is a very complex task. This research is focused on the last stage of character recognition in which separated characters are treated as objects to be identified. This stage consists of pre-processing, feature extraction and classification.

The pre-processing prepares a character in such a way that a representative feature vector may be extracted. In the classification stage a classifier (e.g. classical, statistical, fuzzy logic or neural network) employing certain knowledge and basing on the extracted feature vector assigns the character to a particular class. The knowledge base of the classifier comprises formerly generated feature vectors of characters from the learning set, which membership classes are known.

In order to recognize a character correctly, the closest vector to the character's vector from among learning set vectors must be found (minimal distance classifier

L. Polkowski and A. Skowron (Eds.): RSCTC'98, LNAI 1424, pp. 601–604, 1998.

- used mostly in the research). Although this method is relatively simple and fast, it may cause mistakes and thus often so called k-nearest neighbor classifier is used. It improves the recognition results by a few percentage points, but the general tendency stays the same.

3   Elements of the Rough Sets Theory

Due to the increasing size of databases and huge amount of information stored in them, there occurred a necessity of developing efficient methods of acquiring the most essential and useful knowledge out of databases. This problem was among the most important ones in the field of modern information systems. In 1982 Z. Pawlak proposed in [1] a new theory of reasoning about data called the Rough Sets Theory.

3.1   Decision Tables

In this approach knowledge is represented by a set of data organized in a table called an information system. Rows of the table refer to objects (e.g. characters), and columns to their attributes (features). Reduction of knowledge means deleting unnecessary and superfluous attributes and leaving only the most important ones.

Let DT be a consistent decision table (one without objects with identical condition attributes and different decision attributes [1]), C be a set of condition attributes, D be a set of decision attributes and $\alpha \in$ C. An attribute a is dispensable in the decision table DT if the decision table DT is also consistent without the attribute α; otherwise the attribute α is indispensable in the decision table DT. A decision table DT is called independent if all the attributes $\alpha \in$ C are indispensable in the decision table DT. The subset of attributes R $\subseteq$ C is called a reduct of C in the decision table DT if the new decision table $\langle$U,R $\cup$ D,V,f$\rangle$ is consistent and independent.

Any subset satisfying the above except for being independent also defines the knowledge but is superfluous. These subsets (later referred to as dependent subsets) are of great importance in the conducted research.

3.2   Discernibility Matrix

The process of verification if a given subset is a reduct or dependent subset is a process involving a huge number of comparisons and therefore is very time-consuming.

Notions of discernibility matrix and function introduced by Rauszer and Skowron in [2] make the search time much shorter. The discernibility matrix stores only information whether the corresponding attributes of any pair of objects are different or not. This information can be stored on single bits grouped in bytes. Hence, there is much less data to be compared than previously and the processing time is shorter.

---

[1] in other words: one with identical attributes belonging to different classes

4 Research and Conclusions

4.1 Reducts

All the tests were performed on one set of objects: printed numbers (0-9) based on MS Windows true-type fonts. Each number was stored on a monochrome 48x48 pixels bitmap. On the whole 2330 characters were tested, 233 of each kind. 155 characters belonged to the teaching set, the other 78 to the testing set.

At first a few types of feature vectors were tested and the best one was established. It comprised 64 integer features. An information system then built was being reduced repeatedly. Initially one 8-element reduct was found. The length of this reduct was only 12,5feature vector's length. According to the reduct's definition it allowed perfect recognition of all the objects from the learning set. The testing set, however, was recognized with much lower quality, because the reduct was optimally chosen for the known objects from the learning set and no others. After further examinations the lowest (59,7%), the highest (72,4%) and the average (66,6%) recognition quality for 8-element reducts were found. These results clearly indicate that the reduct is unsuitable for the recognition of characters from outside the learning set. The classification quality in this case is much lower (ca. 30 percent points on the average) than when all the features are used. The only method of increasing the recognition quality was utilization of other(longer)dependent subsets.

4.2 Dependent Subsets

Due to unimaginably big number of all possible subsets of the whole attribute set of the information system ($2^{64} \approx 1.8 * 10^{19}$) verifying every subset is virtually impossible. Therefore a random search algorithm was used. Having tested thousands of subsets some conclusions of statistical nature were reached. As one can notice (Table.1), some of the dependent subsets give better recognition results than the full feature vector!

Table 1. Recognition results of the testing set for dependent subsets of a given length

Length of subset	13	19	26	32	38	45	51	58	64
Minimal quality	70,5	71,5	79,5	83,6	86,8	87,4	89,7	91,3	93,3
Maximal quality	70,5	71,5	79,5	83,6	86,8	87,4	89,7	91,3	93,3
Average quality	**72,5**	**80,1**	**85,3**	**87,8**	**89,9**	**91,1**	**92,2**	**92,8**	**93,3**

They should be regarded as the sets of features with the least important features eliminated. Unfortunately there is no analytical method of finding them. In practical applications one must rely only on statistical results (much less optimistic).

As a matter of fact over 27% of 58-element dependent subsets give better recognition results than the full feature vector, but this number decreases quickly as the vector's length drops (it amounts 5% for 51- element subsets and less for the remaining ones).

The shortest dependent subset found giving recognition results not worse than the full feature vector had 38 elements (ca. 60% of the full vector's length). Graphical interpretation of this subset is shown in Fig. 1a. For comparison, in Fig. 1b the worst dependent subset found of the same length is shown. The two subsets have 14 different elements.

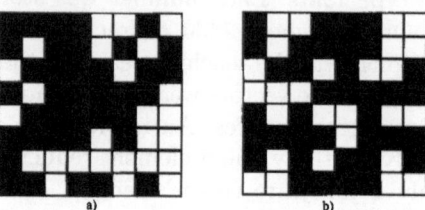

a)   b)

Fig. 1. The distribution of features for a) the best and b) the worst 38-element reduct (white squares represent the features that were neglected in the process of recognition)

4.3   Conclusions

On the ground of the conducted research it can be stated that:

- it is enough to use 5-15% of the knowledge from an information system in order to recognize all the characters from this system perfectly. In other words, up to 95% of data in an information system is dispensable.
- it is possible to find a reduct that will give the same recognition results for characters from outside the information system as the full length vector. In this case ca. 40-50% of data in an information system is dispensable.

Generally one can notice that the more features a single character has the shorter (in percent, not absolutely) the minimal reduct is. Also less (in percent, not absolutely) data is required to provide the same recognition quality for characters outside the information system. The obtained results are quite interesting. They mean substantial (even 10 times in the first case) reduction of recognition time and hardware requirements, which leads directly to economical savings.

References

1. Pawlak, Z.: Rough Sets - Theoretical Aspects of Reasoning about Data. Kluwer Academic Publishers, 1991
2. Meng-Chieh, L.: Software System For Intelligent Data Processing And Discovering Based On The Fuzzy-Rough Sets Theory. Thesis, San Diego State University, 1995
3. Lichon, J.: Classical Hand-written Latin Character Recognition Algorithms. Master's thesis, Warsaw University of Technology, Warszawa (Poland), 1995 (in Polish)
4. Czajewski, W., Zuraw, J.: Application of the Rough Sets Theory to Optical Character Recognition. Master's thesis, Warsaw University of Technology, Warszawa (Poland), 1997 (in Polish)

ROSE - Software Implementation
of the Rough Set Theory

Bartłomiej Prędki, Roman Słowiński, Jerzy Stefanowski
Robert Susmaga, Szymon Wilk

Institute of Computing Science
Poznan University of Technology
Piotrowo 3A, 60-965 Poznan, Poland

[Abstract.] This paper briefly describes ROSE software package. It is an interactive, modular system designed for analysis and knowledge discovery based on rough set theory in 32-bit operating systems on PC computers. It implements classical rough set theory as well as its extension based on variable precision model. It includes generation of decision rules for classification systems and knowledge discovery.

1 Introduction

ROSE (Rough Set Data Explorer) is a modular software system implementing basic elements of the rough set theory and rule discovery techniques. It has been created at the Laboratory of Intelligent Decision Support Systems of the Institute of Computing Science in Poznan, basing on fourteen-year experience in rough set based knowledge discovery and decision analysis.

All computations are based on rough set fundamentals introduced by Z. Pawlak [6]. One of implemented extensions applies the variable precision rough set model defined by W. Ziarko [14]. It is particularly useful in analysis of data sets with large boundary regions.

The ROSE system is a successor of RoughDAS and RoughClass systems [3][5][10]. RoughDAS is historically one of the first successful implementations of the rough set theory, which has been used in many real life applications. Due to limitations of RoughDAS, especially its incapability to make full use of currently available computers, there was a need to design and implement new software.

ROSE started as several independent modules that were later put together in one system. First we were motivated to create computational engine working on more powerful computers (e.g. UNIX workstations), allowing faster analysis of large data sets. Then we came to the point of creating user friendly interface, where Microsoft Windows was chosen as our basic platform. So the modules can be separately redesigned and recompiled without much interference from user's point of view. The only component that is strictly platform dependent is graphical user interface (GUI). All this guarantees that the system can be easily adapted for future operating systems and platforms.

L. Polkowski and A. Skowron (Eds.): RSCTC'98, LNAI 1424, pp. 605–608, 1998.
© Springer-Verlag Berlin Heidelberg 1998

2   ROSE system

The program ROSE is an interactive software system running under 32-bit GUI operating systems (Windows 95/NT 4.0) on PC compatible machines. The core modules were written in C++ programming language (standard ANSI), while the interface modules were developed using Borland C++ (with Object Windows libraries) and Borland Delphi.

The system consists of a graphical user interface (GUI) and a set of separate computational modules. The modules are platform independent and can be recompiled for different targets including UNIX machines. GUI acts as an overlay on all computational modules. So it is quite easy to add new modules to the ROSE system and that is an important characteristic. This guarantees greater expandability of the system in the future.

ROSE is designed to be easy in use, point and click, menu-driven, user friendly tool for exploration and data analysis. It is meant as well for experts as for occasional users who want to do the data analysis. System communicates with users using dialog windows and all the results are represented in the environment. Data can be edited using spreadsheet like interface.

3   Input/output data

ROSE accepts input data in form of a table called an information table in which rows correspond to objects (cases, observations, etc.) and columns correspond to attributes (features, characteristics). The attributes are divided into disjoint sets of condition attributes (e.g. results of particular tests or experiments) and decision attributes (expressing the partition of objects into classes, i.e. their classification). The data is stored in a plain text file according to a defined syntax (Information System File, ISF). ROSE can also import data stored by its predecessor RoughDAS and export to several other formats (including files accepted by the system LERS or C4.5).

ISF file specification allows for long attribute names (up to 30 alphanumerical characters) and string values of attributes (such as 'high', 'low') aside real and integer values. Because it is plain text file it can be transferred between different operating systems without modifications. It is also easy to edit and verify correctness of data contained in the file.

The file format has an open structure. It is divided into sections and it is possible to add some new sections so far undefined for further use. The user can decide to ignore some of the attributes just by changing the qualification of attributes.

Except visualization in GUI, all results are also written to plain text files, so they are readable also outside the system, and easily converted to other file formats.

4   Features

Features currently offered by computational modules include:

- data validation and preprocessing,
- automatic discretization of continuously-valued attributes according to Fayyad & Irani method[1] as well as user-driven discretization,
- qualitative estimation of the ability of the condition attributes to approximate the objects' classification, using either standard rough set model or variable precision model extension,
- finding the core of attributes as well as looking for reducts in the information table (either all reducts or a population of reducts of predetermined size) using several methods such as algorithm by S. Romanski[7] and modified algorithm by A. Skowron [8],
- examining the relative significance of a given attribute for the classification of objects, by observing the changes in the quality of classification,
- reducing superfluous attributes and selecting the most significant attributes for the classification of objects; there are available several techniques that support the choice o subsets of attributes ensuring a satisfactory quality of the classification (e.g. the technique of adding the most discriminatory attributes to the core),
- inducing decision rules using either the LEM2 algorithm [2] or the Explore algorithm [4][13],
- postprocessing of induced rules, e.g. pruning; looking for interesting rules according to the user defined queries [4],
- applying the decision rules to classify new objects using different techniques of rule matching, in particular an original approach based on valued closeness relation [3][9] ,
- evaluation of the sets of decision rules using k-fold cross validation techniques.

It will be quite easy to add new modules to the system due to its open architecture.

5   Final Remarks

In the near future we plan to add new capabilities to our system, such as: incremental reduct generation, incremental rule generation, working with incomplete information tables, working with similarity relations for rough approximations, working with dominance relations for rough approximation of multicriteria classification problems, working with dominance relations and pairwise comparison tables for rough approximation of multicriteria choice and ranking problems. These functionalities are based on recent research results of the team members.

The ROSE system and its predecessor RoughDAS have been applied to many real-life data sets. The references to these applications are given, e.g. in [5]. Some of the main fields of applications include: medicine, pharmacy, technical diagnostics, finance and management science, image and signal analysis, geology, software project evaluation.

References

1. U.M. Fayyad, K.B. Irani. On the Handling of Continuous-Valued Attributes in Decision Tree Generation, Machine Learning, Vol 8, 1992, 87–102.
2. J.W. Grzymala-Busse. LERS - a system for learning from examples based on rough sets. In R. Slowinski, (ed.) Intelligent Decision Support, Kluwer Academic Publishers, 1992, 3–18.
3. R. Mienko, R. Slowinski, J. Stefanowski. Rule Classifier Based on Valued Closeness Relation: ROUGHCLASS version 2.0, ICS Research Report RA-95/002, Poznan University of Technology, Poznan, April 1995.
4. R. Mienko, J. Stefanowski, D. Vanderpooten. Discovery-Oriented Induction of Decision Rules. Cahier du Lamsade no. 141, Paris, Univeriste de Paris Dauphine, spetembre 1996.
5. R. Mienko, R.Slowinski, J. Stefanowski, R. Susmaga. RoughFamily - software implementation of rough set based data analysis and rule discovery techniques. In Tsumoto S. (ed.) Proceedings of the Fourth International Workshop on Rough Sets, Fuzzy Sets and Machine Discovery, Tokyo Nov. 6-8 1996, 437–440.
6. Z. Pawlak Rough Sets. Theoretical Aspects of Reasoning About Data, Kluwer Academic Publishers, Dordrecht, 1991.
7. S. Romanski. Operation on families of sets for exhaustive search, given a monotonic function. In W. Beeri, C. Schmidt, N. Doyle (eds.), Proceedings of the 3rd Int. Conference on Data and Knowledge Bases, Jerusalem 1988, 310–322.
8. A. Skowron, Rauszer C.. The discernibility matrices and functions in information systems in: Slowinski R. (ed.) Intelligent Decision Support. Handbook of Applications and Advances of the Rough Sets Theory. Kluwer Academic Publishers, 1992, 331–362.
9. R. Slowinski. Rough sets learning of preferential attitude in multi-criteria decision making. In Komorowski J., Ras Z.W. (eds.), Proc. of Int. Sump. on Methodologies for Intelligent Systems, Springer Verlag LNAI 689, 1993, 642–651.
10. R. Slowinski, J. Stefanowski. 'RoughDAS' and 'RoughClass' software implementations of the rough set approach. In R. Slowinski (ed.) Intelligent Decision Support. Handbook of Applications and Advances of the Rough Sets Theory. Kluwer Academic Publishers, 1992, 445–456.
11. R. Slowinski, J. Stefanowski. Rough classification with valued closeness relation. In Didey E. et al. (eds.), New Approaches in Classification and Data Analysis, Springer Verlag, 1993, 482–489.
12. R. Slowinski, J. Stefanowski. Rough set reasoning about uncertain data. Fundamenta Informaticae, 27 (2-3), 1996, 229–244.
13. J. Stefanowski. On rough set based approaches to induction of decision rules. In A. Skowron, L. Polkowski (eds.), Rough Set in Knowledge Discovery, 1998.
14. W. Ziarko. Analysis of Uncertain Information in The Framework of Variable Precision Rough Sets. Foundations of Computing And Decision Sciences Vol 18 (1993) No. 3-4, 381–396.

Rough Sets and Bayesian Methods Applied to Cancer Detection

Roman W. Swiniarski

San Diego State University
Department of Mathematical and Computer Sciences
5500 Campanile Drive, San Diego, California 92182-7720, U.S.A.

Abstract. The paper describes an application of soft computing methods of rough sets and Bayesian inference to a breast cancer detection using electro-potentials. The statistical principal component analysis (PCA) and the rough sets methods were applied for feature extraction, reduction and selection. The quadratic discriminant was applied as a classifier for a breast cancer detection.

1   Introduction

An investigation of a breast model indicates variations in epithelial electropotentials that may occur in the area of abnormally proliferating cells in vicinity of neoplazm. From observation that altered skin surface electropotentials may be caused by presence of underlying abnormal proliferation, and idea of cancer detection by measurements and recognition has been formed (Crowe and Faupel 1996; Long et al. 1996; Davies 1996; Dixon 1996).

This technique is based on the processing and recognition of electropotentials (EPs) measured by an array of sensors in the suspicious regions of a breast (Crowe and Faupel 1996; Long et al. 1996).

The cancerous tissue influences metabolic, chemical, ionic and thus electromagnetical processes of a breast (Crowe and Faupel 1996; Long et al. 1996; Davies 1996; Dixon 1996).

These complex processes in a healthy and cancerous breast can be modeled based on the classical findings in the system and model theories. A discovery of differences of EPs readings for a healthy and a cancerous breast gives an outstanding chance to design a device for a breast cancer detection using non invasive EPs measurements (Crowe and Faupel 1996; Long et al. 1996; Davies 1996; Dixon 1996).

It is generally known how cells develop and stabilize (maintain) an ionic gradient (difference) which results in electrical potential gradient across a cell membrane (Davies, 1994). In a breast the epithelial cells are arranged as a line ducts or lobules. They have specialized apical and basolateral domains in order to provide basal absorption, secretion and milk production during lactation phase. The

L. Polkowski and A. Skowron (Eds.): RSCTC'98, LNAI 1424, pp. 609–616, 1998.
© Springer-Verlag Berlin Heidelberg 1998

membranes of considered cells have different preambilities and transport (transition) functions. They have different electrical potentials in respect to each other, which results in transepithelial electrical potential.

It has been observed that the proliferation is increased in cancer cells and proliferation becomes disregulated in in the surrounding breast terminal ductal tabular units (Davies, 1994). Field (subarea) cancerization results in field depolarization. This eventually extends in a penumbra measurable on a skin surface, for example as electro-potential. Different proliferation in healthy and cancerous breast cause different depolarizations and thus resulting difference in electropotentials on a skin surface. This means that there is ability to indirectly measure the altered transepithelial electrical potentials using skin electrodes measuring electropotentials. This leads to use of comparative measurements of electropotentials on a skin surface for a healthy and cancerous breast as feature to non invasive cancer diagnosis.

We discuss an application of soft computing methods of rough sets, principal components analysis and Bayesian inference to data mining and a breast cancer detection using electro-potentials.

2   Data
Measurement and processing of breast electropotentials

Electro potentials EP measurements (signals)

For each patient the measuring of electrical potentials (EPs) on breast skin spots was carried out by the designed system (Long, et al., 1996; Dixon, 1996; Davies, 1994). The array of 8 sensors were glued to each breast with one reference sensor located on left and right palm. The $(2 \times (2 \times 8))$ electro potential signals from $N_{sensors} = 16$ sensors located on breasts were simultaneously measured and record. Total number of measurements in one time moment was 32 (for L and R reference signal). There were two sets of measurements collected simultaneously: left palm and right palm relative. Only signals referring to the left palm were considered in recognition task. The gathered data were like follows. For each jth patient ($j = 1, 2, \cdots, N$), with one (left) palm reference, there were 16 time-series simultaneously measured and recorded for each ith sensor. Each time-series of jth patient for ith sensor contained raw electropotentials measured in $N_{EP} = 150$ time moments (with a sampling time $\Delta t = 0.2\ sec$), forming a data set

$$Patient^{j,i} = x^{j,i}(1), x^{j,i}(2), \cdots, x^{j,i}(N_{EP}), j = 1, 2, \cdots, N, i = 1, 2, \cdots, N_{sensors}$$
(1)

Measurements may contain disturbances, even though, that it is required that measuring device is properly designed, calibrated, with insulation of leads and electronics (but not insulated sensors), without internal noise or oscillations.

Sources of disturbances:

1. Not accurate experiment procedure (sequence, conditions).
2. Noisy environment (disturbing external devices).
3. Patient move, talk, sneezing, coughing, etc).
4. Loosing connection of sensor with the skin due to rapid patient move.
5. Deodorant used by patient.

Preprocessing of raw EP data

Due to measurement disturbances and time dependent dynamics of sensor signals, the raw EP signals are preprocessed. The preprocessing includes filtering and averaging combined with outlier removal. Each sensed spot of breast is initially represented by a time-series of 150 noisy raw EP signals. The two stage averaging process, combined with outliers removal, was used to find representative EP for a sensor. The final average of time-series (after outlier removal) is a representation of a EP in measured skin spot.

The resulting average $\bar{x}^{j,i}$ was considered as representative electropotential for a spot related to the attached ith sensors for jth patient. The collection of all averaged $N_{sensors}$ electropotentials for jth patient can form an averaged EP pattern

$$\bar{\mathbf{x}}^j = [\bar{x}^{j,1}, \bar{x}^{j,2}, \cdots, \bar{x}^{j,N_{sensors}}]^T \tag{2}$$

Scaling of the averaged EP data

Some data sets may include data which should be a subject of scaling and normalization. These operations are suggested in the front of some processing stages (for example predictor design). This is needed to remove biases. For example, the usual goal is to classify the same type of signals or data and classification must be for example invariant to the shift of attribute values.

We have scaled each attribute x separately according to the the standard normalization formula. **Data sets and patterns**

The averaged EP data set used for classification, contained EP patterns with only 14 elements (7 electropotentials for each breast, with one unsufficient removed) and the corresponding class (bening - class 0, malignant class 1). The EP data contains cases for each jth patient $(\bar{\mathbf{x}}^j, class^j)$ with $14 - elements$ raw EP patterns $\bar{\mathbf{x}}^i$ and associated class. The corresponding patients' IDs and age Were stored in a separate file.

Features. Differentials

The initially preprocessed EP data contains averaged EP patterns including 14 electropotentials (averaged) from both breasts. From these patterns created features have been formed assuming that these features will be more relevant for classification. Furthermore some of them may have close medical interpretation used in medical diagnosis. The all concept of cancer recognition is based on observation that electropotentials in specific spots on cancerous breast will be different (larger) that corresponding electropotentials on the healthy breast.

Thus, differences between corresponding electropotentials can be considered as major detection features. However, this may depend on data type and classifier design. The original EP data carries on the same information as differentials. Creation of new features is a kind of transformation $\mathbf{y}^j = F(\bar{\mathbf{x}}^j)$ of $N_{sensors} - dimensional$ raw EP patterns $\bar{\mathbf{x}}^j = [\bar{x}^{j,1}, \bar{x}^{j,2}, \cdots, \bar{x}^{j,N_{sensors}}]^T$ into $N_{differentials} = n - dimensional$ differential space patterns

$$\mathbf{y}^j = [y^{j,1}, y^{j,2}, \cdots, y^{j,n}]^T \tag{3}$$

Differentials are created feature from the original averaged EP data set containing $14 - element$ EP patterns. Differentials are linear combinations of EPs features and other defined differentials. There are *within breast, between breast* or *mirror* types of differentials. We can list some examples of differentials: *symptomatic breast differentials, symptomatic breast high*, etc. (Long, et al., 1996; Dixon, 1996; Davies, 1994). Some of differentials have medical meanings. For example: a measure of local depolarization for the symptomatic breast (estimated by from differences between sensors EPs in the symptomatic breast).

The considered data set with differential contains cases ($\mathbf{y}^j, class^j$) where 28 element pattern contains *age code* $x^{j,1}$ and 27 differential features $x^{j,2}, x^{j,3}, \cdots, x^{j,28}$. The numerical code for an age was defined as: age 1: $age < 35$ code 1; age 2: $35 < age < 65$ code 2; age 3: $age > 65$ code 3.

Feature extraction, reduction and optimal selection by principal component analysis (PCA)

Previous study has concerned mostly classification of 28 selected differentials derived from the averaged EPs. Experiments and research show that decorrelating and orthogonalizing transformations may provide better representation of data patterns with stronger discernibility and classificatory power. Additionally the reduction of data, preserving classification accuracy, based on Occam razor and Rissanen minimum description length law, may lead to better generalization of designed classifier for unseen future patterns. We applied Karhunen-Loéve transformation KLT (resulting from the Principal Component Analysis PCA) for $28 - elements$ differential patterns (with an age code) as feature extraction and reduction. This transformation allows us to discover most expressive pattern features and hidden (latent) variables.

An application of rough sets for feature reduction, classifiability and dependency discovery

We used the method of rough sets (Pawlak, 1991) for selection features from patterns obtained as a result of the KLT transformation. Rough sets allow us to find from the data so called reducts sets. Each reduct is a satisfactory set of features allowing proper classification of the training set (without loosing a classification accuracy). Since for a given data set rough sets may generate several reducts, we provided a technique of selecting the best reduct guaranteeing best generalization based on statistical processing. The final classifiers was designed for reduced data for best pattern features defined by selected best reduct. Additionally, rough sets allow to find most important pattern features defined by so called core, being a set intersection of all reducts for a given data.

Classification

The goal of classification is to classify the patients, being preliminary recognized by physical examination as having a cancer, into two distinct categories (classes)

– benign (category 0)
– malignant (category 1)

From the medical point of view it is important to obtain especially maximal rate of confidence in detecting of the malignant cancer. The designed classifier has to be tested and their performance evaluated. This will give an information about the accuracy of recognition of new objects. We will discuss some of classifier's performance evaluation criteria in the next section.

Classification accuracy measures

Assume that a test set T_{test} (not used for a training) consists of N_{test} cases (patients) with $N_{test,M}$ - true malignant cases, $N_{test,B}$ - true benign cases, with $N_{test,M} + N_{test,B} = +N_{test}$.

Assume that a classifier recognized correctly $N_{test,M,correc}$ malignant cases and $N_{test,B,correc}$ bening cases. This means that following number of malignant cases $N_{test,M,nocorrec} = N_{test,M} - N_{test,M,correc}$ and bening cases $N_{test,B,nocorrec} = N_{test,B} - N_{test,B,correc}$ were recognized not correctly. The total accuracy of classification can be measured by

$$J_{class} = \frac{\text{number of all correctly classified cases}}{\text{number of all cases}} \times 100\% = \qquad (4)$$

$$\frac{N_{test,M,correc} + N_{test,B,correc}}{N_{test}} \times 100\%$$

To evaluate the malignant classification accuracy the sensitivity (malignant) measure is defined

$$Jsens = \frac{\text{number of correctly classified malignants cases}}{\text{number of true malignant cases}} \times 100\% = \qquad (5)$$

$$\frac{N_{test,M,correc}}{N_{test,M}} \times 100$$

To evaluate the bening classification accuracy the specificity (bening) measure is defined as

$$J_{spec} = \frac{\text{number of correctly classified bening cases}}{\text{number of true bening cases}} \times 100\% =$$

$$\frac{N_{test,B,correc}}{N_{test,B}} \times 100\% \qquad (6)$$

The results of the Bayesian quadratic discriminant based classifier

First, we have considered the quadratic Gaussian discriminant based classifier. We have assumed of multivariate normal Gaussian distribution of the feature vector $\mathbf{x}$ within each class. The vector form of the Gaussian distribution of the probability density function $p(\mathbf{x}|C_i)$, for the feature vector $\mathbf{x}$ within a class C_i, is given by the equation:

$$p(\mathbf{x}|c_i) = \frac{1}{(2\pi)^{\frac{n}{2}}|\Sigma_i|^{\frac{1}{2}}} exp\left[-\frac{1}{2}(\mathbf{x}-\boldsymbol{\mu}_i)^T\Sigma_i^{-1}(\mathbf{x}-\boldsymbol{\mu}_i)\right] \tag{7}$$

where $\boldsymbol{\mu}_i$ is the mean vector of the ith class feature vector, Σ_i is the ith class feature vector covariance matrix, $|\Sigma_i|$ is the determinant of the covariance matrix, and n is the dimension of the feature vector $\mathbf{x}$. Based on the Bayes' optimal decision rule the following *a quadratic discriminant* can be derived

$$d_i(\mathbf{x}) = ln\frac{1}{(2\pi)^{\frac{n}{2}}|\Sigma_i|^{\frac{1}{2}}} exp\left[-\frac{1}{2}(\mathbf{x}-\boldsymbol{\mu}_i)^T\Sigma_i^{-1}(\mathbf{x}-\boldsymbol{\mu}_i)\right] + ln\,P(c_i),\ i = 1, 2, \cdots, l \tag{8}$$

Additional derivations give the following discriminant function

$$d_i(\mathbf{x}) = -\frac{1}{2}ln\,|\Sigma_i| - \frac{1}{2}(\mathbf{x}-\boldsymbol{\mu}_i)^T\Sigma_i^{-1}(\mathbf{x}-\boldsymbol{\mu}_i) + ln\,P(c_i),\quad i = 1, 2, \cdots, l \tag{9}$$

The quadratic Gaussian discriminant classifier was designed and experiments were performed for the training and test sets containing patterns in the differentials feature space. In the classifier design we decided consideration of the unknown category. The unknown category was detected when the absolute difference between discriminant for the benign and the malignant is less than a given threshold (from the range [0,1]). Malignant cases were favored by the factor wm We have found a proper value for wm as 0.9. The following data sets were analyzed:

1. The training set T_{tra}, for designing the discriminant parameters, containing 615 cases with patterns constituted with the age code plus 27 differentials: 469 cases of class 0 (benign), 146 cases of class 1 (malignant).
2. The test set T_{test}, for the classifier performance evaluation, contained 1301 cases, with patterns containing the age code plus 27 differentials: 968 class 0 (benign), 333 class 1 (malignant).

The results of three experiments are shown in Table 1. Experiment 1 - malignant class favorized by the weight $wm = 0.9$, with threshold $thu = 1.0$. Experiment 2 - malignant class favorized by the weight $wm = 0.9$, with threshold $thu = 0.8$. Experiment 3 - malignant class favorized by the weight $wm = 0.9$, $thu = 0.5$.

Table 1 Results of differential classification by Gaussian quadratic discriminant based classifier)

Experiment	All misclassified %	Sensitivity %	Specificity %	Unknown %
1	31.83	93.70	26.45	14.30
2	32.21	93.12	24.38	9.76
3	33.39	92.53	21.38	6.81

Results of Gaussian quadratic discriminant classifier with 28–dimensional differential pattern transformed into PCA space

The Gaussian quadratic discriminant classifier was also designed for the transformed by KLT and reduced patterns. The $28 - dimensional$ patterns $\mathbf{y}$, in the differential feature, were transformed into the principal component space using the Karhunen-Loéve transformation (KLT) derived from the training set. First, the full size $28 - element$ PCA feature patterns were obtained as a result of KLT transformation. Then the reduction of PCA feature pattern to dimension $m \leq n$ was provided. We present results with $28 - dimensional$ differential pattern transformed into $m = 12$ dimensional patterns in the PCA space. The reduced feature vector $\mathbf{x}_{PCA}$ contained the first 12 principal components (from the total number of $n = 28$ PCAs of the transformed pattern).

The following data sets were used:

1. The training set $T_{tra,PCA}$, for designing of the discriminant parameters, contained 615 cases with patterns elements being the first 12 PCAs of transformed original training set (containing differentials): 469 class 0 (benign), 146 class 1 (malignant).
2. The test set $T_{test,PCA}$, used for the performance evaluation, contained 1301 cases, with pattern's elements being the first 12 PCAs of the transformed original test set (containing differentials): 968 class 0 (benign), 333 class 1 (malignant).

The results of two experiments are shown in Table 2. Experiment 1 - both classes equally treated. Experiment 2 - malignant class favorized.

Table 2 Results of quadratic Gaussian classifier for PCA features.

Experiment	All misclassified %	Sensitivity %	Specificity %
1	34.43	81.93	18.02
2	35.71	94.62	15.74

Results of classification with the KLT transformation of differential patterns with the rough sets based reduction

The rough sets method was applied for the training set $T_{tra,PCA}$ containing 615 cases. Each case contained the transformed (from differentials) by KLT full size ($n = 28$) patterns and a corresponding class. For the discretized training data set the rough sets method was applied, and the reducts set and a core were computed. The results of Gaussian quadratic discriminant classifier for selected $7-element$ reduct (with corresponding reduced PCA feature patterns) are presented in Table 3.

We present results of two experiments: experiment 1 - both classes equally treated, experiment 2 - malignant class favorized.

Table 3 Results of quadratic Gaussian classifier for PCA features reduced by rough sets.

Experiment	All misclassified %	Sensitivity %	Specificity %
1	38.89	82.89	17.11
2	36.65	92.45	17.55

3 Conclusion

The classification experiments show the possibility of a breast cancer detection using electropotentials. However, the initial results show that this type of data is difficult to classify. The application of the principal component analysis improved classification accuracy for the breast cancer detection using quadratic Gaussian discriminant classifier. Additionally, the rough sets method allowed to reduce the feature number without substantial decreasing of the classification accuracy.

References

1. J. P. Crowe, and M. L. Faupel. 1996. Use non-directed (screening) arrays in the evaluation of symptomatic and asymptomatic breast potentials. In J. M. Dixon (Editor), *Electropotentials in the Clinical Assessments of Breast Neoplasia*. Springer 1996.
   R. J. Davies. 1996. Underlying mechanism involved in surface electrical potential measurements for the diagnosis of breast cancer. An electrophysiological approach to cancer diagnosis. In J. M. Dixon (Editor), *Electropotentials in the Clinical Assessments of Breast Neoplasia*. Springer 1996.
2. J. M. Dixon (Editor), *Electropotentials in the Clinical Assessments of Breast Neoplasia*. Springer 1996.
3. D. M. Long, Jr., M. L. Faupel, Yu-Sheng Hsu, J. A. Escobar, J. P. Michel, L. F. Mittag, r. M. Mitten, B. L. Witt, W. C. Herrick, and R. E. Keefe. 1996. Initial preclinical experiments with the electropotential differential diagnosis of mammary cancer. In J. M. Dixon (Editor), *Electropotentials in the Clinical Assessments of Breast Neoplasia*. Springer 1996.
4. A. A. Marino, I. G. Iliev, M. A. Schwalke, E. Gonzalez, K. C. Marler, and C. A. Flanagan. 1994. Association between cell membrane potential and breast cancer. *Cancer Research*, 15:82-89, 1994.
5. Z. Pawlak. *Rough sets: Theoretical aspects of reasoning about data*. Kluwer Academic Publishers, Dordrecht 1991.

Rough Sets and Neural Networks Application to Handwritten Character Recognition by Complex Zernike Moments

Roman W. Swiniarski

San Diego State University
Department of Mathematical and Computer Sciences
5500 Campanile Drive , San Diego, California 92182-7720, U.S.A.

Abstract. The paper presents a hand-written character recognition by the data mining and knowledge discovery software system RoughNeuralLab. In recognition experiments the Zernike moments were applied as the extracted features of character images. For further feature reduction the rough set theory method was applied as a front end of neural network. Eventually the error backpropagation neural network classifiers were designed for the reduced feature subsets.

1 Introduction

The contemporary concern about soft computing and knowledge discovery has put forward applications of neural processing and useful extensions of elementary set theory such as rough sets (Pawlak, 1991; Skowron, 1990). Basically, rough sets embody the idea of indiscernibility between objects in a set. Neural networks, provide soft computation, like error backpropagation, with generalization abilities for processing unseen instances. We combined both ideas together, and implemented them as a rough-neural software system RoughNeuralLab (Swiniarski, 1995) for data mining and pattern recognition. The system is especially suitable for processing images and data sets in form of decision tables. In this software system, rough sets were utilized to reduce the size of a knowledge base without losing valuable information due to the process of reduction, while neural networks are used for classification. By applying the rough set theory, unimportant knowledge in a decision table can be eliminated and useful information (reducts) can be obtained. The reduced decision table, with reduced features of a pattern, can be used for a classifier design.

The paper is organized as follows. In Section 2 we introduce *Zernike moments* as a robust method of feature extraction in terms of rotation invariancy and ability of image reconstruction via the inverse transform. Then we outline the ideas and notation used in the rough set theory, in knowledge reduction. Eventually, we discuss the software system RoughNeuralLab (Swiniarski, 1995) and its usage for data mining and classification. The paper concludes with a discussion of experiments with the described system for the handwritten character recognition.

L. Polkowski and A. Skowron (Eds.): RSCTC'98, LNAI 1424, pp. 617–624, 1998.
© Springer-Verlag Berlin Heidelberg 1998

2   Feature extraction by complex Zernike moments

2.1   Introduction

A robust recognition system must be able to recognize an image regardless of its orientation, size and position. In other words, *rotation-, scale-,* and *translation-invariancy* properties for the extracted features are desirable. Historically, Hu (1962) first introduced the use of image moment invariants for two-dimensional pattern recognition applications in 1961. He derived a set of invariant moments which has the desirable properties of being invariant under image rotation, scaling, and translation. However, the question of what is gained by including higher order moments in the analysis of an image has not been addressed.

Teague (1980) has suggested the notion of orthogonal moments to recover the image from moments based on the theory of orthogonal polynomials, and has introduced *Zernike moments.* This approach is usually simple and in a straight-forward manner allows moment invariants to be constructed of an arbitrarily high order. In addition, *Zernike moments* possess a useful rotation invariance property. Rotating the image does not change the magnitudes of the moments.

Since the defined features by means of Zernike moments are only rotation invariant, to obtain the scale and translation invariance, an image must be normalized via *image normalization.* Before giving the idea of the image normalization process we briefly describe *translation, scaling,* and *rotations.*

A gray-scale spatial domain image can be defined as

$$\{f(x,y) \in \{0,1,...,255\} : x = 0,1,...,M-1; y = 0,1,...,N-1\}, \tag{1}$$

where x is a column index, y a row index, M a number of columns, N a number of rows, and $f(x,y)$ the pixel value at location (x,y).

The proposed Zernike moments are only rotationally invariant, but the considered images have scale and translation differences as well. Therefore, prior to extraction of Zernike moments, these images should be normalized in respect to scaling and translation.

Translation invariance can be achieved by moving the origin to the center of an image. In order to get the centroid location of an image, **general moments** (or **regular moments**) have been utilized. General moments are defined as

$$m_{pq} = \int_{-\infty}^{\infty} \int_{-\infty}^{\infty} x^p y^q f(x,y)\,dxdy \tag{2}$$

where m_{pq} is the $(p+q)$th order moment of the continuous image function $f(x,y)$. Since we are dealing with digital images, the integrals can be replaced by summations. Given a two-dimensional $M \times N$ image, m_{pq} becomes

$$m_{pq} = \sum_{x=0}^{M-1} \sum_{y=0}^{N-1} x^p y^q f(x,y) \tag{3}$$

To keep the dynamic range of m_{pq} consistent for any different size of images, the $M \times N$ image plane should be first mapped onto a square defined by $x \in [\text{-}1,$

+1], $y \in [-1, +1]$. This implies that grid locations will no longer be integers but will have real values in the [-1, +1] range. This changes the definition of m_{pq} to

$$m_{pq} = \sum_{x=-1}^{+1} \sum_{y=-1}^{+1} x^p y^q f(x, y) \qquad (4)$$

Now we can find the centroid location of an image by the general moment. According to Zernike (1934), the coordinates of the image centroid $(\bar{x}, \bar{y})$ are

$$\bar{x} = \frac{m_{10}}{m_{00}}; \quad \bar{y} = \frac{m_{01}}{m_{00}} \qquad (5)$$

To achieve a translation invariancy, Hu (1980) suggested that we can transform the image into a new one whose first order moments, m_{01} and m_{10}, are both equal to zero. This is done by transforming the original image into the $f(x + \bar{x}, y + \bar{y})$ image, where $\bar{x}$ and $\bar{y}$ are the centroid locations of the original image. Let $g(x, y)$ be the translated image, the new image function becomes

$$g(x, y) = f(x + \bar{x}, y + \bar{y}) \qquad (6)$$

Scale invariancy is accomplished by enlarging or reducing each image such that its zero order moment, m_{00}, is set equal to a predetermined value β. We achieve this by transforming the original image function $f(x, y)$ into a new function $f(x/a, y/a)$, with the scaling factor a, where

$$a = \sqrt{\frac{\beta}{m_{00}}} \qquad (7)$$

Let $g(x, y)$ be the scaled image. After scale normalization, we get

$$g(x, y) = f(\frac{x}{a}, \frac{y}{a}) \qquad (8)$$

Hence,

$$g(x, y) = f(\frac{x}{a} + \bar{x}, \frac{y}{a} + \bar{y}) \qquad (9)$$

with $(\bar{x}, \bar{y})$ the centroid of $f(x, y)$ and $a = \sqrt{\frac{\beta}{m_{00}}}$, β a predetermined value, normalizes a function with respect to scale and translation.

2.2   Zernike moments

Zernike (1934) introduced a set of complex polynomials which form a complete orthogonal set over the interior of unit circle, i.e., $x^2 + y^2 = 1$. Let the set of these polynomials be denoted by $V_{nl}(x, y)$. The form of these polynomials is:

$$V_{nl}(x, y) = V_{nl}(\rho \sin\theta, \rho \cos\theta) = V_{nl}(\rho, \theta) = R_{nl}(\rho) exp(il\theta) \qquad (10)$$

where

n Positive integer or zero.

l Positive or negative integers, subject to constraints $n - |l| = $ even, $|l| \leq n$.

i The complex number i.

ρ Length of vector from origin to (x, y) pixel.

θ Angle between vector ρ and x-axis in counterclockwise direction.

Radial polynomials $R_{nl}(\rho)$ are defined as

$$R_{nl}(\rho) = \sum_{s=0}^{\frac{n-|l|}{2}} \frac{(-1)^s[(n-s)!]\rho^{n-2s}}{s!(\frac{n+|l|}{2} - s)!(\frac{n-|l|}{2} - s)!} \tag{11}$$

These polynomials are orthogonal.

Zernike moments are the projection of the image function onto these orthogonal basis functions. The Zernike moments of order n with repetition l for a continuous image function $f(x, y)$ is

$$A_{nl} = \frac{n+1}{\pi} \int \int_{x^2+y^2 \leq 1} f(x, y)[V_{nl}(\rho, \theta)]^* dxdy = (A_{n,-l})^* \tag{12}$$

For a digital image:

$$A_{nl} = \frac{n+1}{\pi} \sum_x \sum_y f(x, y) V_{nl}^*(\rho, \theta); \quad x^2 + y^2 \leq 1 \tag{13}$$

$$\begin{bmatrix} C_{nl} \\ S_{nl} \end{bmatrix} = \frac{2n+2}{\pi} \int \int_{x^2+y^2 \leq 1} f(x, y) R_{nl}(\rho) \begin{bmatrix} \cos l\theta \\ -\sin l\theta \end{bmatrix} dxdy \tag{14}$$

or

$$C_{nl} = 2 \, Re \, (A_{nl})$$
$$= \frac{2n+2}{\pi} \int \int_{x^2+y^2 \leq 1} f(x, y) R_{nl}(\rho) \, \cos l\theta \, dxdy \tag{15}$$

$$S_{nl} = -2 \, Im \, (A_{nl})$$
$$= \frac{-2n-2}{\pi} \int \int_{x^2+y^2 \leq 1} f(x, y) R_{nl}(\rho) \, \sin l\theta \, dxdy \tag{16}$$

and for $l = 0$

$$C_{n0} = A_{n0} = \frac{1}{\pi} \int \int_{x^2+y^2 \leq 1} f(x, y) R_{n0}(\rho) dxdy$$
$$S_{n0} = 0 \tag{17}$$

For a digital image, when $l \neq 0$

$$\begin{bmatrix} C_{nl} \\ S_{nl} \end{bmatrix} = \frac{2n+2}{\pi} \sum_{x=-1}^{+1} \sum_{y=-1}^{+1} f(x, y) R_{nl}(\rho) \begin{bmatrix} \cos l\theta \\ -\sin l\theta \end{bmatrix} \tag{18}$$

and when $l = 0$

$$C_{n0} = A_{n0} = \frac{1}{\pi} \sum_{x=-1}^{+1} \sum_{x=-1}^{+1} f(x, y) R_{n0}(\rho)$$

$$S_{n0} = 0 \qquad (19)$$

3   Rough sets in knowledge reduction

The rough set theory has been developed for knowledge discovery in experimental data sets. Rough set techniques reduce the computational complexity of learning processes and eliminate irrelevant attributes making knowledge discovery more efficient. Rough sets were introduced by Zdzisław Pawlak to provide a systematic framework for studying imprecise and incomplete knowledge. The rough sets theory based on the concept of an *upper* and a *lower approximation* of a set, the *approximation space* and probabilistic and deterministic models of sets. The rough sets theory deals with information represented by a table called an information system. This table consists of objects (or cases) and attributes. The entries in the table are the categorical values of the features and possibly categories. We use the standard notation (see Pawlak (1991) i.e. in particular we denote by $IND(S)$ the indiscernibility relation of the information system S; by $[x]_A$ the equivalence class defined by $IND(A)$; by $\underline{A}X$, $\bar{A}X$ the lower and upper approximation of X with respect to A; by X_i the i-th decision class of the decision table DT.

The **accuracy** of an approximation of set X by the set of attributes A is defined as follows:

$$\alpha_A(X) = \frac{card\ \underline{A}X}{card\ \bar{A}X} \qquad (20)$$

The **rough** (**A-rough**) membership function of the set X is defined as:

$$\mu_X^A(X) = \frac{card([x]_A \cap X)}{card([x]_A)} \qquad (21)$$

The **accuracy of approximation of classification** $\Upsilon = X_1, X_2, ..., X_n$ by the set of attributes A is defined as follows:

$$\alpha_A(\Upsilon) = \frac{\Sigma_{i=1}^n card\ (\underline{A}X_i)}{\Sigma_{i=1}^n card\ (\bar{A}X_i)} \qquad (22)$$

The process of finding a smaller set of attributes than original one with same classificatory power as original set is called *attribute reduction*. A **reduct** is the essential part of an information system i.e. a minimal attribute subset discerning all objects discernible by the original information system. A **core** is a common part of all reducts. The set of all reducts in S with the attribute set A is denoted by $RED(A)$.

Given a decision table S with condition and decision attributes $Q = C \cup D$, for a given set of condition attributes $A \subset C$ we can define a positive region A $POS_A(D)$ in the relation $IND(D)$, as

$$POS_A(D) = \bigcup \{\underline{A}X | X \in IND(D)\} \tag{23}$$

The positive region $POS_A(D)$ contains all objects in U which can be classified without error (ideally) into distinct classes defined by $IND(D)$ based only on information in the relation $IND(A)$.

The cardinality (size) of the A-positive region of B is used to define a measure (a degree) $\gamma_A(B)$ of dependency of the set of attributes B on A:

$$\gamma_A(B) = \frac{card(POS_A(B))}{card(U)} \tag{24}$$

We say that the set of attributes B depend on the set of attributes A in a degree $\gamma_A(B)$.

One can find a *reduct* of condition attributes relative to the decision attributes by removing *superfluous* condition attributes, without losing the classification power of the reduced decision table. The core of decision table is the intersection of all its relative reducts. It consists of all indispensable condition attributes.

3.1   Classification results

In numerical classification experiments, we used the character image database which is a collection of digits selected from hand-written Zip Codes of 49 different writers from National Institute of Standards and Technology, formerly National Bureau of Standards. This database is a subset of "NIST Special Database 1". These characters have been isolated and specially normalized to be 32×32 pixel images. From this image database, 500 images were selected as our training set and 200 images (different from the 500 images) were randomly selected as our test set.

In the numerous numerical experiments, the image data sets were divided into two sets, the training image data set and the testing image data set. The images of the training data set are the prototypes of the images that we intend to train in the system while the images of the testing data set are randomly selected from images of the database.

For generating the training and test sets, from the training image data set, there were several selection options including: image thinning, moments extraction, reduct computation, etc.

We tested the recognition system with two testing image data set. The original testing image data set is randomly selected from the images database. The second testing image data set contains contaminated images by injecting some noise to original images.

We considered 500 data images for training because of the limited system capacity. The training image data set was transformed into the decision table set by several ways described in Table 1. One of the input data sets was not thinned, another was thinned before the general moments were applied and the other was thinned after the general moments had been extracted. Each of these input data sets was the source of Zernike moments and the system generates a decision table from the Zernike moments by two selections, the domain and the range. We select 5 as the domain and 1.0 as the range by our experience. Among the attributes of the decision table, several attributes are selected as the final input data set by one of the reduct sets. So we select two reduct sets for getting the two final input data sets. One of the selected reduct sets was 1st reduct set, that has the attributes showing more then 40% in all reduct sets, while another set is consisted of the smallist number of attributes among all reduct sets. The error backpropagation neural networks were trained with one hidden layer or two hidden layers. Therefore, the system is trained by 12 different types of input data sets or neural network selections in Table 1.

Table 1. The user's selection options for the training sets

	The options of training sets		
Option	Thinning	Reduct Sets	Hidden layers
Step	Preprocessing		Classifier
Substep	Thinning	Rough Sets	Neural Networks
Selections	no, before or after General Moments		1,2
System 1	no	1st	1
System 2	before	1st	1
System 3	after	1st	1
System 4	no	2nd	1
System 5	before	9th	1
System 6	after	3rd	1
System 7	no	1st	2
System 8	before	1st	2
System 9	after	1st	2
System 10	no	2^{cd}	2
System 11	before	9st	2
System 12	after	3rd	2

The testing image data set, with 200 images, was divided into two sets, one was the original testing image data set and the other contained injected 3% of global noise (injected noise on any pixels of the original image).

For the systems 1,3,4,6,7,9,10 and 12, the value 12 was selected as the experimental order α and the value 320 as the experimental pixel number β. For the system 2,5,8 and 11, 12 was selected as the experimental order α and 65 as the experimental pixel number β. For the test set 1, the best classification result of 84.0% was obtained for the system 9 with thinned images (after the general moments were extracted and applied for transformation) and the neural network with 2 hidden layers. The best result of classification of 79.5% for the testing set 2 with injected noise to images was obtained for the system 7 with the training set of unthinned images and the neural network with 2 hidden layers.

Conclusions. The objective of the research was to build an image classification system by using neural networks based on invariant Zernike moments for feature extraction and rough set tools for feature reduction. In the best experiment setting we obtained 84% classification accuracy for handwritten digits recognition task. Through implementation of the rough sets theory, we could significantly reduce the number of information about images and improve the task running time to one-fourth of that for the regular system.

References

1. M.-K. Hu. 1962. Visual pattern recognition by moment invariants. IRE Trans. on Inform. Theory, vol. IT-8, pp. 179–187.
2. R. Swinirski.1994. Zernike Moment Feature For Shape Recognition. Internal report, SDSU.
3. A. Khotanzad, Y. H. Hong. 1990. Invariant image recognition by Zernike moments. IEEE Trans. on Pattern Anal. Machine Intell., vol. 12., no. 5., pp. 489–497.
4. A. Khotanzad, J.-H. Lu. 1990. Classification of invariant image representations using a neural network. IEEE Trans. on Acoust., Speech, and Signal Processing, vol. 38, no. 6, pp. 1028–1038.
5. M. Teague. 1980 Image analysis via the general theory of moments. J. Opt. Soc. Amer., vol. 70, no. 8, pp. 920-930.
6. F. Zernike. 1934. Physical, vol.. 1, p. 689.
7. T. Nguyen, R. Swiniarski, A. Skowron, J. Bazan, K. Thagarajan. 1994. Applications of rough sets, neural networks and maximum likelihood for texture classification based on singular value decomposition. Proceedings of Third International Workshop on Rough Sets and Soft Computing. San Jose, U.S.A., November 10-12, pp. 332–339.
8. Pawlak, Z. 1991. Rough sets – Theoretical aspects of reasoning about data, Kluwer, Dordrecht.
9. Skowron, A. 1990. The rough sets theory and evidence theory. Fundamenta Informaticae vol.13, pp. 245–262.
10. Swiniarski, R. 1995. RoughNeuralLab. Software package developed at San Diego State University. San Diego.
11. Swiniarski, R. 1993. Introduction to rough sets. Materials of The International Short Course Neural Networks. Fuzzy and Rough Systems. Theory and Applications. April 2, 1993. San Diego State University.
12. Swiniarski, R. 1996. Rough sets expert system for on–line prediction of volleyball game progress for US Olympic team. in: B.D. Czejdo, I.I. Est, B. Shirazi, B. Trousse (eds.), Proceedings of the Third Biennial European Joint Conference on Engineering Systems Design and Analysis, July 1–4, Montpellier, France, pp. 15–20.
13. Swiniarski, R. 1996. Rough sets expert system for robust texture classification based on 2D fast Fourier transformation spectral features. in: S. Tsumoto, S. Kobayashi, T. Yokomori, H. Tanaka and A. Nakamura (eds.), *The fourth International Workshop on Rough Sets, Fuzzy Sets, and Machine Discovery, PROCEEDINGS (RS96FD)*, November 6-8, The University of Tokyo, pp. 419–425.

Author Index